Ref.: Mineral Nutrition of Higher Plants
by Horst Marschner

Corrigendum

Page	error	correction
55, para.1, line 3	chelator	chelate
55, para.2, line 2	(Fig.2.21)	(Fig.2.24)
116, Fig. 5.1	Catalase	Peroxidase
137, Fig.5.11	formula Zeatin:	
	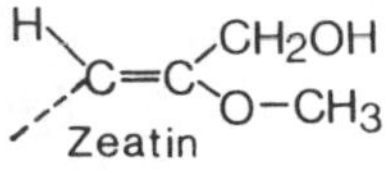 H, C=C, CH_2OH, $O-CH_3$, Zeatin	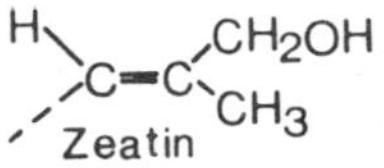 H, C=C, CH_2OH, CH_3, Zeatin
137, Fig.5.11	formula Abscisic acid:	
	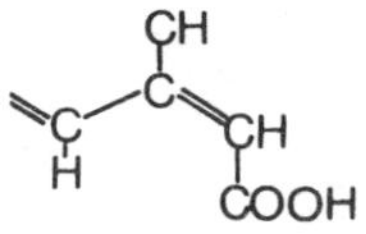 CH, C, CH, H, COOH	CH_3, C, CH, H, COOH
251, Fig.8.22	direction of arrow	
	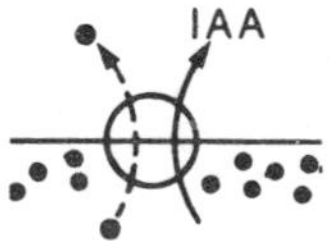 IAA	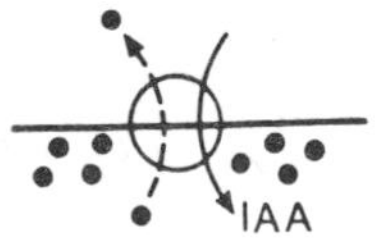 IAA
281, para.2, line 15	are catalase, peroxidase, and ...	are peroxidase, and ...
459, last line	Banksiaa	Banksia
519, Table 16.19, legend line 3	Asvantages	Advantages
520, Table 16.20, legend line 1	Nutritonal	Nutritional
659	Methemoblobemia	Methemoglobemia

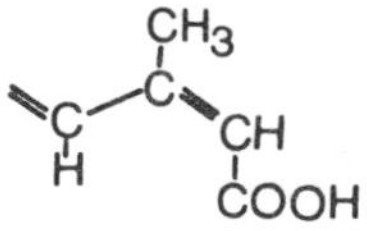

Mineral Nutrition of Higher Plants

We apologise for a minor printing error in the title of this book as shown on the front cover.

The correct title is MINERAL NUTRITION OF HIGHER PLANTS, as given on the title page.

The error will be corrected in future printings.

Academic Press

Mineral Nutrition of Higher Plants

Horst Marschner

Institute of Plant Nutrition
University of Hohenheim
Federal Republic of Germany

Academic Press
Harcourt Brace Jovanovich, Publishers
London Orlando San Diego New York Austin
Boston Sydney Tokyo Toronto

ACADEMIC PRESS INC. (LONDON) LTD.
24/28 Oval Road, London NW1 7DX

United States Edition Published by
ACADEMIC PRESS INC.
Orlando, Florida 32887

British Library Cataloguing in Publication Data
Marschner, Horst
Mineral nutrition of higher plants.
1. Crops 2. Soil mineralogy 3. Plants
— Nutrition
I. Title
631.5 SB91

ISBN 0-12-473540-1
ISBN 0-12-473541-X (Pbk)

Typeset and printed by
W. & G. Baird Ltd., The Greystone Press,
Antrim, Northern Ireland.

Preface

Mineral nutrients are essential for plant growth and development. Mineral nutrition of plants is thus an area of fundamental importance for both basic and applied science. Impressive progress has been made during the last decades in our understanding of the mechanisms of nutrient uptake and their functions in plant metabolism; at the same time, there have also been advances in increasing crop yields by the supply of mineral nutrients through fertilizer application. It is the main aim of this textbook to present the principles of the mineral nutrition of plants, based on our current knowledge. Although emphasis is placed on crop plants, examples are also presented from noncultivated plants including lower plants in cases where these examples are considered more suitable for demonstrating certain principles of mineral nutrition, either at a cellular level or as particular mechanisms of adaptation to adverse chemical soil conditions.

Plant nutrition as a subject is closely related to other disciplines such as soil science, plant physiology and biochemistry. In this book, mineral nutrients in soils are treated only to the extent considered necessary for an understanding of how plant roots acquire mineral nutrients from soils, or how roots modify the chemical soil properties at the soil-root-interface. Fundamental processes of plant physiology and biochemistry, such as photosynthesis and respiration, are treated mainly from the viewpoint of how, and to what extent, they are affected or regulated by mineral nutrients. Crop physiology is included as an area of fundamental practical importance for agriculture and horticulture, with particular reference to source-sink relationships as affected by mineral nutrients and phytohormones.

Mineral nutrition of plants covers a wide field. It is therefore not possible to treat all aspects with the detail they deserve. In this book, certain aspects are covered in more detail, either because they have recently become particularly important to our understanding of mineral nutrition, or because many advances have been made in a particular area in the last decade. Naturally, personal research interests and evaluation are also factors which have influenced selection. Particular emphasis is placed on short and long-distance transport of mineral elements, on source-sink relationships, and on plant-soil relationships. It is also the intention of this book to enable the reader to become better acquainted with the mechanisms of adaptation of plants to adverse chemical soil conditions. The genetical basis of mineral nutrition is therefore stressed, as well as the possibilities and limitations of "fitting crop plant to soils", especially in the tropics and subtropics.

I have written this textbook for graduate students and researchers in the various fields of agricultural, biological and environmental sciences, who already have a profound knowledge of plant physiology, biochemistry and soil science. Instead of extensive explanations of basic processes, emphasis is placed on representative examples—tables, figures, schematic presentations—illustrating the various aspects of mineral nutrition. In a textbook of such wide scope, generalizations cannot be avoided, but relevant literature is cited for further and more detailed studies. In the literature, preference has been given to more recent publications. Nevertheless, representative examples of classical contributions are also cited in the various sections.

Although this book is written by one person, it is nevertheless the product of cooperation at various levels. My interest in plant nutrition and my scientific career in this field are due to the inspirations of Dr. G. Michael. The book as it is presented here would not have been accomplished without the excellent support of two colleagues, Dr. V. Römheld and Mr. Ernest A. Kirkby. I am very much indebted to both of them. Dr. V. Römheld not only prepared the drawings but also gave highly valuable advice regarding the arrangement of tables and improvements to the text. My old friend Ernest A. Kirkby corrected the English and improved the first draft considerably by valuable suggestions and stimulating criticism. My colleagues in the institute, Dr. P. Martin, Dr. W. J. Horst and Dr. B. Sattelmacher helped me greatly, both by valuable discussions in various subject areas treated in this book and by keeping me free for some time from teaching and administrative responsibilities. Many colleagues were kind enough to supply me with their original photographs, as indicated in the legend of the corresponding figures.

The preparation of such a manuscript requires skilful technical assistance. I would especially like to thank Mrs. H. Hoderlein for typing the manuscript.

Last but not least, I have to thank my family for encouraging me to write the book and for their assistance and patience throughout this time-consuming process.

Stuttgart-Hohenheim — Horst Marschner
August 1985

Contents

Glossary of Plant Species

Botanical names of plant species frequently cited in this book.

Alfalfa	*Medicago sativa* L.
Aubergine	*Solanum melongena* L.
Barley	*Hordeum vulgare* L.
Bean	*Phaseolus vulgaris* L.
Brussels sprouts	*Brassica oleracea gemmifera* L.
Cabbage	*Brassica oleracea* L.
Carrot	*Daucus carota* L. subsp. *sativus* (Hoffm.) Arcang.
Cassava	*Manihot esculenta* Crantz
Castor bean	*Ricinus communis* L.
Celery	*Apium graveolens* L. var. *rapaceum* (Mill.)
Cotton	*Gossypium hirsutum*
Cowpea	*Vigna unguiculata* L.
Cucumber	*Cucumis sativus* L.
Faba bean	*Vicia faba* L.
Grapevine	*Vitis vinifera* L. subsp. *vinifera*
Leek	*Allium porrum* L.
Lentil	*Lens culinaris* L.
Lettuce	*Lactuca sativa* L.
Maize	*Zea mays* L.
Oat	*Avena sativa* L.
Onion	*Allium cepa* L.
Pea	*Pisum sativum* L.
Peanut	*Arachis hypogaea* L.
Pigeon pea	*Cajanus cajan* L. Huth.
Potato	*Solanum tuberosum* L.
Pumpkin	*Cucurbita pepo* L.
Rape	*Brassica napus* L. var. *napus*
Red beet	*Beta vulgaris* L. subsp. *vulgaris* var. *conditiva* Alef.
Red clover	*Trifolium pratense* L.
Red pepper	*Capsicum annuum* L.
Rice	*Oriza sativa* L.
Rye	*Secale cereale* L.
Ryegrass	*Lolium perenne* L.
Soybean	*Glycine max* (L.) Merr.
Sugar beet	*Beta vulgaris* L. subsp. *vulgaris*
Sugarcane	*Saccharum officinarum* L.

Sorghum	*Sorghum bicolor* (L.) Moench
Spinach	*Spinacea oleracea* L.
Squash	*Cucurbita maxima* Duch.
Subterranean clover	*Trifolium subterraneum* L.
Sunflower	*Helianthus annuus* L.
Sweet potato	*Ipomea batatas* L. Lam.
Taniers	*Xanthosoma* sp.
Tea plant	*Thea sinensis* L.
Timothy	*Phleum pratense* L.
Tobacco	*Nicotiana tabacum* L.
Tomato	*Lycopersicon esculentum* L.
Turnip	*Brassica rapa* var. *rapa* L.
Welsh onion	*Allium fistulosum* L.
Wheat	*Triticum aestivum* L.
White clover	*Trifolium repens* L.
White mustard	*Sinapis alba* L.
Winter radish	*Raphanus sativus* L. var. *niger* (Mill.) S. Kerner
Yams	*Dioscorea alata* L.

Part I

Nutritional Physiology

1

Introduction, Definition, and Classification of Mineral Nutrients

The beneficial effect of adding mineral elements (e.g., plant ash or lime) to soils to improve plant growth has been known in agriculture for more than 2,000 years. Nevertheless, even 150 years ago it was still a matter of scientific controversy as to whether mineral elements function as nutrients for plant growth. It was mainly to the credit of Justus von Liebig (1803–1873) that the scattered information concerning the importance of mineral elements for plant growth was compiled and summarized and that the mineral nutrition of plants was established as a scientific discipline. These achievements led to a rapid increase in the use of mineral fertilizers. By the end of the nineteenth century, especially in Europe, large amounts of potash, superphosphate, and, later, inorganic nitrogen were used in agriculture and horticulture to improve plant growth.

Liebig's conclusion that the mineral elements nitrogen, sulfur, phosphorus, potassium, calcium, magnesium, silicon, sodium, and iron are essential for plant growth was arrived at by observation and speculation rather than by precise experimentation. The fact that the "mineral element theory" was based on this unsound foundation was one of the reasons for the large number of studies undertaken at the end of the nineteenth century. From these and other extensive investigations on the mineral composition of different plant species growing on various soils, it was realized as early as the beginning of this century that neither the presence nor the concentration of a mineral element in a plant is a criterion for essentiality. Plants have a limited capability for the selective uptake of those mineral elements which are essential for their growth. They also take up mineral elements which are not necessary for growth and which may even be toxic.

The mineral composition of plants growing in soils cannot therefore be used to establish whether a mineral element is essential. Once this fact was appreciated, both water and sand culture experiments were carried out in which particular mineral elements were omitted. These techniques made possible a more precise characterization of the essentiality of mineral elements and led to a better understanding of their role in plant metabolism.

Progress in this research was closely related to the development of analytical chemistry, particularly in the purification of chemicals and methods of estimation. This relationship is reflected in the time scale of the discovery of the essentiality of micronutrients (Table 1.1).

Table 1.1
Discovery of the Essentiality of Micronutrients for Higher Plants

Element	Year	Discovered by
Iron	1860	J. Sachs
Manganese	1922	J. S. McHague
Boron	1923	K. Warington
Zinc	1926	A. L. Sommer and C. B. Lipman
Copper	1931	C. B. Lipman and G. MacKinney
Molybdenum	1938	D. I. Arnon and P. R. Stout
Chlorine	1954	T. C. Broyer *et al.*

The term *essential mineral element* (or mineral nutrient) was proposed by Arnon and Stout (1939). These authors concluded that, for an element to be considered essential, three criteria must be met:

1. A given plant must be unable to complete its life cycle in the absence of the mineral element.
2. The function of the element must not be replaceable by another mineral element.
3. The element must be directly involved in plant metabolism—for example, as a component of an essential plant constituent such as an enzyme—or it must be required for a distinct metabolic step such as an enzyme reaction.

According to this definition those mineral elements which compensate for the toxic effects of other elements or which simply replace mineral nutrients in some of their less specific functions, such as the maintenance of osmotic pressure, are not essential, but can be described as "beneficial" elements (Chapter 10).

It is still difficult to generalize when discussing which mineral elements are essential for plant growth. This is particularly obvious when higher and lower plants are compared (Table 1.2). For higher plants the essentiality of 13 mineral elements is well established, although the known requirement for chlorine is as yet restricted to a limited number of plant species.

Because of gradual but continuous improvements in analytical techniques, especially in the purification of chemicals, this list might well be extended to include mineral elements that are essential only in very low concentrations in plants (i.e., that act as micronutrients). This holds true in

Table 1.2
Essentiality of Mineral Elements for Higher and Lower Plants

Classification	Element	Higher plants	Lower plants
Macronutrient	N, P, S, K, Mg, Ca	+	+ (Exception: Ca for fungi)
Micronutrient	Fe, Mn, Zn, Cu, B, Mo, Cl	+	+ (Exception: B for fungi)
Micronutrient and "beneficial" element	Na, Si, Co,	±	±
	I, V	−	±

particular for sodium and silicon, which are abundant in the biosphere. The essentiality of these two mineral elements has been established for some higher plant species (Chapter 10). Most micronutrients are predominantly constituents of enzyme molecules and are thus essential only in small amounts. In contrast, the macronutrients either are constituents of organic compounds, such as proteins and nucleic acids, or act as osmotica. These differences in function are reflected in the average concentrations of mineral nutrients in plant shoots that are sufficient for adequate growth (Table 1.3). The values can vary considerably depending on plant species, plant age, and concentration of other mineral elements. This aspect is discussed in Chapters 8 to 10.

Table 1.3
Average Concentrations of Mineral Nutrients in Plant Shoot Dry Matter that are Sufficient for Adequate Growth[a]

Element	Abbreviation	μmol/g dry wt	mg/kg (ppm)	%	Relative number of atoms
Molybdenum	Mo	0·001	0·1	—	1
Copper	Cu	0·10	6	—	100
Zinc	Zn	0·30	20	—	300
Manganese	Mn	1·0	50	—	1,000
Iron	Fe	2·0	100	—	2,000
Boron	B	2·0	20	—	2,000
Chlorine	Cl	3·0	100	—	3,000
Sulfur	S	30	—	0·1	30,000
Phosphorus	P	60	—	0·2	60,000
Magnesium	Mg	80	—	0·2	80,000
Calcium	Ca	125	—	0·5	125,000
Potassium	K	250	—	1·0	250,000
Nitrogen	N	1,000	—	1·5	1,000,000

[a]From Epstein (1965).

2

Ion Uptake Mechanisms of Individual Cells and Roots: Short-Distance Transport

2.1 General

As a rule there is a great discrepancy between the mineral nutrient concentration in the soil or nutrient solution, on the one hand, and the mineral nutrient requirement of plants, on the other. Furthermore, these substrates may contain high concentrations of mineral elements not needed for plant growth. The uptake mechanisms of plants must therefore be selective. This selectivity can be demonstrated particularly well in algal cells (Table 2.1), where the external and internal (cell sap) solutions are separated by only two membranes: the plasma membrane and the tonoplast.

Table 2.1
Relationship between Ion Concentration in the Substrate and in the Cell Sap of *Nitella* and *Valonia*[a]

	Nitella concentration (m*M*)			*Valonia* concentration (m*M*)		
Ion	A, Pond water	B, Cell sap	Ratio B/A	A, Seawater	B, Cell sap	Ratio B/A
Potassium	0·05	54	1080	12	500	42
Sodium	0·22	10	45	498	90	0·18
Calcium	0·78	10	13	12	2	0·17
Chloride	0·93	91	98	580	597	1

[a]Modified from Hoagland (1948).

In *Nitella* the concentration of potassium, sodium, calcium, and chloride ions is higher in the cell sap than in the pond water, but the concentration ratio differs considerably among the ions. In *Valonia* growing in highly saline seawater, on the other hand, only potassium is much more concentrated in the cell sap, whereas the sodium and calcium concentrations remain at a lower level in the cell sap than in the seawater.

Although usually less dramatic, selectivity of ion uptake is also a typical feature of higher plants. When plants are grown in a nutrient solution of limited volume, the external concentration changes within a few days (Table 2.2). The concentrations of potassium, phosphate, and nitrate decline markedly, whereas those of sodium and sulfate might even increase, indicating that water is taken up faster than either of these two ions. Uptake rates, especially for potassium and calcium, differ between the two plant species (maize and bean). The ion concentration in the root press sap is generally higher than that in the nutrient solution; this is most evident in the case of potassium, nitrate, and phosphate.

Table 2.2
Changes in the Ion Concentration of the External (Nutrient) Solution and in the Root Press Sap of Maize and Bean

	External concentration (m*M*)			Concentration in the root press sap (m*M*)	
		After 4 days[a]			
Ion	Initial	Maize	Bean	Maize	Bean
Potassium	2·00	0·14	0·67	160	84
Calcium	1·00	0·94	0·59	3	10
Sodium	0·32	0·51	0·58	0·6	6
Phosphate	0·25	0·06	0·09	6	12
Nitrate	2·00	0·13	0·07	38	35
Sulfate	0·67	0·61	0·81	14	6

[a]No replacement of water lost through transpiration.

The results obtained from both lower and higher plants demonstrate that ion uptake is characterized by the following:

1. Selectivity. Certain mineral elements are taken up preferentially, while others are discriminated against or nearly excluded.
2. Accumulation. The concentration of mineral elements can be much higher in the plant cell sap than in the external solution.
3. Genotype. There are distinct differences among plant species in ion uptake characteristics.

These results pose many questions. For example, how do individual cells and higher plants regulate ion uptake? Is ion uptake solely a reflection of demand, or are ions that have no function in plant metabolism or that are toxic also taken up? Before the regulation of ion uptake on a cellular level and the role of cell membranes in this process are discussed in more detail, the pathway of solutes (charged ions, molecules) from the soil or nutrient solution into the root tissue is described. This description is necessary for two reasons: (1) In higher terrestrial plants the roots are the principal sites of solute uptake, and (2) the walls of root cells can interact with solutes and thus may facilitate or restrict further movement to the uptake sites at the plasma membrane of individual root cells.

2.2 Pathway of Solutes from the External Solution into the Roots

2.2.1 Influx into the Apparent Free Space

Movement of low-molecular-weight solutes (e.g., ions, organic acids, and amino acids) by diffusion or mass flow is not restricted to the external surface of the roots, that is, the rhizodermal cells (Fig. 2.1). The cell walls and water-filled intercellular spaces of the root cortex are also accessible to these solutes from the external solution.

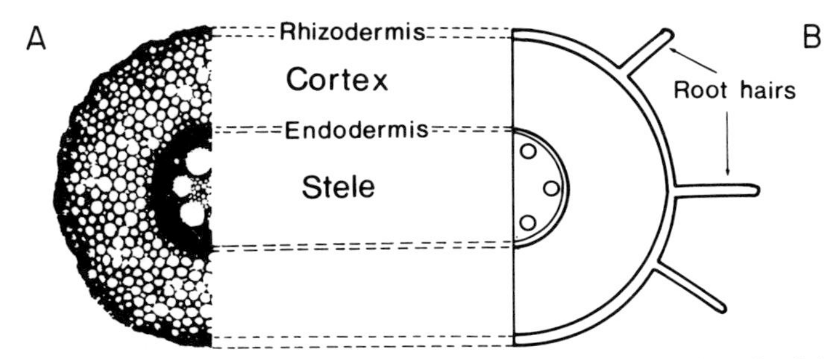

Fig. 2.1 A. Cross section of a differentiated root zone of maize. B. Schematic representation of cross section.

This solute movement into the roots is a nonmetabolic, passive process, the main barriers to which are the plasma membrane of the individual cortex cells and the endodermis, the innermost layer of cells of the cortex (Fig. 2.1). In the radial and transverse walls of the endodermis, hydrophobic incrustations (suberin)—the Casparian strip—constitute an effective barrier against passive solute movement into the stele.

The volume of root tissue available for passive solute movement—the *free space*—represents ~10% of the total volume of young roots. The presence of this free space means that there is a substantial internal root surface enabling the individual cortex cells to take up solutes directly from the external solution. The extent of solute diffusion into the free space depends on various factors such as solute concentration and root hair formation. When the external concentration is low and there is extensive root hair formation, the uptake of solutes such as potassium and phosphorus is limited mainly to the rhizodermal cell layer (Vakhmistrov, 1967). This is particularly true for roots growing in soil (see Section 2.10).

Primary cell walls consist of a network of cellulose, hemicellulose (including pectins), and glycoprotein. This network contains pores, the so-called interfibrillar and intermicellar spaces, which differ in size. For root hair cells of radish a maximum diameter of 3·5 to 3·8 nm (35–38 Å) has been calculated; maximum values for plant cell walls are in the range of 5·0 nm (Carpita *et al.*, 1979). The dimensions of hydrated ions such as K^+ and Ca^{2+} are small compared with the diameter of these pores, as shown in the following tabulation; thus, the pores themselves should not restrict the movement of these ions within the free space.

	Diameter (nm)
Rhizodermal cell wall (maize; Fig. 2.3)	500–3000
Cortical cell wall (maize)	100–200
Pores in cell wall	<5·0
Sucrose	1·0
Hydrated ions	
K^+	0·66
Ca^{2+}	0·82

In contrast to mineral nutrients and low-molecular-weight organic solutes, high-molecular-weight solutes (e.g., metal chelates, fulvic acids, and toxins) or viruses and other pathogens are either severely restricted or prevented by the diameter of pores from entering the free space of root cells.

In this network, a variable proportion of the pectins consist of polygalacturonic acid, originating mainly from the middle lamella. In the free space of the roots, therefore, the carboxylic groups ($R \cdot COO^-$) act as cation exchangers. Thus cations can accumulate in a nonmetabolic step in the free space, whereas anions are "repelled" (Fig. 2.2).

Because of these negative charges in the cell walls of the apoplast (the cell wall continuum in plant tissue), Hope and Stevens (1952) introduced the term *apparent free space* (AFS). This comprises the *water free space* (WFS), which is freely accessible to ions, and the *Donnan free space* (DFS), where cation exchange and anion repulsion take place (Fig. 2.2). Ion distribution within the DFS is the typical Donnan distribution which occurs in soils at the surfaces of negatively charged clay particles. Divalent cations such as Ca^{2+} are therefore preferentially bound to these cation-exchange sites. Plant species differ considerably in their cation-exchange capacity (CEC), that is, in the number of cation-exchange sites (fixed anions; $R \cdot COO^-$), in their cell walls, as shown in Table 2.3.

As a rule, the CEC of dicotyledonous species is much higher than that of monocotyledonous species. In intact roots, the CEC is much lower than the values shown in Table 2.3; only the exchange sites of the AFS (and not of the

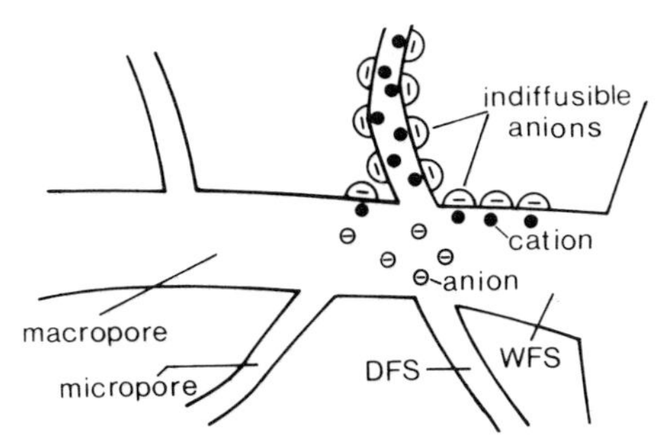

Fig. 2.2 Schematic diagram of the pore system of the apparent free space. DFS, Donnan free space; WFS, water free space.

whole root) are directly accessible to cations from the external solution. Nevertheless, the differences shown are typical of those that exist among plant species.

This exchange adsorption in the AFS of the apoplast is not an essential step for ion uptake or transport through the plasma membrane into the cytoplasm. Nevertheless, the preferential binding of di- and polyvalent cations increases their concentration in the apoplast of the roots and thus in the vicinity of the active uptake sites at the plasma membrane. As a result of this indirect effect, a positive correlation can be observed between CEC and the ratio of Ca^{2+} to K^{+} contents in different plant species (Crooke and Knight, 1962). The importance of cation binding in the AFS for uptake and subsequent shoot transport is also indicated by experiments with the same plant species but with different binding forms of a divalent cation such as zinc (Table 2.4). When zinc is supplied in the form of an inorganic salt (i.e., as free Zn^{2+}), the zinc content not only of the roots but also of the shoots is several times higher than when zinc is supplied as a chelate (ZnEDTA), that is, without substantial binding of the solute in the AFS. In addition,

Table 2.3
Cation-Exchange Capacity of Root Dry Matter of Different Plant Species[a]

Plant species	Cation-exchange capacity (meq/100 g dry wt)
Wheat	23
Maize	29
Bean	54
Tomato	62

[a]Based on Keller and Deuel (1957).

Table 2.4
Uptake and Translocation of Zinc by Barley Plants[a]

Zinc supplied as[b]	Rate of uptake and translocation (μg Zn/g dry wt per 24 hr)	
	Roots	Shoots
$ZnSO_4$	4598	305
ZnEDTA	45	35

[a]Based on Barber and Lee (1974).
[b]Concentration of zinc in nutrient solution: 1 mg/liter.

restricted permeation of the chelated zinc within the pores of the AFS may be a contributing factor.

With heavy-metal cations in particular, the binding in the AFS can be quite specific. Copper, for example, may be bound in a nonionic form (coordinative binding) to nitrogen-containing groups of either glycoproteins or proteins of ectoenzymes, such as phosphatases or peroxidases, in the cell wall (Harrison *et al.*, 1979; van Cutsem and Gillet, 1982). This cation binding in the apoplast can contribute significantly to the total cation content of roots, as shown by studies of the uptake of polyvalent cations such as copper, zinc, and iron. This is also demonstrated by the data in Table 2.4. As long as the exchangeable fraction of Zn^{2+} in the apoplast of roots has not been removed (removal can be quite difficult), such data cannot be interpreted as evidence that, compared with the chelated form, the translocation of Zn^{2+} to the shoots is inhibited by immobilization within the cytoplasm or vacuoles of the root tissue.

2.2.2 Passage into the Cytoplasm and the Vacuole

Despite some selectivity for cation binding in the cell wall (Section 2.2.1), the main sites of selectivity in the uptake of cations and anions as well as solutes in general are located in the *plasma membrane* of individual cells. The plasma membrane is an effective barrier against the diffusion of solutes either into the cytoplasm (influx) or from the cytoplasm into the cell walls and the external solution (efflux). The plasma membrane is also the principal site of active transport in either direction. The other main barrier to diffusion is the *tonoplast* (vacuolar membrane). In most fully differentiated cells the vacuole comprises more than 90% of the cell volume (Leigh *et al.*, 1981; also see Fig. 2.3) and is the main compartment for ion accumulation.

It can be readily demonstrated that the plasma membrane and the

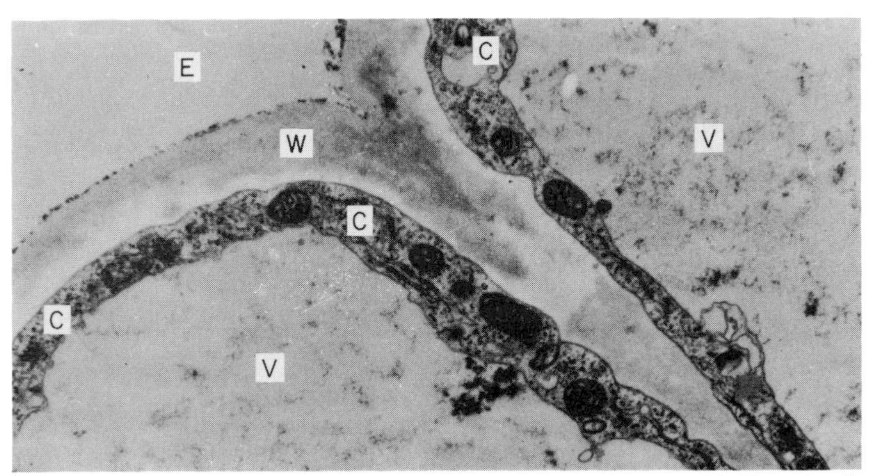

Fig. 2.3 Cross section of two rhizodermal cells of a maize root. V, Vacuole; C, cytoplasm; W, cell wall; E, external solution. (Courtesy of C. Hecht-Buchholz.)

tonoplast function as effective barriers to diffusion and exchange of ions by short-term experiments on cation uptake, for example, of K^+ and Ca^{2+} (Fig. 2.4). Most of the Ca^{2+} (^{45}Ca) taken up within 30 min (influx) is still readily exchangeable (efflux) and is almost certainly located in the AFS. In contrast, only a minor fraction of the K^+ (^{42}K) is readily exchangeable within this 30-min period, most of the K^+ having already been transported across the membranes into the cytoplasm and vacuoles ("inner space").

Although the plasma membrane and the tonoplast are the main biomembranes directly involved in solute uptake and transport in roots, it must be kept in mind that compartmentation by biomembranes is a general prere-

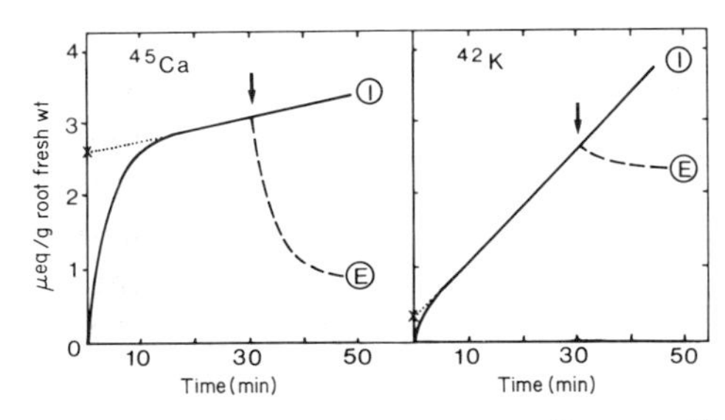

Fig. 2.4 Time course of influx (I) and efflux (E) of ^{45}Ca and ^{42}K in isolated barley roots. After 30 min (arrow) some of the roots were transferred to solutions of nonlabeled Ca^{2+} and K^+. The proportion of the exchangeable fraction in the apparent free space is calculated by extrapolation to zero time (×).

quisite for living systems. Solute transport into organelles such as mitochondria and chloroplasts therefore must also be regulated by membranes which separate these organelles from the surrounding cytoplasm. An example of solute transport across the outer chloroplast membrane is given in Section 8.4 for phosphorus and sugars.

The capability of biomembranes for solute transport and its regulation is closely related to their chemical composition and molecular structure. Before the mechanisms of solute transport across membranes are discussed in more detail (Sections 2.4 and 2.5), it is therefore appropriate to consider some fundamental aspects of the composition and structure of biomembranes.

2.3 Structure and Composition of Membranes

The capacity of plant cell membranes to regulate solute uptake has fascinated botanists since the nineteenth century. At that time the technical possibilities for experimentation were rather restricted. Nevertheless, in the early twentieth century, investigators, using various solutes, dyes, and sugars and by studying plasmolysis and deplasmolysis of cells with the light microscope, were able to elucidate some basic facts of solute permeation across the plasma membrane and tonoplast. From comparisons of membrane permeation of nonelectrolytes of various diameters, the *Ultrafiltertheorie* has been developed. According to this theory, cell membranes have pores of certain sizes, and the rate of permeation of molecules decreases as the pore diameter increases. The validity of this theory was confirmed, at least in principle, by Zimmermann and Steudle (1970), as shown in Table 2.5.

It is evident from these data that, in addition to the cell walls (Section 2.2), the cell membranes are effective barriers to solutes of higher molecular weight. Most synthetic chelators such as EDTA (see also Table 2.4) are also thus restricted in their permeation through membranes. It is possible, therefore, to use high-molecular-weight organic solutes such as polyethyleneglycol at high external concentrations as effective osmotica in order to induce water stress in plants.

Even some of the earlier studies demonstrated that some solutes penetrate membranes much faster than would be predicted on the basis of their molecular size. These molecules are characterized by high solubility in organic solvents; that is, they have lipophilic properties. It was therefore concluded that membranes have lipophilic pores which allow rapid permeation of even the larger lipophilic molecules. This view has now been discredited by experimental evidence showing that simple solution and

Table 2.5
Reflection Coefficient δ of Some Nonelectrolytes at the Cell Membranes of *Valonia utricularis*[a]

Compound	δ[b]	Molecule radius (nm)
Raffinose	1·00	0·61
Sucrose	1·00	0·53
Glucose	0·95	0·44
Glycerol	0·81	0·27
Urea	0·76	0·20

[a]Based on Zimmermann and Steudle (1970).
[b]1·00 indicates that the membranes are impermeable to the solute; 0 indicates that the membranes are freely permeable to the solute.

diffusion of these solutes through the lipid core of the membranes are responsible for the faster permeation.

In 1935 Danielli and Davson put forward a *unit-membrane model* which presented biological membranes as a unit consisting of two lipid molecular layers in which the hydrophobic tails of the fatty acids are oriented inward. The outer surfaces of both lipid layers are hydrophilic (consist, e.g., of phospholipids) and are covered by a protein layer. Such a unit membrane with continuous lipid layers would certainly be an effective barrier to polar solutes, including inorganic ions. It would not, however, explain the active transport of solutes, particularly polar ones, through membranes. This membrane model was later replaced, therefore, by hydrophilic models in which protein-coated pores were included in the lipid layer.

On the basis of electron microscopic studies in combination with freeze-etching of tissues and cells, Branton (1969), and later Singer (1972), presented a *fluid-membrane model* consisting of a protein matrix and two phospholipid layers with the hydrophilic, charged head regions (amino, phosphate, and carboxyl) oriented toward the membrane surfaces. Protein molecules can be attached (extrinsic proteins), for example, by electrostatic binding to the surfaces as membrane-bound enzymes. Other proteins may be integrated into the membranes (intrinsic proteins) or even traverse the membranes to form "protein channels" (transport proteins). These channels can be considered the hydrophilic pores through which polar solutes such as ions are transported. This model is shown in Fig. 2.5.

Membranes are typically composed of two main classes of compounds: proteins and lipids. Carbohydrates comprise only a minor fraction of membranes. The relative abundance of proteins and lipids can be quite variable depending on whether the membrane is a plasma, mitochondrial, or chloroplast membrane (Clarkson, 1977). Three polar lipids represent the

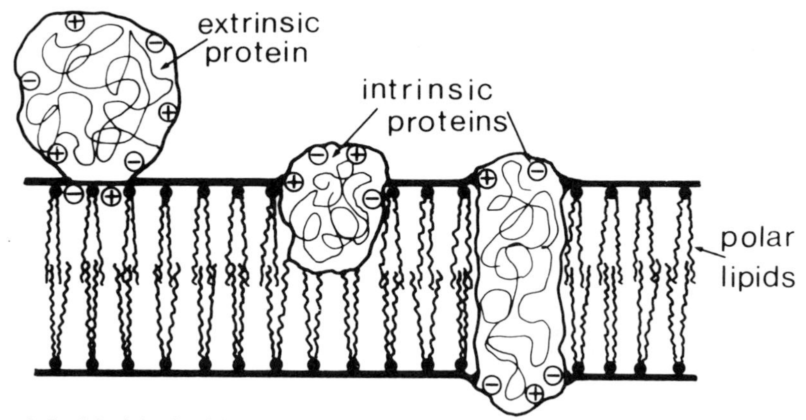

Fig. 2.5 Model of a biomembrane with polar lipids and with either extrinsic or intrinsic, integrated proteins. The latter can cross the membrane to form "protein channels".

major lipid components of membranes: phospholipids, glucolipids, and, less abundant, sulfolipids; examples of these are the following:

R^1, R^2 — CH_2—O—P(=O)(O^-)—O—CH_2—CH_2—$N^+(CH_3)_3$; CH; CH_2

Phosphatidylcholine (lecithin)

R^1, R^2 — CH_2—O— (ring: O, CH_2OH, H, OH, H, OH, H, OH, H); CH; CH_2

Monogalactosyl diglyceride

(Long chain polyunsaturated fatty acids)

R^1, R^2 — CH_2—O— (ring: O, CH_2—S(=O)(=O)—O^-, H, H, H, OH, OH, OH, H); CH; CH_2

Sulfoquinovosyl diglyceride

In polar lipids, long-chain fatty acids represent the hydrophobic tail oriented toward the inside of the membrane (Fig. 2.5). Variation in both the length and degree of unsaturation (i.e., number of double bonds) of the

Table 2.6
Fatty Acid Composition[a] of Roots of Various Plant Species Grown at 25°C[b]

Fatty acid	Chain length	Melting point (°C)	Plant species		
			Bean	Barley	Sugar beet
Palmitic acid	C_{16}	+62·8	25	22	18
Stearic acid	C_{18}	+70·1	4	3	1
Oleic acid	$C_{18:1}$[c]	+13·0	3	3	4
Linoleic acid	$C_{18:2}$[c]	− 5·5	27	38	47
Linolenic acid	$C_{18:3}$[c]	−11·1	31	17	8

[a]Percentage of total fatty acids.
[b]Based on Stuiver *et al.* (1978).
[c]Numeral to the right of the colon indicates the number of double bonds.

chain influences the melting point, as shown in Table 2.6. Generally, highly unsaturated fatty acids predominate, particularly in plants that grow in cold climates, indicating that these fatty acids have an important role in the maintenance of high membrane fluidity even at low temperatures.

Another important group of membrane lipids consists of sterols, especially cholesterol in animals and β-sistosterol in higher plants. Sterols may

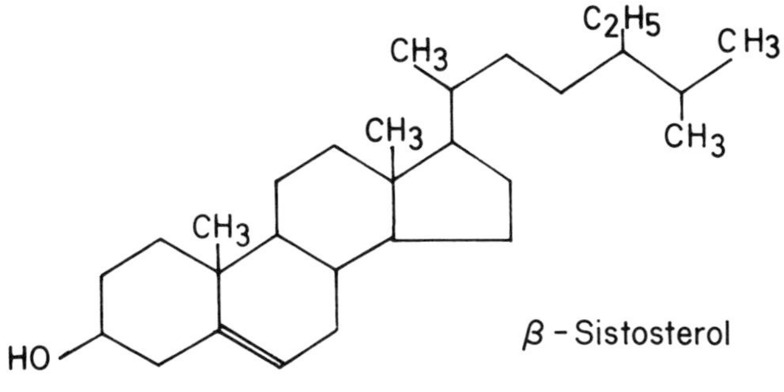

β-Sistosterol

have primarily a structural role in membranes. In the plasma membrane the proportion of sterols can be as high as that of the phospholipids, namely, 15–40% of the total lipids (Clarkson, 1977).

The lipid composition of membranes not only differs characteristically among plant species (Table 2.6); it is also strongly affected by environmental factors. In leaves, for example, distinct annual variations in the levels of sterols occur (Westerman and Roddick, 1981). In many instances changes in lipid composition reflect the adaptation of a plant to its environment through the adjustment of membrane properties. Lowering of the temperature from

25 to 10°C during the growth of wheat changes the fatty acid composition of the roots such that highly unsaturated fatty acids predominate; the proportion of linoleic acid (18 : 2) decreases from 50 to 33%, whereas that of linolenic acid (18 : 3) increases from 21 to 39% of the total fatty acids (Ashworth *et al.*, 1981). Such a change shifts the freezing point (the transition temperature) of membranes to a lower temperature and may thus be of importance for the maintenance of membrane functions even at lower temperatures. During acclimatization to low temperatures an increase in the amount of phospholipids (Yoshida, 1984) and of certain sterols in the membranes can also be observed (Uemura and Yoshida, 1984). As shown by Singh and Paleg (1984) low temperatures increase the proportion of phosphatidylcholine in the membrane. An increase in the responsiveness of membranes to gibberellic acid at low temperatures may be related to these changes, since phospholipids probably act as receptors for phytohormones such as gibberellic acid (Singh and Paleg, 1984).

There also exist distinct correlations between ion selectivity and lipid composition of membranes. This seems to be the case with sterols (Douglas and Walker, 1983) and has been clearly demonstrated for chloride transport and galactolipids in membranes (Section 16.6). Interestingly, the plant species bean, barley, and sugar beet differ not only in the fatty acid composition of root membranes (Table 2.6) but also considerably in the uptake of sodium (Section 10.2).

Alterations in the lipid composition of root membranes are also typical responses to changes in the mineral nutrient supply or exposure to salinity (Kuiper, 1980). In soybean roots changes in calcium and nitrogen supply affect the ratio of saturated to unsaturated fatty acids as well as the uptake rate of certain herbicides (Rivera and Penner, 1978). An increase in membrane permeability can be observed in roots suffering from phosphorus deficiency (Ratnayake *et al.*, 1978) and zinc deficiency (Welch *et al.*, 1982). In the case of phosphorus deficiency, a shortage of phospholipids in the membranes has been assumed to be the responsible factor. In the case of zinc deficiency, autoxidation in the membranes of highly unsaturated fatty acids is presumably involved in membrane leakiness (Section 9.4). Of the mineral nutrients, calcium plays the most direct role in the maintenance of membrane integrity, a function which is discussed in Section 2.5.2.

The dynamic nature of membranes is clearly demonstrated also by the rapid incorporation of externally supplied membrane constituents (e.g., phospholipids) into the membrane structure. The incorporation of some externally supplied compounds, however, renders membranes more sensitive to injury. The incorporation of antibiotics such as nystatin induces the formation of pores ("holes") in the membranes and a corresponding rapid leakage of low-molecular-weight solutes such as potassium. Monocarbonic acids, such as acetic acid and butyric acid, also induce membrane injury. The undissociated species of these acids are readily taken up and lead to a sharp rise in membrane leakiness, as indicated by the leakage of potassium and nitrate from the root tissue (Lee, 1977). The capacity of monocarbonic acids to induce membrane leakiness increases with a lowering of the external pH ($R{\cdot}COO^- + H^+ \rightarrow R{\cdot}CCOH$) and with an increase in the chain length of the acids [C_2 (acetic acid) $\rightarrow$ C_8 (caprylic acid)] and hence with an increase in their lipophilicity.

These undissociated carbonic acids obviously induce membrane leakiness by changing the fatty acid composition of the membranes (Table 2.7). Even

Table 2.7
Effect of Acetic Acid on the Fatty Acid Composition of Barley Roots Treated with 10 m*M* Potassium Acetate[a]

Treatment	pH	Percentage of total fatty acids as: Linoleic acid	Linolenic acid	Ratio Linoleic acid/ Linolenic acid
Acetate	7	50·6	18·6	2·7
	5	46·3	10·5	4·4
Water	7	50·8	18·6	2·7
	5	49·8	18·1	2·7

[a]Duration of treatment: 6 hr. Based on Jackson and St. John (1980).

within a few hours, the proportion of polyunsaturated fatty acids, and particularly of linolenic acid, decreases markedly when the concentration of undissociated acetic acid in the external solution is raised (pH 5). The effects of monocarbonic acids on the membrane permeability of roots have considerable ecological importance, since these acids accumulate in waterlogged soils (Section 16.4).

2.4 Solute Transport across Membranes

2.4.1 Carrier-Mediated Transport

Intact membranes are effective barriers to the passage of ions and uncharged molecules. On the other hand, they are also the sites of selectivity and transport against the concentration gradient of solutes. In the experiment recorded in Table 2.2, for example, the potassium concentration in maize root sap (which is approximately equal to the potassium concentration of the vacuoles) rose to 80 times higher than that in the external solution. In contrast, the sodium concentration in the root sap remained lower than that in the external solution. It is generally agreed that such selectivity and accumulation require specific binding sites, *carriers*, which bind to ions such as potassium or uncharged molecules and transport them across the membrane, as shown in Fig. 2.6A. Although transport against a concentration gradient does not necessarily resemble active transport in a thermodynamic sense (see Section 2.4.2), energy is nonetheless required, either directly or indirectly. A direct coupling of carrier-mediated selective ion transport and the consumption of energy-rich phosphates in the form of ATP is shown in Fig. 2.6B. The ATP is regenerated from ADP + P_i via respiration (oxidative phosphorylation; see Section 8.4).

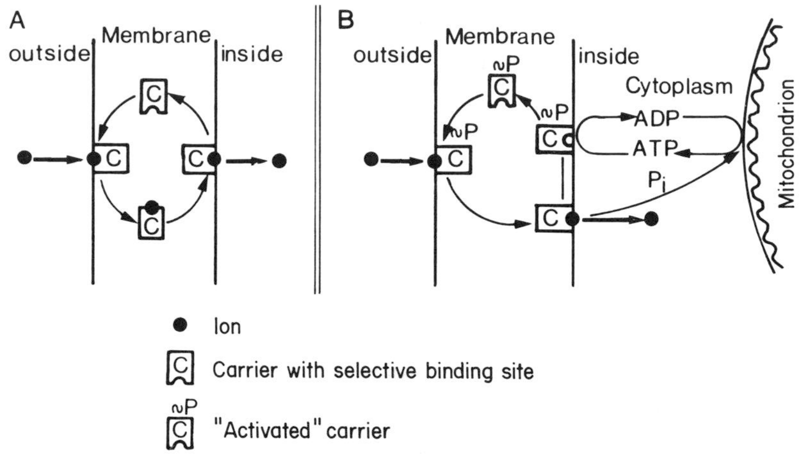

Fig. 2.6 Model of carrier-mediated membrane transport of ions (A) and coupling between energy expenditure (oxidative phosphorylation) and ion transport (B).

In Fig. 2.6B the energy-requiring step is the activation of the carrier. This is not necessarily so, however, since energy-requiring steps could also occur at the binding of the ion on the external surface, at the release of the ion from the internal surface, or during the transport of the ion through the membrane by the carrier. In any case such a model requires a membrane with highly dynamic properties. This is in accordance with the fluid-membrane model discussed earlier.

Strong support for the involvement of ATP in carrier-mediated ion transport was first presented by Fisher *et al.* (1970). Studying the uptake of K^+ by roots of various plant species, these workers demonstrated a close relationship between K^+ uptake and ATPase activity (Fig. 2.7). Of course, such a correlation does not answer the question as to which step of the model (Fig. 2.6B) involves ATP. It is, however, well established that Mg-ATPases (Section 8.5) of the plasma membrane are strongly stimulated by K^+ (Fisher *et al.*, 1970; Briskin and Poole, 1983) and that ions such as K^+, when added to the external solution, trigger their own transport across the plasma membrane.

Higher ion transport rates are not necessarily coupled with higher respiration rates. For example, when the supply of carbohydrate from the shoots exceeds the amount of carbohydrate required by the roots for growth and uptake processes, a relatively large proportion of these carbohydrates are oxidized in a nonphosphorylating mitochondrial electron transport chain (Chapter 5). Depending on the mineral nutrient supply, more than 50% of the total respiration in roots may take place via this "alternative pathway", which acts as an "energy overflow". If this shift in respiratory pathway is taken into account, a requirement of one molecule of ATP per ion transported across the plasma membrane is calculated (Lambers *et al.*, 1981b).

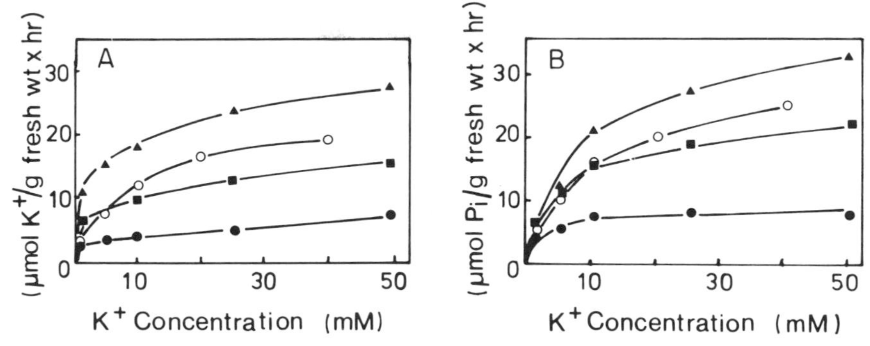

Fig. 2.7 (A) Potassium ion uptake (influx) and (B) ATPase activity (ATP → ADP $+P_i$) in isolated roots of different plant species. Key: ▲, barley; ○, oat; ■, wheat, ●, maize. (After Fisher *et al.*, 1970).

The whole concept of a carrier mechanism is based on conformational changes of specific macromolecules integrated into a membrane matrix. Only protein macromolecules appear to provide enough structural possibilities to explain the wide range of specific binding capacities for inorganic ions that have been identified. Such examples are a sulfate-binding protein (sulfate permease, isolated from microorganisms; Pardee, 1967) and the phosphate translocator in the outer chloroplast membrane (Section 8.4).

A somewhat different model has been developed by Mueller and Rudin (1967) for K^+ transport through membranes. This transport is facilitated, by several orders of magnitude, by cyclic antibiotics such as valinomycin, which are presumably incorporated in membranes in the form of ion-specific "channels". Specific proteins in combination with other membrane constituents such as phospholipids and sterols probably act as the carriers in intact membranes. This would explain the difficulties involved in even the apparent impossibility of isolating these carriers.

2.4.1.1 The Kinetics

The carrier-mediated process of ion transport through membranes as indicated by the model in Fig. 2.6 is subject to control by saturation kinetics, assuming that the number of carriers (binding sites) in the membrane is limited. Epstein and his group (Epstein and Hagen, 1952; Elzam and Epstein, 1965) regarded the kinetics of ion transport through membranes of plant cells as formally equivalent to the relationship between an enzyme and its substrate, using terms for enzymology. Comparing the carrier to an enzyme molecule and the ion to the substrate for the enzyme, the transport rate of an ion is dependent on the following two factors:

V_{max} A capacity factor denoting the maximal transport rate when all available carrier sites are loaded, that is, the maximal transport rate

K_m The Michaelis constant, equal to the substrate ion concentration giving half the maximal transport rate

At a given ion concentration in the substrate, C_S, the rate of transport v is given by

$$v = \frac{V_{max}C_S}{K_m + C_S}$$

When the external concentration of ions is low (mechanism I), the uptake isotherm is well described by this Michaelis–Menten equation. An example is given in Fig. 2.8 for K^+ uptake by barley roots. It is evident from this figure that the uptake isotherm of K^+ is the same whether the source of K^+ is KCl or K_2SO_4. As we shall see, however, when the substrate concentrations are higher, the accompanying anion has an effect on the uptake rate of the cation and vice versa. The K_m value reflects the affinity of the carrier sites for the ion, just as in enzymatic reactions it indicates the affinity of the enzyme for the substrate.

With very low external concentrations the net uptake of ions ceases before the ions are completely depleted. This is either the result of a limited affinity of the carrier sites for the ions or the result of an equilibrium being established between influx and efflux. This C_{min} concentration differs among ions. In maize, Barber (1979) found C_{min} values of 2 μM for potassium and 0·2 μM for phosphorus. In barley, the corresponding C_{min} values are about 1 μM and $<0{\cdot}1$ μM respectively (Drew *et al.*, 1984). The C_{min} concentration is an important factor in ion uptake from the soil, because it is the lowest concentration at which roots can extract an ion in a soil solution. It therefore determines diffusion gradients in the rhizosphere (Section 13.4.2.2).

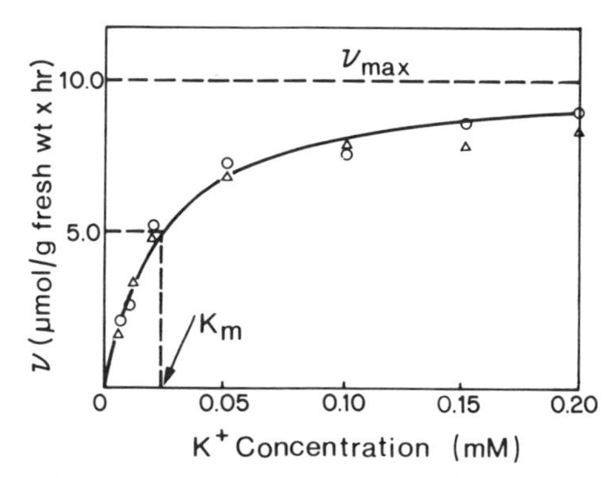

Fig. 2.8 Rate of K^+ uptake (v) as a function of the external concentration of KCl (○) or K_2SO_4 (△); K_m = 0·023 m*M*. (After Epstein, 1972.)

The original concept of a single carrier-mediated mechanism of ion transport (one carrier system for each ion) did not sufficiently describe the kinetics of uptake when wide concentration ranges were tested. At K^+ concentrations above 1 m*M*, for example, the kinetics differ considerably from those at lower concentrations (Epstein *et al.*, 1963). The selectivity of the binding sites is lower (Na^+ competes with K^+) and the accompanying anion has an effect on the uptake rate (Section 2.5). This led to the hypothesis of a *dual mechanism*, that is, the operation of two separate uptake systems for each ion. System I has a higher selectivity and is located at the plasma membrane. Whether system II is also located at the plasma membrane or at the tonoplast was matter of contention between Epstein and his group and Laties and his group. For further details see Epstein (1972).

According to Nissen *et al.* (1980) the uptake isotherm, when considered over a wide concentration range, does not provide evidence for two separate systems, but is "multiphasic." They argue that the uptake of each ion is mediated by only one carrier system and that the affinity of the system for an ion depends on the external concentration of the ion. This model, which has characteristics that are similar to the properties of allosteric enzymes, has been questioned, however, for several reasons. Borstlap (1983), for example, considers the multiphasic patterns an artificial interpretation of the uptake isotherms. In more recent years it has been recognized that a formal application of the carrier model is too mechanistic and that it overlooks important aspects of self-regulation in ion uptake. Depending on the nutritional status or age of a plant, K_m and V_{max} for any particular ion can vary considerably. This aspect is discussed in more detail in Section 2.5.4.

2.4.2 Active and Passive Transport: Electrogenic Pumps

In the preceding section on carrier-mediated transport it was more or less implied that solute transport across membranes is an active process. This, however, is not necessarily so. Solutes may be more concentrated on one side of the membrane (i.e., they may possess more free energy) and thus diffuse from a higher to a lower chemical potential (Fig. 2.9). This "downhill" transport across the membrane is, in thermodynamic terms, a passive transport. It takes place as diffusion either across the lipid phase, with the aid of carriers, or across aqueous pores. In cells, such downhill transport of ions across the plasma membrane may be maintained by a lowering of the ion activity in the cytoplasm, for example, due to adsorption at charged groups (e.g., $R \cdot COO^-$ or $R \cdot NH_3^+$) or to incorporation into organic structures (e.g., phosphate into nucleic acids). This is particularly true in meristematic tissues (e.g., root tips). In such cells cations and anions can accumulate without active transport. In this case the metabolic coupling of the transport is indirect.

In contrast, membrane transport against the gradient of potential energy ("uphill") must be linked to an energy-consuming mechanism, a "pump" in the membrane (Fig. 2.9). To determine whether an ion is actively transported across a membrane, however, both the activity or concentration of the ion on either side of the membrane (i.e., the chemical potential gradient) and the electrical potential gradient (i.e., differences in millivolts) across the membrane must be known. By means of small glass microelectrodes inserted into the vacuoles, a strongly negative electrical potential can be measured between the cell sap and the external solution (Fig. 2.10). The first

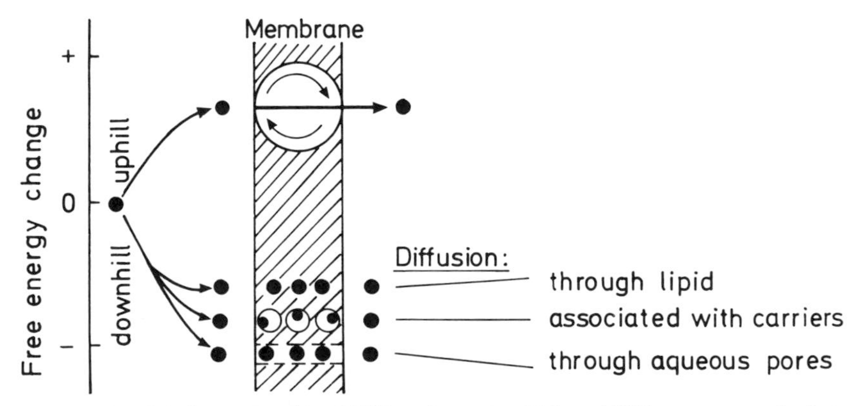

Fig. 2.9 Models for active ("uphill") and passive ("downhill") transport of solutes (•) across a membrane. (Modified from Clarkson, 1977.)

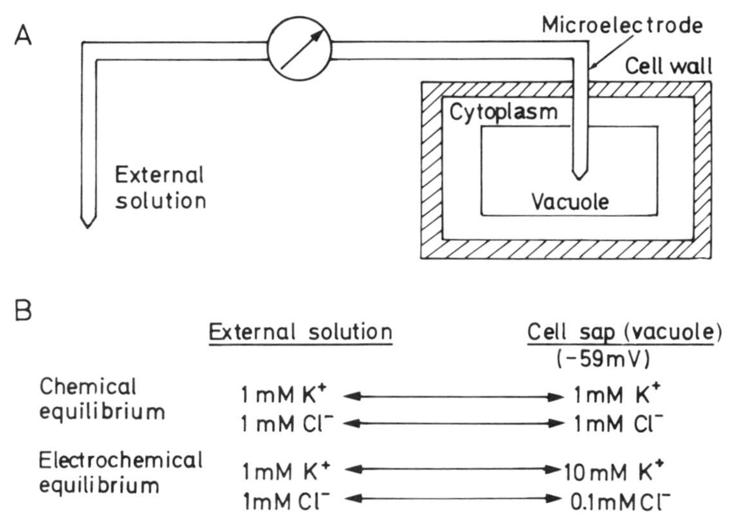

Fig. 2.10 A. Schematic presentation of the system for measuring electropotentials in plant cells. B. Example of the calculation of ion distribution at chemical and electrochemical equilibrium assuming an electropotential of −59 mV.

measurements of this kind were made in cells of giant algae such as *Chara*, where a strongly negative electrical potential of between −100 and −200 mV was found.

The same method was used by Higinbotham *et al.* (1967) and Glass and Dunlop (1979) to demonstrate the existence of similar electrical potential gradients in cells of higher plants. The concentration at which cations and anions on either side of a membrane are in electrochemical equilibrium or at which ions in the external solution are in the equilibrium with those in the vacuole can be calculated according to the Nernst equation:

$$E\ (\text{mV}) = -59 \log \frac{\text{concentration inside (vacuole)}}{\text{concentration outside (external solution)}}$$

According to this equation, at a negative electropotential of −59 mV, monovalent cations such as K^+ or anions such as Cl^- would be in electrochemical equilibrium if their concentration in the vacuole were 10 times higher (K^+) or 10 times lower (Cl^-) than in the external solution (Fig. 2.10). For divalent cations or anions the difference between chemical and electrochemical equilibrium differs by even more than the factor of 100. In cells of higher plants the electrical potential differences between vacuoles and the external solution are generally higher than −59 mV (Table 2.8). Thus, as a rule, in terms of electrophysiology, only anion uptake into the vacuoles would always require an active transport process. This is indicated in Table 2.8 in the differences between the anion concentrations in the vacuoles based on calculations according to the electrochemical equilibrium and the anion concentrations in the vacuoles actually found in the experiment.

Table 2.8
Experimentally Determined and Calculated Ion Concentration (m*M*) According to the Electrical Potential Differences in Roots of Pea and Oat[a]

	Pea roots (−110 mV)		Oat roots (−84 mV)	
Ion	Experimental	Calculated	Experimental	Calculated
Potassium	75	74	66	27
Sodium	8	74	3	27
Calcium	2	10,800	3	1,400
Chloride	7	0·014	3	0·038
Nitrate	28	0·027	56	0·076

[a]Composition of the external solution: 1 m*M* KCl, 1 m*M* $Ca(NO_3)_2$, and 1 m*M* NaH_2PO_4. Based on Higinbotham *et al.* (1967).

In this example, the only cation that would require active transport for its uptake by oat roots is K^+. At lower external K^+ concentrations (<0·5 m*M*, operation of system I of ion uptake), active K^+ transport is usually required (Cheeseman and Hanson, 1979). For Na^+ and Ca^{2+} in particular, the equilibrium concentration in the cell sap (i.e., the calculated ones) would be much higher than that experimentally found in the steady state (Table 2.8). A possible explanation for this discrepancy is that the plasma membrane strongly restricts permeation by these ions or that the ions are pumped (transported) back into the external solution. For Na^+ such an efflux pump at the plasma membrane of root cells seems very likely (Jeschke and Jambor, 1981). For Ca^{2+} an active extrusion (Ca^{2+}-efflux pump) at the plasma membrane of root cells has also been suggested (Marmé, 1983). Since the Ca^{2+} concentration in soil solutions is usually higher than 1 m*M*, a Ca^{2+}-extrusion pump to prevent Ca^{2+} transport along the electrochemical gradient (e.g., Table 2.8) would require considerable energy. It is likely, therefore, that physiochemical factors such as the size and charge of Ca^{2+} (Section 2.5.1) strongly restrict permeation along the electrochemical potential gradient across the plasma membrane.

The formation and maintenance of electropotentials across membranes require *electrogenic pumps*, which transport an ion in one direction only, without an accompanying ion of the opposite charge (e.g., H^+ without HCO_3^-) and without transport of an ion of the same charge in the opposite direction (e.g., no H^+/K^+ exchange). It is now generally accepted that the electrogenic pumps in higher plant cells are mainly proton (H^+) pumps that create a negative electropotential across cell membranes (Pitman *et al.*, 1975; Spanwich, 1981). These pumps are mainly membrane-bound ATPases (H^+-ATPases) which extrude H^+ from the cytoplasm. The "proton motive force" (Poole, 1978) generated by the extrusion of protons provides the driving force for the transport not only of cations and anions (Fig. 2.11) but also of amino acids and sugars (Chapter 5).

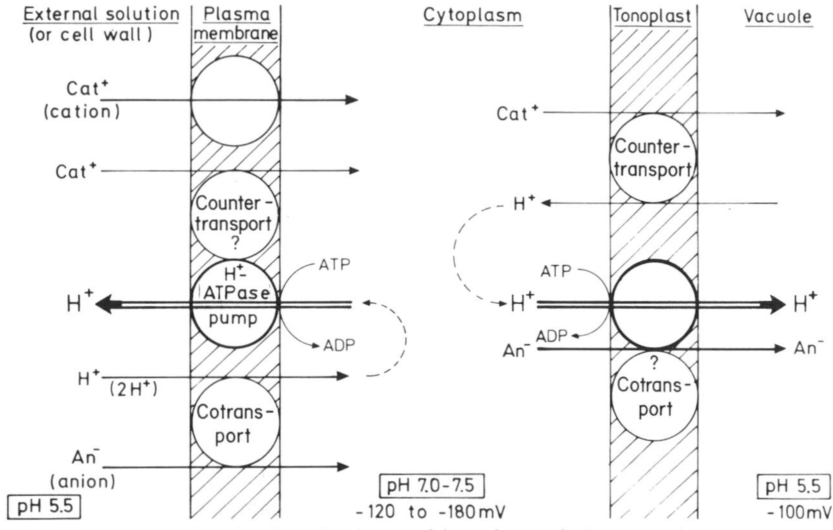

Fig. 2.11 Models for the functioning and locations of electrogenic proton pumps (H^+-ATPases) in plant cells. (*Left*) Plasma membrane H^+-ATPase: H^+ extrusion from the cytoplasm; stimulated by monovalent cations, insensitive to anions. (*Right*) Tonoplast H^+-ATPase: H^+ transport into the vacuole; sensitive to anions (stimulated by Cl^-, inhibited by NO_3^-); relatively insensitive to cations. (*Right:* Based on Sze, 1984.)

The existence of electrogenic proton pumps at the plasma membrane has been well established for several years. More recently, firm evidence has been presented that an H^+-ATPase is also located at the tonoplast and transports protons from the cytoplasm into the vacuole (Komor *et al.*, 1982; Bennett *et al.*, 1984). These two classes of electrogenic proton pump are important for the maintenance of the cytoplasmic pH in the range of 7·0 to 7·5 (Martin *et al.*, 1982) by means of proton extrusion (plasma membrane) and proton transport (tonoplast), as shown in Fig. 2.11.

According to this model, the transport of cations and anions across the plasma membrane into the cytoplasm can be driven by the H^+-ATPase in various ways. A direct stoichiometric 1:1 coupling of H^+ efflux and cation influx in a process known as *countertransport* (*antiport*) is less likely, since the ratio of H^+ efflux to K^+ influx can be considerably lower than 1 (Glass and Siddiqui, 1982). Most experimental evidence supports the view that the electrical potential gradient maintained by the H^+-efflux pump provides the driving force for the downhill transport of cations across the plasma membrane. Specific structures (carriers) in the plasma membrane regulate the transport rate and selectivity of downhill cation transport.

Although for anions the electrical potential gradient (Table 2.8) seems to imply the necessity of active transport in the thermodynamic sense (uphill), anion transport can also occur as *cotransport* (*symport*) with protons from the free space into the cytoplasm (Fig. 2.11). For protons, a steep electrochemical potential gradient exists across the plasma membrane toward the cytoplasm. This gradient is composed of a chemical component (pH difference) and an electrical component (negative potential). Evidence for proton–anion cotransport at the plasma membrane has been presented, for example, in barley roots for chloride (Jacoby and Rudich, 1980) and in *Lemna* for phosphate (Ulrich-Eberius *et al.*, 1981). Measurements of changes in pH and in electropotentials during uptake indicate that in this cotransport more than one proton, probably two, is transported with a monovalent phosphate anion (Ulrich and Novacky, 1981), and probably three protons with a divalent sulfate anion (Lass and Ulrich-Eberius, 1984).

However, there are other views on the mechanism of anion transport across the plasma membrane. According to Lin (1979), phosphate uptake by corn roots is mediated by OH^-/phosphate (P_i) countertransport in which the downhill transport of OH^- from the high electrochemical potential in the cytoplasm (high pH and strong negative charge) into the free space is coupled with a countertransport of anions into the cytoplasm.

Although our knowledge of transport at the tonoplast is much more limited than our understanding of transport at the plasma membrane, the existence of another ATP-driven proton pump (H^+-ATPase) at the tonoplast is now well established (Fig. 2.11). This proton pump is probably coupled to anion transport into the vacuole with a stoichiometry of two protons transported per molecule of ATP hydrolyzed (Bennett and Spanswick, 1984) and a proton/anion cotransport ratio of 1 : 1 (Struve *et al.*, 1985). Cation transport into the vacuoles is probably mediated by countertransport driven by the electrochemical gradient for protons from the vacuole across the tonoplast into the cytoplasm (Fig. 2.11).

The two classes of ATP-driven proton pumps differ not only in location (plasma membrane versus tonoplast) but also in their sensitivity to cations and anions. The plasma membrane H^+-ATPase is stimulated by monovalent cations in the order $K^+ > NH_4 > Na^+$ and is relatively insensitive to anions (O'Neill and Spanswick, 1984). An example of the stimulation by K^+ is given in Fig. 2.7. Interestingly, the plasma membrane H^+-ATPase is also stimulated by auxin (Cleland, 1982) and particularly fusicoccin, a fungal toxin (Ballio *et al.*, 1981). Both substances lead to acidification of the free space, an increase in the negative membrane potential (Cleland and Rayle, 1977), and a corresponding increase in the uptake of cations such as K^+ and Rb^+ (Stout *et al.*, 1978).

In contrast to the plasma membrane H^+-ATPase, the tonoplast H^+-ATPase is largely insensitive to monovalent cations (Bennett *et al.*, 1984) but is stimulated by most anions, particularly chloride (Table 2.9). Compared with chloride stimulation, the stimulation by sulfate is small. It is interesting that the tonoplast H^+-ATPase is strongly inhibited by nitrate (Table 2.9). This difference in anion sensitivity might have important consequences for the rate of radial transport of anions and cations in the symplast across the root and the subsequent release to the xylem (Section 2.7).

Table 2.9
Effect of Monovalent Salts on Tonoplast ATPase Activity[a]

Salt added (3 mM $MgSO_4$ + 50 mM monovalent salts)	ATPase activity	
	Specific activity (μmol P_i/mg[b] × hr)	Salt stimulation (%)
$MgSO_4$	19·9	—
KCl	36·2	82
NaCl	38·3	92
K_2SO_4	23·2	16
KNO_3	10·2	−49

[a]Based on Bennett *et al.* (1984).
[b]Membrane preparation.

According to our current knowledge, the two classes of proton pumps are present in all vacuolated cells. In leaf tissue the plasma membrane H^+-ATPase is light dependent (Gepstein, 1982) and may be responsible for the light-stimulated uptake of, for example, potassium into leaf tissue. In leaves of plants with Crassulacean Acid Metabolism (CAM), nocturnal malic acid accumulation (Chapter 5) is characterized by the stoichiometric active transport of two protons and the cotransport of a divalent malate anion ($^-OOC{\cdot}CH_2{\cdot}CHOH{\cdot}COO^-$) into the vacuole (Struve *et al.*, 1985). The degree of stimulation of tonoplast H^+-ATPase by malate is similar to its stimulation by chloride (Jochem *et al.*, 1984).

In conclusion, the main driving forces of membrane transport of solutes in general and of ions in particular are H^+-ATPases, which create an electrical potential gradient across membranes. Cation and anion transport is then a gradient-dependent or a coupled transport. The rate of cation and anion transport is determined not only by the electrical and chemical potential gradients, however, but also by the physiochemical properties of ions and their affinity to carriers in the membranes (see the following section).

2.5 Characteristics of Ion Uptake by Roots

2.5.1 Role of Physicochemical Properties of Ions and Root Metabolism

During their passage from the external solution into the roots, ions interact with the negatively charged groups in the AFS of the cell walls (Section 2.2). Although these interactions may influence subsequent uptake into the inner space (the cytoplasm and vacuoles), the characteristics of ion uptake by roots are determined primarily by the transport across membranes, the plasma membrane in particular. Despite the dynamic properties of membranes (Section 2.3), there are some physicochemical properties of ions and other solutes, especially ion diameter and valency, which generally determine their rates of membrane transport.

2.5.1.1 Ion Diameter

For ions with the same valency there is often a negative correlation between the uptake rate and the ion radius. An example of this for monovalent cations is given in Table 2.10. The inverse relationship between ion radius and uptake rate is observed only when Li^+, Na^+, and K^+ are compared. Despite its smaller diameter, Cs^+ is taken up at a much lower rate than K^+. Obviously, factors other than ion diameter are involved in the regulation of uptake; one of these is the affinity of membrane-bound carriers for ions of a given valency.

Table 2.10
Relationship between Ion Radius and Uptake of Alkali Cations[a]

Cation	Ion radius[b] (nm)	Uptake rate (μmol/g fresh wt × 3 hr)
Lithium	0·38	2
Sodium	0·36	15
Potassium	0·33	26
Cesium	0·31	12

[a]Cations were supplied at pH 6·0 as bromide salts, 5m*M*. Based on Jacobson *et al*. (1960).
[b]Data from Conway (1981).

2.5.1.2 Molecule versus Ion Uptake and the Role of Valency

Membrane constituents, particularly phospho- and sulfolipids and proteins, contain electrically charged groups, and ions interact with these groups. As a general rule, the strength of this interaction increases in the following order:

$$\text{Uncharged molecules} < Cat^{+}, An^{-} < Cat^{2+}, An^{2+} < Cat^{3+}, An^{3-}$$

Conversely, the uptake rate often decreases in this order. The increase in the diameter of a hydrated ion with valency is certainly an additional factor responsible for this order. A few examples will illustrate this general rule.

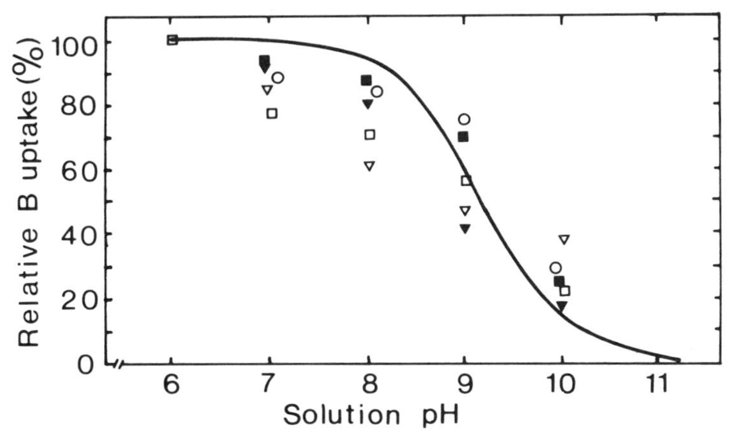

Fig. 2.12 Relative uptake of boron as a function of the external solution pH. Uptake at pH 6 = 100 at each supply concentration. Solid line: percentage of undissociated H_3BO_3. Key: ▽, 1·0 ppm B; □, 2·5 ppm B; ○, 5·0 ppm B; ▼, 7·5 ppm B; ■, 10·0 ppm B. Reproduced from Oertli and Grgurevic, 1975, by permission of the American Society of Agronomy.

As shown in Fig. 2.12 the uptake rate of boron falls dramatically when the external pH is increased. This pattern of behavior is closely related to the shift in the ratio of boric acid to borate anion. It is obvious that cell membranes are much more permeable to the uncharged solute species, the boric acid molecule, than to the borate anion. This is in complete agreement with the results obtained with other weak acids such as acetic acid (Table 2.7). The ratio of dissociated to undissociated species of acids—and thus the solution pH—also determines the membrane permeation of certain phytohormones such as abscisic acid (ABA); only the undissociated species readily permeates cell membranes, a factor which is closely related to the action of ABA in cells (Kaiser and Hartung, 1981; Cowan *et al.*, 1982).

In the case of weak acids, the high membrane permeation of the uncharged species is certainly not merely the result of less interaction with the charged groups in the membranes. In addition, greater permeation through the lipid phase of the membranes might be involved, as might downhill transport along the pH gradient across the plasma membrane (Fig. 2.11), which also reflects a corresponding chemical gradient toward the cytoplasm for the uncharged species [e.g., H_3BO_3 (outside)→ $H_2BO_3^-$ (cytoplasm)].

A special case of pH-dependent uptake of a molecular species is observed when ammonium nitrogen is supplied (Section 2.5.2). At high external pH the uptake increases sharply, probably owing to an increase in the proportion of the molecular species (NH_3 and NH_4OH).

The influence of valency on the uptake rate is particularly obvious when two ionic species of the same mineral element occur together in the external solution. In the pH range of 4 to 8, for example, phosphate is present as

$$H_2PO_4^- \underset{+H^+}{\overset{-H^+}{\rightleftharpoons}} HPO_4^{2-}$$

Consequently, at low pH $H_2PO_4^-$ dominates, whereas the reverse is true at high pH. As shown in Fig. 2.13, in the pH range of 8·5 to 5·5 there is a striking positive correlation between the proportion of $H_2PO_4^-$ in the external solution and the uptake rate of phosphate. In contrast, there is a much smaller effect on the uptake rate of sulfate, since in this pH range only the divalent anion SO_4^{2-} occurs.

It is, however, an oversimplification to explain the increase in phosphate uptake upon lowering the pH from 8·5 to 5·5 only by the shift in valency. Moreover, the sulfate uptake increases to some degree. As we shall see later (Section 2.5.2) pH-induced changes in membrane potential and in proton–anion cotransport also contribute to this enhanced anion uptake.

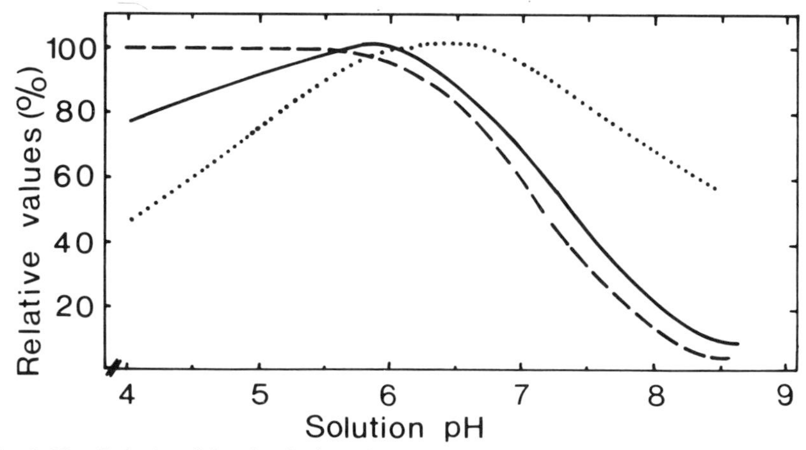

Fig. 2.13 Relationship of solution pH, proportion of $H_2PO_4^-$ (– – –), and uptake rate of phosphate (——) and sulfate (· · · ·) by bean plants; relative values. (After Hendrix, 1967.)

2.5.1.3 Metabolic Activity

In order for ions and other solutes to accumulate against a concentration gradient, an expenditure of energy is required, either directly or indirectly. The main source of energy in nonphotosynthesizing cells and tissues (including roots) is respiration. Thus, all factors which affect respiration may also influence ion accumulation. A few examples will demonstrate this connection.

Oxygen. As oxygen tension decreases, the uptake of ions such as potassium and phosphate falls, particularly at very low oxygen tensions (Table 2.11). Consequently, nutrient deficiency is one of the factors which may restrict plant growth in poorly aerated substrates (e.g., waterlogged soils; Chapter 16).

Table 2.11
Effect of Oxygen Partial Pressure around Roots on Uptake of Potassium and Phosphate by Barley Plants[a]

Oxygen partial pressure (%)	Uptake[b]	
	Potassium	Phosphate
20	100	100
5	75	56
0·5	37	30

[a]Based on Hopkins *et al.* (1950).
[b]Data represent relative values.

Temperature. Whereas physical processes such as exchange adsorption of cations in the AFS are only slightly affected by temperature (Q_{10}* = 1·1–1·2), chemical reactions are much more temperature dependent. An increase in temperature of 10°C usually enhances chemical reactions by a factor of 2 ($Q_{10} = 2$). For enzyme reactions, Q_{10} values considerably higher than 2 are quite often observed. Also, for the uptake of ions such as potassium, Q_{10} is often much higher than 2, at least within the physiological temperature range (Fig. 2.14). A comparison of the Q_{10} values for ion uptake and respiration reveals that ion uptake is more temperature dependent, especially below 10°C. This may possibly indicate that restricted ion uptake at low temperature is primarily the result of low membrane fluidity and a correspondingly higher "membrane resistance" (Section 2.3).

*Q_{10}, or quotient$_{10}$ refers to the change in the rate of a reaction or process (i.e., rate of membrane transport) imposed by a change in the temperature of 10°C.

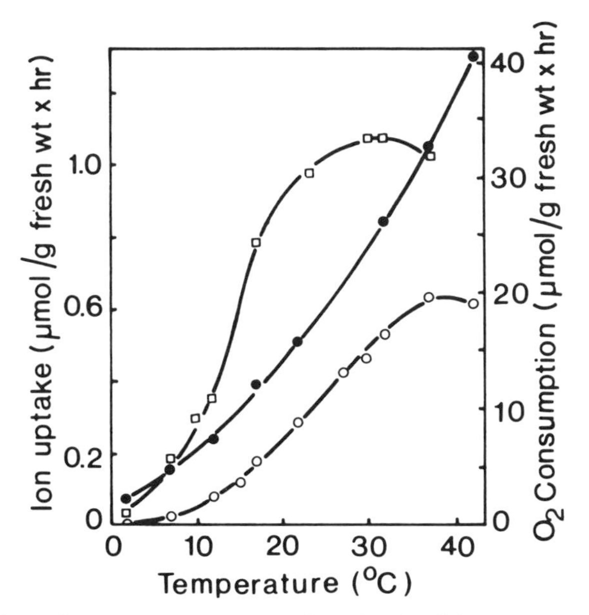

Fig. 2.14 Effect of temperature on rates of respiration (●) and uptake of phosphorus (○) and potassium (□) (supply of 0·25 m*M* potassium and 0·25 m*M* phosphorus) by maize root segments. (After Bravo and Uribe, 1981.)

The differences in temperature dependency of the uptake rates of potassium and phosphate are quite striking (Fig. 2.14). This might be interpreted as indicating a closer coupling between enzyme reactions and uptake of potassium as compared with uptake of phosphate. In the temperature range between 14 and 5°C the uptake rates of nitrate are also much more temperature sensitive than those of ammonium (Clarkson and Warner, 1979). These differences, however, are thought to be caused mainly by the differences between nitrate and ammonium metabolism within the cells and corresponding feedback control of the uptake rate (Chapter 8). One should use results of short-term studies of temperature effects on ion uptake with reservation when drawing conclusions concerning ion uptake in long-term experiments, where temperature effects on root growth, root morphology, and root/shoot ratio can become prominent factors in the regulation of ion uptake (Cumbus and Nye, 1982). Phosphorus uptake rates might then be depressed at low temperatures by inhibition of root elongation growth rather than by lower uptake rates per unit root length (Moorby and Nye, 1984).

Carbohydrates. The main energy substrates for respiration are carbohydrates. Therefore, in roots and other nonphotosynthesizing tissues, under conditions of limited carbohydrate supply from a "source" (e.g., leaves) a close correlation can often be found between the carbohydrate content and the uptake of ions (Table 2.12).

Table 2.12
Carbohydrate Content and Uptake Rate of Potassium by Isolated Barley Roots[a]

Carbohydrate content (mg glucose eq/g[b] fresh wt)	Potassium uptake rate (nmol/g fresh wt × hr)
3·0	49
1·9	36
1·5	18
1·3	9

[a]Recalculated from Mengel (1962).
[b]Amount of total carbohydrates (sugars, starch . . .) expressed in glucose units (equivalents).

Light. In photosynthesizing cells and tissue, close correlations exist among light, photosynthesis, and ion uptake. Green algae show distinct diurnal patterns of ion uptake (Kondo, 1982), and in submerged vascular plants such as *Valisneria* the action spectra for photosynthesis and for chloride uptake are nearly identical (van Lookeren-Campagne, 1957). A direct effect of light on ion uptake can also be demonstrated in green leaf tissue of terrestrial plants, where 2,4-DNP—a classical inhibitor of mitochondrial (oxidative) phosphorylation (i.e., ATP formation)—inhibits potassium uptake in the dark much more than in the light (Chapter 4). Obviously, in the light part of the energy required for ion uptake is supplied directly from photophosphorylation, that is, independently of respiration (Jeschke, 1972).

Measurements of ion uptake by roots of higher terrestrial plants show that the short-term effects of light are usually small. Under certain conditions, however, distinct diurnal patterns in ion uptake by roots can be observed—for example, for nitrate uptake in pepper (Pearson and Steer, 1977) and in ryegrass (Clement *et al.*, 1978b). The much higher nitrate uptake rates in the daytime are probably a reflection of fluctuations in nitrate reductase activity in the roots (Radin, 1978; Breteler and Nissen, 1982). Nitrate reduction requires reducing equivalents (NADH) generated by the breakdown of carbohydrates during respiration (Section 8.1), and the rate of nitrate reduction has an indirect effect on nitrate uptake (Section 2.5.5).

The close interrelationship of light availability, carbohydrate supply to the roots, root respiration, and ion uptake by roots is of particular ecological

Table 2.13
Effect of Shading the Basal Leaves of Rice Plants on Roots[a]

Treatment	Root dry weight (g/plant)	Root respiration ($\mu l\ O_2$/g dry wt)	Phosphorus uptake/g root dry wt (relative values)
Control	2·46	0·174	100
Shading of leaves	1·70	0·062	32

[a]Period of shading: 6 days.

importance for dense plant stands when the light supply to the basal leaves is limited, since the basal leaves are the main source of carbohydrates for the roots. Table 2.13 illustrates this interrelationship.

2.5.2. Interactions between Ions

In the preceding sections, for the sake of simplicity the transport of a particular ion was treated as a singular process, regulated only by the physicochemical properties of the ion and the metabolic activities of the cells. In reality, however, in the external solution (soil or nutrient solution) both cations and anions are present in different concentrations and states. Various interactions between ions during their uptake are therefore to be expected. Some of these general interactions are discussed in this section.

2.5.2.1 Competition

In general, ion transfer from the external solution to the cytoplasm requires binding of the ions at more or less specific sites (carriers) on the surface of the plasma membrane (Fig. 2.9). Thus competition between ions of the same electrical charge can be expected, assuming that the number of binding sites is small in relation to the number or concentration of competing ions. Such competition occurs particularly between ions with similar physicochemical properties (valency and ion diameter), as shown in Table 2.14 for alkali cations.

As one would expect, ^{42}K uptake is decreased by about half when an equivalent concentration of nonlabeled K^+ is added to the external solution. Since the radius of hydrated Rb^+ is similar to that of hydrated K^+, the depressing effect of Rb^+ on ^{42}K uptake is similar to that of K^+. Although Rb^+ cannot replace K^+ in its functions in plant metabolism, the binding sites (carriers) at the plasma membrane of root cells do not seem to distinguish between these two cations. Radioactive rubidium (^{86}Rb) is therefore often

Table 2.14
Effect of Equivalent Addition of Other Cations on Uptake of ^{42}K and ^{136}Cs by Isolated Barley Roots[a]

Treatment	^{42}K Uptake	^{136}Cs Uptake
Control	100[b]	100[c]
+ Sodium	94	84
+ Potassium	54	20
+ Rubidium	56	20
+ Cesium	97	54
+ Calcium	129	118

[a]Duration of experiment: 2 hr; concentration of added cations: 1mEq/liter. Relative values.
[b]Treatment with 1 meq/liter ^{42}K only.
[c]Treatment with 1 meq/liter ^{136}Cs only.

used as a tracer for K^+ uptake studies, although this can give misleading results under certain circumstances (Behl and Jeschke, 1982). From the data in Table 2.14 it is further evident that in the low concentration range the competing effects of Na^+ or Cs^+ on ^{42}K uptake are negligible and that Ca^{2+} even stimulates ^{42}K uptake, an effect which is discussed later. The uptake of ^{136}Cs is inhibited more than proportionally by K^+ and Rb^+, indicating that both cations have a higher affinity for the binding sites than does Cs^+.

Among monovalent cations the competition between K^+ and NH_4^+ is difficult to explain simply by competition for binding sites at the plasma membrane (Table 2.15). Whereas NH_4^+ is quite effective in competing with K^+, the converse (inhibition of NH_4^+ uptake by K^+) is not observed. This seems quite surprising, but Mengel *et al.* (1976) obtained similar results with rice. These authors assumed that at least a substantial proportion of ammonium nitrogen is not taken up in the form of NH_4^+, but that also NH_3

Table 2.15
Interaction between Uptake of NH_4^+ and K^+ in Maize Roots[a,b]

	Contents in roots (μmol/g fresh wt)			
	NH_4^- Nitrogen		Potassium	
$(NH_4)_2SO_4$ (m*M*)	$-K^+$	$+K^+$	$-K^+$	$+K^+$
0	6·9	6·7	8·2	53·7
0·15	7·3	7·1	6·7	48·4
0·50	17·1	13·5	8·9	41·1
5·00	29·4	31·5	9·3	27·1

[a]Based on Rufty *et al.* (1982a).
[b]Duration of the experiment: 8 hr; +K indicates addition of 0·15 m*M* K^+; calcium concentration constant at 0·15 m*M*.

permeates the plasma membrane after deprotonation, leaving H^+ in the external solution. Studies with lower plants indicate that deprotonation before uptake (i.e., permeation as NH_3) may become increasingly important at higher substrate concentrations of NH_4^+ (Bertl *et al.*, 1984). Inhibition of uptake of K^+ and other cations by NH_4^+ is then merely a reflection of competition for negative charges within the cells, that is, of cation–anion relationships (Section 2.5.3).

Table 2.16
Effect of K^+ and Ca^{2+} on the Uptake of Labeled Mg^{2+} (^{28}Mg) by Barley Seedlings[a]

	Mg^{2+} Uptake (μeq Mg^{2+}/10 g fresh wt × 8 hr)		
	$MgCl_2$	$MgCl_2 + CaSO_4$	$MgCl_2 + CaSO_4 + KCl$
Roots	165	115	15
Shoots	88	25	6·5

[a]Concentration of each cation: 0·25 meq-liter. Based on Schimansky (1981).

Of the mineral nutrients that are taken up as cations, the affinity of the highly hydrated Mg^{2+} for binding sites at the root plasma membrane seems to be particularly low. Other cations, Mn^{2+} (Heenan and Campbell, 1981) and Ca^{2+} in particular, therefore compete quite effectively with Mg^{2+}, and the uptake rate of Mg^{2+} is thus strongly depressed (Table 2.16). This strong competition is in agreement with observations of magnesium deficiency induced in crop plants by extensive application of potassium and calcium fertilizers.

Competition and limited selectivity of binding sites at the root plasma membranes are also observed for anions. Representative examples are competition between SO_4^{2-} and MoO_4^{2-} (Trobisch, 1966; Pasricha *et al.*, 1977) and between SO_4^{2-} and SeO_4^{2-} (Leggett and Epstein, 1956). When equimolar concentrations of SO_4^{2-} and SeO_4^{2-} are supplied to barley roots, uptake of SO_4^{2-} is preferred by a factor of only 1·4, whereas this factor increases to 3·0 when sulfur and selenium are incorporated into the protein fraction of the roots (Ferrari and Renosto, 1972). This is a further indication that selectivity during uptake is not as well expressed as selectivity in subsequent metabolic processes.

Another distinct type of anion competition occurs between Cl^- and NO_3^-. The net influx of NO_3^- is decreased by Cl^-, and the Cl^- already accumulated in the vacuoles seems to be particularly effective in this respect (Cram, 1973). In intensive crop production, Cl^- competition can be used to decrease the NO_3^- content of such plant species as spinach which tend to accumulate a large amount of NO_3^- and to use it mainly as an osmoticum (Section 8.2). In saline substrates the competing effect of Cl^- and NO_3^-

uptake has implications for crop production (Bernal *et al.*, 1974). In such substrates the anion competition can be used to decrease remarkably the uptake and contents of Cl^- in plants through an increase in substrate NO_3^- concentration (Table 2.17).

Table 2.17
Relationship between NO_3^- Supply and Cl^- Content in Soybean Leaves[a]

Supply (meq/liter)		Content in leaves
Cl^-	NO_3^-	(meq Cl/100 g dry wt)
10	1·25	90
10	2·50	51
10	5·00	34
10	7·00	19

[a]Based on Weigel *et al.* (1974).

The examples of strong competition between K^+ and Rb^+ and between SO_4^{2-} and SeO_4^{2-} demonstrate that the selectivity of the binding sites in root plasma membranes is not a reflection of the role of a given mineral element in plant metabolism, but merely a reflection of the physicochemical similarities between ions that are plant nutrients (e.g., K^+ and SO_4^{2-}) and ions that have no function in metabolism (Rb^+ and SeO_4^{2-}). Plants are unable to exclude the unneeded ions from uptake (*Exklusionsunvermögen*). This aspect has important practical implications in, for example, the channeling of certain heavy metals into the food chain via their uptake by plants (Marschner, 1983).

2.5.2.2 Role of pH

The competition between H^+ and other cations and between OH^- and other anions is of general importance for plant mineral nutrition. Because pH values below 7 are more common in soil solutions (at least in humid climates) than are higher values, competition between H^+ and cations has attracted more attention than the competition between OH^- and anions. A typical pattern of the influence of external pH on the uptake rate of a cation is shown in Fig. 2.15 for K^+. As the H^+ concentration increases (i.e., the pH falls), in the absence of Ca^{2+} the net uptake of K^+ sharply declines; below pH^4 (i.e., above $10^{-4} M H^+$), there is a net loss of K^+ from the roots. This effect of pH could be explained, at least in the pH range of 7 to 4, by competition between H^+ and K^+ for binding sites at the plasma membrane. It is more likely, however, that according to the model shown in Fig. 2.11, at high substrate H^+ concentrations the efficiency of the H^+-efflux pump at the plasma membrane decreases and the downhill transport of H^+ into the cytoplasm is enhanced. Evidence for the latter assumption is the observation that the electropotentials of root cells decreases from −150 mV at pH 6 to −100 mV at pH 4 (Dunlop and Bowling, 1978). In agreement with this, cation uptake in general is inhibited at low substrate pH.

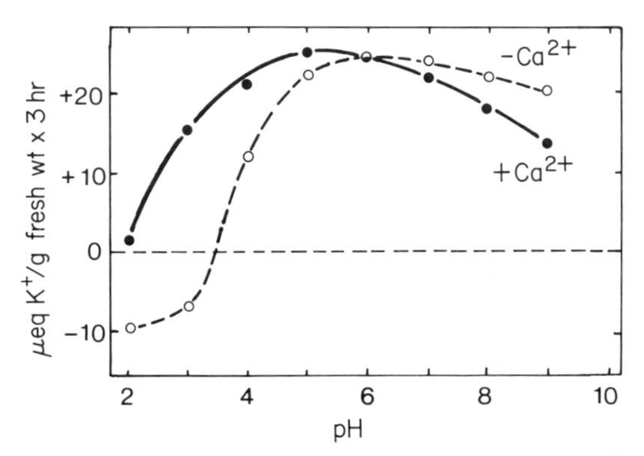

Fig. 2.15 Net uptake of K^+ from 5 m*M* KBr by barley roots as a function of the pH of the external solution and the Ca^{2+} supply (±5 m*M*). (Modified from Jacobson *et al.*, 1960.)

As shown in Fig. 2.15, in the absence of Ca^{2+} at pH < 4 a net efflux of K^+ was induced, indicating that the function of the plasma membrane as a barrier to ion diffusion had been impaired. The addition of Ca^{2+} was quite effective in reducing the H^+-induced depression both in net K^+ uptake and in net K^+ efflux.

The inhibition of cation uptake at low pH can also be demonstrated by long-term experiments (Fig. 2.16). In contrast to cation uptake, the uptake of anions is either not affected or slightly stimulated at low pH. These differences in pH effects on cation and anion uptake have been confirmed by other authors in, for example, rice (Zsoldos and Haunold, 1982) and soybean (Rufty *et al.*, 1982b). In the latter case, a decrease in the pH from 6·1 to 5·1 resulted in an increase in the ratio of anion to cation uptake from about 1·0 to 1·25. The distinct stimulation of anion uptake at low pH is in agreement with the model of proton–anion cotransport across the plasma membrane into the cytoplasm (Fig. 2.11) enhanced by the downhill transport of H^+ along its electrochemical potential gradient.

The effect of pH on nitrogen uptake depends on whether the nitrogen is supplied as NH_4^+ or NO_3^- (Michael *et al.*, 1965; Zsoldos and Haunold,

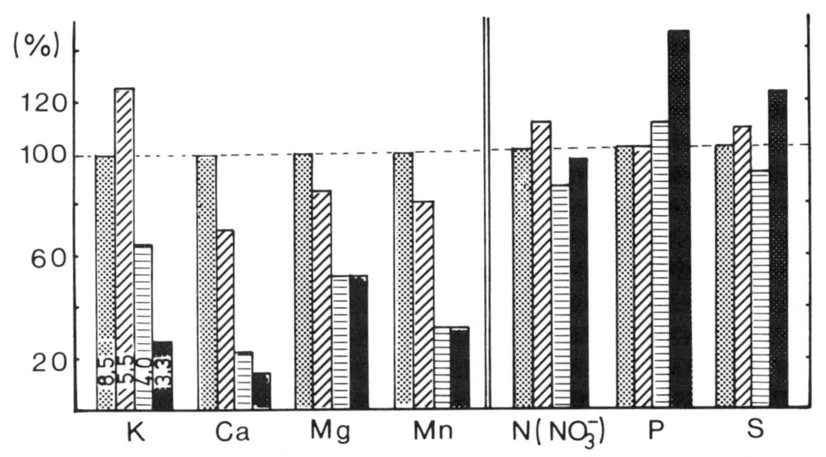

Fig. 2.16 Effect of pH of external solution on mineral element content (percentage of dry weight) of shoots of bean (*Phaseolus vulgaris*); pH 8·5, 5·5, 4·0, and 3·3, respectively, as indicated in the columns for potassium. Relative values, pH 8·5 = 100. (Data recalculated from Islam *et al.*, 1980.)

1982). As expected, lowering the pH from 7 to 4 decreases the uptake of the cation (NH_4^+) but increases the uptake of the anion (NO_3^-).

There is substantial evidence that, at low substrate pH and with corresponding inhibition of net extrusion of H^+, the pH of the cytoplasm is lowered (Smith and Walker, 1976). Analysis of the root sap indicates that this lowering of cellular pH is enhanced when at low substrate pH NH_4^+ is supplied instead of NO_3^- (Findenegg *et al.*, 1982). This effect of NH_4^+ on cellular pH is probably the result, at least in part, of the inhibitory effect of NH_4^+ on the uptake of metal cations such as K^+, Mg^{2+}, and Ca^{2+}.

2.5.2.3 Ion Synergism and the Role of Ca^{2+}

Synergism, like competition, is a feature of ion interaction during uptake. Stimulation of cation uptake by anions and vice versa is often observed and is mainly a reflection of the necessity of maintaining charge balance within the cells (Section 2.5.3). Synergism in uptake can also be the result of an increase in metabolic activity of the roots when mineral nutrients are supplied after a period of deprivation. In long-term experiments involving different growth rates, when "concentration" or "dilution" of mineral nutrients in the dry matter plays an important role, the interpretation of mutual effects of ions during uptake is rather difficult and should be undertaken with care.

The best example of ion synergism is Ca^{2+}-stimulated cation and anion uptake, first discovered by Viets (1944) and known as the *Viets effect*. Examples have already been given (Table 2.14). As shown in Fig. 2.15, the stimulation of K^+ uptake by Ca^{2+} increases with decreasing pH, indicating that Ca^{2+} counteracts the negative effect of high H^+ concentrations on K^+ uptake. Similarly, the stimulation of K^+ uptake by Ca^{2+} declines at higher pH and Ca^{2+} even inhibits the K^+ uptake, probably because of cation competition. The transition point between stimulation and inhibition does not occur at a fixed pH, but depends, for example, on the plant species (Volz and Jacobson, 1974). Furthermore, the nature of the interaction between Ca^{2+} and K^+ varies according to the nutritional status of the roots. In roots with high K^+ content ("high salt"), the stimulation of K^+ uptake by Ca^{2+} at low pH is due mainly to a decrease in K^+ efflux. In contrast, in roots with a low K^+ content ("low salt"), Ca^{2+} stimulates K^+ uptake by an increase in influx (Table 2.18). This Ca^{2+} stimulation is also apparent for Cl^- influx. Similar stimulation has also been shown for SO_4^{2-} (Leggett and Epstein, 1956) and other anion species, such as Cl^- (Table 2.18).

Table 2.18
Effect of Ca^{2+} on the Rates of K^+ and Cl^- Uptake in "Low-Salt" Barley Roots

External solution (m*M*)	Uptake rate (μeq/g dry wt × 2 hr)			
	K^+ Influx	K^+ Net uptake	Cl^- Influx	Cl^- Net uptake
0·1 KCl	116 ± 3	117 ± 6	35 ± 1	34 ± 4
0·1 KCl + 1·0 $CaSO_4$	137 ± 2	140 ± 7	53 ± 3	52 ± 4

The effect of Ca^{2+} on the flux of ions through membranes is related to its role in maintaining membrane integrity and stability. Calcium enables membranes to function as barriers against uncontrolled permeation processes. As a divalent cation, it presumably reacts with the negatively charged phosphate groups of the phospholipids in membranes and thus stabilizes the membranes (Section 8.6). Calcium has a stabilizing effect on membranes even under N_2 treatment and at low temperatures, indicating that it influences the physicochemical properties of membranes. Specific membrane effects of Ca^{2+} can also be demonstrated in isolated plasma membranes; Ca^{2+} treatment increases the membrane diameter from 103 Å (its value in distilled water) to 114 Å (Morré and Bracker, 1976).

Because Ca^{2+} can be removed from the binding sites of membranes, by chelators, for example (van Steveninck, 1965), or can be exchanged by high concentrations of H^+ and of metal cations (e.g., aluminum), the concentration of Ca^{2+} in the external solution necessary for the maintenance of membrane integrity and selectivity depends on the pH and the concentration of competing cations. In saline substrates, for example, Ca^{2+} can substantially improve the growth of plants, particularly those species that are sensitive to high Na^+ concentrations in the tissue (Section 16.6). This Ca^{2+} effect is mainly the result of a shift in the K^+/Na^+ uptake ratio in favor of K^+, as shown in Table 2.19.

Table 2.19
Effect of Ca^{2+} on the K^+/Na^+ Selectivity of Roots

External solution, NaCl + KCl (10 meq/liter each)	Uptake rate (μeq/g fresh wt × 4 hr)					
	Maize			Sugar beet		
	Na^+	K^+	Na^++K^+	Na^+	K^+	Na^++K^+
− Calcium	9·0	11·0	20·0	18·8	8·3	27·1
+ Calcium (1 mEq $CaCl_2$/liter)	5·9	15·0	20·9	15·4	10·7	26·1

It is generally accepted that an important mechanism for the selectivity of K^+/Na^+ uptake in plant roots is a K^+-stimulated Na^+-efflux pump located at the plasma membrane (Jeschke and Jambor, 1981). A K^+, Na^+-ATPase in the plasma membrane seems to be responsible for the coupling of both the K^+ influx and the Na^+ efflux (Lindberg, 1980). Such coupled fluxes rely on intact membranes. In the absence of Ca^{2+} the plasma membrane is more permeable ("leaky") and thus influx and efflux are not as well controlled. In the presence of Ca^{2+} the influx is shifted in favor of the (less hydrated) K^+ and, simultaneously, the Na^+ efflux is enhanced. This countertransport of K^+ and Na^+ explains the close relationship between the selectivity of K^+/Na^+ uptake and, for example, root respiration or carbohydrate supply.

Studying ion uptake in the absence of external Ca^{2+} is therefore a questionable approach and is justified only when the role of Ca^{2+} in the process is to be demonstrated.

2.5.3 Cation–Anion Relationship

Because cation uptake and anion uptake are regulated differently (Fig. 2.11), direct interactions between them do not necessarily occur. For instance, at low external concentrations the uptake rate of a cation is not affected by the accompanying anion and vice versa, as shown in Table 2.20 for K^+ and Cl^-. At high external concentrations, however, ions with lower uptake rates (SO_4^{2-} and Ca^{2+}) depress the uptake rate of K^+ and Cl^- considerably.

Table 2.20
Effect of the Accompanying Ion on the Rate of K^+ and Cl^- Uptake by Maize Plants[a]

Concentration (mEq/liter)	Uptake rate (μeq/g fresh wt × hr) K⁺ from KCl	K⁺ from K_2SO_4	Cl⁻ from KCl	Cl⁻ from $CaCl_2$
0·2	1·6	1·6	0·8	0·7
2·0	2·7	1·9	2·0	1·0
20·0	5·7	2·2	4·3	2·1

[a]Recalculated from Lüttge and Laties (1966).

Different uptake rates of cations and anions require compensation of electrical charges both within the cells and in the external solution. Obviously, at high external concentrations, charge compensation within the cells becomes a limiting factor for the uptake of K^+ and Cl^- when accompanied by SO_4^{2-} and Ca^{2+}, respectively. Under these conditions, nonspecific competition between ions of the same charge can also occur. For example, cations such as K^+ which rapidly permeate the plasma membrane depress the uptake rate of slower cations such as Ca^{2+} or Mg^{2+}, not by specific carrier competition at the plasma membrane, but by nonspecific competition for "native" anions in the cytoplasm or the vacuole.

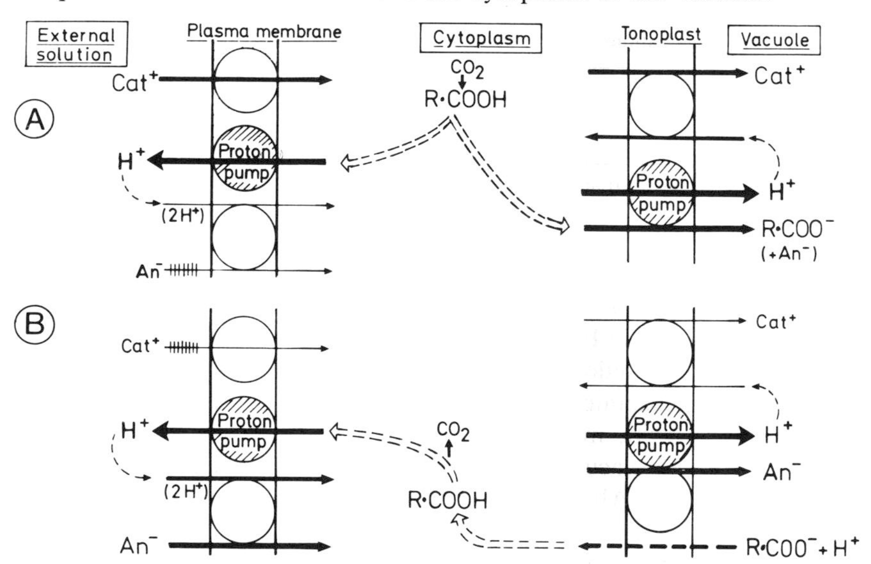

Fig. 2.17 Model for charge compensation and internal pH stabilization at different ratios of cation–anion uptake from the external solution. A. Excessive uptake of cations (Cat^+) when, for example, K_2SO_4 is supplied. B. Excessive uptake of anions (An^-) when, for example, $CaCl_2$ is supplied. For further details see Fig. 2.11.

According to the model shown in Fig. 2.17A, excessive cation uptake leads to an increase in the pH of the cytosol (i.e., the matrix of the cytoplasm). However, the cytosol pH has to be maintained within a narrow range ("pH stat," according to Davies, 1973). Thus the excess of cations is compensated for by enhanced synthesis of organic acids, with subsequent transport of the organic acid anions ($R{\cdot}COO^-$) and the cations into the vacuole. In contrast, excessive anion uptake (Fig. 2.17B) decreases the cytosol pH due, for example, to proton–anion cotransport across the plasma membrane. Maintenance of a high cytosol pH requires enhanced decarboxylation of organic acids supplied from the storage pool (i.e., the vacuoles).

The role of organic acids in maintaining cation–anion balance in plant tissues in response to mineral nutrient uptake was appreciated by Ulrich as early as 1941 and was later investigated in more detail by Jacobson (1955). The close relationship between cation–anion uptake and net changes in organic acid content is shown in Table 2.21 for barley roots. The net changes in organic acid content are related to different rates of $^{14}CO_2$ fixation ("dark fixation") in root tissue.

Table 2.21

Relationship between Cation–Anion Uptake and Organic Acid Content in Isolated Barley Roots[a]

External solution (mEq/liter)	Uptake (μeq/g fresh wt)		Change in organic acid content (μeq/g fresh wt)	$^{14}CO_2$ Fixation (relative)
	Cations	Anions		
K_2SO_4	17	1	+15·1	145
KCl	28	29	− 0·2	100
$CaCl_2$	1	15	− 9·7	60

[a]Based on Hiatt (1967a,b) and Hiatt and Hendricks (1967).

In the cytoplasm an equilibrium exists between CO_2 fixation (carboxylation) and decarboxylation. This equilibrium is regulated by the pH of the cytosol (Hiatt, 1967b). The main reactions involved in the process are shown schematically in Fig. 2.18. An increase in pH activates the enzyme PEP carboxylase [reaction (1)], and both the rate of CO_2 fixation and the synthesis of oxaloacetate are increased. After oxaloacetate is reduced to malate by the enzyme malate dehydrogenase, the malate can be directly translocated into the vacuoles [reaction (2)], where it acts as a counterion for cations (Fig. 2.17A). Alternatively, the malate can be incorporated into the cytoplasmic pool of the organic acids of the Krebs cycle, and another organic acid from this pool (e.g., citric acid) can be translocated into the vacuole, where it acts as a counterion for cations [reaction (3)]. If, however, anions are taken up in excess and thus anion–proton cotransport predominates (Fig. 2.17B), the pH of the cytoplasm decreases and the "malic enzyme"

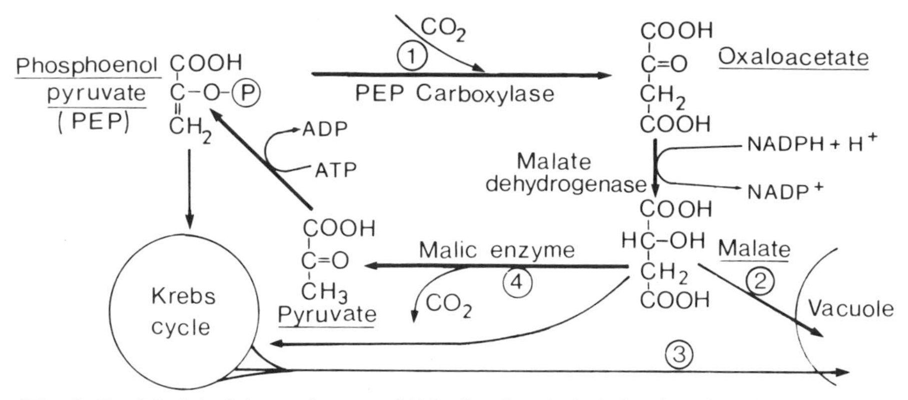

Fig. 2.18 Model of the pathways of CO_2 fixation ("dark fixation") and decarboxylation. Reactions (1)–(4) are explained in the text.

[reaction (4)] is activated, leading to decarboxylation of malate and the production of CO_2. As a result of these reactions, the cation–anion equilibrium and the pH of the cells are stabilized.

The cation/anion uptake ratio affects not only the content of organic acid anions in the cells but also the pH of the external solution (Fig. 2.17). Excess anion uptake increases the external pH, whereas excess cation uptake decreases the external pH. In the experiment recorded in Table 2.21, when K_2SO_4 was supplied, the net efflux of H^+ was 4·3 μeq per gram of root fresh weight per 2 hr, leading to a decrease in pH from 5·60 to 5·12 (Hiatt and Hendricks, 1967).

About 70% of the cations and anions taken up by plants are represented by either nitrate or ammonium (Jungk, 1970). The form in which nitrogen is supplied thus has a distinct effect on both the organic acid content of plants and the external pH (Kirkby and Mengel, 1967, 1970). The effect of the form of nitrogen on external pH is shown in Fig. 2.19.

The increase in external pH upon nitrate uptake is the result not only of preferential anion uptake but also of nitrate reduction in plants, which produce OH^- equivalents according to the following equation:

$$NO_3^- + 8e^- + 8H^+ \rightarrow NH_3 + 2H_2O + OH^-$$

If this process takes place preferentially in the roots, these OH^- equivalents can either neutralize H^+ within the cytoplasm or be released into the external solution. In either case, the external pH increases. Nitrate reduction in the shoots, on the other hand, is often, but not necessarily (Chapter 8), correlated with a stoichiometric increase in organic acid anions in the shoots (Kirkby and Knight, 1977). The form of nitrogen supply therefore

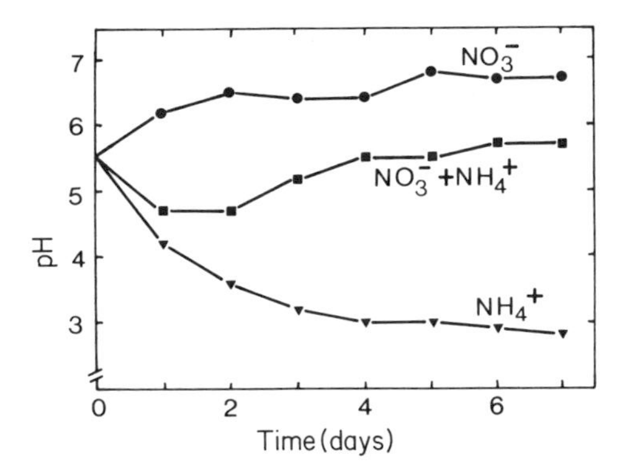

Fig. 2.19 Time course of external solution pH when sorghum plants were supplied with only NO_3^-, only NH_4^+, or both at a ratio of 8 NO_3^- to 1 NH_4^+. Total nitrogen concentration, 300 mg/liter. (Redrawn from Clark, 1982b, by courtesy of Marcel Dekker, Inc.).

Table 2.22
Influence of the Form of Nitrogen Supply on the Cation–Anion Balance in the Leaves of White Mustard Plants[a]

Form of N supply	Cations					Anions					
	Ca^{2+}	Mg^{2+}	K^+	Na^+	Total	NO_3^-	$H_2PO_4^-$	SO_4^{2-}	Cl^-	Organic acids	Total
NO_3^-	107	28	81	5	221	1	26	25	25	162	239
NH_4^+	72	22	40	7	141	1	25	25	31	54	136

[a]Data are expressed as milliequivalents per 100 g dry wt. Based on Kirkby (1968).

considerably affects both the mineral composition and the organic acid content of plants (Table 2.22).

The proportion of nitrate that is reduced in the roots and shoots varies considerably among plant species (Pate, 1973) and even among cultivars of a species such as soybean (Hunter *et al.*, 1982). The corresponding differences in the pH of the external solution, the consequences for ion uptake, and, particularly, the consequences for the availability of mineral nutrients at the soil–root interface (rhizosphere) are discussed in Chapter 15. In the special case of the cation–anion relationship in legumes symbiotically supplied with nitrogen, the H^+ release by roots in exchange for cations is of the same order of magnitude as for NH_4^+ nutrition. Legumes which rely on N_2 fixation are therefore characterized by both low rhizosphere pH and by some particularities in nutrient mobilization (Chapter 15).

2.5.4. External Concentration

As discussed in Section 2.4.1, in the low concentration range the uptake rate of ions such as K^+ is governed by saturation kinetics (Fig. 2.8). With an increase in the external concentration, however, the uptake rate further increases, though by a much lower rate. In this high concentration range (e.g., >1 m*M* K^+) the uptake is less selective (e.g., competition between K^+ and $Na^+ + Ca^{2+}$) and is less sensitive to metabolic inhibitors such as 2,4-DNP (Barber, 1972) but the effects of the accompanying ion (e.g., Table 2.20) and of the transpiration rate of the plants are much greater (Chapter 3). For further details on the interpretation as to whether the kinetics is a reflection of a dual-mechanism process (one highly specific—mechanism I—and another less specific—mechanism II) or resembles a multiphasic system, the reader is referred to Epstein (1972).

The ecological relevance of many uptake studies with roots, however, can be questioned if, for example, concentrations of K^+ and phosphate greater than 10 and 1 m*M*, respectively, are used; the same holds true for micronutrient concentrations greater than 1 m*M*.

If we exclude extremely high external concentrations of a given mineral element and consider the concentration range commonly found in soil solutions, the concentration-dependent uptake differs in a typical manner among mineral elements. An example of these differences is shown in Fig. 2.20 for K^+ and Na^+. In terms of the uptake kinetics (Fig. 2.8), they reflect differences in the affinity of the binding sites on the plasma membrane of root cells for K^+ and Na^+, namely, a high affinity for K^+ (low K_m value) and a low affinity for Na^+ (high K_m value). The uptake isotherm of phosphate is

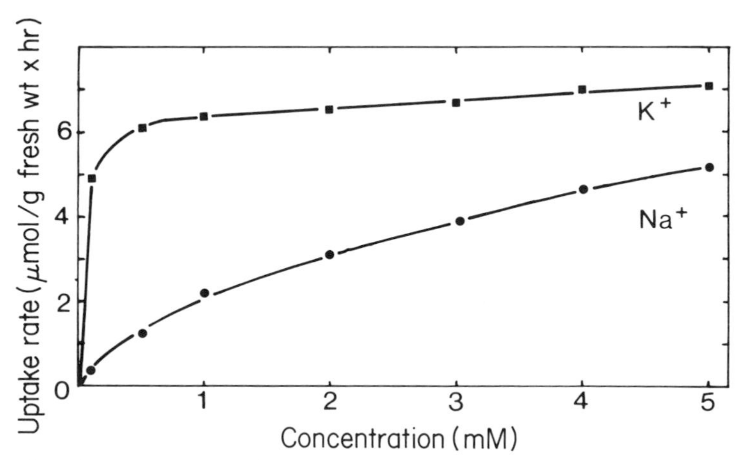

Fig. 2.20 Uptake isotherm of K^+ and Na^+ in isolated barley roots with an increasing supply of KCl + NaCl in the presence of 0·5 m*M* Ca^{2+}

similar to that shown for K^+, whereas the uptake isotherms for Ca^{2+} and Mg^{2+} are similar to that for Na^+.

What is the ecological relevance of these differences in uptake isotherms? Compared with the requirement for optimal growth, concentrations in soil solutions are usually low for K^+ (<1 m*M*) and phosphate (<0·1 m*M*); on the other hand, concentrations of Ca^{2+} and Mg^{2+} are often considerably higher (Chapter 13). To satisfy the different requirements for these nutrients, plants have binding sites on the plasma membrane of root cells which differ in affinity (K_m values) for the various mineral nutrients. In agreement with this, when the supply of K^+ (e.g., in nutrient solutions) is constant, optimal growth can be obtained at K^+ concentrations below 1·0 m*M*, that is, in the concentration range of the high affinity mechanism (mechanism I) (Johansen *et al.*, 1968). With a few exceptions (e.g., in saline substrates) the external concentrations of the other macronutrients in the range of mechanism I are also sufficient for optimal growth of most plant species (Asher and Edwards, 1983). As we shall see in Chapter 12, in long-term experiments the uptake at higher external concentrations reflects what is known as *luxury uptake*. Under field conditions during ontogeny, however, luxury uptake in preceding periods can be important because it provides an internal reserve pool in periods of high demand or interrupted root supply.

The relationship between the external concentration and the uptake rate of a particular mineral nutrient also depends on the duration of an experiment, as is clearly demonstrated in Fig. 2.21. Whereas in the long-term

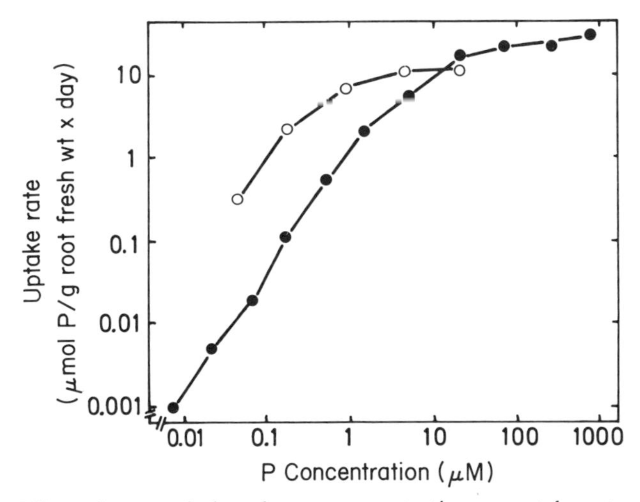

Fig. 2.21 Effect of external phosphorus concentration on uptake rate of phosphorus by eight plant species during 4 weeks of growth (○——○) and of barley in 24 hr (●——●). (After Loneragan and Asher, 1967.) (Copyright 1967 The Williams & Wilkins Co., Baltimore).

experiment the phosphorus uptake rate levels off at an external concentration of $\sim 5\,\mu M$, in the short-term experiment the uptake rate continues to increase even at considerably higher external concentrations. The main reason for this discrepancy is the difference in the phosphorus nutritional status of the plants. In short-term studies with isolated roots or intact young plants, preculture usually takes place either in very dilute nutrient solutions or in $CaSO_4$ solution (Epstein, 1972). In these "low-salt" plants or roots the internal concentration of nutrients is quite low. After nutrients are supplied, the uptake rates are very high; they continue to increase, even in the high concentration range, at least for several hours or a few days. Thereafter, as the internal concentration rises, the uptake rate declines.

2.5.5 Internal Concentration and Nutritional Status

The relationship between the external concentration and the uptake rate of a given ion is not characterized by fixed values for K_m or V_{max}. The relationship is rather variable, depending in particular on the internal concentration of the mineral nutrient, that is, on the nutritional status of the plant (e.g., Fig. 2.21). The mode of pretreatment of plants and their growth rate (i.e., their requirement for nutrients) (Pitman, 1972b; Clement *et al.*, 1978a) are important regulation factors—feedback mechanisms—in ion uptake. In general, as the internal concentration of a particular ion

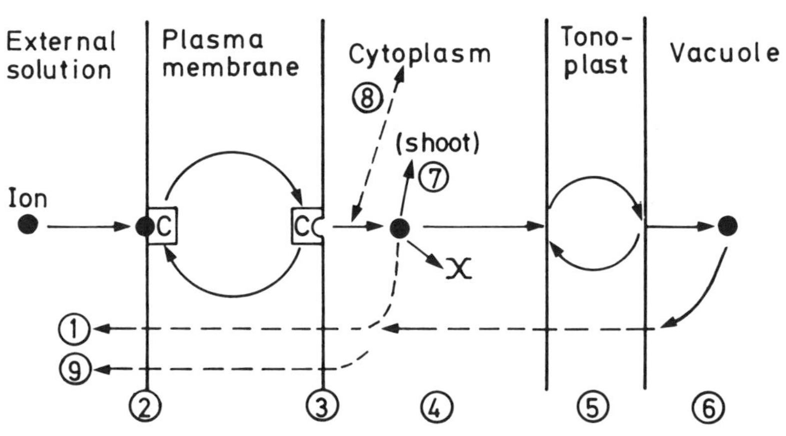

Fig. 2.22 Model of possible feedback mechanisms controlling ion uptake by the internal concentration (nutritional status). (1) Ion efflux; (2) number of binding sites (carriers); (3) unloading site; (4) transformation, incorporation; (5) transport through the tonoplast; (6) vacuolar concentration; (7) xylem and phloem loading, long-distance transport; (8) feedback regulation from the shoot; (9) enhanced efflux of H^+ and/or organic compounds.

increases, its uptake rate declines and vice versa. Various mechanisms [(1) to (9), Fig. 2.22] bring about this regulation.

On the basis of studies of isolated roots or storage tissue, the main regulatory factor is considered to be an increase in efflux (1) (Mengel and Helal, 1967). Such an increase with increasing internal concentration and in the presence of Ca^{2+} can readily be demonstrated for potassium (Mengel and Schneider, 1965), nitrate (Morgan *et al.*, 1973; Breteler and Nissen, 1982) and phosphate (Schjørring and Jensen, 1984). Although efflux might contribute to the decrease in uptake rate, it is certainly not the main mechanism, since at high internal concentrations the rate of influx is significantly lower (Johansen *et al.*, 1970; Lee, 1982). Because ion uptake and transport require energy, one cannot rule out the possibility that a decline in carbohydrate content is responsible for the decline in influx rates at high internal concentrations. This would be an unspecific response, however, and would also cause a corresponding decline in the uptake rate of other ions. The feedback regulation in either direction is rather specific for a particular mineral nutrient. For example, in phosphorus-deficient barley plants, only the influx rate of phosphate is increased. There is no effect on nitrate, chloride, or sulfate influx (Lee, 1982). This stimulation of influx was found to be due to an increase in V_{max} rather than in K_m, indicating that it was the capacity for uptake of a particular nutrient, possibly related to the number of binding sites, rather than the affinity which had been increased.

Evidence of the effects of nutritional status on the number of specific binding sites at the plasma membrane [mechanism (2) Fig. 2.22] has been reported for sulfate (Table 2.23). For the "initial" uptake, a fraction of $^{35}SO_4^{2-}$ was characterized, which could be readily exchanged and was most likely located on the external surface of the plasma membrane. In terms of regulation, such a response could be interpreted as a specifically increased uptake rate at low internal concentrations.

A decline in the influx rate of K^+ by a factor of 3 to 6 as the internal concentration was increased was found to be correlated with an increase in

Table 2.23
Initial Sulfate Uptake by Wheat Roots after Pretreatment with Different Sulfate Concentrations[a]

Pretreatment (m*M* sulfate)	Initial uptake of $^{35}SO_4^{2-}$ subsequently supplied (nmol/g fresh wt)
0·50	32
0·25	40
0·05	92
0·0005	197

[a]Based on Persson (1969).

Table 2.24
Relationship between Content and Influx of Potassium in Barley Roots[a]

K Content (μmol/g fresh wt)	K^+ Influx (μmol/g fresh wt × hr)
20·9	3·05
32·1	2·72
47·9	2·16
57·8	1·61

[a]From Glass and Dunlop (1979).

the K_m value from 1 to 15–20 μM (Wild *et al.*, 1979). Negative feedback mechanisms induced primarily at the unloading step [i.e., the release of the ion from the carrier (3)] via allosteric inhibition have been discussed as the possible cause of this decrease in influx rate of K^+ (Table 2.24).

A similar negative feedback mechanism controlling phosphorus influx may develop within a few hours. This control is presumably exerted by inorganic phosphorus at the unloading step [(3), Fig. 2.22] (Lefebvre and Glass, 1982). Feedback mechanisms at unloading sites cannot, however, be restricted to the plasma membrane, but must include step (5) at the tonoplast. Because both transport processes are acting in series, it is difficult to differentiate between them. Cram (1983) suggested that the feedback control of net sulfate uptake is located primarily at the tonoplast, a view which is also supported for phosphorus by Lee and Ratcliffe (1983). In pea root tips the uptake rate of phosphorus was closely correlated to the concentration of inorganic phosphorus (P_i) in the vacuoles, whereas there was no correlation to the phosphorus concentration in the cytoplasm.

In intact plants, the feedback mechanism regulating phosphorus uptake can be delayed for several days, because rapid phosphorus translocation into the shoot [step (7), Fig. 2.22] prevents a marked increase in phosphorus concentration within the roots (Table 2.25).

The resupply of phosphorus after a period of deficiency can therefore lead to a greatly increased phosphorus content in the shoot and also to phosphorus toxicity, as shown by Green and Warder (1973). Although such rapid changes in phosphorus supply are unlikely to occur in soil culture, in nutrient solution culture this factor should be considered, especially after the replacement of solutions.

For those mineral nutrients that are metabolized in the cells (e.g., sulfate and nitrate, which are reduced), the uptake rate can also be regulated by a feedback mechanism induced by metabolites [mechanism (4), Fig. 2.22]. This is a well-documented phenomenon in sulfate uptake which is severely inhibited by the sulfur-containing amino acids cysteine and methionine

Table 2.25
Contents of Phosphorus in Barley Plants after Growth without Phosphorus or Resupply of Phosphorus[a]

	Phosphorus content (μmol P/g dry wt)[b]		
	8 days − P[c]	7 days −P +1 day +P[d]	7 days −P +3 days +P[e]
Shoot total	49 (20)	151 (61)	412 (176)
Youngest leaves	26 (5)	684 (141)	1647 (483)
Roots	43 (24)	86 (48)	169 (94)

[a]Based on Clarkson and Scattergood (1982).
[b]Numerals in parentheses are relative values; 100 represents control with continuous phosphorus supply of 150 μMP throughout the experiment.
[c]Eight days of growth without phosphorus.
[d]Seven days of growth without phosphorus and 1 day of growth upon addition of phosphorus (150μM).
[e]Seven days of growth without phosphorus and 3 days of growth upon addition of phosphorus (150μM).

(Brunold and Schmidt, 1978). For nitrate, the time course of the feedback mechanism can be more complex. The uptake rate of plants without pretreatment with nitrate is distinctly lower in the first few hours than the rate of pretreated plants, until induction of nitrate reductase in the root cells (Jackson *et al.*, 1972) or of a nitrate carrier in the plasma membrane of root cells (Deane-Drummond, 1984). Afterwards nitrate accumulates in the vacuoles of root cells and/or is transported into the shoots, until negative feedback control (regression) by reduced nitrogen (4), (8) or by accumulated nitrate (6) depresses the uptake rate. Negative feedback control is well known from enzyme kinetics; it is imposed by inhibitory effects of the endproducts of a reaction.

Negative feedback control of uptake rate by a high vacuolar nitrate concentration is not necessarily specific for nitrate. It is also observed for chloride uptake (F. A. Smith, 1973). This rather unspecific control by a high vacuolar concentration (6) is thought to be a turgor-regulated process (Glass, 1983) involving the modificaton of membrane structure due to membrane compressibility (Enoch and Glinka, 1981). Turgor-regulated ion uptake in the tissue of glycophytes is different from the process in halophytes (Cram, 1980), an effect which might be related to differences in salt tolerance (Chapter 16, Section 16.6).

The relationship between influx rate and the internal concentration of a particular mineral nutrient cannot always be explained satisfactorily by consideration of the root alone. Positive and negative feedback control by the shoots can also affect the uptake rate of the root [step (8), Fig. 2.22]. A more general and unspecific effect might be induced by carbohydrate transport into the roots (Pitman, 1972b) or by growth of the shoot, increasing the demand for and thus the translocation rate of mineral nutrients into the shoot. However, there must also be more specific feedback mechanisms. For example, the uptake rates of phosphorus (de Jager, 1979, Drew *et al.*, 1984) and potassium (Table 2.26) are more closely related to the corresponding contents of these nutrients in the shoots than to their contents in the roots.

Table 2.26
Rate of Potassium Uptake by Maize Roots in Relation to the Potassium Contents of the Roots and Shoot[a]

K^+ Uptake (pmol/cm × sec)	K Content (% dry wt)	
	Roots	Shoot
15·8	5·85	8·00
28·0	5·55	6·45
33·8	4·99	4·35
36·8	5·51	4·13

[a]From Barber (1979).

The mode of action by which the shoot specifically controls uptake by the roots is not yet known. Retranslocation into the roots of phloem-mobile mineral elements such as potassium and phosphorus and control by phytohormones (Jeschke, 1982) have been proposed as possible feedback control mechanisms (Marschner, 1983; de Jager, 1984). Retranslocation of organic compounds such as amino acids and amides is likely to be involved in the feedback control of nitrogen uptake rates (Simpson *et al.*, 1982).

An unusual response is found in the rate of uptake of cadmium by tomato roots (Petit *et al.*, 1978). As the cadmium content in the roots increases, the uptake rate also increased. This might reflect the induction of synthesis of compounds such as metallothioneins, which are proteins with a specific binding affinity for heavy metals (Section 9.3). A similar mechanism might also be responsible for the surprising differences in the rate of uptake of copper in copper-sufficient and copper-deficient plants: After copper is supplied to deficient plants, the uptake rate is much lower than in plants with a sufficient copper content (Jarvis and Robson, 1982).

A well-studied regulatory mechanism for iron uptake which depends on iron nutritional status exists in most dicotyledonous and in monocotyledonous plant species, except of the grasses. With iron deficiency these plant species increase the "reducing capacity" of their roots. This is reflected in an enhanced reduction at the root surface of Fe(III) to Fe(II) when an Fe(III) chelate such as Fe-EDTA is supplied (Chaney *et al.*, 1972). The rate of uptake and translocation of iron to the shoot increases sharply as the reducing capacity increases. This enhanced iron reduction is associated with a splitting of the chelator and takes place mainly at the plasma membrane of the rhizodermal cells. In iron-deficient plants, these cells develop a distinct cell wall labyrinth and can be characterized as rhizodermal transfer cells (Section 9.1). This reduction of Fe(III) to Fe(II) is drastically depressed when the integrity of the plasma membrane is impaired (Barrett-Lennard *et al.*, 1983), an observation which supports the idea of the involvement of a plasma membrane–bound enzyme (reductase, according to Chaney *et al.*, 1972, and Bienfait *et al.*, 1982). There is, however, some evidence that phenolic compounds might also be involved in this process. Iron-deficient roots of many dicotyledonous species accumulate phenolics, which under certain conditions are released into the external solution in relatively large amounts (Olsen *et al.*, 1981; Fig. 2.23), a process resembling regulation mechanism (9) in Fig. 2.22. Phenolics are very effective in the mobilization of iron from inorganic Fe(III) compounds by complexation and/or reduction.

Furthermore, iron-deficient roots of these plant species extrude protons at a much higher rate, which acidifies the external solution (Fig. 2.21). When the iron supply is suboptimal—when, for example, it contains a low concentration of Fe(III) chelates or sparingly soluble inorganic Fe(III) compounds [e.g., Fe(III) hydroxide]—the increase in the reducing capacity of the roots, combined with the acidification and the release of phenolics, leads to enhanced mobilization and uptake of iron. On recovery of the plants from

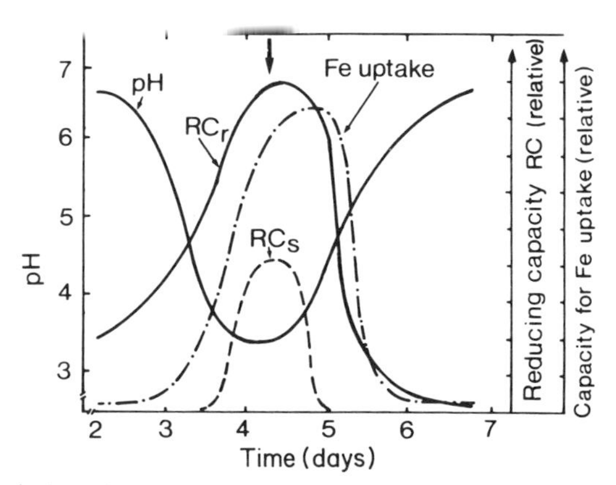

Fig. 2.23 Root-induced changes in the external solution (pH; reducing substances, RC_s) and in root properties (reducing capacity, RC_r; capacity for iron uptake) in sunflower in response to iron deficiency and resupply of iron (↓)). [Data compiled from Marschner *et al.* (1974, 1978) and Römheld and Marschner (1981a).]

iron deficiency, reducing capacity, proton efflux, and iron uptake drop to control levels within 1 day. When the supply of iron is suboptimal, these iron-deficiency-induced changes in root physiology (and in root anatomy; see Section 9.1) are reflected in periodic changes in both the iron nutritional status of the plants and corresponding changes in the pH of the substrate. A model for this regulation mechanism is shown in Fig. 2.24.

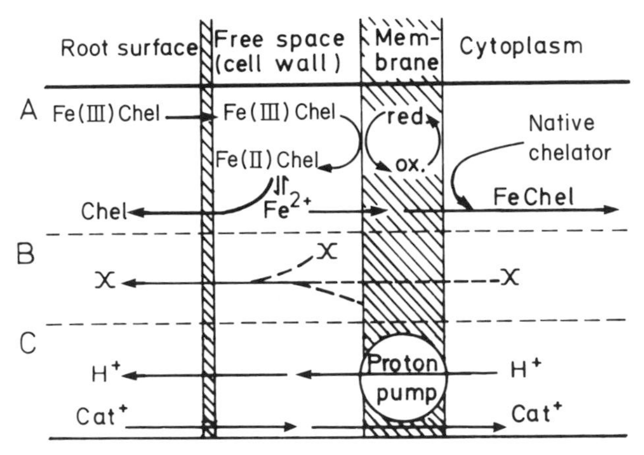

Fig. 2.24 Model of iron deficiency–induced regulation mechanisms in roots of "Fe-efficient" dicots for enhancement of iron uptake. A. Increase in reducing capacity. B. Release of phenolic compounds. C. Enhanced proton efflux.

This sophisticated mechanism for the regulation of iron uptake is sensitive to high bicarbonate concentrations and is not element specific. When iron deficient, the roots of these plant species readily reduce other compounds such as manganese oxides (MnO_2). Interestingly, gramineceous species (grasses) respond quite differently to iron deficiency, by increasing sharply the release of nonproteinogenic amino acids ("phytosiderophores") from the roots (Section 9.1). These amino acids mobilize sparingly soluble inorganic Fe(III) in the rhizosphere by the formation of Fe(III) chelates. Most likely, these chelates are taken up readily by the roots of grasses. The ecological advantages and disadvantages of the different mechanisms involved in the response of roots to iron deficiency are discussed in Section 16.5.

The regulatory mechanisms discussed in this section demonstrate that plants strive by means of feedback control to maintain fairly constant internal concentrations of mineral nutrients over a wide range of external concentrations. It is important, therefore, not to overestimate the relevance of short-term studies on uptake isotherms and corresponding calculations on K_m and V_{max} to the mineral nutrition of plants over long periods. In addition, changes in root growth and morphology are also quite common responses to suboptimal internal concentrations of mineral nutrients (Chapter 16).

2.6 Ion Uptake along the Roots

Growing roots vary both anatomically and physiologically along their longitudinal axes (Fig. 2.25). The rates of ion uptake are different at different zones along the root. This has to be kept in mind when models for mechanisms of "the" behavior of root tissue and root cells are based on uptake studies with isolated roots or roots of intact plants. In the apical zone, nonvacuolated cells dominate; they differ in many respects from the vacuolated cells in the basal zones. For example, the apical root zones are characterized by higher rates of fermentation (Ramshorn, 1958), in many instances higher cation-exchange capacity (Crooke *et al.*, 1960), higher K^+/Na^+ selectivity (Jeschke and Stelter, 1976), and negligible effects of the accompanying cation on anion uptake, even in high concentration ranges (Torii and Laties, 1966).

In general, there is a tendency for the rate of ion uptake per unit root length to decline as the distance from the apex increases. A similar tendency is observed in water uptake along the roots (Sanderson, 1983). Three factors are mainly responsible for this decline: (a) an increase in suberin formation in the rhizodermis (Clarkson *et al.*, 1978b), (b) the formation of the

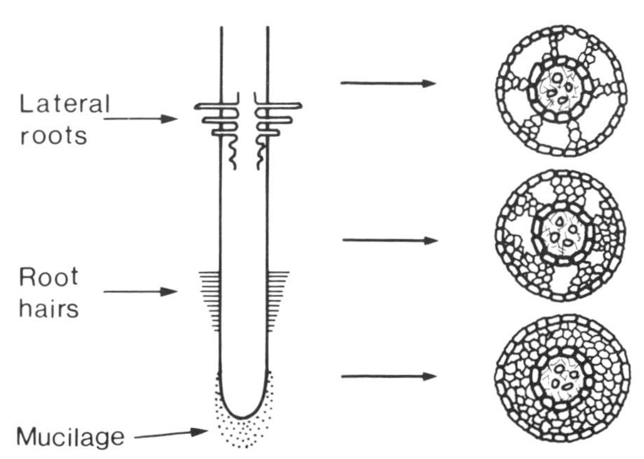

Fig. 2.25 Schematic representation of typical anatomical changes along the axis of a maize seminal root. In basal zones there is degeneration of cortical cells and formation of tertiary endodermis with passage cells.

secondary and tertiary endodermis and corresponding inhibition of radial transport into the stele, and (c) the partial degeneration of cortical cells and formation of cavities in the cortex, sometimes referred to as *aerenchyma* formation (Fig. 2.25). Aerenchyma formation is also observed in well-aerated solutions and is enhanced by nitrogen deficiency and ethylene treatment. The mechanism of formation and the role of aerenchyma in adaptation to waterlogged soils is discussed in Section 16.4.

Despite these anatomical changes, basal root zones still have a considerable capacity for ion uptake, even in woody species (Atkinson and Wilson, 1979). In maize roots, the capacity for uptake and translocation of ions per unit root length may still be surprisingly high in basal zones, indicating that the strands of cells bridging the cortex maintain sufficient ion transport capacity from the rhizodermis up to the endodermis (Drew *et al.*, 1980). The transport of ions and water across the tertiary endodermis takes place preferentially in nonlignified passage cells (Fig. 2.25). These cells are of particular importance for the radial transport of solutes in older roots of perennials (P. J. Kramer, 1983). Nevertheless, the gradient in anatomy and physiology along the axis of an individual root reflects a shift in function from uptake to the long-distance transport of solutes such as water and mineral nutrients.

The ion uptake gradient along the root axes varies between nodal and seminal roots (Robards *et al.*, 1973) and also among ions. For calcium and magnesium (Ferguson and Clarkson, 1976) the decline in uptake is much sharper than for potassium (Marschner and Richter, 1973) or phosphorus (Ferguson and Clarkson, 1975). The gradient also depends on the nutritional status of a plant. With iron-deficient dicotyledonous species, for example, the apical, but not the basal, root zones increase both their reduction and uptake of iron by a factor of up to 100 (Römheld and Marschner, 1981b). In contrast, in phosphorus-deficient plants, the basal root zones increase their rate of phosphorus uptake much more than the apical zones (Table 2.27).

These differences in ion uptake rate along the roots do not necessarily reflect the contributions of these root zones to nutrient transport to the shoot. Apical root zones have a higher nutrient requirement for growth (cell division and elongation) than the differentiated cells in the basal zones. Thus, despite their high uptake rate, the contribution of apical zones to translocation to the shoot can be quite low (Table 2.28).

The apical root zones not only transport less potassium toward the shoot; they also compete with the shoot for potassium taken up in the basal zones; that is, they act as a "sink" for potassium (Table 2.28). In contrast, accumulation and transport of calcium to the shoot are extensive in the apical zones. This contribution of the apical zones to calcium transport can

Table 2.27
Effect of Phosphorus Nutritional Status on the Rate of Phosphorus Uptake by Various Root Zones of Barley Plants[a]

	Root zone (distance from root tip, cm)		
Pretreatment for 9 days	1	2	3
With phosphorus	2019	1558	970
Without phosphorus	3150	4500	4613

[a]Uptake rate expressed as picomoles per cubic millimeter of root segment in 24 hr. Based on Clarkson *et al.* (1978a).

Table 2.28
Uptake and Translocation of Potassium (^{42}K) and Calcium (^{45}Ca) Supplied to Different Zones of the Seminal Roots of Maize[a]

		Root zone supplied (distance from tip, cm)		
Nutrient	Accumulation and translocation	0–3	6–9	12–15
Potassium	Translocation to the shoot	3·8	14·6	15·6
	Accumulation in the zone of supply	11·5	3·8	1·9
	Translocation to the root tip	—	4·3	2·0
	Total	15·2	22·7	15·9
Calcium	Translocation to the shoot	2·4	2·2	2·4
	Accumulation in the zone of supply	4·1	1·6	0·4
	Translocation to the root tip	—	—	—
	Total	6·5	3·8	2·8

[a]Data expressed as microequivalents per 12 plants in 24 hr. Based on Marschner and Richter (1973).

be even higher in other plant species (Robards *et al.*, 1973), which is of particular importance for the calcium nutrition of plants (Section 8.6).

2.7 Radial Transport across the Roots

There are two parallel pathways of solute movement across the cortex toward the stele: one passing through the extracellular space, or apoplast (cell walls and intercellular spaces; see Section 2.2), and another passing from cell to cell in the symplast through the plasmodesmata which bypass the vacuoles (Fig. 2.26). Transport through the apoplast is terminated at the endodermis by the Casparian strip in the walls of the endodermal cells. This strip is suberized and completely surrounds each cell; it forms a barrier in the apoplast between the cortex and the stele (Section 2.2). Because of this barrier, the transport of solutes from the cortex into the stele and of assimilates from the stele into the cortex must in principle occur through the endodermal cells. There are, however, some "leaky" zones in the barrier; they are found in the apical meristems where the differentiation of the various cell types is not yet complete and in basal zones where lateral roots penetrate the endodermis. Lateral roots develop from the pericycle in the stele. Radial transport in the apoplast from the external solution into the stele might thus be of importance in these zones (Queen, 1967), particularly for calcium and magnesium (Ferguson and Clarkson, 1976) and aluminum (Rasmussen, 1968) because of their limited mobility within the symplast.

The relative importance of the two pathways—apoplasmic and symplasmic—within the cortex depends on such factors as external ion concentration, root hair formation, and suberization of the rhizodermal cells. At low external concentrations the uptake and entry of ions into the symplast seem to be restricted mainly to the rhizodermal cell layer (Vakhmistrov, 1967; Grunwald *et al.*, 1979). In agreement with this, uptake studies with labeled mineral nutrients show that their concentration in the rhizodermal cells is often much higher than in the cortex cells (Läuchli *et al.*, 1971; Chino, 1979).

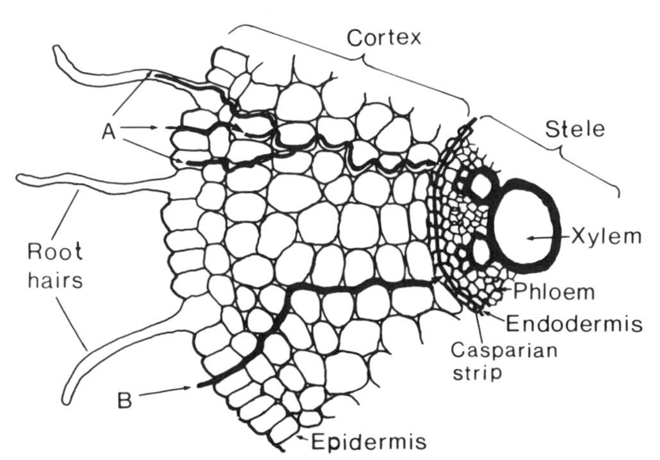

Fig. 2.26 Part of transsection of a maize root showing the symplastic (A) and apoplastic (B) pathway of ion transport across the root.

Ion movement into the apoplast might be restricted in the basal root zones by suberization of the rhizodermal cells (Clarkson *et al.*, 1978b) and in the apical root zones by mucilage, which often forms a thick layer at the external surface of the rhizodermal cell walls (Oades, 1978). Polyvalent cations, such as aluminum, might accumulate in this mucilage (Section 16.3), indicating that it acts as a barrier to the passive movement of solutes into this incompletely differentiated root zone (Rovira *et al.*, 1979).

Radial transport in the symplast from cell to cell requires bridges, called *plasmodesmata*, across the cell walls that connect the cytoplasm of neighboring cells. Helder and Boerma (1969) found an average of 20,000 plasmodesmata per cell in the endodermis of young barley roots. In the tertiary (liquified) endodermis of older sections of barley roots, there are far fewer plasmodesmata, but there is a sufficient number of these to permit considerable radial transport of both water and ions through the endodermis (Clarkson *et al.*, 1971).

In evaluating the role of the two pathways in roots, the number of plasmodesmata in the rhizodermal cells is of particular interest. As shown in Table 2.29 rhizodermal cells which have developed into root hairs have more plasmodesmata than the remaining rhizodermal cells. The relatively small number of plasmodesmata in *Raphanus* raises the question as to whether the root hairs are of major importance for symplasmic radial transport in this species. However, not only the number but also the diameter of the individual plasmodesmata must be taken into account (Tyree, 1970).

Table 2.29
Intracellular K^+ Activity and Number of Plasmodesmata in Tangential Walls of Hair and Hairless Cells of the Root Epidermis[a]

			Number of plasmodesmata	
Plant species	Cell type	K^+ Activity (mM)	Per μm^2	Per cell junction
Trianea bogotensis	Hair	133	2·06	10,419
	Hairless	74	0·11	693
Raphanus sativus	Hair	129	0·16	273
	Hairless	124	0·07	150

[a]From Vakhmistrov (1981).

The mechanism of symplasmic transport of potassium seems to be chiefly diffusional. Inhibition of cytoplasmic streaming has no effect on the radial transport of potassium (Glass and Dunlop, 1979). On the other hand, in the radial transport of phosphorus, various metabolic steps, such as esterifi-

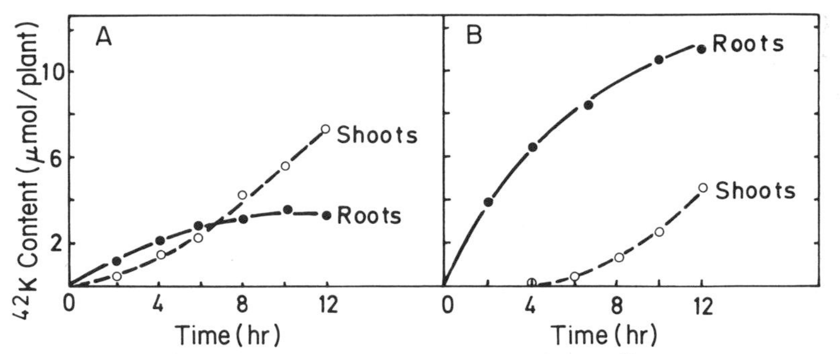

Fig. 2.27 Accumulation and translocation rate of K^+ (^{42}K) from 1 m*M* KCl (+0·5 m*M* $CaSO_4$) in barley plants (A) after preculture with 1 m*M* KCl or (B) without KCl.

cation, seem to be involved (Loughman, 1966). This should lead to a close coupling between metabolic activity of cortical cells and radial transport of phosphorus.

At low internal concentrations ("low-salt" roots) acumulation in the vacuoles competes very effectively with symplasmic transport (Hooymans, 1976). In short-term studies this competition is reflected in a typical delay in the translocation of ions from the roots to the shoots of plants which originally had low internal concentrations of the ion being investigated (Fig. 2.27). As a result of this competition, when the supply of a nutrient is suboptimal the roots usually have higher contents of the particular nutrient than the shoot ("restricted translocation"). In long-term studies, this regulation mechanism is in part responsible for a shift in the relative growth rates of roots and shoots in favor of the roots (Chapter 14).

Exchange between ions in the vacuoles of the individual cortex cells and ions in the symplast takes place through the tonoplast. For potassium this exchange is quite rapid (Hooymans, 1976), whereas for sodium (Wallace *et al.*, 1972) or sulfate (Marschner and Michael, 1960) it is very slow.

2.8 Mechanism of Ion Release into the Xylem

After radial transport in the symplast into the stele, most of the ions are released into the xylem. Although in apical root zones there are some metaxylem vessels containing cytoplasm, most of the xylem vessels are free of cytoplasm and therefore represent an apoplasmic space. Release of ions and water within the stele into the xylem vessels therefore involves a retransfer from the symplast into the apoplast. The question arises as to whether this release is a passive process ("leakage") or an active secretion.

Crafts and Broyer (1938) postulated uphill transport in the symplast across the cortex to the endodermal cells and, in the stele, a "leakage" into the xylem. This idea was supported by Laties and Budd (1964), who showed that stelar tissue isolated from corn roots is highly permeable and incapable of active ion uptake. Furthermore, in intact roots the oxygen tension is distinctly lower in the stele than in the cortex (Fiscus and Kramer, 1970), a factor which might in part be responsible for the higher permeability of the stele.

In contrast to this hypothesis of "leakage," there is increasing evidence that ions are secreted into the xylem. For example, the potassium concentration in stelar tissue can be much higher than that in the cortex (Läuchli *et al.*, 1971). The idea of secretion is also supported by the results of experiments with metabolic inhibitors, particularly inhibitors of protein synthesis (Läuchli *et al.*, 1978). These compounds, an example being cycloheximide, depress the release of ions into the xylem but not their accumulation in the root cells (Läuchli *et al.*, 1973). On the basis of these observations a two-pump model has been developed in which active transport sites are located at the outer surface of the symplasm in the rhizodermis and cortex and at the symplasm–xylem interface in the stele (Pitman, 1972a; Läuchli, 1976a; see Fig. 2.28). In this model the xylem parenchyma cells play a key role in ion secretion. The discovery of transfer cell–like structures (Pate and Gunning, 1972) in xylem parenchyma cells of roots (Läuchli, 1976a) lends support to this view.

The model is not fully supported by electrophysiological studies, however (Dunlop, 1974; Bowling, 1981). According to these studies, ion movement from the symplast into the xylem appears to occur along the electrochemical gradient. Regardless of these discrepancies, however, it can be concluded that in most instances the final step of ion transfer from the symplast into the xylem is a carrier-mediated step which is regulated in a different way than loading into the symplast of rhizodermis and cortex. Although not neces-

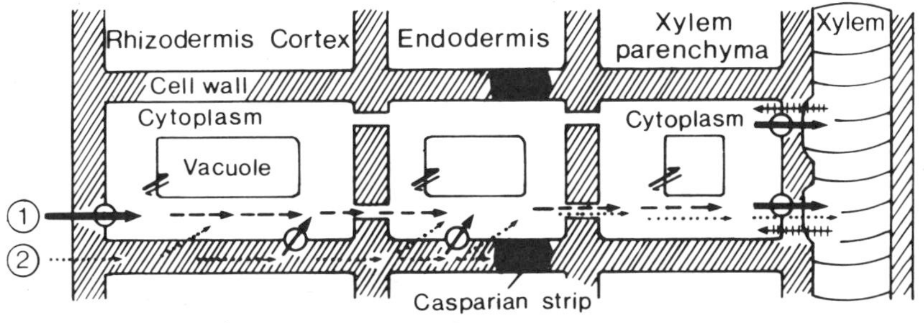

Fig. 2.28 Model for symplasmic (1) and apoplasmic (2) pathways of radial transport of ions across the root into the xylem. Key: ⊖→, active transport; ↚, resorption. (Modified from Läuchli, 1976a.)

sarily an active membrane transport against the electrochemical gradient (Section 2.4.2), this transport fulfills the criteria for an active, metabolic-coupled process. It enables the plant to regulate the long-distance transport of ions to the shoot via the rate and selectivity of release or by reabsorption of certain ions such as sodium (Chapter 3). Ion transfer from symplast to apoplast and vice versa might occur directly at the xylem parenchyma cells or at other stelar cells (Bowling, 1981). In any case, as a result of this transfer, the ion concentration in the apoplast of the stele is usually much higher than that in the apoplast of the cortex, and the Casparian strip must prevent diffusion of ions along this gradient.

2.9 Factors Affecting Ion Release into the Xylem: Guttation and Exudation

The permeability of plant membranes to water is much higher than that to ions. Plant cells or roots therefore behave as osmometers. Ion release into the apoplast of the stele increases both the osmotic potential and the water potential (they become more negative), and a corresponding net flux of water from the external solution is induced according to the model in Fig. 2.28. As a result of this water flux, both osmotic and water potentials decrease, whereas hydrostatic pressure, or the turgor pressure of individual cells, increases. This hydrostatic pressure in the stele induces a volume flow of solutes (water, ions, and molecules) in the xylem vessels toward the shoot. Because of this "root pressure", droplets are sometimes released on the tips and margins of leaves, a process known as guttation. This is particularly apparent in seedlings and young plants at night and in the early morning (under conditions of high relative humidity and low transpiration). Exudation from the stumps of cut plants (e.g., freshly mown grass) is also the result of root pressure.

Root pressure and the corresponding volume flow in the xylem are of particular importance for the long-distance transport of mineral nutrients at low transpiration rates, such as might be found in seedlings and at night. This is particularly true for calcium transport into low-transpiring organs such as fruits (Chapter 6). Volume and composition data on the xylem exudate of decapitated plants are also important for understanding root metabolism and activity, particularly in relation to the mineral nutrition of the whole plant. For example, the nitrogen fractions (nitrate and organic nitrogen) of the xylem exudate are indicators of the capacity of roots to reduce nitrate (Ezeta and Jackson, 1975), and in legumes they also indicate the intensity of rhizobial N_2 fixation (Pate *et al.*, 1980), even under field conditions (Streeter, 1979). From xylem exudate analysis, models for the carbon and nitrogen economy of plants can be developed (Section 8.2). Furthermore, the composition of xylem sap reflects the capability of roots for complexing heavy-metal cations (Chapter 3), and sap composition data can be used to calculate the recirculation of mineral nutrients between shoots and roots. For these reasons, studies of ion release into the xylem and root pressure make important contributions to our understanding of mineral nutrition and its regulation in higher plants.

Ion release into the xylem requires at least one, and presumably two, active membrane transport steps. Root pressure therefore depends on external and internal factors in much the same way as does the accumulation of ions in the root tissue. There are, however, distinct differences between these processes, since the release of ions into the xylem involves several additional factors. For technical reasons, it is difficult to measure directly the release of ions into the xylem. Most experimental evidence on this final step of radial transport across the roots is based on studies of xylem exudate from individual roots or root systems, although this includes a step of the long-distance transport of solutes (Chapter 3).

2.9.1 External Concentration

An increase in the external ionic concentration leads to an increase in the concentration of ions in the xylem exudate. However, the relative concentration falls as the external concentration is increased (Table 2.30). This concentration gradient ("concentration factor") between the external solution and the xylem exudate decreases and can even fall below 1 in the case of calcium.

The exudation volume flow shows a somewhat different pattern and is maximal at 1·0 m*M* external concentration (Table 2.30). At 0·1 m*M* this flow is limited by the ion concentration in the xylem. In contrast, at 10·0 m*M*, the flow is limited by the water availability (i.e., the high water potential in the

Table 2.30
Relationship between External Concentration, Exudate Concentration, and Exudate Volume Flow in Decapitated Sunflower Plants

External solution $KNO_3 + CaCl_2$ (m*M* each)	Exudate (m*M*)			"Concentration factor"			Exudation volume flow (ml/4 hr)
	K^+	Ca^{2+}	NO_3^-	K^+	Ca^{2+}	NO_3^-	
0·1	7·3	2·8	7·4	73	28	74	4·0
1·0	10·0	3·2	10·7	10	3·2	10·1	4·5
10·0	16·6	4·2	10·3	1·7	0·4	1·0	1·6

external solution) and the small concentration gradient between the external solution and the xylem. The increase in the exudate concentration of the mineral nutrients, with the rise in external concentration from 1·0 to 10·0 m*M*, does not compensate for the decrease in the exudation volume flow. Thus, in contrast to the accumulation in roots (hyperbolic function of the external concentration; see e.g., Fig. 2.8), the rate of root pressure–driven transport of mineral nutrients can decline at high external concentrations.

2.9.2 Temperature

An increase in the root temperature has a much greater effect on the exudation volume flow than on the ion concentration in the exudate (Table 2.31). This is consistent with the expectation that a root behaves as an osmometer: Temperature determines the rate of ion release into the xylem,

Table 2.31
Temperature Effect on Exudation Volume Flow and on Potassium and Calcium Concentration in the Exudate of Decapitated Maize Plants[a]

Temperature (°C)	Exudation volume flow (ml/4 hr)	Exudate concentration (m*M*)		Ratio K^+/Ca^{2+}
		K^+	Ca^{2+}	
8	5·3	13·4	1·5	8·9
18	21·9	15·2	1·0	15·2
28	31·7	19·6	0·8	24·5

[a]Concentration of KNO_3 and $CaCl_2$ in the external solution: 1m*M* each.

and water moves accordingly along the water potential gradient. There are, however, distinct differences between a root and a simple osmometer. An increase in the root temperature results in an increase in the potassium concentration but a decrease in the calcium concentration of the exudate. This shift in the potassium/calcium ratio might reflect changes in membrane selectivity with temperature. Similar shifts in the potassium/calcium translocation ratio are also observed at different soil temperatures (Walker, 1969). This temperature effect could have important implications for the calcium nutrition of plants and may explain the enhancement of calcium deficiency symptoms in lettuce at elevated root temperatures, despite a slight increase in the calcium concentration of the leaf tissue (Collier and Tibbitts, 1984).

2.9.3 Respiration

As with ion accumulation (Table 2.11), the rate of release of ions into the xylem is closely related to root respiration (Table 2.32). A lack of oxygen strongly depresses the exudation volume flow but not the concentrations of potassium and calcium in the exudate. In contrast to the effect of temperature (Table 2.31), inhibition of respiration slows down the rate but does not affect the K^+/Ca^{2+} transport ratio.

Table 2.32
Effect of Root Respiration on Exudation Volume Flow and Ion Concentration in the Exudate of Decapitated Maize Plants[a]

Treatment[b]	Exudation volume flow (ml/3 hr)	Exudate concentration (m*M*) K^+	Ca^{2+}
O_2	26·5	16·6	1·8
N_2	5·7	15·2	1·7

[a]Concentration of KNO_3 and $CaCl_2$ in the external solution: 0·5 m*M* each.
[b]Respiration treatment consisted of bubbling oxygen or nitrogen through the external (nutrient) solution.

2.9.4 Accompanying Ion

As in ion accumulation in root cells, maintenance of the cation–anion balance is necessary in the release of ions into the xylem (Anderson and Collins, 1969). However, the accompanying ion seems to affect the transport rate even at low external concentrations (Cooil, 1974). As shown in Table 2.33, when KNO_3 is supplied, the exudation flow rate is almost twice

Table 2.33
Flow Rate and Ion Concentration in the Xylem Exudate of Wheat Seedlings[a]

Parameter	Treatment KNO_3	K_2SO_4
Exudation flow rate (μl/hr × 50 plants)	372	180
Ion concentration (μEq/ml)		
Potassium	23·3	24·5
Calcium	9·1	9·5
Nitrate	18·1	0·0
Sulfate	0·2	0·8
Organic acids	9·6	25·8

[a]Seedlings were supplied with either KNO_3 (1 m*M*) or K_2SO_4 (0·5 m*M*) in the presence of 0·2 m*M* $CaSO_4$. From Triplett *et al.* (1980).

as high as the flow rate when an equivalent concentration of K_2SO_4 is added. Since the potassium concentration in the exudate is similar in both treatments, the transport rate of potassium supplied as K_2SO_4 is only about half the rate of potassium supplied as KNO_3.

In contrast to the potassium concentration, the concentrations of nitrate and sulfate in the exudate exhibit a large difference (18·1 and 0·8 μEq/ml, respectively) between the treatments. The corresponding difference in negative charges in the exudate is approximately compensated for by elevated concentrations of organic acid anions. In the K_2SO_4 treatment, however, the rate-limiting factor is probably the capacity of the roots to maintain the cation–anion balance by organic acid synthesis; this leads to a decrease in the rate of potassium and calcium release into the xylem and a corresponding decrease in exudation flow rate. The differences between nitrate, on the one hand, and sulfate and organic acid anions on the other, in stimulating the proton pump at the tonoplast (Section 2.4.2) might contribute to the differences in the release of cations into the xylem.

2.9.5 Carbohydrate Status of the Roots

The release of ions into the xylem and the corresponding changes in root pressure are closely related to the carbohydrate status of the roots (Louwerse, 1967). The length of photoperiod affects the carbohydrate

Table 2.34
Relationship between Photoperiod, Carbohydrate Content of Roots, and Uptake and Translocation of Potassium in Decapitated Maize Plants[a]

	Photoperiod (hr)	
	12/12[b]	24/0
Carbohydrate in roots (mg)	122 (48)	328 (226)[c]
Total potassium uptake (meq)	1·3	5·0
Potassium translocation in exudation volume flow (meq)	1·0	3·5
Exudation volume flow ml/8 hr	30·3	88·5
Relative decline in flow rate within 8 hr (%)	60	12

[a]Data per 12 plants.
[b]Hours of light/hours of darkness. This pre-treatment with different day lengths was for one day (i.e., the day prior to decapitation).
[c]Numbers in parentheses denote carbohydrate content after 8 hr (decapitation).

status of the roots and correspondingly the rate and duration of exudation volume flow after decapitation (Table 2.34). Both the uptake and translocation rate of potassium in roots with a high carbohydrate content are considerably greater than in roots that are low in carbohydrate. The higher translocation rate is closely related to the exudation volume flow. In roots with a low carbohydrate content, reserves are rapidly depleted after decapitation and there is a corresponding decline in the rate of exudation volume flow within 8 hr. This depletion of carbohydrates in the roots of decapitated plants is one of the factors which limits studies on exudation volume flow.

The release of ions into the xylem and exudation volume flow can show distinct diurnal fluctuations that are not necessarily related to the supply of carbohydrates from the shoots (Vakhmistrov and Ali-Zade, 1974). Moreover, other substances, such as organic acids and phytohormones, are retranslocated from the shoots into the roots, where they affect, for example, the rate of nitrate reduction (Deane-Drummond *et al.*, 1979) and the root electropotential (Kelday and Bowling, 1980). In addition, the concentration of ions at the collecting sites can be different from that at the sites of ion release into the xylem. Finally, the phloem can also play an important role in long-distance transport from the roots to the shoots. Thus, results of xylem exudate analysis of decapitated plants should be interpreted with care in relation to the root to shoot transport of mineral elements in intact plants. These and other aspects of long-distance transport are discussed in the following chapter.

3

Long-Distance Transport in the Xylem and Phloem and Its Regulation

3.1 General

The long-distance transport of solutes—mineral elements and low-molecular-weight organic compounds—takes place in the vascular system of the xylem and phloem, water being the transporting agent. Long-distance transport from the roots to the shoots occurs predominantly in the nonliving xylem vessels. This xylem transport is driven by the gradient in hydrostatic pressure (root pressure) and by the gradient in the water potential. The gradient in water potential between roots and shoots is usually quite steep during the day when the stomata are open. It follows the pattern: atmosphere $\gg$ leaf cells $>$ xylem sap $>$ root cells $>$ external solution. Solute flow in the xylem from the roots to the shoots is therefore unidirectional (Fig. 3.1).

In contrast, long-distance transport in the phloem with its living sieve tubes is bidirectional. The direction of transport is determined by the nutritional requirements of the various plant organs or tissues and occurs, therefore, from source to sink (Chapter 5). Also in the roots, mineral elements can enter the phloem and thus be translocated bidirectionally. The translocation of different mineral elements taken up by a particular zone of the root varies markedly during long-distance transport from the zone of supply, as shown in Table 3.1 for maize seedlings. For the reasons already mentioned, long-distance transport from the zone of supply to the root tip must take place in the phloem. Whereas ^{45}Ca is rapidly translocated into the shoot, the translocation of ^{22}Na toward the shoot is severely restricted. The steep basipetal gradient in the ^{22}Na content of the root sections reflects resorption by the surrounding root tissue and is a typical feature of so-called natrophobic plant species (Chapter 10). Some ^{22}Na has also been translocated via the phloem to the root tip. In contrast, ^{42}K is quite mobile both in the xylem and in the phloem, and a markedly high proportion of the potassium taken up in more basal root zones is translocated via the phloem toward the root tip, which acts as a sink for this mineral nutrient.

During long-distance transport mineral elements and organic solutes are

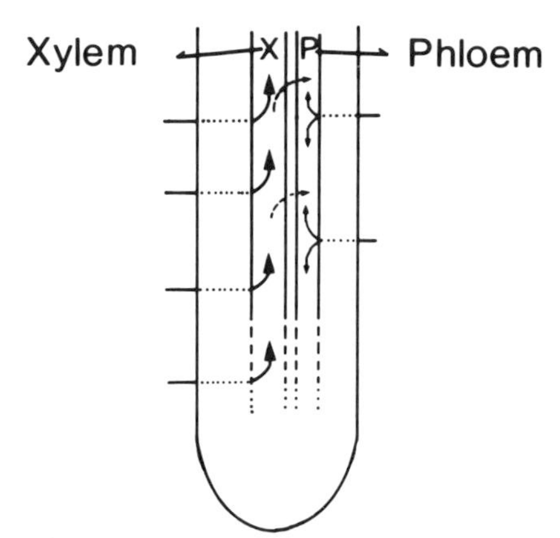

Fig. 3.1 Direction of long-distance transport of mineral elements in roots.

transferred between the phloem and xylem by extensive exchange processes, referred to as loading and unloading. The transfer is mediated by specific cells called *transfer cells* (Pate and Gunning, 1972). Despite this interchange, mineral nutrients, such as phosphorus, supplied to only one part of the root system (lateral or seminal roots) are transported preferentially to those parts of the shoots that have direct vascular connections with

Table 3.1
Accumulation and Long-Distance Transport of ^{45}Ca, ^{22}Na and ^{42}K in Maize Seedlings[a,b]

	Content (μeq/12 plants × 24 hr)		
Plant part	^{45}Ca	^{22}Na	^{42}K
Shoot	2·20	0·01	9·07
Endosperm	0·18	0·04	2·38
24–27 cm root	0·01	0·06	0·35
21–24 cm root	0·01	0·09	0·85
18–21 cm root	0·01	0·18	1·30
15–18 cm root	0·01	0·46	1·58
12–15 cm zone of supply	0·40	1·28	1·93
9–12 cm root	0	0·03	0·40
6–9 cm root	0	0·02	0·38
3–6 cm root	0	0·02	0·45
0–3 cm root	0	0·01	0·75
Total	2·82	2·20	19·44

[a]Based on Marschner and Richter (1973).
[b]Each seedling was supplied with 1 mEq/liter of labeled nutrient solution to the root zone 12–15 cm from the root tip. The remainder of the root system was supplied with the same solution in which the nutrients were not labeled.

particular root zones (Stryker *et al.*, 1974). This distribution pattern is especially important for the mineral nutrition of trees that are supplied with fertilizer in a localized area of the root system.

3.2 Xylem Transport

3.2.1 Mechanism

Although the mechanism of solute transport in the xylem sap is predominantly one of mass flow in the nonliving xylem vessels (i.e. in the apoplast), important interactions between solutes and both the cell walls of the vessels and the surrounding xylem parenchyma cells take place. The major interactions are exchange adsorption of polyvalent cations and reabsorption of mineral elements and the release (excretion) of organic compounds by surrounding living cells (xylem parenchyma and phloem).

3.2.1.1 Exchange Adsorption

The interactions between cations and the negatively charged groups in the cell walls of the xylem vessels are similar to those in the AFS of the root cortex (Fig. 2.1). The long-distance transport of cations in the xylem can be compared to ion movement in a cation exchanger with a corresponding decline in the translocation rate of cations such as Ca^{2+} (Bell and Biddulph, 1963) and Cd^{2+} (Petit and Geijn, 1978) relative to that of water (Thomas, 1967) or anions such as phosphate (Ferguson and Bollard, 1976). This cation-exchange adsorption is not restricted to the xylem vessels, in addition the cell walls of the surrounding tissue take part in these exchange reactions (Geijn and Petit, 1979; Wolterbeek *et al.* 1984).

The degree of retardation of cation translocation depends on the valency of the cation ($Ca^2 > K^+$), its concentration and activity, the presence of other competing cations and of complexing agents (Jacoby, 1967; Isermann, 1978; McGrath and Robson, 1984), the charge density of the negative groups (dicots > monocots), the diameter of the xylem vessels and the pH of the xylem sap. On average, the pH of xylem sap is 5·5.

An example of cation exchange in long-distance transport through the stem is shown in Table 3.2. When only $^{45}CaCl_2$ is supplied to the cut stem of derooted bean plants, the acropetal transport of ^{45}Ca is severely retarded. The addition of other cations strongly facilitates acropetal ^{45}Ca transport, the effect being similar to that seen with the exudate (xylem sap) of decapitated bean plants.

In intact plants, the role of cation-exchange adsorption in long-distance

Table 3.2
Effect of Other Cations and Root Exudates on the Long-Distance Transport of ^{45}Ca within 24 Hours in Stems of Derooted Bean[a]

Plant part	Cut ends of stem base supplied with		
	$^{45}CaCl_2$ only	$^{45}CaCl_2$ plus Ca^{2+}, Mg^{2+}, K^+, and Na^+	$^{45}CaCl_2$ plus root exudate
Primary leaves	0·04	4·7	1·8
12–18 cm stem	7	19	11
8–12 cm stem	28	56	40
4–8 cm stem	84	57	61
0–4 cm stem	159	81	81

[a]Quantity of ^{45}Ca transport expressed as microequivalents per gram dry weight. Based on Jacoby (1967).

transport through the xylem is therefore quite variable. For example, even organic acids which influence ion activity might be important for calcium translocation (Bradfield, 1976). They play a significant part in the transport of heavy metals such as iron (Tiffin, 1970) and, together with amino acids, in the movement of other heavy-metal cations such as copper and cadmium (White *et al.*, 1981a,b). Accordingly, the translocation rate in the stem of copper (Smeulders and Geijn, 1983), zinc (McGrath and Robson, 1984) and calcium (Isermann, 1978) can be greatly enhanced by synthetic chelators.

Exchange adsorption reduces the speed of the long-distance transport of polyvalent cations such as calcium. It is, however, a mechanism for the regulation of the distribution of these cations within the shoot, independent of the transpiration rate of the various shoot organs and tissues (see Section 3.3.9).

3.2.1.2 Resorption

Solutes are resorbed from the xylem (apoplast) into the living cells (cytoplasm and vacuole) along the pathway of the xylem sap from the roots to the leaves. The concentration and composition of the xylem sap change, therefore, along the pathway. In nodulated legumes, for example, the concentrations of amides and amino acids in the xylem sap decrease with increasing path length (Pate *et al.*, 1964). In plants grown in dilute nutrient solutions, the ion concentration in the xylem sap declines sharply from the roots to the tops of the plants, and the water released by guttation is virtually free of salts (Klepper and Kaufmann, 1966). On the other hand, with high external concentrations of mineral elements such as boron and silicon, the resorption of water from the xylem can be higher than that of the ions. These mineral elements therefore accumulate at the end of the xylem vessels and are either released by guttation, as in the case of boron (Oertli, 1962), or concentrated in the cell walls of the leaf epidermis, as in the case of silicon (Section 3.2.3). Necrosis on the tips or margins of leaves is a reflection of this insufficient resorption of certain mineral elements (e.g., boron and chlorine) during long-distance transport in the xylem.

Resorption from the xylem sap can be the result either of accumulation (storage or transformation) in individual parenchyma cells or of transient accumulation in specialized cells called the xylem parenchyma transfer cells. These cells are of particular importance in the stem for the transfer of mineral elements and organic compounds from the xylem to the phloem (Kuo *et al.*, 1980).

In some plant species, the resorption of certain mineral elements from the xylem sap is very pronounced and can have important consequences for the mineral nutrition of these plants. This is most evident in so-called natrophobic plant species (Section 10.2). In these plant species (e.g., bean), Na^+ is retained mainly in the roots and lower stem, whereas in natrophilic species (e.g., sugar beet) translocation into the leaves readily occurs (Fig. 3.2).

This restricted upward Na^+ translocation is caused by selective Na^+ accumulation in the xylem parenchyma cells of the roots (Kramer *et al.*, 1977) and stems (Rains, 1969). There is considerable evidence (Yeo *et al.*, 1977) that the xylem parenchyma transfer cells play a key role not only in this resorption but also in the phloem loading of Na^+ with its subsequent translocation to the roots (see Section 3.3.4). Winter (1982) demonstrated that concentrations of both Na^+ and Cl^- in the leaf blades of *Trifolium*

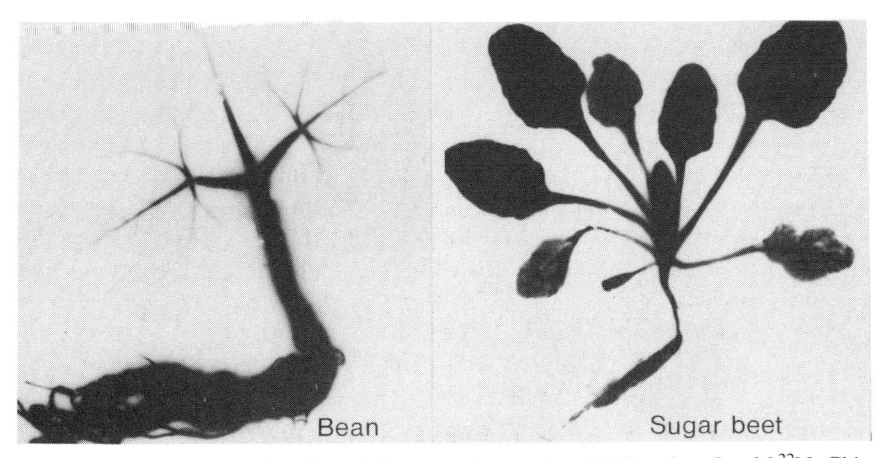

Fig. 3.2 Distribution of sodium in bean and sugar beet 24 hr after 5 mM $^{22}NaCl$ is supplied to the roots. Autoradiogram.

Table 3.3
Effect of Sodium Fertilizer on the Sodium Content of Roots and Shoots of Pasture Plants[a]

	Na Content (% dry wt)			
	Without Na fertilizer		With Na fertilizer	
Plant species	Roots	Shoots	Roots	Shoots
Lolium perenne	0·03	0·26	0·06	1·16
Phleum pratense	0·10	0·04	0·28	0·38
Trifolium repens	0·27	0·22	0·77	1·96
Trifolium hybridum	0·45	0·03	0·77	0·22

[a]Based on Saalbach and Aigner (1970).

alexandrinum were quite low as long as the xylem parenchyma cells in the petioles of leaf blades were intact.

Resorption of Na^+ from the xylem sap is therefore an effective mechanism of restricting translocation to the leaf blades. This mechanism, however, is not necessarily advantageous for the salt tolerance of plants (see Section 16.6) and is also a disadvantage in forage plants. For animal nutrition the sodium content of the forage should be at least 0·2%. As shown in Table 3.3, in *Lolium perenne* and *Trifolium repens*, Na^+ is readily translocated to the shoots, whereas in *Phleum pratense* and *Trifolium hybridum* this translocation is rather restricted. It is evident that for increasing the sodium content of forage the selection of suitable plant species is more important than the application of sodium fertilizers.

Resorption from the xylem sap can also be a determining factor in the distribution of micronutrients in plants. In certain species, such as bean and sunflower, molybdenum is preferentially accumulated in the xylem parenchyma of the roots and stems. In these species a steep gradient occurs in the molybdenum concentrations from the roots to the leaves (Table 3.4). In contrast, in other species, such as tomato, molybdenum is readily translo-

Table 3.4
Distribution of Molybdenum in Bean and Tomato Plants Supplied with Molybdenum in the Nutrient Solution[a]

	Molybdenum content (mg/g dry wt)	
Plant parts	Bean	Tomato
Leaves	85	325
Stems	210	123
Roots	1030	470

[a]Concentration of molybdenum in solution: 4 mg per liter. Based on Hecht-Buchholz (1973).

cated from the roots to the leaves. In agreement with this finding, when the molybdenum supply in the nutrient medium is high, toxicity occurs much earlier in tomato than in bean or sunflower (Hecht-Buchholz, 1973).

3.2.1.3 Release or Secretion

The composition of the xylem sap along the transport pathway can also be changed by the release or secretion of solutes from the surrounding cells. For example, in nonlegumes supplied with nitrate, the nitrate concentration in the xylem sap decreases as the path length increases, whereas the concentration of organic nitrogen, glutamine in particular, increases (Pate *et al.*, 1964). In nodulated legumes (where N_2 fixation occurs), on the other hand, the ratio of amides to amino acids is shifted in favor of the amino acids (Pate *et al.*, 1979).

Besides these specific aspects of nitrogen translocation, the release or secretion of mineral nutrients from the xylem parenchyma (and stem tissue in general) is of major importance for the maintenance of a continuous nutrient supply to the growing parts of the shoots. In periods of ample supply to the roots, mineral nutrients are resorbed from the xylem sap, whereas in periods of insufficient root supply they are released into the xylem sap. Changes in the potassium and nitrate contents of the stem base reflect this functioning of the tissues along the xylem in response to changes in the nutritional status of a plant. From this information a rapid test for nitrate in the stem base has been developed as a basis for recommending nitrogen fertilizer application (Chapter 8).

3.2.2 Effect of Transpiration Rate on Uptake and Translocation

The rate of water flux across the root (short-distance transport) and in the xylem vessels (long-distance transport) is determined by the root pressure and the rate of transpiration. An increase in the transpiration rate enhances both the uptake and the translocation of mineral elements in the xylem. This enhancement can be achieved in various ways, as shown in Fig. 3.3. An increase in mass flow of the external solution into the apparent free space of the cortex (Fig. 3.3C) can be of particular importance for soil-grown plants (see Chapter 15).

The effect of transpiration on the uptake and translocation rate of mineral elements depends predominantly on the following factors:

1. *Plant age.* In seedlings and young plants with a low leaf surface area, the effect of transpiration is extremely small; water uptake and transport to the

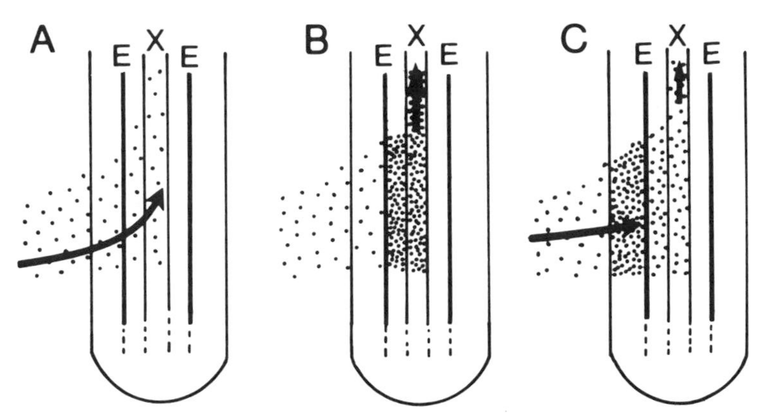

Fig. 3.3 Model for the enhancement effects of high transpiration rates on the uptake and translocation of mineral elements in roots. A. "Passive" transport through the apoplast into the stele. B. More rapid removal of mineral elements released in the xylem vessels (Emmert, 1972). C. Increase in the mass flow of the external solution into the apparent free space of the cortex, favoring active uptake into the symplasm. E, Endodermis; X, xylem; arrow, water flux (A to C see text).

shoots are determined mainly by the root pressure. As the age and size of the plants increase, the relative importance of the transpiration rate, particularly for the translocation of mineral elements increases rapidly.

2. *Time of day.* In leaves more than 90% of the total transpiration is stomatal. During the light period, transpiration rates and thus the enhancement of uptake and translocation of mineral elements are much higher than during the dark period. Short-term transient falls in the translocation rates of mineral elements at the onset of the dark period reflect the change from transpiration–mediated to root pressure–mediated xylem volume flow (Crossett, 1968). A consistent decline in uptake and translocation rates of mineral elements in the dark period is sometimes caused by a shortage of carbohydrates in the roots.

A particular situation exists in nodulated legumes in the diurnal pattern of fixed nitrogen. The sharp decrease in transpiration–driven xylem volume flow during the dark period is compensated for by a sharp increase in the concentration of fixed nitrogen (as ureides, see Chapter 7) in the xylem sap, thus keeping the total xylem transport rate of fixed nitrogen constant throughout the light/dark cycle (Rainbird *et al.*, 1983).

3. *External concentration.* It is well known that an increase in the concentration of mineral elements in the nutrient medium enhances the effect of transpiration rate on the uptake and translocation of mineral elements. This is most likely the result of the factors A and C shown in Fig. 3.3. Usually,

Table 3.5
Effect of Transpiration Rate of Sugar Beet Plants on Uptake and Translocation of Potassium and Sodium from Nutrient Solutions[a,b]

	Potassium		Sodium	
External concentration (m*M*)	Low transpiration	High transpiration	Low transpiration	High transpiration
	Uptake rate (μmol/plant × 4 hr)			
$1\ K^+ + 1\ Na^+$	4·6	4·9	8·4	11·2
$10\ K^+ + 10\ Na^+$	10·3	11·0	12·0	19·1
	Translocation rate (μmol/plant × 4 hr)			
$1\ K^+ + 1\ Na^+$	2·9	3·0	2·0	3·9
$10\ K^+ + 10\ Na^+$	6·5	7·0	3·4	8·1

[a]Based on Marschner and Schafarczyk (1967) and W. Schafarczyk (unpublished).
[b]Transpiration in relative values: low transpiration = 100; high transpiration = 650.

translocation rates are more sensitive to transpiration than are uptake rates, as shown for potassium and sodium in Table 3.5. The effect of transpiration on potassium is quite small in comparison with the effect on sodium. This difference corresponds to the differences in the uptake isotherms of these elements at increasing external concentrations (Fig. 2.20).

4. *Internal concentration.* The effect of transpiration also depends on the internal concentration, that is, the nutritional status of a plant. Transpiration usually enhances the uptake rate of a mineral nutrient to a greater degree in plants with high internal concentrations. This is merely an indirect effect related to the lower metabolically controlled uptake rates at high internal concentrations. Assuming a certain "passive" passage of mineral elements across the roots in all plants, this component is relatively more important in plants with lower rates of metabolically controlled uptake.

5. *Type of mineral element.* Under otherwise comparable conditions (e.g., plant age and external concentration), a typical ranking exists in the effect of transpiration rates on the uptake of mineral elements (Table 3.5). In addition, transpiration enhances the uptake and translocation of uncharged molecules to a greater extent than that of ions, a feature which is related to the corresponding differences in membrane permeation (Fig. 2.9). There is a close relationship between the transpiration rate and uptake rates of anionic dyes (e.g., light green) and certain herbicides (Shone *et al.,* 1973). The uptake and translocation of mineral elements as molecules is of greatest significance in the case of boron (Fig. 2.12) and silicon (monosilisic acid; Jones and Handreck, 1965). The close correlation between transpiration and the uptake of silicon is shown for oat plants in Table 3.6.

Table 3.6
Calculated and Measured Silicon Uptake in Relation to Transpiration (Water Consumption) of Oat Plants[a]

Harvest after days	Transpiration (ml/plant)	Measured uptake (mg/plant)	Calculated Si uptake[b] (mg/plant)
44	67	3·4	3·6
58	175	9·4	9·4
82	910	50·0	49·1
109	2785	156·0	150·0

[a]From Jones and Handreck (1965).
[b]Silicon concentration in the soil solution: 54 mg/liter.

There is perfect agreement between the silicon content measured in plants and that predicted from the transpiration values (water loss times silicon concentration in the soil solution). It is possible, therefore, to use the silicon content of field-grown wheat plants as a parameter for calculating the water consumption of plants during ontogeny (Hutton and Norrish, 1974).

Despite these close correlations, roots are not freely permeable to the radial transport of silicon. There is a certain filtration effect (Jones and Handreck, 1969), most likely at the endodermis, where heavy deposition of silicon occurs in the cell walls of field-grown plants (Bennett, 1982). Although transpiration can also affect the rates of uptake of other mineral elements, such as calcium (Section 3.7) and cadmium (Hardiman and Banin, 1982), this effect is small compared with the effect on silicon or boron uptake. In general, the effect of transpiration is much more pronounced in the distribution of mineral elements within the shoot and its organs.

3.2.3 Effect of Transpiration Rate on Distribution within the Shoot

The long-distance transport of a mineral element exclusively in the xylem should be expected to give a distinct distribution pattern in the shoot organs that depends on both transpiration rates (e.g., milliliters per gram dry weight each day) and duration of transpiration (e.g., age of the organ). Both the distribution and content of silicon accurately reflect the loss of water from the various organs. The silicon content increases with leaf age and is particularly high in spikelets of cereals such as barley. Even within a certain tissue, the silicon distribution resembles perfectly the pathway of transpiration flow in the apoplast. Silicon is deposited in the walls of the epidermis cells and in the cell walls of inflorescent bracts of rice grains (Soni and Parry, 1973) or in the pericarp and outer aleurone layer of grass seeds such as *Setaria italica* (Hodson and Parry, 1982).

The distribution of boron is also related to the loss of water from the shoot

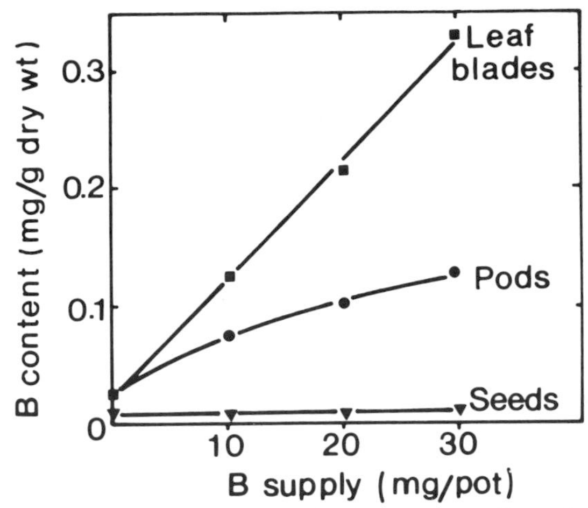

Fig. 3.4 Effect of increasing boron application to the soil on the distribution of boron in the shoots of rape. (Recalculated from Gerath *et al.*, 1975.)

organs (Michael *et al.*, 1969), as shown by the boron distribution in shoots of rape in response to an increasing boron supply (Fig. 3.4). The typical gradient in the transpiration rates among the shoot organs (leaves > pods ≫ seeds) corresponds to the gradient in boron content.

Even within a particular leaf, an excessive supply of boron creates a steep gradient in the boron content: petioles < middle of the leaf blade < leaf tip (Oertli and Roth, 1969). Necrosis on the margins or leaf tips is therefore a typical symptom of boron toxicity (Fig. 3.5). In salt-affected plants the

Fig. 3.5 Boron toxicity in the leaves of lentil. (*Left*) Control; (*right*) boron toxicity.

visible symptoms of toxicity (e.g., by chloride) are often quite similar, reflecting the transpiration-mediated distribution pattern within the shoot and its organs.

Frequently, a close positive correlation is observed (Wiersum, 1966) between calcium distribution and the transpiration rates of shoot organs. This is shown, for example, by the low calcium content of fleshy fruits (<0·1% calcium dry wt) as compared with that of the leaves (3–5% calcium dry wt) in the same plant. A lowering of the transpiration rate further decreases the calcium content of fruits (Table 3.7). The effect of transpiration on magnesium is much lower than its effect on calcium, and that on potassium is negligible.

Despite the correlations shown in Table 3.7, the interactions are much more complex between the rates of water and calcium influx into a plant organ (Section 3.3.9).

Table 3.7
Effect of Transpiration Rates of the Shoots of Red Pepper during Fruit Growth on the Mineral Element Content of the Fruits[a]

Transpiration rate (relative)	Mineral element content (mg/g dry wt)			Fruit dry wt (g/fruit)
	K	Mg	Ca	
100	91·0	3·0	2·75	0·62
35	88·0	2·4	1·45	0·69

[a]From Mix and Marschner (1976b).

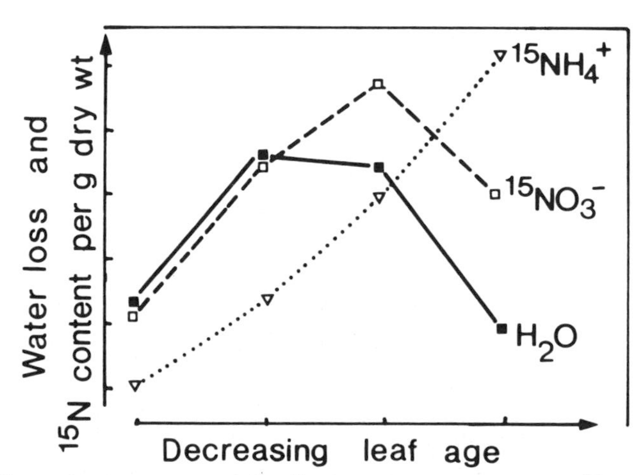

Fig. 3.6 Transpiration rates and distribution of labeled nitrogen (^{15}N) in different leaves of bean after $^{15}NO_3^-$ and $^{15}NH_4^+$ are supplied to the root. (Redrawn from Martin, 1971.)

The influence of transpiration on the distribution differs not only among mineral elements but also among the various forms of the same element, as shown in Fig. 3.6 for nitrogen. Whereas the distribution within the shoot of ^{15}N from nitrate follows the transpiration pattern quite closely, the distribution of ^{15}N from ammonium is independent of the transpiration rates of the leaves and is translocated preferentially to the shoot apex, which acts as a sink for reduced nitrogen.

The fact that transpiration rates are higher and the leaf water potentials are lower in mistletoe than in the host plant presumably explains why xylem parasites such as *Loranthus* can compete effectively with the host for mineral nutrients, nitrogen in particular, in the xylem fluid (Schulze *et al.*, 1984).

3.3 Phloem Transport

3.3.1 Principles and Phloem Anatomy

Long-distance transport in the phloem takes place in living cells. The phloem consists of a complex of various cell types: sieve tube elements, companion cells, and parenchyma cells (Fig. 3.7). Some of these individual sieve tube elements are stretched end to end in a long series, forming the sieve tubes which are connected by conspicuous pores (inset, Fig. 3.7) called sieve plate pores. The sieve tubes are highly specialized vascular systems for the long-distance transport of solutes. The sieve tube cells contain a thin layer of cytoplasm, which forms transcellular filaments (the so-called P-protein) that pass through the sieve plate pores. The anatomical features of long-distance transport in the sieve tube across the sieve plate pores are similar to those of short-distance transport in the symplasm across the plasmodesmata.

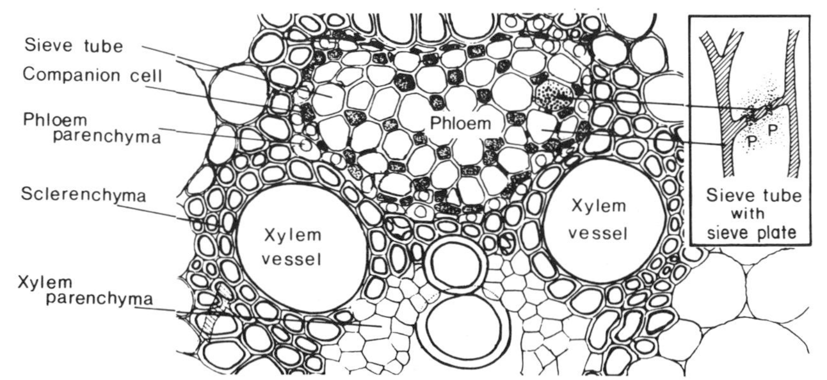

Fig. 3.7 Cross-sectional area of a vascular bundle from the stem of maize. Inset: sieve tube with sieve plate pores and "P-protein." (From Eschrich, 1976.)

In most plant species the sieve plate pores are lined with callose, a highly hydrated polysaccharide. There is good evidence that this callose can swell rapidly and fill the pores, thus blocking long-distance transport in the sieve tubes. This process of plugging is induced by such factors as heat treatment or mechanical pertubation of the stem (Jaffe *et al.*, 1985), as well as by mechanical injury of the sieve tubes, such as incision. Incision causes a sudden fall in the high internal pressure of the sieve tubes (>10 bars), which presumably triggers the plugging of the sieve tube plates. Considering the role of sieve tubes as food-connecting channels, this process can be thought of as performing the same function as a "security valve" that prevents "bleeding" when the system is injured. For studies on long-distance transport this plugging mechanism is both an advantage and a disadvantage. It is an advantage in that, very soon after decapitation of a plant, only xylem exudate is obtained at the stump of the root or stem; it is a disadvantage, in that, with a few exceptions—for example, the inflorescent stalks of certain palm tree species—it is very difficult to collect phloem exudate and thus to conduct extended studies on the mineral element composition of the phloem sap. There are some plant species (e.g., *Ricinus* and *Lupinus* spp.) from which small amounts of phloem exudate can be collected relatively easily by careful incision. However, with the incision technique there is always a possibility of contaminating the phloem sap by cut parenchyma cells and by substances from the apoplast. Another method is to use aphids. In the process of feeding, an aphid inserts its stylet into the phloem tissue; if the insect is snipped off, the stylet remains in the tissue. The high internal pressure within the sieve tubes forces the phloem sap out of the open end of the stylet. This technique, of course, is very troublesome, and the amounts of exudate obtained are quite small. For these reasons our knowledge of long-distance transport based on phloem sap analysis is rather limited, particularly for mineral elements.

3.3.2 Composition of the Phloem Sap

Phloem exudate analysis indicates that phloem sap has a high pH (7–8) and high concentrations of solids, on average 15–25% dry matter. The main component is usually sucrose, which comprises more than 90% of the solids. In addition, there are fairly high concentrations of organic acids and organically bound nitrogen, particularly in the form of amino acids and amides; nitrate is not detectable in phloem exudates. Of the mineral elements, potassium is usually present in by far the highest concentration. It is followed by phosphorus, magnesium, and sulfur, the latter mainly in the reduced form (glutathione > methionine ≫ cysteine; Rennenberg *et al.*,

1979). The calcium concentration is always very low. Reliable data on micronutrients are very rare.

In a comprehensive study, Hocking (1980) analyzed both the phloem and xylem exudate of the same *Nicotiana* plant (Table 3.8). With the exception of calcium, the concentrations of all the solids were several times greater in the phloem exudate than in the xylem exudate. These data are in fairly good agreement with those obtained from analyses of the stems of *Ricinus* (Hall and Baker, 1972), the leaf sheaths of rice (Chino *et al.*, 1982), and the peduncles of lupins (Pate *et al.*, 1974).

Table 3.8
Comparison of the Levels of Organic and Inorganic Solutes in the Phloem and Xylem Exudates of *Nicotiana glauca*[a]

Substance	Phloem exudate (stem incision) pH 7·8–8·0 (μg/ml)[b]	Xylem exudate (tracheal) pH 5·6–5·9 (μg/ml)[b]	Concentration ratio phloem/xylem
Dry matter	170–196[c]	1·1–1·2[c]	155–163[c]
Sucrose	155–168[c]	ND	—
Reducing sugars	Absent	NA	—
Amino compounds	10,808·0	283·0	38·2
Nitrate	ND	NA	—
Ammonium	45·3	9·7	4·7
Potassium	3,673·0	204·3	18·0
Phosphorus	434·6	68·1	6·4
Chloride	486·4	63·8	7·6
Sulfur	138·9	43·3	3·2
Calcium	83·3	189·2	0·44
Magnesium	104·3	33·8	3·1
Sodium	116·3	46·2	2·5
Iron	9·4	0·60	15·7
Zinc	15·9	1·47	10·8
Manganese	0·87	0·23	3·8
Copper	1·20	0·11	10·9

[a]From Hocking (1980).
[b]ND, Not present in a detectable amount; NA, data not available.
[c]Milligrams per milliliter.

In Hocking's study (Table 3.8) organic acids were not analyzed. Nonetheless, these acids and a whole range of other organic compounds (vitamins, ATP etc.) are found in phloem exudates. A comparison of the concentrations of inorganic cations and anions in the phloem reveals a large excess of cations. This is compensated for mainly by amino acids (Hall and Baker, 1972; Mengel and Haeder, 1977). For a comprehensive review of the composition of the phloem, the reader is referred to Ziegler (1975).

3.3.3 Mobility in the Phloem

There are only a few mineral nutrients (boron, molybdenum, and nitrate nitrogen) which have not yet been found in reasonable concentrations in the phloem exudate. The question arises, however, as to whether the phloem exudate does in fact, perfectly reflect the *in vivo* mobility of mineral elements. Another approach to the study of phloem mobility is the use of radioactive-labeled elements as shown in a representative example for phosphorus (^{32}P) and sodium (^{22}Na) in Fig. 3.8. The long-distance transport of the labeled elements was followed after application to the tips of a leaf blade (Fig. 3.8). Due to the gradient in xylem water potential, retranslocation from the leaf tip and out of the treated leaf must take place in the phloem.

On the basis of studies such as this, a classification scheme has been devised (Table 3.9). From this classification, two profound discrepancies are apparent: the immobility of calcium despite its occurrence in the phloem

Table 3.9
Mobility of Mineral Elements in the Phloem[a]

Mobile	Intermediate	Immobile
Potassium	Iron	Lithium
Rubidium	Manganese	Calcium
Sodium	Zinc	Strontium
Magnesium	Copper	Barium
Phosphorus	Molybdenum	Boron
Sulfur		
Chlorine		

[a]From Bukovac and Wittwer (1957).

exudate (see Table 3.8) and the immobility of boron. The contradictory results on the phloem mobility of calcium are discussed in Section 3.3.9. In studies both with labeled boron (Martini and Thellier, 1980) and on the time course of boron translocation into developing fruits such as peanut (Campbell *et al.*, 1975) and apple (Goor and Lune, 1980), boron seems to be sufficiently phloem mobile, in order to meet the requirement of growing fruits. The limiting factor in the long-distance transport of boron is most likely the high permeability of the sieve tube plasma membrane to boron and a correspondingly high leakage of boron out of the sieve tubes (Oertli and Richardson, 1970; Raven, 1980). In vascular bundles with a high xylem volume flux rate (e.g., in the stem) the probability of long-distance transport of boron in the phloem in the opposite direction (from shoots to roots) is thus quite small (Oertli and Richardson, 1970).

3.3.4 Direction of the Phloem Transport

In contrast to xylem transport which is undirectional (for exceptions, see Section 3.7), phloem transport is bidirectional, from source to sink. In green plants the source (phloem loading sites) of photosynthates, which represent the main component of the phloem sap, are the leaves, and the sinks (phloem unloading sites) are the roots, shoot apices, fruits, and seeds. The sources of the mineral elements, which in principle are subject to the same bidirectional transport, are: (a) the apoplast of the stele in the roots, (b) the xylem in the stem and leaves, and (c) the leaf tissue, particularly during senescence (remobilization). The main sinks for mineral nutrients are the shoot apices, fruits, and seeds.

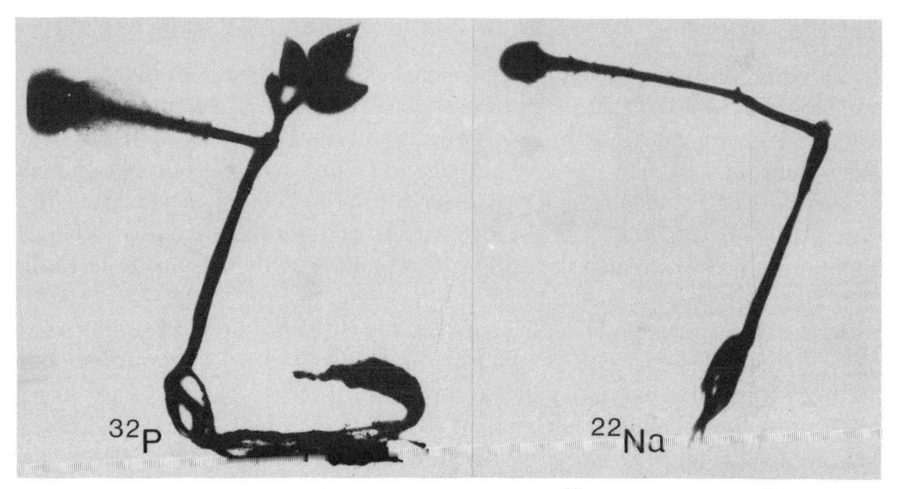

Fig. 3.8 Retranslocation of labeled phosphorus (^{32}P) and sodium (^{22}Na) after application to the tip of a primary leaf of bean. Autoradiogram, 24 hr after application.

This source-to-sink transport of mineral elements in the phloem can readily be demonstrated after the application of labeled phosphorus or sodium to the leaves (Fig. 3.8). For phosphorus the major sinks are the shoot apex and the roots, whereas transport from the treated mature primary leaf to the other primary leaf is negligible. The translocation to the root is somewhat surprising since the plants were well supplied with (nonlabeled phosphorus) in the nutrient solution. Possible reasons for this retranslocation are discussed later. The direction of ^{22}Na retranslocation is quite different from that of phosphorus: It is nearly exclusively confined basipetally to the roots, and there is considerable efflux into the external medium (Table 3.10).

Table 3.10
Distribution of ^{22}Na and ^{36}Cl in Plant Species 48 Hours after Application of Labeled NaCl to the Leaves[a]

	Treated leaf		Remainder of the shoot		Roots		External solution (efflux)	
Plant species	^{22}Na	^{36}Cl	^{22}Na	^{36}Cl	^{22}Na	^{36}Cl	^{22}Na	^{36}Cl
Sugar beet	92	86	7	13	1	1	0	0
Maize	69	45	29	28	1	6	1	1
Bean	52	91	29	7	5	1	14	1

[a]Relative values (percentage of total amount applied). Based on Lessani and Marschner (1978).

The direction of long-distance transport in the phloem can therefore be explained only in part by the source–sink relationship. There must be other mechanisms involved that prevent the accumulation of toxic concentrations of certain mineral elements such as sodium and chloride in the leaves of plants growing in saline substrates (Walker *et al.*, 1981; Winter, 1982). Both the intensity and direction of retranslocation of mineral elements such as sodium and chloride also depend on the plant species, as shown in Table 3.10.

In the salt-tolerant species (sugar beet) most of the applied sodium and chloride remains in the leaf, whereas in the species less tolerant to salt (maize) a much greater degree of retranslocation is observed with some preference for chloride, even back into the roots. In the salt-sensitive species (bean), only sodium is retranslocated preferentially to the roots and released from the root base (Fig. 3.8) into the external solution. This discrepancy between the retranslocation of sodium and that of chloride in bean explains the usually low sodium levels and high chloride levels in the shoots of this species and the corresponding chloride toxicity in plants grown on substrates with NaCl (Section 16.6).

3.3.5 Transfer between the Xylem and Phloem

In the vascular bundles, phloem and xylem are separated by only a few cells (Fig. 3.7). In the regulation of long-distance transport, an exchange of solutes between the two conducting systems is very important. From the concentration differences shown in Table 3.8 it is evident that a transfer from phloem to xylem can occur downhill, that is, as "leakage" through the plasma membrane of the sieve tubes if a sufficient concentration gradient exists. In contrast, for most organic and inorganic solutes a transfer from xylem to phloem is an uphill transport against a steep concentration gradient; that is, it is most likely an active transport process. This transfer of mineral elements can take place all along the pathway from root to shoot. There is convincing evidence that the stem plays an important role in this xylem-to-phloem transfer (McNeil, 1980), most likely via transfer cells (Kuo *et al.*, 1980, Fig. 3.9). Xylem-to-phloem transfer is of particular importance for the mineral nutrition of plants, because xylem transport is directed mainly to the sites (organs) of highest transpiration, which are usually not the sites of highest demand for mineral nutrients. In stems of graminaceous species (e.g., cereals) the nodes are the sites of intensive xylem-to-phloem transfer for mineral nutrients such as potassium (Haeder and Beringer, 1984b).

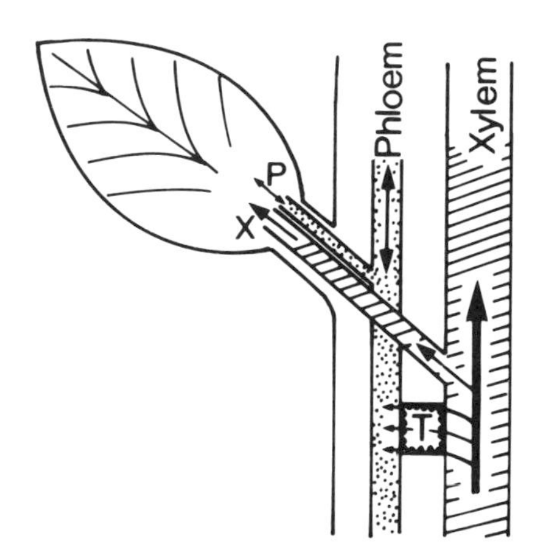

Fig. 3.9 Long-distance transport in xylem (X) and phloem (P) in a stem with a connected leaf, and xylem-to-phloem transfer mediated by a transfer cell (T).

No information is available on the opposite process, phloem-to-xylem transfer, except in the case of wheat stems after anthesis (Martin, 1982); retranslocation in the phloem from the flag leaf to the stem is followed by a considerable release of phosphorus, magnesium, and nitrogen, but not potassium, into the xylem with the subsequent transport of these mineral elements through the xylem into the ears. This particular mechanism may at least in part be responsible for the relatively low content of potassium in the ears and grains of cereals, which is caused by restricted import, but not enhanced export, of potassium (Martin and Platz, 1982; Haeder and Beringer, 1984a).

3.4 Relative Importance of Phloem and Xylem for the Long-Distance Transport of Mineral Elements

The relative importance of the phloem and xylem for the long-distance transport of mineral elements depends on various factors, including the kind of mineral element and the developmental stage of an organ. Nitrate is not detectable in the phloem exudate: its long-distance transport is confined to the xylem. In the reduced form however, nitrogen is readily transported in the phloem (e.g., as amino acids). A large proportion of the total amount of calcium required by growing organs has to be transported in the xylem (Section 3.3.9). The relative importance of the phloem and xylem for the transport of mineral elements that are readily mobile in both conducting systems, varies considerably. This is shown in Table 3.11 for the transport of potassium in barley leaves.

Table 3.11
Rate of Potassium Influx via Xylem and Phloem in Barley Leaves of Different Ages[a]

	Leaf age		
Influx	Young (area and dry wt increasing)	Intermediate (maximum area but dry wt increasing)	Old (maximum area and dry wt)
Via xylem	2·0	2·7	1·9
Via phloem	1·3	0·7	−1·6
Net uptake	3·3	3·4	0·3

[a]Influx rate expressed as micromoles per day. Based on Greenway and Pitman (1965).

During leaf expansion potassium influx takes place in both the xylem and phloem. The phloem influx of potassium declines sharply at leaf maturation and is reversed at later stages, becoming phloem export. Even at these later stages there is still a high xylem influx of potassium. Bidirectional long-distance transport in an organ is not limited to potassium, but can also be demonstrated, for example, for phosphorus (Greenway and Gunn, 1966). A lack of change in the net contents of phloem-mobile mineral elements in fully expanded leaves is therefore a reflection either of cessation of the influx or, more often, of an equilibrium between influx and efflux (retranslocation).

A major difficulty in making a quantitative assessment of solute transport in xylem and phloem is the need to determine not only the concentrations of solutes but also the velocity of transport and the cross-sectional area of the conducting vessels, according to the formula

$$\underset{(\text{g/cm}^2 \times \text{hr})}{\text{Specific mass transfer}} = \underset{(\text{cm/hr})}{\text{velocity}} \times \underset{(\text{mg/ml})}{\text{concentration}}$$

The velocity of transport in the xylem and phloem varies enormously, of course. On average, velocities between 10 and 100 cm/hr are often found, and there is a tendency for rates in the phloem to be considerably lower than those in the xylem. In fruit stalks of lupins, the maximal velocities were 22 cm/hr in the phloem and 147 cm/hr in the xylem (Pate *et al.*, 1978).

3.5 Circulation of Mineral Nutrients between Shoots and Roots

Phloem-mobile mineral nutrients (including reduced nitrogen) are retranslocated from the shoots to the roots even when the roots are supplied with these nutrients from the external solution. This retranslocation is somewhat surprising, because under these conditions the roots are the source and the shoots are the sink for the mineral nutrients. It is obvious that this view is an oversimplification of the source–sink concept. A certain proportion of the phloem-mobile mineral nutrients circulate from the shoots to the roots and back again. This could simply be the consequence of the mechanism (mass flow, see Section 5.4) and the direction of the phloem transport governed by the sugar transport from source (leaves) to sink (roots). There is, however, substantial evidence that specific mechanisms are involved for mineral nutrients. Simpson *et al.* (1982) supplied the roots of barley plants with nitrate nitrogen and found that, of the nitrogen translocated in the xylem to the shoots (100%), up to 79% was retranslocated in the phloem as reduced nitrogen back to the roots; of this 79%, ~21% was incorporated into the root tissue and the remainder translocated back in the xylem to the shoots. A limited capacity of the roots for nitrate reduction could be one of the reasons for this retranslocation of reduced nitrogen.

Retranslocation of potassium from shoots to roots can make up as much as 20% of the root-to-shoot transport (Armstrong and Kirkby, 1979a) and might be related, at least in part, to the role of potassium as a counterion for nitrate transport in the xylem (cation–anion balance). According to Ben-Zioni *et al.* (1971), after nitrate reduction in the shoot and the corresponding synthesis of organic acids for charge balance, potassium and organic acid anions (mainly malate) are retranslocated via the phloem to the roots, where potassium can act again as a countercation for nitrate transport. This circulation (Fig. 3.10) would be quite an economic mechanism for avoiding excessive organic acid storage in the leaf vacuoles as a consequence of nitrate reduction (Chapter 8). Furthermore, potassium could act more than once as a countcrion in the long-distance transport of nitrate.

Regardless of the physiological reasons for retranslocation of mineral nutrients from the shoots to the roots, this process provides a specific means of regulating uptake rates by the roots. The concentration of mineral

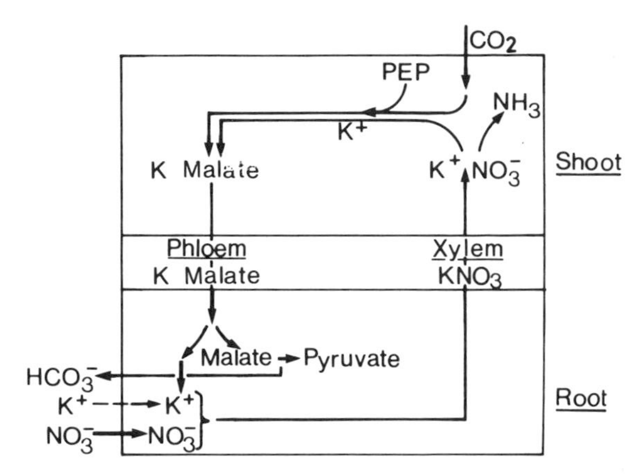

Fig. 3.10 Model for the circulation of potassium between roots and shoots in relation to nitrate and malate transport (PEP, phosphoenol pyruvate). [Based on Ben-Zioni *et al.* (1971) and Kirkby and Knight (1977).]

nutrients in the phloem varies according to the nutritional status of a plant, as has been shown for potassium (Mengel and Haeder, 1977). It is quite likely that a decline in the shoot concentration of a particular mineral nutrient and a corresponding decline in the retranslocation rate to the roots constitute a "signal" for increasing the rate of uptake of this mineral nutrient (Fig. 2.20).

3.6 Remobilization of Mineral Nutrients

Influx and efflux of mineral nutrients can occur simultaneously in plant organs such as leaves. Efflux (or retranslocation) therefore results in a decrease in net content only if the rate of efflux exceeds the rate of influx. Determinations of this dynamic relationship are complicated and require the use of either radioisotopes or separate collections of xylem and phloem exudates. It is therefore common to determine only the time course of net changes in the mineral element composition of various plant parts, leaves in particular. In order to avoid confusion, this decrease in net content (or, more precisely, in the amount per organ) is denoted by the term *remobilization*.

Remobilization of mineral nutrients is important during the ontogenesis of a plant in the following stages: seed germination; periods of insufficient supply to the roots during vegetative growth; reproduction; and, in perennials, the period before leaf drop. For reviews on remobilization the reader is referred to Pate (1975) and Hill (1980).

3.6.1 Seed Germination

During the germination of seeds (or storage organs such as tubers) mineral nutrients with the exception of calcium (Helms and David, 1973), are remobilized within the seed tissue and translocated in the phloem and/or xylem to the developing roots and shoots. As a result, seedlings will grow at least several days without an external supply of mineral nutrients.

3.6.2 Insufficient and/or Interrupted Supply

During vegetative growth nutrient supply to the roots is often either permanently insufficient (as in the case of low soil content) or temporarily interrupted (when, for example, there is a lack or excess of soil moisture). Remobilization of mineral nutrients from mature leaves to areas of new growth is thus essential for the completion of the life cycle of plants under these environmental conditions. The extent to which remobilization takes place, however, differs among mineral nutrients and is reflected in the distribution of visible deficiency symptoms in the plants. Deficiency symptoms in older leaves reflect high rates of remobilization, whereas those in young leaves and apical meristems reflect insufficient remobilization. Obviously, in the latter cases only a relatively small fraction of the mineral nutrients can be remobilized from fully expanded leaves, and the additional demand of areas of new growth cannot be transformed into a "signal" strong enough to bring about extensive remobilization of these mineral nutrients. Table 3.12 summarizes the characteristic differences in remobilization according to this criterion—distribution of deficiency symptoms. This general classification is quite helpful as a first step in the identification of nutrient deficiencies in various crop species under field conditions (Chapter 12).

3.6.3 Reproductive Stage

Remobilization of mineral nutrients is particularly important during the formation of seeds, fruits, and storage organs. In this growth stage the carbohydrate supply to the roots, and thus root activity and nutrient uptake, generally decrease rapidly. The mineral nutrient content of vegetative parts,

Table 3.12
Characteristic Differences between Distribution of Visible Deficiency Symptoms and Degree of Remobilization

Mineral nutrient	Part of plant in which deficiency symptoms occur predominantly	Retranslocation
Nitrogen, potassium magnesium and phosphorus	Old leaves	Very good
Sulfur	Young leaves	Insufficient
Iron, zinc, copper and molybdenum	Young leaves	Very low
Boron and calcium	Young leaves and apical meristems	Extremely low or nil

therefore, quite often declines sharply during the reproductive stage (Fig. 3.11).

The extent of this remobilization depends on various factors, including (a) the specific requirements of the seeds and fruits, (b) the mineral nutrient level in the vegetative parts, (c) the ratio between vegetative mass (source size) and number and size of seeds or fruits (sink size), and (d) the nutrient uptake rate by the roots during the reproductive stage. Cereal grains, for example, are characterized by a high content of nitrogen and a low content of potassium, magnesium, and calcium, whereas fleshy fruits (e.g., tomatoes) or storage organs (e.g., potato tubers) have a high content of potassium but a relatively low content of nitrogen and phosphorus.

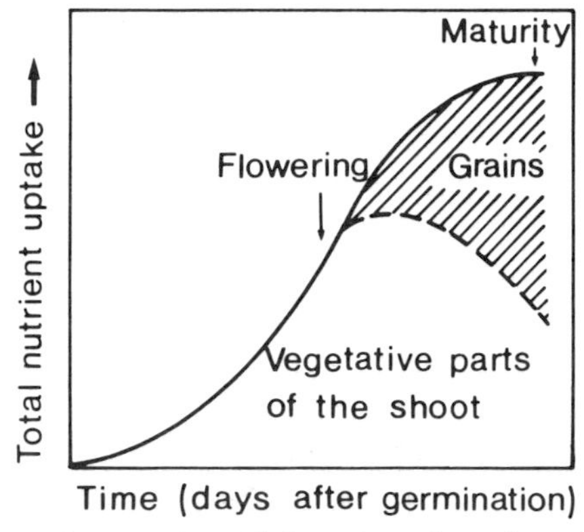

Fig. 3.11 Schematic representation of the mineral nutrient distribution in cereal plants during ontogeny.

A typical example of the differences in the degree of remobilization of these mineral nutrients from the vegetative shoots is shown in Table 3.13 for pea plants grown under field conditions. The percentage of remobilization of nitrogen and phosphorus is very high, whereas there is a lack of remobilization of magnesium and calcium; instead, a net increase in these nutrients takes place. In this particular case it should be kept in mind that part of the nitrogen demand of the seeds of legumes is met by N_2 fixation after flowering and that some leakage from the leaves is probably involved in the relatively high data for remobilization of potassium (Chapter 4). High rates of calcium uptake by the root after anthesis and negligible remobilization and a corresponding net increase in calcium in the vegetative shoots are typical features of field-grown plants, however.

Table 3.13
Remobilization of Mineral Nutrients in a Pea Crop between Flowering and Ripening[a]

	Mineral nutrients in leaves and stems (kg/ha)				
	N	P	K	Mg	Ca
Harvest time					
June 8 (flowering)	64	7	53	5	31
June 22	87	10	66	8	60
July 1	60	7	61	8	69
July 12 (ripening)	32	3	46	9	76
Percentage of increase or decrease after June 22	−63	−73	−30	+10	+21
Percentage in seeds in relation to the total content of shoots	76	82	29	26	4

[a]Based on Garz (1966).

The remobilization of highly phloem-mobile mineral nutrients can lead to such a rapid decline in their content in the vegetative shoots that severe deficiency symptoms and even leaf drop occur during the reproductive phase. Under these conditions, plants behave as "self-destructing" systems. An interesting example of this is the remobilization of potassium in two tomato cultivars (Table 3.14). The cultivar VF-13L was developed for mechanical harvesting. It is characterized by a heavy fruit load combined with an early and uniform maturation. Severe potassium-deficiency symptoms during fruit ripening occur in this cultivar even in plants growing in soils with high potassium availability. Obviously, in this genotype a particularly

Table 3.14
Potassium Content of the Petioles of Two Tomato Cultivars at Various Stages of Growth[a]

Cultivar	Third cluster, full bloom	First cluster, mature green	First cluster, fruit pink	50% fruits ripe
VFN-8	5·30	6·83	3·48	0·97
VF-13L	5·24	5·86	1·80	0·40[b]

[a]Potassium content expressed as percentage of dry weight. Based on Lingle and Lorenz (1969).
[b]Severe potassium-deficiency symptoms on the leaves.

strong sink competition for carbohydrates between fruits and roots causes a rapid decline in the root uptake of potassium in the period of high potassium demand for fruit growth. This is an instructive example of a specific yield limitation induced by a mineral nutrient (Chapter 6) and also demonstrates some of the physiological limitations of plant breeding for higher yields.

Remobilization is highly selective for mineral nutrients. This selectivity and the corresponding discrimination against mineral elements which are either not essential or required only at very low levels is quite impressive, as shown in Table 3.15 for barley grown in saline substrates. In the vegetative shoots the content of potassium is lower than that of sodium and chloride. During remobilization, however, potassium is highly preferred and the ratio of the three mineral elements is reversed in the ears. An additional step in the selection of nutrients takes place before their entry into the grains.

During the reproductive stage the degree of remobilization of micronutrients and of calcium is often astonishingly high compared with that during vegetative growth. In lupins (*Lupinus albus*), for example, up to 50% of the micronutrients and 18% of the calcium that originally accumulated in the leaves were retranslocated to the fruits (Hocking and Pate, 1978). The extent of remobilization of micronutrients strongly depends on their con-

Table 3.15
Mineral Element Content of Barley (cv. Palladium) Grown in a Saline Substrate[a]

Plant part	Content (μmol/g dry wt)		
	K	Na	Cl
Vegetative shoot	0·22	2·27	1·52
Rachis, glume, awn	0·56	0·42	0·43
Grain	0·13	0·04	0·04

[a]Based on Greenway (1962).
[b]Substrate contained 6mM K^+ and 125 mM Na^+ (as NaCl).

centration in the fully expanded leaves (Loneragan *et al.*, 1976). During grain development in wheat, for example, leaves with a high copper content lost more than 70% of their copper, whereas leaves of copper-deficient plants lost less than 20% (Hill *et al.*, 1978). This relationship between levels in leaves and degree of remobilization is in contrast to that of highly mobile mineral nutrients, such as nitrogen and potassium, of which a much higher percentage is remobilized in deficient plants.

The extent of remobilization of the micronutrients copper and zinc—but not manganese (Nable and Loneragan, 1984) is also closely related to leaf senescence. This is reflected, for example, in a close positive correlation that exists between the remobilization of nitrogen and that of copper (Fig. 3.12). Enhancement of senescence by shading is associated with more rapid remobilization of both nitrogen and copper; in copper-deficient plants most of the copper can then be remobilized. Nitrogen deficiency, like shading, also enhances the copper remobilization (Hill *et al.*, 1978). The same is true for zinc (Hill *et al.*, 1979b). These relationships may be in part responsible for the results of field experiments showing a particularly high copper demand in plants supplied with high levels of nitrogen fertilizers and a corresponding delay in leaf senescence (Chapter 6).

The relatively high remobilization rates of micronutrients during fruit development are presumably the result of fruit-induced leaf senescence. The development of sulfur-deficiency symptoms in either old or young leaves depends on the level of nitrogen supply (Loneragan *et al.*, 1976) and is most likely also related to leaf senescence.

Remobilization of mineral nutrients requires several steps: (a) mobilization within individual leaf cells, (b) short-distance transport in the sym-

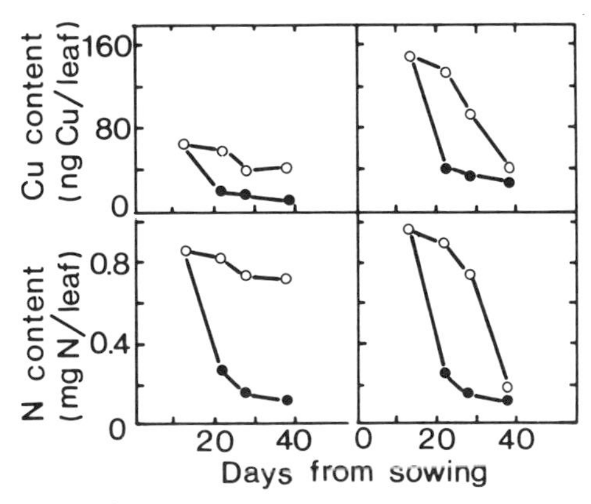

Fig. 3.12 Efects of copper supply and shading on the copper and nitrogen content of the oldest leaf of wheat. Key: ○——○, unshaded; ●——●, shaded. (From Hill *et al.*, 1979a.)

plasm to the phloem, (c) phloem loading, and (d) phloem transport. Discrepancies between high phloem mobility (Table 3.9) and low rates of remobilization (Table 3.12) are most likely caused by insufficient mobilization within the leaf cells. A large proportion of micronutrients are incorporated into organic structures, particularly chloroplasts. The enhanced breakdown of these structures during senescene is therefore associated with a marked increase in remobilization of some of these mineral nutrients.

The extent of remobilization of mineral nutrients is attracting increasing attention in connection with the selection and breeding of genotypes of high "nutrient efficiency". Genotypes that grow well on soils of low nutrient availability not only may have a higher rate of uptake and translocation of a particular mineral nutrient but may also exhibit higher nutrient efficiency at the cellular level (compartmentation, binding stage, etc.), including high rates of remobilization from older to younger leaves, seeds, and storage organs.

3.6.4 Period before Leaf Drop (Perennials)

Remobilization of mineral nutrients (except calcium) from the leaves to the woody parts is a typical feature of perennial species before leaf drop. Remobilization is closely related to the discoloration of leaves in the autumn. During this period, typical visible deficiency symptoms are often observed, indicating that during the growing period there might have been a latent deficiency of a particular mineral nutrient. In plants growing on saline substrates, preferential remobilization of certain mineral nutrients (Table 3.15) often gives rise to toxicity symptoms in leaf margins, indicating a further shift toward extreme ionic imbalance before leaf drop.

3.7 Long-distance Transport of Calcium: Xylem versus Phloem Transport

Calcium nutrition in general, and the long-distance transport of calcium in particular, are attracting increasing attention owing to the widespread occurrence of calcium deficiency and the so-called calcium-related disorders, such as tipburn in lettuce, blossom end rot in tomato, and bitter pit in apple. It is now well established that these disorders are caused primarily by the limited capability of plants to regulate calcium distribution internally in relation to the demand of low-transpiring organs, especially fast-growing leaves, tubers, and fruits. For comprehensive reviews on this subject, see Bangerth (1979), Kirkby (1979), and Marschner (1983).

From xylem and phloem exudate analysis, it is predicted that plant parts or organs supplied predominantly via the phloem should have a very low calcium but a high potassium content. There is a discrepancy, however, between the high potassium/calcium ratios that exist in phloem exudate and the relatively high calcium requirement of meristems and expanding tissues, particularly of dicotyledonous species. Furthermore, phloem sap analysis might overestimate the phloem mobility of calcium. Nevertheless, evidence has been presented in intact plants that a low rate of calcium influx via the phloem into growing tissue such as fruits can occur (Tromp, 1979). But for several reasons (high pH and high phosphate concentration) the concentration of free calcium (Ca^{2+}) in the phloem sap has to be very low (Raven, 1977). There is general agreement that, in order for the requirements of growing tissue to be met, most of the calcium has to be either translocated via the xylem into the tissue (Hill, 1980) or taken up directly from the external solution (soil solution), as in the case of developing fruits of peanut (Hallock and Garren, 1968) and potato tubers (Krauss and Marschner, 1975).

To increase the calcium content of growing fruits, increasing the transpiration rates of the fruits is more effective than increasing the calcium supply in the nutrient medium (Table 3.16). As expected, highly phloem mobile potassium is not affected by these treatments. Furthermore, there is a distinct negative correlation between the growth rate and the calcium content of growing fruits, whereas the potassium content, again, is unaffected (Table 3.16). Higher growth rates are based on higher solute influx via the phloem and thus correlated with higher potassium, but a very low influx of calcium. In addition, in organs with low transpiration rates, such as fruits, a high phloem solute influx either strongly depresses or even reverses the direction of the xylem volume flow (Mix and Marschner, 1976); thus, high growth rates have a corresponding negative effect on the calcium influx and content of fruits such as pepper (Table 3.16), tomato (Wiersum, 1966), and apple (Wieneke and Führ, 1973).

High transpiration rates of the whole shoot, however, often decrease rather than increase the calcium influx into low-transpiring organs such as rosettes of cauliflower (Krug *et al.*, 1972) or the inner leaves of cabbage (Palzkill *et al.*, 1976). Under these conditions the xylem volume flow is directed to the high-transpiring outer leaves at the expense of the inner leaves or the rosettes. Inhibition of transpiration (by high relative humidity or during the dark period) usually favors the direction of the xylem volume flow toward low-transpiring organs. Diurnal shrinking (during the light period) and swelling (during the dark period) of the rosettes of cauliflower

Table 3.16
Effect of Environmental Factors and Growth Rate on Calcium and Potassium Content of Red Pepper Fruits[a]

Treatment and growth rate	Content in fruits (μmol/g dry wt)	
	Ca	K
Calcium supply to the roots (m*M*)		
0·5 m*M*	26·9	1315
5·0 m*M*	33·2	1228
Relative humidity in the fruit environment (%)		
90%	32·7	1892
40%	55·4	1918
Growth rate of the fruits (mg dry wt/day)		
20·1	28·2	1772
29·9	20·7	1846
38·5	17·2	1813

[a]From Marschner (1983).

(Krug *et al.*, 1972) or the head of cabbage (Wiebe *et al.*, 1977) are closely correlated with corresponding changes in the xylem flow–directed calcium influx into various parts of the shoots.

Under conditions of low transpiration, the rate of xylem volume flow from the roots to the shoots is determined by the root pressure. The influx of water and calcium via the xylem into low-transpiring organs therefore depends on the root pressure. Water availability in the rooting medium, particularly during the dark period, is thus crucial for the long-distance transport of calcium into low-transpiring organs. In agreement with this, high osmotic potential of the soil solution (e.g., soil salinity) decreases both root pressure and calcium influx into young leaves or fruits and induces calcium-deficiency symptoms (Ende *et al.*, 1975; Palzkill *et al.*, 1976; Bradfield and Guttridge, 1984). These relationships are shown in Table 3.17 for expanding strawberry leaves. High root pressure, as indicated by the intensity of guttation, is closely correlated with an increased concentration of calcium in expanding leaves and either the absence of, or only mild symptoms of, calcium deficiency (tip necrosis). Magnesium, which is highly phloem mobile, is only slightly affected by root pressure. On the other hand, as has been shown in Section 2.9.2, an increase in root pressure caused by elevated root temperatures affects the potassium/calcium ratio in the xylem and might thereby enhance calcium deficiency symptoms.

The rate of calcium influx into an organ via the xylem is regulated by yet another factor. As shown above, exchange adsorption plays an important role in the long-distance transport of calcium through the xylem. Removal of calcium from these exchange sites (binding sites) at the end of the transport

Table 3.17
Relationship between Root Pressure (Guttation) and Calcium Transport into Expanding Strawberry Leaves[a]

Concentration of the nutrient solution[b]		Guttation (relative) (0–3)[c]	Tip necrosis (0–5)[d]	Content (μg/leaf)	
Day	Night			Ca	Mg
Concentrated	Concentrated	0·3	3·0	7	57
Concentrated	Diluted	2·4	0·3	25	77
Diluted	Concentrated	0·8	1·3	16	74
Diluted	Diluted	2·3	0·0	62	78

[a]Based on Guttridge *et al.* (1981).
[b]Root pressure varied by the concentration of the nutrient solution: Concentrated = 6·5 atm; diluted = 1·6 atm.
[c]0 = none; 3 = high.
[d]0 = none; 5 = very severe.

system (at the sink) enhances the transport rate and vice versa. Xylem transport rates of water and calcium are therefore not directly related. In leaves, for example, the calcium influx rate per unit of transpired water is high during expansion and decreases sharply after maturation (Koontz and Foote, 1966; Marschner and Ossenberg-Neuhaus, 1977). The same shift can be observed during the growth of fruits (Mix and Marschner, 1976a). New exchange sites in the apoplast (mainly cell wall $R{\cdot}COO^-$) of growing tissue act as sinks for calcium and are mainly responsible for the preferential long-distance transport of calcium into the apical meristems, expanding leaves, or developing fruits (Marschner and Ossenberg-Neuhaus, 1977). This may explain why preferential calcium transport in the dark period into the apical mersitems is not limited to intact plants (with root pressure), but is also observed in isolated shoots, although at a much lower level of calcium influx (Geijn and Smeulders, 1981). It should be kept in mind that xylem volume flow into expanding tissue is conceivable without transpiration or root pressure, since water is taken up during cell expansion (Hill, 1980).

Our knowledge of calcium transport is still rather limited, especially in terms of its regulation in the zone of transition from long- to short-distance transport in the apoplast of growing tissue. Nevertheless, the direct connection between the long-distance transport of calcium and the water economy of plants and their organs and the indirect connection with the phloem/xylem flux ratio into an organ make it possible to control the calcium distribution to some extent. In the long run, to overcome calcium-related disorders, it will certainly be necessary to make more use of genotypical differences within species with regard to the uptake, transport, and distribution of calcium on both the tissue and cellular level.

The role of the formation of new binding sites in the apoplast for the

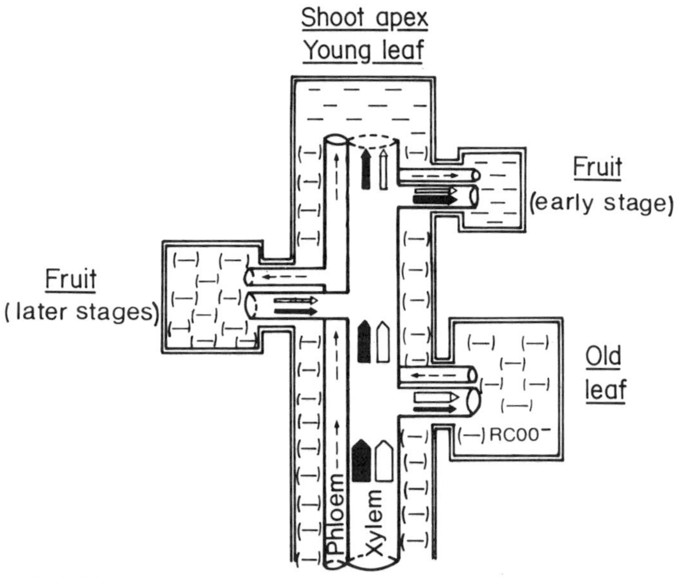

Fig. 3.13 Model for the long-distance transport of calcium in the shoot (➡ ➡ ➡, decreasing rate) and mass flow of water in the xylem (⇨ ⇨ ⇨, decreasing rate). Note fixed negative charges in the cell walls (−) and formation of new exchange sites (negative charges) in the cell walls (−).

preferential transport of calcium to the shoot apex can be demonstrated by the application of TIBA (triiodobenzoic acid) to sunflower plants which results in a decrease in both the cation-exchange capacity and calcium content of the shoot apex but which has no effect on the content of other mineral nutrients (Marschner and Ossenberg-Neuhaus, 1977). In agreement with this, TIBA application leads to a decrease in the calcium content of apple fruits and an increase in the incidence of bitter pit (Lüdders and Fischer-Bölükbasi, 1980). Whether auxins, which are synthesized mainly in apical meristems, seeds, and developing fruits, are involved directly or indirectly in the action of these tissues as sinks for calcium is still an open question (Marschner, 1983). There is experimental evidence indicating a couple of acropetal calcium transport with a basipetal auxin transport (Bangerth, 1976; Guzman and Dela Fuente, 1984). Figure 3.13 presents a simplified model of the long-distance transport of calcium in relation to xylem and phloem water flow rates and the formation of new binding sites. Obviously, the mechanism for the regulation of calcium transport is more complex, or at least more likely to be affected by external and internal factors, than the regulation of highly phloem mobile mineral nutrients such as potassium.

4

Uptake of Mineral Elements by Leaves and Other Aerial Plant Parts

4.1 Uptake of Gases through Stomata

In terrestrial plants the stomata are the sites of exchange of gases (CO_2, O_2) with the atmosphere. Mineral nutrients in the form of gases, such as SO_2, NH_3 and NO_x, also enter the leaves predominantly through the stomata. This has been demonstrated with labeled SO_2 ($^{35}SO_2$), which is rapidly metabolized (reduced) and incorporated into organic compounds (Weigl and Ziegler, 1962). When sulfur was supplied to the shoots exclusively as SO_2, the growth rate of plants was comparable to that when sulfur was supplied as sulphate (SO_4^{2-}) to the roots (Table 4.1)

Table 4.1
Dry Matter Production and Sulfur Content of Tobacco after Exposure of SO_2 to the Shoots or Supply of SO_4^{2-} to the Roots[a,b]

	Dry weight (mg/plant)			Sulfur content (mg S/g dry wt)		
Plant part	No sulfur supply	SO_2	SO_4^{2-}	No sulfur supply	SO_2	SO_4^{2-}
Leaves	0·8	2·0	2·0	1·5	11·4	7·4
Roots	0·4	0·6	0·6	1·9	1·9	4·9

[a]Based on Faller (1972).
[b]Shoots were exposed to 1·5 mg SO_2 per cubic meter for 3 weeks. The concentration of SO_4^{2-} supplied to the roots was 80 mg/liter.

In the same study (Faller, 1972), NH_3 and nitrogenous gases were supplied to the leaves and their uptake and incorporation measured. The promotion of growth by these gases was demonstrated. Only rarely, however, does the uptake of these gases through the leaves make any real contribution to growth under field conditions. It may occur with SO_2 uptake by leaves in a sulfur-deficient soil or with NH_3 uptake in pastures. At night a steep concentration gradient of NH_3 can occur within a pasture canopy from the base (soil surface) to the atmosphere above to the canopy (the free atmosphere above the vegetation); during the day, however, the NH_3 concentration within the canopy drops to a very low level, indicating substantial NH_3 uptake through the stomata (Lemon and Houtte, 1980). Daily uptake rates of NH_3 by leaves in a pasture have been calculated to be between 100 and 450 g nitrogen per hectare (Cowling and Lockyer, 1981). In highly industrialized areas characterized by high levels of air pollution, plant growth may be retarded by excessive leaf uptake of gases such as SO_2 and various nitrogen oxides (e.g. NO, N_2O) and their reaction products (Mohr, 1983). At high concentrations, SO_2 can competitively inhibit the binding of CO_2 at the active center of ribulosebisphosphate carboxylase, the key enzyme in the C_3 pathway of CO_2 fixation. This effect might explain, at least in part, why in certain plant species such as *Picea abies* long-term treatment with SO_2 results in a decline in net photosynthesis at SO_2 concentrations as low as 0·05 mg/liter (Keller, 1981). The metabolic pathway of SO_2 incorporation in leaves is discussed in more detail in Section 8.3.

4.2 Uptake of Solutes

4.2.1 Structure and Function of the Cuticular Layer

In aquatic plants the leaves are the sites of mineral nutrient uptake. In terrestrial plants the uptake of solutes by the surfaces of leaves and other aerial parts is severely restricted by the external wall of the epidermal cells. This wall is covered by a layer of wax and cutin (a condensation product of C_{18} hydroxy fatty acids with semihydrophilic properties) and contains pectin, hemicellulose ("hemisubstances") and cellulose. A model of the external wall of an epidermal leaf is shown in Fig. 4.1.

A distinct gradient occurs within this structure from the hydrophobic external surface to the hydrophilic internal cell wall. The major function of the hydrophobic surface layer is to protect the leaf from excessive water loss by transpiration. The control of water economy in terrestrial higher plants by the stomata is dependent on a very low water permeability of the remaining surfaces of the plant. The Casparian strip in the endodermis of the roots acts as a similar "waterproof" barrier to block the movement of free solutes in the apoplast. It has to be borne in mind that mineral elements entering the leaves via the xylem are released into the apoplast of leaf tissue before uptake by individual leaf cells (Pitman *et al.*, 1974). The other main function of the hydrophobic surface layer of leaf surfaces is therefore

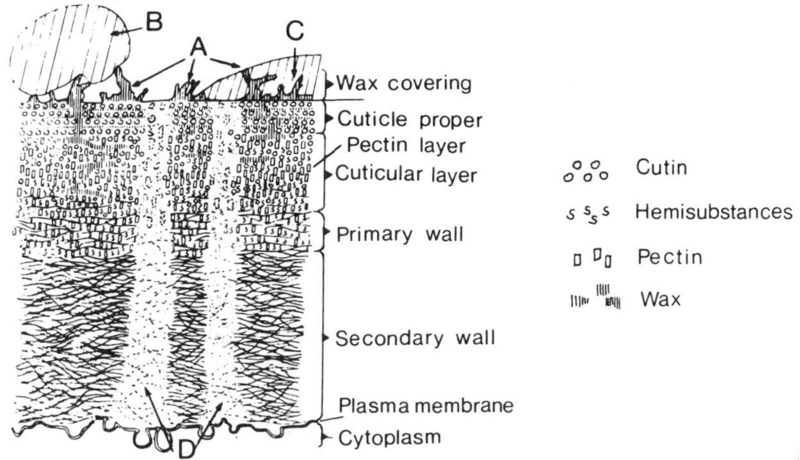

Fig. 4.1 Simplified scheme of the outer wall of a leaf epidermal cell. A. Wax rodlets. B. Water droplet. C. Water droplet with detergent. D. Ectodesmata. (Reproduced from Franke, 1967, with permission by Annual Reviews Inc.)

protection against excessive leaching of inorganic and organic solutes from the leaves by rain (see Section 4.4). The relative importance of these two functions depends on climatic conditions (arid regions versus humid tropics).

The functioning of the cuticle layer as a weak cation exchanger is attributable to the negative charge of the pectic material and nonesterified cutin polymers. A distinct gradient occurs from low to high charge density, from the external surface toward the cell walls (Yamada *et al.*, 1964). Ion penetration across the cuticle is therefore favored along this gradient, an important factor for both uptake from foliar sprays (Section 4.3) and losses by leaching (Section 4.4).

4.2.2 Role of Ectodesmata

The movement of solutes across the cuticular layer takes place in cavities (or channels), the so-called *ectodesmata* (see Fig. 4.1). The ectodesmata are definitely of a nonplasmatic nature (Franke, 1967; Schönherr and Bukovac, 1970) and serve as pathways for cuticular or peristomatal transpiration, which is particularly extensive in the cell wall system between the guard cells and subsidiary cells (Maier-Maercker, 1979). This explains the commonly observed positive correlation between the number or distribution of stomata (e.g., between the upper and lower leaf surface) or both and the intensity of mineral element uptake from foliar sprays (Franke, 1975; Levy and Horesh, 1984).

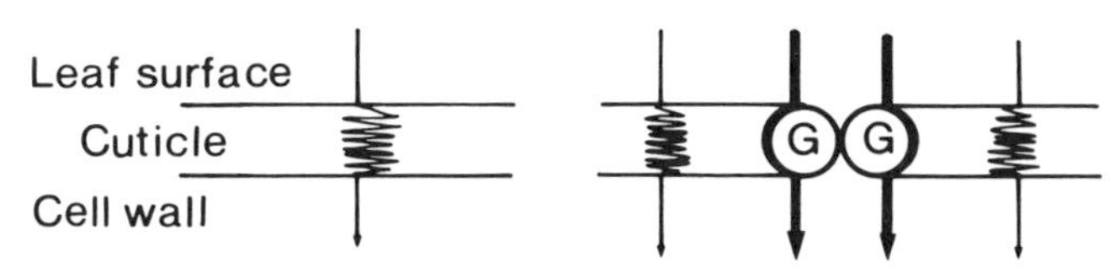

Fig. 4.2 Schematic representation of solute penetration across the cuticular layer of leaf epidermal cells (G, guard cell).

The differences in resistance to solute penetration at various parts of the cuticle are shown schematically in Fig. 4.2. It is unlikely that direct penetration of solutes from the leaf surface through open stomata into the leaf tissue plays an important role, because a cuticular layer (the internal cuticle) also covers the surface of the guard cells in stomatal cavities. Furthermore, ion uptake rates from foliar sprays are usually higher at night, when the stomata are closed, than during the day, when the stomata are open.

4.2.3 Role of External and Internal Factors

Leaf cells, like roots cells, take up mineral elements from the apoplast. Leaf uptake is thus similarly affected by external factors, such as mineral nutrient concentration and valency as well as temperature, and by internal factors, such as metabolic activity. The particularly high rates of urea uptake by leaves compared with those of ammonium or nitrate nitrogen uptake are in agreement with the principles of permeation in the AFS and across membranes (small uncharged molecules are taken up to a greater extent than ions), as discussed for root cells in Chapter 2. For a given external concentration of mineral elements, the rates of uptake by intact leaves are, of course, much lower than the rates of uptake by roots, since the cuticular layer severely restricts diffusion to the plasma membranes, the uptake sites. The thickness of the cuticular layer differs widely between plant species and is also affected by environmental factors; this is particularly evident in comparisons of plants grown under shaded and non-shaded conditions (Takeoka *et al.*, 1983).

The rate at which leaves take up mineral nutrients supplied to their surfaces also depends on the nutritional status of a plant, as shown for phosphorus in Table 4.2. The rate of uptake by the leaves of phosphorus-deficient plants was twice as high as that of control plants well supplied with phosphorus via the roots. In addition, much more phosphorus was translocated out of the leaf, particularly to the roots, in the deficient plants.

Rates of mineral element uptake by leaves usually decline with leaf age. Several factors are responsible for this decline, including a decrease in

Table 4.2
Foliar Absorption and Translocation of Labeled Phosphate by Barley Plants[a,b]

	Rate of absorption and translocation (μmol P/g leaf dry wt × hr)	
	Control plants	Phosphorus-deficient plants
Uptake by treated leaf	5·29 ± 0·54	9·92 ± 2·17
Translocation from treated leaf	2·00 ± 0·25	5·96 ± 1·08
Translocation to roots	0·63 ± 0·04	4·38 ± 0·42

[a]From Clarkson and Scattergood (1982).
[b][^{32}P] Phosphate was supplied to the mature leaf. Duration of experiment: 3 days.

metabolic activity (sink activity), an increase in membrane permeability (i.e., an accompanying increase in ion concentration in the apoplast), and an increase in the thickness of the cuticle.

In contrast to ion uptake by root cells, uptake by green leaf cells is directly stimulated by light. This can be demonstrated either with intact leaves after vacuum infiltration of the solutes (MacDonald *et al.*, 1975) or with leaf segments (Rains, 1968), where the resistance to solute penetration by the cuticle is minimized (Table 4.3).

During the light period, not only is the potassium uptake rate higher, but the type of energy coupling is different, as shown by the effect of 2,4-DNP, an inhibitor of oxidative phosphorylation (mitochondrial ATP synthesis). Part of the ATP necessary for the active uptake of potassium is obviously supplied by photophosphorylation in the chloroplasts.

Mineral element uptake by intact leaves from foliar sprays, however, is often either not stimulated by light or even depressed as a result of indirect light effects. During the daytime, as the ambient temperature increases there is usually a decrease in relative humidity, leading to more rapid water

Table 4.3
Effect of Light and Inhibitor (2,4-DNP) on Potassium Uptake by Maize Leaf Segments[a]

	K Uptake (μmol/g × hr)	
	Darkness	Light
Treatment		
Control	2·3	3·7
10^{-5} *M* 2,4-DNP	0·2	2·0
Percent inhibition	91	46

[a]Potassium supply 0·1 m*M* KCl. Based on Rains (1968).

evaporation from the foliar sprays and thus more rapid drying of the sprays at the leaf surface. Differences in rates of magnesium uptake by apple leaves from various salts [$MgCl_2 \gg Mg(NO_3)_2 > MgSO_4$] applied during light and dark periods were related exclusively to differences in the solubility and hygroscopicity of these salts (Allen, 1960).

Rates of mineral element uptake by leaves increase as a hyperbolic function of increasing external concentration (Chapter 2). At low concentrations of macronutrients such as potassium and phosphorus, the uptake rates through leaf surfaces are particularly low. Relatively high internal concentrations of these ions in the apoplast of the leaf tissue—for example, up to 5 mEq potassium per liter in barley plants (Pitman *et al.*, 1974)—severely restrict the penetration of ions from the leaf surface into the apoplast. This situation is different in deficient plants (Table 4.2) or when micronutrients are applied to the leaves.

4.3 Foliar Application of Mineral Nutrients

4.3.1 General

The foliar application of mineral nutrients by means of sprays offers a method of supplying nutrients to higher plants more rapidly than methods involving root application. The supply is more temporary, however, and the method creates several problems:

1. Low penetration rates, particularly in leaves with thick cuticles (e.g., citrus and coffee)
2. Run-off from the hydrophobic surfaces
3. Washing off by rain
4. Rapid drying of spray solutions
5. Limited rates of retranslocation of certain mineral nutrients such as calcium from the sites of uptake (mainly the mature leaves) to other plant parts
6. Limited amounts of macronutrients which can be supplied by one foliar spray (on average, 1% × 400 liters/ha, an exception being urea, can comprise 10% of a spray)
7. Leaf damage (necrosis and "burning").

Leaf damage by high nutrient concentrations is a serious practical problem and is mainly the result of local nutrient imbalance in the leaf tissue rather than osmotic effects, as shown in Table 4.4. At high urea concentration leaf damage is quite severe but can be overcome by simultaneous spraying with sucrose, despite the further increase in the osmotic potential of the foliar spray.

Table 4.4
Effect of Foliar Application of Urea and Sucrose on Leaf Damage in Soybean[a]

Treatment ($\mu g/cm^2$)		Damaged area (% of total leaf area)
Urea	Sucrose	
159	0	0
478	0	25
478	909	15
478	2726	3

[a]Based on Barel and Black (1979).

Leaf damage is much less severe when the spray solution pH is low (Neumann *et al.*, 1983). The addition of silicon-based surfactants seems to be a means of decreasing leaf damage and simultaneously increasing the efficiency of sprays, particularly in leaves with thick cuticles (Horesh and Levy, 1981).

4.3.2 Practical Importance of Foliar Application of Mineral Nutrients

Despite the drawbacks of supplying nutrients to plants by means of foliar application, the technique has great practical utility under certain conditions.

4.3.2.1 Low Nutrient Availability in Soils

In calcareous soils, for example, iron availability is very low and iron deficiency ("lime chlorosis") widespread. Foliar spraying is much more efficient than the soil application of expensive iron chelates (Horesh and Levy, 1981) and is also a method of alleviating manganese toxicity (Moraghan, 1979). In soils with a high pH and large content of organic matter, manganese deficiency can be overcome effectively by foliar sprays containing manganese (Farley and Draycott, 1978). In acid mineral soils, molybdenum is strongly fixed; thus foliar sprays with molybdenum are much more effective than soil applications in increasing the molybdenum content of, for example, maize kernels (Weir *et al.*, 1976).

4.3.2.2 Dry Topsoil

In semiarid regions, a lack of available water in the topsoil and a corresponding decline in nutrient availability during the growing season are common phenomena. Even though water may still be available in the subsoil, mineral

Table 4.5
Effects of Soil and Foliar Application of Copper ($CuSO_4 \cdot 5H_2O$) on Growth Parameters and Grain Yield in Wheat[a]

Treatment	Ears/m²	Grains/ear	Grain yield (g dry wt/m²)
No application	37·0	0·14	0·03
Soil application (kg copper sulfate per ha)			
2·5	28·8	2·3	1·0
10·0	58·5	2·9	2·3
Foliar application (2%; 2 kg copper sulfate per ha)			
Once at stem extension	63·8	17·1	14·0
Once at stem extension and once at booting stage	127·4	52·7	79·7

[a]Based on Grundon (1980).

nutrition becomes the growth-limiting factor. Under these conditions, soil application of nutrients is much less effective than foliar application, as demonstrated in Table 4.5 by the results of an experiment in which copper was applied to wheat under field conditions in a semiarid region of Australia.

4.3.2.3 Decrease in Root Activity during the Reproductive Stage

As a result of sink competition for carbohydrates, root activity and thus nutrient uptake by the roots decline with the onset of the reproductive stage. Foliar sprays containing nutrients can compensate for this decline (Trobisch and Schilling, 1970). In legumes that rely on symbiotic N_2 fixation, sink competition for carbohydrates between developing seeds and root nodules causes a marked decrease in the rate of N_2 fixation (Chapter 7). Foliar application of nitrogenous compounds (e.g., urea) can therefore be quite effective in increasing the seed yields of these plants (Garcia and Hanway, 1976). It is not possible, however, to make generalizations concerning the beneficial effect of nitrogen-containing foliar sprays on legumes (Neumann, 1982).

4.3.2.4 Increase in Protein Content of Cereal Seeds

In cereals such as wheat the protein content of the seeds and thus their quality for certain purposes (e.g., baking, animals feeding) can be increased quite readily by the foliar application of nitrogen at later stages of growth. Nitrogen supplied at these stages is rapidly retranslocated or remobilized from the leaves and directly transported to the developing grains (Chapter 8).

4.3.2.5 Increase in Calcium Content of Fruits

As shown in Chapter 3, calcium-related disorders are widespread in certain plant species. Due to the limited phloem mobility of calcium, foliar sprays are not very effective and have to be reapplied several times during the growing season. Nevertheless, some decrease, for example, in bitter pit in apple can be achieved by the repeated application of calcium sprays, particularly if the surfaces of developing fruits are sprayed directly (Schumacher and Frankenhauser, 1968).

4.3.3 Foliar Uptake and Irrigation Methods

Foliar uptake of mineral elements can also occur as a negative side effect of sprinkler irrigation with saline water (Table 4.6).

Table 4.6
Effects of Sprinkler and Drip Irrigation with Saline Water on the Mineral Element Content of Leaves of Chili (*Capsicum frutescens* L.)[a]

	Mineral content of leaves (meq/1100 g dry wt)					
	Chloride		Sodium		Potassium	
Salt content of the water	Sprinkler	Drip	Sprinkler	Drip	Sprinkler	Drip
Low	110	20	20	1	101	118
Medium	121	51	26	1	97	121
High	165	76	48	1	86	113

[a]Based on Bernstein and François (1975).

Sprinkler irrigation leads to a greater increase in the contents of chloride and sodium in the leaves than does drip irrigation (to the soil surface only), indicating substantial, direct leaf uptake from the irrigation water. The levels of these two mineral elements in the leaves therefore become toxic quite rapidly when saline water is used for sprinkler irrigation (François and Clark, 1979). The effect of the two irrigation methods on the potassium content of leaves reflects another phenomenon. With sprinkler irrigation the potassium content of the leaves is lower and declines with increasing salt content of the irrigation water. This pattern indicates leaching of potassium from the leaves.

4.4 Leaching of Mineral Elements from Leaves

4.4.1 Causes and Mechanisms

Leaching can be defined as the removal of substances from the aerial parts of plants by the action of aqueous solutions such as rain, irrigation, dew, and fog (Tukey, 1970). These substances are (a) excreted actively to the external surfaces (e.g., salt excretion by salt glands), (b) excreted by guttation (root pressure), (c) lost from damaged leaf areas, or (d) leached from the apoplast of intact leaf tissue. In the following discussion only the latter two possibilities are considered, since they are of general ecological importance.

Mechanical leaf damage (e.g., by the wind) is quite common, tips and margins being especially susceptible. These areas of the leaf tissue are often particularly high in certain mineral elements when supply of these minerals is excessive (Chapter 3). In tomato plants with a high chloride supply, for example, the ion is preferentially translocated to the leaf margin and into the leaf hairs, so that mechanical damage leads to a loss of between 10 and 70% of the total chloride taken up during the growing period (Chhabra *et al.*, 1977). Stress conditions other than mechanical damage, such as prolonged darkness, water shortage, and high temperatures, can also increase the leaching rate of mineral nutrients from leaves (Tukey and Morgan, 1963). Comparable stress is imposed on leaves or needles by air pollutants such as ozone when combined with nitrogenous gases. This leads to more rapid leaf senescence and modification of the chemical composition of the cuticle or both. Enhanced leaching of organic solutes and mineral nutrients from aerial parts can therefore be considered to be among the damaging effects of air pollutants on perennials.

The leaching rate generally increases with leaf age (Wetselaar and Farquhar, 1980). With the onset of senescence the permeability ("leakiness") of biomembranes increases and there is a corresponding rise in the concentrations of organic and inorganic solutes in the apoplast of the leaf tissue. The resulting steep concentration gradient across the cuticle favors leaching by runoff water at the leaf surfaces. Under field conditions in wetland rice, for example, the maximum amount of nitrogen or potassium in the shoots is reached at anthesis and declines thereafter during the rainy season by about 30% when the plants reach maturity (Tanaka and Navasero, 1964). Losses through leaching under field conditions have to be considered in studies on the time course of nutrient uptake and retranslocation. Presumably, the losses are largely responsible for the generally lower mineral element contents in leaves of plants grown under field conditions compared with those in leaves of plants grown indoors.

Losses of ammonia (Farquhar *et al.*, 1979) and volatile organic compounds through the stomata amounting to as much as 15 kg nitrogen per hectare within 100 days have been found (Silva and Stutte, 1981). The losses of ammonia are closely related to leaf senescence, and the losses of volatile organic compounds to transpiration rates. Failure to recognize these losses may lead to an overestimate of losses of nitrogen from the soil by denitrification, leaching, and ammonia volatilization (Wetselaar and Farquhar, 1980).

4.4.2 Ecological Importance of Leaching from Leaves

Leaching from leaves and other aerial parts is important both for the ecosystem and for individual plants. This is particularly true of perennials. The amount of a mineral element that is leached depends on the type of mineral element and the amount and intensity of rainfall. It would appear that low but continuous rainfall leads to greater losses through leaching than high rainfall. Net losses by leaching also depend on the mineral element composition of the rainwater, as shown in Table 4.7 for spruce. These data as a first approximation may be considered to represent conditions in temperate climates with relatively low levels of air pollution.

Table 4.7
Mineral Element Supply by Rainwater and Mineral Element Leaching from Needles (Canopy Drop) of Spruce (*Picea abies*)[a]

	Na	K	Ca	Mn	N	S	P
Rainwater	23·2	2·4	10·2	0·04	7·3	26·7	0·05
Canopy drop	13·4	13·4	11·9	0·42	10·0	21·9	0·74
Leaching	—	9·0	1·7	0·38	2·7	—	0·69
(Uptake +)	(+9·8)	—	—	—	—	(+4·8)	—

[a]Data are annual values expressed as kilograms per hectare. Based on Fassbender (1977).

The losses of potassium and nitrogen are particularly high, but the losses of the micronutrient manganese, though lower, might be of more importance for the mineral nutrition of plants on soils with an excessive supply of this mineral element. In contrast, there is some net uptake of sodium and sulfur by the needles, because the concentration gradient of these two mineral elements is reversed (going from the external surface toward the apoplast of the needles).

In tropical rain forests the amounts of mineral elements leached from the canopy are, as expected, several orders of magnitude higher, and the annual values, expressed in kilograms per hectare, are a follows: potassium, 100–200; nitrogen, 12–60; magnesium, 18–45; calcium, 25–29; and phosphorus, 4–10 (Nye and Greenland, 1960; Bernhard–Reversat, 1975). The magnitude of this leaching is similar to the nutrient supply from the throughfall (litter) on the soil surface and is thus an important component in the recycling of mineral elements, particularly in ecosystems with low amounts of available nutrients in the soil, e.g., in highly weathered tropical soils. Reabsorption of leached mineral nutrients also offers a possibility for plants to supply the sites of demand (e.g., new growth) with mineral nutrients whose retranslocation is very limited or absent (e.g., calcium and manganese). Leaching is presumably also important for the removal from the leaves of certain mineral elements, such as aluminum or manganese, when they are present in toxic concentrations. Leaching of mineral elements and organic compounds such as phenolics, organic acids, and amino acids (Tyagi and Chauhan, 1982) can affect other plant species within the canopy as well as soil microorganisms.

5

Yield and the Source–Sink Relationship

5.1 General

More than 90% of plant dry matter consists of organic compounds such as cellulose, starch, lipids, and proteins. The total dry matter production of plants, the *biological yield*, is therefore in the first instance directly related to photosynthesis, the primary process of synthesis of organic compounds in green plants. In crop plants, however, yield is usually defined by the dry matter production of those plant organs for which particular crops are cultivated and harvested (e.g., grains and tubers). For the latter the term *economic yield* may be used (Barnett and Pearce, 1983). Thus, in many crop plants not only the total dry matter production is important but also the partitioning of the dry matter, the so-called *harvest index*. The partitioning of the photosynthates and the source–sink relationship and its controlling mechanisms are therefore of crucial importance in crop production.

In this chapter some principles of photosynthesis are discussed, as are the related processes of photophosphorylation and photorespiration, and examples of the direct involvement of mineral nutrients are given. In higher plants the main sites of photosynthesis—the source (mature green leaves)—and the sites of consumption and storage—the sink (roots, shoot apices, seeds, and fruits)—are separate from one another. The long-distance transport of photosynthates in the phloem from source to sink is therefore essential to growth and plant yield. Thus, it is necessary to have a basic understanding of the processes of phloem loading, phloem transport, and phloem unloading at the sink sites and the regulation of these processes, particularly in relation to the action of phytohormones. Finally, the source–sink relationship and the question of whether yield can be limited by source or sink are discussed.

5.2 Photosynthesis and Related Processes

5.2.1 Photosynthetic Energy Flow and Photophosphorylation

The conversion of light energy to chemical energy is brought about by a flow of electrons through pigment systems. In the chloroplasts these pigment systems are embedded in thylakoid membranes in a distinct structural arrangement. Often, the thylakoid membranes are stacked into piles (see Fig. 5.2) which appear as grains or "grana" in the light microscope. The principles involved in the process of electron flow are illustrated in Fig. 5.1. Light energy is absorbed by two pigment systems: photosystem I (PS I) and photosystem II (PS II). In each of these photosystems between 400 and 500 individual chlorophyll molecules and accessory pigments (e.g., carotinoids) act as centers for trapping light energy (photons), with an "antenna" chlorophyll absorbing at 680 nm (PS II) and 700 nm (PS I). In both photosystems the absorption of light energy induces the emission and uphill transport of two electrons against the electrical gradients, from 0·8 to −0·1 V in PS II and from 0·46 to −0·44 V in PS I. The electrons required for this process are derived from the photolysis of water, mediated by PS II. In higher plants PS II and PS I act in series (Z scheme). At the end of the uphill transport chain, the electrons are accepted by an unknown compound Z and transferred to ferredoxin, the first stable redox compound. Ferredoxin in its reduced form has a high negative potential (−0·43 V) and is able to reduce $NADP^+$ (nicotinamide adenine dinucleotide phosphate), as well as other compounds (see below). Several mineral nutrients are directly involved in this photosynthetic electron transport chain (Fig. 5.1). In PS II and PS I chlorophyll molecules with their central magnesium atom absorb photons, thereby

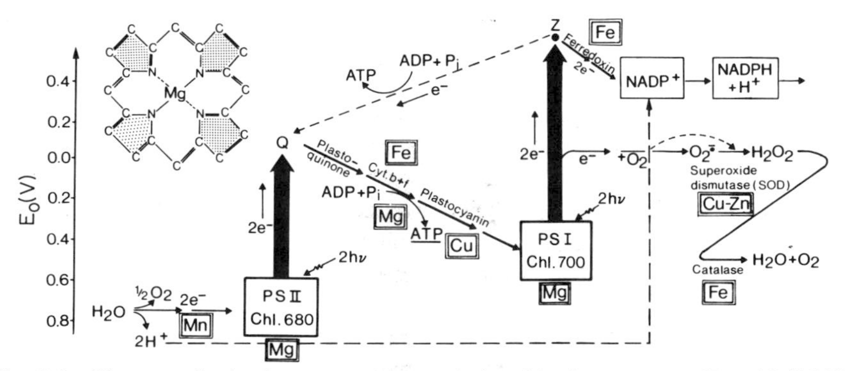

Fig. 5.1 Photosynthetic electron transport chain with photosystems II and I (PS II and PS I) and photophosphorylation (ATP formation). Q, Quencher; Z, unknown compound; Cyt., cytochrome. (*Inset*) Section of the prophyrin structure of chlorophyll with the central magnesium atom.

initiating the electron flow. The photolysis of water is mediated by a manganese-containing enzyme (Cheniae and Martin, 1969). Cytochromes with their central iron atom in the heme configuration mediate the electron flow between PS II and PS I. One of the electron acceptors is plastocyanin, a copper-containing coenzyme. Finally, the iron atoms in ferredoxin are bound to sulfur and act as transmitters of electrons from the so-called Z compound to NADP$^+$, which is reduced to NADPH. Reduced ferredoxin in the chloroplasts can also function as an electron donor for other acceptors. The ferredoxin-mediated reduction of nitrite (NO_2^-) and of sulfate is of particular importance for the mineral nutrition of plants:

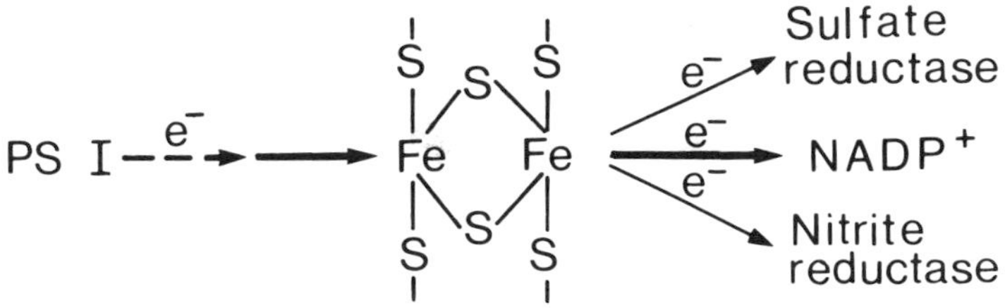

PS I $\xrightarrow{e^-}$ Fe–S cluster (Fe, Fe, S) $\xrightarrow{e^-}$ Sulfate reductase; $\xrightarrow{e^-}$ NADP$^+$; $\xrightarrow{e^-}$ Nitrite reductase

Thus both nitrite and sulfate compete within the chloroplasts with NADP$^+$ for reduction. In leaves the rates of reduction of nitrite and sulfate are much higher in periods of light (Chapter 8). This light coupling is also an example of a regulatory principle, since photosynthesis supplies the structures (carbon skeletons) required for the incorporation of reduced nitrogen (—NH_2) and sulfur (—SH) into organic compounds such as amino acids.

A downhill flow of electrons occurs from quencher molecule Q in PS II to to PS I (Fig. 5.1). This step in the electron transport chain provides the energy for the synthesis of ATP by so-called noncyclic photophosphorylation. Another route of ATP synthesis (cycle photophosphorylation) involves downhill electron flow from reducing agent Z to the original donator Q.

According to the well-established chemiosmotic hypothesis of Mitchell (1966), the coupling of electron flow and ATP synthesis requires charge separation by membranes (thylakoid or mitochondrial membranes) and a proton gradient (pH gradient) across these membranes. Photophosphorylation supplies energy in the form of ATP for the various steps of CO_2 fixation and carbohydrate synthesis, which proceed within the chloroplasts, as shown in Section 5.2.2.

There is increasing evidence (Woolhouse, 1978) that PS I can also act as electron donor for O_2 forming the superoxide radical $O_2^{\cdot -}$ (Fig. 5.1). In order to prevent damage to the photosynthetic apparatus, these highly reactive radicals have to be converted to hydrogen peroxide (H_2O_2) by the enzyme superoxide dismutase (SOD). This reaction is of particular interest, because SOD contains copper and zinc and presumably also manganese (Chapter 9). The reaction product of SOD, the H_2O_2 molecule, must also be removed, and photorespiration (Section 5.2.4) might be essential to this removal (Woolhouse, 1978).

5.2.2 Carbon Dioxide Fixation and Reduction

In order to utilize the energy stored during the light reaction (as NADPH and ATP) for CO_2 reduction and the synthesis of other compounds within the chloroplasts, a CO_2 acceptor is necessary. The outstanding work of Calvin and his group in the 1950s has provided most of our current knowledge of this process. Carbon dioxide is incorporated (assimilated) by the carboxylation of ribulose bisphosphate (RuBP), a C_5 compound:

CH₂-O-(P) | C-OH ‖ C-OH | HC-OH | CH₂-O-(P) —— $+CO_2$ $+H_2O$ / RuBP Carboxylase / Mg ——→ CH₂-O-(P) | HC-OH | COOH ; COOH | HC-OH | CH₂-O-(P)

Ribulose bisulphate (RuBP) 2 × Phosphoglycerate (PGA)

The final reaction product consists of two molecules of the C_3 compound phosphoglycerate (PGA); hence this route to CO_2 incorporation is referred to as the C_3 pathway. The enzyme RuBP carboxylase, which mediates the CO_2 incorporation, is strongly activated by Mg^{2+}. In a series of further steps PGA is reduced to glyceraldehyde 3-phosphate (GAP), a reaction using NADPH and ATP supplied from the light reaction of photosynthesis.

The principles of CO_2 fixation according to the C_3 pathway in chloroplasts are illustrated in Fig. 5.2. The enzymes responsible for CO_2 fixation and carbohydrate synthesis are located in the stroma of the chloroplasts, whereas NADPH and ATP are supplied from the thylakoids. The CO_2 acceptor RuBP has to be regenerated in the Calvin–Benson cycle. The remaining carbohydrates are either utilized for transient starch formation in the chloroplasts or released as C_3 compounds into the cytoplasm for further synthesis of mono- and disaccharides. The rate of release of C_3 compounds from the chloroplasts is controlled by the concentration of inorganic phosphate (P_i) in the cytoplasm; P_i therefore has a strong regulatory effect on the ratio of starch accumulation to sugar release from the chloroplasts (Chapter 8).

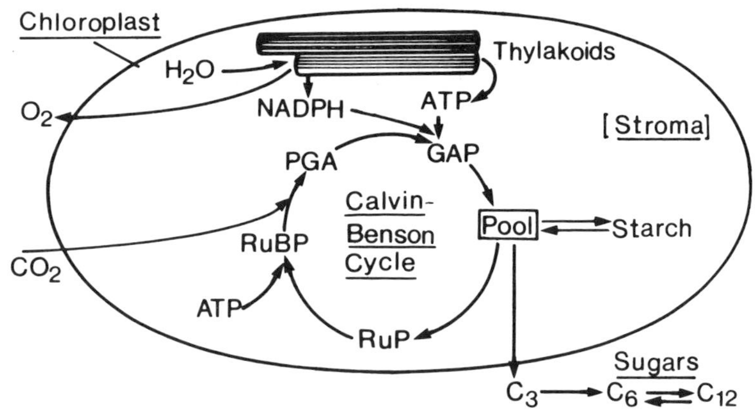

Fig. 5.2 Simplified scheme of CO_2 fixation and carbohydrate synthesis according to the Calvin–Benson cycle in C_3 plants. (Modified from Larcher, 1980.)

5.2.3 C_4 Pathway of Photosynthesis and Crassulacean Acid Metabolism

The incorporation of CO_2 into organic compounds is not restricted to the C_3 pathway just described. It has already been shown (Section 2.9) that an imbalance of cation–anion uptake in roots in favor of cations has to be compensated for by the incorporation of CO_2 and formation of organic acids. The same pathway of CO_2 incorporation occurs in the chloroplasts of certain plant species. The main outline of this pathway, discovered by Hatch and Slack (1970), is as follows:

$$\underset{\text{Phosphoenol pyruvate (PEP)}}{\begin{array}{c} COOH \\ | \\ C{-}O{-}\text{Ⓟ} \\ \| \\ CH_2 \end{array}} \xrightarrow[\text{PEP-carboxylase}]{+CO_2} \underset{\text{Oxaloacetate}}{\begin{array}{c} COOH \\ | \\ C{=}O \\ | \\ CH_2 \\ | \\ COOH \end{array}} \xrightarrow[\text{Malate dehydrogenase}]{NADPH \;\; NADP^+} \underset{\text{Malate}}{\begin{array}{c} COOH \\ | \\ HC{-}OH \\ | \\ CH_2 \\ | \\ COOH \end{array}}$$

Phosphoenol pyruvate (PEP) acts as CO_2 acceptor, forming oxaloacetate, which is reduced to malate. The products of this CO_2 incorporation are C_4 dicarbonic acids. Plant species with this C_4 pathway are therefore classified as C_4 plants. For a comprehensive review of the C_3 and C_4 pathways the reader is referred to Edwards and Walker (1983).

In the chloroplasts, however, incorporation of CO_2 into malate is only a transition stage. In a subsequent step the malic enzyme decomposes malate to pyruvate (C_3) and CO_2. This CO_2 is again fixed by RuBP and channeled into the Calvin–Benson cycle. Thus in C_4 plants the final fixation and

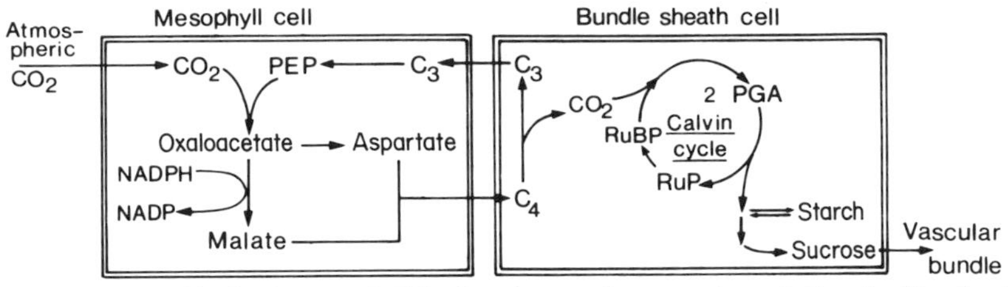

Fig. 5.3 Simplified scheme of CO_2 fixation and compartmentation in C_4 plants.

reduction of CO_2 is identical to that in C_3 plants. In C_4 plants, however, there is a characteristic spatial separation of these two forms of CO_2 fixation, as shown in Fig. 5.3.

The primary fixation of CO_2 in C_4 plants takes place in the chloroplasts of the mesophyll cells (Fig. 5.3). Depending on the plant species (Ray and Black, 1979) either malate or the amino acid aspartic acid (aspartate) is translocated to the bundle sheath cells. Here the CO_2 is released and is subsequently fixed and reduced in the bundle sheath chloroplasts by the C_3 pathway. The remaining C_3 compounds are retranslocated to the mesophyll cells to act again as CO_2 acceptors (shuttle system). In most C_4 species the two cell types are arranged in the so-called Kranz-type leaf anatomy. The minor veins of the vascular bundles are surrounded by bundle sheath cells, forming a Kranz, or wreath. The bundle sheath cells are in turn surrounded by a layer of large mesophyll cells. Furthermore, in C_4 species the chloroplasts are dimorphic, those in the bundle sheath cells being larger and possesing grana that are not as well developed as those of the mesophyll. Starch accumulation is also much more abundant in the bundle sheath chloroplasts.

The type of photosynthesis that takes place in C_4 plants, which is obviously more complicated than that in C_3 plants, usually occurs in plant species of tropical origin with high photosynthetic rates and a large production of dry matter (e.g., sugarcane, sorghum, maize, and various Chenopodiaceae). The C_4 mechanism enables plants for various reasons to utilize both CO_2 and water more efficiently. The strong affinity of PEP for CO_2 results in the maintenance of a low CO_2 partial pressure in the intercellular spaces of the leaf tissue. This is true even at high temperatures (above 25°C), where endogenous respiratory CO_2 production is correspondingly high. The temperature optimum of PEP carboxylase is also much higher than that of RuBP carboxylase (Treharne and Cooper, 1969). This explains, at least in part, why at high light intensities and at high temperatures C_4 plants have considerably lower CO_2 compensation points, that is, levels of CO_2 at which CO_2 consumption and CO_2 production (respiration) are in equilibrium, than C_3 plants (0–20 and 50–100 ppm CO_2, respectively).

The greater efficiency of water use by C_4 plants than by C_3 plants is also related to the lower endogenous CO_2 partial pressure and the correspondingly steeper CO_2 gradient from the ambient atmosphere through the open stomata into the leaf tissue. In C_4 plants there is a relatively greater inward diffusion of CO_2 through the stomata (expressed in terms of units of water vapor lost), which can be utilized for photosynthesis and dry matter production. In addition, when stomata partially close in response to water deficit, the decrease in CO_2 influx is less in C_4 than in C_3 plants, because the internal recycling of CO_2 maintains a lower CO_2 concentration in the leaf tissue of C_4 plants. Correspondingly, the relative efficiency of water use (grams transpired water per gram dry matter produced) is around 300 in C_4 species compared with more than 600 in C_3 species (Woolhouse, 1978).

It is not possible, however, to divide plants strictly into C_3 or C_4 categories. In C_3 species a substantial proportion of the CO_2 fixation occurs via the PEP carboxylase pathway, particularly in the reproductive organs, such as developing wheat grains (Wirth *et al.*, 1977) and legume fruits (Atkins *et al.*, 1977). Such observations might indicate a shift in the photosynthetic pathway toward a more efficient use of CO_2, even if the anatomical structures of typical C_4 plants are lacking.

Fixation of CO_2 via the PEP carboxylase pathway is also a characteristic feature of plant species in certain families, such as Crassulaceae and Bromeliaceae, which are particularly well adapted to dry habitats. These plants are mostly succulent; that is, they have a low surface area per unit of fresh weight. These plant species are characterized by their so-called crassulacean acid metabolism (CAM) and differ from C_4 species in a number of features: (a) The stomata of CAM species are open at night. (b) Carbon dioxide enters the leaves and is fixed by PEP carboxylase in the cytoplasm with subsequent reduction to malic acid, which is stored in the vacuoles during the night. (c) During the day the stomata are closed and malic acid is released from the vacuoles; after decarboxylation, CO_2 is fixed and reduced in the chloroplasts following the C_3 pathway. In contrast to the spatial separation of the two steps of CO_2 fixation in C_4 species, the separation of the three steps of CO_2 fixation in CAM species is temporal (*diurnal acid rhythm*). CAM species generally also have lower growth rates than C_4 species. The combination of CAM and succulence is of particular advantage for adaptation to dry habitats and/or high soil salinity.

5.2.4 Photorespiration

In C_3 species, light stimulates not only the incorporation of CO_2 but also its liberation, a process enhanced by the presence of O_2. This light-driven efflux of CO_2 (photorespiration) proceeds simultaneously with the net influx of

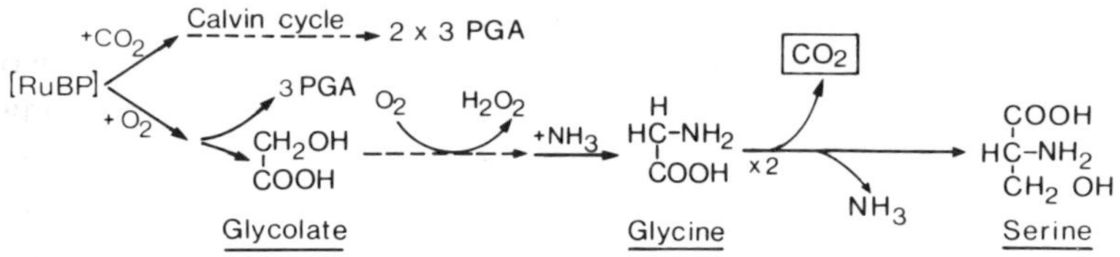

Fig. 5.4 Photorespiration, glycolate pathway, and synthesis of the amino acids glycine and serine.

CO_2. At high temperatures in particular, the rate of CO_2 efflux increases relative to the rate of influx, resulting in a decline in net photosynthesis. The principle of the reactions involved in photorespiration are shown in Fig. 5.4.

Bowes and Ogren (1972) discovered that RuBP carboxylase can also react with O_2; that is, it can behave as an oxygenase (RuBP oxygenase). The reaction depends on the prevailing partial pressure of the two substrates, CO_2 and O_2, which compete at the sites of carboxylation reaction. Photorespiration is therefore stimulated by low CO_2 concentrations, which are the ultimate consequence of CO_2 fixation in light. In the oxygenase reaction (Fig. 5.4) the C_5 compound RuBP is split into 3-PGA (C_3) and glycolate, the first compound of the "glycolate pathway." Glycolate is released from the chloroplasts into the cytoplasm and transferred to certain microbodies (peroxisomes) in which glycolate acts as acceptor for NH_3, forming the amino acid glycine. After the translocation of glycine into the mitochondria, two molecules are converted to the amino acid serine with simultaneous liberation of CO_2 (photorespiration). In this reaction ammonia is also liberated (Fig. 5.4). In order to prevent both ammonia toxicity and losses by volatilization, reassimilation of ammonia is required via the formation of glutamine from glutamate (Chapter 8). This "photorespiratory nitrogen cycle" has been reviewed by Wallsgrove *et al.* (1983).

Although similar patterns of photorespiration cannot be dismissed as being absent in C_4 species, the actual rates of photorespiration are very low. There are two possible explanations for this. Either photorespiratory CO_2 in the C_4 species is fixed by cytoplasmic PEP carboxylase as it passes through the mesophyll cells and is recycled internally (Woolhouse, 1978), or photorespiration is depressed by the liberation of CO_2 in the bundle sheath chloroplasts, where the higher CO_2 partial pressure inhibits photorespiration directly (Edwards and Walker, 1983). The lower intracellular CO_2 partial pressure that would result from either one or both of these processes can be considered the main factor responsible for the lower CO_2 compensation point and higher net photosynthetic rates of C_4 species.

Photorespiration should not be considered, however, merely from the viewpoint of net CO_2 fixation. Photorespiration is an important pathway of amino acid synthesis in leaf cells (Fig. 5.4). Light-induced enhancement of glycolate synthesis, on the one hand, and ferredoxin-mediated nitrite reduction (Fig. 5.1) and thus NH_3 formation, on the other hand, are synchronous steps in this pathway. According to Woolhouse (1978), O_2 consumption in photorespiration might also play a vital role in the stabilization of the photosynthetic apparatus against damage by free superoxide radicals formed in PS I (Fig. 5.1).

5.3 Respiration and Oxidative Phosphorylation

In nongreen tissue (e.g., roots, seeds, and tubers) or in green tissue during the dark period, respiratory carbohydrate decomposition is the main source of energy required for energy-consuming processes such as synthesis and transport. As shown in Fig. 5.5 the major steps of respiration are the

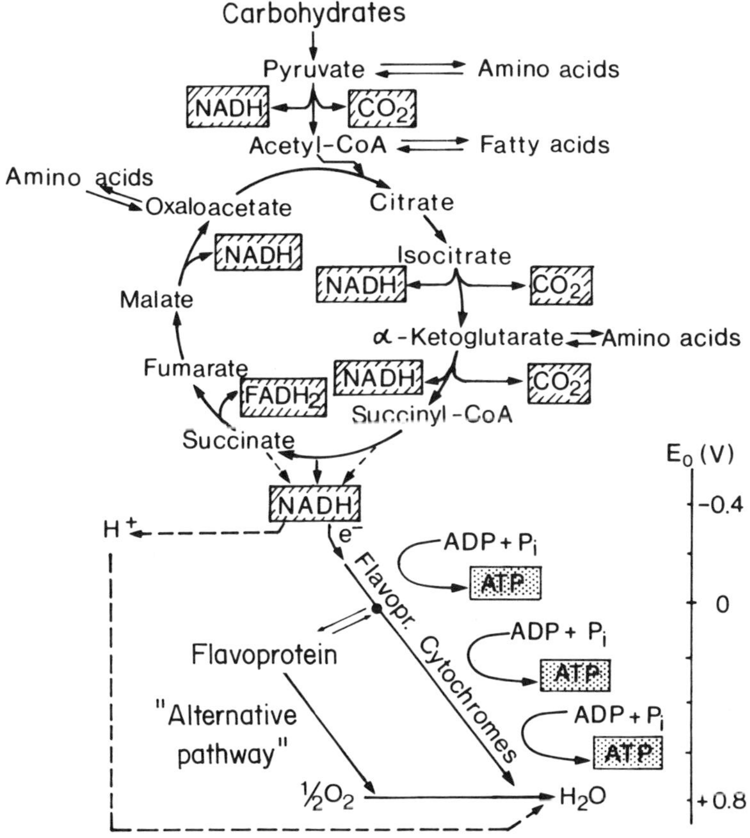

Fig. 5.5 Scheme of the tricarbonic acid cycle (Krebs cycle) and electron transport chain with oxidative phosphorylation and "alternative pathway" (or cyanide-insensitive respiration).

decarboxylation reactions of pyruvate and of organic acids in the Krebs cycle (also called the tricarbonic acid cycle or citric acid cycle) and the oxidation of NADH (reduced nicotinamide dinucleotide). The downhill flow of electrons from the donor NADH to the acceptor O_2 takes place in the respiratory chain in the mitochondrial membranes and supplies the energy needed for ATP synthesis (oxidative phosphorylation). The principles of ATP synthesis in the mitochondria are the same as those of ATP synthesis in the thylakoids of the chloroplasts, namely, charge separation by a membrane with a corresponding proton (pH) gradient across the membrane, constituting a pump for ATP synthesis.

The NADH synthesized in the decarboxylation reactions represents a universal reducing agent in nongreen tissues and is therefore also required for various synthetic processes involving reduction, such as amino acid and fatty acid synthesis. Furthermore, the various intermediates of carbohydrate decomposition are essential structures (carbon skeletons) for the synthesis of, for example, amino acids and fatty acids. The rate of respiration is therefore regulated not only by environmental factors such as temperature or by energy requirements (e.g., ATP for ion uptake in the roots), but also by the demand for reducing equivalents and intermediates.

The fact that the respiration is dependent on the requirement of ATP and NADH implies the presence of a feedback reaction in the form of end-product inhibition if the requirement is low. Depending on the prevailing pathway of synthesis (e.g., lipid or protein synthesis) or other energy-consuming processes (e.g., transport), the requirement of intermediates such as organic acids, amino acids or sugars for NADH (reducing agent) or for ATP (activating agent) can vary over a wide range. This variable demand is met by metabolic "by-passes", of which the "alternative pathway" is particularly interesting (Fig. 5.5). Depending upon plant species, developmental stage and plant organ, the proportion of this alternative pathway can vary between 32–64% and 17–80% of total respiration in the roots and leaves, respectively (Lambers *et al.*, 1983). In accordance with the shift in the demand for intermediates, in wheat roots the proportion of the alternative pathway also depends on the form of nitrogen supply; the proportion is very high with ammonium supply but merely negligible with nitrate supply (Barneix *et al.*, 1984). As we shall see later (Section 5.7), in fast-growing tissues the demand for intermediates in the synthesis of organic structures is especially high compared with the demand for NADH or ATP. A large proportion of the electrons from NADH thus bypass the cytochromes in the respiratory chain and are directly transferred from a flavoprotein to oxygen. As a consequence, less ATP is synthesized per molecule of NADH oxidized. Although this alternative pathway is less efficient in energy conversion than the cytochrome pathway, it has an important function in the regulation of metabolism and growth (Lambers, 1982). On the other hand, it enables nongreen plant cells or tissues to increase ATP synthesis without increasing respiration or carbohydrate consumption, an aspect which demonstrates the difficulty of correlation respiration rates with active ATP-mediated processes, such as ion transport across biomembranes.

5.4 Phloem Transport of Assimilates and Its Regulation

5.4.1 Phloem Loading of Assimilates

In young expanding leaves, most or all assimilates produced during photosynthesis (photosynthates) are required for growth and energy supply. Therefore, in their early growth stages green leaves, like nongreen tissues, act as sinks and require assimilates supplied by a source via the phloem. The major sources of assimilates are fully expanded leaves.

The first step in supplying young leaves with assimilates from the source leaves is short-distance transport of the assimilates from individual leaf cells to the vascular bundles. Although symplastic transport may also be important, transport in the apoplast of the mesophyll tissues seems to dominate (Geiger, 1975). In the apoplast near the minor veins, relatively high sucrose concentrations have been detected. These are ~20 m*M* in sugar beet leaves (Sovonick *et al.*, 1974) and ~7 m*M* in maize leaves (Heyser *et al.*, 1978). Sucrose is most likely released into the apoplast by leakage along the concentration gradient. In contrast, transport into the sieve tubes has to be active, since it is directed against a concentration gradient. The sucrose concentration in the phloem sap is very high, being in the range 100–200 m*M* (Section 3.3). The uphill transport (phloem loading) of sucrose takes place mainly in the minor veins, transfer cells playing a major role (Pate and Gunning, 1972). The process is thus very similar to the phloem loading of mineral nutrients (Chapter 3). Phloem loading of sucrose in the minor veins of a source leaf can be demonstrated quite impressively if ^{14}C-labeled sucrose is infiltrated into the leaf (Fig. 5.6).

Recent years have seen a considerable increase in our knowledge of the mechanism of phloem loading, particularly of sucrose, the dominant sugar in the phloem sap of most plant species. Phloem loading of sucrose has been reviewed by Giaquinta (1983). In 1977 Giaquinta as well as Malek and Baker (1977) and Komor *et al.* (1977) observed that sucrose loading into the phloem is pH dependent and postulated a sucrose–H^+ cotransport at the plasma membrane of sieve element cells. On the basis of these observations and further studies by other groups, a model of phloem loading of sucrose has been developed (Fig. 5.7).

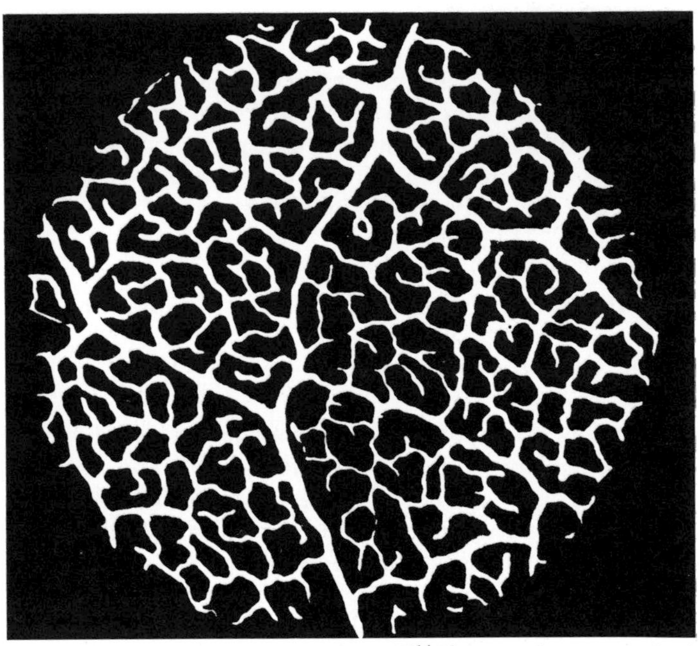

Fig. 5.6 Autoradiograph of phloem loading of [^{14}C]sucrose into source leaf tissue of bean. Sucrose concentration, 1 m*M*; accumulation period, 30 min. White areas = minor veins with ^{14}C. (From Giaquinta and Geiger, 1977.)

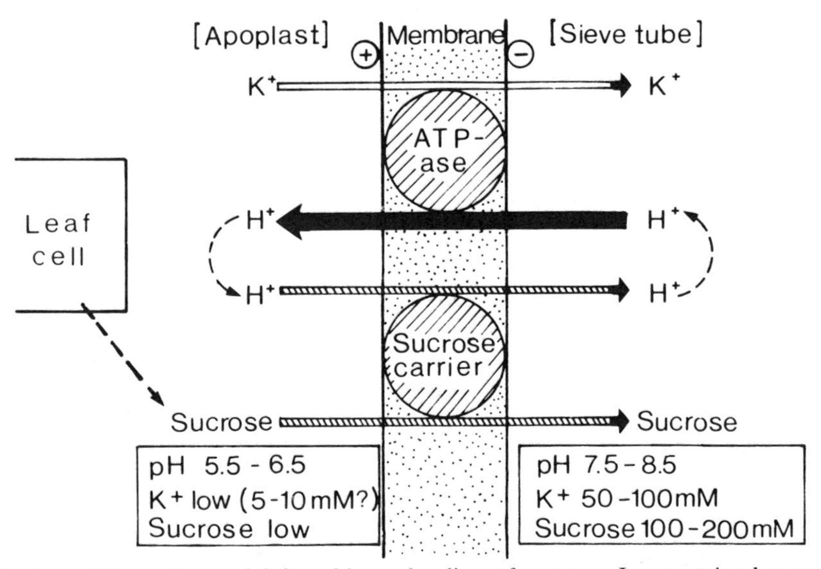

Fig. 5.7 Schematic model for phloem loading of sucrose. Large striped arrows indicate sucrose–proton cotransport. (Based on Baker, 1978; Giaquinta, 1980.)

In this model the cotransport of sucrose and H^+ into the sieve tubes is mediated by a carrier system in the plasma membrane of the sieve tube cells. The driving force for this cotransport is a separate electrogenic proton pump (H^+-efflux pump), which is likely a membrane-bound ATPase. This pump is comparable to the H^+-efflux pump in the plasma membrane of root cells (Section 2.5). In both cases the pump creates a gradient in pH and in electrical potential across the plasma membrane. In sieve tubes a negative potential of −155 mV has been found (Wright and Fisher, 1981), and the pH differs by ~2 units between the phloem sap (7·5–8·5) and the apoplast (5·5–6·5) of leaf tissue. A steep transmembrane electrochemical proton gradient thus exists. The movement of protons along this gradient from the apoplast into the sieve tube cells is obviously coupled by a carrier-mediated cotransport of sucrose. Further evidence supporting this mechanism is the inhibition of sucrose loading by high pH (6) in the apoplast (Delrot and Bonnemain, 1981) and the transient pH increase in the apoplast (i.e., H^+ consumption) at the onset of sucrose loading (Komor *et al.*, 1980).

The model also postulates a direct coupling of H^+ efflux and K^+ influx (H^+/K^+ antiport), which could account for the high K^+ concentrations in the phloem sap (Chapter 3). The role of K^+ in the phloem loading of sucrose, however, is not clear. Low K^+ concentrations in the apoplast might stimulate then H^+-efflux pump and thus facilitate sucrose loading; high K^+ concentrations are most likely inhibitory, because a high K^+ influx should depolarize the membrane and thus decrease the H^+ gradient (Giaquinta, 1980; Ho and Baker, 1982).

Although K^+ does not seem to be essential for sucrose–H^+ cotransport (Cho and Komor, 1980), enhancement by K^+ of both phloem loading and transport of sucrose is well established (Mengel and Haeder, 1977; Doman and Geiger, 1979; Peel and Rogers, 1982). The activation of ATPase by K^+ might be involved in this effect (Giaquinta, 1980; Ho and Baker, 1982). Potassium, however, can also facilitate sucrose loading indirectly by increasing the sucrose concentration in the apoplast of the leaf tissue (Doman and Geiger, 1979). Potassium and Na^+, unlike Ca^{2+}, considerably increase the rate of sucrose efflux from leaf cells into the apoplast, as shown in Table 5.1. In contrast to the efflux of sucrose, the efflux of reducing sugars (mainly glucose and fructose) is very low and is hardly affected by the cations. The close positive correlation between the efflux of sucrose and K^+ from mesophyll cells of wheat and tobacco has led to the idea of a K^+–sucrose cotransport at the plasma membrane of leaf cells (Huber and Moreland, 1981). It well might be that the higher activity of sucrose-P-synthase, the key enzyme of sucrose synthesis (Section 5.5) also contributes to the higher export rates in leaves well supplied with potassium (Huber, 1984).

Judging from the high amino acid concentrations in the phloem sap

Table 5.1
Effect of Cations on the Efflux of Sugars from Sugar Beet Leaf Segments[a,b]

	Control[c]	Ca^{2+}	K^+	Na^+
Reducing sugars				
Leaves	5·11	4·10	3·70	3·80
Efflux	—	0·04	0·09	0·14
Sucrose				
Leaves	2·23	2·70	2·00	2·60
Efflux	—	0·10	1·02	0·85

[a]Based on Hawker *et al.* (1974).
[b]Segments were incubated for 8 hr in the light in solutions of 12·5 m*M* K_2SO_4, 12·5 m*M* Na_2SO_4, or 10 m*M* $CaSO_4$. Sugar efflux is expressed as milligrams sugar per gram fresh weight.
[c]Zero time.

(Chapter 3) the loading of amino acids may be of similar importance. In soybean leaves, both the loading of sucrose and the loading of amino acids are energy dependent, although loading may in each instance be achieved by different carriers (Servaites *et al.*, 1979). Evidence for an H^+–amino acid cotransport in phloem loading is given in Table 5.2. In accordance with the model of phloem loading already discussed (Fig. 5.7), a high pH in the apoplast also depresses phloem loading of amino acids (Table 5.2). Furthermore, in this process K^+ is much more effective than Na^+.

Table 5.2
Effect of pH and Potassium or Sodium in the Apoplast on the Loading and Transport of [^{14}C] Alanine in the Phloem of Castor Bean Petioles[a]

Treatment		[^{14}C] Alanine in the Phloem Exudate (^{14}C counts/ml)
Ion	pH	
K^+	5	27,840
	8	8,441
Na^+	5	13,760
	8	4,090

[a]Based on Baker *et al.* (1980).

5.4.2 Mechanism of Phloem Transport

The principles of transport in the sieve tubes, the anatomy of the phloem, and the transport direction (from source to sink) have been discussed in Chapter 3 in connection with the long-distance transport of mineral nutrients. The solute flow rate in the sieve tubes and the direction of flow are, however, primarily determined not by mineral nutrients but by photosynthates supplied by the source and required at the sink. As early as 1930 Münch proposed a *pressure flow hypothesis* (*Druckstromtheorie*) based on the principle of the osmometer. He suggested that solutes such as sucrose are concentrated in the phloem of leaves and that water is sucked into the phloem, creating a positive internal pressure; this pressure induces a mass flow in the phloem to the sites of lower positive pressure caused by the removal of solutes from the phloem. Flow rate and direction of flow are therefore closely related to the removal of solutes from the phloem, that is, from release or *unloading* at the sink. This type of pressure-driven mass flow in the phloem differs from that in the xylem (root pressure; Section 2.12) in three important ways. (a) Organic compounds are the dominant solutes in the phloem sap, (b) transport takes place in living cells (sieve tubes), and (c) the unloading of solutes at the sink plays an important role.

In accordance with the pressure flow hypothesis, the sucrose concentration in the phloem (sieve tubes) of soybean decreases from 336 m*M* in the leaves to 155 m*M* in the roots, and there is a corresponding decrease in the pressure potential from 6·0 to 1·8 bars (Fisher, 1978). It can also be demonstrated that phloem loading in the leaves is associated with lateral water transport in the leaves toward the phloem (Minchin and Thorpe, 1982). Water availability in the leaves is therefore a determining factor in the solute flow rate in the sieve tubes (Smith and Milburn, 1980).

Although sucrose and other organic compounds are the main solutes in the sieve tubes, potassium can contribute substantially to the total osmotic potential in the sieve tubes and thus to the volume flow rate (Table 5.3). In plants well supplied with K^+, the concentration of K^+ and the osmotic potential of the phloem sap, and particularly the volume flow rate (exudation rate), are all higher than in plants supplied with a lower level of K^+. Sucrose concentration in the phloem sap remains more or less unaffected,

Table 5.3
Effect of Potassium Supply to Castor Bean Plants on the Composition of Phloem Sap and Rate of Phloem Sap Exudation[a]

	Potassium supply in the growth medium	
	0·4 m*M*	1·0 m*M*
Phloem sap concentration (m*M*)		
Potassium	47	66
Sucrose	228	238
Osmotic potential (bars)	12·5	14·5
Exudation rate (ml/3 hr)	1·35	2·49

[a]Based on Mengel and Haeder (1977).

and a high K^+ supply increases the transport rate of sucrose in the phloem by a factor of ~2. There could be several reasons for this enhancement of the volume flow rate by K^+, including higher rates of sucrose synthesis (Section 5.4.1), higher rates of release into the apoplast (Table 5.1), enhancement of phloem loading (Fig. 5.8), or direct osmotic effects of K^+ within the sieve tubes.

The pressure flow hypothesis has been criticized by Spanner and his group (Spanner, 1975), who are of the opinion that active transport of solutes at the sieve plates of individual sieve tube cells takes place along the phloem from source to sink. The decrease or even cessation of phloem transport by localized cooling or anoxia of petioles or stems (Geiger, 1975) indeed seems to disprove the pressure flow hypothesis. However, it has been demonstrated by Giaquinta and Geiger (1977) that the inhibition of phloem transport by these treatments is most likely due to severe damage of the fine structure of sieve tube cells, particularly at the sieve plates. A generalization from this result, however, may be questioned in view of recent findings showing that the inhibitory effects of anoxia, or low temperatures, on phloem transport can be immediately overcome after replacement of N_2 by O_2, or raising the temperature around the stem or petiole (Fensom *et al.*, 1984; Goeschl *et al.*, 1984). Although these latter results may indicate the involvement of active processes within the sieve tubes, the pressure flow can still be considered as the most acceptable mechanism of phloem transport.

5.4.3 Phloem Unloading and Sucrose Storage

The process of release of solutes from the sieve tubes into the surrounding tissue at the sink sites is not well understood. Although cooling of sink tissue has no immediate effect on the phloem transport rate into the tissue (Chamberlain and Spanner, 1978), other, more indirect evidence supports the involvement of an active process in phloem unloading (Jenner, 1980a; Bel and Patrick, 1985). Passive release or leakage of solutes from the sieve tubes could be facilitated by an increase in the permeability of the sieve tube plasma membranes. The phytohormone abscisic acid (ABA) seems to be involved in the unloading of sucrose (Tanner, 1980; Schussler *et al.*, 1984); even low concentrations of ABA increase the rate of sucrose efflux from phloem tissue (Vreugdenhil, 1983). The induction of a localized increase in the membrane permeability of the phloem cells of the host seems to be the mechanism by which stem parasites such as *Cuscuta europea* act as sinks, acquiring the assimilates and mineral nutrients they require for growth (Wolswinkel, 1978; Wolswinkel *et al.*, 1984).

Unloading of sucrose into the apoplast of the sink tissue is also certainly facilitated by the concentration gradient created by the consumption of sucrose during growth (*growth sink*) and the storage of sucrose (*storage sink*), either as sucrose (e.g., in sugar beet) or as starch (e.g., in cereal grains and tubers). In young growing tissues (growth sinks), high activities of acid invertase can be observed in the apoplast. This enzyme hydrolyzes sucrose to form monosaccharides; that is, it maintains a low sucrose concentration in the apoplast and is most likely involved in the phloem unloading of sucrose in stems of sugarcane (Hawker and Hatch, 1965), in the apical zones of growing roots (Eschrich, 1980) and other growth sinks (Eschrich, 1984), including the storage roots of sugar beet during the early stages of growth (Giaquinta, 1979). Similarly, in storage sinks accumulating starch, the unloading of sucrose is enhanced by high rates of sucrose consumption and a correspondingly steep gradient in sucrose concentration from phloem to the apoplast of storage cells (Jenner, 1980a). There is also evidence that a turgor-sensitive component is involved in this sucrose unloading: sucrose unloading is inhibited not only by high concentrations of sucrose in the apoplast but also by other slowly permeating osmotica such as mannitol (Patrick, 1984).

The situation is different, however, during sucrose accumulation in the individual storage cells of sugar beet roots. Sucrose released into the apoplast is not hydrolyzed before uptake into the storage cells (Stein and Willenbrink, 1976; Giaquinta, 1979). The mechanism of this release is not yet known. The sucrose concentration in the apoplast of the storage tissues of sugar beet can be astonishingly high (~60 m*M*) and is in equilibrium with the concentration in the cytoplasm, but not with the very high concentrations (500 m*M*) in the vacuoles (Saftner *et al.*, 1983). The only active step in sucrose accumulation in the storage cells of sugar beet seems to take place at the tonoplast (Saftner *et al.*, 1983).

The accumulation of sucrose in the storage cells of sugar beet roots is stimulated by K^+ (Fig. 5.8). Sodium has an even greater stimulatory effect on sucrose accumulation according to Saftner and Wyse (1980) and Willenbrink *et al.* (1984). These authors suggest that this stimulation by alkali cation takes place at the tonoplast and that the mechanism has features in common with the mechanism operating at the phloem loading sites (Fig. 5.7). An impressive example of the direct role of mineral nutrients in sucrose transport into the vacuoles is given in Table 5.4.

Even after isolation, the vacuoles of storage cells maintain their capacity for sucrose accumulation. The accumulation depends on Mg^{2+} and is further stimulated by K^+. This strongly supports the view that a membrane-bound Mg·ATPase of the type shown in Fig. 5.7 is also responsible for sucrose transport into the vacuoles of storage cells. The activation of Mg·ATPases by K^+ is a well-known phenomenon in other membrane transport processes, such as ion transport at the plasma membrane of root cells (Chapter 2).

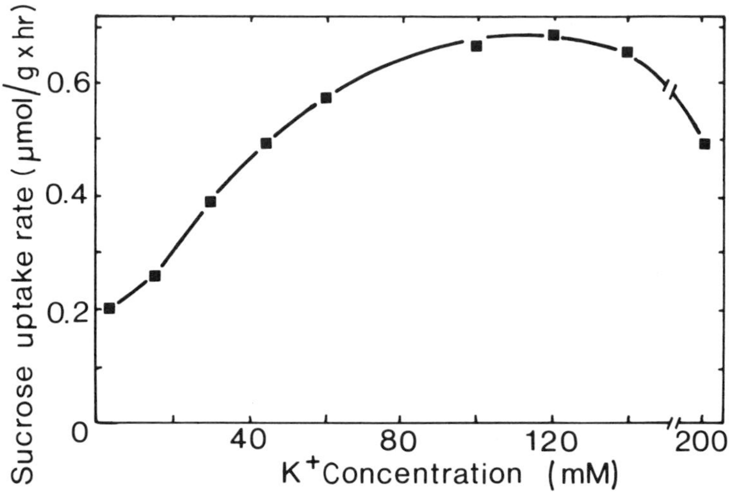

Fig. 5.8 Effect of potassium concentration on sucrose uptake rates by slices of sugar beet storage roots. Sucrose concentration, 40 m*M*. (From Saftner and Wyse (1980.)

Table 5.4
Effect of Magnesium and Potassium on Sucrose Transport into Vacuoles Isolated from Red Beet Tissue[a]

Mg^{2+}	K^+	Uptake rate of sucrose (nmol/unit β-Cyanin × hr)
–	–	4·9
+	–	42·3
+	+	55·3

[a]Based on Doll *et al.* (1979).

5.5 Shift in the Sink–Source Relationship

5.5.1 Effect of Leaf Maturation on Source Function

During its life cycle each leaf undergoes a shift in which its function as a sink changes to that of a source for both mineral nutrients and photosynthates. For mineral nutrients, this shift is correlated with a change in the prevailing long-distance transport in the phloem and xylem (Section 3.4). The long-distance transport of sugars such as sucrose, however, is restricted to the phloem, and the shift from sink to source of each leaf has to be regulated by a corresponding shift from phloem unloading (import) to phloem loading (export). As shown in Fig. 5.9, this transition occurs in sugar beet at 40 to 50% leaf expansion, when about half of the net photosynthetic capacity has been reached.

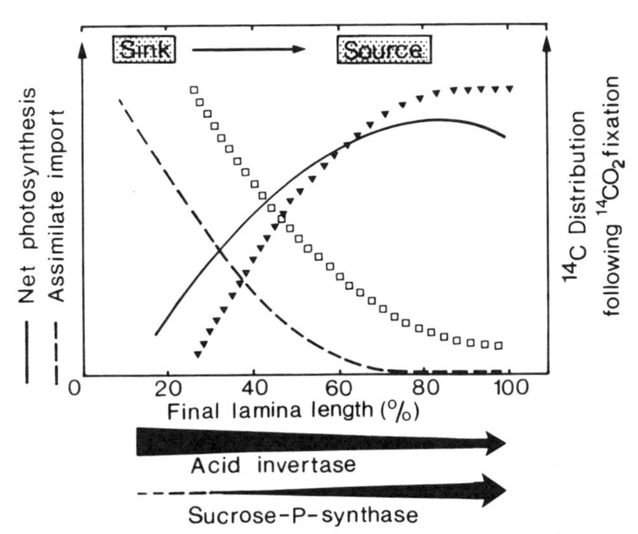

Fig. 5.9 Relationship between assimilate import, net photosynthesis, rate of sugar synthesis (▼, sucrose; □, glucose + fructose), and enzyme activities during maturation of sugar beet leaves. (Based on Giaquinta, 1978.)

During leaf maturation a typical shift also occurs in the incorporation of carbon into sugars, as can be demonstrated by supplying $^{14}CO_2$ to leaves of different age (Fig. 5.9). The shift in favor of sucrose synthesis is closely correlated with changes in the enzyme activities associated with carbohydrate metabolism in the leaves, namely, a decrease in acid invertase activity (sucrose hydrolysis) and a sharp increase in sucrose-P-synthase activity (sucrose synthesis). In very young leaves, sucrose-P-synthase is virtually absent. In plants where sucrose is the dominant sugar in the phloem sap, the functioning of a leaf as a source therefore relies on the induction and activity of this sucrose-synthesizing enzyme. Results similar to those obtained with sugar beet leaves were found with soybean leaves during maturation (Silvius *et al.*, 1978). In apple, on the other hand, the alcohol sorbitol is the dominant form of organic carbon in the phloem sap. Correspondingly, in apple leaves the activity of the enzyme responsible for sorbitol synthesis (aldose-6-P reductase) is absent in young leaves but increases markedly during leaf maturation (Loescher *et al.*, 1982).

Our knowledge of the mechanisms by which the import and export of mineral nutrients are regulated during leaf maturation is, in contrast to our understanding of the mechanism for photosynthates, quite poor. Considering both the mechanism of phloem transport (solute mass flow) and the average composition of phloem sap (Chapter 3), it is clear that, when the rate of sucrose import into a sink leaf is high, the corresponding phloem

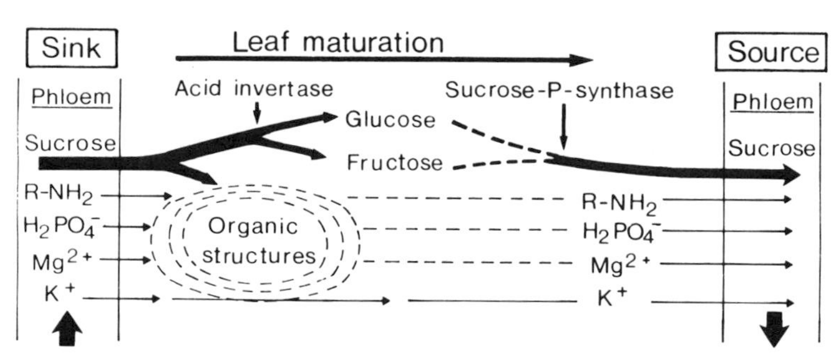

Fig. 5.10 Schematic representation of the shift from import to export of assimilates and mineral nutrients during leaf maturation and the sink–source transition.

import rate not only of mineral nutrients such as potassium and phosphorus, but also of amino compounds, should also be high (Fig. 5.10). It may be supposed that phloem unloading of these solutes, whether active or passive, is also maintained by the requirement for these solutes in growth processes. It has been shown by Schilling and Trobisch (1971), however, that preferential transport from source to sink can also be observed when a nonproteinogenic amino acid (α-aminobutyric acid) is supplied to a source leaf. This amino acid accumulates in the sink in the soluble nitrogen fraction, indicating that it is not the sink demand for a particular amino acid that regulates its source-to-sink transport but rather the direction of mass flow in the phloem and the unloading of other solutes, probably sucrose.

With the onset of leaf maturation and the capacity for sucrose synthesis, the leaf becomes a new source of phloem sap as loading of sucrose begins and an increase in the flow rate in the phloem from the leaf is induced. This means, of course, that the export of other solutes in the phloem such as mineral nutrients and amino compounds can also increase. As discussed previously (Chapter 3) for phloem mobile mineral nutrients such as potassium and phosphorus, import via the xylem and export via the phloem can be in equilibrium in mature leaves. The degree to which mature leaves also act as a source of mineral nutrients depends, however, not only on the rate of photosynthate export but also on the nutrient content of the source leaf and the demand of the sink.

5.5.2 Leaf Senescence

With the beginning of leaf senescence the rates of both photosynthesis (Fig. 5.9) and export of sugars from the leaf decline. Along with this membrane permeability increases (Poovaiah, 1979) and compartmentation is correspondingly affected. Among the other changes that occur, proteolytic enzymes such as acid proteases, previously sequestered in the vacuoles, are released into the cytoplasm and induce rapid breakdown of the proteins in the cytoplasm and chloroplasts. The fall in chlorophyll content (chlorosis) is, of course, the visible symptom of senescence. The composition of the phloem sap exported from senescing leaves changes as well; the sugar concentration declines and that of low-molecular-weight organic nitrogen compounds (Pate and Atkins, 1983) and phloem mobile mineral nutrients increases.

Leaf senescence can be induced when leaves are maintained in the dark, an effect which is enhanced if they are also detached. In detached leaves of *Tropaeolum* kept in darkness and sprayed with distilled water only, there is a rapid breakdown of chlorophyll and protein in the leaf blades within 6 days (Table 5.5). Senescence and enhanced phloem export of soluble nitrogen, potassium, and phosphorus take place, as indicated by the accumulation of

Table 5.5
Inhibitory Effect of Kinetin on the Senescence of Isolated *Tropaeolum majus* Leaves in Darkness[a,b]

Treatment of leaf blades	Content in leaf blades		Content in petioles (base)			
	Chlorophyll	Protein N	Total N	Carbohydrates	K	P
Zero time	100	100	100	100	100	100
After 6 days						
+ Distilled water	53	57	323	90	135	217
+ Kinetin	98	87	95	47	107	117

[a]Based on Allinger *et al.* (1969).
[b]Leaf blades were sprayed daily with either distilled water or kinetin. Relative values.

these constituents at the base of the petioles. Senescence and export from the detached leaves, however, can be inhibited almost totally by spraying of the leaves with kinetin, a synthetic analogue of the phytohormone cytokinin.

Richmond and Lang (1957) and Mothes (1960) were the first to demonstrate this role of cytokinin in delaying leaf senescence. It is interesting that "green islands" can often be observed in blades of senescing leaves of perennials in the autumn. These green islands are usually areas of fungal infection or of parasitic insect attack and are particularly high in cytokinins (Engelbrecht *et al.*, 1969; Abou and Volk, 1971). The production of cytokinin-like substances by parasites appears to be an elegant mechanism by which the parasites maintain the function of the leaf as a source at the site of infection.

5.6 Role of Phytohormones in the Regulation of the Sink–Source Relationship

5.6.1 General

Phytohormones play a dominant role in the regulation of the growth and development of higher plants. This is reflected, for example, in their effect on the sink–source relationship and plant yield. Both synthesis and the action of phytohormones are affected by environmental factors, such as the mineral nutrient supply. It is necessary, therefore, to bear in mind that some effects of mineral nutrients on plant growth and yield are most likely caused primarily by their influence on the phytohormone balance in the plant. Some examples of these effects are given in the following sections and in Chapter 6.

By definition phytohormones are substances for which sites of synthesis and sites of action are separate. Transport either from cell to cell or from organ to organ is therefore necessary. Phytohormones are translocated in both the phloem (Weiler and Ziegler, 1981) and the xylem. The prevailing direction of transport depends on the type of phytohormone (e.g., whether they are synthesized in roots or in the shoots) and the developmental stage of the plant. Each phytohormone has a broad action spectrum; that is, the same phytohormone can affect or regulate various processes depending on its concentration and the conditions at the sites of action—the receptor sites (see following section).

5.6.2 Structure, Sites of Biosynthesis, and Main Actions of Phytohormones

The following four groups of phytohormones are of particular importance: cytokinins (CYT), gibberellins (GA), auxins (IAA), and abscisic acid (ABA). Their molecular structures (including that of ethylene) are shown in Fig. 5.11, and some of their major characteristics are summarized in Table 5.6. There is a general tendency for CYT, GA, and IAA to activate growth and developmental processes, whereas ABA has a more antagonistic effect. The synthesis of the "stress hormone" ABA occurs in rapid response to environmental factors such as a deficiency of water or nitrogen. Some of the actions of ABA—for example, its enhancement of membrane permeability (e.g., stomatal closure)—also occur very rapidly.

There is often a poor correlation between the concentrations of endogenous phytohormones, as determined by bioassays or chemical methods, and

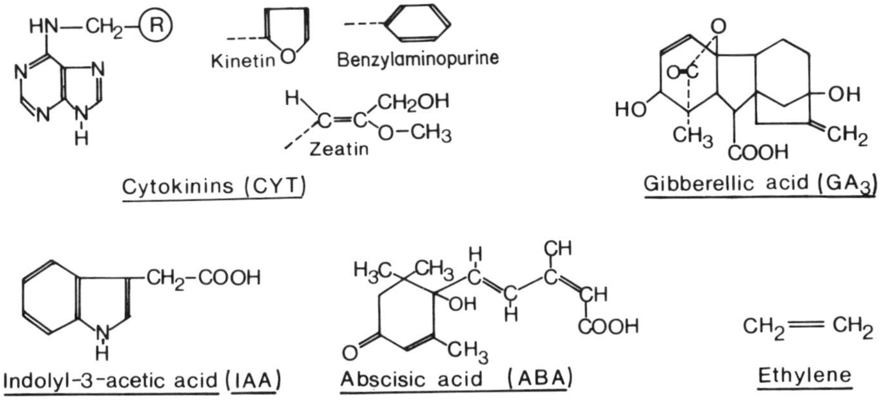

Cytokinins (CYT) — Kinetin, Benzylaminopurine, Zeatin

Gibberellic acid (GA_3)

Indolyl-3-acetic acid (IAA)

Abscisic acid (ABA)

$CH_2=CH_2$ Ethylene

Fig. 5.11 Molecular structures of the main phytohormones.

the actions of phytohormones in plants. For example, high GA concentrations are found in certain dwarf mutants. The expected actions of applied phytohormones are also often very much at variance with their real actions in plants. The concept of a direct regulatory role of phytohormones in growth and development has therefore been criticized for good reasons (Trewavas, 1981). The reasons for the poor correlation between phytohormone action and cellular concentration are outlined in Fig. 5.12. Usually only a fraction of the total phytohormones are physiologically active, the remainder being at least temporarily inactivated either by chemical binding (e.g., CYT as zeatine ribotide or riboside, Sattelmacher and Marschner, 1978b; Koda, 1982) or by compartmentation (e.g., ABA in chloroplasts, Hartung *et al.*, 1981; or in root cell vacuoles; Behl and Hartung, 1984). The main reason for the poor correlations—or unexpected effects—however, is the requirement for receptors at the sites of phytohormone action. During cell and tissue differentiation and organ maturation not only the response (sensitivity) to given phytohormones changes; the type of action can also be different. This is well documented, for example, by the declining stimulation of RNA synthesis by CYT with increasing leaf age (Naito *et al.*, 1981).

For some phytohormone actions, transmitters or *secondary messengers* are also required at the sites of action (Fig. 5.12). The most well known secondary messenger is ethylene ($CH_2=CH_2$). There is good evidence that the enhancement of senescence by ABA is mediated by ABA-stimulated ethylene synthesis (Ronen and Mayak, 1981). Increasing attention has been paid to polyamines as secondary messengers (Dai *et al.*, 1982). The polyamines putrescine ($NH_2CH_2CH_2CH_2NH_2$) and spermidine are as effective as CYT in delaying leaf senescence (Altman, 1982). In young fast-growing tissues the polyamine concentrations are very high (e.g., in developing bean fruits, >200 μg/g fresh wt; Palavan and Galston, 1982), and their synthesis is

Table 5.6
Pathway and Main Sites of Biosynthesis and Some Major Functions of Phytohormones

Cytokinins (CYT)
- Biosynthesis
  - Purine derivatives (adenine), presumably formed during degradation of transfer RNA
- Main sites of biosynthesis
  - Root meristems; to some extent shoot meristems (e.g., embryo of seeds); prevailing long-distance transport via xylem from roots to shoot
- Functions
  - Cell division and expansion, stimulation of RNA and protein synthesis, induction of enzymes, delay in senescence, apical dominance

Gibberellins (GA)
- Biosynthesis
  - From mevalonic acid to the gibbane carbon skeleton; more than 60 gibberellins with this basic structure have been found, gibberellic acid (GA_3) being one of the main gibberellins
- Main sites of biosynthesis
  - Mainly expanding leaves and shoot apex; also other parts of shoots, including fruits and seeds and, presumably, roots
- Functions
  - Cell division and expansion, cessation of dormancy of buds and seeds, induction of flowers and enzyme synthesis (especially of hydrolases)
- Antagonists/inhibitors
  - Chlorocholine chloride (CCC), Ancymidol

Auxins (IAA)
- Biosynthesis
  - Indol derivatives of the amino acid tryptophan, the most prominent being IAA ("auxin")
- Main sites of biosynthesis
  - Meristems or young expanding tissues; in dicots mainly the apical meristems; prevailing direction of long-distance transport basipetal in the phloem
- Functions
  - Cell expansion and division in cambial tissues, apical dominance, induction and activation of enzymes (e.g., H^+ pump)
- Antagonists/inhibitors
  - Coumarins, TIBA, 2,4-D

Abscisic acid (ABA)
- Biosynthesis
  - Terpenoids; presumably also a degradation product of xanthophylls.
- Main sites of biosynthesis
  - Fully differentiated tissues of shoots and roots
- Functions
  - Favors abscission of leaves and fruits and enhances or induces dormancy ("dormin") of seeds and buds; inhibits DNA synthesis, activates ribonucleases, increases membrane permeability
- Antagonists
  - Fusicoccin, IAA, CYT, GA

enhanced by GA application (Dai *et al.*, 1982). In developing soybean seeds, both the concentration and composition of the polyamines change dramatically with time in the cotyledons and embryo; these changes may be related to one of the proposed functions of the polyamines, the cellular pH stabilization (Lin *et al.*, 1984). Polyamines are also of particular interest in

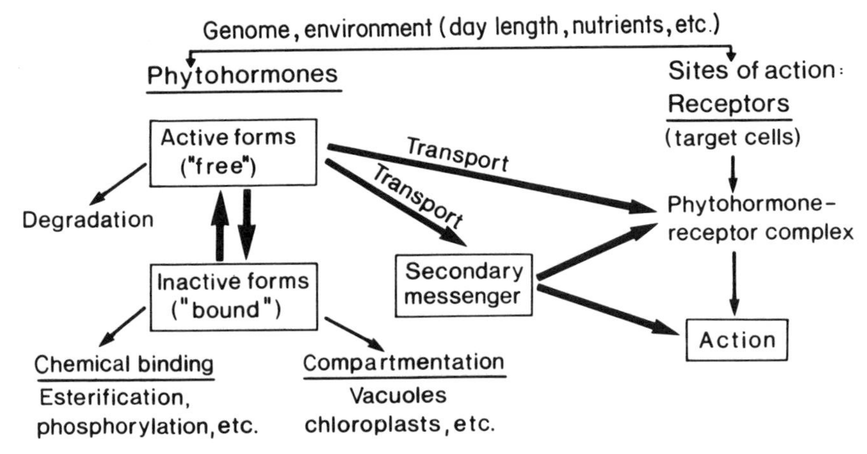

Fig. 5.12. Possible relationships between levels and activities of phytohormones, receptors, and the action of phytohormones.

relation to the mineral nutrition of plants; for example they accumulate in cases of potassium deficiency (Klein *et al.*, 1979). The stimulation of flower induction by ammonium nutrition in particular (Chapter 6) might be related to polyamine accumulation.

From the viewpoint of crop physiology and crop production, there is an urgent need for a better understanding of the actions of phytohormones and of the endogenous and environmental factors which affect the synthesis of both phytohormones and secondary messengers as well as the sensitivity of the receptor sites. The use of *bioregulators*, that is, synthetic plant hormones such as kinetin or growth inhibitors or retardants such as CCC (chlorocholine chloride) and TIBA (2,3,5-triiodobenzoic acid), for the regulation of vegetative and reproductive growth, as well as of senesence and abscission, is a very promising future approach in crop production. Although in some instances bioregulators have been successfully used in crop production (Jung, 1980), their application on a wide scale is still limited by the difficulty of predicting their effects with sufficient accuracy. Quite often plant responses are lacking, or even negative effects are obtained. Given the mode of action shown in Fig. 5.12 it is not surprising that when bioregulators are applied there is a large degree of variability in the results.

5.6.3 Phytohormones and Sink Action

In 1950 Nitsch convincingly demonstrated the role of IAA in the sink action of developing strawberry fruits. Removal of the seeds from the developing fruits resulted in the immediate cessation of fruit growth. Application of

Table 5.7
Effects of Seed or Fruit Removal and Hormone Application on the Accumulation of Leaf-Applied ^{32}P in the Peduncles of Bean[a]

Treatment	^{32}P (cpm)
Control (intact fruit)	373
Seeds removed	189
Fruit removed	34
Fruit removed and cut end treated with	
Lanoline	6 (320)[b]
Kinetin	20
IAA	235 (5520)[b]
IAA + kinetin	471

[a]Based on Seth and Wareing (1967).
[b]Numbers in parentheses indicate counts per minute (cpm) ^{14}C from $^{14}CO_2$ applied to a mature leaf.

IAA to the seedless fruits replaced the sink action of the seeds and restored the growth rate of the fruits. This indicates that solute volume flow via the phloem into developing strawberry fruits is mediated by IAA produced in the seeds. Phytohormones are also involved in the role of tissues as sinks for mineral nutrients, as shown in Table 5.7 for phosphorus in bean plants. Removal of the seeds and, especially, the fruit drastically reduces the accumulation of ^{32}P in the peduncles. This can be restored to some extent, however, by IAA application to the cut end of the stump. When IAA is applied in combination with kinetin, the accumulation of ^{32}P in the peduncle is even greater than in the control, indicating that the action of the fruit as a sink for ^{32}P can be stimulated by treatment with phytohormones. IAA similarly enhances the accumulation of ^{14}C in the peduncle after exposure of a mature leaf to $^{14}CO_2$.

An increase in sink activity at sites of phytohormone application has also been demonstrated by long-term experiments with intact plants (Table 5.8). Foliar sprays containing kinetin and, especially, those containing GA greatly

Table 5.8
Effect of Foliar Sprays on the Growth of Carrot Plants[a]

Spray	(g dry wt/plant)			Ratio shoot/root
	Shoot	Root	Total	
H_2O	3·2	10·9	14·1	0·29
Kinetin	7·3	8·8	16·1	0·83
GA	9·9	5·7	15·6	1·74
CCC	2·8	10·8	13·6	0·26

[a]Sprays were applied once per week for 7 weeks. From Linser *et al.* (1974).

increase the shoot growth of carrot plants, but largely at the expense of the growth of the storage root. This is a typical example of both sink competition between shoot and root and effects of phytohormones on the sink strength of tissues and organs. CCC inhibits shoot growth without affecting the storage root growth; that is, it supports the sink strength of storage roots.

Although direct applications of IAA or GA to storage organs such as roots of winter radish can increase their growth rate (Starck *et al.*, 1980), the results presented in Table 5.8 demonstrate one of the difficulties involved in the practical use of bioregulators for increasing the economic yield. The situation might be somewhat different in plant species such as potato where the tuber yield is quite often limited by the source capacity and where foliar application of phytohormones can therefore have favorable effects on tuber yield (Ahmed and Sagar, 1981).

In crop species such as food legumes or cereals where the seeds represent the yield, the deliberate application of bioregulators for the sake of obtaining direct yield increases is even more complicated and less predictable. Examples such as that shown in Table 5.9 are therefore an exception given the limited state of our knowledge on phytohormones.

Table 5.9
Effect of Foliar Sprays Containing GA on Pod and Seed Number and on Seed Yield in Faba Bean[a]

Treatment[b]	Pod number	Seed number	Seed yield (g)
Control	25·3	81·0	32·4
+GA	31·8	107·0	45·5

[a]From Belucci *et al.* (1982)
[b]Plants were treated with GA at the six-leaf stage. Data indicate amounts per plant.

Interestingly, the yield increase obtained upon GA application was mainly the result of an increased number of pods and seeds per plant. Faba beans are well known for the high proportion of flowers which abort. The application of GA decreases this abortion. However, GA can also have negative effects on flowering; GA application to citrus buds, for example, caused a decrease in the number of flowers through degeneration of the flower primordia (Guardiola *et al.*, 1982).

A large number of experiments involving bioregulators have been performed with cereals. Except for the widespread use of CCC and other growth retardants and their predictable effects, the results of applying bioregulators, including phytohormones, in order to obtain direct yield increases have not been very promising (Michael and Beringer, 1980). This

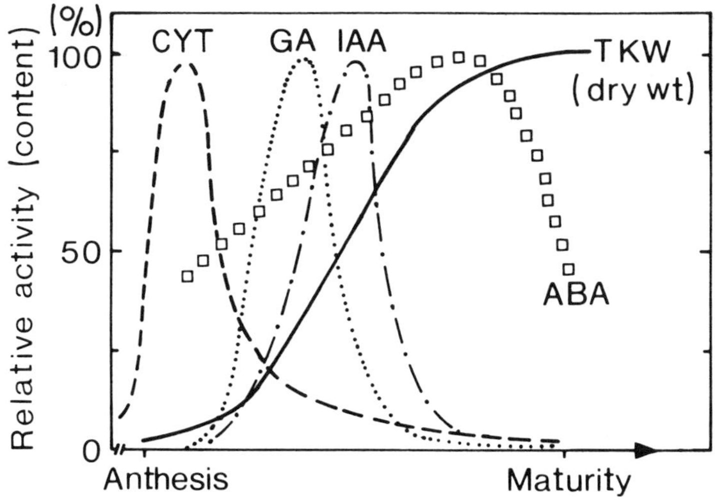

Fig. 5.13 Tentative patterns of phytohormone activities and contents in cereal grains during grain development. CYT, cytokinins; GA, gibberellins; IAA, auxins; TKW, thousand kernel weight; ABA, abscisic acid. Relative values; maximum = 100. (Data compiled from Rademacher, 1978; Radley, 1978; Michael and Beringer, 1980; Mounla *et al.*, 1980; and Jameson *et al.*, 1982.)

is not surprising since the endogenous level of various phytohormones during seed development is characterized by distinct patterns (Fig. 5.13) which might correspond to their period of action.

Maximum CYT activity is reached as soon as a few days after anthesis (Jameson *et al.*, 1982) and obviously coincides with the maximum rate of cell division. In contrast, ABA activity increases much later and reaches its maximum during the period of rapid decline in the rate of dry matter accumulation; the peak in ABA activity is also correlated with enhanced water loss and corresponding desiccation of the grains. The maxima of GA and IAA activity are reached when rates of dry matter accumulation are highest, that is, when the rates of both sink activity and phloem unloading are highest. Causal interpretations of the relationships between sink activity and average values for phytohormone levels in grains and seeds are further complicated by the differences in the level of individual phytohormones between the tissues. In soybeans, for example, the levels of ABA and IAA vary dramatically and independently between embryo, cotyledons and testa during seed development (Hein *et al.*, 1984a).

There is a well-established positive correlation between the final grain weight and the number of endosperm cells (Singh and Jenner, 1982; Schacherer and Beringer, 1984) as well as the length of the grain-filling period (days between anthesis and maturity). In other words, grain weight depends on the capacity of the grain to store carbohydrates (i.e., on the number of storage cells), on the length of the storage period, and on the intensity of storage. In agreement with this, the single grain weight can be increased by application of CYT to the roots shortly before anthesis (Herzog and Geisler, 1977) and decreased by elevated ABA levels, induced, for example, by high leaf temperatures during the grain-filling period (Goldbach and Michael, 1976) and correspondingly enhanced leaf senescence (Al-Khatib and Paulsen, 1984).

In principle these distinct patterns of endogenous phytohormone level should also be observed in fleshy fruits such as tomatoes (Desai and Chism, 1978) and grapes (Alleweldt *et al.*, 1975). However, what proportion of phytohormones are synthesized in the fruits or seeds themselves and what proportion are derived from source leaves or roots is an open question. There is convincing evidence that most of CYT derives from the roots (Varga and Bruinsma, 1974; Michael and Beringer, 1980; Carmi and Van Staden, 1983). Most of the ABA is obviously translocated from source leaves via the phloem into sinks, such as cereal grains (Goldbach and Goldbach, 1977). No information is available on the relationship between import and the synthesis of GA and IAA in seeds and fruits.

Reports on the effect of ABA on sink activity are contradictory. From its main actions (Table 5.6), one would expect ABA to have an inhibitory effect on sink activity (King and Patrick, 1982). Therefore, reports of stimulatory effects of ABA on assimilate transport into sinks such as cereal grains (Tietz *et al.*, 1981), grapes (Düring and Alleweldt, 1980) and roots (Karmoker and Steveninck, 1979) seem, at first, difficult to explain. However, it must be remembered that, by increasing membrane permeability, ABA probably favors the unloading of sucrose in the sink. In the early stages of grain development ABA derived from the source is rapidly degraded in the sink, whereas in later stages the rate of degradation decreases (Goldbach *et al.*, 1977), leading to an increase in ABA level in the grains (Fig. 5.13). Since ABA increases membrane permeability to water (Glinka and Reinhold, 1971), desiccation of the grains will be enhanced and compartmentation negatively affected; both processes are expressions of tissue senescence. These various effects of ABA in the sink illustrate the complexity of the regulation of sink activity by phytohormones, even without consideration of the receptor sites. They also reflect the difficulty of increasing the sink activity of storage organs through the application of bioregulators.

5.6.4 Effect of Environmental Factors on the Endogenous Level of Phytohormones

The synthesis, activity, and degradation of phytohormones are affected by such environmental factors as temperature, day length, and water and nutrient supply. Because some of these factors can be varied relatively easily by agronomical practices (e.g., the application of fertilizers), growth and development and ultimately economic yield may be manipulated indirectly via the manipulation of endogenous levels of phytohormones. In the following discussion attention is directed toward the mineral nutrients and their effect on phytohormone levels.

It is generally agreed that most of the CYT is synthesized in the roots (Van Staden and Davey, 1979). There is thus a close relationship between root growth in general, the number of root meristems (the sites of CYT synthesis), and CYT production in the roots in particular (Forsyth and Van Staden, 1981). Of the mineral nutrients, nitrogen has the most prominent influence on both root growth and the production and export of CYT to the shoots (Wagner and Michael, 1971). Because CYT is exported mainly in the xylem, collecting xylem exudate is a simple method of obtaining information on this nitrogen effect, as shown in Table 5.10 for potato plants. When the nitrogen

Table 5.10
Effect of the Supply of Nitrogen to the Roots of Potato Plants on the Export of CYT from the Roots[a]

Plant age at zero time[b]	CYT exported ng/plant × 24 hr[c]	
(days)	+N	−N
0	196	196
3	420	26
6	561	17
9[d]	—	132

[a]Based on Sattelmacher and Marschner (1978a).
[b]30 days after sprouting.
[c]The amount of CYT exported (over a 24 hr period) is determined by levels in the xylem exudate. +N, Continuous nitrogen supply; −N, interrupted nitrogen supply.
[d]Restoration of nitrogen supply after 6 days without nitrogen.

supply is continuous, CYT export increases with plant age, whereas when the supply is interrupted, the roots respond rapidly with a drastic decrease in CYT export. After the nitrogen supply is restored, CYT export is rapidly enhanced.

The synthesis and export of CYT are also affected by phosphorus and potassium supply, although this effect is somewhat less prominent than in the case of nitrogen (Table 5.11). Results similar to those obtained with sunflower have been obtained with perennials (Horgan and Wareing, 1980).

Table 5.11
Nutrient Supply and CYT Content of Root and Leaves of Sunflower Plants Grown in Nutrient Solution with Deficient Levels of Nutrients[a]

Treatment (15 days)	CYT (kinetin equivalents, μg/kg fresh wt)	
	Roots	Leaves
Control	2·38	3·36
1/10 N[b]	0·94	1·06
1/10 P	1·06	1·28
1/10 K	1·06	2·02

[a]From Salama and Wareing (1979).
[b]Indicates proportion of nutrients in relation to fully concentrated control solution.

Although the possibility that these mineral nutrients have a direct effect on the biosynthesis of CYT cannot be dismissed, it is much more likely that they act indirectly via root growth and the induction of root primordia (see above). The close positive correlation between the number of root primordia and leaf area in tomato plants (Richards, 1981) is presumably related to CYT production.

A deficiency of water or nitrogen strongly enhances the synthesis of ABA in plants. An example of the effect of nitrogen is shown in Table 5.12. In plants well supplied with nitrogen, the ABA content of young leaves is somewhat higher than that of fully expanded or old leaves, reflecting phloem transport of ABA from the older (source) to the young (sink) leaves

Table 5.12
Relationship between Nitrogen Supply and the ABA Content of Sunflower Plants[a]

Plant part	Plant grown in nutrient solution	
	With nitrogen	Without nitrogen (7 days)
Leaves		
Old	8·1	29·8
Fully expanded	6·8	21·0
Young	13·5	24·0
Stem	2·5	4·9

[a]Content of ABA expressed as micrograms per gram fresh weight. Based on Goldbach *et al.* (1975).

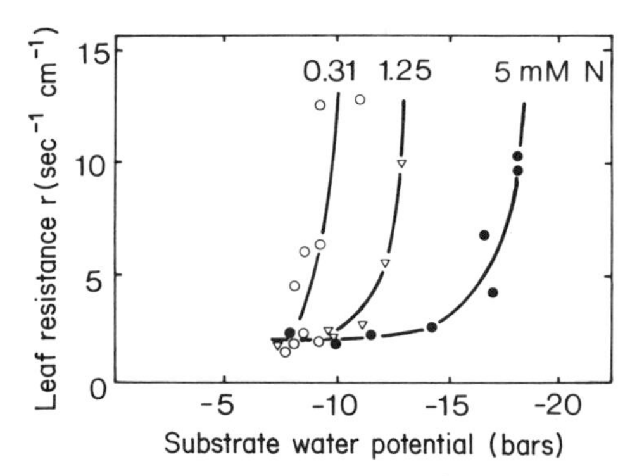

Fig. 5.14 Relationship between nitrogen supply (m*M* nitrate nitrogen), leaf resistance to water vapor diffusion, and substrate water potential in cotton plants. (Based on Radin and Ackerson, 1981.)

(Zeevaart and Boyer, 1984). Under conditions of nitrogen deficiency, the ABA content increases sharply in all parts of the shoots. In potato plants this response can be observed within 3 days, not only in the shoots but also in the roots and xylem exudate (Krauss, 1978a).

These effects of nitrogen supply on ABA levels have important consequences for the water balance of plants. Under conditions of water deficiency (e.g., in dry soils or soils with a high salt content) elevated ABA levels in the leaves favor stomatal closure and thus prevent excessive water loss (Fig. 5.14).

As one would expect, when plants are nitrogen deficient or are supplied with suboptimal amounts of nitrogen, they respond to a shortage of available water in the substrate (i.e., an increase in substrate water potential) by a more rapid stomatal closure (indicated by an increase in leaf resistance to water vapor diffusion) than do plants well supplied with nitrogen (Fig. 5.14). This faster stomatal response is not caused exclusively by higher ABA levels, however; most likely lower CYT levels in the plants are also responsible (Radin *et al.*, 1982). It is well documented that CYT and ABA have opposite effects on stomatal aperture (Radin *et al.*, 1982). The higher drought resistance of low-nitrogen plants (Radin and Parker, 1979) is therefore the result not only of morphological changes in root growth (Chapter 14) or leaf anatomy (e.g., smaller leaf blades; Radin and Parker, 1978) but also of physiological changes such as an increase in the ABA/CYT ratio. It has to be kept in mind, however, that elevated ABA levels in plants can also have negative effects on sink activity (e.g., in cereal grains; see above) and on fertilization (i.e., the formation of a sink; see Chapter 6).

Somewhat similar relationships to those described for nitrogen and stomatal response can be observed for phosphorus (Radin, 1984). In phosphorus-deficient cotton plants, more ABA is accumulated in the leaves in response to water stress than in phosphorus-sufficient plants: in the deficient plants, the stomata close at leaf water potentials of approximately −12 bars, compared to −16 bars in the sufficient plants. In the deficient plants, the sensitivity of the stomata to ABA is also increased, an effect which is blocked by kinetin. These results are seen as the phosphorus nutritional status influencing the balance between ABA and CYT, or the partitioning of ABA between active and inactive pools. A decrease in the hydraulic conductance of the roots of phosphorus-deficient plants may contribute to the faster stomatal closure in response to water stress (Radin and Eidenbock, 1984).

The levels of GA can also be modulated by environmental factors, particularly, by the day length. The GA levels in plants grown under long-day conditions are much higher than those in plants grown under short-day conditions (Railton and Wareing, 1973). Of the mineral nutrients, nitrogen again has the most prominent effect on GA levels. In potato plants, for example, an interruption in the supply of nitrogen induces a sharp drop in the GA levels of the shoots, associated with a sharp increase in the ABA levels (Krauss and Marschner, 1982). After restoration of the nitrogen supply, the levels of GA and ABA respond quite rapidly in the opposite direction, GA increasing and ABA falling. Comparable changes in GA and ABA levels induced by the nitrogen supply can also be observed in the tubers of potato plants (Krauss, 1978b) where the changes are correlated with distinct differences in tuber growth pattern (Chapter 6).

The effects of nitrogen on GA levels are presumably indirect. The main sites of GA synthesis are the shoot apex and the expanding leaves. Environmental factors which favor the growth rate of the shoots, therefore, also indirectly favor GA synthesis. The decrease in GA levels that occurs when the supply of nitrogen is interrupted is presumably indirectly caused by the decrease in CYT export from the roots and a corresponding lack of CYT for the maintenance of a high growth rate of the shoot apex and the young leaves, the main sites of GA synthesis. These relationships are summarized in Fig. 5.15. The immediate cessation of shoot growth observed after the supply of nitrogen to the roots is withheld (Krauss and Marschner, 1982) and the failure to restore shoot growth under these conditions by the application of nitrogen to the leaves (Sattelmacher and Marschner, 1979) support these views.

The examples just given demonstrate that a change in the supply of nutrients to the roots, nitrogen in particular, can markedly modulate not only the levels but also the balance of phytohormones in plants. The application of nitrogen fertilizers can therefore affect growth and develop-

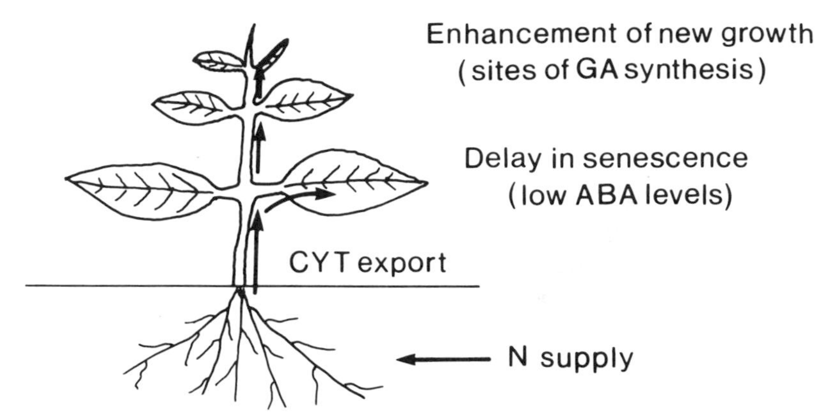

Fig. 5.15 Model for some effects of nitrogen supply on the phytohormone balance in plants.

ment not only directly (supplying nitrogen as a constituent of protein) but also indirectly by changing the phytohormone balance. These changes are also reflected in plant morphology. In cereals to which high levels of nitrogen have been applied, for example, the elongation of the stems is enhanced and the potential danger of lodging increases. In order to counteract these effects in cereals, high levels of nitrogen fertilizer are often applied in combination with growth retardants such as CCC which depress GA synthesis.

5.7 Source and Sink Limitations on Growth Rate and Yield

The growth rate of the shoot apex, fruits, and storage organs can be limited either by the supply of assimilates from the source leaves (*source limitation*) or by a limited capacity of the sink itself (*sink limitation*). Sink limitation can be related to low rates of phloem unloading or cell division, a small number of storage cells, or a decline in assimilate conversion rate (e.g., sugars to starch). In the following examples, both types of limitation are considered and possible mechanisms of feedback control from sink to source are discussed.

During vegetative growth in plants with a large number of source leaves, the growth rate is often limited by the capacity of the young leaves (sinks) to utilize the photosynthetic potential of the source. In mustard plants (Table 5.13), the removal of four leaves (leaves 3–6) led to an approximate doubling of both the rate of photosynthesis and the export of assimilates from the remaining source leaf (leaf 2). This demonstrates that, in the intact plant with all the leaves present, the potential photosynthetic capacity of leaf

Table 5.13
Effect of Removal of Source Leaves of White Mustard on the Photosynthesis and Assimilate Export of a Remaining Source Leaf (no. 2)[a]

Treatment	Photosynthetic rate of leaf no. 2 (%)	^{14}C Export from leaf no. 2 (%)
Control	100	36
Source leaves removed (nos. 3–6)	187	62

[a]Based on Römer (1971).

2 was not being fully utilized. The situation is certainly different in such plant species as cabbage or lettuce, where there is a large total number of leaves but only a small number of source leaves.

Sink limitation can be demonstrated quite readily during the reproductive growth stage if the major sinks—the fruits, seeds, and storage organs—are removed. In aubergine plants, for example, the rate of net photosynthesis declines much more during the light period in plants from which the fruits have been removed. At the beginning of the light period the carbohydrate content of the leaves in plants without fruits is still higher than that of plants with fruits (Claussen and Biller, 1977). Similar responses in photosynthesis and carbohydrate accumulation in leaves can be demonstrated in potato plants after removal of the tubers (Nösberger and Humphries, 1965).

During vegetative growth, sink limitation is not necessarily reflected in low rates of photosynthesis. A higher proportion of carbohydrates can be respired via the "alternative pathway" (i.e., the nonphosphorylating electron transport chain; see Fig. 5.5) in the sink tissue. This provides another mechanism for maintaining a high rate of assimilate influx into sink tissue. In early stages of growth (*structural growth*) of storage tissue, such as storage roots, the requirement for amino compounds and mineral nutrients is high compared with the demand for sugars. The surplus of sugars imported via mass flow into the storage tissue is then respired as an *energy overflow* (Lambers, 1982). An example of this is shown in Table 5.14 for the storage roots of carrots. Rapid growth of the storage root and sugar storage at the expense of lateral root growth started at about day 25. Thereafter the growth rate of the roots (i.e., of the main sink) was maintained at a similarly high level, despite the sharp decline in the rate of net photosynthesis and in photosynthate export to the roots. This apparent discrepancy is the result of a shift in the respiratory pathway. Before sugar storage, respiration in the tap roots was high and the alternative pathway contributed 49% to the root respiration (compared with 57% for cytochrome respiration). With the

F

Table 5.14
Photosynthate Production and Utilization in Shoots and Roots of Carrot[a,b]

	Plant age (days)		
	18–25	25–32	32–39
Shoot			
Photosynthesis	555	381	236
Respiration	119	55	35
Growth	236	193	85
Export to roots	200	133	116
Roots			
Alternative respiration	49	9	4
Cytochrome respiration	57	40	28
Growth	93	84	84

[a]From Steingröver (1981).
[b]Shoot data expressed as milligrams organic carbon (CH_2O) per shoot × day); root data expressed as percentages calculated from daily produced photosynthate.

onset of sugar storage the respiration rate decreased, the decrease being particularly pronounced in the alternative respiratory pathway. It can be concluded that alternative respiration consumes only those sugars that are not required for energy metabolism and growth. Another type of energy overflow is observed in aphids sucking phloem sap. In this case, however, the surplus of sugars not required for aphid growth and energy metabolism is excreted as honeydew.

On the other hand, it is not uncommon for source limitation to occur during the phase of rapid growth and photosynthate storage in fruits and seeds. In potato, for example, a reduction of the leaf area (removal of leaves) by 50% leads to ~50% reduction in the tuber growth rate (Engels, 1983). As discussed later (Chapter 6), however, the results of such drastic manipulations of the source–sink relationship should be interpreted with caution. Nevertheless, source limitation in potato can also be demonstrated in a non-destructive way: Cooling (to 8°C) of half the tubers on a plant causes growth cessation of the cooled tubers and a corresponding double of the growth rate of the untreated tubers on the same plant (Engels, 1983).

The limitation of growth rate by source or sink also depends, of course, on several environmental factors affecting the rate of photosynthesis. In cereal plants during the phase of rapid grain growth, removal of some leaves brings about an increase in the rate of photosynthesis of the remaining leaves only at high light intensities and high ambient CO_2 concentrations, that is, under optimal conditions for photosynthesis (Römer, 1971; Martinez-Carrasco and Thorne, 1979).

Another factor determining source–sink limitations is the source size (leaf area) in relation to the sink size (e.g., number of grains or tubers per plant). This was demonstrated for three maize genotypes, which were subjected to defoliation treatment (25% reduction of the leaf area) 2 weeks after 50% silking (Barnett and Pearce, 1983). Defoliation hardly affected grain weight in the genotype with a relatively large source size and a large source/sink ratio. In contrast, defoliation of the other two genotypes with a smaller source/sink ratio reduced grain yield. In all three genotypes defoliation reduced the stalk weight as a consequence of the mobilization of nonstructural carbohydrates stored in the stalk (Barnett and Pearce, 1983). This reflects the role of the stem as a transient storage pool for photosynthates during vegetative growth and its importance as an additional source of photosynthates during the reproductive stage (Spiertz and Ellen, 1978; Setter and Meller, 1984).

In most plant species several sinks compete for photosynthates. The major sinks are the (vegetative) roots, shoot apices, reproductive organs, and storage roots and tubers. As a rule in the reproductive stage the fruits, seeds, and storage organs are the dominant sinks. Under conditions of sink limitation, competition is reflected in a decrease in the growth rate of the other sinks. The mechanism of sink competition is not yet clear. According to a simple model higher rates of phloem unloading in the dominant sink (perhaps because of higher phytohormone activity) induce higher rates of solute mass flow in the phloem toward this sink at the expense of the other sinks. There is evidence, however, for more sophisticated regulation mechanisms in which phytohormones are produced in the dominant sink and then translocated to the other sinks in order to suppress competition. The latter type of *correlative inhibition* seems to exist in aubergine plants between fruits and roots (Claussen and Biller, 1976) and also between the individual grains of a wheat ear (Radley, 1978).

Sink competition between reproductive organs and roots is also of importance for the uptake of water and mineral nutrients. In cereals, for example, a considerable proportion of the nitrogen translocated to the developing grains is taken up by the roots during the grain-filling period (Spiertz and Ellen, 1978). In agreement with this, removal of some spikelets from wheat ears (i.e., a decrease in sink size) results in an increase in the nitrogen content of the remaining grains but does not affect the grain weight (Jenner, 1980a).

In perennials with indeterminate vegetative growth during the reproductive phase, sink competition by the developing fruits can be quite dramatic, as in the case of citrus trees (Table 5.15). With increasing fruit load the growth of vegetative shoots and roots is drastically reduced. Simultaneously the total dry weight per plant decreases, indicating that more is affected by

Table 5.15
Effect of Fruit Load on Dry Matter Production and Distribution and on Water Consumption of *Citrus madurensis* Lour[a]

	No. of fruits per plant		
	0	50	100
Dry weight (g/plant)			
Fruits	—	134	175
Vegetative shoots and flowers	457	305	118
Roots	68	49	17
Total dry wt	525	488	310
Water transpired			
Liters/plant	91	90	59
Liters/kg leaf dry wt	370	520	1030

[a]Based on Lenz and Döring (1975).

the fruit load than the distribution of photosynthates. The effect of increasing fruit load on water consumption per plant is much lower than that on dry weight. The transpiration rate per unit leaf dry weight increases by nearly a factor of 3. The close positive correlations between the number of fruits per plant and the transpiration rate per unit leaf area is a general phenomenon (Lenz, 1970). Plants with a heavy fruit load are therefore more sensitive to inadequate supplies of water and mineral nutrients (Lenz, 1970), because their shoots place a higher demand on their small root system than do the shoots of plants without fruits or with only a small number of fruits.

In nodulated legumes characterized by N_2 fixation, the root nodules represent an additional sink for carbohydrates supplied from the leaves. As shown in Table 5.16 the removal of source leaves leads to a decrease in both nodule growth and N_2 fixation, whereas the removal of flowers and pods (competing sinks) results in an increase in both nodule weight and N_2 fixation to values that are higher than those of untreated control plants.

Sink competition in legumes between fruits (seeds) and root nodules for carbohydrates from the source leaves often leads to a sharp decline in N_2 fixation rates at the onset of a rapid fruit growth. During this period, however, the demand of the fruits for organic nitrogen is also very high. An insufficient supply of fixed N_2 from the root nodules is then compensated for by enhanced mobilization of nitrogen from the leaves, an effect which can accelerate senescence and cause "self-destruction" of the system (Sinclair and de Wit, 1976). These aspects are discussed in Chapter 7.

In this section, we have presented several examples of the ways in which the sink may influence the rate of photosynthesis and transpiration of source

Table 5.16
Effect of Defoliation or Removal of Flowers and Developing Pods on Nodule Weight and Nitrogen Content of Soybean Plants[a]

Treatment	Dry weight of root nodules (mg/plant)	Nitrogen (mg/plant)
Control	298	475
Defoliation	176	266
Removal of flowers and pods	430	548

[a]Data are for plants harvested after 60 days of growth. Based on Bethlenfalvay *et al.* (1978).

leaves. How these feedback mechanisms are regulated is an important question. When the demand for photosynthates at the sink sites is low (sink limitation), sugars and starch accumulate in the source leaves and the rate of photosynthesis declines. This decline is closely correlated with either partial or total closure of the stomata (Hall and Milthorpe, 1978). It is thought that certain intermediates of carbohydrate metabolism, rather than the level of sugars or starch per se, are responsible for this stomatal response (Hall and Milthorpe, 1978; Claussen and Lenz, 1979).

On the other hand, when the demand of the sink is high, the stomata are wide open and the rates of both photosynthesis and transpiration are high. Evidence has been presented that ABA and CYT are involved in this regulation. Because the transport rates of sucrose and ABA from the source leaves to the sink are positively correlated, when the sink demand for photosynthates is high the ABA concentrations in the source leaves are low and the stomata are fully open (Setter *et al.*, 1980). The high transpiration rates of the source leaves in turn enhance the xylem volume flow into the leaves not only of water but also of the CYT synthesized in the roots (Carmi and Koller, 1979), a mechanism which is also of importance for delaying leaf senescence (Neumann and Stein, 1984).

The role of transpiration rate on the partitioning of root-derived CYT into the various shoot organs is well documented (Monselise *et al.*, 1978). Since stomatal opening is enhanced by CYT and stomatal closure is enhanced by ABA, changes in the ABA/CYT ratio by different rates of import and export seem to be an interesting hypothesis for feedback control from the sink. However, additional control mechanisms might also exist, such as "remote" hormonal signals from the developing fruits to the source leaves (Hein *et al.*, 1979). There is evidence that increased rates of auxin export in the phloem from the developing fruits to the source leaves may also act as such a feedback signal (Hein *et al.*, 1984b).

5.7.1 Concluding Remarks

In this section sink and source limitations in relation to yield have been discussed on a single-plant basis. Under field conditions, however, yield is determined by dry or fresh weight per unit surface area (e.g., hectares), that is, the sum of the yield of the individual plants. The number of plants and the number of reproductive or storage organs per unit surface (the plant density) are therefore important yield components. As plant density increases, however, competition for light and thus source limitation become more important. This explains why under field conditions there is often a close correlation in wheat between grain yield and the integral of leaf area duration (LAD, see Section 6.2) from ear emergence to maturity (Evans *et al.*, 1975).

6

Mineral Nutrition and Yield Response

6.1 General

As the supply or availability of growth factors such as light, CO_2, water, and mineral nutrients increases, the growth rate and yield increase, although with diminishing returns. For mineral nutrients this relationship was formulated mathematically by Mitscherlich as a *law of diminishing yield increment* (Mitscherlich, 1954; Boguslawski, 1958). According to this law, the yield response curves for a particular mineral nutrient are asymptotic; when the supply of one mineral nutrient (or growth factor) is increased, other mineral nutrients (or growth factors) or the genetic potential of crop plants become limiting factors. Typical yield response curves for mineral nutrients are shown in Fig. 6.1. The slopes of the three curves differ. Micronutrients have the steepest and nitrogen the flattest slope, if the nutrient supply is expressed in the same weight units. The slopes reflect the different demands of plants for particular mineral nutrients.

In contrast to the assumptions made by Mitscherlich, the slope of the yield response curve for a particular mineral nutrient cannot be described by a constant factor, nor is the curve asymptotic. When there is an abundant supply of nutrients, a point of inversion is obtained, as shown for micronu-

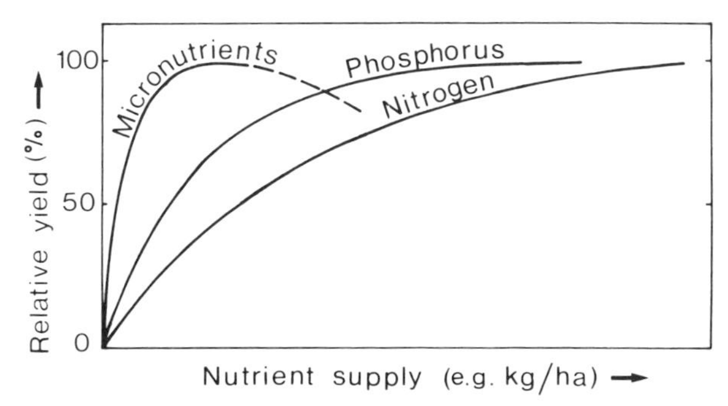

Fig. 6.1 Yield response curves for nitrogen, phosphorus, and micronutrients.

trients in Fig. 6.1. This inversion point exists for other mineral nutrients such as nitrogen (e.g., in the case of grain yield depression by lodging in cereals) and is caused by a number of factors such as the toxicity of a nutrient per se or the induced deficiency of another nutrient. The effects of an ample supply of nitrogen on the phytohormone level and thus on development processes can also be the cause of yield depressions. Furthermore, distinct deviations from typical yield response curves (Fig. 6.1) can be obtained when mineral nutrients such as copper and boron are supplied in very low quantities to a severely deficient soil, in which case the seed set is either prevented or severely inhibited (Section 6.3).

An example of the effect of interaction among mineral nutrients on yield is given in Fig. 6.2. At the lowest potassium level, the response to increasing nitrogen supply is small and at high nitrogen supply yield depression is severe. Under field conditions, however, yield depressions caused by excessive nutrient supply are usually less severe.

Yield response curves differ between grain and straw, particularly at higher potassium levels (Fig. 6.2). In contrast to the straw yield, the grain yield levels off when the nitrogen supply is high, reflecting sink limitation (e.g., small grain number per ear), sink competition (e.g., enhanced formation of tillers), or source limitation (e.g., mutual shading of leaves).

Yield response curves are strongly modulated by interactions between

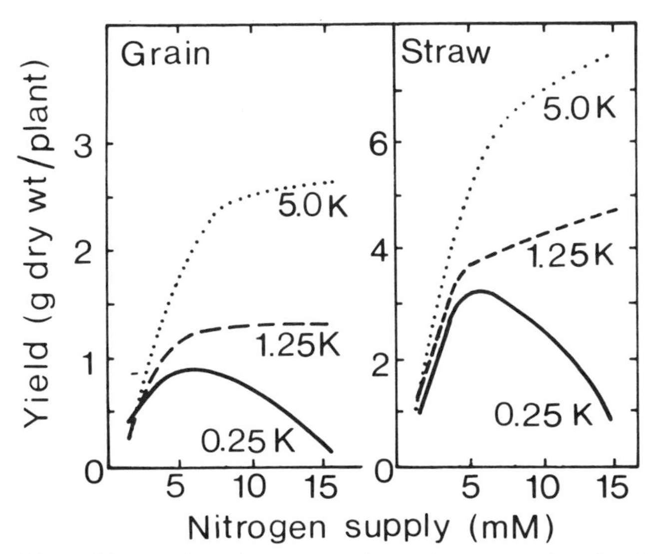

Fig. 6.2 Effect of increasing nitrogen supply at three potassium levels (mM) on grain and straw yield of barley grown in water culture. (Reproduced from MacLeod, 1969, by permission of the American Society of Agronomy.)

mineral nutrients and other growth factors. Under field conditions the interactions between water availability and nitrogen supply are of particular importance. In maize, for example, with increasing nitrogen supply and different soil moisture levels, the grain yield response curves obtained (Shimshi, 1969) are similar to those shown for different potassium levels (Fig. 6.2). The depressions in yield that accompany a large supply of nitrogen in combination with low soil moisture levels are presumably caused by the negative effects of nitrogen on the stomatal response to water deficiency (Chapter 5) as well as by the higher water consumption of vegetative growth and the correspondingly higher water stress in critical periods of grain formation.

Yield response curves can differ not only between vegetative and reproductive organs (Fig. 6.2) but also between the yield components of harvested products. In most crops, both quantity (e.g., dry matter yield in tons per hectare) and quality (e.g., concentration of sugars or protein) are important yield components. As shown schematically in Fig. 6.3 maximum quality can be obtained either before [curve (1)] or after [curve (2)] the maximum dry matter yield has been reached, or both yield components can have a synchronous pattern [curve (3)]. Examples of the behavior described by curve (1) are nitrate accumulation in spinach and sucrose accumulation in sugar beet with increasing levels of nitrogen fertilizer. Examples of curve (2)

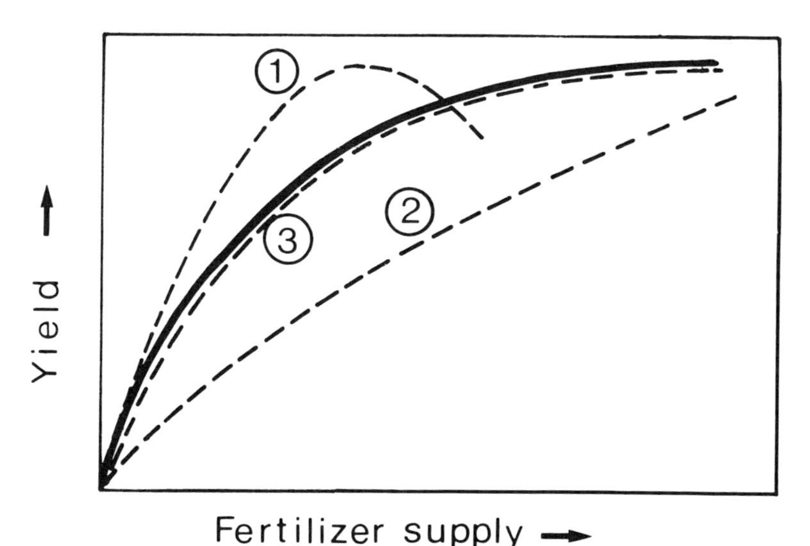

Fig. 6.3 Schematic representation of yield response curves of harvested products. Key: ——, quantitative yield (e.g., dry matter per hectare); – – –, qualitative yield (e.g., content of sugar, protein, and mineral elements).

are the change in protein content of cereals, grains or forage plants with increasing supply of nitrogen fertilizer and the change in concentrations of certain mineral elements (e.g., magnesium and sodium) in forage plants with increasing mineral fertilizer supply.

6.2 Leaf Area Index and Net Photosynthesis

Positive yield response curves are the result of different individual processes, such as an increase in plant density and net photosynthesis per unit leaf area (i.e., effects at the source) or an increase in fruit and seed number (i.e., effects at the sink). In this section, processes that affect the source site are considered.

Generally, the density of a crop population is expressed in terms of the *leaf area index* (LAI), which is defined as the leaf area of plants per unit area of soil (Watson, 1952). For example, an LAI of 5 means that there is a 5 m^2 leaf area of plants growing on a soil area of 1 m^2. As a rule the crop yield increases until an optimal value in the range of 4 to 8 is reached, the exact value depending on plant species, light intensity, leaf shape, leaf angle and other factors (Mengel and Kirkby, 1982). At a high LAI, mutual shading usually becomes the main limiting factor. When the water supply is limited, however, water stress and corresponding negative effects, particularly at the sink sites (see Section 6.3), can decrease the optimal LAI to values far below those resulting from mutual shading.

When the nutrient supply is suboptimal, the leaf growth rate, and thus the LAI, can be limited by low rates of net photosynthesis and/or insufficient cell expansion. In the case of phosphorus deficiency, the small size and dark green color of the leaf blades are the result of insufficient cell expansion and a correspondingly larger number of cells per surface area (Hecht-Buchholz, 1967). In plants suffering from nitrogen deficiency the leaf cells are also smaller (Radin and Parker, 1979), this effect being the result of a decrease in hydraulic conductivity, which causes a water deficit in expanding leaf blades (Radin and Boyer, 1982). These interactions have already been discussed in relation to root pressure (Chapter 3). The effect of nitrogen deficiency on leaf expansion differs among plant species (Table 6.1). In cereals (monocotyledons) cell expansion is inhibited to the same extent during the day and at night. In dicotyledons, however, the inhibition is much more severe during the daytime. This difference in response is related to morphological differences among species and corresponding differences in competition for the water available for transpiration and for cell expansion. In dicotyledons, cell expansion occurs in leaf blades which are exposed to the atmosphere and therefore experience a high rate of transpiration during the daytime. In cereals, however, cell expansion occurs at the base of the leaf blade. This zone is protected from the atmosphere by the sheath of the preceding leaf, so that little transpiration occurs from this zone of elongation. In contrast to leaf expansion, net photosynthesis per unit leaf area is depressed to a similar extent in both groups of plants by nitrogen deficiency. Thus the rate of dry matter production is more efficient in cereals than in dicotyledons when nitrogen is limiting (Radin, 1983).

Table 6.1
Inhibition of Leaf Growth by Nitrogen Deficiency in Different Plant Species[a]

	Average growth inhibition (%)	
Plant species	Day	Night
Cereals (wheat, barley, maize, sorghum)	16	18
Dicotyledons (sunflower, cotton, soybean, radish)	53	8

[a]Based on Radin (1983).

Similar results to those shown in Table 6.1 for the nitrogen effect in dicotyledons have been obtained for phosphorus in cotton plants (Radin and Eidenbock, 1984). It was found that phosphorus deficiency severely inhibited leaf growth rate only during the daytime, and had very little effect at night. These day/night differences were primarily a response to limited water availability for cell expansion during the daytime, caused by low hydraulic conductance of the root system due to phosphorus deficiency. The mode of this phosphorus effect is not clear, but a decrease in root cell wall elasticity is assumed by the authors to be one of the possible explanations.

Mineral nutrition can also affect net photosynthesis in various ways (Nátr, 1975; Barker, 1979). As summarized in Table 6.2, mineral nutrients are required for various processes related to the formation and function of chloroplasts. In green leaf cells, for example, up to 75% of the total organic nitrogen is located in the chloroplasts, mainly as enzyme protein. A deficiency of mineral nutrients that are directly involved in protein or chlorophyll synthesis therefore results in the formation of chloroplasts with low photosynthetic efficiency. The same holds true when there is a deficiency of a mineral nutrient which is directly involved in the electron transport chain and in photophosphorylation (Fig. 5.1). A deficiency of almost any of the individual mineral nutrients therefore also leads to a decrease in the photosynthetic activity of isolated chloroplasts (Spencer and Possingham,

Table 6.2
Examples of the Direct and Indirect Roles of Mineral Nutrients in Photosynthesis

	Mineral nutrients	
Process	Constituents of organic structure	Activators of enzymes, osmoregulation
Chloroplast formation		
Protein synthesis	N, S	Mg, Zn, Fe, K (Mn)[a]
Chlorophyll synthesis	N, Mg	Fe
Electron transport chain		
PS II + I, photophosphorylation	Mg, Fe, Cu, S, P	Mg, Mn (K)
CO_2 Fixation	—	Mg (K, Zn)
Stomatal movement	—	K (Cl)
Starch synthesis, sugar transport	P	Mg, P (K)

[a]Mineral nutrients are in parentheses where their role is based mainly on indirect evidence.

1960) and a change in the fine structure of chloroplasts in a more or less specific manner (Vesk *et al.*, 1966; Hecht-Buchholz, 1972). A mineral nutrient deficiency can also depress net photosynthesis by influencing the CO_2 fixation reaction (Table 6.2) and entry of CO_2 through the stomata. Finally, starch synthesis in the chloroplasts and transport of sugars across the chloroplast envelope into the cytoplasm are directly controlled by the concentration of inorganic phosphate (Heldt *et al.*, 1977). These functions of mineral nutrients in photosynthesis are discussed in more detail in Chapters 8 and 9.

The close positive correlations often observed between the rate of net photosynthesis and the mineral nutrient content of leaves (Nátr, 1975) can also be demonstrated in mature leaf blades when the root supply of mineral nutrients such as nitrogen is withheld (Fig. 6.4). As nitrogen becomes increasingly deficient, the light response curve of net photosynthesis in the mature leaf declines to a very low level. Simultaneously stomatal resistance increases (Nevins and Loomis, 1970). The changes in light response curves with nitrogen deficiency are not necessarily simply reflections of a decrease in the efficiency of the photosynthetic apparatus (e.g., leaf senescence and retranslocation of nitrogen), however, but can also be caused by a lower demand for photosynthates at the sink sites (Chapter 5). In agreement with this, under otherwise optimal growth conditions nutrient deficiency often induces lower growth rates during the vegetative phase, despite the accumulation of carbohydrates in source leaves.

In the phase of photosynthate storage in reproductive organs, roots, or tubers, not only the current LAI and net photosynthesis are important for the final yield; the *leaf area duration* (LAD), that is, the length of time in which the source leaves supply photosynthates to sink sites, such as potato

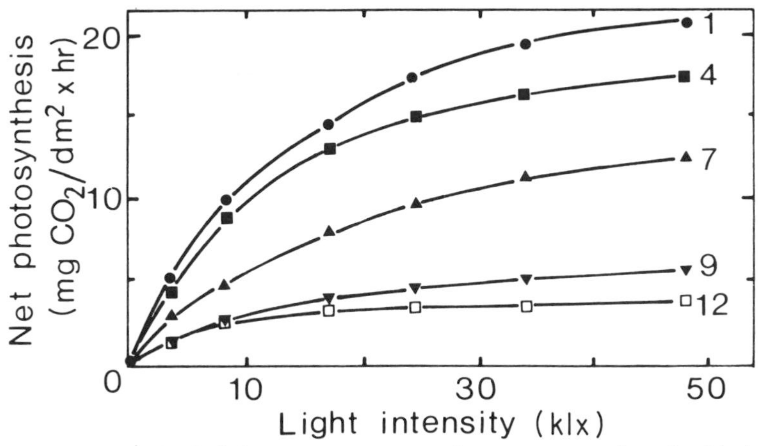

Fig. 6.4 Photosynthetic light response curves of a mature sugar beet leaf 1, 4, 7, 9, and 12 days after the plant was transferred to nitrogen-free nutrient solution. (Reproduced from Nevins and Loomis, 1970, by permission of the Crop Science Society of America.)

tubers (Gunasena and Harris, 1971) and wheat grains (Evans *et al.*, 1975), also affects the yield. Examples of the role of mineral nutrients as limiting factors for the LAD and thus the final yield are described in Section 6.4.

6.3 Mineral Nutrient Supply, Sink Formation, and Sink Activity

6.3.1 General

In crop species where fruits, seeds, and tubers represent the yield, the effects of mineral nutrient supply on yield response curves are often a reflection of sink limitations imposed by either a deficiency or an excessive supply of mineral nutrients during certain critical periods of plant development, including flower induction, fertilization, and tuber initiation. These effects can be either direct (as in the case of nutrient deficiency) or indirect (e.g., effects on the levels of photosynthates or phytohormones).

6.3.2 Flower Initiation

In apple trees, flower formation is affected to a much greater extent by the time and/or form of nitrogen application than by the level of nitrogen supply; the application of urea-containing foliar sprays during the period of flower bud differentiation results in a marked increase in the number of

Table 6.3
Effect of Ammonium versus Nitrate Supply during the Period of Flower Bud Differentiation on the Flowering of Jonathan Apple Trees in the Following Growing Season[a]

Treatment	No. of buds per tree		Percentage of flower buds
	Emerged buds	Flower buds	
Continuous nitrate N supply	35·5	12·8	38·7
Nitrate N supply interrupted by 8 weeks of ammonium N supply	33·3	21·2	63·7
Nitrate N supply interrupted by 24 hr of ammonium N supply	38·8	30·3	77·5

[a]From Grasmanis and Edwards (1974).

flowers produced in the following year and is an effective procedure for smoothing yield fluctuations from year to year (Lüdders and Bünemann, 1970). As shown in Table 6.3 nitrogen supplied in the form of ammonium is much more effective in flower induction than is nitrate nitrogen. Surprisingly, even a 24-hr "pulse" application of ammonium to the roots is more effective than an 8-week treatment.

Since the trees in this study had an ample supply of nitrogen throughout the growing season, it is unlikely that these effects on flower initiation (i.e., on development processes) are related to the nutritional role of nitrogen. Although higher concentrations of certain amino acids and amides (arginine and asparagine) have been found in apple trees supplied with ammonium, it can be assumed that it is not these amino compounds, but compounds such as phytohormones that are responsible for the enhancement effect on flower initiation. This is supported by results of Buban *et al.* (1978) demonstrating that CYT export from the roots to the shoots is higher with ammonium than with nitrate (Table 6.4).

Table 6.4
Effect of Nitrogen Supply on CYT Concentration in Xylem Sap of Apple Root Stocks and on Shoot Growth

Treatment	Zeatin concentration 1 day after start of nitrogen supply (μg/ml)	Shoot growth after 4 months	
		No. of new spurs (<5 cm)	Total length (cm) of the new shoots (>5 cm)
Control (N supply withheld)	0·05	12	16
Ammonium N	1·95	17	34
Nitrate N	0·82	13	48

[a]Based on Buban *et al.* (1978).

The form of nitrogen supply also affects subsequent shoot growth (Table 6.4). Nitrate promotes elongation more than ammonium, whereas only ammonium nutrition increases the number of new spurs (the flower-bearing parts of the shoots). Although the promotion of flower morphogenesis by CYT is well established for various plant species (Bruinsma, 1977; Herzog, 1981), there is a possibility that the enhancement of flower initiation by ammonium involves other compounds (e.g., polyamines) as secondary messengers (Chapter 5).

Flower formation in tomato (Menary and Van Staden, 1976) and wheat (Rahman and Wilson, 1977) is also positively correlated with the phosphorus supply. In apple trees the number of flowers per tree is almost linearly related to the phosphorus content of the leaves (Bould and Parfitt, 1973). The positive correlations between the number of flowers and CYT level in tomato (Menary and Van Staden, 1976), on the one hand, and between the phosphorus supply and the CYT level, on the other (Dhillon, 1978; Horgan and Wareing, 1980) provide strong evidence that CYT contributes to the enhancement effect of phosphorus on flower formation. Basically similar conclusions were drawn from the effects of potassium on flower formation in *Solanum sisymbrifolium* (Wakhloo, 1975a,b). Low potassium levels in the leaves were correlated with a high proportion of sterile female flowers. This sterility did not occur in plants either with high or with low potassium levels when the plants had been sprayed with CYT. Effects of potassium supply on CYT levels in plants are documented (Chapter 5).

The effects of mineral nutrient supply on flower formation described above are most likely caused by changes in the phytohormone balance and are expressions of the regulation of the sink–source relationship (e.g., the ratio of sink size to source size and source capacity) that is necessary to ensure proper completion of the reproductive phase even under conditions of limited supply of mineral nutrients. This internal regulation can be modulated in cereals by the late application of nitrogen fertilizers before anthesis in order to increase the number of grains per ear (Herzog, 1981). Similarly, in sunflowers the number of seeds per plant was increased several times by elevated nitrogen supply prior to flower initiation, but hardly affected by the same treatment after flower initiation (Steer *et al.*, 1984). It has to be kept in mind, however, that the seed number per plant can also be increased by a high supply of sucrose prior to flower initiation, as it has been demonstrated by Waters *et al.* (1984) in wheat. Mineral nutrients may therefore also exert their influence on flower initiation and seed number by increasing the rate of photosynthesis and export of sucrose to these sink sites, as can similarly be demonstrated by increasing light intensity (Stockman *et al.*, 1983).

6.3.3 Fertilization

The number of seeds and/or fruits per plant can also be directly affected by mineral nutrient supply. This is clearly the case with copper and boron. In cereals in particular, copper deficiency affects the reproductive phase (Table 6.5). When the copper deficiency is severe, no grains are produced even though the straw yield is quite high owing to enhanced tiller formation (loss of apical dominance of the main stem). As the copper supply is increased, grain yield rises sharply, whereas the straw yield is only slightly enhanced.

Table 6.5
Effect of Increasing Copper Supply in Wheat (cv. Chatilerma) Grown on Copper-Deficient Soil[a]

	Copper supply (mg/pot)			
	0	0·1	0·4	2·0
Number of tillers	22	15	13	10
Straw yield (g)	7·7	9·0	10·3	10·9
Grain yield (g)	0·0	0·5	3·5	11·8

[a]Total number of plants per pot: 4. Based on Nambiar (1976).

These results provide an informative example of both sink limitation on yield and deviation from the typical response curve (Fig. 6.1) between grain yield and mineral nutrient supply.

The primary causes of failure in grain set in copper-deficient plants are inhibition of anther formation, the production of a much smaller number of pollen grains per anther, and particularly the nonviability of the pollen (Graham, 1975). That nonviability of the pollen is the main cause of failure in grain set has been confirmed by cross-pollination experiments (Table 6.6).

Table 6.6
Grain Set by Cross-Pollination between Copper-Deficient and Copper-Sufficient Wheat Plants[a]

Cross				Total no.	
Female	×	Male	No. of grains set	Florets	Heads
−Cu		−Cu (selfed)[b]	0	76	3
+Cu		−Cu	2	76	3
−Cu		+Cu	47	157	7
+Cu		+Cu (selfed)	86	86	3

[a]Reprinted from Graham (1975). Copyright 1975 MacMillan Journals Limited.
[b]Self-pollination.

The critical period in copper-deficient wheat plants is the early booting stage at the onset of pollen formation (microsporogenesis). The synchronous meiotic division of a large number of pollen mother cells may create a high localized demand for copper not available in sufficient amounts in deficient plants (Graham, 1975). In these plants, inhibited lignification of the anther cell walls may also be involved in the lack of fertilization (Chapter 9). In copper-deficient soils, especially in dry soils, the lack of fertilization can be a major yield-limiting factor, but it can be overcome by the application of foliar sprays containing copper salts (Chapter 4).

The production and viability of pollen are also affected by molybdenum. In maize, a decrease in the molybdenum content of pollen was correlated with a decrease in the number of pollen grains per anther as well as a decrease in the size and viability of the pollen grains (Section 9.5). As yet no information is available on the extent to which molybdenum deficiency also depresses fertilization and grain set.

Boron is another mineral nutrient that affects fertilization. Boron is essential for pollen tube growth (Section 9.6), a role reflected under conditions of boron deficiency by a decrease in the number of grains per head in rice (Garg *et al.*, 1979) or even a total lack of fertilization in barley (Simojoki, 1972). The failure of seed formation in maize suffering from boron deficiency is caused by the nonreceptiveness of the silks to the pollen (Vaughan, 1977). As the level of boron nutrition increases, vegetative growth, including the structural growth of the silks, either is not affected or is even somewhat depressed (Fig. 6.5). In contrast, grain formation is absent in severely deficient plants but increases dramatically when the boron supply

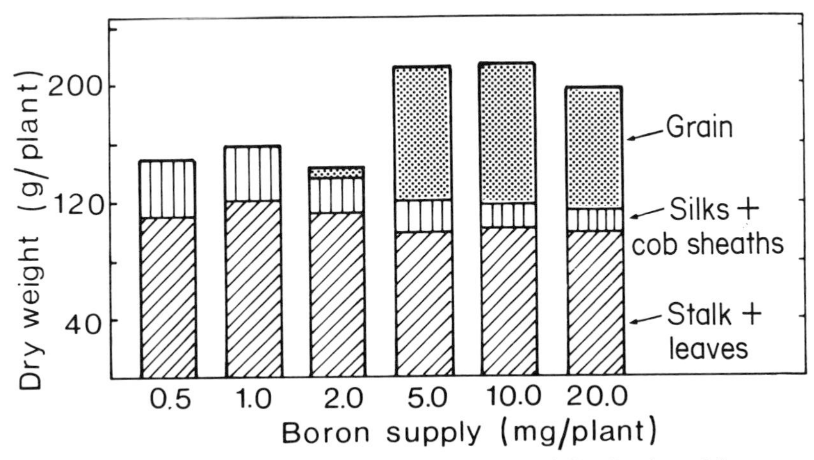

Fig. 6.5 Effect of boron supply on the production and distribution of dry matter in maize plants. (Based on Vaughan, 1977.)

is adequate. Obviously there is a minimum boron requirement, which is in the range of 3 mg boron per maize plant for fertilization and grain set. Figure 6.5 provides another example of both strict sink limitation induced by mineral nutrient deficiency and a yield response curve quite different from the typical curve.

Water deficiency (water stress) a few days before anthesis (during which meiosis of the pollen mother cells takes place) is another environmental factor which reduces grain set. As shown in Table 6.7 this effect is related to the accumulation of ABA in the ears. The negative effect of applying ABA to the well-watered control plants was similar to the effect of water stress on fertilization.

Table 6.7
Relationship between Water Stress, ABA, and Seed Set in Wheat[a]

Treatment	Fertile spikelets (%)	ABA in ears (μg/g fresh wt)
Control	68	35
Water stress[b]	44	111
Control + ABA[b]	37	—

[a]Based on Morgan (1980).
[b]Application during meiosis of pollen mother cells.

Both water stress and ABA application during meiosis of pollen mother cells cause malformation of the pollen grains without affecting female fertility (Morgan, 1980; Saini and Aspinall, 1982). The inhibitory effect of ABA on fertilization is restricted to a short period before anthesis (Saini and Aspinall, 1982). From these results it can be concluded that mineral nutrient deficiency can depress grain set and grain yield in a similar manner, provided that this deficiency occurs during the critical period of pollen formation, and that the deficiency is correlated with an accumulation of ABA (e.g., nitrogen or potassium deficiency).

6.3.4 Flower and Seed Development

In certain plant species, such as soybean, drop of flowers and developing pods is a major yield-limiting factor. Nitrogen deficiency during the flowering period enhances flower drop and depresses seed yield correspondingly (Streeter, 1978). An ample quantity of nitrogen during this critical phase, supplied either via higher basal nitrogen dressing or by late application of nitrogen, is therefore quite effective in reducing flower and pod drop and in increasing final seed yield in soybean (Brevedan *et al.*, 1977, 1978).

Although the physiological reasons for both the flower and pod drop and the decline in drop produced by nitrogen application are not yet known in detail, it is certain that phytohormones, especially ABA, are involved. The inhibitory effect of an ample nitrogen supply both on ABA accumulation (Chapter 5) and on flower and pod drop and the specific effects of ABA on the formation of abscission layers support this assumption.

Premature ripening of fruits and seeds imposed by water or nutrient deficiency is another yield-limiting factor. In this case it is not the number of grains but the weight (size) of a single grain or fruit that is low. There is substantial evidence that elevated ABA levels are also involved in premature ripening. An example of this is shown in Table 6.8 for potassium-deficient wheat plants. In these plants, and particularly 4–6 weeks after

Table 6.8
Effect of Potassium Supply to Wheat on ABA Content and Weight of Grains[a]

K Supply	ABA Content (ng/grain), days after anthesis				Days from anthesis to full ripening	Weight of a single grain (mg)
	28	35	38	44		
Low (deficient)	7·7	13·4	16·5	2·2	46	16·0
High	3·7	4·4	ND[b]	9·4	75	34·4

[a]Based on Haeder and Beringer (1981).
[b]ND, Not determined.

anthesis, the levels of ABA in the grains are much higher than those in the grains of plants well supplied with potassium. Correspondingly, the grain-filling period in potassium-deficient plants is much shorter and the weight of a single grain at maturity is lower than that in control plants. As has been shown before (Chapter 5), high ABA levels in grains coincide with a sharp decline in the sink activity of grains. It is quite likely that the elevated ABA levels in the flag leaves of potassium-deficient wheat plants (Haeder and Beringer, 1981) and a correspondingly higher ABA import to the developing grains are responsible for the premature ripening. Similar causal relationships can be assumed to occur between premature ripening and water and/or nitrogen deficiency during the grain-filled period.

6.3.5 Tuberization and Tuber Growth Rate

In root and tuber crops such as sugar beet or potato, the induction and growth rate of the storage organ are strongly influenced by the environmental factors. In contrast to crop species in which seeds and fruits are the main storage sinks, root and tuber crops exhibit a distinct sink competition between vegetative shoot growth and storage tissue growth for fairly long periods after the onset of storage growth. This competition is particularly evident in so-called indeterminate genotypes—for example, in potato (Kleinkopf *et al.*, 1981). In general, environmental factors (e.g., high nitrogen supply) with pronounced favourable effects on vegetative shoot growth delay the initiation of the storage process and decrease growth rate and photosynthate accumulation in storage organs—for example, sugar beets (Forster, 1970) and potatoes (Ivins and Bremner, 1964; Gunasena and Harris, 1971).

A large and continuous supply of nitrogen to the roots of potatoes delays or even prevents tuberization (Krauss and Marschner, 1971). After tuberization the tuber growth rate is also drastically reduced by a large nitrogen supply, whereas the growth rate of the vegetative shoot is enhanced. The effect of nitrogen supply on tuber growth rate is illustrated in Table 6.9. Resumption of the tuber growth rate to normal levels after interruption of the nitrogen supply indicates that sink competition between the vegetative shoot and tubers can readily be manipulated by the nitrogen supply.

Table 6.9
Growth Rate of Potato Tubers in Relation to Nitrate Supply to the Roots of Potato Plants[a]

Nitrate concentration (mEq/liter)	Nitrate uptake (meq/day × plant)	Tuber growth rate (cm^3/day × plant)
1·5	1.18	3·24
3·5	2·10	4·06
7·0	6·04	0·44
Nitrogen supply withheld for 6 days	—	3·89

[a]From Krauss and Marschner (1971).

In potato, cessation of tuber growth caused by a sudden increase in nitrogen supply to the roots induces "regrowth" of the tubers, that is, the formation of stolons on the tuber apex (Krauss and Marschner, 1976, 1982). Interruption and resupply of nitrogen, therefore, can result in the production of chain-like tubers or so-called secondary growth (Fig. 6.6). After a temporary cessation of growth, resumption of the normal growth rate is usually restricted to a certain area of the tubers (meristems or "eyes"), leading to typical malformations and knobby tubers, which are often observed under field conditions after transient drought periods.

These effects of nitrogen supply are brought about by nitrogen-induced changes in the phytohormone balance both in the vegetative shoots and in the tubers. As already shown (Chapter 5), an interruption of the nitrogen supply results in a decrease both in CYT export from roots to shoots and in

Fig. 6.6 Secondary growth and malformation of potato tubers induced by alternating high and low nitrogen supply to the roots. (From Krauss, 1980.)

the sink strength and growth rate of the vegetative shoot. A corresponding increase in the ABA/GA ratio of the shoots seems to trigger tuberization. In agreement with this, tuberization can also be induced by the application of either ABA or the GA antagonist CCC (Krauss and Marschner, 1976) or by the removal of the shoot apices, the main sites of GA synthesis (Hammes and Beyers, 1973). On the other hand, the regrowth of tubers induced by a sudden increase in the nitrogen supply is correlated with a decrease in the ABA/GA ratio not only in the vegetative shoots but also in the tubers, where the GA level increases by a factor of 2 but the ABA level drops to less than 5% of that in normal growing tubers (Krauss, 1978b).

6.4 Mineral Nutrition and the Sink–Source Relationships

In root and tuber crops, unlike grain crops, the sink–source relationship is quite labile even after the onset of the storage process. This has to be considered, for example, in the application of nitrogen fertilizer to potato. On the one hand, a large nitrogen supply is important for shoot growth and for obtaining an LAI between 4 and 6, the value required for high tuber yields (Kleinkopf *et al.*, 1981; Dwelle *et al.*, 1981). On the other hand, a large supply of nitrogen delays either tuberization or the onset of the linear phase of tuber growth. These interactions are demonstrated in Fig. 6.7. The advantage of earlier tuberization obtained by supplying a low level of nitrogen is offset by a low LAI and earlier leaf senescence, that is a short LAD and a correspondingly lower tuber yield. When the nitrogen supply is large, both LAI and LAD, and thus final tuber yield, are much higher. It is also evident, however, that higher tuber yield induced by a large nitrogen supply can be realized only when the vegetation period is sufficiently long, that is, in

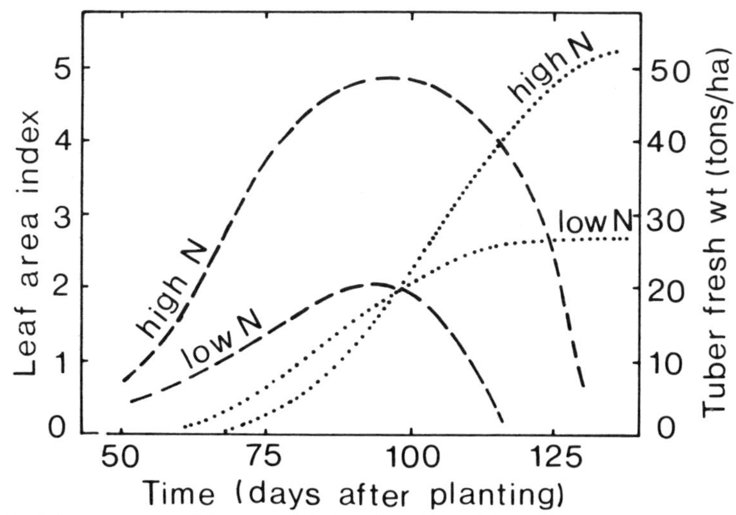

Fig. 6.7 Time course of leaf area index (---) and fresh weight of potato tubers (· · ·) at two levels of nitrogen supply. (Based on Ivins and Bremner, 1964, and Kleinkopf *et al.*, 1981.)

the absence of early frost (Clutterbuck and Simpson, 1978) or in the absence of severe water deficiency.

The early decline in LAI when the supply of nitrogen is low (Fig. 6.7) indicates that the final tuber yield is limited by the source. The question arises as to the reasons for this source limitation. In potato plants at maturity, between 60 and 80% of the total nitrogen is located in the tubers (Kleinkopf *et al.*, 1981). Therefore, when the nitrogen supply is low, exhaustion of nitrogen in the source leaves presumably plays a key role in leaf senescence and in the termination of tuber growth.

Competition for nitrogen rather than for carbohydrates supplied from the source leaves can also be the main limiting factor for seed yield in mustard and rape plants (Trobisch and Schilling, 1970; Schilling and Trobisch, 1970). In mustard plants the developing seeds and leaves compete for nitrogen, and seed set, seed growth, and final seed yield are determined primarily by the size of the nitrogen pool in the vegetative parts. In crucifers, the flower differentiation at the auxiliary stems occurs after the onset of flowering of the main stem and is strongly dependent on the availability of nitrogen during this period. Additional application of nitrogen at the onset of flowering therefore leads to an increase in seed number and yield (Fig. 6.8).

The example with mustard plants in Fig. 6.8 demonstrates that source limitation can be imposed by nitrogen rather than carbohydrates. This aspect has also to be considered in source–sink manipulations. Removing source leaves from a plant is a common procedure for evaluating source

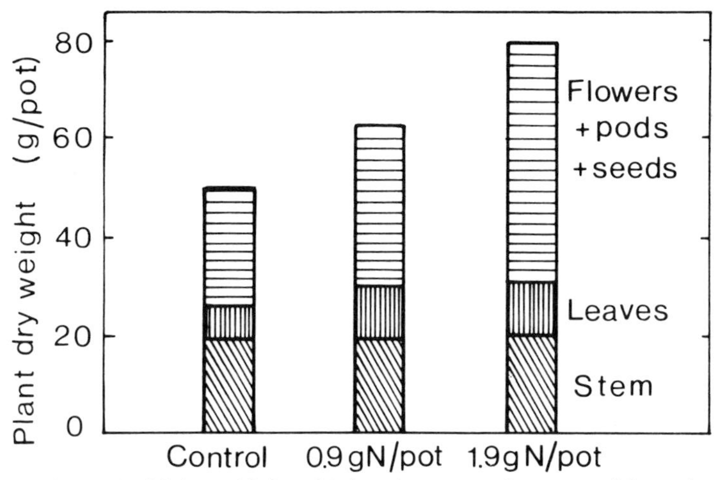

Fig. 6.8 Effect of addition of 0·9 and 1·9 g nitrogen at the onset of flowering on the total dry weight and dry weight distribution in shoots of white mustard plants. (Based on Trobisch and Schilling, 1970.)

limitation in photosynthesis (Chapter 5). Of course, when source leaves are removed, nitrogen and other mineral nutrients are also removed. Therefore, shading of source leaves has a different effect on the reduction of seed yield than does leaf removal. Shading the source leaves of mustard plants reduced the seed yield by only 20%, whereas removal of these leaves depressed the yield by 50% (Trobisch and Schilling, 1969).

In principle, similar to the example shown above for nitrogen, each other mineral nutrient can become the dominant factor inducing source limitation on final yield of seeds, fruits and tubers, provided this mineral nutrient can be readily retranslocated from the source (Section 3.6). The probability that such a limitation will exist depends on such factors as the availability of a given nutrient in the soil, the concentration and amount of the nutrient (source size) in the vegetative shoot, the specific demand of the sink for the nutrient, and the growth rate of the sink. In mature cereal plants, for example, as much as 80% of the total amount of nitrogen or phosphorus is located in the grains, compared to less than 20% of the total potassium. Thus, in cereal plants where there is a suboptimal supply of the three mineral nutrients during the vegetative stage, source limitation during grain filling is most likely induced by nitrogen or phosphorus, but not by potassium. On the other hand, in fleshy fruits or tubers, the potassium concentrations are very high (2–3% of the dry weight), and at maturity most of the potassium is located in the fruits or tubers. Potassium-induced source limitation is therefore more likely in this type of crop.

An instructive example of potassium-induced limitations for two tomato cultivars has been given in Table 3.14. The cultivar VF-13L was developed for mechanical harvesting, and ripening is therefore more rapid and uniform than that of the conventional cultivar VFN-8. In both cultivars, the fruits act as a strong sink for potassium from the leaves, causing a sharp decline in potassium concentrations in the petioles at later stages of fruit growth. The decline was more rapid and intensive in the cultivar VF-13L, however, and was correlated with severe potassium-deficiency symptoms in the leaf blades. This deficiency could not be prevented by a large supply of potassium to the roots (Lingle and Lorenz, 1969), presumably because of the preferential carbohydrate supply from the source leaves to the fruits (dominant sink) and the corresponding decline in both root growth and root activity (Chapter 5).

These examples illustrate the role of mineral nutrients as yield-limiting factors when fruits, seeds, or other organs are the dominant sink sites and mineral nutrient uptake by the roots is declining. Progress in selecting and breeding for genotypes with a high *harvest index* (ratio of economic yield to total dry matter) and short periods of fruit growth or ripening (e.g., the filling period in cereals) might be severely restricted, not because of the limited capacity of the source to supply carbohydrates, but rather because of the limited amount of mineral nutrients such as potassium, nitrogen, phosphorus, and magnesium that are available for retranslocation from source to sink.

7

Nitrogen Fixation

7.1 General

Total world biological dinitrogen (N_2) fixation is assumed to be about three to four times higher than the current industrial fixation of atmospheric N_2 (5×10^7 tons/year; Werner, 1980). The conversion of the inert N_2 molecule to combined nitrogen (NH_3; NO_3^-, etc.) which can be utilized as a mineral nutrient is brought about either by reduction to ammonia (NH_3) or oxidation to nitrate (NO_3^-). This conversion, also referred to as fixation, is highly energy consuming. In both industrial and biological conversion the reaction $N_2 \rightarrow 2NH_3$ dominates. In industrial fixation, N_2 is catalytically reduced to NH_3 by reaction with hydrogen (produced, e.g., from natural gas) in the Haber–Bosch process ($N_2 + 3H_2 \rightarrow 2NH_3$) under conditions of high temperature and pressure. The increase in both the costs of fossil energy and the worldwide demand for nitrogen fertilizer in food production are major reasons for renewed interest in biological N_2 fixation as an alternative or at least a supplement to the use of chemical nitrogen fertilizer.

7.2 Biological Nitrogen-Fixing System

The capacity for biological N_2 fixation is restricted to prokaryotes, namely, bacteria and blue-green algae (Cyanophyceae). According to our present knowledge, some species in 11 of the 47 bacterial families and some species in the 8 Cyanophyceae families are capable of N_2 fixation (Werner, 1980). Some of these, such as blue-green algae, are free living in terrestrial and marine ecosystems. Others live in associations in the rhizosphere of host plants; for higher plants the most important of these are N_2-fixing microorganisms living in symbiosis, for example, with legumes or nodulated nonlegumes. The major N_2-fixing systems in terrestrial ecosystems are summarized in Fig. 7.1.

Distinct differences in both energy source and fixation capability exist among the three N_2-fixing systems. On average, symbiotic systems have the

System of N_2 fixation ($N_2 \rightarrow NH_3$) and microorganisms involved	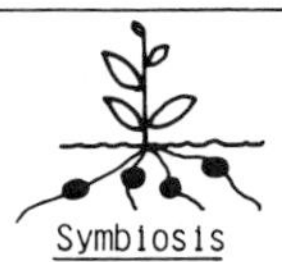 Symbiosis (e.g. Rhizobium, actynomycetes)	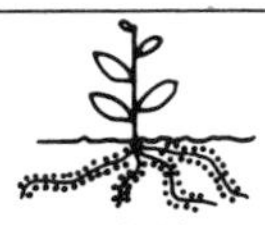 Associations (e.g. Azospirillum, Azotobacter paspali)	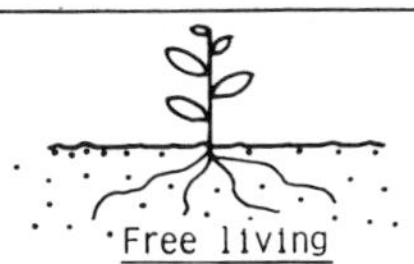 Free living (e.g. Azotobacter, Klebsiella, Rhodospirillum)	
Energy source (organic carbon)	Sucrose (and other carbohydrates from the host plant)	Root exudates from the host plant	Heterotroph: Plant residues	Autotroph: Photosynthesis
Fixation capabilities (kg N/ha·year)	Legumes: 57 - 600 Nodulated non-legumes: 2 - 300	12 - 313	0.1 - 0.5	25

Fig. 7.1 Type, energy source, and fixation capabilities of biological N_2 fixation systems in soils. Data for fixation capabilities are from Evans and Barber (1977). By courtesy of K. Isermann.

highest fixation capability, because N_2-fixing microorganisms are supplied directly from the host plant with carbohydrates as energy source for N_2 fixation. Nodulated legumes, such as soybean, faba bean (Fig. 7.2), clover, and alfalfa, in symbiosis with *Rhizobium* are among the most prominent N_2-fixing systems in agriculture, but most of the nitrogen input in forests and woodlands is provided by a variety of nodulated nonleguminous symbiotic systems between *Actinomycetes* (genus *Frankia*) and higher plants such as *Alnus* or *Ceanothus*.

Some N_2-fixing systems with high host–microorganisms specificity do not develop nodules; instead, the root surface or the intercellular spaces of the cortex cells are the habitat of N_2-fixing microorganisms (Fig. 7.1). These are

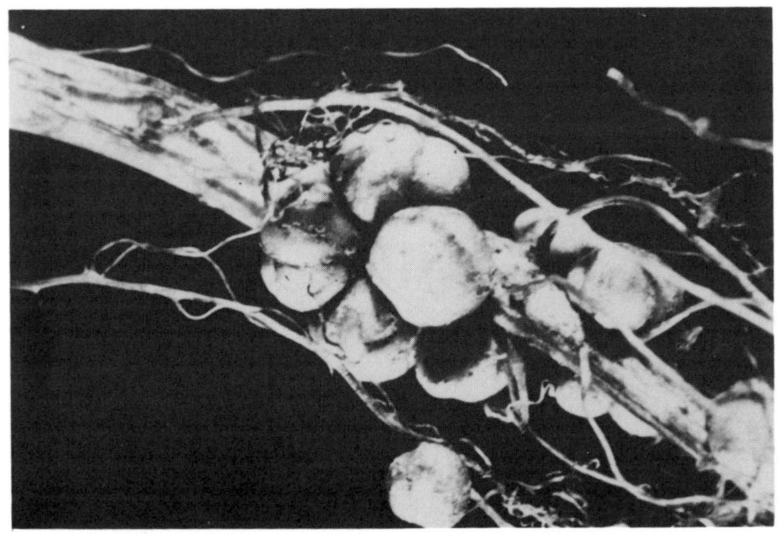

Fig. 7.2 Section of a nodulated root of faba bean.

called *rhizosphere associations* (Döbereiner, 1983). In certain grasses the N_2 fixation capability of these associations can be very high and has led to exciting speculations about the possibility of providing nitrogen for those crop species (see Section 7.5). In the case of free-living microorganisms in the soil or at the soil surface, usually the capability for N_2 fixation is severely restricted by substrate limitation (an insufficient amount of organic residues in the soil) unless the organisms are carbon autotrophic, as are blue-green algae.

7.3 Biochemistry of Nitrogen Fixation

Because industrial N_2 fixation requires both high temperature and high pressure, the question arises as to how this reaction can proceed in living cells growing at low temperatures and atmospheric pressure.

In all N_2-fixing microorganisms both the energy requirement of 355 kJ/mol NH_3 and the principal steps of the $N_2 \rightarrow 2NH_3$ reaction are the same (Fig. 7.3). The enzyme complex referred to as nitrogenase, in combination with the necessary reactants, catalyzes the reduction of N_2 to NH_3. Nitrogenase is unique to N_2-fixing microorganisms and has been found, for example, in aerobic and anaerobic bacteria, blue-green algae, and root nodules of legumes. Nitrogenase consists of two oxygen-sensitive, nonheme iron proteins (Fig. 7.3). One has a molecular weight of ~222,000 and contains iron and molybdenum in a ratio of 24–36 : 2 per molecule; the other has a molecular weight of ~60,000 (Bothe *et al.*, 1983). This smaller component is free of molybdenum and contains four atoms each of iron and acid-labile sulfur per molecule (4Fe–4S clusters; Bothe *et al.*, 1983). The smaller component provides the electrons for the reduction of N_2 by the larger molybdenum–iron protein.

For the nitrogenase reaction, energy in the form of a reductant and as ATP is essential (Fig. 7.3). Energy from ATP and electrons from the electron carrier (usually ferredoxin; see Chapter 5) induce a conformational change in the iron protein and convert it to a powerful reductant capable of

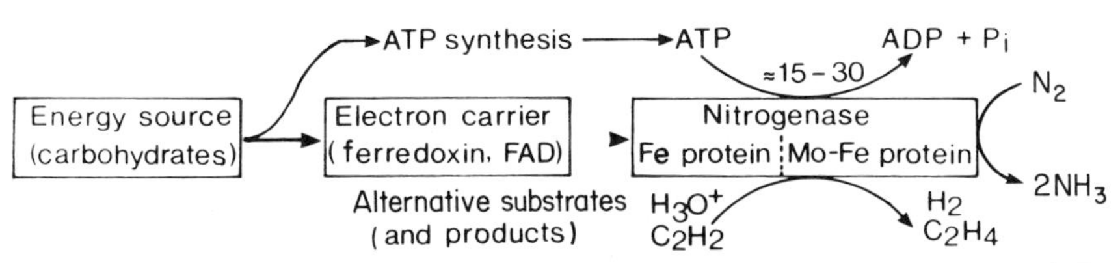

Fig. 7.3 Scheme illustrating the energy supply and principal reactions of the nitrogenase system. (Based on Evans and Barber, 1977.)

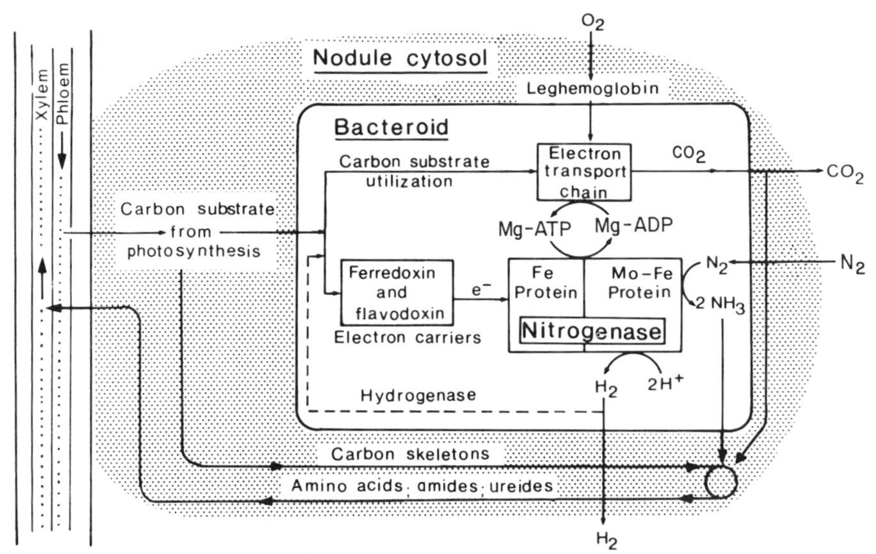

Fig. 7.4 Diagram of the relationship between nitrogenase and related reactions in the bacteroids and the cytosol of legume nodules. (Modified from Evans and Barber, 1977; copyright 1984 by the AAAS.)

transferring electrons to the molybdenum–iron protein, which in turn reduces N_2. Reduction of N_2 to NH_3 requires between 15 and 30 ATP molecules per N_2 molecule reduced (Shanmugam *et al.*, 1978). Several other substrates, including protons and acetylene (C_2H_2), can function in the nitrogenase reaction as alternatives to N_2 (Fig. 7.3). They therefore compete with N_2 for the electrons from nitrogenase and considerably decrease the efficiency (utilization of energy) of N_2 fixation (Fig. 7.4). In *Rhizobium* between 30 and 60% of the energy supplied to the nitrogenase is released as H_2 (Schubert *et al.*, 1978). *Rhizobium* strains are also capable of splitting the H_2 by hydrogenase, however, thus recycling the electrons for subsequent N_2 reduction. This recycling can be important for the efficiency of the N_2 reduction. A higher recycling during the dark period obviously permits the maintenance of similar rates of N_2 fixation and of the root export of fixed nitrogen, despite a distinct drop in the carbohydrate level of the nodules (Rainbird *et al.*, 1983). Selection of *Rhizobium* strains with a generally higher recycling of electrons from H_2 may be important for higher efficiency of the N_2 fixation (Schubert *et al.*, 1978), although this potential should not be overestimated (Truelsen and Wyndaele, 1984). The utilization of alternative substrates in the nitrogenase reaction, on the other hand, offers an excellent tool for studies on nitrogenase activity and N_2 fixation. Ethylene (C_2H_4) can be readily measured in very low concentrations by means of gas chromatography. Supplying acetylene and then determining the rate of ethylene evolution (acetylene reduction assay) is a relatively simple method and has been responsible for the enormous progress made in the past decade in research on biological N_2 fixation (Bothe *et al.*, 1983; Döbereiner, 1983; Quispel, 1983).

The nitrogenase reaction is severely inhibited by O_2, and at high O_2 concentrations the enzyme is irreversibly destroyed. On the other hand, there is a large requirement for energy in the form of ATP, which is produced mainly during respiration (Evans and Barber, 1977). Nitrogen-fixing microorganisms have developed the following different strategies to overcome the difficulty of O_2 inhibition:

1. Living under anaerobic conditions (e.g., *Clostridium*)
2. Respiratory protection by consumption of most of the O_2 through excessive respiration (e.g., *Azotobacter*)
3. Living in colonies covered by slime sheets, which restrict O_2 diffusion
4. Controlling O_2 diffusion enzymatically by leghemoglobin (e.g., in the root nodules of legumes).

The large capacity for N_2 fixation in the nodules of legumes is also the result of other factors (Fig. 7.4). The host plant delivers sucrose via the phloem to the nodules as an energy substrate. The oxidative degradation of the carbon substrate produces ATP and reductants, and there is a considerable release of respiratory CO_2. Leghemoglobin regulates the transport of O_2 from outside the nodules to the bacteroids in order to maintain an exceedingly low O_2 concentration on the bacteroid surface. The N_2 fixation product, NH_3, is released into the plant cytosol, where amides, amino acids, and ureides (e.g., allantoin; see Chapter 8) are synthesized. These compounds are excreted by transfer cells (Pate and Gunning, 1972), predominantly into the root xylem, and transported to the shoots.

Evidence has recently been presented that recycling of CO_2 and synthesis of organic acids via the PEP carboxylase pathway (Section 2.5.3) play an important role in the carbon economy of nodules (Fig. 7.4). In the nodule fresh weight, the PEP carboxylase activity is more than ten times higher than in the fresh weight of roots or leaves (Deroche *et al.*, 1983). About 2% of the soluble protein of nodule tissue consists of this enzyme (Vance and Stade, 1984). The fixation of respiratory CO_2 in the nodules delivers up to 25% of the carbon skeleton (as organic acids) required for the incorporation of fixed NH_3 and its translocation as amino acids and amides to the host plant (Vance *et al.*, 1983).

The enzymes responsible for primary NH_3 assimilation in the nodule cytosol are glutamine synthetase and glutamate synthase (Meeks *et al.*, 1978; Werner *et al.*, 1980). The same enzymes are also responsible for NH_3

assimilation in the roots and shoots of the host (Chapter 8). Nitrogenase is severely inhibited by end products (Shanmugam *et al.*, 1978); high concentrations of NH_3, glutamine, and glutamate also repress the "N_2-fixing genes" (*nif* genes). This has important implications for the application of nitrogen fertilizer to the legumes and other N_2-fixing systems (Section 7.4.4).

The concentration of leghemoglobin, a red-colored enzyme with a central iron atom in the porphyrin ring (identical to cytochromes; see Section 8.7), is closely but not linearly correlated with the N_2-fixing capacity of root nodules (Werner *et al.*, 1981). The synthesis of leghemoglobin requires cobalt, although iron rather than cobalt is the metal component of the enzyme. The function of cobalt is presumably related to the involvement of vitamin B_{12} in the synthesis of leghemoglobin (Section 10.4).

7.4 Symbiotic Systems

7.4.1 General

With respect to the carbon supply required for N_2 fixation, two types of symbiotic systems exist:

I. Nodulated legumes and nodulated nonlegumes
II. Symbioses with blue-green algae

In system I the N_2-fixing microorganisms are either *Rhizobium* (in legumes) or actinomycetes (in nonlegumes, e.g., *Alnus* or *Casuarina*) living in nodules on the host roots and supplied directly with carbohydrates from the host. In system II the fixing bacteria meet their carbon (energy) requirement for N_2 fixation at least in part by their own photosynthesis. Examples of system II are lichens (fungi and blue-green algae of the genus *Nostoc*) and the freshwater fern *Azolla* in symbiosis with the blue-green alga *Anabaena azolla*. The water fern *Azolla* has long been appreciated, particularly in China, for its contribution to the nitrogen balance in paddy soils, which is assumed to be on average 50–80 kg fixed N_2 per hectare each year (Bothe *et al.*, 1983). From long-term field studies even higher values (79–103 kg fixed N_2 per hectare each year) have been calculated (App *et al.*, 1984). In other agricultural production systems, however, the main suppliers of fixed N_2 are nodulated legumes. The following discussion of symbiotic systems therefore focuses on legumes.

7.4.2 Infection and Host Specificity

The initial basis for our understanding of the symbiotic relationship between legumes and *Rhizobium* species was provided by Hellriegel and Wilfarth in

Table 7.1
Examples of Host Preference among *Rhizobium* Species[a]

Rhizobium species	Preferred host genus
R. leguminosarum	*Pisum, Vicia, Lens; Cicer*
R. trifolii	*Trifolium*
R. phaseoli	*Phaseolus*
R. meliloti	*Medicago, Melilotus*
R. japonicum	*Glycine*
R. lupini	*Lupinus, Lotus, Ornithopus*

[a]Based on Quispel (1983).

1888, but the favorable effect of legumes on subsequent crops in crop rotation had been exploited in ancient agriculture. The first step in the colonization of legume roots with *Rhizobium* involves recognition of the host plant. Most probably, sugar-binding proteins (lectins) are responsible for this recognition. They are located at the external surface of root cells, mainly root hairs, and cross-react with, and bind, carbohydrates on the surface of the appropriate *Rhizobium* strain (Quiespel, 1983). Different *Rhizobium* species have various degrees of host specificity or host preference (Table 7.1). Therefore, if legumes are introduced into soils in which the appropriate *Rhizobium* species is not present, efficient N_2 fixation requires seed inoculation with the appropriate species. Infection starts with the penetration of the bacteria into the root cells, preferentially the root hairs. In peanut, for example, *Rhizobium* infection takes place only in genotypes with abundant root hair formation (Nambiar *et al.*, 1983), and in bean plants *Rhizobium* infection is closely correlated with root hair length (Franco and Munns, 1982b). Infected root hairs are curled, most likely as a result of auxin production by the bacteria (Munns, 1968b). This step in the infection process is quite sensitive to unfavorable environmental conditions, such as salinity (Singleton *et al.*, 1982) or soil acidity (Section 7.4.4). Inside the infected root cell, the bacteria are enclosed by a thread which grows into the cortical tissue of the roots (Quispel, 1983). The infected cortical cell proliferates, and nodule growth begins (Fig. 7.5).

During nodule growth, the bacteria are transformed into bacteroids, which are several times larger than the original bacteria. The transformation is closely related to the synthesis of hemoglobin, nitrogenase and other enzymes required for N_2 fixation. Between infection and the onset of N_2 fixation there exists a period of 3 to 5 weeks, during which time carbohydrates, mineral nutrients, and amino acids are supplied by the host without benefit to the host ("parasitism"). Legumes therefore rely on a sufficient supply of combined nitrogen during this period in order to establish a source leaf area large enough to supply photosynthates to the growing nodules.

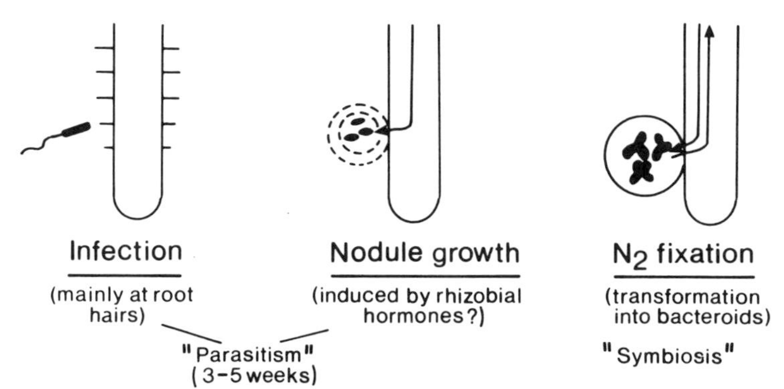

Fig. 7.5 Schematic representation of the establishment of an N_2-fixing system in the roots of legumes.

7.4.3 Fluctuations in Nitrogen Fixation Rates and the Energetics of Nitrogen Fixation

There is a close correlation between the N_2 fixation rate and the supply of carbohydrates to the root nodules. This is reflected, for example, in diurnal changes in N_2 fixation rates (Minchin and Pate, 1974). If nodulated legume plants are kept for several days under low light, not only does the N_2 fixation rate decrease, but an increasing proportion of the root nodules are transformed from the red, active, hemoglobin-containing types to green, nonfixing types (Haystead and Sprent, 1981). In mixed stands of legumes with nonlegumes, the competing power of legumes is therefore closely related to the light interception and light intensity.

In annual grain legumes, but to a lesser extent in pasture legumes such as white clover (Haystead and Sprent, 1981), a characteristic pattern occurs in the N_2 fixation rate during ontogenesis, as shown in Fig. 7.6 for cowpea. The maximum rate of N_2 fixation is reached at the beginning of the flowering period and is followed by a rapid decline. After some delay, a decline in the rate of net photosynthesis also occurs (see Peoples *et al.*, 1983). The decline in N_2 fixation rate at the beginning of flowering is an expression of sink competition for photosynthates between the developing pods and the root nodules; this decline in N_2 fixation can thus be alleviated by the removal of flowers and pods (Chapter 5).

Considering the high energy requirement of N_2 fixation, the close correlation between the rates of net photysynthesis and N_2 fixation is to be expected. When symbiosis is involved, the legume plant has to pay a price in the form of an additional supply of carbohydrate for nodule growth and nodule respiration. Respiration rates per unit dry weight of nodulated

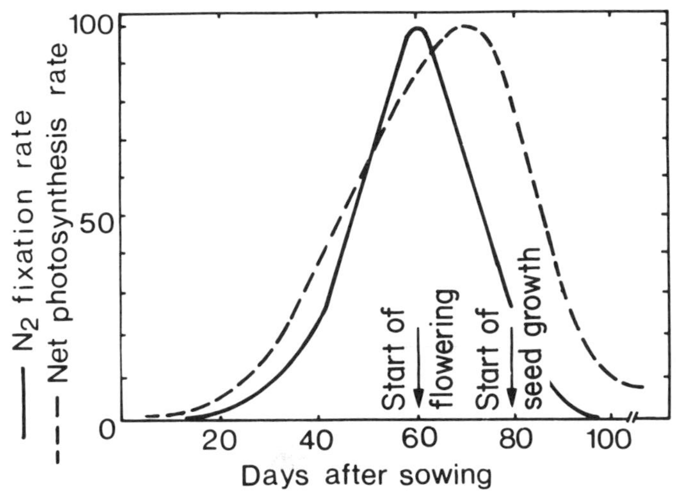

Fig. 7.6 Changes in rates of net photosynthesis and N_2 fixation during growth of cowpea. Relative values. (Based on Herridge and Pate, 1977.)

legume roots are about twice as high as the rates of nonnodulated legume roots (Ryle *et al.*, 1979). On average, over the whole growing period in cowpea ~10% of the carbon from net photosynthesis is translocated to the nodules. About half of this is consumed by nodule respiration, the remainder being utilized for the synthesis of amino compounds in the nodules which are returned to the shoot (Herridge and Pate, 1977). These average values for sink competition for photosynthates during the growing period are somewhat misleading because during the period of maximum N_2 fixation (Fig. 7.6) ~50% of the photosynthates are translocated to nodulated roots, where 78% are respired, 10% used for growth, and only 12% returned as amino compounds to the shoot (Pate and Herridge, 1978). In view of the recent findings on recycling of CO_2 in the nodules (Section 7.3) calculations which only consider the net loss of CO_2 from nodules might require some reconsideration. This also holds true for what is discussed in the following paragraph.

On the basis of "balance sheets" the requirement for carbohydrates during N_2 fixation in legumes has been calculated to be between 4 and 10 mg per milligram nitrogen (Pate and Herridge, 1978; Pate *et al.*, 1979). Assuming an average value of 7 mg, a crop of nodulated legume plants fixing 150 kg nitrogen would require an additional net photosynthesis of ~1 ton of carbohydrates compared with a similar crop supplied with combined nitrogen. Such calculations, however, do not take into consideration the energy requirement for assimilating inorganic nitrogen in the form of nitrate. Nitrate reduction (Chapter 8) also has a high energy requirement. Thus in plant species such as lupins in which nitrate is preferentially reduced in the roots, 10·2 mg carbohydrate per unit assimilated nitrogen is required when the plants are supplied solely by N_2 fixation, whereas the corresponding value for nitrate assimilation is 8·1 mg carbohydrate (Pate *et al.*, 1979). In plant species in which the main site of nitrate reduction is the leaves, photosynthetic reductants can be directly utilized for nitrate reduction (Chapter 8). In these plant species the differences in carbohydrate requirements for nitrogen assimilation are thus larger; that is, the costs of N_2 fixation are relatively higher. In legumes nodulated roots respire an average of 11 to 13% more of the net photosynthate than do nonnodulated roots (Ryle *et al.*, 1979). Whether this additional demand for respiration becomes limiting for shoot growth depends on a number of factors in the sink–source relationship (Chapter 5). At least in annual grain legumes at the onset of flowering, the additional carbohydrate requirement for N_2 fixation becomes critical for reproductive growth. At this stage the sink competition by the nodules is depressed quite rapidly even before the decline in the rate of net photosynthesis (Fig. 7.6). Possible mechanisms for the regulation of sink competition have been discussed in Chapter 5.

The rate of decline after flowering and pod set in N_2 fixation differs among plant species. This decline is more rapid, for example, in cowpea (Fig. 7.6; Peoples *et al.*, 1983) than in soybean, where distinct differences among cultivars also exist in the rate of decline (Spaeth and Sinclair, 1983). Thus the beneficial effects of late application of nitrogen fertilizer on legume seed yield depend on the genotype, being high, for example, in pea but negligible in faba bean (Schilling, 1983).

7.4.4 Effects of Mineral Nutrients

In legumes N_2 fixation can be affected indirectly or directly by other mineral nutrients. In many legume species, nodulation is severely inhibited in acid soils. Various factors, such as an inadequate calcium supply, are responsible for this phenomenon. As shown in Fig. 7.7, root infection and nodule initiation have a much higher calcium requirement than the root and shoot growth of the host plant.

Additional experiments of Lowther and Loneragan (1968) demonstrated that after nodule initiation further nodule growth was not affected by a decrease in the calcium concentration, indicating that only the first step of infection is highly sensitive to calcium supply. A similar negative effect on nodulation can be brought about in soybean by lowering the pH from 5·5 to 5·0, that is, by increasing the H^+ concentration with a simultaneous decrease in root hair length (Franco and Munns, 1982a,b). Therefore, by liming acid

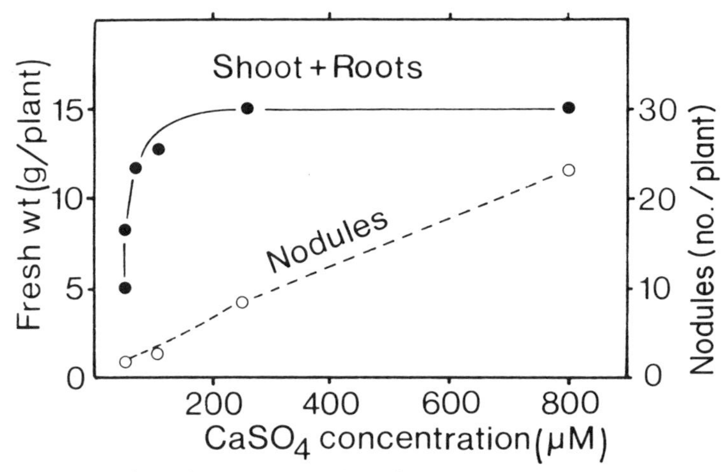

Fig. 7.7 Effect of calcium concentrations in nutrient solution (pH 5·0) on fresh weight (●) and nodule number (○) in subterranean clover. (Based on Lowther and Loneragan, 1968.)

soils it is possible to increase nodulation in, for example, *Medicago* species (Robson and Loneragan, 1970a) and soybean (Sartain and Kamprath, 1975). Low rates of nodulation in acid soils can also be caused by the effects of aluminum on root morphology or on the survival of certain *Rhizobium* strains in acid soils (Section 16.3.3).

The level of phosphorus supply, like that of calcium supply, affects nodulation and nodule growth much more than root or shoot growth (Table 7.2). In soybean the minimum concentration of phosphorus required for nodulation is ~0·5 μg/liter in the external solution. In particular, an increase in the concentration from 200 to 500 μg/liter leads to a relatively greater increase in the nodule dry weight than in the shoot and, especially, the root dry weight.

Table 7.2
Relationship between Phosphorus Supply and Dry Weight of Roots and Nodules in Soybean Plants[a]

P Supply (μg/liter)	Dry weight (g/plant)		
	Roots	Nodules	Shoots
0·5	0·60	0·07	1·21
20	0·76	0·10	1·55
50	0·79	0·16	1·86
200	1·23	0·35	4·22
500	1·35	0·64	6·57

[a]Based on Cassman *et al.* (1980).

When legumes dependent on symbiotic nitrogen receive an inadequate supply of phosphorus, they therefore also suffer from nitrogen deficiency. Under these conditions, nitrogen-deficiency symptoms are dominant and can be alleviated by the application of phosphorus fertilizers. Correspondingly, in soybean grown in a soil low in available phosphorus, a similar yield increase was obtained with mineral fertilizer supply in the combination of either 90 kgP/0 kg N or 0 kg P/120 kg N per hectare (Dadson and Acquaah, 1984).The high specific phosphorus requirement for nodulation is responsible, at least in part, for the interesting interactions in certain legume species between the infection of roots with mycorrhizas and nodulation. With plants growing in phosphorus-deficient soil, nodulation could be induced either by the application of phosphorus fertilizer or by root infection with mycorrhizas (Section 15.7).

The requirement for molybdenum, the metal component of nitrogenase, explains the occurrence of nitrogen-deficiency symptoms in legumes growing on soils low in available molybdenum. Acid mineral soils of the humid tropics are particularly low in molybdenum. Under these conditions, seed pelleting and soil treatment with molybdenum are common and effective methods of obtaining high rates of N_2 fixation by legumes (Section 9.5).

Because cobalt is required for the synthesis of leghemoglobin (see Section 7.3), there are close correlations in legumes between cobalt supply, N_2 fixation, and the leghemoglobin content of the nodules (Table 7.3). In lupins, the application of cobalt increases not only the nitrogen content but also the number and shape of the bacteroids in the nodules, indicating that cobalt also affects rhizobial cell division (Chatel *et al.*, 1978).

Table 7.3
Cobalt Requirement for N_2 Fixation in Symbiotically Grown Alfalfa Plants[a]

Treatment	Shoot dry wt (g/pot)	N_2 Fixation (mg/pot)	Color of nodules	Leghemoglobin (mg/cm^3 nodule)
−Co	3·5	0	White	Traces
+Co	4·5	30·5	Pink	0·73

[a]Based on Wilson and Reisenauer (1963).

Of the mineral nutrients, combined nitrogen both in soils and in plants has the most prominent influence on N_2 fixation in legumes. This influence can be stimulating or depressing, depending on the level of nitrogen supplied. As shown schematically in Fig. 7.8, increasing the supply of combined nitrogen results in an asymptotic increase in total nitrogen per plant or per unit surface area (e.g., hectares). The contribution of N_2 fixation to the total nitrogen per plant is increased by moderate levels of combined nitrogen but

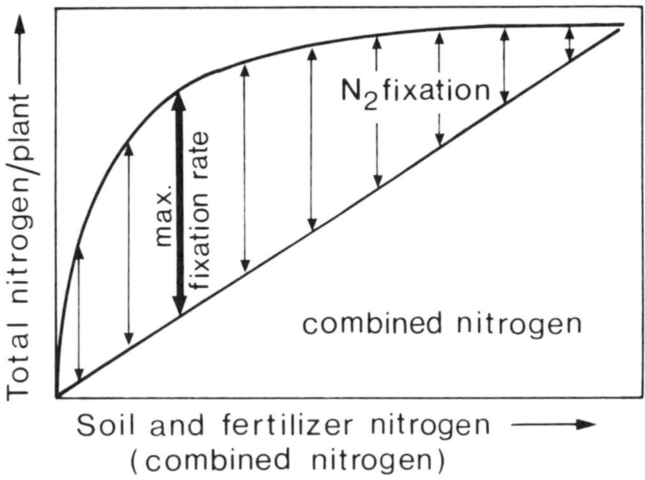

Fig. 7.8 Simplified scheme of the relationship between N_2 fixation and nitrogen uptake from soil and fertilizer in nodulated legumes.

declines at high levels, reflecting the depression of N_2 fixation by the high levels of either soil or fertilizer nitrogen.

The enhancing effect of low levels of combined nitrogen on N_2 fixation in legumes is related to the lag phase between infection and the onset of N_2 fixation (Fig. 7.5). A shortage of nitrogen in the host plant during this lag phase is detrimental to the formation of a source leaf area that is sufficiently large to supply the photosynthates needed for nodule growth and activity. The highest nodule activity (N_2 fixation) is therefore obtained when combined nitrogen either from soil reserves or from fertilizers is available in amounts that are sufficient for vigorous plant growth during the first weeks of legume establishment. This beneficial effect is demonstrated in Table 7.4 at a treatment level of 25 kg nitrogen (applied as nitrate fertilizer).

Table 7.4
Influence of Nitrogen Fertilizer Supply on Nitrogenase Activity and Shoot and Root Growth in Bean[a]

Nitrate N fertilizer supply (kg/ha)	Nitrogenase activity (μmol C_2H_4 produced/ plant × hr)		N in shoots (%), day 49	Dry weight (g/plant), day 49	
	35 days	49 days		Shoot and roots	Nodules
0	1·13	0·19	1·54	2·53	0·18
25	2·26	0·33	1·82	3·35	0·28
50	0·60	0·10	1·67	3·65	0·13
100	0·14	0·03	1·69	4·35	0·11

[a]Based on Sundstrom *et al.* (1982).

When the levels of combined nitrogen increase, nitrogenase activity, as an expression of the N_2 fixation rate, declines drastically (Table 7.4). Shoot growth, however, continues to increase, indicating the shift from symbiotic to inorganic nitrogen nutrition. The highest nitrogen content in the shoot dry weight coincides with the highest nitrogenase activity but not with the highest dry weight of the plants. Evidently, under these experimental conditions the dry matter production was source-limited (Chapter 5) and the additional supply of photosynthates required for N_2 fixation restricted the plant growth.

As the nitrate supply increases, the nodule weight correspondingly decreases, but the effect is much less noticeable than with nitrogenase activity (Table 7.4). The detrimental effects of high levels of combined nitrogen on nitrogenase activity are well known in N_2-fixing systems and are caused by several feedback mechanisms for N_2 fixation control induced directly or indirectly by combined nitrogen. As shown schematically in Fig. 7.9 nitrogenase activity can be directly repressed by ammonia (NH_3) or the primary amino compounds glutamine and glutamate synthesized in the nodules. Ammonium fertilizer therefore directly represses nitrogenase activity. Nitrate reduced in the nodules (Streeter, 1982) would have a similar effect. In addition, in nodules the activity of the nitrate reductase is much higher than that of the nitrite reductase, leading to accumulation of nitrite, NO_2^- (Becana *et al.*, 1985). This accumulation of toxic nitrite concentrations in the nodules might lead to a further decrease in nitrogenase activity if the

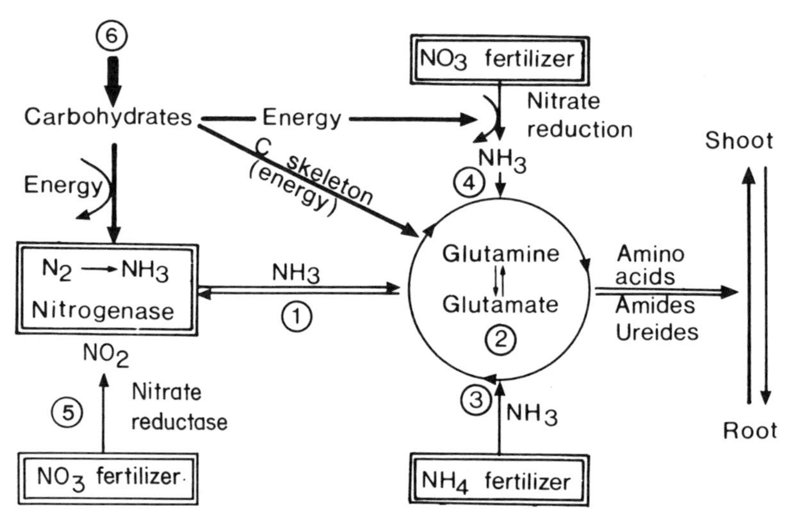

Fig. 7.9 Schematic representation of some major inhibitory effects of nitrogen fertilizer on N_2 fixation. Steps (1) to (5): feedback control, repression or inhibition of nitrogenase; Step (6): feedback control by the carbohydrate supply from the shoot.

external supply of nitrate is large (Streeter, 1982). A decrease in nitrogenase activity when the nitrate supply is large can also be brought about by competition for photosynthates, particularly in those legume species which preferentially reduce nitrate in the roots. High external concentrations of combined nitrogen even inhibit root infection with *Rhizobium* (Carroll and Gresshoff, 1983); glutamic acid, present in many soils at fairly high concentrations, is particularly effective in this respect (Iruthayathas *et al.*, 1983). Interestingly, the inhibitory effects of nitrate differ remarkably between legume species. As shown by Harper and Gibson (1984) high nitrate supply inhibits nodulation much more in soybean than in chickpea or lupins, whereas the nitrogenase activity is severely inhibited in lupins and chickpea but only slightly affected in subterranean clover. Thus, there may be quite different results from mineral fertilizer supply on the rate of N_2 fixation, depending upon plant species and time period of high nitrate availability during the growing season.

Besides these direct feedback mechanisms induced by combined nitrogen, there is also indirect control of the N_2 fixation rate via the carbohydrate supply from the shoots [Fig. 7.9, step (6)]. When the level of nitrogen nutrition is high (the nitrogen being supplied in the form of either nitrate or ammonium), a considerable proportion of carbohydrate is required for amino acid and protein synthesis, so that correspondingly less carbohydrate is available for transport to the nodules.

These various feedback control mechanisms enable legume plants to regulate N_2 fixation rates according to the demand for combined nitrogen and to restrict the expenditure of additional respiratory costs for N_2 fixation to those conditions in which the external supply of combined nitrogen is the growth-limiting factor.

7.5 Free-Living and Associative Nitrogen-Fixing Microorganisms

7.5.1 Free-Living Nitrogen-Fixing Microorganisms

Bacteria that are capable of fixing N_2, such as *Clostridium pasteurianum* (isolated by Winogradsky in 1893) and *Azotobacter chroococcum* (isolated by Beijerinck in 1901), are present in most soils. Their contribution to the nitrogen balance is considered to be very small, the annual value being on average less than 1 kg nitrogen per hectare (see Fig. 7.1; Bothe *et al.*, 1983). The main limiting factor in N_2 fixation by these free-living bacteria is the availability of carbon (in the form of organic residues), for which they have to compete with other carbon-heterotrophic microorganisms. Furthermore, for several reasons (see Section 7.4.3) the efficiency of N_2 fixation by free-living bacteria is much lower than that of symbiotic systems. In free-living bacteria between 50 and 400 mg carbon are utilized per milligram N_2 fixed (Mulder, 1975).

The situation is different with carbon-autotrophic (photosynthetic) blue-green algae (*Cyanobacteria*) living at the soil surface or in shallow water (wetland rice fields). From long-term field experiments in Rothamsted (Great Britain) their annual contribution to the nitrogen balance was calculated to be between 13 and 28 kg nitrogen per hectare (Witty *et al.*, 1979). In wetland rice fields the annual N_2 fixation rates of blue-green algae are assumed to be 30–50 kg nitrogen per hectare, as long as the availability of combined nitrogen is maintained at a fairly low level (Stewart *et al.*, 1975). More recent calculations considerably exceed these values of N_2 fixation rates (Section 7.4.1).

Free-living N_2-fixing microorganisms are also abundant on leaf surfaces, especially on the leaves of plants growing in the humid tropics (Ruinen, 1975). The extent to which these microorganisms in the *phyllosphere* (the leaves of a canopy) can be classified as free-living components or as associative components of N_2-fixing systems (see Section 7.5.2) is not clear. As yet no specificity in plant–bacteria interactions has been demonstrated in the phyllosphere. These bacteria consist of both carbon-autotrophic and carbon-heterotrophic types, the latter relying on photosynthates leached out of leaves (Ruinen, 1975). In tropical ecosystems, N_2 fixation in the phyllosphere is assumed to make a substantial contribution to the total input of combined nitrogen (Ruinen, 1975; Döbereiner, 1983).

There is some prospect of utilizing N_2 fixation in the phyllosphere in tropical argicultural systems. As shown in Table 7.5, spraying leaves several times with N_2-fixing bacteria substantially increases the nitrogenase activity of the leaves, the grain yield, and the nitrogen content of the plants.

Similar results were obtained with rice and wheat by Sen Gupta *et al.* (1982), who found that up to 50 kg nitrogen (as chemical nitrogen fertilizer) could be saved per hectare for each crop. It should be remembered,

Table 7.5
Effect of Leaf Spraying with N_2-Fixing Bacteria versus Urea Application on Rice[a]

Treatment	Nitrogenase activity of the leaves	Grain yield	Nitrogen content	
			Straw	Grain
Control	100	100	100	100
+ *Mycobacterium flavum*	560	211	175	160
+ *Klebsiella pneumoniae*	524	182	180	169
+ Urea	0	318	236	195

[a]Relative values. Based on Nandi and Sen (1981).

however, that the nitrogenase activity of the N_2-fixing bacteria is also repressed by combined nitrogen. This type of N_2 fixation therefore usually makes a major contribution only when plants are severely nitrogen deficient or when the yield potential is not being fully utilized (Table 7.5; Sen Gupta *et al.*, 1982).

7.5.2 Rhizosphere Associations

A considerable proportion of the carbon fixed during photosynthesis in higher plants is released into the rhizosphere in the form of root exudates or decaying root cells (Section 15.6). The population density of soil microorganisms, including N_2-fixing bacteria, in the rhizosphere is thus several times higher than in the bulk soil (Chapter 15). In temperate zones N_2-fixing bacteria of the genus *Enterobacter* dominate in the rhizosphere of most higher plants (Jagnow, 1983). In some tropical grasses (e.g., sugarcane, sorghum, and maize) N_2-fixing bacteria of the genus *Azospirillum* or *Azotobacter* dominate and invade not only the rhizosphere but also the intercellular spaces of the root cortex and thus have much better access to root exudates (Döbereiner and Day, 1975; Smith *et al.*, 1976). Malate seems to be a root exudate that especially favors growth and N_2 fixation of associative microorganisms (Jagnow, 1979).

Azospirillum can even be found within root cells (Pohlman and McColl, 1982) or in the stems of grasses (De-Polli *et al.*, 1982). This demonstrates that the division of N_2-fixing systems into free living, associative or symbiotic does not fully reflect the various transitional features existing in nature. Rhizosphere associations are characterized by high host specificity. The most prominent associations are *Beijerinckia* and sugarcane; *Azotobacter paspali* and bahia grass (*Paspalum notatum*); and *Azospirillum* and guinea grass (*Panicum maximum*), pear millet (*Pennisetum glaucum*), or maize (Döbereiner, 1983). The host specificity can be extremely developed: Among the 31 cultivars of tropical bahia grass, only one (Batatais) produces root exudates which specifically stimulate growth of *Azotobacter paspali* (Döbereiner, 1983). Similar results have been obtained with wheat and *Azospirillum brasilense* (Millet *et al.*, 1984). Out of 20 wheat cultivars only two responded to inoculation with a yield increase. These genotypical differences demonstrate how difficult it is to generalize in this area of bacteria/plant interaction even within a species. The following examples illustrate this further.

There are quite conflicting reports on the amounts of N_2 fixed by these rhizosphere associations, varying from a few kilograms up to several hundred kilograms of nitrogen per hectare each year. The very high N_2 fixation rates are based on short-term studies of nitrogenase activity (C_2H_4 evolution) using isolated roots or roots in undisturbed soil cores. Extrapolations from these short-term studies to an annual N_2 fixation rate per hectare must be interpreted with great caution (see below). There is general agreement that under tropical conditions the combination of high soil temperature, low O_2 tension in the rhizosphere, high irradiation, and high rates of net photosynthesis favors root exudation as well as growth and N_2 fixation of the rhizosphere associations (Cohen *et al.*, 1980). The N_2 fixation rates of these associations in grass species with C_4 photosynthesis (e.g., sugarcane and maize) growing in the humid tropics are therefore usually higher than those in C_3 species growing in temperate zones (Ehleringer, 1978). In barley, for example, the N_2 fixation of rhizosphere associations may reach only 5–10 kg nitrogen per hectare for a growing season (Idris *et al.*, 1981). Nevertheless, in temperate zones also inoculation of wheat with *Azospirillum lipoferum* might substantially increase yield and nitrogen content in the grains, and is thus considered as a promising future approach to save mineral nitrogen (Mertens and Hess, 1984). However, as we shall see later, this stimulatory effect of inoculation is not necessarily only due to the improvement of nitrogen nutrition by N_2 fixation.

Other important factors determining the N_2 fixation rates of rhizosphere association are the availability of combined nitrogen and low O_2 tension. The highest rate of nitrogenase activity is usually obtained in the absence of combined nitrogen (Table 7.6). With increasing nitrogen supply the nitrogenase activity is drastically depressed, but the shoot growth of the host plant (maize) increases. Similar results have been obtained with tall oat-grass, *Arrhenatherum elatior* (Martin *et al.*, 1984).

In field experiments, however, the highest rates of nitrogenase activity are obtained when the supply of combined nitrogen is moderate (Kapulnik *et al.*, 1981b; Vinther, 1982). This is in agreement with the results in symbiotic systems (Section 7.4.4), as the supply of photosynthates becomes the limiting factor in severely nitrogen-deficient host plants. As in symbiotic systems, in rhizosphere associations distinct diurnal (Trolldenier, 1977) and

Table 7.6
Effect of Combined Nitrogen (NH_4NO_3) on the Growth of Maize Inoculated with *Azospirillum*[a]

NH_4NO_3 Concentration (g/liter)	Nitrogenase activity (nmol C_2H_4/plant × hr)	Shoot dry weight (g/plant)
0	200	0·49
0·04	156	0·97
0·08	10	1·84
0·16	0	2·93

[a]From Cohen *et al.* (1980).

seasonal fluctuations occur in N_2 fixation rates (Sims and Dunigan, 1984). Nitrogenase activity is low during the first weeks of host plant growth. Activity increases sharply at the booting stage (Kapulnik *et al.*, 1981; Vinther, 1982) and reaches its peak at flowering, as in rice (Sano *et al.*, 1981). Measuring nitrogenase activity only once in the growing period can therefore lead to substantial errors in the calculation of total N_2 fixation rates for the whole growing period. Rhizosphere associations seem to be capable of making use of other carbon sources, as indicated by a distinct stimulation of N_2 fixation in *Azospirillum* associations with sorghum (Pal and Malik, 1981) and maize (Meshram and Shende, 1982) by the application of farm manure.

Because of the high sensitivity of nitrogenase to O_2 (Section 7.3), the N_2 fixation rates of rhizosphere associations are favored by low O_2 tension. High soil moisture therefore enhances N_2 fixation rates (Werner *et al.*, 1981), a factor which presumably contributes to the high N_2 fixation rates of rhizosphere associations in the humid tropics. The exceptionally high N_2 fixation rates in the rhizosphere of flooded (wetland) rice can be explained by the combination of low O_2 tension, high soil temperature, and high rate of root exudation (Trolldenier, 1982; Rao and Rao, 1984). The majority of N_2-fixing bacteria in the rhizosphere of wetland rice are aerobic species (Balandreau *et al.*, 1975), and optimal conditions for N_2 fixation exist at the oxygenated zone around the roots of intact rice plants (Section 15.3). A rapid decline in nitrogenase activity in response to shading or decapitation of rice is most likely mainly a reflection of a decrease in O_2 transport from the shoots to the roots and the rhizosphere (Berkum and Sloger, 1982). Under field conditions between 3 and 63 kg nitrogen per hectare each year can be attributed to N_2 fixation in the rhizosphere of rice (Yoshida and Ancajas, 1973). In the rhizosphere of rice, unlike that of maize or of *Paspalum*, specific associations do not appear to occur; therefore, the term *rhizosphere* associations cannot, strictly speaking, be applied to this N_2-fixing system. Nevertheless, there is also a considerable transfer of fixed nitrogen from the bacteria to the host rice plant, which can be observed within a few days (Yoshida and Yoneyama, 1980).

The N_2 fixation rate in the rhizosphere of rice also depends on the availability of combined nitrogen as well as carbon other than that supplied from root exudates. As shown in Table 7.7 the N_2 fixation rate is increased by the application of straw, whereas high doses of combined nitrogen have the reverse effect. Accumulation of phenolics in the soil during anaerobic decomposition might be involved in the stimulating effect of straw. It has been shown by Krotzky *et al.* (1983) that phenolics can dramatically increase the nitrogenase activity in the rhizosphere of graminaceous species.

The stimulation of growth of grasses by certain associative N_2-fixing

Table 7.7
Fixation of $^{15}N_2$ in Rhizosphere Soil of Wetland Rice

Treatment	$^{15}N_2$ Fixed (mg/kg soil per 30 days)
Nonrhizosphere soil (bulk soil)	5·2
Rhizosphere soil	22·2
Rhizosphere soil + 6 tons straw/ha	29·4
Rhizosphere soil + 80 kg fertilizer N/ha	16·5

[a]From Charyulu *et al.* (1981).

microorganisms may also be due to effects other than an improvement in the nitrogen nutrition of the host plant. Bacteria of the genus *Azospirillum*, for example, also secrete plant hormones such as auxins and gibberellins (Tien *et al.*, 1979; Inbal and Feldman, 1982) which favor root growth and especially, root hair formation (Martin and Glatzle, 1982; Martin *et al.*, 1984) and affect the development of the host plant. This is reflected, for example, in earlier heading and an increase in spikelet number (Kapulnik *et al.*, 1981a; Millet and Feldman, 1984). In agreement with this, in pot and field experiments inoculation with *Azospirillum* stimulates the growth and grain yield of wheat when supplied with both small and large quantities of

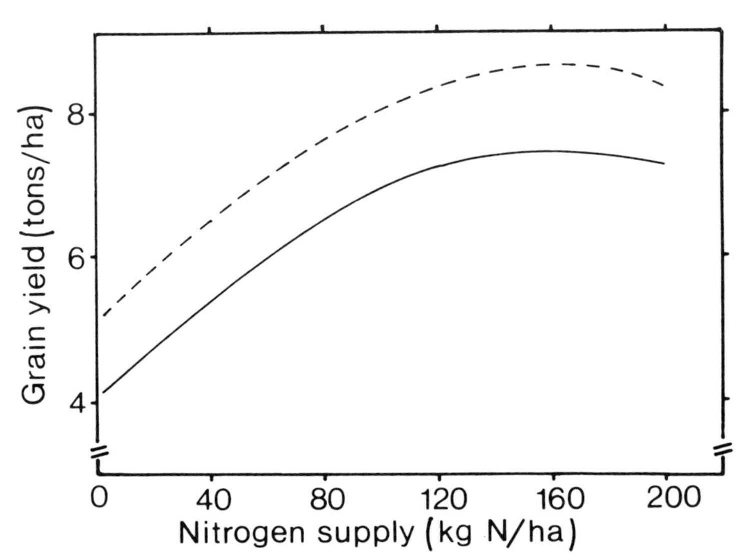

Fig. 7.10 Effect of level of nitrogen fertilizer and inoculation (foliar sprays) during tillering with a suspension of *Azospirillum* on the yield of 10 winter wheat cultivars (——, control; – – –, inoculated). (Based on Rynders and Vlassak, 1982.)

nitrogen fertilizers (Avivi and Feldman, 1982) and under high yielding conditions, as shown in Fig. 7.10.

The yield increase observed upon inoculation with *Azospirillum* was manifested in a greater number of ears per unit surface area, the contribution by N_2 fixation being considered to be small (Rynders and Vlassak, 1982; see also Yahalom *et al.*, 1984). This inoculation effect opens new areas of research on the use of *Azospirillum* as "biofertilizers" in intensive cropping under temperate climatic conditions. The probability of obtaining predictable grain yield increases by inoculation with this bacterium should not be overestimated, however (Rynders and Vlassak, 1982), when hormonal effects are the main contributing factors (Chapter 5).

7.6 Outlook

Future developments in the use of both the N_2-fixing capability and the phytohormone production of associative bacteria are focused on plant breeding, selection, and genetic engineering. In maize, for example, as in *Paspalum notatum* (Döbereiner, 1983), distinct genetic variability exists in rhizosphere associations. The capacity of maize to support the association with *Azospirillum* can be increased through screening and transferred to other germplasm (Ela *et al.*, 1982). This approach is both a complementary and an alternative way of achieving the long-term goal of expression of genetically engineered bacterial N_2 fixation. In this way, for example, *nif* genes from N_2-fixing bacteria are transferred to nonfixing bacteria species dominant in the rhizosphere of many grass species (Kleeberger and Klingmüller, 1980) in order to establish new associations between higher plants and N_2-fixing microorganisms, such as between wheat and *Rhizobium* (Hess, 1981). Although speculations such as these are exciting, the carbon requirement has to be kept in mind. Increasing the capacity of a plant for N_2 fixation is closely associated with a corresponding increase in the demand for photosynthates for the competing (N_2-fixing) sink. In future agricultural systems based on better use of N_2 fixation by rhizosphere associations, therefore, the N_2-fixing process will not replace the use of nitrogenous fertilizers. It might, however, allow a reduction in their usage without dramatic loss in yield.

8

Functions of Mineral Nutrients: Macronutrients

8.1 Classification and Principles of Action of Mineral Nutrients

By definition, mineral nutrients have specific and essential functions in plant metabolism. Depending on how great the growth requirement for a given nutrient, the nutrient is referred to as either a *macronutrient* or a *micronutrient*. Another classification, based on physicochemical properties, divides nutrients, into *metals* (potassium, calcium, magnesium, iron, manganese, zinc, copper, molybdenum) and *nonmetals* (nitrogen, sulfur, phosphorus, boron, chlorine). Both classifications are inadequate since each mineral nutrient can perform a variety of functions, and some of these functions are only loosely correlated to either quantity of requirement or physicochemical properties. A mineral nutrient can function as a constituent of an organic structure, as an activator of enzyme reactions, or as a charge carrier and osmoregulator. In this book the more common classification of macro- and micronutrients is preferred.

The main functions of mineral nutrients such as nitrogen, sulfur, and phosphorus that serve as constituents of proteins and nucleic acids are quite evident and readily described. Other mineral nutrients, such as magnesium and the micronutrients (except chlorine), may function as constituents of organic structures, predominantly of enzyme molecules, where they are either directly or indirectly involved in the catalytic function of the enzymes. Potassium, and presumably chlorine, are the only mineral nutrients that are not constituents of organic structures. They function mainly in osmoregulation (e.g., in vacuoles), the maintenance of electrochemical equilibria in cells and their compartments and the regulation of enzyme activities. Naturally, because of their low concentrations, micronutrients do not play a direct role in either osmoregulation or the maintenance of electrochemical equilibria.

The different types of functions that mineral elements perform in enzyme reactions require further comment. Nitrogen and sulfur are integral constituents of protein structure and thus of *apoenzymes* (Fig. 8.1). However, in the case of only a few enzymes can an apoenzyme by itself catalyze a

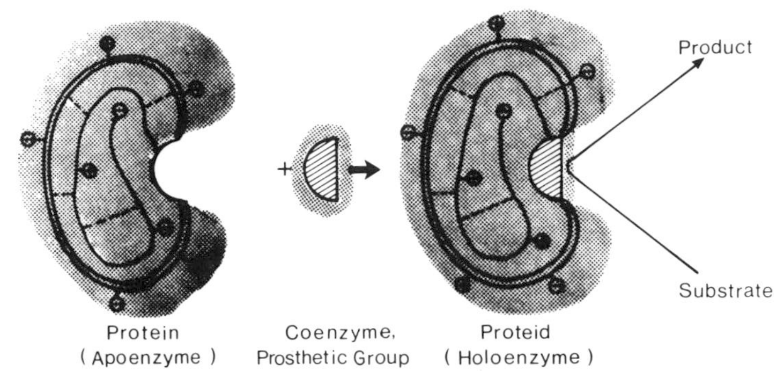

Fig. 8.1 Schematic representation of the components of an enzyme molecule. Shaded area: hydration shells of water molecules (cluster).

reaction. For the catalytic reaction of a majority of enzymes, a nonprotein component is required, namely, a coenzyme, a prosthetic group, or a metal component. The difference between coenzymes and prosthetic groups is primarily a matter of convention. Typical coenzymes are ATP and FAD; typical prosthetic groups are chlorophyll, cytochromes, and nitrogenase, in which a metal acts as the functional group. In several enzymes the prosthetic group resembles the metal component only. Examples of this are given in Chapter 9. Most of the metal atoms integrated into metalloproteins are transition metals, which perform their catalytic function through a change in valency. This is the case for iron in cytochromes, copper in plastocyanin, and molybdenum in nitrogenase. In some enzymes, however, the metal performs its catalytic function by forming an enzyme–substrate–metal complex (e.g., magnesium in ATPase). Detailed information on the enzyme function of heavy metals can be found in Sandmann and Böger (1983).

Mineral nutrients can have a dominant role in enzyme reactions without being integral components of enzyme structure. Potassium is a typical example of a mineral nutrient that exerts its regulatory function by changing the conformation of the protein component of the enzyme. Proteins are charged macromolecules which are highly hydrated in living, metabolically active cells (Fig. 8.1). Because of intramolecular hydrogen bonding, water molecules form partial but nonpermanent associations ("structures" or "clusters") and thus have a stabilizing effect on protein conformation. Dissolved ions, including mineral nutrients, alter the physical properties of the solvent water through the formation of hydration shells around the ion, as well as the properties of the protein molecule through interactions, particularly with the charged groups of the macromolecule (electrostatic interaction). The hydration, stability, and conformation of enzymes or other biopolymers (e.g., membranes) are therefore affected not only by temperature and pH but also by the type (cation or anion, and their valencies) and concentration of the mineral nutrients. The conformation (spatial orientation) of an enzyme is again a crucial factor for both the affinity between the active centre of the enzyme and the substrate (K_m value) and the turnover rate of the enzyme (V_{max}; see Fig. 2.12). Potassium, as the major cytoplasmic cation, has a prominent effect on the conformation of enzymes and thus regulates the activities of a large number of enzymes (Evans and Wildes, 1971). For details on the interactions between inorganic ions and enzymes, see Wyn Jones and Pollard (1983). Typical examples of the various functions of macronutrients will be given in this chapter, while micronutrients are dealt with in Chapter 9. Mineral elements that either replace certain mineral nutrients in some of their functions (e.g., sodium replacing potassium) or stimulate growth by other means are discussed in Chapter 10.

8.2 Nitrogen

8.2.1 Assimilation of Nitrogen

Nitrate and ammonium are the major sources of inorganic nitrogen taken up by the roots of higher plants. Most of the ammonium has to be incorporated into organic compounds in the roots (see Section 8.2.1.2), whereas nitrate is mobile in the xylem and can also be stored in the vacuoles of roots, shoots, and storage organs. Nitrate accumulation in vacuoles can be of considerable importance for cation–anion balance (Section 2.5.3) and for osmoregulation, particularly in so-called "nitrophilic" species such as *Chenopodium album* and *Urtica dioica* (Smirnoff and Stewart, 1985). However, in order to be incorporated into organic structures and to fulfil its essential functions as a plant nutrient, nitrate has to be reduced to ammonia. The importance of the reduction and assimilation of nitrate for plant life is similar to that of the reduction and assimilation of CO_2 in photosynthesis.

8.2.1.1 Nitrate Assimilation and Reduction

Mechanism. The currently accepted pathway of nitrate reduction in higher as well as in lower plants is as follows:

$$NO_3^- + 8H^+ + 8e^- \rightarrow NH_3 + 2H_2O + OH^-$$

Some bacteria use nitrate as an electron acceptor under anaerobic conditions ("nitrate respiration") and produce nitrogenous gases (N_2 and NO_x), a process which causes a considerable loss of combined nitrogen from soils by denitrification. The reduction of nitrate to ammonia is mediated by two separate enzymes: *nitrate reductase*, which reduces nitrate to nitrite; and *nitrite reductase*, which reduces nitrite to ammonia (Fig. 8.2).

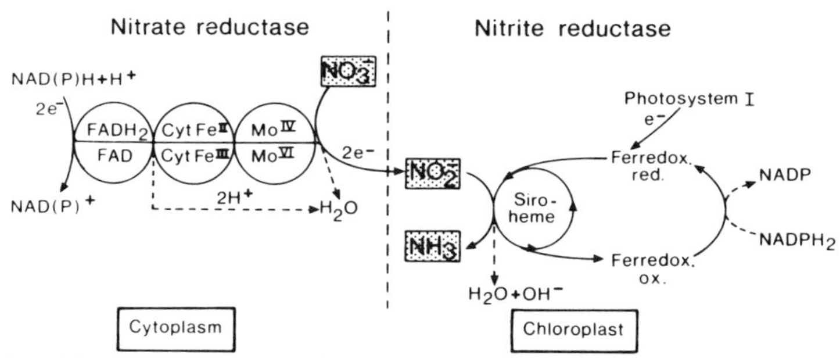

Fig. 8.2 Schematic representation of the sequence of nitrate assimilation in leaf cells. (Based on Guerrero *et al.*, 1981, and Beevers and Hageman, 1983.)

Nitrate reductase is a complex enzyme with a molecular weight of ~200,000 in higher plants and up to 500,000 in lower plants. It contains several prosthetic groups, including FAD, cytochrome, and molybdenum. It is localized in the cytoplasm of higher plant cells and requires either NADH or NADPH as an electron donor. It is assumed that during the reduction electrons are directly transferred from molybdenum to nitrate (Guerrero *et al.*, 1981). In lower as well as in higher plants, nitrate reductase has a half-life of only a few hours (Beevers and Hageman, 1983; Ullrich, 1983). It is present only in low levels in plants not receiving nitrate (Timpo and Neyra, 1983), but it can be induced within a few hours by the addition of nitrate (Oaks *et al.*, 1972) and also by cytokinins (Hänisch ten Cate and Breteler, 1982; Gzik and Günther, 1984). Nitrate reductase activity can be inhibited or completely suppressed by ammonium (Ullrich, 1983) and certain amino acids (Oaks *et al.*, 1977) or amides (Breteler and Smit, 1974). In roots the inhibitory effect of ammonium seems to be due partly to acidification, which is related to ammonium uptake (Mengel *et al.*, 1983).

As would be expected, nitrate reductase activity is very low in molybdenum-deficient plants (Table 8.1). Incubation of deficient leaf segments

Table 8.1
Effect of Pretreatment with Molybdenum on Nitrate Reductase Activity in Wheat Leaf Segments[a]

Molybdenum supply during plant growth (μg/plant)	Pretreatment of Leaf Segments (μg Mo/liter)	Nitrate reductase activity (μmol NO_2^-/g fresh wt) after	
		24 hr	70 hr
0·005	0	0·2	0·3
0·005	100	2·8	4·2
5·0	0	—	8·0
5·0	100	—	8·2

[a]From Randall (1969).

in solutions containing molybdenum markedly increases enzyme activity. The enzyme can even be reactivated *in vitro* if the molybdenum-free apoenzyme is treated with molybdenum-containing complexes (Notton and Hewitt, 1979). The striking differences in the nitrate reductase activities of molybdenum-deficient and molybdenum-sufficient plants and the rapid response to molybdenum supply can be used to determine the molybdenum nutritional status of plants (Randall, 1969; Witt and Jungk, 1977; see also Chapter 12).

In higher plants, *nitrite reductase* (Fig. 8.2) converts nitrite to ammonia, without release of free intermediates. The enzyme has a low molecular weight and is associated in leaves with chloroplasts and in roots most likely with proplastids (Beevers and Hageman, 1983). In green leaves, the electron donor is reduced ferredoxin, generated in the light by photosystem I (Chapter 5) and in the dark via respiration. In root tissues ferredoxin is absent, and another, as yet unknown compound may serve as an electron carrier between NADPH and nitrite reductase.

Despite the spatial separation of nitrate reductase and nitrite reductase (Fig. 8.2), as a rule, nitrite rarely accumulates in intact plants under normal conditions, presumably since nitrite reductase is present at much higher levels than nitrate reductase. Root nodules of legumes are obviously exceptions to this rule (Section 7.4.4). Certain herbicides such as diuron strongly and selectively inhibit nitrite reductase in leaves and correspondingly increase nitrite concentration in the tissue (Peirson and Elliott, 1981).

In C_4 plants, mesophyll and bundle sheath cells differ in their functions not only in CO_2 assimilation (Fig. 5.3) but also in nitrate assimilation. Both nitrate reductase and nitrite reductase are localized in the mesophyll cells and are virtually absent in the bundle of sheath cells (Neyra and Hageman, 1978). This has led to the interesting speculation (Moore and Black, 1979) that the higher nitrogen efficiency (percentage of nitrogen in dry matter per matter production) in C_4 plants is due to a "division of labor" whereby mesophyll cells utilize light energy for nitrate reduction and bundle sheath cells utilize light energy for CO_2 reduction.

Localization in Roots and Shoots. In most plant species both roots and shoots are capable of nitrate reduction. The proportion of reduction carried out in each location depends on various factors, including the level of nitrate supply, the plant species, the plant age, and has important consequences for the mineral nutrition and carbon economy of plants. In general, when the external nitrate supply is low, a high proportion of nitrate is reduced in the roots. With an increasing supply of nitrate, the capacity for nitrate reduction in the roots becomes a limiting factor and an increasing proportion of the total nitrogen is translocated to the shoots in the form of nitrate (Fig. 8.3).

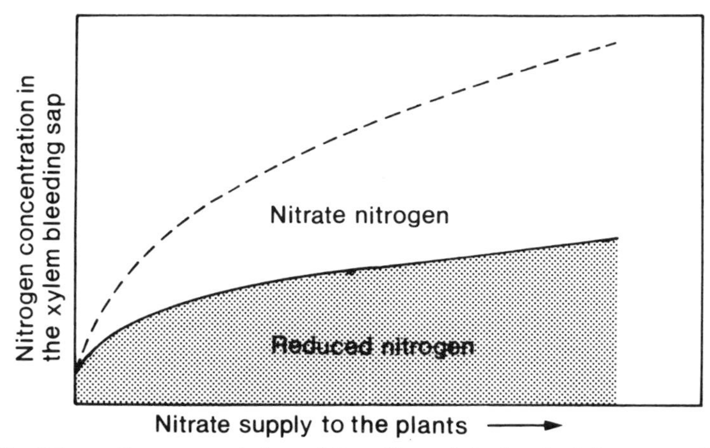

Fig. 8.3 Schematic representation of the effect of the level of nitrate supply in the rooting medium of noninoculated field pea (*Pisum arvensis* L.) on the nitrogen compounds in the xylem sap of decapitated plants. (Data recalculated from Wallace and Pate, 1965.)

The large carbohydrate requirement for nitrate reduction in the roots (Chapter 7) is certainly one of the factors limiting the capacity of roots for nitrate reduction (Minotti and Jackson, 1970). This capacity also differs among cultivars of a species such as soybean (Hunter *et al.*, 1982), and the difference is even more striking among plant species. In the latter case the capacity for nitrate reduction in the roots of woody species is usually very high (Smirnoff and Stewart, 1985) and decreases among annual species in the following sequence: rape > barley > sunflower > maize ≫ *Xanthium* (Pate, 1973). In legumes the proportion of nitrate reduction in the roots as compared to the leaves decreases in the following sequence: lupins (*L. angustifolius*) ≫ chickpea > pea > subterranean clover ≫ soybean (Harper and Gibson, 1984). In leaves, but not in roots, nitrate reduction and CO_2 reduction compete for reductants and ATP from photosynthesis. This competition may have important ecological consequences for the adaptation of plants to low-light and high-light conditions (Smirnoff and Stewart, 1985). In a given species the proportion of nitrate reduced in the roots increases with temperature (Theodorides and Pearson, 1982) and plant age (Hunter *et al.*, 1982). The uptake rate of the accompanying cation also affects this proportion. With potassium as the accompanying cation, translocation of both potassium and nitrate to the shoots is rapid; correspondingly, nitrate reduction in the roots is relatively low. In contrast, when calcium or sodium is the accompanying cation, nitrate reduction in the roots is considerably higher (Blevins *et al.*, 1978; Rufty *et al.*, 1981).

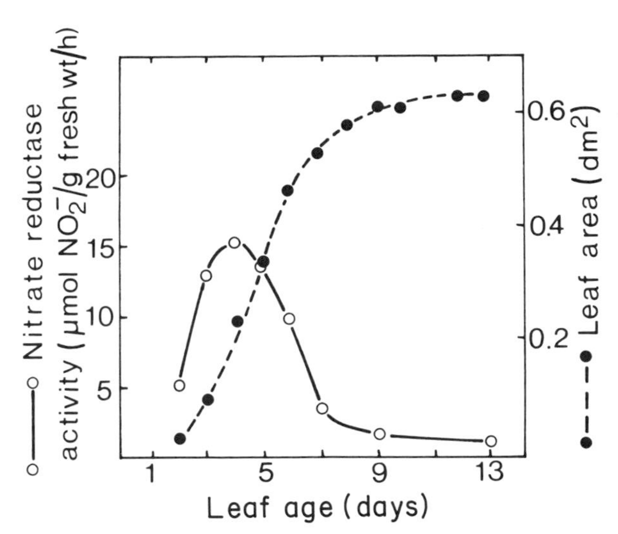

Fig. 8.4 Time course of nitrate reductase activity and leaf area development during the ontogeny of the first trifoliate leaf of soybean. (Modified from Santoro and Magalhaes, 1983.)

Leaf Age. During the ontogenesis of an individual leaf, a typical pattern is observed in nitrate reductase activity (Fig. 8.4). Maximum activity occurs when the rate of leaf expansion is maximal. Thereafter, the activity declines rapidly. Thus, in fully expanded leaves, nitrate reductase activity is usually very low, and often the nitrate levels are correspondingly high (Egmond and Breteler, 1972; Santoro and Magalhaes, 1983). In roots, nitrate reductase activity is high in the expanding cells of the apical zones and declines rapidly towards the basal root zones (Dudel and Kohl, 1974).

Because nitrate is phloem immobile (Chapter 3), high nitrate levels in fully expanded leaves are of limited use for the nitrogen metabolism of plants. Furthermore, in the individual cells nitrate is stored nearly exclusively in the vacuoles (Martinoia *et al.*, 1981). The release of nitrate from the vacuole into the cytoplasm can become a rate limiting step for nitrate reduction (Martin, 1973; Rufty *et al.*, 1982c), and thus for the utilization of stored nitrate nitrogen in growth processes (Clement *et al.*, 1979). Interruption of the nitrate supply to the roots may therefore lead to a drop in both nitrate reductase activity in the leaves and shoot growth rate, despite a still high nitrate content in the shoot (Blom-Zandstra and Lampe, 1983). These results have important consequences for the timing of nitrate fertilizer supply.

Light. In green leaves a close correlation exists between light intensity and nitrate reduction. For example, there is a distinct diurnal pattern of reduc-

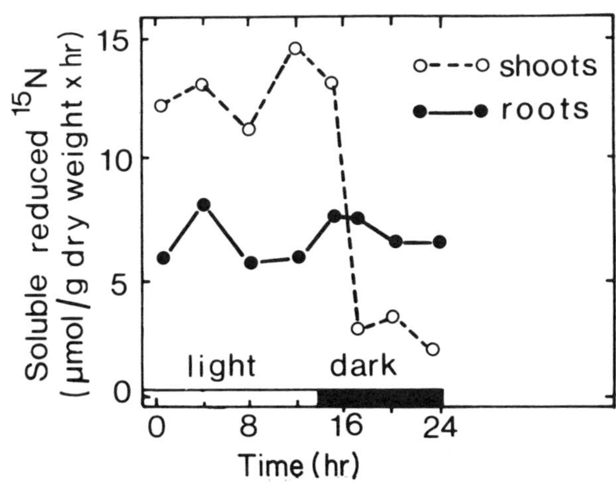

Fig. 8.5 Accumulation of soluble reduced ^{15}N in maize during a 24-hr period of $^{15}NO_3$ supply to the roots. (Based on Pearson *et al.*, 1981.)

tion in the shoots but not in the roots (Fig. 8.5). The daytime proportion of nitrate reduction in shoots and roots therefore differs from the proportion at night (Aslam and Huffaker, 1982).

According to the mechanism of nitrate reduction, this light effect reflects fluctuations in the carbohydrate level (Aslam and Huffaker, 1984) and in the corresponding supply of reducing equivalents (ferredoxin and NADPH). In addition, however, leaves of plants grown under low light usually—although not always (Steingröver *et al.*, 1982)—have lower levels of nitrate reductase; enzyme activity increases again after the plants are transferred to conditions of high light intensity (Beevers and Hageman, 1983). Light also affects the stability of the enzyme (Lillo and Henriksen, 1984). The rate of nitrate reduction in leaves is thus affected by light in various ways. Plants cultivated permanently under low-light conditions (e.g., in greenhouses) may contain nitrate concentrations that are several times higher than those of plants grown under high-light conditions (e.g., in an open field during the summer). This is particularly evident in certain vegetables such as spinach and other members of the Chenopodiaceae which have a high preference for nitrate accumulation in the shoots and which obviously use nitrate in vacuoles for osmoregulation. In these species, under low-light conditions, nitrate concentrations in the leaves can reach more than 6000 mg/kg fresh wt, that is, ~100 m*M* (Wedler, 1980).

In addition to a number of other agronomic procedures (see Section 8.2.4) the nitrate content can be kept low by the proper selection of harvest time and harvested organs. As shown in Table 8.2, the nitrate content of petioles of spinach, like that of other nitrate-storing plant species, is several times

Table 8.2
Time Course of Nitrate Content in Spinach Leaves during the Light Period from 9:00 to 18:00[a]

	Concentration of nitrate N (mg/kg fresh wt)	
Time of day	Leaf blade	Petioles
8:30	228·2	830·2
Light: 9:30	166·6	725·1
Light: 13:30	100·8	546·0
Light: 17:30	91·0	504·0
18:30	106·4	578·2

[a]Steingröver *et al.* (1982).

higher than that of the leaf blades, the main sites of nitrate reduction. Furthermore, in both tissues the nitrate content decreases considerably during the light period.

Nitrate Assimilation and Osmoregulation. In those plant species where most or all nitrate assimilation occurs in the shoots, organic acid anions are synthesized in the cytoplasm and stored in the vacuole (Fig. 8.6) in order to maintain both cation–anion balance and intracellular pH. This might lead to osmotic problems if nitrate reduction were to continue after the termination of leaf cell expansion (Raven and Smith, 1976). However, several mechanisms exist for the removal of excess osmotic solutes from the shoot tissue:

1. Precipitation of excess solutes in an osmotically inactive form. Synthesis of oxalic acid for charge compensation in nitrate reduction and precipitation as calcium oxalate are common in plants, including sugar beet (Egmond and Breteler, 1972).

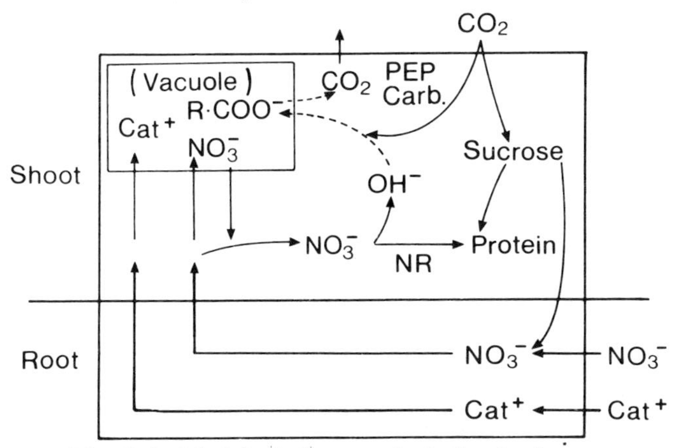

Fig. 8.6 Model of nitrate assimilation in shoots.

2. Retranslocation of reduced nitrogen (amino acids and amides) together with phloem-mobile cations, such as potassium and magnesium, to areas of new growth.

3. Retranslocation of organic acid anions, preferentially malate, together with potassium into the roots and release of an anion (OH^- or HCO_3^-) after decarboxylation. In exchange for the released anions, nitrate can be taken up without cations, because the endogenous potassium acts as a counter-ion for long-distance transport, a mechanism discussed in Chapter 3 and well documented by Kirkby and Armstrong (1980), among others.

8.2.1.2 Assimilation of Ammonium

Whereas nitrate can be stored in vacuoles without detrimental effect, ammonium and in particular its equilibrium partner ammonia

$$NH_3 \text{ (dissolved in water)} \rightleftharpoons NH_4^+ + OH^-$$

are toxic at quite low concentrations. The formation of amino acids, amides and related compounds is the main pathway of detoxification of either ammonium ions taken up by the roots or ammonia derived from nitrate reduction or N_2 fixation.

The principal steps in the assimilation of ammonium ions supplied to the roots are uptake into the root cells and incorporation into amino acids and amides with a simultaneous release of protons for charge compensation (Fig. 8.7). The permeation of ammonia across the plasma membrane, with proton liberation occurring before the permeation, has been discussed as an alternative model (Mengel *et al.*, 1976).

From both experimental findings (Martin, 1970) and theoretical considerations (Raven and Smith, 1976) it appears that nearly all of the assimilated ammonia is translocated to the shoots as amino acids, amides, and related compounds for further utilization. Ammonium assimilation in roots has a large carbohydrate requirement because of the need for carbon

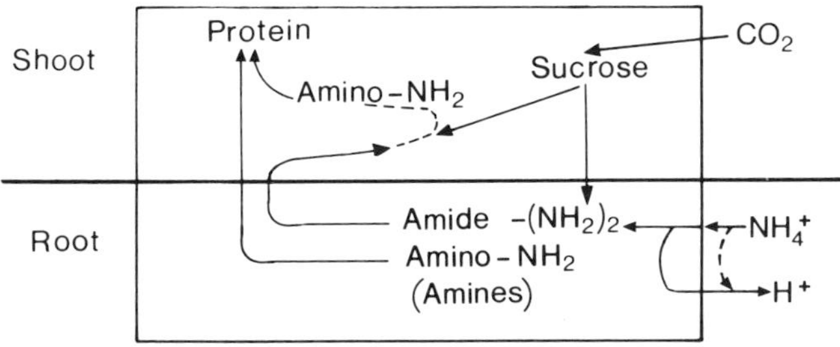

Fig. 8.7 Model of ammonium assimilation in roots. (Based on Raven and Smith, 1976.)

skeletons in the synthesis of amino acids and amides. The same is true in roots assimilating ammonia from nitrate reduction or N_2 fixation. This root to shoot transport of assimilated ammonia takes place predominantly or exclusively in the xylem. In order to minimize the carbon loss caused by nitrogen transport, nitrogen-rich compounds (N/C ratio $>0{\cdot}4$) carry the bulk of the assimilated nitrogen leaving the roots (Wallace and Pate, 1965; Streeter, 1979). One, and rarely two or more, of the following compounds dominate in the xylem exudate of the roots: the amides glutamine (2N/5C) and asparagine (2N/4C); the amino acid arginine (4N/6C); and the ureide allantoin (4N/4C). In agreement with this model of carbon economy, in phloem transport to developing fruits, which are nonphotosynthetic sinks, amino acids with an N/C ratio of $>0{\cdot}4$ are the predominant transport forms of nitrogen (Pate, 1973).

Table 8.3
Low-Molecular-Weight Organic Nitrogen Compounds Which Are Important Storage and Long-Distance Transport Forms

Compound	Plant family
Glutamine, asparagine	Gramineae
Glutamine	Ranunculaceae
Asparagine	Fagaceae
Arginine, glutamine	Rosaceae
Proline, allantoin	Papilionaceae
Betaine	Chenopodiaceae

Allantoin (ureide)

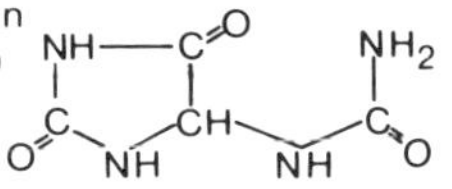

Betaine

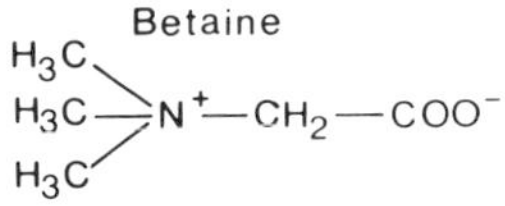

$(H_3C)_3N^+—CH_2—COO^-$

The low-molecular-weight organic nitrogen compounds used predominantly for long-distance transport or for storage in individual cells differ among plant families (Table 8.3). In legumes in general and in soybean in particular, the majority of the fixed nitrogen transported in the xylem of nodulated roots is incorporated into the ureides allantoin and allantoic acid (Layzell and LaRue, 1982).

Despite the different sites of ammonia assimilation (roots, root nodules, and leaves) the key enzymes involved are in each case glutamine synthetase and glutamate synthase (Fig. 8.8). Both enzymes have been found in roots, in chloroplasts, and in N_2-fixing microorganisms, and convincing evidence exists that assimilation of most if not all ammonia derived from ammonium uptake, N_2 fixation (Bothe *et al.*, 1983), nitrate reduction, and photorespiration (Chapter 5) is mediated by the glutamine synthase–glutamate synthase pathway (Skokut *et al.*, 1978; Fentem *et al.*, 1983).

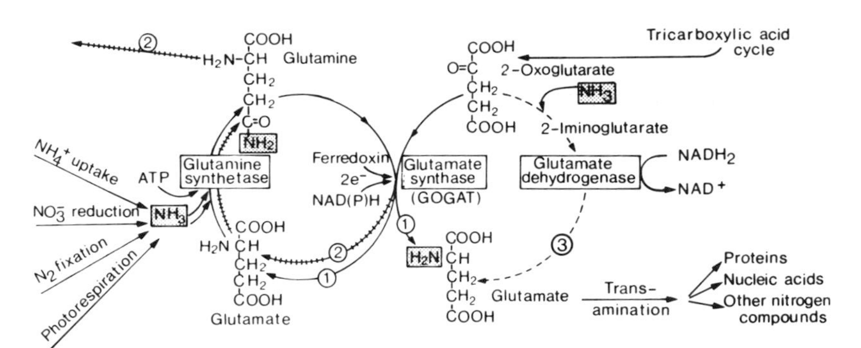

Fig. 8.8 Model of ammonia assimilation pathways. (1,2) Glutamine synthetase–glutamate synthase pathway, with low NH_3 supply (1) and with high NH_3 supply (2). (3) Glutamate dehydrogenase pathway. GOGAT, Glutamine-oxoglutamate aminotransferase.

In this pathway the amino acid glutamate acts as the acceptor for ammonia, and the amide glutamine is formed (Fig. 8.8). Glutamine synthetase has a very high affinity for ammonia (low K_m value) and is thus capable of incorporating ammonia even if present in very low concentrations. The latter is activated by high pH and high concentrations of both magnesium and ATP, and all three factors are increased in the chloroplast stroma upon illumination (see Section 8.5.4)

In chloroplasts, light-stimulated nitrate reduction and enhanced ammonia assimilation are therefore efficiently coordinated to prevent ammonia levels from becoming so elevated that they uncouple photophosphorylation (Krogmann *et al.*, 1959). Ammonia toxicity may be related to the rapid permeation of ammonia across biomembranes. For example, ammonia, but not ammonium (NH_4^+), diffuses rapidly across the outer membranes of chloroplasts (Heber *et al.*, 1974).

The other enzyme in ammonia assimilation, glutamate synthase (GOGAT), catalyzes the transfer of the amide group (—NH_2) from glutamine to 2-oxoglutarate, the latter of which is a product of the tricarboxylic acid cycle (Fig. 8.8). For this reaction either reduced ferredoxin (from photosystem I) or NAD(P)H (from respiration) is required. The reaction results in the production of two molecules of glutamate, one of which is required for the maintenance of the ammonia assimilation cycle and the other of which can be utilized for biosynthesis of proteins, for example. As an alternative, when the ammonia supply is large, both glutamate molecules can act as an ammonia acceptor, and one molecule of glutamine leaves the cycle [pathway (2), Fig. 8.8)].

Another enzyme, glutamate dehydrogenase (Fig. 8.8), used to be considered the key enzyme in ammonia assimilation. However, it is localized principally in the mitochondria of root and leaf cells and has a low affinity for ammonia (high K_m value), which is inconsistent with the need to maintain a very low intracellular ammonia concentration.

8.2.2 Amino acid and Protein Biosynthesis

The organically bound nitrogen of glutamate and glutamine can be utilized for the synthesis of other amides, as well as of ureides, amino acids, and high-molecular-weight compounds such as proteins. Up to 20 different amino acids are required for protein synthesis. The peptide chain of each protein has a genetically fixed amino acid sequence (see below). The carbon skeletons for these different amino acids are derived mainly from intermediates of photosynthesis (3-phosphoglycerate), glycolysis (3-phosphoglycerate and pyruvate), and the tricarboxylic acid cycle (oxaloacetate and 2-oxoglutarate). The transfer of the amino group from amino acids to other carbon skeletons—the transamination reaction—is catalyzed by aminotransferases, which are also referred to as transaminases:

$$\underset{\text{Amino acid (A)}}{H\text{-}\overset{NH_2}{\underset{R_A}{C}}\text{—}COOH} + \underset{\text{2-Oxo acid (B)}}{\overset{O}{\underset{R_B}{\overset{\|}{C}}}\text{—}COOH} \underset{\text{transferases}}{\overset{\text{amino-}}{\rightleftharpoons}} \underset{\text{2-Oxo acid (A)}}{\overset{O}{\underset{R_A}{\overset{\|}{C}}}\text{—}COOH} + \underset{\text{Amino acid (B)}}{H\text{—}\overset{NH_2}{\underset{R_B}{C}}\text{—}COOH}$$

The prosthetic group of the transaminases is pyridoxal phosphate, a vitamin B_6 derivative. Higher plants contain a complete set of transaminases capable of shuttling amino groups between appropriate acceptors. Monogastric animals and humans rely on an external supply in the diet of both the transaminase prosthetic group (i.e., on vitamin B_6) and certain amino acids which cannot be synthesized and which therefore are "essential" in the diet (e.g., valine, leucine, lysine, methionine, and tryptophan).

In protein synthesis the individual amino acids are coupled by peptide bonds ($R_1 \cdot CO—NH—R_2$) in a condensation reaction:

$$\text{Amino acid} \underset{+H_2O}{\overset{-H_2O}{\rightleftharpoons}} \text{dipeptide} \underset{+nH_2O}{\overset{-H_2O}{\rightleftharpoons}} \text{polypeptide/protein}$$

Proteins are polypeptides formed from more than 100 individual amino acids, and their sequence is determined by the genetic information carried by the deoxyribonucleic acid (DNA) molecules (Fig. 8.9). Expression of the genetic information involves the production of messenger ribonucleic acid (mRNA) and transfer ribonucleic acid (tRNA) molecules. Whereas most of the DNA is located in the nucleus of the cell, most of the RNA is located in the cytoplasm ribosomes, the sites of protein synthesis. Ribosomes are ribonucleoprotein particles composed of subunits with a highly ordered

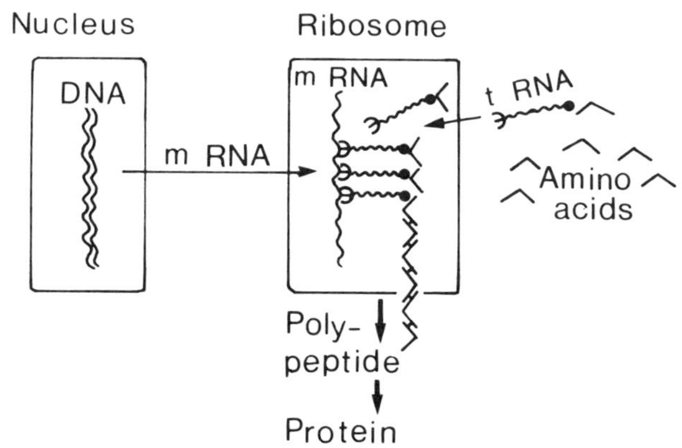

Fig. 8.9 Main steps of protein biosynthesis.

structure. Ribosomes are essential for the several distinct steps of protein biosynthesis: (a) amino acid activation and aminoacyl-tRNA synthesis; (b) peptide chain initiation and elongation; and (c) chain termination and peptide and ribosome release from the polysome complex. Mineral nutrients have an important role in these processes. The maintenance of the highly ordered ribosome structure requires divalent cations, Mg^{2+} in particular. Magnesium is also necessary for the activation of amino acids by ATP, and potassium seems to be involved in the step of chain elongation. Zinc is the metal component of RNA polymerase, and iron is required in some way for ribosome integrity. These effects of individual mineral nutrients are discussed in more detail in later sections.

The biosynthesis and breakdown of proteins occur simultaneously in cells and tissues. The equilibrium between these two processes is determined by various factors, including the stage of development (e.g., leaf age), the source–sink relationship, and the nutritional status of a plant. Either directly or indirectly, phytohormones play a dominant role in the regulation of this equilibrium. Thus so too do those mineral nutrients that strongly affect phytohormone balance—nitrogen, for example. During leaf senescence the rapid decline in net protein content is mediated by the *de novo* synthesis of certain enzymes (endopeptidases) which enhance the fomation of low-molecular-weight organic nitrogen compounds from protein structures (Weckenmann and Martin, 1981).

8.2.3 Role of Low-Molecular-Weight Organic Nitrogen Compounds

In higher plants low-molecular-weight organic nitrogen compounds (Fig. 8.10) not only act as intermediates between the assimilation of inorganic

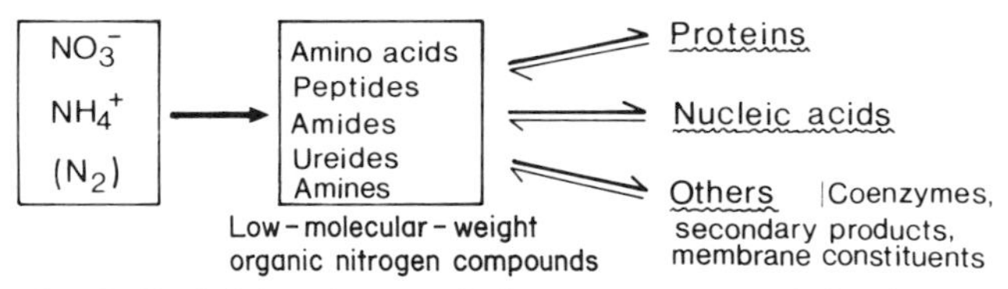

Fig. 8.10 Major classes of nitrogen compounds in plants.

nitrogen and the synthesis of high-molecular-weight compounds. They are also important for the nitrogen economy of plants for another reason. In contrast to lower plants, animals, and humans, higher plants are not capable of excreting substantial amounts of organically bound nitrogen, for example, as urea. They have to store this nitrogen in a nontoxic form, for example, as urea derivatives such as allantoin. Nor are higher plants capable of reoxidizing organically bound nitrogen to nitrate, which would be a safe storage form, for example, in periods of enhanced protein degradation.

Some other functions of low-molecular-weight organic nitrogen compounds should be mentioned. For example, the tripeptide glutathione functions in the redox system of chloroplasts and in the long-distance transport of reduced sulfur in the phloem (Section 8.3). Several antibiotics such as valinomycin are low-molecular-weight polypeptides. There is increasing evidence that some low-molecular-weight organic nitrogen compounds are involved in the long-distance transport of certain heavy-metal cations in the xylem. In xylem exudates of sunflower roots, a high proportion of manganese seems to be complexed to amino acid-carbohydrate compounds of molecular weight below 1500 (Höfner, 1970), and in xylem exudates of maize roots manganese is preferentially bound to peptides (Ebeid and Kutacek, 1979). Copper is transported in the xylem exclusively in chelated form (Graham, 1979), being bound most likely to asparagine, histidine, and glutamine (White *et al.*, 1981b). The level and form of nitrogen supply are therefore likely to have an important effect on the transport of certain heavy-metal cations via the xylem.

Low-molecular-weight organic nitrogen compounds are also precursors of amine synthesis, which is mediated, for example, by the decarboxylation of an amino acid:

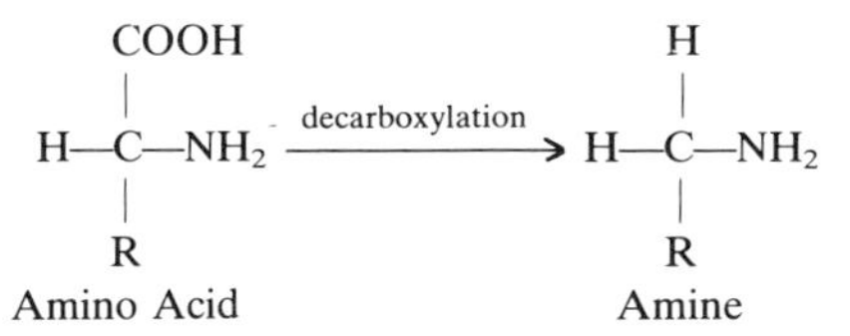

Amines are components of the lipid fraction of biomembranes; the amine component ethanolamine, for example, is synthesized by decarboxylation of the amino acid serine. Di- and polyamines have been attracting increasing attention as secondary messengers in the action of phytohormones (Chapter 5). Polyamines also have protective effects on certain enzymes. For example, 1,3-diaminopropane, a normal constituent of leaves, delays senescence of isolated leaves by inhibiting ethylene formation (Shih *et al.*, 1982). The accumulation of di- and polyamines under conditions of potassium deficiency is well documented and is also closely related to ammonium nutrition. In both cases the diamine putrescine dominates; this compound is formed from the amino acid arginine:

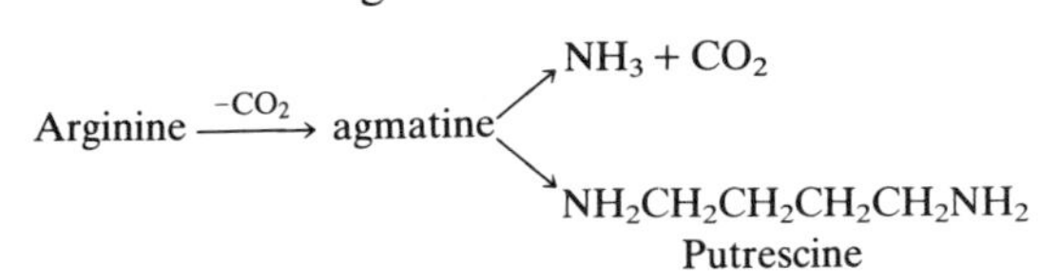

It is assumed that these basic amines have important functions in both cation–anion balance and stabilization of the cellular pH.

Low-molecular-weight organic nitrogen compounds are also involved in osmoregulation in higher plants. Under salt or water stress, the synthesis of such compounds as proline and betaine (also referred to as glycine betaine) is greatly enhanced. These "compatible solutes" accumulate preferentially in the cytoplasm to counteract the osmotic perturbation caused by high vacuolar concentrations of inorganic ions such as chloride and sodium, which are incompatible with cytoplasmatic metabolism (Gorham *et al.*, 1980, 1981). Glycine betaine protects enzymes from inactivation by high NaCl concentrations (Fig. 16.16) and membranes against heat destabilization (Jolivet *et al.*, 1982). Low-molecular-weight organic nitrogen compounds are therefore important for the adaptation of plants to saline substrates (Section 16.6).

8.2.4 Ammonium versus Nitrate Nutrition

Whether plants should be supplied with nitrate- or ammonium-based fertilizers, eventually combined with nitrification inhibitors (Sahrawat, 1980), is a matter of great practical importance. Nitrification inhibitors may include naturally occurring compounds such as neem cake, obtained from Karanja trees, *Pougamia glabra* (Thomas and Prasad, 1983), or synthetic ones, such as N-Serve or dicyanodiamide (Amberger and Gutser, 1978). The subject of ammonium versus nitrate nutrition has been reviewed by Haynes and Goh (1978) and Kirkby (1981). As a general pattern of genotypical differences among plant species, *calcifuges*— or plants adapted to acid soils—and plants adapted to low soil redox potential (e.g., wetland rice) have a preference for ammonium (Ismunadji and Dijkshoorn, 1971). In contrast, *calcicoles*— or plants with preference to calcareous, high pH soils—utilize nitrate preferentially (Kirkby, 1967).

For any given plant species, the uptake and utilization of ammonium is greater than that of nitrate at low temperatures (Clarkson and Warner, 1979). In some cases, highest growth rates are obtained with a combination of both ammonium and nitrate (Gashaw and Mugwira, 1981) or with ammonium only (Sommer and Six, 1982a). Specific effects of ammonium on growth and development have already been discussed in relation to phytohormone balance (Chapter 5) and are most likely also responsible for the stimulation of barley yield under field conditions (Sommer and Six, 1982b).

Contradictory effects of the two nitrogen sources on plant growth are to be expected for other reasons, such as their different effects on cation–anion balance (Kirkby, 1968; Kurvits and Kirkby, 1980), on root-induced pH changes in the rhizosphere (Chapter 15), and on energy metabolism (Middleton and Smith, 1979). Ammonium generally inhibits cation uptake and can depress growth by inducing a deficiency of magnesium (Manolakis and Lüdders, 1977) or calcium (Pill *et al.*, 1978).

Ammonium, unlike nitrate, increases root respiration (Matsumoto and Tamura, 1981). This effect can be attributed both to enhancement of root exudation and hence stimulation of bacterial growth (Trolldenier and Rheinbaben, 1981a) and to the necessity for increasing substrate flux rate in the tricarboxylic acid cycle in order to supply more carbon skeletons for ammonia assimilation (Section 5.3). This higher demand of carbon skeletons is also reflected in lower levels of soluble carbohydrates in roots supplied with ammonium compared to nitrate (Talouizte *et al.*, 1984).

Growth inhibition by ammonium nutrition is closely related to the fall in substrate pH (Findenegg *et al.*, 1982) imposed by ammonium uptake (Figs. 2.19 and 8.7). At low substrate pH, ammonium uptake is not as depressed as the uptake of other cations, which further increases the cation–anion imbalance. When low substrate pH is induced by ammonium nutrition the pH of the root press sap is also distinctly lower (Findenegg *et al.*, 1982), and under these conditions a drastic increase in the di- and polyamine content of roots is observed (Smith and Wilshire, 1975). Upon stabilization of the rhizosphere pH by $CaCO_3$, the di- and polyamine levels remain fairly low with ammonium nutrition. Putrescine formation in roots can be increased either by ammonium supply without pH stabilization or by disruption of the potassium supply (Table 8.4). A combination of both increases the putrescine level by a factor of more than 80 compared with the level in nitrate-fed plants well supplied with potassium. Because the enzyme systems involved in putrescine synthesis are stimulated by low pH (Smith and Sinclair, 1967) the increase in putrescine level in ammonium-fed plants can be regarded as

Table 8.4
Effect of Ammonium and Potassium Supply on Root Growth and Root Putrescine Content in Pea Plants[a]

Treatment			
Nitrogen supply (NO_3^-/NH_4^+)	Potassium supply	Dry weight of the roots (g)	Putrescine (μmol/g dry wt)
100:0	$+K^+$	134	0·6
	$-K^+$	113	10·3
50:50	$+K^+$	92	24·5
	$-K^+$	80	50·5

[a]From Klein *et al.* (1979).

the response of the roots to low cellular pH and as part of a homeostatic mechanism for maintaining cellular pH by the synthesis of basic compounds (Klein *et al.*, 1979).

Accordingly, ammonium nutrition should be less effective in putrescine formation at higher substrate pH. This is indeed the case, unless the pH reaches values where the concentration of ammonia in the aqueous solution are high. As shown in Table 8.5 ammonia concentrations as low as 0·16 m*M* in the root medium are high enough to depress severely both CO_2 assimilation and transpiration. Maximal growth rates were obtained with the combination of both nitrogen sources—ammonium and nitrate—and maintenance of the substrate pH at 6·50.

Under field conditions in well-buffered soils within the pH range 5–7, these detrimental side effects of ammonium nutrition are of minor importance. However, in soils with either very low cation–exchange capacity or with pH values below 5 and above 7·5, ammonium nutrition can be detrimental to growth.

Table 8.5
Effect of pH and Nitrogen Source in the Nutrient Solution on the Assimilation and Transpiration Rate of Cucumber Plants[a]

	Nitrogen source (m*M*)				
pH	Nitrate N	Ammonium N	Ammonia[b]	Assimilation rate (mg $CO_2/dm^2 \times hr$)	Transpiration rate (mg $H_2O/dm^2 \times hr$)
6·50	3	0	0	6·16	2·00
7·75	3	0	0	6·55	2·18
6·50	3	5	0·01	6·60	1·80
7·75	3	5	0·16	4·48	1·39

[a]Based on Schenk and Wehrmann (1979).
[b]Calculated NH_3 concentration in the aqueous solution.

Urea, another nitrogen source, can be classified according to its effect on plant metabolism and growth somewhere between nitrate and ammonium (Kirkby, 1967; Kirkby and Mengel, 1967). Urea can be taken up directly by the roots or aerial parts (Chapter 4); after being taken up by the roots (Hartel, 1977), it is rapidly hydrolyzed by the enzyme urease either within the roots (e.g., in soybean) or after translocation to the shoots (e.g., in maize). In soils the hydrolysis of urea usually takes place before root uptake.

8.2.5 Nitrogen Supply, Plant Growth, and Plant Composition

Depending on the plant species, development stage, and organ, the nitrogen content required for optimal growth varies between 2 and 5% of the plant dry weight. When the supply is suboptimal, growth is retarded; nitrogen is mobilized in mature leaves and retranslocated to areas of new growth. Typical nitrogen-deficiency symptoms, such as enhanced senescence of older leaves, can be seen. An increase in the nitrogen supply not only delays senescence and stimulates growth but also changes plant morphology in a typical manner (Fig. 8.11), particularly if the nitrogen availability is high in the rooting medium during the early growth. Shoot elongation is enhanced and root elongation inhibited (Klemm, 1966), a shift which is unfavorable for nutrient acquisition and water uptake in later stages.

Typical nitrogen-induced changes in leaf morphology are shown in Table 8.6 for rice. The length, width and area of the leaf blades increase, but the thickness decreases. In addition, the leaves become increasingly droopy, an effect that interferes with light interception.

In cereals the enhancement of stem elongation by nitrogen increases the susceptibility to lodging. This change in shoot morphology is less distinct

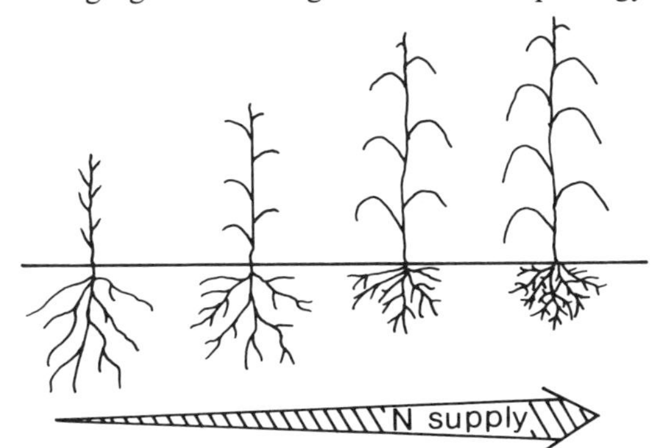

Fig. 8.11 Schematic representation of the effect of increasing levels of nitrogen supply to the roots during early growth stages on the root and shoot growth of cereal plants.

Table 8.6
Effect of Increasing Nitrogen Supply as NH_4NO_3 on leaves of Rice[a]

	Leaf blade			
N Supply (mg/liter)	Length (cm)	Width (cm)	Area (cm^2)	Thickness (mg/cm^2)
5	49·0	0·89	30·6	4·9
20	56·1	1·13	47·8	4·1
200	60·3	1·25	56·1	3·8

[a]Based on Yoshida *et al.* (1969).

with ammonium than with nitrate nutrition (Sommer and Six, 1982b) and is presumably related to nitrogen-induced changes in the phytohormone balance (Chapter 5). In cereals this side effect of nitrogen on stem elongation can become the dominant yield limiting factor when high doses of nitrogen fertilizers are applied. Efforts have thus been made in the last couple of decades to shorten stem length of cereals by other means. Short stem length, which is obtained by breeding in most high-yielding cultivars, can also be induced by the application of growth retardants such as chlorocholine chloride (CCC; trade names: Chlormequat or Cycocel), either solely (Mayr, 1986) or in combination with other growth retardants (Kühn *et al.*, 1977), counteracting the negative side effects of a large nitrogen supply (Table 8.7). The increase in grain yield obtained by CCC application is mainly, but not exclusively, the result of its counteractive effect on lodging. Other effects on development (tillering, flowering, etc.) are involved, as would be expected from the application of bioregulators that interfere with the balance of phytohormone balance in plants. Nitrogen alters plant

Table 8.7
Interacting Effects of Nitrogen Fertilizer Supply and Growth Regulation by Chlorocholine Chloride (CCC) on Lodging and Grain Yield of Winter Wheat[a]

N Supply (kg/ha)	Degree of lodging[b]		Grain yield (dt/ha)[c]	
	−CCC	+CCC	−CCC	+CCC
0	2·4	1·0	39·7	41·8
80	4·8	1·2	47·1	51·3
120	5·8	1·8	46·7	51·3
160	6·3	1·7	48·0	53·1

[a]Based on Jung and Sturm (1966).
[b]1 = no lodging; 9 = total lodging.
[c]dt = decitons.

Table 8.8
Effect of Increasing Nitrogen Supply as NH_4NO_3 on Dry Matter Production and Composition of Ryegrass[a]

	Nitrogen supply (g/pot)			
	0·5	1·5	1·5	2·0
Dry matter (g/pot)	14·9	23·2	26·2	26·0
Composition (% dry wt)				
Total nitrogen	2·0	2·8	3·6	4·2
Sucrose	7·7	7·3	7·1	6·3
Polyfructosane	10·0	4·3	1·8	1·1
Starch	6·1	3·4	2·1	1·4
Cellulose	14·4	13·9	13·9	17·6

[a]From Hehl and Mengel (1972).

composition much more than any other mineral nutrient. For example, as shown in Table 8.8, the dry matter production of ryegrass is increased by nitrogen in a typical yield response curve. Simultaneously, the total nitrogen content increases, but the contents of the two main storage carbohydrates in grasses, polyfructosane and starch, decrease drastically.

Whether these changes in plant composition represent an increase or decrease in quality depends on the further use of the plant material. The increase in total nitrogen content has to be interpreted with care. Total nitrogen and "crude protein" (total nitrogen content multiplied by a certain factor, usually 6·25, gives the total crude protein content) are the sum of both protein and soluble nitrogen, the latter including, for example, amino acids and amides as well as nitrate. In general the proportion of soluble nitrogen increases with elevated levels of nitrogen supply and is higher in leaves and storage organs with high water content, but low in grains and seeds.

The shift in plant composition with increasing nitrogen supply (Table 8.8) reflects a competition for photosynthates among the various metabolic pathways. This competition is modulated by internal and external factors. Figure 8.12 summarizes the general pattern of variable nitrogen supply in relation to the composition of vegetative shoots. When the nitrogen supply is suboptimal, ammonia assimilation increases both the protein content and leaf growth and correspondingly the leaf area index (LAI). As long as the increase in LAI is correlated with an increase in net photosynthesis, the requirement of carbon skeletons for ammonia assimilation does not substantially depress other biosynthetic pathways related to carbohydrates (sugars, starch, cellulose, etc.), storage lipids, or oils. In this nitrogen concentration range, the plant composition does not change substantially, but the total production of plant constituents per unit surface area (e.g., per

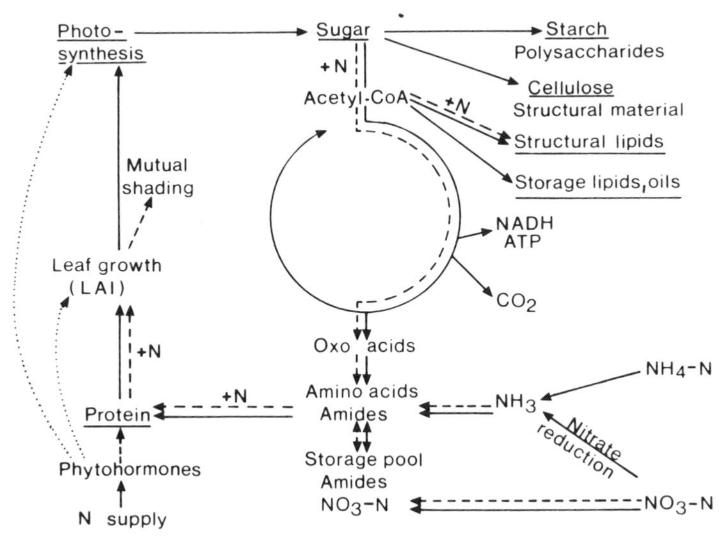

Fig. 8.12 Model of the effects of nitrogen supply on leaf growth and on various plant constituents. Key: →, suboptimal to optimal nitrogen supply; – – →, high to excessive nitrogen supply.

hectare) increases. As the nitrogen supply is further increased, however, a higher proportion of the assimilated nitrogen is sequestered in storage pools, as amides, for example, and owing to mutual shading a further increase in LAI has no effect on the rates of net photosynthesis. Above this transition point, additional ammonia assimilation merely increases the nitrogen content at the expense of the content of most of the other major plant constituents. An interpretation of the effects of nitrogen on the growth and composition of plants requires, of course, a consideration of the effects of nitrogen supply on the phytohormone balance (Chapter 6).

In contrast to the content of storage lipids and oils, the content of lipids in green leaves is closely related to the nitrogen supply (Beringer, 1966). This apparent contradication reflects a different type of regulation (Fig. 8.12). In leaves the majority of lipids are galactolipids, acting as structural components of the chloroplasts (Beringer, 1966). Correspondingly, an enhancement of protein synthesis and chloroplast formation leads to an increase in the lipid content of leaves, as well as to an increase in chloroplast constituents such as chlorophyll and carotine, the provitamin A (Schulze, 1957). An increase in the protein content of vegetables or cereal grains is often positively correlated with an increase in the content of B vitamins: riboflavine, thiamine, and nicotinic acid (Dressel and Jung, 1979), indicating that a high proportion of this protein represents enzyme proteins in which these B vitamins act as a prosthetic group (Section 8.1). The enhancing effect of nitrogen supplied to the roots on the formation and content of nicotine in tobacco (Schmid, 1967) is also indirect. The alkaloid nicotine not only is structurally closely related to cytokinins (adenine derivates) but is also synthesized only in the apical zones of growing roots. The enhancement of root growth in tobacco by nitrogen therefore increases the synthesis and export to the shoots of both cytokinins and nicotine.

For various reasons, increasing concern is being expressed about elevated nitrate levels in certain plant species and plant organs. Elevated nitrate levels in plants at harvest are uneconomical in relation to nitrogen utilization and are also undesirable nutritionally. Sometimes nitrite is formed from nitrate during either the storage or processing of vegetables. Infants fed these nitrite-containing foods run the serious risk of developing methemoglobemia. In addition, nitrite increases the risk of formation of the carcinogenic nitrosamine (Tannenbaum *et al.*, 1978; Heyns, 1979). High nitrate levels (>1–2% in the dry weight) in forage can be toxic to ruminants (Prins, 1983) and in canned vegetables high nitrate levels cause detinning of containers within a few months of storage (Farrow *et al.*, 1971). Effective procedures for eliminating high nitrate levels in plants at harvest are the partial replacement of nitrate by chloride (Alt and Stüwe, 1982), adjustment of the nitrogen supply to the prevailing light conditions during the growing period, appropriate selection of harvest time and harvested organs (Section 8.2.1.1), better control of the nitrogen supply from both soil reserves and fertilizers, and, in general, utilization of the knowledge available on nitrate accumulation in plants (Breimer, 1982).

In cereals a high grain protein content is necessary for processing (bread-baking quality) and for nutrition. An increasing nitrogen supply to the roots, however, affects the nitrogen content of grains mainly indirectly via retranslocation from vegetative growth. The complications can be overcome by the late application of nitrogen. When there is sufficient rainfall during the growing period, nitrogen fertilizer can be applied until anthesis. Most of this nitrogen bypasses the leaves and is directly translocated as amides and amino acids from the roots to the developing grains (Michael *et al.*, 1960). In wheat or barley, however, the corresponding increase in the protein content of the grains is correlated with a decrease in the content of the essential amino acid lysine in the grain protein, that is, with a decrease in the nutritional quality of the protein. As shown in Table 8.9 the decrease in the lysine content of grain protein brought about by the late application of nitrogen results from a shift in the proportions of the various grain protein fractions, mainly in favor of the endosperm protein prolamine, which has a very low lysine content. In agreement with the regulation of the synthesis of individual proteins (albumin, prolamine, etc.) according to the genetic code (Fig. 8.9), the proportion of lysine in the various protein fractions is not altered by the nitrogen supply.

Table 8.9
Effect of Additional Nitrogen Supply as NH_4NO_3 at Anthesis on Wheat Grains[a]

	Percentage of Lysine N of total Protein N		Amount of protein fractions (mg protein/10 g grain dry wt)	
Protein fraction	9·5[b]	19·1[c]	9·5[b]	19·1[c]
Albumin	4·05	4·00	13·6	13·3
Globulin	2·20	2·25	12·5	16·7
Prolamine	0·40	0·40	41·6	123·0
Glutelin	1·90	1·85	36·5	86·1

[a]From Ewald (1964).
[b]Percentage of crude protein, in the dry weight control.
[c]Percentage of crude protein in the dry weight additional nitrogen supply.

A similiar relationship occurs in barley between nitrogen supply, protein content, and lysine content. In oat and rice, on the other hand, the main endosperm storage protein is glutelin. Increasing the grain protein content in these species by late nitrogen application therefore has only negligible effects on lysine content and protein quality.

8.3 Sulfur

8.3.1 General

Although atmospheric SO_2 is taken up and utilized by the aerial parts of higher plants, the most important source of sulfur is sulfate taken up by the roots. In the physiological pH range, the divalent anion SO_4^{2-} is taken up by the roots at relatively low rates, and long-distance transport of sulfate is confined mainly to the xylem (Chapter 2). In several respects, sulfur assimilation resembles nitrate assimilation. For example, reduction is necessary for the incorporation of sulfur into amino acids, proteins, and coenzymes, and in green leaves ferredoxin is the reductant for sulfate. Unlike nitrate nitrogen, however, sulfate can also be utilized without reduction and incorporated into essential organic structures such as sulfolipids in membranes or polysaccharides such as agar. Also in contrast to nitrogen, reduced sulfur can be oxidized again in higher plants, a process that is most evident in the rapid protein degradation that takes place during leaf senescence (Mothes, 1939). In this oxidation reaction the reduced sulfur of cysteine is converted to sulfate (Sekiya *et al.*, 1982a), the "safest" storage form of sulfur in plants.

8.3.2 Sulfate Assimilation and Reduction

It is now generally accepted that in higher plants and in green algae, the first step of sulfur assimilation is the activation of the sulfate ion by ATP (Fig. 8.13). In this reaction the enzyme ATP sulfurylase catalyzes the replacement of two phosphate groups of the ATP by the sulfuryl group, which leads to the formation of adenosine phosphosulfate (APS) and pyrophosphate (Fig. 8.13). The synthesis of sulfate esters [pathway (1)] requires further activation by ATP to form phosphoadenosine phosphosulfate (PAPS).

For sulfate reduction [pathway (2)] the sulfuryl group of APS is transferred to the thiol (—SH) of a carrier, probably the tripeptide glutathione with a cysteine residue serving as the reactive —SH group (Glut—SH). The subsequent reduction,

$$R{-}S{-}SO_3^- \rightarrow R{-}S{-}SH$$

is mediated by ferredoxin, and the newly formed —SH group is transferred to acetylserine, which is split into acetate and the amino acid cysteine. Cysteine, the first stable product of the assimilatory sulfate reduction, acts as a precursor for the synthesis of all other organic compounds containing reduced sulfur, as well as for other biosynthetic pathways, such as the formation of the secondary messenger ethylene (Shih *et al.*, 1982).

Assimilatory sulfate reduction is regulated at various levels. High concentrations of cysteine inhibit APS sulfotransferase (Brunold and Schmidt, 1978), the enzyme catalyzing the transfer APS → R—S—SO_3^- (Fig. 8.13), whereas ammonium nutrition (compared to nitrate nutrition) increases the

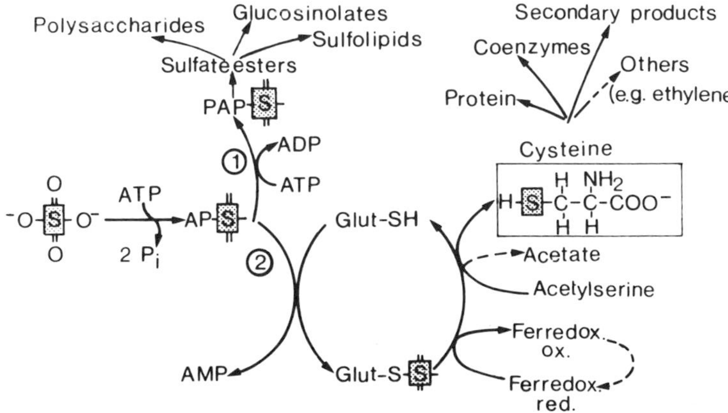

Fig. 8.13 Pathways of sulfur assimilation in higher plants and green algae. (1) Synthesis of sulfate esters; (2) sulfate reduction according to the APS pathway. (Based on Schiff, 1983.)

activity of this enzyme (Brunold, 1981; Brunold and Suter, 1984), indicating a higher demand for S-containing amino acids in connection with the ammonium assimilation. At high cellular levels of either cysteine (Sekiya *et al.*, 1982a) or SO_2 (Sekiya *et al.*, 1982b), the evolution of hydrogen sulfide (H_2S) from green cells is strongly enhanced by light. The light-dependent SO_2 reduction coupled with H_2S release from green leaves of up to 10% of the absorbed SO_2 is considered an important mechanism for the detoxification of SO_2 in leaves and needles (Sekiya *et al.*, 1982b). This type of sulfate reduction may be considered a modification of the dissimilatory sulfate reduction pathway in prokaryotic anaerobes such as *Desulfovibrio* which use sulfate as an oxidant in the formation of ATP and sulfide during respiration (Schiff, 1983).

In higher plants the enzymes of the assimilatory sulfate reduction are localized in the chloroplasts (Fankhauser and Brunold, 1978) but can also be found, at a much lower level, in plastids of roots (Fankhauser and Brunold, 1979). In C_4 plants the bundle sheath chloroplasts are the main sites of sulfate assimilation (Gerwick and Black, 1979; Schmutz and Brunold, 1984), whereas the mesophyll chloroplasts are the sites of nitrate assimilation. In general, sulfate reduction is several times higher in green leaves than in roots, and in leaves the reaction is strongly stimulated by light (Willenbrink, 1964; Fankhauser and Brunold, 1978). Because of the requirement for ferredoxin as a reductant for sulfate this light enhancement is to be expected. The enhancement of sulfate reduction by light might also be related to higher levels of serine (acetylserine; Fig. 8.13) synthesized during photorespiration (Fig. 5.4). Excessive sulfate reduction in the leaves leads to the export of reduced sulfate in the phloem, mainly as glutathione (Rennenberg *et al.*, 1979). During leaf development, the pattern of sulfate reduction is similar to that of nitrate reduction; that is, it is maximal during leaf expansion but declines rapidly after leaf maturation (Schmutz and Brunold, 1982).

8.3.3 Metabolic Functions of Sulfur

Sulfur is a constituent of the amino acids cysteine and methionine and hence of proteins. Both of these amino acids are precursors of other sulfur-containing compounds such as coenzymes and secondary plant products. Sulfur is a structural constituent of these compounds (e.g., $R^1—C—S—C—R^2$) or acts as a functional group (e.g., R—SH) directly involved in metabolic reactions. Under conditions of sulfur deficiency, protein synthesis is inhibited. In green leaves the majority of the protein is located in the chloroplasts where the chlorophyll molecules comprise prosthetic groups of the chromoproteid complex. Thus in sulfur-deficient plants

Table 8.10
Effect of Sulfur Deficiency on Leaf Composition in Tomato[a]

	Content in leaves (mg/100g dry wt)			Sulfur content of protein (μg/mg protein)	
Treatment	Chlorophyll	Protein	Starch	Cytoplasm	Chloroplast
Control ($+SO_4^{2-}$)	5·8	48·0	2·8	13·5	6·5
S Deficiency	0·9	3·5	27·0	3·8	5·2

[a]Based on Willenbrink (1967).

the chlorophyll content declines also (Table 8.10). In contrast, starch may accumulate as a consequence either of impaired carbohydrate metabolism at the sites of production (the source) or of low demand at the sink sites (growth inhibition). In response to sulfur deficiency proteins of low sulfur content are synthesized, which is most evident in the cytoplasm of leaf cells (Table 8.10). Changes in protein composition brought about by changes in the nutrient supply have already been discussed in relation to the nitrogen supply (Section 8.2).

Sulfur is a structural constituent of several coenzymes and prosthetic groups, such as ferredoxin, biotin (vitamin H), and thiamine pyrophosphate (vitamin B_1). Sulfur deficiency impairs the synthesis of these compounds.

Sulfur plays a key role in redox reactions, according to

$$(R^1)\text{—SH} + \text{HS—}(R^2) \underset{+2H}{\overset{-2H}{\rightleftharpoons}} (R^1)\text{—S—S—}(R^2)$$

in which (R^1) and (R^2) represent two molecules of cysteine forming the dipeptide cystine or two molecules of the tripeptide glutathione, with the cysteine residue (cysteinyl moiety) as the reactive group (Fig. 8.13). Because it is highly water soluble glutathione is considered to comprise a major redox system in the cytosol and the chloroplasts (Rennenberg, 1982).

The formation of a disulfide bond between two adjacent cysteine residues in polypeptide chains is of fundamental importance for the tertiary structure and thus the function of enzyme proteins. This bond may form a permanent (covalent) cross-link between polypeptide chains or a reversible dipeptide bridge. During dehydration, the number of disulfide bonds in proteins increases at the expense of the —SH groups, and this shift is associated with protein aggregation and denaturation (Tomati and Galli, 1979). The protection of —SH groups in proteins from the formation of disulfide bridges is considered to be of great importance for providing cellular resistance to dehydration (caused by drought and heat) and frost damage (Levitt, 1980).

Low-molecular-weight proteins ($MW < 10{,}000$) that are high in sulfur and heavy-metal cations such as copper, zinc, and cadmium can be isolated from a variety of living tissues, including those of higher plants. In these proteins, referred to as *metallothioneins*, cysteine residues represent up to one-third of the total amino acids (Weser *et al.*, 1973; Rauser, 1984), and the —SH groups of these residues form strong complexes with certain heavy-metal cations. It has been shown that the synthesis of metallothioneins in roots and shoots can be induced by heavy metals such as cadmium (Petit *et al.*, 1978; Wagner and Trotter, 1982), and copper (Rauser, 1983). These findings support the view that metallothioneins are important for eliminating excessive amounts of copper, cadmium, and perhaps also zinc in order to prevent the irreversible binding of these cations to the functional —SH groups of enzymes.

In many enzymes and coenzymes, such as ureas, APS sulfotransferase and coenzyme A, the —SH groups act as functional groups in the enzyme reaction. In the glycolic pathway, for example, decarboxylation of pyruvate and the formation of acetyl coenzyme A are catalyzed by a multienzyme complex involving three sulfur-containing coenzymes: thiamine pyrophosphate (TPP), the sulfhydryl–disulfide redox system of lipoic acid, and the sulfhydryl group of coenzyme A:

Pyruvate (CH_3–C=O–COOH) —[TPP, Lipoic acid; CoA-SH (Multienzyme complex)]→ CO_2 + H-C(H)(CH_3)-S-CoA → Fatty acids (CO_2, Biotin, $+Mn^{2+}$) or Tricarboxylic acid cycle

The acetyl group (—CH_2—CH_3) of the coenzyme A is then transferred to the tricarboxylic acid cycle or to the pathway of fatty acid synthesis. The coupling of C_2 units for the synthesis of long-chain fatty acids requires transient carboxylation, which is mediated by the sulfur-containing coenzyme biotin and activated by manganese.

Volatile compounds such as isothiocyanates and sulfoxides with sulfur as a structural constituent are mainly responsible for the characteristic odor of such plant species as onion, garlic, and mustard. Of these volatile compounds, mustard oils of the Cruciferae are of particular agricultural importance. Mustard oils accumulate in intact cells predominantly as nonvolatile glucosides (glucosinolates) containing sulfur both as a sulfhydryl and as a sulfo group:

Glucosinolate: R–CH_2–C(S–Glucose)=N–O–SO_2–O^-

R = CH_2=CH–CH_2– Sinigrin (*Brassica nigra*)

R = HO–C_6H_4–CH_2– Glucosinalbin (*Sinapis alba*)

—Myrosinase (− Glucose, − Sulfate)→ R–CH_2–N=C=S Isothiocyanates (e.g. mustard oil)

The side chain R of mustard oils varies among plant species. Hydrolysis of the glucosinsolates, catalyzed by the enzyme myrosinase, leads finally to volatile compounds with either reduced sulfur (e.g., mustard oil) or partially oxidized sulfur (e.g., sulfoxides, allicin of onions). The hydrolysis is greatly enhanced by mechanical damage to the cells. Although the role of these compounds is not fully understood, they definitely act as phytocids and protect plants from animals.

Sulfur in its nonreduced form is a component of sulfolipids and is thus a structural constituent of all biological membranes. In sulfolipids the sulfo group is coupled by an ester bond to a C_6 sugar, for example, glucose:

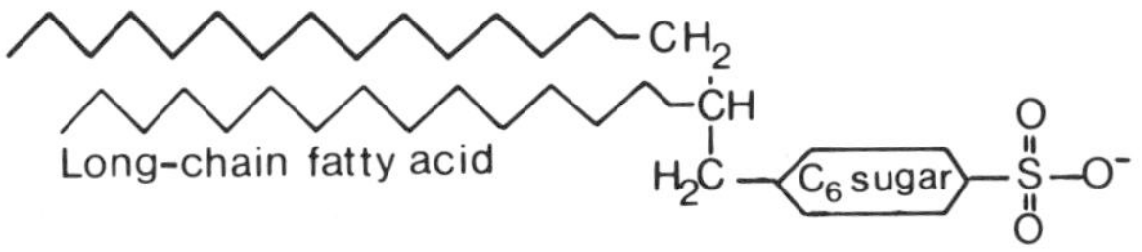

Long-chain fatty acid — CH_2 — CH — H_2C — C_6 sugar — $S(=O)_2$—O^-

Sulfolipids are particularly abundant in the thylakoid membranes of chloroplasts (Heise and Jacobi, 1973). There is evidence that sulfolipids are also involved in the regulation of ion transport across other biomembranes. The sulfolipid level in roots and the salt tolerance of plants are positively correlated, for example (Erdei *et al.*, 1980; Stuiver *et al.*, 1981).

8.3.4 Sulfur Supply, Plant Growth, and Plant Composition

The sulfur requirement for optimal growth varies between 0·2 and 0·5% of the dry weight of plants. Among the families of crop plants, the requirement increases in the order Gramineae $<$ Leguminosae $<$ Cruciferae and is also reflected in corresponding differences in the sulfur content (percentage of dry weight) of their seeds: 0·18–0·19, 0·25–0·3, and 1·1–1·7, respectively (Deloch, 1960). The sulfur content of protein also varies considerably both among the protein fractions of individual cells (Table 8.10) and among plant species. On average, proteins from legumes contain less sulfur than proteins from cereals, the N/S ratios being 40 : 1 and 30 : 1, respectively (Dijkshoorn and Wijk, 1967).

Under conditions of sulfur deficiency, inhibition of protein synthesis is correlated with an accumulation of soluble organic nitrogen and nitrate (Table 8.11). Amides are usually present in higher concentration in the soluble organic nitrogen fraction (Freney *et al.*, 1978). The sulfate content is extremely low in deficient plants and increases markedly when the sulfate supply is sufficient for optimal growth. The sulfate content of plants is therefore a more sensitive indicator of sulfur nutritional status than the total sulfur content, the best indicator being the proportion of sulfate sulfur in the total sulfur content (Freney *et al.*, 1978).

Table 8.11
Effect of Sulfate Concentration in the Nutrient Solution on Plant Fresh Weight and Sulfur and Nitrogen Content of Cotton Leaves[a]

		Sulfur or Nitrogen (% dry wt)				
Supply (mg SO_4^{2-}/liter)	Leaf dry wt (g/plant)	Sulfate S	Organic S	Nitrate N	Soluble organic N	Protein N
0·1	1·1	0·003	0·11	1·39	2·23	0·96
1·0	2·4	0·003	0·12	1·37	2·21	1·28
10·0	3·4	0·009	0·17	0·06	1·19	2·56
50·0	4·7	0·10	0·26	0·00	0·51	3·25
200·0	4·7	0·36	0·25	0·10	0·45	3·20

[a]Based on Ergle and Eaton (1951).

Inhibition of protein synthesis during sulfur deficiency leads to chlorosis, just as it does during nitrogen deficiency. Unlike nitrogen, however, sulfur is much more uniformly distributed between old and new leaves and its content is similarly affected in old and young leaves by the level of sulfate supply (Freney *et al.*, 1978). Furthermore, the distribution of sulfur in sulfur-deficient plants is also affected by the nitrogen supply; sulfur-deficiency symptoms may occur either in young (ample nitrogen) or in old (low nitrogen) leaves (Robson and Pitman, 1983), indicating that the extent of sulfur retranslocation from older leaves depends on the rate of nitrogen deficiency-induced leaf senescence, a relationship which is also evident for the micronutrients copper and zinc (Section 3.6). In legumes, during the early stages of sulfur deficiency, nitrogenase activity in the root nodules is much more depressed than photosynthesis (DeBoer and Duke, 1982); symptoms of sulfur deficiency in symbiotically grown legumes are therefore indistinguishable from nitrogen-deficiency symptoms (Anderson and Spencer, 1950).

The decrease in the protein content of sulfur-deficient plants is correlated with the preferential synthesis of proteins with lower proportions of methionine and cysteine but higher proportions of other amino acids such as arginine and aspartic acid (Table 8.12).

The lower sulfur content of proteins influences nutritional quality considerably: Methionine is an essential amino acid in human nutrition and often a limiting factor in diets in which seeds are a major source of protein (Arora and Luchra, 1970). Furthermore, a decrease in the cysteine content of cereal grains reduces the baking quality of flour, since disulfide bridging during dough preparation is responsible for the polymerization of the glutelin fraction (Ewart, 1978).

The concentration of glucosinolates and their volatile hydrolytic products

Table 8.12
Effect of Sulfur Fertilization on the Amino Acid Composition of Endosperm Protein from Wheat[a]

	Amino acid concentration (nmol/16 g protein N)	
Amino acid	Control[b]	Sulfur deficiency[c]
Methionine	11	5
Cysteine	21	7
Arginine	27	34
Aspartic acid	33	93

[a]Based on Wrigley *et al.* (1980).
[b]0·25% total S dry wt.
[c]0·10% total S dry wt.

is closely related to the sulfate supply. Their concentration in plants can be increased beyond the level at which sulfate supply affects growth (Table 8.13). From the qualitative viewpoint this increase can be considered favorable (e.g., because it enhances the taste of vegetables, making them spicier) or unfavorable (e.g., because it decreases acceptability as animal feed).

In highly industrialized areas the sulfur requirement of plants is met to a substantial degree by atmospheric SO_2 pollution. However, in forest trees a permanently high uptake of SO_2 through stomata may cause severe damage, accentuated by acidification of the soil ("acid rain"). On the other hand in rural areas, particularly in the humid tropics and in highly leached soils, sulfur deficiency in crop production is quite common. Under these conditions, the application of fertilizers containing nitrogen in the form of urea are ineffective unless sulfur is applied simultaneously (Wang *et al.*, 1976).

Table 8.13
Effect of Sulfate Supply on Yield and Mustard Oil Content of the Shoots of *Brassica juncea*[a]

Sulfate supply (mg S/pot)	Shoots (g fresh wt)	Mustard oil content (mg/100 g fresh wt)
1·5	80	2·8
15·0	208	8·1
45·0	285	30·7
405·0	261	53·1
1215·0	275	52·1

[a]Based on Marquard *et al.* (1968).

8.4 Phosphorus

8.4.1 General

Unlike nitrate and sulfate, phosphate is not reduced in plants but remains in its highest oxidized form. After uptake—at physiological pH mainly as $H_2PO_4^-$—either it remains as inorganic phosphate (P_i) or it is esterified through a hydroxyl group to a carbon chain (C—O—P) as a simple phosphate ester (e.g., sugar phosphate) or attached to another phosphate by the energy-rich pyrophosphate bond Ⓟ ~ Ⓟ (e.g., in ATP). The rates of exchange between P_i and the Ⓟ in ester and the pyrophosphate bond are very high. For example, P_i taken up by roots is incorporated within a few minutes into organic Ⓟ but thereafter is released again as P_i into the xylem. Another type of phosphate bond is characterized by the relative high stability of its diester state (C—Ⓟ—C). In this association phosphate forms a bridging group connecting units to more complex or macromolecular structures.

8.4.2 Phosphorus as a Structural Element

The function of phosphorus as a constituent of macromolecular structures is most prominent in nucleic acids, which, as units of the DNA molecule, are the carriers of genetic information and, as units of RNA, are the structures responsible for the translation of the genetic information. In both DNA and RNA, phosphate forms a bridge between ribonucleoside units to form macromolecules:

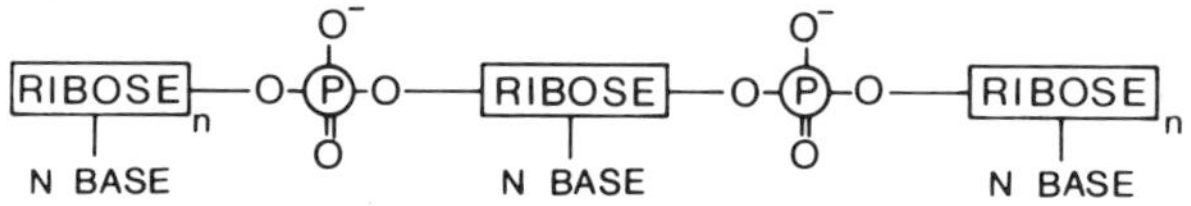

```
                   O⁻                              O⁻
                   |                               |
[RIBOSE]n—O—Ⓟ—O—[RIBOSE]—O—Ⓟ—O—[RIBOSE]n
    |              ‖          |                    ‖          |
 N BASE            O       N BASE                  O       N BASE
```

(Section of an RNA molecule)

Phosphorus is responsible for the strongly acidic nature of nucleic acids and thus for the exceptionally high cation concentration in DNA and RNA structures. The proportion of phosphorus in nucleic acids to total organically bound phosphorus differs among tissues and cells; it is high in meristems and low in storage tissues.

The bridging form of phosphorus diester is also abundant in the phospholipids of biomembranes. There it forms a bridge between a diglyceride and another molecule (amino acid, amine, or alcohol). In biomembranes the amine choline is often the dominant partner, forming phosphatidylcholine (lecithin):

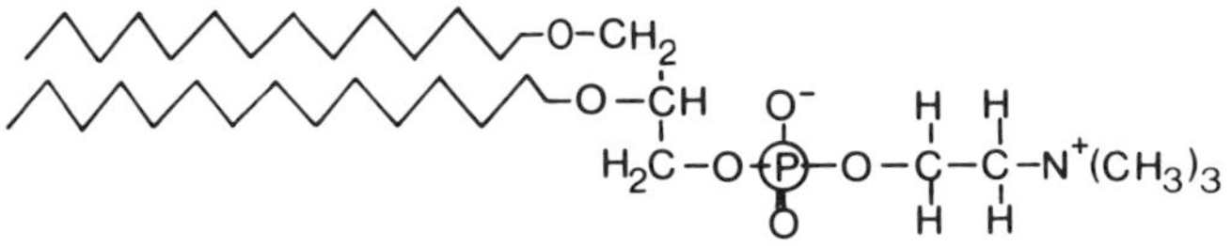

$$
\begin{array}{l}
\text{(fatty acid chain)}-O-CH_2 \\
\text{(fatty acid chain)}-O-CH \quad O^- \quad\;\; H \;\; H \\
\qquad\qquad\qquad\quad H_2C-O-\text{Ⓟ}-O-C-C-N^+(CH_3)_3 \\
\qquad\qquad\qquad\qquad\qquad\;\; O \qquad\;\; H \;\; H
\end{array}
$$

The functions of phospholipids (and also of sulfolipids; see Section 8.3) are related to their molecular structure. There is a lipophilic region (consisting of two long-chain fatty acid moieties) and a hydrophilic region in one molecule; at a lipid–water interface, the molecules are oriented so that the boundary layer is stabilized (Fig. 2.5). The electrical charges of the hydrophilic region play an important role in the interactions between biomembrane surfaces and ions in the surrounding medium.

8.4.3 Role in Energy Transfer

Although present in cells only in low concentrations, phosphate esters (C—Ⓟ) and energy-rich phosphates (Ⓟ ~ Ⓟ) represent the metabolic machinery of the cells. Up to 50 individual esters formed from phosphate and sugars and alcohols have been identified, about 10 of which, including glucose 6-phosphate and phosphoglyceroaldehyde, are present in relatively high concentrations in cells. The common structure of phosphate esters is as follows:

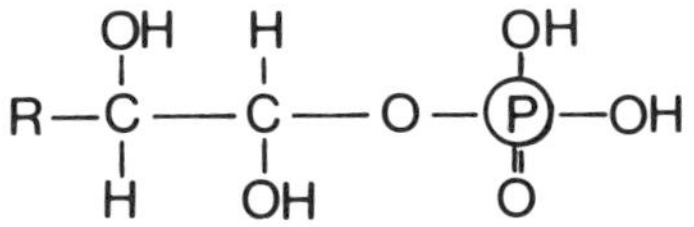

$$
\begin{array}{l}
\quad\;\; OH \quad\; H \qquad\quad\;\; OH \\
R-C-C-O-\text{Ⓟ}-OH \\
\quad\;\; H \quad\;\; OH \qquad\quad\;\; O
\end{array}
$$

Most phosphate esters are intermediates in metabolic pathways of biosynthesis and degradation. Their function and formation are directly related to the energy metabolism of the cells and to energy-rich phosphates. The energy required, for example, for biosynthesis or for ion uptake is supplied by an energy-rich intermediate or coenzyme, principally ATP:

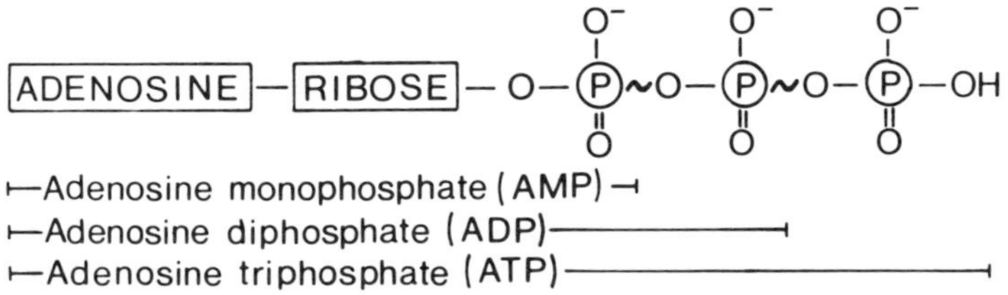

$$
\text{ADENOSINE}-\text{RIBOSE}-O-\underset{O}{\overset{O^-}{\text{Ⓟ}}}\sim O-\underset{O}{\overset{O^-}{\text{Ⓟ}}}\sim O-\underset{O}{\overset{O^-}{\text{Ⓟ}}}-OH
$$

⊢Adenosine monophosphate (AMP)⊣
⊢Adenosine diphosphate (ADP)⊣
⊢Adenosine triphosphate (ATP)⊣

Energy liberated during glycolysis, respiration, or photosynthesis is utilized for the synthesis of the energy-rich pyrophosphate bond, and on hydrolysis of this bond ~30 kJ per mole ATP are released. This energy can be transmitted with the phosphoryl group in a phosphorylation reaction to another compound, which results in the activation (priming reaction) of this compound:

Adenosine–P~P~Ⓟ [ATP] → HO–[R]

Adenosine–P~P [ADP] ← Ⓟ–O–[R] - - - - - →

In some phosphorylation reactions the substrate remains attached to the adenosine moiety (e.g., ADP-glucose or adenosine phosphosulfate; see Section 8.3.2). ATP is the principal energy-rich phosphate required for starch synthesis. The energy-rich pyrophosphate bonds of ATP can also be transmitted to other coenzymes which differ from ATP only in the nitrogen base, for example, uridine triphosphate (UTP) and guanosine triphosphate (GTP), which are required for the synthesis of sucrose and cellulose, respectively.

In actively metabolizing cells, energy-rich phosphates are characterized by extremely high rates of turnover. From pulse labeling experiments with ^{32}P, the turnover rates of various phosphorus compounds can be calculated, as shown in Table 8.14. It is impressive that such a small amount of ATP can satisfy the energy requirement of plant cells. It has been calculated, for example, that 1 g of actively metabolizing maize root tips synthesize about 5 g ATP per day (Pradet and Raymond, 1983). The amount of phospholipids and RNA synthesized is much higher, but they are much more stable compounds and have a relatively low rate of synthesis (Table 8.14).

In vacuolated cells of higher plants two major phosphate pools exist. In the "metabolic pool", represented by the cytoplasm and including the chloroplasts, phosphate esters dominate, whereas in the "nonmetabolic pool," or the vacuole, P_i is the dominant fraction. In a plant with an adequate phosphorus supply, ~85–95% of the total P_i is located in the

Table 8.14
Turnover Times and Rates of Synthesis of Organic Phosphate Fractions in *Spirodela*[a]

Phosphorus fraction	Amount (nmol/g fresh wt)	Turnover time (min)	Synthesis rate (nmol P/g fresh wt × min)
ATP	170	0.5	340
Glucose 6-phosphate	670	7	95
Phospholipids	2700	130	20
RNA	4900	2800	2
DNA	560	2800	0.2

[a]Based on Bieleski and Ferguson (1983).

vacuoles (Bieleski, 1968; Bieleski and Ferguson, 1983). With interruption of the phosphorus supply the P_i concentration in the vacuoles decreases sharply, whereas the P_i concentration in the cytosol (metabolic compartment) decreases only from about 6 m*M* to somewhat less than 3 m*M* (Rebeille *et al.*, 1984). Release of P_i from the vacuoles is usually slow and may limit growth when the external supply is interrupted.

8.4.4 Regulatory Role of Inorganic Phosphate

In addition to its role in the nonmetabolic pool, inorganic phosphate has various essential functions in the metabolic pool. In many enzyme reactions, P_i is either a substrate or an end product (e.g., ATP $\rightarrow$ ADP + P_i). Furthermore, P_i controls some key enzyme reactions. Compartmentation of P_i is therefore essential for the regulation of metabolic pathways in the cytoplasm and chloroplasts. In fruit tissue of tomato, for example, P_i released from the vacuoles into the cytoplasm can stimulate phosphofructokinase activity (Woodrow and Rowan, 1979); the latter is the key enzyme in the regulation of substrate flux into the glycolytic pathway. Thus an increase in the release of P_i from vacuoles can initiate the respiratory burst correlated with fruit ripening (Woodrow and Rowan, 1979). Delayed fruit ripening in phosphorus-deficient tomato plants (Pandita and Andrew, 1976) may be related to this function of P_i.

A fascinating control function of P_i in photosynthesis and carbohydrate metabolism has been discovered. A relatively low external concentration of P_i depresses the rate of starch synthesis in isolated chloroplasts to an insignificant level, despite stimulation of the total carbon fixation (Table 8.15). At higher P_i concentrations the total carbon fixation is also depressed. In the stroma of the chloroplasts, the corresponding concentration of P_i required for severe inhibition of starch synthesis is ~5 m*M* (Heldt *et al.*, 1977).

Table 8.15
Total Carbon Fixation and Carbon Incorporation into Starch in Isolated Spinach Chloroplasts as a Function of the External Phosphate (P_i) Concentration[a]

	P Concentration in the external solution (m*M*)				
	0·15	0·65	1·0	2·0	3·5
Total carbon fixation	11·0	13·9	10·9	7·1	3·3
Incorporation into starch	2·6	0·4	0·1	0·1	0·1

[a]Data represent micrograms carbon per milligram chlorophyll after 8 min of light exposure. Based on Heldt *et al.* (1977).

The inhibition of starch synthesis by P_i is caused by two separately regulated mechanisms located in the chloroplasts. The key enzyme of starch synthesis in chloroplasts, ADP-glucose pyrophosphorylase [pathway (1), Fig. 8.14], is allosterically inhibited by P_i and stimulated by triosephosphates. The ratio of P_i to triosephosphates therefore determines the rate of starch synthesis in chloroplasts (Heldt *et al.*, 1977; Portis, 1982), at high ratios the enzyme being "switched off." The other mechanism regulated by P_i is the release of triosephosphates (glyceraldehyde 3-phosphate and dihydroxyacetone phosphate), the main products of CO_2 fixation leaving the chloroplasts. This release is controlled by a phosphate translocator, a specific carrier located in the inner membrane of the chloroplast envelope [pathway (2), Fig. 8.14] and facilitating the counterexchange $P_i \rightleftharpoons$ triosephosphate (Heldt *et al.*, 1977; Flügge *et al.*, 1980). In this way the net uptake of P_i into the chloroplasts regulates the release of photosynthates from the chloroplast. High P_i concentrations therefore deplete the stroma triosephosphate metabolites, which serve as both substrates for, and activators of, starch synthesis. Thus, inhibition of starch synthesis by high P_i concentrations is also the result of substrate depletion.

Carbon dioxide fixation in the Calvin cycle is a process in which five-sixths of the carboxylation products are required in the stroma to regenerate the CO_2 acceptor ribulose bisphosphate (RuBP). Excessive export of triosephosphates induced by high P_i concentrations leads to the depletion of these metabolites which are required for the regeneration of RuBP (Fig. 8.14). High external P_i concentrations therefore also inhibit total CO_2

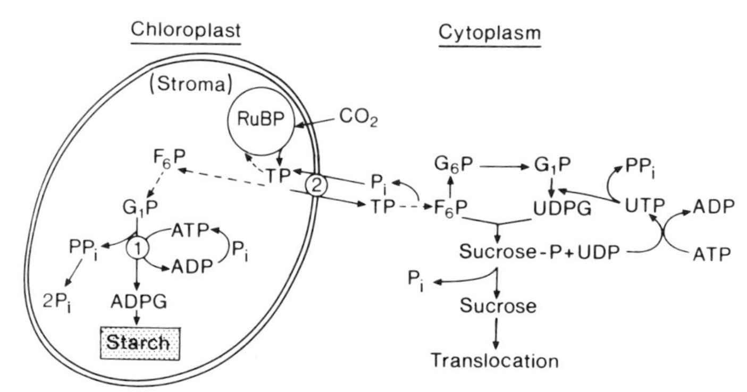

Fig. 8.14 Involvement and regulatory role of phosphate (P) in starch synthesis and carbohydrate transport in a leaf cell. (1) ADP-Glucose pyrophosphorylase: regulates the rate of starch synthesis; inhibited by P_i and stimulated by PGA. (2) Phosphate translocator: regulates the release of photosynthates from chloroplasts; enhanced by P_i. TP, Triosephosphate (3-phosphoglycerate, PGA); F_6P, fructose 6-phosphate; G_6P, glucose 6-phosphate. (Based on Walker, 1980.)

fixation (Table 8.15; Flügge *et al.*, 1980). This regulatory function of P_i is also reflected in the leaves of intact phosphorus-deficient plants, where an extremely large accumulation of starch can be observed in the chloroplasts (Ariovich and Cresswell, 1983) and starch is insufficiently remobilized from the leaves at night or during reproductive growth (Giaquinta and Quebedeaux, 1980).

Carbohydrate metabolism in leaves and sucrose translocation are also affected by P_i, at least indirectly. The priming reactions in the synthesis of hexoses and sucrose require energy-rich phosphates (ATP and UTP). The ATP requirement is also high for the sucrose–proton cotransport in phloem loading (Chapter 5). The simplified model in Fig. 8.14 summarizes the reactions of carbohydrate metabolism in which P_i has a regulating function or in which energy-rich phosphates and the corresponding phosphate esters are involved. Because of these functions, strict compartmentation and regulation of the P_i level in the metabolic pool are essential for carbohydrate metabolism in leaf cells.

In principle, similar P_i regulation of starch synthesis should occur in the amyloplasts of storage cells. ADP-glucose pyrophosphorylase is also the key enzyme in the regulation of starch synthesis in potato tubers (Sowokinos, 1981) and cassava storage roots (Hawker and Smith, 1982); when isolated from these storage tissues the enzyme is severely inhibited by P_i. In contrast, starch accumulation in the endosperm of wheat grains is not affected by high P_i levels (Rijven and Gifford, 1983), which may indicate that these cells have a particularly large capability for effective P_i compartmentation.

The storage of phosphate in cells as inorganic polyphosphates is widespread among lower plants (bacteria, fungi, and green algae). It has also been found in higher plants, for example, in apple leaves (Schmidt and Buban, 1971) and in various species of Australian heath plants (Jeffrey, 1968). Polyphosphates synthesized by plants are linear polymers of P_i (>500 molecules) with pyrophosphate linkages energetically equivalent to ATP. Polyphosphates may therefore function as energy storage compounds and as compounds controlling the P_i level in the metabolic pool of the cells. They certainly also function as cation exchangers, for example, for potassium in *Chlorella* (Peverly *et al.*, 1978) and for calcium in mycorrhizal fungi (Strullu *et al.*, 1982). Polyphosphate formation in the hyphae of mycorrhizal fungi is attracting much attention because of its proposed role in the phosphate nutrition of higher plants living symbiotically with mycorrhiza. Hyphae take up P_i from the soil solution and synthesize polyphosphates; these may be the main forms in which phosphate is transported in the hyphae (Cox *et al.*, 1980), although phosphorus is subsequently transferred from the hyphae to the host root cells, most likely as P_i (Marx *et al.*, 1982).

8.4.5 Phosphorus Fractions and the Role of Phytate

The amount of phosphorus being supplied to a plant affects the various phosphorus fractions in a typical manner. With an increase in the supply from a suboptimal to an optimal level, all the phosphorus fractions in the leaves increase (Table 8.16). Above this level, only the P_i increases, reflecting the storage function of P_i in highly vacuolated tissues.

Table 8.16
Effect of Phosphorus Supply on Phosphorus Fractions of Tobacco Leaves[a]

		Phosphorus fraction (mg P/100 g leaf dry wt)			
P Supply (mg/liter)	Leaf dry wt (g/leaf)	Lipid	Nucleic acid	Ester	Inorganic
2	0·82	32	74	36	33
6	1·08	83	134	91	83
8	1·10	89	133	104	123
20	1·06	91	142	109	338

[a]Based on Kakie (1969).

In grains and seeds, however, the level of P_i at maturity is very low and most of the phosphate is present as phytate. In these organs phytate is usually the only phosphorus fraction to be affected by a change in phosphate supply (Michael *et al.*, 1980). During the early stages of development, the phytate level is low, but it rises sharply during the period of rapid starch synthesis. In contrast, the level of P_i during this period is generally low and further declines during rapid phytate formation (Fig. 8.15).

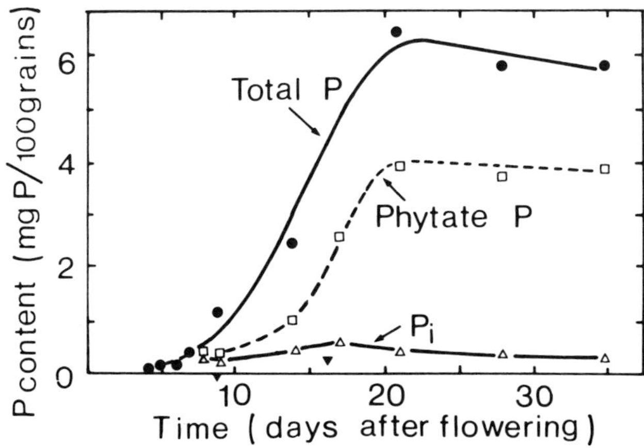

Fig. 8.15 Time course of content of inorganic phosphorus (P_i) and phytate phosphorus in rice grains during grain development. (Based on Ogawa *et al.*, 1979b.)

Phytates are salts of phytic acid, which is hexainositol phosphoric acid. Phytic acid is synthesized from the cyclic alcohol myoinositol by esterification of the hydroxyl groups with phosphoryl groups:

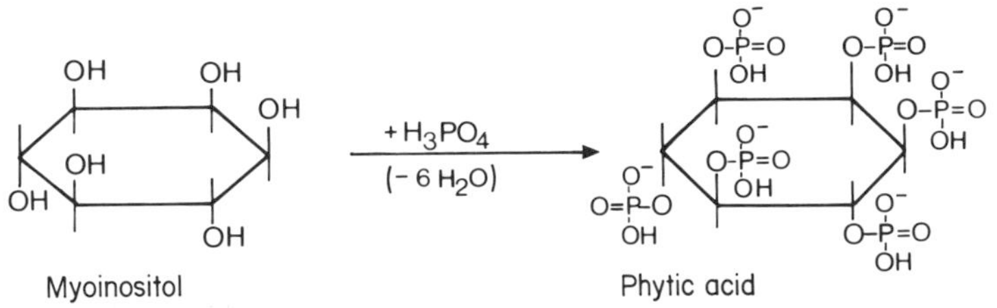

$$\text{Myoinositol} \xrightarrow[(-6\,H_2O)]{+H_3PO_4} \text{Phytic acid}$$

The sparingly soluble calcium–magnesium salt of phytic acid is termed phytin. Phytic acid also has high affinity for zinc and iron. In legume seeds and cereal grains the main phytates are the potassium–magnesium salts (Lott and Buttrose, 1978; Ogawa *et al.*, 1979a), phytin being virtually absent, Phytate phosphorus makes up ~50% of the total phosphorus in legume seeds, 60–70% in cereal grains, and 86% in wheat mill bran (Lolas *et al.*, 1976). Most of the phytate is localized in the aleurone layer in cereals, in the germ in maize, or in the cotyledons in legumes. The phytates are incorporated as discrete particles ("globoid crystals") in the protein matrix of large protein bodies (Lott and Buttrose, 1978; Ogawa *et al.*, 1979b). The occurrence of phytates is not restricted to grains and seeds but can also be demonstrated, for example, in potato tubers, where it accounts for 15–30% of the total phosphorus (Quick and Li, 1976).

Some phosphorus is associated with the starch fraction and is incorporated into the starch grains. In cereals only a small proportion is involved, but in potato tubers up to 40% of the total phosphorus may be incorporated in starch. Starch-bound phosphorus may reflect another type of compartmentation of P_i and control of its concentration at the sites of starch synthesis. It could also act as a source of phosphorus for sugar export from the amyloplasts during the sprouting of tubers.

Phytates are presumably involved in the regulation of starch synthesis during grain filling or tuber growth. The synthesis of phytate and a decrease in the P_i level within the grains are closely related (Fig. 8.15; Michael *et al.*, 1980). In addition, in grains and seeds in the final stage of the filling period, with the onset of desiccation phytic acid may act as a major cation trap that eliminates excessive cellular concentrations of potassium and magnesium.

The function of phytate in seed germination is quite obvious. In the early stages of seedling growth, the embryo has a large requirement for mineral nutrients, including magnesium (necessary for phosphorylation and protein synthesis), potassium (required for cell expansion), and phosphorus (incor-

Table 8.17
Changes in Phosphorus Fractions of Rice Seeds during Germination[a]

Germination hours	Phosphorus fraction (mg P/g dry wt)				
	Phytate	Lipid	Inorganic	Ester	RNA + DNA
0	2·67	0·43	0·24	0·078	0·058
24	1·48	1·19	0·64	0·102	0·048
48	1·06	1·54	0·89	0·110	0·077
72	0·80	1·71	0·86	0·124	0·116

[a]From Mukherji *et al.* (1971).

porated in membrane lipids and nucleic acids). In agreement with this, digestion of the globoid crystals is one of the earliest observable changes in the protein bodies of cotyledon during germination (Lott and Vollmer, 1973). Degradation of phytates, catalyzed by phytases, leads to a rapid decline in phytate-bound phosphorus (Table 8.17).

Within the first 24 hr most of the phosphorus released from phytate is incorporated into phospholipids, indicating membrane reconstitution, which is essential for compartmentation and thus for the regulation of metabolic processes within the cells. An increase in P_i and phosphate ester levels reflects the onset of intensive respiration, phosphorylation, and related processes. The degradation of phytate continues with time, and finally the levels of DNA and RNA phosphorus increase, indicating enhancement of cell division and net protein synthesis. The rate of phytate degradation is also controlled by P_i; high levels of P_i repress the synthesis of phytase (Sartirana and Bianchetti, 1967).

Phytates have attracted considerable attention among nutritionists. These compounds interfere with intestinal resorption of mineral elements, especially zinc but also iron and calcium. They thereby cause nutritional deficiencies in both monogastric animals (Welch *et al.*, 1974) and humans, especially children (Hambidge and Walravens, 1976). Of a given supply, the amount of zinc resorbed by the intestine is determined by the zinc/phytate ratio in the diet (Lantzsch *et al.*, 1981). In man, on cereal diets, zinc deficiency results from both the low zinc content of the grains and the consumption of phytate-rich unleavened whole-wheat bread (Reinhold *et al.*, 1973). This unfavorable effect of phytates has led human nutritionists to take the questionable step of classifying phytates as phytotoxicants (Overleas, 1973). This problem can be alleviated by zinc supplementation in the diet, by an increase in the zinc/phytate ratio in seeds and grains through the application of zinc fertilizers (Peck *et al.*, 1980), and perhaps by the selection of genotypes with lower phytate contents (Griffiths and Thomas, 1981).

8.4.6 Phosphorus Supply, Plant Growth, and Plant Composition

The phosphorus requirement for optimal growth is in the range of 0·3 to 0·5% of the plant dry weight during the vegetative stages of growth. Plants suffering from phosphorus deficiency exhibit retarded growth, and often a reddish coloration occurs because of enhanced anthocyanin formation. Phosphorus-deficient plants also often have a darker green color than do normal plants. Under conditions of phosphorus deficiency, cell and leaf expansion are more retarded than chlorophyll formation, the chlorophyll content per unit leaf area therefore being higher (Hecht-Buchholz, 1967), but the photosynthetic efficiency per unit of chlorophyll is much lower (Tombesi *et al.*, 1969). Inhibition of leaf cell expansion is particularly expressed during the daytime and caused by decreased root hydraulic conductance in the phosphorus-deficient plants (Section 5.6.4; Radin and Eidenbock, 1984).

Because of the functions of phosphorus in the growth and metabolism of plants, deficiency leads to a general reduction of most metabolic processes, including cell division and expansion, respiration, and photosynthesis (Terry and Ulrich, 1973). The regulatory function of P_i in photosynthesis and carbohydrate metabolism of leaves (Fig. 8.14) can be considered to be one of the major factors limiting growth, particularly during the reproductive stage. The level of phosphorus supply during this period regulates the starch/sucrose ratio in the source leaves and the partitioning of photosynthates between the source leaves and the reproductive organs (Giaquinta and Quebedeaux, 1980). Presumably, this effect of phosphorus on partitioning is also responsible, in part, for the insufficient photo-synthate supply to nodulated roots of phosphorus-deficient legumes and the occurrence of nitrogen deficiency as a dominant symptom in N_2-fixing legumes receiving deficient levels of phosphorus.

Any interpretation of the effects of phosphorus supply on plant growth and development must take into consideration the effects of phosphorus on the phytohormone balance (Chapter 5). This is particularly true of the relationship between phosphorus deficiency and a decrease in the number of flowers (Bould and Parfitt, 1973) and delay in flower initiation (Rossiter, 1978).

8.5 Magnesium

8.5.1 General

Magnesium is a small and strongly electropositive divalent cation with a hydrated ionic radius of 0·428 nm and a very high hydration energy of

1908 J/mol. Its rate of uptake can be strongly depressed by other cations, such as K^+, NH_4^+ (Kurvits and Kirkby, 1980), Ca^{2+}, and Mn^{2+} (Heenan and Campbell, 1981), as well as by H^+, that is, by low pH (Chapter 2). Magnesium deficiency induced by competing cations is thus a farily widespread phenomenon, although a decrease in the magnesium content of leaves (e.g., as a result of a large potassium supply) does not necessarily lead to a decrease, and may even lead to an increase in the magnesium content of fruits or storage tubers (Kirkby and Mengel, 1976).

The functions of Mg^{2+} in plants are related to its mobility within the cells, its capacity to interact with strongly nucleophilic ligands (e.g., phosphoryl groups) through ionic bonding, and to act as a bridging element and/or form complexes of different stabilities. Although most bonds involving Mg^{2+} are ionic, some are partially covalent, as in the chlorophyll molecule. Magnesium forms ternary complexes with enzymes in which bridging cations are required for establishing a precise geometry between enzyme and substrate (Clarkson and Hanson, 1980). A high proportion of the total Mg^{2+} is involved in the regulation of cellular pH and the cation–anion balance.

8.5.2 Chlorophyll Synthesis and Cellular pH Control

A major function of Mg^{2+}, and certainly its most familiar function, is its role as the central atom of the chlorophyll molecule (Fig. 5.1). In green leaves, however, only a small proportion of the total Mg^{2+} is bound to chlorophyll, and even under severe magnesium deficiency this value does not exceed ~25% (Table 8.18). About 70% of the total Mg^{2+} can be readily extracted with water.

When the supply of Mg^{2+} is optimal for growth, between 10 and 20% of the total Mg^{2+} of leaves is localized in the chloroplasts, less than half of it bound to chlorophyll; a similar proportion of K^+ (10–20%) is localized in the

Table 8.18
Effect of Magnesium Deficiency on Fully Expanded Oat Leaves[a]

Magnesium supply	Chlorophyll (mg/g dry wt)	Magnesium content (mg/g dry wt)	Percentage of the total magnesium bound to chlorophyll
$+Mg^{2+}$	10·4	5·1	10
$-Mg^{2+}$	4·5	1·0	24

[a]Based on Michael (1941).

chloroplasts. The proportion of Ca^{2+}, however, is usually much lower (Mix and Marschner, 1974). The cytoplasm, as the other compartment of the metabolic pool (see Section 8.4), may contain another 10–20% of the total Mg^{2+} and K^+. When supplied at high levels, these cations are stored in the vacuoles (the storage pool), there being a corresponding shift in the distribution among the cell compartments. The main compartments differ considerably in size. In cells of mature leaf tissue, ~15% of the whole cell volume is occupied by the chloroplasts, the cytoplasm, and the apparent free space (~5% each), the remaining 85% by the vacuole (Cowan *et al.*, 1982). Thus a proportion of 20% of the leaf Mg^{2+} localized in the chloroplasts, which occupy only 5% of the total volume, indicates a very high Mg^{2+} concentration in this compartment. High concentrations of Mg^{2+} and K^+ are required in the chloroplasts and cytoplasm to maintain a high pH between 6·5 and 7·5, compared to the much lower vacuolar pH of 5·0 to 6·0. The influence of pH on protein structure and hence enzyme activity is dependent on the strict regulation of the pH of the metabolic pool, and cations such as Mg^{2+}, K^+, and, to a limited extent, Ca^{2+} (see Section 8.6) perform this regulatory function (Smith and Raven, 1979). In the metabolic pool Mg^{2+} is also necessary for the neutralization of organic acids, the phosphoryl group of phospholipids, and particularly nucleic acids.

8.5.3 Protein Synthesis

Magnesium also has an essential function as a bridging element for the aggregation of ribosome subunits (Cammarano *et al.*, 1972), a process that is necessary for protein synthesis. When the level of Mg^{2+} is deficient, or in the presence of excessive levels of K^+ (Sperrazza and Spremulli, 1983), the subunits dissociate and protein synthesis ceases. Magnesium is also required for RNA polymerases and hence for the formation of RNA in the nucleus. This latter role might be related to both bridging between individual DNA strains and neutralization of the acid proteins of the nuclear matrix (Wunderlich, 1978).

As shown in Fig. 8.16, in response to magnesium deficiency, the net synthesis of RNA immediately stops; synthesis resumes rapidly after the addition of Mg^{2+}. In contrast, protein synthesis remains unaffected for more than 5 hr, but thereafter it rapidly declines. The requirement for Mg^{2+} in protein synthesis can also be directly demonstrated in chloroplasts (Table 8.19).

In leaf cells at least 25% of the total protein is localized in chloroplasts. This explains why a deficiency of magnesium particularly affects the size, structure, and function of chloroplasts, including the electron transfer in

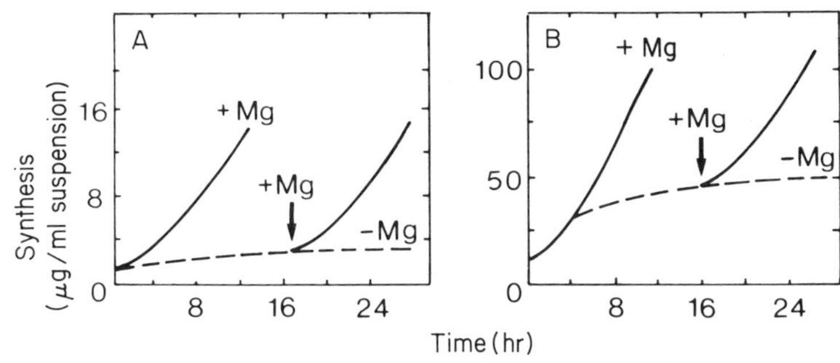

Fig. 8.16 Effect of magnesium supply on (A) RNA and (B) protein synthesis in *Chlorella pyrenoidosa* suspension culture. (Based on Galling, 1963.)

Table 8.19
Magnesium Requirement for the Incorporation of ^{14}C [Leucine] into the Protein Fraction of Isolated Wheat Chloroplasts[a]

Magnesium supply (μM)	^{14}C Incorporation (cpm/mg chlorophyll)	Relative value
0	412	11·5
0·25	688	19·5
2·50	3550	100·0

[a]Based on Bamji and Jagendorf (1966).

photosystem II (McSwain *et al.*, 1976). Lower chlorophyll contents in magnesium-deficient leaves are therefore caused by inhibited protein synthesis rather than a lack of Mg^{2+} for the synthesis of chlorophyll molecules (see Table 8.18). This also explains why in magnesium-deficient plants the other plastid pigments are often affected in the same way as chlorophyll (Table 8.20). Regardless of this decline in the concentration of chloroplast pigments, starch accumulates in magnesium-deficient chloroplasts (Vesk *et al.*, 1966) and is mainly responsible for the increase in the dry matter content of magnesium-deficient leaves (Table 8.20).

Table 8.20
Magnesium Deficiency-Induced Changes in Plastid Pigments and Leaf Dry Matter in Rape[a]

Treatment	Chlorophyll (*a* and *b*) (mg/g fresh wt)	Carotenoids (mg/g fresh wt)	Leaf dry matter (%)
Control	2·33	0·21	13·6
Magnesium-deficient	1·33	0·11	17·7

[a]Based on Baszynski *et al.* (1980).

8.5.4 Ezyme Activation and Energy Transfer

There is a long list of enzyme reactions which require or are promoted by Mg^{2+}. Most of these can be categorized by the general type of reaction to which they conform, such as the transfer of phosphate (e.g., phosphatases and ATPases) or of carboxyl groups (e.g., carboxylase). The central role of ATP in energy metabolism had been discussed in Section 8.4. The substrate for most ATPases is Mg·ATP:

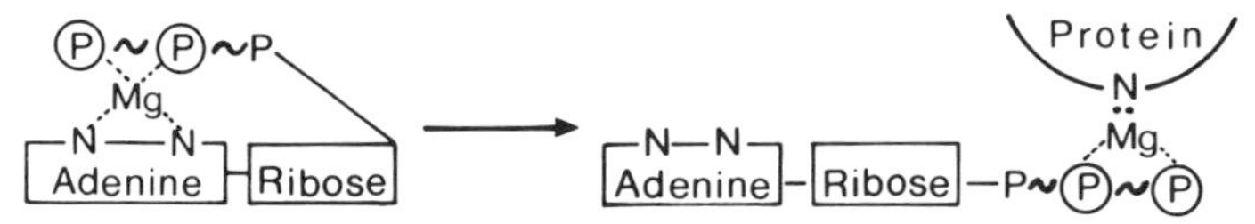

Magnesium is preferentially bound to nitrogen bases and phosphoryl groups. In the case of ATP, an Mg·ATP complex is formed with reasonable stability above pH 6 and in which most of the negative charges have been neutralized. This complex can be utilized by the active sites of ATPases for the transfer of the energy-rich phosphoryl group (Balke and Hodges, 1975). An example of the Mg^{2+} requirement of membrane-bound ATPases is shown in Fig. 8.17. It is evident that Mg·ATP rather than ATP is the substrate for plasma membrane–bound ATPase in corn roots. Maximal activity requires the presence of both Mg^{2+} and K^+. A similar requirement

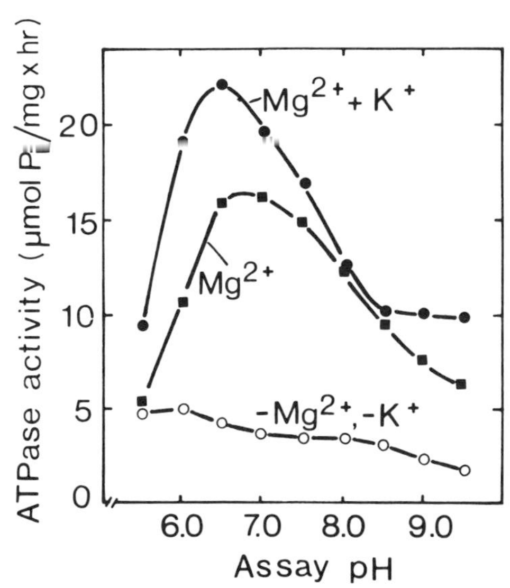

Fig. 8.17 Effect of pH, magnesium (3 mM), and potassium (50 mM) on the ATPase activity of the plasma membrane fraction of corn roots. (Based on Leonard and Hotchkiss, 1976.)

Table 8.21
Effect of Cations in the Incubation Medium on the Photophosphorylation of Isolated Pea Chloroplasts[a]

Cation in incubation medium[b]	Photophosphorylation rate (μmol ATP formed/mg chlorophyll $\times$ hr)
None	12·3
5 mM Mg^{2+}	34·3
5 mM Ca^{2+}	4·3

[a]Based on Lin and Nobel (1971).
[b]Incubation medium contained ADP, P_i, and cation indicated.

for Mg^{2+} and additional stimulation by K^+ can be demonstrated for ATP-dependent sucrose transport into vacuoles of the storage roots of beet (Chapter 6).

The synthesis of ATP (phosphorylation: ADP + $P_i \rightarrow$ ATP) has an absolute requirement for Mg^{2+} as a bridging component between ADP and the enzyme. As shown in Table 8.21 ATP synthesis in isolated chloroplasts is increased considerably by external supplies of Mg^{2+}. Because the endogenous Mg^{2+} content of chloroplasts is relatively high even in the control (no added cations), the additional Mg^{2+} supply can only further stimulate photophosphorylation. The addition of Ca^{2+} severely inhibits photophosphorylation; that is, a low Ca^{2+} concentration also has to be maintained *in vivo* within the chloroplasts at the sites of photophosphorylation.

Another key reaction of Mg^{2+} is the modulation of RuBP carboxylase in the stroma of chloroplasts. The activity of this enzyme is highly dependent on both Mg^{2+} and pH (Fig. 8.18A). Binding of Mg^{2+} to the enzyme increases its affinity K_m for the substrate CO_2 and the turnover rate V_{max} (Sugiyama *et al.*, 1968). Magnesium also shifts the pH optimum of the reaction towards the physiological range (below 8). In chloroplasts the light-triggered activa-

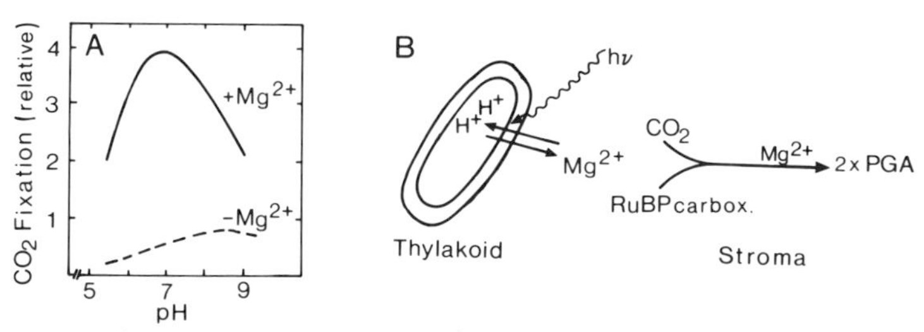

Fig. 8.18 A. Activation of ribulose-1,5-bisphosphate carboxylase from spinach leaves by magnesium. (Redrawn from Sugiyama *et al.*, 1969). B. Model for light-induced magnesium transport from the thylakoids into the stroma of chloroplasts with subsequent enhancement of CO_2 fixation.

tion of RuBP carboxylase is related to increases both in pH and in the Mg^{2+} concentration of the stroma. As shown in Fig. 8.18B, upon illumination, protons move rapidly from the stroma into the thylakoids and the stromal pH increases from ~6·0–6·5 to ~7·5–8·0. The efflux of protons from the stroma is counterbalanced by an influx of Mg^{2+} from the thylakoid spaces. This light-triggered reaction increases the Mg^{2+} concentration of the stroma from ~2 m*M* in the dark to 4 m*M* in the light (Portis and Heldt, 1976; Portis, 1981). Changes of this magnitude in both pH and Mg^{2+} concentration are sufficient to increase the activity of the RuBP carboxylase and also of other stromal enzymes which depend on higher Mg^{2+} concentrations and which have a pH optimum above 6.

One of the key enzymes with a high Mg^{2+} requirement and high pH optimum is fructose–1,6-diphosphatase, which regulates assimilate partitioning between starch synthesis and triosephosphate export in the chloroplasts (Baièr and Latzko, 1976). Another key enzyme with similar Mg^{2+} requirements is glutamate synthetase (O'Neal and Joy, 1974). A light-induced increase in nitrite reduction and thus NH_3 production requires a simultaneous increase in the activity of enzymes such as glutamate synthetase regulating ammonia assimilation within the chloroplasts. Thus the model of regulation for CO_2 fixation and reduction (Fig. 8.18B) also holds true, in principle, for nitrite reduction and ammonia assimilation. A light-induced increase in stromal pH also causes an influx of abscisic acid from the cytoplasm into the stroma (Cowan *et al.*, 1982). The importance of this change is not yet clear.

Only a relatively small proportion of the total plant Mg^{2+} is required for the various functions of Mg^{2+} in the chloroplasts and cytoplasm. In vacuolated cells a majority of the Mg^{2+} cations act as counterions for organic acid anions and inorganic anions stored in the vacuoles and for pectates in the middle lamella of cell walls. In general the proportion of wall-bound Mg^{2+} is lower than that of Ca^{2+}; it may increase from a low percentage in calcium-sufficient plants to more than 25% in calcium-deficient plants (Kirkby and Mengel, 1976).

8.5.5 Magnesium Supply, Plant Growth, and Plant Composition

On average, the magnesium requirement for optimal plant growth is ~0·5% of the dry weight of the vegetative parts. Chlorosis of fully expanded leaves is the most obvious visible symptom of magnesium deficiency. In accordance with the function of Mg^{2+} in protein synthesis, the proportion of protein nitrogen is depressed and that of nonprotein nitrogen increased in magnesium-deficient leaves. As calculated on both unit leaf area and unit chlorphyll, the rate of photosynthesis is lower in magnesium-deficient plants than in normal plants, as is the respiration rate (Bottrill *et al.*, 1970; Terry and Ulrich, 1974). The appearance of slight and transient magnesium-deficiency symptoms during the vegetative growth stage, however, is not necessarily associated with a depression of the final yield unless irreversible changes, such as a reduction in grain number per ear in cereals, occur (Forster, 1980).

The accumulation of starch in the leaves of magnesium-deficient plants is mainly responsible for the higher dry matter content of these leaves (Table 8.20), indicating that photosynthesis per se is less impaired than starch degradation in the chloroplasts, sugar transport within the cells, and/or phloem loading of sucrose. All these metabolic processes have a high requirement for energy-rich phosphates and thus also for Mg^{2+} for energy transfer. Impaired transport of photosynthates from source leaves to sinks such as roots, fruits, or storage tubers is therefore a main consequence of magnesium deficiency. This effect of Mg^{2+} on photosynthate partitioning is shown in Table 8.22 for subterranean clover. Root growth is affected by magnesium deficiency to a much greater extent than shoot growth,

Table 8.22
Effect of Magnesium or Potassium Deficiency on Growth and Dry Matter Distribution of Subterranean Clover[a]

		Relative distribution of dry weight		
Treatment	Dry weight (g/plant)	Lamina	Petioles	Roots
Control	0·85	42	29	29
$-Mg^{2+}$	0·50	53	28	19
$-K^{+}$	0·64	44	27	29

[a]Based on Bouma *et al.* (1979).

particularly lamina growth, leading to an increase in the shoot/root ratio. In contrast, potassium deficiency affects different plant parts similarly. In nodulated legumes, magnesium deficiency may be assumed to have a corresponding detrimental effect on carbohydrate supply to the root nodules and thus on the N_2 fixation rate.

When magnesium is deficient, there is a decrease in the starch content of storage tissues such as potato tubers (Werner, 1959) and in the single-grain weight of cereals (Beringer and Forster, 1981). These effects result mainly from impaired assimilate partitioning. In cereal grains, however, Mg^{2+} might play an additional role in the regulation of starch synthesis through its effect on the level of P_i and formation of Mg-K-phytate. As has been shown in Section 8.4.4, high P_i levels inhibit starch synthesis. In magnesium-deficient wheat grains, twice as much phosphorus remained as P_i, and there was a correspondingly smaller proportion of phytate phosphorus, compared to the grains adequately supplied with magnesium (Beringer and Forster, 1981).

Increasing the Mg^{2+} supply beyond the growth-limiting level results in additional Mg^{2+} being stored, predominantly in the vacuoles as inorganic salts. In most instances, elevated magnesium contents improve the nutritional quality of plants. For example, inadequate contents in forage grasses are mainly responsible for grass tetany (Grunes *et al.*, 1970). In intensive agriculture there is increasing concern about the tendency toward a decreasing magnesium content in crop plants (Arzet, 1972) and a corresponding increase in magnesium deficiency-related disorders such as grape stem rot—*Stiellähme* (Bergmann, 1983). Insufficient Mg^{2+} intake in the human diet leading to a magnesium-deficiency syndrome has also attracted considerable attention.

8.6 Calcium

8.6.1 General

Calcium is a relatively large divalent cation with a hydrated ionic radius of 0·412 nm and a hydration energy of 1577 J/mol. It readily enters the apoplast and is bound in an exchangeable form to cell walls and at the exterior surface of the plasma membrane. Its rate of uptake into the cytoplasm is severely restricted and seems to be only loosely coupled to metabolic processes. The mobility of Ca^{2+} from cell to cell and in the phloem is very low, and it is the only mineral nutrient other than, possibly, boron which functions mainly outside the cytoplasm in the apoplast. Most of its activity is related to its capacity for coordination, by which it provides stable but reversible intermolecular linkages, predominantly in the cell walls and at the plasma membrane. These Ca^{2+}-mediated linkages respond to local changes in environmental conditions and are part of the control mechanism for growth and developmental processes. Calcium is a non-toxic mineral nutrient, even in high concentration, and is very effective in detoxifying high concentrations of other mineral elements in plants. For a comprehensive review of the function of calcium as a plant nutrient, see Hanson (1984) and Kirkby and Pilbeam (1984).

8.6.2 Cell Wall Stabilization

In contrast to the other macronutrients, a high proportion of the total Ca^{2+} in plant tissue is located in the cell walls (apoplast). This unique distribution

Table 8.23
Relationship between Calcium Supply and Proportion of Total Calcium in Various Binding Forms in Young Sugar Beet Plants[a]

Binding form of calcium	Calcium supply (mEq/liter)	
	0·33	5·0
Water soluble	27	19
Pectate	51	31
Phosphate	17	19
Oxalate	4	25
Residue	1	6

[a]Based on Mostafa and Ulrich (1976).

is the result of an abundance of binding sites for Ca^{2+} in the cell walls as well as the restricted transport of Ca^{2+} across the plasma membrane into the cytoplasm. In the middle lamella, it is bound to $R{\cdot}COO^-$ groups of polygalacturonic acids (pectins) in a more or less readily exchangeable form (Chapter 2). In dicotyledons, which have a large cation-exchange capacity, and particularly when the level of Ca^{2+} supply is low, up to 50% of the total Ca^{2+} can be bound as pectates (Table 8.23; Armstrong and Kirkby, 1979b). In the storage tissue of apple fruits, the cell wall-bound fraction of Ca^{2+} can make up as much as 90% of the total (Faust and Klein, 1974).

In sugar beet plants well supplied with Ca^{2+}, the proportion in the pectate fraction decreases mainly at the expense of the oxalate fraction (Table 8.23). Calcium oxalate and most of the calcium phosphate are precipitated in the vacuoles. Moreover, most of the water-soluble Ca^{2+} is in the vacuoles and is accompanied by organic acid anions (e.g., malate) and inorganic ions (e.g., nitrate). For reasons to be discussed later, the Ca^{2+} concentration has to be extremely low in the cytosol.

The typical distribution of Ca^{2+} in cells of fully expanded tissue is shown in Fig. 8.19. There are two distinct areas in the cell wall with high Ca^{2+} concentrations: the middle lamella and the exterior surface of the plasma

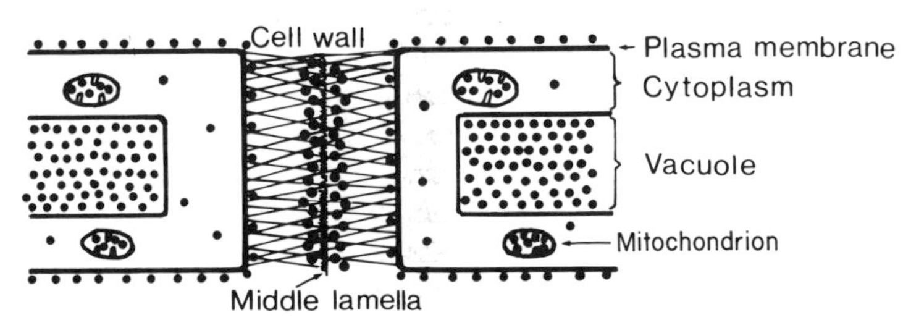

Fig. 8.19 Schematic representation of two adjacent cells with a typical distribution of Ca^{2+} (•).

Table 8.24
Effect of Calcium on the Hydrolysis of Sodium Pectate by Polygalacturonase[a]

Ca^{2+} Concentration (mg/liter)	Amount of galacturonic acid released (μmol/4 hr)
0	3·5
40	2·5
200	0·6
400	0·2

[a]Based on Corden (1965).

membrane. In both areas Ca^{2+} has essential structural functions, namely, the regulation of membrane permeability and related processes (Section 8.6.4) and the strengthening of the cell walls, respectively. The degradation of pectates is mediated by polygalacturonase, which is drastically inhibited by high Ca^{2+} concentrations (Table 8.24). In agreement with this, in calcium-deficient tissue polygalacturonase activity is increased (Konno *et al.*, 1984), and a typical symptom of calcium deficiency is the disintegration of cell walls and the collapse of the affected tissues, such as the petioles and upper parts of the stems (Bussler, 1963).

In the leaves of plants receiving high levels of Ca^{2+} during growth or grown under conditions of high light intensity, a large proportion of the pectic material exists as calcium pectate. This makes the tissue highly resistant to degradation by polygalacturonase (Cassells and Barlass, 1976). The proportion of calcium pectate in the cell walls is of importance for determining the susceptibility of the tissue to fungal infections (Chapter 11) and for the ripening of fruits. As shown by Rigney and Wills (1981) in experiments with tomato pericarp tissue during fruit development, the Ca^{2+} content of the cell walls increases to the fully grown immature stage, but this is followed by a drop in the content just before the onset of ripening ("softening" of the tissue). Simultaneously a shift in the binding stage of Ca^{2+} occurs in which water-soluble Ca^{2+} is favored over wall-bound Ca^{2+}. This shift is associated with a sharp increase in ethylene formation in the fruit tissue. Increasing the Ca^{2+} content of fruits, for example, by spraying several times with calcium salts during fruit development or by post-harvest dipping in $CaCl_2$ solutions, leads to an increase in the firmness of the fruit (Cooper and Bangerth, 1976) and delays or even prevents fruit ripening (Wills *et al.*, 1977). The importance of Ca^{2+} for fruit ripening can also clearly be seen from a comparison between a normally ripening cultivar (Rutgers) and a nonripening (*rin*) mutant of tomato (Table 8.25). In the *rin* mutant there is an increase in the content of bound Ca^{2+} during fruit maturation, whereas in the Rutgers cultivar the total content remains constant and the

Table 8.25
Calcium Content of the Pericarp Tissue of Nonripening Mutant *rin* and Normal Rutgers Tomatoes at Different Stages of Development[a]

Days after anthesis	Soluble calcium		Bound calcium	
	rin	Rutgers	*rin*	Rutgers
40	299	349	530	562
50	412	602	667	246
60	492	622	1357	291

[a]Calcium content expressed as micrograms per gram dry weight. Based on Poovaiah (1979).

content of bound Ca^{2+} declines. This decline was found to be associated with an increase in polygalacturonase activity (Poovaiah, 1979).

These results strongly support the view that solubilization of Ca^{2+} binding sites in the cell walls and subsequent redistribution of Ca^{2+} within the cells (probably via transport to the vacuoles) depresses polygalacturonase and also activates the ethylene-generating system, which is located in a cell wall–plasma membrane complex, that is, at the membrane–cell wall interface (Mattoo and Lieberman, 1977). Intracellular redistribution of Ca^{2+} is therefore the primary stimulus of, or acts as a secondary messenger (Marmé, 1983) in, the onset of the ripening process of fleshy fruits (Rigney and Wills, 1981).

8.6.3 Cell Extension

In the absence of an exogenous Ca^{2+} supply, root extension ceases within a few hours (Fig. 8.20). This effect is more distinct in a Ca^{2+}-free nutrient

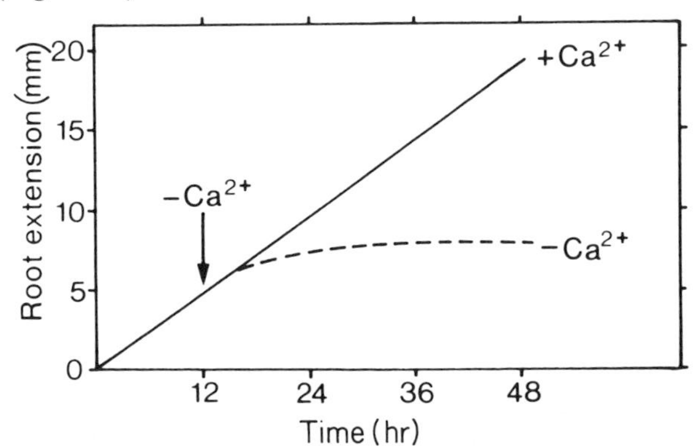

Fig. 8.20 Relationship between the extension of primary roots of bean and the calcium concentration (±2 m*M*) in the nutrient solution. (Based on Marschner and Richter, 1974.)

solution than in distilled water, an observation consistent with the function of Ca^{2+} in counterbalancing the harmful effects of high concentrations of other ions at the plasma membrane. Although Ca^{2+} is also involved in cell division (Burström 1968; Schmit, 1981), the cessation of root growth in the absence of exogenous Ca^{2+} is primarily the result of inhibited cell extension. The role of Ca^{2+} in cell extension is not yet clear; there are indications, however, that it is required for the incorporation of material into the cell walls (Reiss and Herth, 1979).

Pollen tube growth is also dependent on the presence of Ca^{2+} in the growth medium, and the direction of growth of the pollen tube is chemotropically controlled by the extracellular Ca^{2+} gradient (Mascarenhas and Machlis, 1964). The Ca^{2+} level is highest in the growing pollen tip (Reiss and Herth, 1978), where the plasma membrane fuses with the secretory vesicles, which carry structural material for cell wall formation (Reiss and Herth, 1979).

How Ca^{2+}, auxin (IAA), and extension growth are interrelated is still a matter of controversy (Burström, 1968; Hertel, 1983). Auxin is involved in Ca^{2+} transport within plant tissue, and the inhibition of Ca^{2+} transport or a decline in the level of auxin (e.g. by TIBA application; see Chapter 2) induces calcium-deficiency symptoms (for a review, see Bangerth, 1979). Moreover, auxin-induced enhancement of proton excretion, a prerequisite for the wall-loosening process and thus for cell extension, requires exogenous Ca^{2+} (Cohen and Nadler, 1976). At the same time, Ca^{2+} stabilizes the cell walls, an action that is contrary to auxin-induced extension growth (Cleland and Rayle, 1977). Consistent with this is the observation that at the onset of gravistimulated growth of coleoptiles, Ca^{2+} redistribution in the apoplast occurs with Ca^{2+} movement from the lower (extending) region of the coleoptile toward the upper region (Slocum and Roux, 1983). The crucial role of auxin-induced redistribution of Ca^{2+} in gravistimulation has also been demonstrated in roots. Auxin-inhibitors like TIBA inhibit both Ca^{2+} redistribution and gravitropism (Lee *et al.*, 1984). These results reflect the complicated control function of Ca^{2+} in root extension. It should be stressed that in intact plants root extension growth is decreased only slightly by high exogenous Ca^{2+} concentrations.

8.6.4 Membrane Stabilization and Enzyme Modulation

The fundamental role of Ca^{2+} in membrane stability and cell integrity is reflected in various ways. It can be demonstrated most readily by the increased leakage of low-molecular-weight solutes from cells of calcium-deficient tissue (e.g., tomato fruits; Goor, 1968) and, in severely deficient

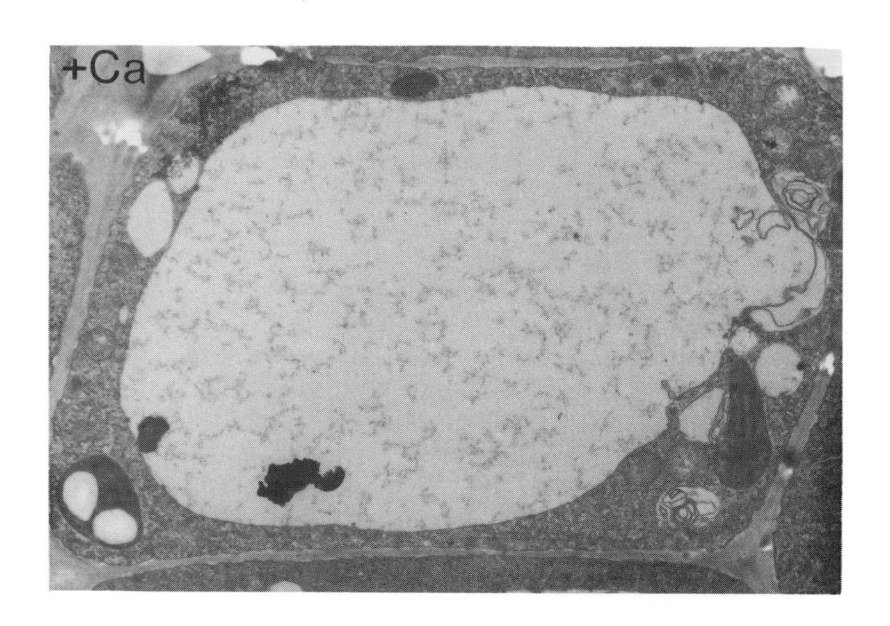

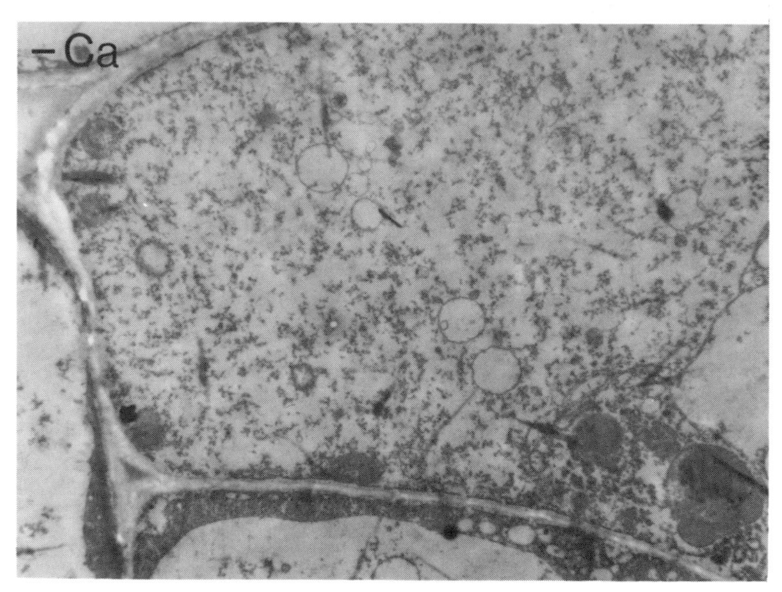

Fig. 8.21 Calcium nutritional status and fine structure of cells of potato sprouts. (Top) Calcium-sufficient. (Bottom) Calcium-deficient; loss of compartmentation. (Courtesy of C. Hecht-Buchholz.)

plants, by a general disintegration of membrane structures (Hecht-Buchholz, 1979) and a loss of cell compartmentation (Fig. 8.21).

In calcium-deficient tissues increased respiration rates are related to enhanced leakage of respiratory substrates from vacuoles to the respiratory enzymes in the cytoplasm (Bangerth *et al.*, 1972). Calcium treatment of deficient tissues therefore decreases respiration rates; it also enhances the net rate of protein synthesis (Faust and Klein, 1974). These features of calcium deficiency are similar to those related to senescence. It is now well established that Ca^{2+}, as well as phytohormones, is involved in the regulation of senescence. Senescence of isolated maize leaves can be deferred, for example, by the addition of cytokinins or Ca^{2+}, and the effects of both substances are additive (Poovaiah and Leopold, 1973).

Calcium stabilizes cell membranes by bridging phosphate and carboxylate groups of phospholipids (Caldwell and Haug, 1981) and proteins, preferentially at membrane surfaces (Legge *et al.*, 1982). There can be an exchange between Ca^{2+} at these binding sites and other cations (e.g., K^+, Na^+, or H^+), although the latter cannot replace Ca^{2+} in its membrane stabilization role; even the divalent cation Mg^{2+} cannot replace Ca^{2+} here (Van Steveninck, 1965). To fulfil its functions at the plasma membrane, therefore, Ca^{2+} must always be present in the external solution, where it regulates the selectivity of ion uptake (Chapter 2) and prevents solute leakage from the cytoplasm. The membrane-protecting effect of Ca^{2+} is most prominent under various stress conditions such as low temperature (Zsoldos and Karvaly, 1978) and anaerobiosis (Table 8.26).

In contrast to Mg^{2+}, which is a strong activator of enzymes (Section 8.6), Ca^{2+} increases the activity of only a few enzymes. These include α-amylase, phospholipases, and ATPases (Wyn Jones and Lunt, 1967). In case of α-amylase, the effect of Ca^{2+} depends on the presence of the phytohormone GA (Jones and Carbonell, 1984) and is more likely a reflection of a Ca^{2+}/calmodulin induced enhancement of synthesis and secretion of this

Table 8.26
Influence of Calcium on Carbohydrate Loss from Cotton Roots[a]

Treatment			Carbohydrate loss
Aeration	Temperature (°C)	Solution	(μg/seedling × 4 hr)
O_2	31	Distilled water	18
O_2	5	Distilled water	57
O_2	5	10^{-5} M Ca^{2+}	7
N_2	31	Distilled water	89
N_2	31	10^{-5} M Ca^{2+}	7

[a]Based on Christiansen *et al.* (1970).

enzyme by the cells (Mitsui *et al.*, 1984). In general Ca^{2+} stimulates membrane-bound enzymes (Rensing and Cornelius, 1980), particularly ATPases at the plasma membrane of roots of certain plant species (Kuiper *et al.*, 1974). Because the activities of many membrane-bound enzymes are modulated by membrane structure, Ca^{2+} presumably enhances the activity of those enzymes even though it is bound to non-catalytic sites at the membranes (Clarkson and Hanson, 1980). On the other hand, the inhibitory effects of Ca^{2+} on various enzymes located in the cytoplasm and in the chloroplasts have been well documented. Hexodiphosphatase (Baier and Latzko, 1976) and PEP carboxylase (Table 8.27) are two examples of enzymes inhibited by Ca^{2+}

Table 8.27
Effect of Calcium on Phosphoenolpyruvate Carboxylase from Leaves of C_4 and C_3 Plants[a]

Type of CO_2 fixation pathway	Plant species	V_{max} (μmol CO_2/g fresh wt × min)	Inhibition of V_{max} by 1 mM Ca^{2+} (%)
C_4	*Amaranthus viridis*	42.4	57
	Atriplex tatarica	31·4	70
C_3	*Spinacea oleracea*	1·1	34
	Stellaria media	0·5	39

[a]From Gavalas and Manetas (1980).

Because PEP carboxylase is the key enzyme for CO_2 fixation in C_4 species (Chapter 5), a very low level of free Ca^{2+} in the chloroplasts is a prerequisite for a high photosynthetic rate. In C_3 species as well, CO_2 fixation ceases abruptly if the concentration of Ca^{2+} in the stroma of the chloroplasts reaches the level of Mg^{2+} (Portis and Heldt, 1976).

8.6.5 Regulation of Intracellular Calcium Distribution

There is general agreement that the level of free Ca^{2+} in the cytoplasm and in the chloroplasts is very low, probably 1 μM or less (Wyn Jones and Pollard, 1983). Such low levels must be maintained in order to prevent precipitation of P_i, competition with Mg^{2+} for binding sites, and inactivation or uncontrolled activation of certain enzymes. The question arises as to how cells can control the cytoplasmic levels of free Ca^{2+}. Some of the possibilities are shown in Fig. 8.22. The plasma membrane presents an effective barrier to Ca^{2+} penetration. Whether additional active Ca^{2+}-efflux pumps exist at the plasma membranes of plant cells is still a matter of speculation.

Increasing evidence has been presented that the Ca^{2+}-binding protein calmodulin, which plays a key role in enzyme regulation in animal cells

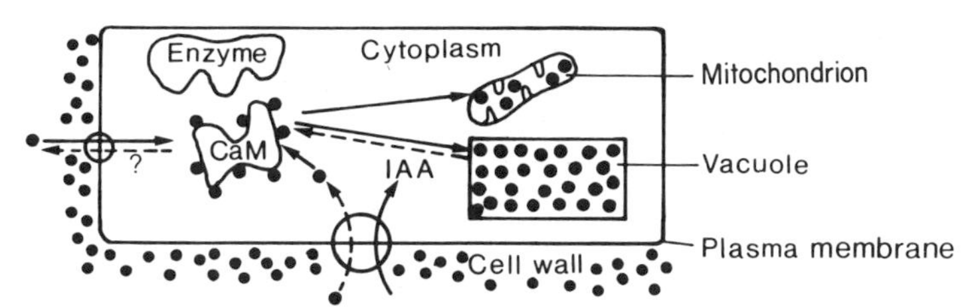

Fig. 8.22 Model of intracellular regulation of calcium (•) concentration via calmodulin (CaM) and auxin (IAA).

(Means and Dedman, 1980), is also important in plant cells for both the regulation of free Ca^{2+} in the cytosol and enzyme activation (Marmé, 1983; Dieter, 1984). Calmodulin is a low-molecular-weight compound (MW $\approx$ 20,000) which binds Ca^{2+} reversibly and with high affinity and selectivity. It activates a number of key enzymes such as phospholipase (Leshem *et al.*, 1984) and NAD kinase, probably by the formation of calcium/calmodulin complexes with the enzymes (Dieter, 1984). It might also control Ca^{2+} transport within the cell as well as mediate Ca^{2+} transfer to the vacuoles (Marmé, 1983). Competition with Ca^{2+} for binding sites on calmodulin has been discussed in relation to aluminum toxicity in barley roots (Siegel and Haug, 1983). Le Gales *et al.* (1980) found high levels of Ca^{2+}-binding proteins (calmodulin?) in the cytoplasm of calcicole plant species but only low levels in calcifuge species. This might reflect a mechanism for the adaptation of plants to calcareous soils (Chapter 16). By such a mechanism the plants may maintain a low level of free Ca^{2+} in the cytoplasm of root cells. There are also striking similarities to another type of adaptation which involves the synthesis of special heavy-metal-binding proteins (Section 8.3).

Whether calmodulin also controls the Ca^{2+} influx at the plasma membrane is not yet clear. Hertel (1983) has offered an interesting hypothesis on the reverse flux of Ca^{2+} and auxin through the plasma membrane mediated by active auxin transport, which opens "gates" for Ca^{2+} influx (Fig. 8.22) along the electrochemical gradient (Chapter 2). This hypothesis would help to explain some of the interactions between the transport and action at the cellular level of both auxin and Ca^{2+}, particularly in relation to cell extension.

Mitochondria are rich in Ca^{2+}, and this mineral nutrient has an important role in their structure and function. The accumulation of Ca^{2+} in mitochondria provides another means of intracellular Ca^{2+} transport and control of Ca^{2+} concentration in the cytosol (Marmé, 1983). The release of Ca^{2+} from mitochondria and from vacuoles, or both, and subsequent changes in the membrane permeability induced by Ca^{2+} are most likely an essential step in the turgor-regulated seismonastic movement of the leaflets of *Mimosa* (Drissche, 1978; see also Section 8.7).

8.6.6 Cation–Anion Balance and Osmoregulation

In vacuolated cells of leaves in particular, a large proportion of Ca^{2+} is localized in the vacuoles, where it contributes to the cation–anion balance by acting as a counterion for inorganic and organic anions (Chapter 2). In plant species which preferentially synthesize oxalate in response to nitrate reduction, the formation of calcium oxalate in the vacuoles helps to maintain a low level of free Ca^{2+} in the cytosol and also in the chloroplasts (Mix and Marschner, 1974). The formation of sparingly soluble calcium oxalate is also important for the osmoregulation of cells and provides a means of salt accumulation in vacuoles of nitrate-fed plants without increasing the osmotic pressure in the vacuoles (Osmond, 1967). In mature sugar beet leaves, for example, up to 90% of the total calcium is bound to oxalate. Calcium oxalate is also abundant in the leaves of many halophytes.

8.6.7 Calcium Supply, Plant Growth, and Plant Composition

The calcium content of plants varies between 0·1 and >5·0% of dry wt depending on the growing conditions, plant species, and plant organ. The calcium requirement for optimum growth is much lower in monocotyledons than in dicotyledons (Loneragan *et al.*, 1968; Loneragan and Snowball, 1969), as shown in Table 8.28. In well-balanced, flowing nutrient solutions with controlled pH, maximal growth rates were obtained at Ca^{2+} supply levels of 2·5 (ryegrass) and 100 μM (tomatoes), i.e., differing by a factor of

Table 8.28
Effect of Calcium Concentration in the Nutrient Solution on Relative Growth Rates of Plants and Calcium Content in the Shoots[a]

	Ca^{2+} Concentration (μM)				
Plant species	0·8	2·5	10	100	1000
Relative growth rate					
Ryegrass	42	100	94	94	93
Tomato	3	19	52	100	80
Calcium content (mg/g dry wt)					
Ryegrass	0·6	0·7	1·5	3·7	10·8
Tomato	2·1	1·3	3·0	12·9	24·9

[a]Based on Loneragan *et al.* (1968) and Loneragan and Snowball (1969).

40. This difference is primarily a reflection of the calcium demand at the tissue level, which is lower in ryegrass (0·7 mg) than in tomato (12·9 mg). Genotypical differences in Ca^{2+} requirement are closely related to the Ca^{2+} binding sites in the cell walls, that is, the cation-exchange capacity. This also explains why the Ca^{2+} requirement of some algal species is in the micronutrient range or even difficult to demonstrate at all (O'Kelley, 1969).

Another factor determining the Ca^{2+} requirement for optimum growth is the concentration of other cations in the external solution (Wyn Jones and Lunt, 1967; Burström, 1968). Because Ca^{2+} is readily replaced by other cations from its binding sites at the exterior surface of the plasma membrane, the Ca^{2+} requirement increases with increasing external concentrations of heavy metals (Wallace *et al.*, 1966), sodium chloride (LaHaye and Epstein, 1971), or protons (Table 8.29). At low compared to high pH the Ca^{2+} concentration in the external solution has to be several times higher in order to counteract the adverse effect of high H^+ concentrations on root elongation. A similar relationship exists between external pH and the Ca^{2+} requirement for nodulation of legumes (Munns, 1970). In order to protect roots against the adverse effects of high concentrations of various other cations in the soil solution, the Ca^{2+} concentration has to be much higher than in balanced flowing nutrient solutions (Asher and Edwards, 1983).

An increase in the concentration of Ca^{2+} in the external solution leads to an increase in the calcium level in the leaves but not necessarily in low-transpiring organs such as fleshy fruits and tubers supplied predominantly via the phloem. Plants have developed mechanisms for restricting the transport of calcium to these organs by maintaining low calcium concentrations in the phloem sap (Chapter 3) or by precipitation of Ca^{2+} as oxalate along the conducting vessels (Liegel, 1970) or in the seed coat (Mix and Marschner, 1976c). Dilution of calcium concentration by growth is another way of maintaining a low Ca^{2+} level, which is necessary in fruits and storage tissues for rapid cell expansion and high membrane permeability (Mix and Marschner, 1976a). High growth rates of low-transpiring organs, however,

Table 8.29
Effects of Calcium Concentration and Solution pH on the Growth Rate of Seminal Roots of Soybean[a]

Ca^{2+} Concentration (mg/liter)	Root growth rate (mm/hr)	
	pH 5·6	pH 4·5
0·05	2·66	0·04
0·50	2·87	1·36
2·50	2·70	2·38

[a]Based on Lund (1970).

enzymes increase the rate of catalytic reactions, V_{max}, and in some cases also the affinity for the substrate, K_m (Evans and Wildes, 1971).

In potassium-deficient plants some gross chemical changes occur, including an accumulation of soluble carbohydrates, a decrease in levels of starch, and an accumulation of soluble nitrogen compounds. It is possible to relate the changes in carbohydrate metabolism to the high K^+ requirement of certain regulatory enzymes, particularly pyruvate kinase and 6-phosphofructokinase (Läuchli and Pflüger, 1978). As shown in Fig. 8.23 the activity of starch synthase is also highly dependent on univalent cations, and of these K^+ is the most effective. The enzyme catalyzes the transfer of glucose to starch molecules:

$$\text{ADP-Glucose} + \text{starch} \rightleftharpoons \text{ADP} + \text{glocosyl-starch}.$$

Potassium similarly activates starch synthase isolated from a variety of other plant species and organs (e.g., leaves, seeds, and tubers), the maximum lying in the range of 50 to 100 m*M* K^+ (Nitsos and Evans, 1969). Higher concentrations, however, may have inhibitory effects (Preusser *et al.*, 1981) and perhaps contribute to the decline in the starch content of potato tubers at high K^+ concentrations (Section 8.7.8).

Another key function of K^+ is the activation of membrane-bound ATPases, which require Mg^{2+} but are further stimulated by K^+ (Section 8.5) and, in some cases, to a similar extent by Na^+ (Fisher and Hodges, 1969; see Section 10.2). The activation of ATPases by K^+ not only facilitates its own

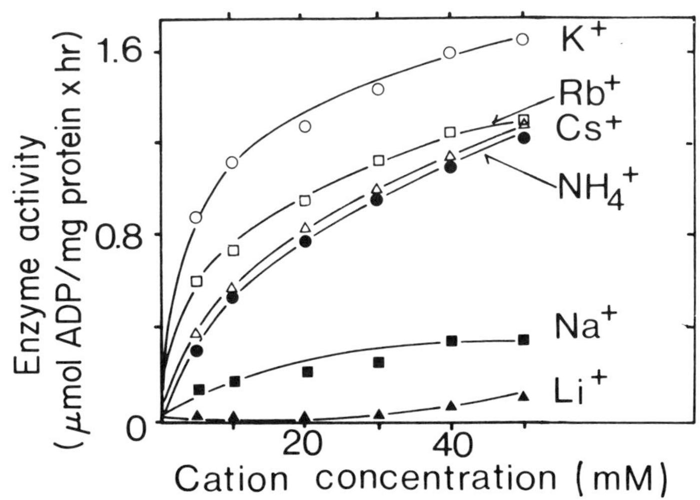

Fig. 8.23 Effect of univalent cations (as chlorides) on the activity of ADP-glucose starch synthase from maize. (Nitsos and Evans, 1969.)

transport from the external solution across the plasma membrane into the root cells (Chapter 2) but also makes K^+ the most important mineral element in cell extension and osmoregulation.

Tissues of potassium-deficient plants exhibit a much higher activity of certain hydrolases such as ß-glucosidase (Amberger, 1954) or of oxidases such as polyphenol oxidase (Welte and Müller, 1966) than do tissues of normal (sufficient) plants. It is not clear whether these changes in enzyme activity are caused by direct or indirect effects of K^+ on the synthesis of the enzymes. An instructive example of indirect effects is the accumulation of the diamine putrescine in potassium-deficient tissues (T. A. Smith, 1973; see Section 8.2.3). The enzymes which catalyze the synthesis of putrescine from arginine are stimulated by low cellular pH. Considering the dominant role of K^+ in the maintenance of high cytoplasmic pH, it appears that enhanced putrescine synthesis is a reflection of a homeostatic mechanism which controls the cytosol pH; in potassium-deficient plants the putrescine level can account for up to 30% of the deficit in K^+ equivalents (Murty *et al.*, 1971).

8.7.3 Protein Synthesis

It is now well established that K^+ is required for protein synthesis in higher plants. In cell-free systems the rate of protein synthesis by ribosomes isolated from wheat germ has an optimum at 130 m*M* K^+ and ~2 m*M* Mg^{2+} (Wyn Jones *et al.*, 1979). It is probable that K^+ is involved in several steps of the translation process, including the binding of tRNA to ribosomes (Evans and Wildes, 1971; Wyn Jones *et al.*, 1979). In green leaves the chloroplasts account for about half of both leaf RNA and leaf protein. In C_3 species the majority of the chloroplast protein is RuBP carboxylase (Section 8.2). Accordingly, the synthesis of this enzyme is particularly impaired when K^+ is deficient and responds rapidly to K^+, as shown in Table 8.31. The maximum activation was obtained at K^+ concentrations in the external solution as low as 1 m*M*. This concentration was obviously sufficient to obtain a ~100-fold higher K^+ concentration in the chloroplasts than is required for high rates of protein synthesis.

The role of K^+ in protein synthesis not only is reflected in the accumulation of soluble nitrogen compounds (e.g., amino acids, amides, and nitrate) in potassium-deficient plants (Mengel and Helal, 1968) but can also be demonstrated directly following the incorporation of ^{15}N-labeled inorganic nitrogen into the protein fraction. For example, within 5 hr in potassium-sufficient and potassium-deficient tobacco plants, 32 and 11%, respectively, of the total ^{15}N taken up had been incorporated into the protein nitrogen (Koch and Mengel, 1974). From studies of Pflüger and Wiedemann

Table 8.31
Effect of Potassium on the Incorporation of [^{14}C] Leucine into RuBP Carboxylase in the Leaves of Potassium-Deficient Alfalfa Plants[a]

Preincubation medium (m*M* KNO_3)	[^{14}C] Leucine incorporation (dpm/mg RuBP carboxylase × 24 hr)
0·0	99
0·01	167
0·10	220
1·00	526
10·00	526
Control (K$^+$-sufficient plants)	656

[a]Leaves were preincubated in light for 20 hr with potassium. From Peoples and Koch (1979).

(1977) it seems highly probable that K$^+$ not only activates nitrate reductase but also is required for the synthesis of this enzyme.

8.7.4 Photosynthesis

In higher plants K$^+$ affects photosynthesis at various levels. In a review Läuchli and Pflüger (1978) stressed the role of K$^+$ as the dominant counter-ion to light-induced H$^+$ flux across the thylakoid membranes and the establishment of the transmembrane pH gradient necessary for the synthesis of ATP (photophosphorylation), in analogy to ATP synthesis in mitochondria.

The role of K$^+$ in CO_2 fixation can be demonstrated most clearly with isolated chloroplasts (Table 8.32). An increase in the external K$^+$ concentration to 100 m*M*, that is, to about the K$^+$ concentration in the cytosol of intact cells, stimulates CO_2 fixation more than threefold. On the other hand,

Table 8.32
Influence of the Antibiotic Valinomycin and Potassium on the Rate of Carbon Dioxide Fixation in Isolated Intact Spinach Chloroplasts[a]

Treatment	Rate of CO_2 fixation (μmol/mg chlorophyll × hr)	Percentage of control
Control	23·3	100
100 m*M* K$^+$	79·2	340
1 μM valinomycin	11·0	47
1 μM valinomycin + 100 m*M* K$^+$	78·4	337

[a]From Pflüger and Cassier (1977).

Table 8.33
Relationship between Potassium Content in Leaves, Carbon Dioxide Exchange, and RuBP Carboxylase Activity in Alfalfa[a]

Leaf potassium (mg/g dry wt)	Stomatal resistance (sec/cm)	Photosynthesis (mg CO_2/ $dm^2 \times hr$)	RuBP Carboxylase activity (μmol CO_2/mg protein $\times$ hr)	Photorespiration (dpm/dm^2)	Dark respiration (mg CO_2/ $cm^2 \times hr$)
12·8	9·3	11·9	1·8	4·0	7·6
19·8	6·8	21·7	4·5	5·9	5·3
38·4	5·9	34·0	6·1	9·0	3·1

[a]From Peoples and Koch (1979).

the ionophore valinomycin, which makes biomembranes "leaky" for passive K^+ flux, severely decreases CO_2 fixation. The effect of valinomycin can be compensated for by high external K^+ concentrations.

The enhanced leakage of K^+ caused by antibiotics also induces severe changes in the fine structure of chloroplasts and proplastids (Marschner and Mix, 1974), which is another indication that K^+ is essential to the structural integrity and thus function of plastids. Accordingly, in the suboptimal concentration range, the potassium content of leaves and the various parameters of CO_2 exchange are closely correlated (Table 8.33). An increase in the leaf potassium content is accompanied by increased rates of photosynthesis, photorespiration, and RuBP carboxylase activity, but a decrease in dark respiration. Enhanced respiration rates are a common feature of potassium deficiency (Bottrill *et al.*, 1970). It is apparent from Table 8.33 that K^+ plays a further role in the photosynthesis of leaves, namely, in stomatal regulation (see Section 8.7.5.2).

8.7.5 Osmoregulation

In Chapter 3 it was shown that a high osmotic potential in the stele of roots is a prerequisite for turgor-pressure-driven solute transport in the xylem and for the water balance of plants. In principle, at the level of individual cells or in certain tissues, the same mechanisms are responsible for cell extension and various types of movement. Potassium, as the most prominent inorganic solute, plays a key role in these processes.

8.7.5.1 Cell Extension

Cell extension involves the formation of a large central vacuole occupying 80–90% of the cell volume. There are two major requirements for cell

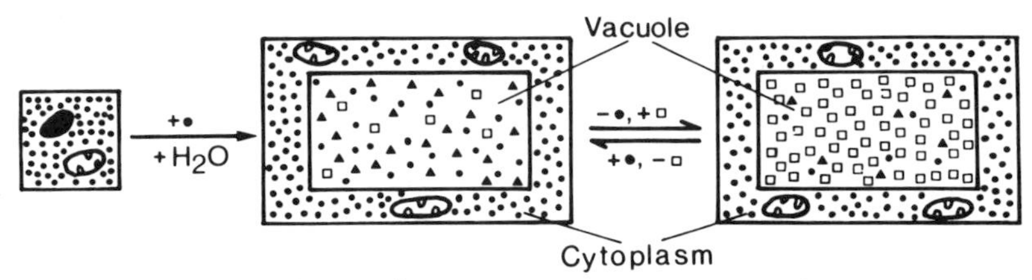

Fig. 8.24 Model of the role of potassium and other solutes in cell extension and osmoregulation. Key: •, K^+; □, reducing sugars; sucrose, Na^+; ▲, organic acid anions.

extension: an increase in cell wall extensibility, probably induced by IAA, and solute accumulation to create an internal osmotic potential (Fig. 8.24).

It is well established that cell extension is the consequence of the accumulation in the cells of K^+, which is required for both stabilizing the pH in the cytoplasm and increasing the osmotic potential in the vacuoles. In *Avena* coleoptiles, IAA-stimulated H^+ efflux is electrochemically balanced by a stoichiometric K^+ influx; in the absence of external K^+, IAA-induced elongation declines and ceases after a few hours (Haschke and Lüttge, 1975). In cucumber cotyledons K^+ supply enhances extension by a factor of ~4 in response to the application of cytokinins (Green and Muir, 1979). Similarly, cell extension in leaves is closely related to their K^+ level. In expanding leaves of bean plants suffering from potassium deficiency, turgor, cell size, and leaf areas were significantly lower than in expanding leaves well supplied with K^+ (Arneke, 1980).

As shown in Table 8.34 the enhancement of stem elongation by gibberellic acid (GA) is also dependent on the K^+ supply. Potassium and GA act synergistically, the highest elongation rate being obtained when both GA and K^+ are applied. The results in Table 8.34 seem to indicate further that K^+ and reducing sugars act in a complementary manner to produce the turgor potential required for cell extension. In the plants with a low K^+

Table 8.34
Effects of Potassium and Gibberellic Acid (GA) on Plant Height and Concentrations of Sugars and Potassium in the Shots of Sunflower Plants[a]

Treatment		Plant height	Concentration (μmol/g fresh wt)		
KCl (m*M*)	GA (mg/liter)	(cm)	Reducing sugars	Sucrose	Potassium
0·5	0	7·0	19·1	5·0	10·2
0·5	100	18·5	38·5	5·4	13·2
5·0	0	11·5	4·6	4·1	86·5
5·0	100	26·0	8·4	2·5	77·8

[a]Based on Guardia and Benlloch (1980).

supply, however, GA-stimulated growth was correlated with a marked increase in K^+ concentration in the elongation zone to a level similar to that of the reducing sugars (Guardia and Benlloch, 1980). These and other data from the literature strongly support the view that K^+, most often associated with organic acid anions (Haschke and Lüttge, 1975), is the main solute required in the vacuoles for cell extension. For maintenance of cell turgor thereafter, however, K^+ can be replaced in the vacuoles by other solutes such as reducing sugars (Fig. 8.24).

Inverse relationships between tissue concentrations of K^+ and reducing sugars are widespread phenomena (Pitman *et al.*, 1971) and can also be observed during the growth of storage tissues. As shown by Steingröver (1983) the osmotic potential of the press sap from the storage roots of carrots remains constant throughout growth. Before sugar storage begins, K^+ and organic acids are the dominant osmotic substances. During sugar storage, however, an increase in the concentration of reducing sugars is compensated for by a corresponding decrease in the concentration of K^+ and organic acids.

8.7.5.2 Stomatal Movement

In most plant species K^+ has the major responsiblity for turgor changes in the guard cells during stomatal movement. An increase in the K^+ concentration in the guard cells results in the uptake of water from the adjacent cells and a corresponding increase in turgor in the guard cells and thus stomatal opening. This has been repeatedly demonstrated (e.g., by Humble and Hsiao, 1970). In Table 8.35 the quantitative relationships are shown for the guard cells of faba bean.

The accumulation of K^+ in the guard cells of open stomata can also be shown by X-ray microprobe analysis (Fig. 8.25). Closure of the stomata in the dark is correlated with K^+ efflux and a corresponding decrease in the osmotic pressure of the guard cells. Abscisic acid inhibits stomatal opening

Table 8.35
Relationship between Stomatal Aperture and Characteristics of Guard Cells of Faba Bean[a]

	Stomatal aperture (μm)	Amount per stoma (10^{-14} g equiv.)		Guard cell volume (10^{-12} liters per stoma)	Guard cell osmotic pressure (bars)
		K^+	Cl^-		
Open stoma	12	424	22	4·8	35
Closed stoma	2	20	0	2·6	19

[a]From Humble and Raschke (1971).

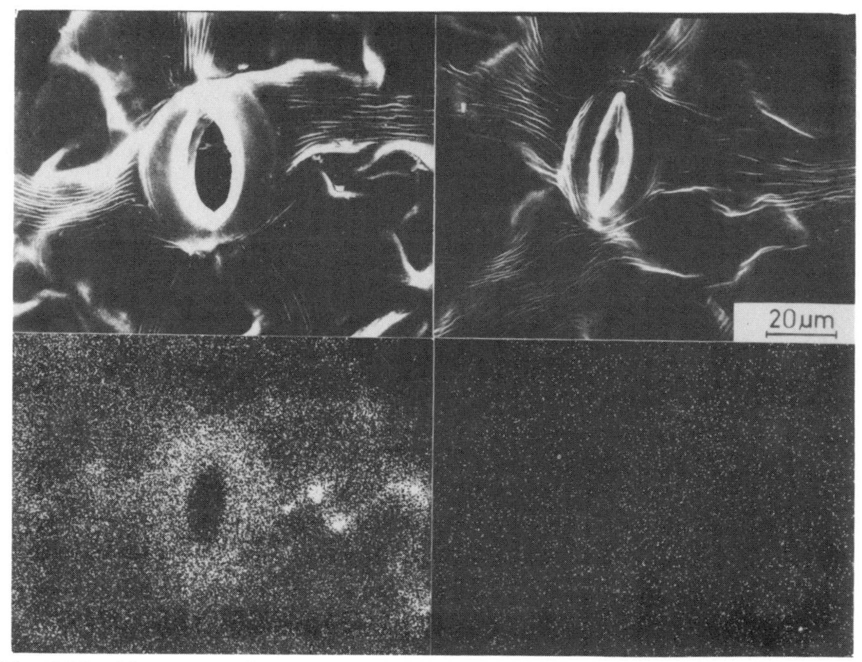

Fig. 8.25 Electron probe analyzer picture (*upper*) and corresponding X-ray microprobe images of the potassium distribution (*lower*) in open and closed stomata of faba bean. (Courtesy of B. Wurster.)

or induces a rapid closure of stomata (Mittelheuser and Van Steveninck, 1971). This effect of ABA is brought about mainly by increased efflux of K^+ from the guard cells (MacRobbie, 1981).

Light-induced accumulation of K^+ in the guard cells is mediated by a membrane-bound H^+-efflux pump, as in other plasma membranes. The energy required for this pump (ATPase) is supplied by photophosphorylation in the guard cells of chloroplasts (Humble and Hsiao, 1970). The accumulation of K^+ in the guard cells has to be balanced by a counteranion, mainly malate or Cl^-, depending on the plant species and availability of Cl^- (Van Kirk and Raschke, 1978; Raschke and Schnabl, 1978). Malate is synthesized within the guard cells via the PEP carboxylase:

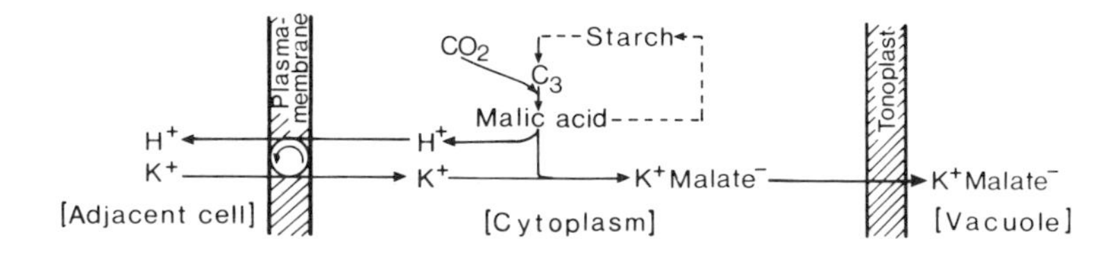

The C_3 compound required for malate synthesis (PEP) is supplied by starch degradation in the guard cell chloroplasts (Schnabl, 1980). In plant species, such as onion, that lack starch in the guard cell chloroplasts, the action of Cl^- as a counterion for K^+ influx might be of fundamental importance, at least for stomatal regulation (Schnabl, 1980), an aspect which deserves further attention in relation to the role of Cl^- as a mineral nutrient.

8.7.5.3 Nyctinastic and Seismonastic Movements

Potassium plays a key role in turgor-regulated movement, not only of individual cells but also of plant organs. In many species such as bean (Kiyosawa, 1979) and *Albizzia* (Scatter *et al.*, 1974), *nyctinastic* (circadian) movements of the leaves are observed; that is, the leaves are open during the day and closed at night. This leaf movement is controlled by K^+-mediated turgor changes in specialized tissues, the motor organs or pulvini. Active K^+ transport into the ventral (lower) motor cell leads to opening; K^+ efflux (leakage) in the opposite direction causes closure. There is evidence that, in analogy to the situation in guard cells, K^+ flux is driven by an H^+ pump in the pulvini (Iglesias and Satter, 1983) and Cl^- acts as a major anion (Kumon and Tsurumi, 1984).

A similar mechanism is responsible for the *seismonastic* reactions of the leaves of *Mimosa pudica*. In response to mechanical stimuli the leaflets fold together within a few seconds and reopen after about 30 min (Campbell and Thomson, 1977). This turgor-regulated response is caused by a redistribution of K^+ within the pulvini (Allen, 1969), a process similar to that which is responsible for circadian movements. Correlated with the leaflet movement in *Mimosa* is a relocation and/or change in the binding of Ca^{2+} within the cells of the pulvini (Toriyama and Jaffe, 1972; Campbell *et al.*, 1979); this most likely is a factor affecting the membrane permeability of K^+ in the pulvini (Campbell and Thompson, 1977).

8.7.6 Phloem Transport

The high K^+ concentrations in the sieve tubes are probably related to the mechanism of phloem loading of sucrose (Chapter 5). Regardless of theoretical considerations of a coupling of sucrose and K^+ transport into the sieve tubes (Malek and Baker, 1977; Giaquinta, 1983), however, it is well established that K^+ in the sieve tubes makes a considerable contribution to the total osmotic pressure and thus to the flow rate of photosynthates from source to sink. The phloem transport towards cereal grains may represent an

Table 8.36
Effect of Potassium Nutritional Status of Sugarcane Plants on the Translocation of ^{14}C-Labeled Photosynthates after Feeding of $^{14}CO_2$ to a Leaf Blade[a]

	^{14}C Distribution (%)			
	90 min		4 hr	
Plant part	+K	−K	+K	−K
Fed leaf blade	54·3	95·4	46·7	73·9
Sheath of fed leaf blade	14·2	3·9	6·8	8·0
Joint of fed leaf and leaves and joint above fed leaf	11·6	0·7	17·0	13·6
Stalk below fed leaf	20·1	<0·1	29·5	4·6

[a]Total label = 100. Based on Hartt (1969).

exception to this rule (Section 3.3.5). In legumes with an adequate (compared with an inadequate) K^+ supply, the root nodules have a greater supply of sugars, which correspondingly increases their rates of N_2 fixation and export of bound nitrogen (Mengel *et al.*, 1974; Collins and Duke, 1981). Moreover, in adequately supplied plants, a much higher proportion of ^{14}C-labeled photosynthates is translocated from the leaves to the storage organs such as potato tubers (Haeder *et al.*, 1973) or storage tissue (stalk) of sugarcane (Table 8.36).

Several factors may account for the lower rates of export of ^{14}C-labeled photosynthates from potassium-deficient leaves. These include the greater requirement for sugars in the osmoregulation of leaves; lower rates of sucrose synthesis, of phloem loading and of flow rates of sucrose in the sieve tubes (Section 5.4.1); and impaired sucrose transport across the tonoplast of the storage cells in the sink tissue. The involvement of K^+ in the last process has been discussed in Chapter 6.

8.7.7 Cation–Anion Balance

From the viewpoint of charge compensation, K^+ is the predominant cation for counterbalancing immobile anions in the cytoplasm, and quite often mobile anions in vacuoles as well as in the xylem and phloem. The statement that "potassium is a cation for anions" (Clarkson and Hanson, 1980) is to a large extent valid, although it insufficiently expresses the active role of K^+ in the cation–anion balance. The accumulation of organic acids is often the consequence of K^+ transport without an accompanying anion into the cytoplasm (e.g., root or guard cells). The role of K^+ in the cation–anion balance is also reflected in nitrate metabolism, in which K^+ often is the predominant counterion for NO_3^- in long-distance transport in the xylem as well as for storage in vacuoles. As a consequence of NO_3^- reduction in leaves, the remaining K^+ requires the stoichiometric synthesis of organic acids for charge balance; part of this newly formed K^+ malate$^-$ may be retranslocated to the roots for subsequent reutilization of K^+ as a counterion for NO_3^- within the root cells and for xylem transport (Chapter 3). In nodulated legumes, this recirculation of K^+ may serve a similar function in the xylem transport of amino acids (Jeschke *et al.*, 1985).

8.7.8 Potassium Supply, Plant Growth, and Composition

The K^+ requirement for optimal plant growth is ~2–5% of the dry weight of the vegetative parts, fleshy fruits, and tubers. In natrophilic species, however, the requirement for K^+ can be much lower (Section 10.2). When K^+ is deficient, growth is retarded, and K^+ is retranslocated from mature leaves and stems, and under conditions of severe deficiency these organs can become chlorotic and necrotic (Bussler, 1964). Lignification of vascular bundles is impaired (Pissarek, 1973), a factor that may also be responsible for the higher susceptibility of potassium-deficient plants to lodging.

When the soil water supply is limited, loss of turgor and wilting are typical symptoms of potassium deficiency. The lower tolerance of potassium-deficient plants to drought is related mainly to (a) the role of K^+ in stomatal regulation, which is a major mechanism controlling the water regime of higher plants, and (b) the role of K^+ as the predominant osmotic solute in the vacuole, maintaining a high tissue water level even under drought conditions. Higher proline levels in plants well supplied with K^+ compared with deficient plants, particularly in response to water deficiency (Mukherjee, 1974), may be an additional aspect of the role of K^+ in drought tolerance. Plants receiving an inadequate supply of K^+ are often more susceptible to frost damage, which, at the cellular level, is related in some respect to water deficiency. An example of this effect is shown in Table 8.37. Frost damage is inversely related to the potassium content of leaves, at least when the increase in potassium is correlated with an increase in tuber yield. Inadequate potassium supply is therefore one factor leading to an increase in the risk of frost damage.

The changes in enzyme activity and organic compounds that take place during potassium deficiency (Section 8.7.1) are in part responsible for the higher susceptibility of plants to fungal attack (Chapter 11). They also affect the nutritional and technological (processing) quality of harvested products. This is most obvious in fleshy fruits and tubers with their high K^+ require-

Table 8.37
Relationship in Potato between Potassium Supply, Tuber Yield, Potassium in Leaves, and Percentage of Leaves Damaged by Frost[a]

Potassium supply (kg/ha)	Tuber yield (tons/ha)	Potassium content of leaves (mg/g dry wt)	Percentage of foliage damaged by frost
0	2·39	24·4	30
42	2·72	27·6	16
84	2·87	30·0	7

[a]Average values of 14 locations. Based on Grewal and Singh (1980).

ment. In tomato fruits, for example, the incidence of so-called ripening disorders ("greenback") increases with inadequate K^+ supply (Lune and Goor, 1977), and in potato tubers a whole range of quality criteria are affected by the K^+ level in the tuber tissue (Table 8.38).

In several cases the relationships between K^+ concentrations and gross changes in tuber tissue (e.g., osmoregulation and cation–anion balance) are quite obvious. In other cases, quality disorders are related directly to the citric acid content and thus only indirectly to K^+. Although differences among cultivars may modify the relationships, they do not eradicate them.

Table 8.38
Effect of Increasing Potassium Concentrations in Potato Tubers on the Composition and Quality of the Tubers

Type of change	Effect of increasing K^+	Responsible mechanism	References[a]
Water content	Increase	Osmoregulation	1,2,3
Reducing sugars	Decrease	Osmoregulation	3
Citric acid	Increase	Cation–anion balance	4
Starch	Decrease	?	1,2,3
Black spot disorder	Decrease	Lower polyphenol oxidase activity?	2,5
Darkening of press sap	Decrease	High citric acid/low polyphenol oxidase	4
Discoloration after cooking	Decrease	High citric acid/low chlorogenic acid	5,6
Storage loss	Decrease	Lower respiration and fungal diseases	2

[a]Key to references: 1, Forster (1981). 2, Mirswa and Ansorge (1981). 3, A. Krauss and H. Marschner (unpublished). 4, Welte and Müller (1966). 5, Vertregt (1968). 6, Hughes and Evans (1969).

By increasing the K^+ supply to plant roots it is relatively easy to increase the potassium content of various organs except grains and seeds, which maintain a relatively constant potassium content of 0·3% of the dry weight. When the K^+ supply is abundant "luxury consumption" of K^+ often occurs, which deserves attention both for its effect on plant composition and for its possible interference with the uptake and physiological availability of Mg^{2+} and Ca^{2+}.

9

Functions of Mineral Nutrients: Micronutrients

9.1 Iron

9.1.1 General

In aerated systems maintained in the physiological pH range, the concentrations of ionic Fe^{3+} and Fe^{2+} are extremely low (10^{-10} *M* or lower; Lindsay, 1974). Chelates of Fe(III) and occasionally of Fe(II) are therefore the dominant forms of soluble iron in soil and nutrient solutions. As a rule Fe(II) is the species taken up. Iron(III), therefore, has to be reduced at the root surface before transport into the cytoplasm (Chaney *et al.*, 1972; Römheld and Marschner, 1983). In grasses, however, uptake of Fe(III) is of major importance (Section 9.1.6). Iron uptake rates are higher in the apical than the basal root zones, particularly when iron is deficient (Clarkson and Sanderson, 1978; Römheld and Marschner, 1981a). In long-distance transport via the xylem, there is a predominance of Fe(III) complexes with citrate (Clark *et al.*, 1973) or peptide–carbohydrate compounds (Höfner, 1970). The high affinity of iron for various ligands (e.g., organic acids or inorganic phosphate) makes it unlikely that ionic Fe^{3+} or Fe^{2+} is of any importance in short- or long-distance transport in plants or that either form is involved in the reactions of iron within the cells. The formation of coordination complexes (chelates) and its action as a reversible oxidation–reduction reaction system [$Fe(II) \rightleftharpoons Fe(III) + e^-$] constitute the major metabolic functions of iron.

9.1.2 Iron-Containing Constituents of Redox Systems

There are two groups of well-defined iron-containing proteins: hemoproteins and iron–sulfur proteins (Sandmann and Böger, 1983).

9.1.2.1 Hemoproteins

The most well known hemoproteins are the cytochromes, which contain a heme iron–porphyrin complex (Fig. 9.1) as a prosthetic group. Cytochromes

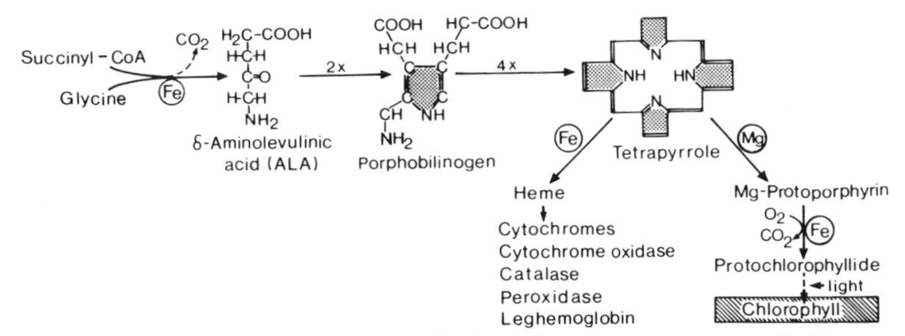

Fig. 9.1 Role of iron in the biosynthesis of heme coenzymes and chlorophyll.

are constituents of the redox systems in chloroplasts (Fig. 5.1) and mitochondria (Fig. 5.5) and, in the form of cytochrome oxidase, participate in the terminal step of the respiratory chain. The role of leghemoglobin in root nodules of legumes in N_2 fixation has been discussed in Chapter 7.

Other heme enzymes are catalase and peroxidases. Under conditions of iron deficiency, the activity of both types of enzyme declines. This is particularly the case for catalase activity in leaves (Table 9.1). The activity of this enzyme is therefore a suitable indicator of the iron nutritional status of plants (Chapter 12).

Catalase facilitates the dismutation of H_2O_2 to water and O_2 according to the equation

$$H_2O_2 \rightarrow H_2O + \frac{1}{2}O_2$$

The enzyme plays a key role in the chloroplasts in cooperation with superoxide dismutase (Fig. 5.1), as well as in photorespiration and the glycolytic pathway (Fig. 5.4).

Peroxidases of various types (isoenzymes) are abundant in plants. They catalyze the following reactions:

$$XH_2 + H_2O_2 \rightarrow X + 2H_2O$$

and

$$XH + XH + H_2O_2 \rightarrow X\text{—}X + 2H_2O$$

Table 9.1
Effect of Iron Supply on Chlorophyll Content and Enzyme Activity in Tomato Leaves

Treatment	Iron in leaves (μg/g fresh wt)	Chlorophyll (mg/g fresh wt)	Enzyme activity (relative)	
			Catalase	Peroxidase
+Fe	18·5	3·52	100	100
−Fe	11·1	0·25	20	56

[a]Based on Machold (1968).

In the second type of reaction, cell wall–bound peroxidases catalyze the polymerization of phenols to lignin. Peroxidases are abundant in the cell walls of the rhizodermis and endodermis of the root (Mueller and Beckman, 1978). There is increasing evidence that the peroxidases in rhizodermal cell walls play a role in the regulation of iron uptake. The reasons for this are as follows. Phenolic compounds, which are precursors of lignin biosynthesis, are released into the apoplast of rhizodermal cells. Lignin synthesis also requires H_2O_2, the formation of which is catalyzed by another peroxidase resulting from oxidation of NADH at the plasma membrane (Mäder, 1977; Mäder and Füssl, 1982). The principles of these reactions are as follows:

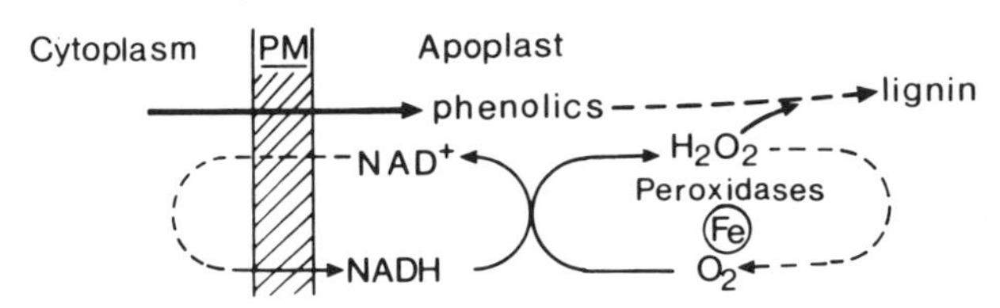

In iron-deficient roots (in contrast to leaves; Table 9.1), the decrease in peroxidase activity is much greater than that of catalase activity. As a consequence, cell wall formation (see Fig. 9.4) and lignification are impaired; phenolics accumulate in the rhizodermis (Römheld and Marschner, 1981a) and are released into the external solution (Olsen *et al.*, 1981). Certain phenolics, such as caffeic acid, are very effective in the chelation and reduction of inorganic Fe(III). With the decrease in peroxidase activity there is a simultaneous decline in the oxidation of NADH or NADPH at the external surface of the plasma membrane, the capacity of the plasma membrane to reduce other substrates such as Fe(III) chelates thereby increasing (Craig and Crane, 1981; Bienfait, 1985). Iron deficiency also inhibits the activity of an isoperoxidase which catalyzes the polymerization of aliphatic and aromatic (phenolic) monomers to suberin (Sijmons *et al.*, 1985). This effect may play an additional role in the impairment of rhizodermal cell wall formation and in the accumulation of certain phenolics in iron-deficient roots.

9.1.2.2 Iron–Sulfur Proteins

In these nonheme proteins the iron is coordinated to the thiol group of cysteine and/or to inorganic sulfur. The most prominent one is ferredoxin, which acts as an electron transmitter in a number of basic metabolic processes according to the principle:

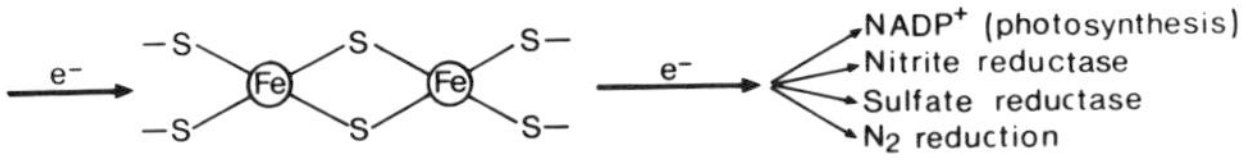

e^- → –S, –S (Fe) S, S (Fe) S–, S– → e^- → $NADP^+$ (photosynthesis); Nitrite reductase; Sulfate reductase; N_2 reduction

In N_2 reduction three different iron–sulfur proteins act in series in the electron transport chain of the nitrogenase complex (Fig. 7.4). Details of the role of ferredoxin are discussed in the relevant sections.

9.1.3 Other Iron-Containing Enzymes

There are a number of less well characterized enzymes in which iron acts either as a metal component in redox reactions or as a bridging element between enzyme and substrate. In iron-deficient plants, the activities of some of these enzymes are impaired (perhaps owing to a lower affinity for iron), which is the main reason for the gross changes in metabolic processes discussed below.

Riboflavin accumulates in the roots of various dicotyledonous species under conditions of iron deficiency. This accumulation is presumably the result of impaired purine metabolism and indicates the iron requirement of xanthine oxidase (Schlee *et al.*, 1968). Relatively large amounts of riboflavin are released from these roots in response to enhanced H^+ secretion (Nagarajah and Ulrich, 1966; Venkat-Raju *et al.*, 1972). The relevance of enhanced riboflavin release to iron mobilization in the rhizosphere and iron uptake is not yet understood.

The accumulation of organic acids in roots and, less consistently, in shoots is a well-documented general phenomenon of iron deficiency (Brown *et al.*, 1971; Venkat-Raju *et al.*, 1972). The predominant organic acids which accumulate are malic and citric acids, as shown in Table 9.2. When the supply of iron is restored, the organic acid level in the roots drops to control levels within ~2 days (Landsberg, 1979).

The accumulation of organic acids in iron-deficient plants may be related directly to a decrease in the activity of aconitase (aconite hydratase). This enzyme catalyzes the isomerization of citrate to isocitrate (Fig. 9.2). It is well established that Fe(II) is a component of the enzyme. It is required for both

Table 9.2
Relationship between Iron Supply, Chlorophyll Content in Leaves, and Organic Acid Content in Roots of Oat[a]

Treatment	Chlorophyll content (relative)	Organic acid content (μg/10 g fresh wt)			
		Malic	Citric	Other	Total
+Fe	100	39	11	23	73
−Fe	12	93	67	78	238

[a]Based on Landsberg (1981).

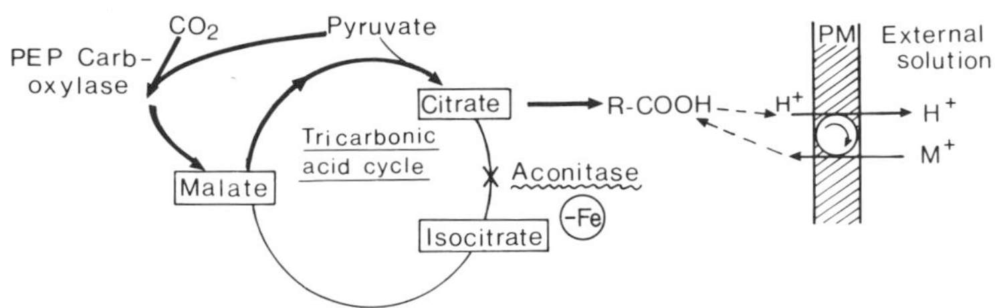

Fig. 9.2 Model of the relationship between a decrease in aconitase activity and the accumulation of organic acids in roots of iron-deficient plants (PM, plasma membrane).

stability and activity (Hsu and Miller, 1968) and is responsible for the spatial orientation of the substrates (citrate and isocitrate). Valency changes are not involved in the reaction (Glusker, 1968).

Inhibition of aconitase should severely impair the tricarbonic acid cycle unless other pathways of organic acid synthesis are activated, most likely the PEP carboxylase pathway (Fig. 9.2). Enhanced rates of CO_2 fixation in leaves (Stocking, 1975) and roots of iron-deficient plants are well documented (Bedri *et al.*, 1960). Elevated levels of organic acids are most likely the source of enhanced H^+ efflux from roots of most iron-deficient dicotyledonous plant species (Landsberg, 1981). In grasses (e.g., maize and oat), however, despite organic acid accumulation (Table 9.2), enhanced H^+ efflux is not observed when nitrate is the souce of nitrogen. There is good evidence that the rates of nitrate reduction are higher in roots of grasses than in dicotyledonous species and, correspondingly, higher rates of OH^- and HCO_3^- efflux are maintained under conditions of iron deficiency, thus counterbalancing the acidification of the substrate by enhanced H^+ efflux (Egmond and Atkas, 1977; Landsberg, 1979).

In young leaves of green plants, a decrease in chlorophyll content (chlorosis) is the most obvious visible symptom of iron deficiency. Various factors are responsible for this decrease, the most direct one being the role of iron in the biosynthesis of chlorophyll (Fig. 9.1). The common precursor of chlorophyll and heme synthesis is δ-aminolevulinic acid (ALA), and the rate of ALA formation is controlled by iron (Chereskin and Castelfranco, 1982; Miller *et al.*, 1982). The incorporation of iron or magnesium as the central atom into tetrapyrrole leads to the formation of heme coenzymes or Mg-protoporphyrin, respectively.

It is well established (Machold and Stephan, 1969) that iron is also required for the formation of protochlorophyllide from Mg-protoporphyrin. Feeding ALA to iron-deficient leaf tissue leads to an increasse in the Mg-protophorphyrin level, whereas the protochlorophyllide and chlorophyll levels remain low compared with the levels in leaf tissue adequately supplied with iron (Spiller *et al.*, 1982).

The enzyme coproporphyrinogen oxidase is an iron protein (Chereskin and Castelfranco, 1982) that catalyzes the oxidative decarboxylation of Mg-protoporphyrin (Vlcek and Gassman, 1979).

9.1.4 Protein Synthesis and Chloroplast Development

The inhibition of chlorophyll formation under conditions of iron deficiency is, at least in part, the result of impaired protein synthesis. The requirement for iron in protein synthesis is reflected in leaves by a drastic decline in the number of ribosomes—the sites of protein synthesis (Lin and Stocking, 1978)—and an increase in the amino acid concentration of chlorotic leaves (Gilfillan and Jones, 1968). A peculiarity of iron deficiency is a greater decline in protein synthesis in the chloroplasts of leaf cells than in the cytoplasm (Table 9.3). In maize leaves suffering from severe iron deficiency, the total protein content was 25% lower than the normal value, whereas the chloroplast protein content was 82% lower (Perur *et al.*, 1961). These differences indicate that the synthesis of certain chloroplast proteins, most likely structural proteins of the grana, is particularly impaired (Machold, 1972; Funkhouser and Price, 1974). In agreement with this, the content of other components of the chromoproteid complex of the chloroplasts also declines when iron is deficient. Examples include the carotenoids (Table 9.3), ferredoxin (Stocking, 1975), and other components of photosystem I (Nishio and Terry, 1983).

Table 9.3
Effect of Iron Deficiency on the Pigment and Protein Content of Tobacco Leaves[a]

Iron supply (chlorosis)	Pigment content (mg/g fresh wt)		Protein content (mg/g fresh wt)	
	Chlorophyll $a + b$	Carotenoids	Chloroplast	Cytoplasm
+Fe	0·98	0·45	8·6	12·8
−Fe (mild)	0·34	0·33	5·0	11·9

[a]Based on Shetty and Miller (1966).

The effect of iron on chloroplast development is summarized in Table 9.4. As the severity of iron deficiency increases (i.e., the chlorophyll content per unit leaf area decreases), protein content per leaf area, leaf cell volume, and number of chloroplasts remain unaffected, whereas the chloroplast volume and the amount of protein per chloroplast decline. In iron-deficient leaves the rate of photosynthesis decreases per unit leaf area but not per unit chlorophyll (Terry, 1980), indicating that the photosynthetic apparatus is still intact.

Table 9.4
Effect of Iron Deficiency on Leaves and Chloroplasts of Sugar Beet[a]

Parameter	µg Chlorophyll/cm² Control, >40	Mild deficiency, 20–40	Severe deficiency, <20
Soluble protein/leaf area (mg/cm^2)	0·57	0·56	0·53
Mean leaf cell volume (10^{-8} cm^3)	2·64	2·78	2·75
No. of chloroplasts/cell	72	77	83
Average chloroplast volume (μm^3)	42	37	21
Protein N/chloroplast (pg)	1·88	1·34	1·24

[a]From Terry (1980).

9.1.5 Localization and Binding State of Iron

In green leaves ~80% of the iron is localized in the chloroplasts regardless of the iron nutritional status (Fig. 9.3). Under conditions of iron deficiency there is a shift in the distribution of iron only within the chloroplasts, in which the lamellar iron content increases at the expense of the stromal iron.

Iron is stored in plant cells in the stroma of plastids as phytoferritin. This is a hollow shell which can store up to 5000 atoms of iron as Fe(III) (Fe content 12–23% dry wt) with the proposed formula $(FeO{\cdot}OH)_8{\cdot}(FeO{\cdot}OPO_3H_2)$, often in well-defined crystalline form (Seckback, 1982). The concentration of phytoferritin is high in dark-grown leaves (up to 50% of the total iron), but it rapidly disappears during regreening (Mark *et al.*, 1981). It is found in chloroplasts of formerly iron deficient plants after the massive uptake of iron that occurs when the supply of iron is restored (Platt-Aloia *et al.*, 1983). The

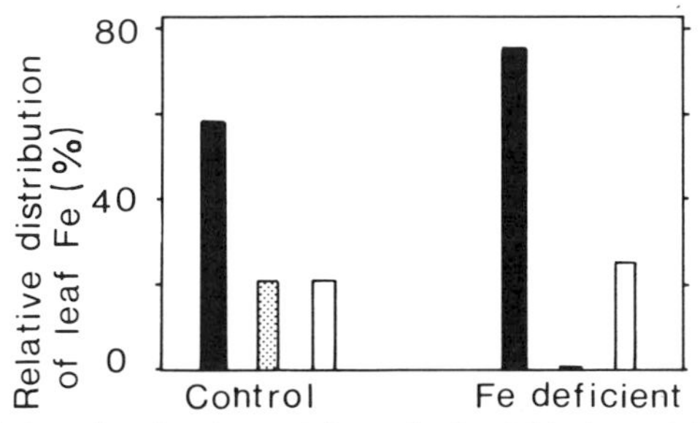

Fig. 9.3 Intracellular distribution of iron in leaf blades of iron-sufficient and iron-deficient sugar beet plants. Solid bars, lamellar iron; stippled bars, stromal iron; open bars, extrachloroplastic iron. (Redrawn from Terry and Low, 1982, by courtesy of Marcel Dekker, Inc.)

localization of phytoferritin is not confined to chloroplasts: it can be detected also in the xylem and phloem (Smith, 1984).

If plants are grown under controlled conditions (e.g., in nutrient solutions), there is a close positive correlation between the total iron content of the leaves and the chlorophyll content when the supply of iron (as chelates) is suboptimal (Römheld and Marschner, 1981a; Terry and Low, 1982). This correlation, however, is often poor or absent in plants grown in calcareous soils, when there is a large supply of phosphorus (Cumbus *et al.*, 1977; DeKock *et al.*, 1979), or when different forms of nitrogen are supplied (Machold, 1967). Under these conditions, the content of iron in chlorotic leaves may be similar to or even higher than that in green leaves. These discrepancies are related in part to the binding state of iron in leaves. Ferrous iron [Fe(II)] is the "physiologically available" form of iron and the fraction which undergoes reversible Fe(III)/Fe(II) oxidoreduction. This has been shown convincingly by direct determinations of the redox state of iron in leaves by Mössbauer spectrometry (Machold *et al.*, 1968). Determination of Fe(II) in leaf extracts (Katyal and Sharma, 1980) or extraction of leaves with diluted acids for characterization of the so-called active iron [presumably mainly the Fe(II) fraction] therefore often considerably improves the correlation between iron and chlorophyll content (DeKock *et al.*, 1979; Mengel and Bübl, 1983). It also seems to be a suitable indicator for characterization of the critical deficiency level (i.e., the level that results in 5–10% growth or yield reduction, see Chapter 12) of iron in upland rice (Katyal and Sharma, 1984).

There is some evidence that the ameliorating effect of ammonium on chlorosis (Farrahi-Aschtiani, 1972; Driessche, 1978) is caused not only by acidification of the rhizosphere (Chapter 15), but also by an increase in the proportion of Fe(II) to Fe(III) in the leaves (Machold, 1967).

The extent of photochemical reduction of Fe(III) to Fe(II) in leaves also depends on the wavelength of light (Brown *et al.*, 1979); short wavelength as well as certain organic acids enhance Fe(III) reduction (Bennett *et al.*, 1982).

9.1.6 Root Responses to Iron Deficiency

Whereas in leaves the major symptom of iron deficiency is inhibition of chloroplast development, in the roots of many plant species (except grasses) the deficiency is associated with distinct morphological changes. These include inhibition of elongation, an increase in the diameter of the apical root zones, abundant root hair formation (Römheld and Marschner, 1981a), and the formation of rhizodermal cells with a distinct wall labyrinth typical of transfer cells (Fig. 9.4). The iron deficiency-induced formation of

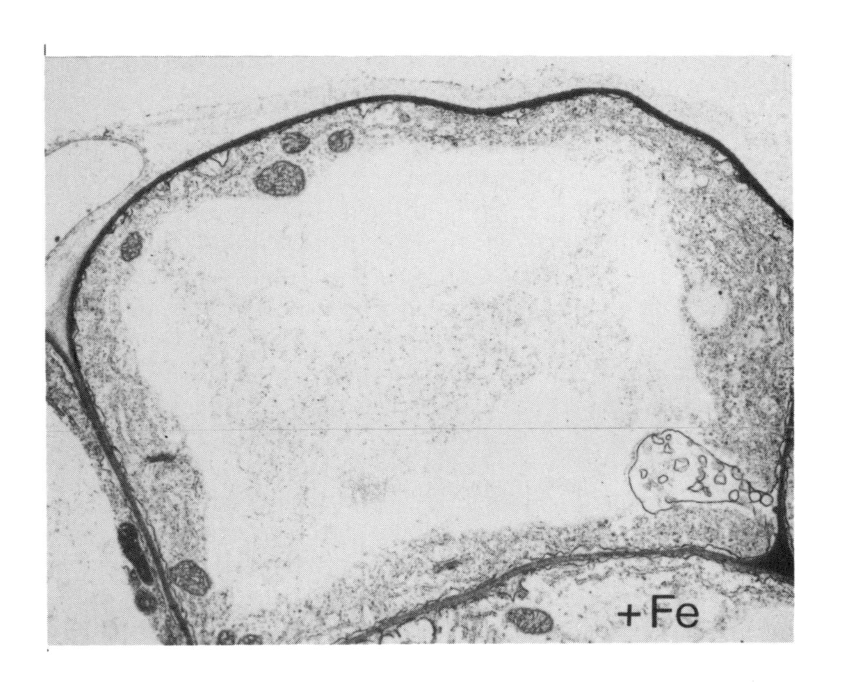

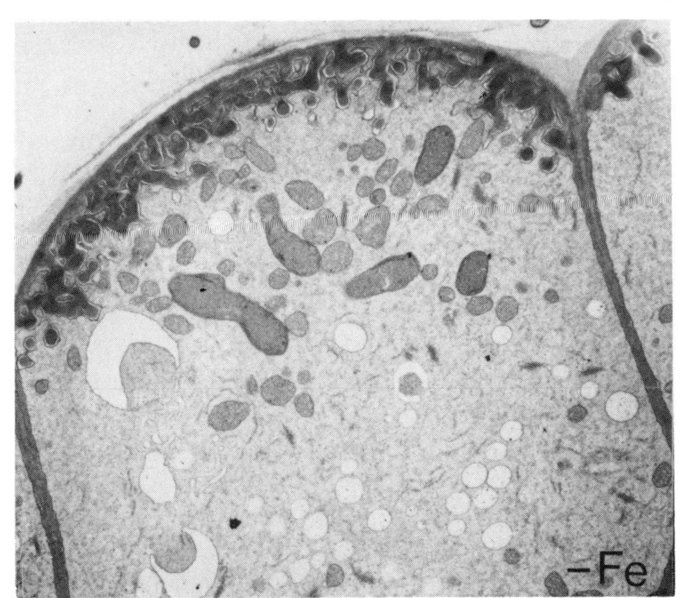

Fig. 9.4 Sections of rhizodermal cells of sunflower. (*Top*) Iron sufficient. (*Bottom*). Iron deficient. (By courtesy of D. Kramer.)

rhizodermal transfer cells (Kramer *et al.*, 1980) is part of a regulation mechanism for enhancing iron uptake and is observed only in plant species that can acidify the substrate (Römheld and Kramer, 1983).

Transfer cells are characterized by high metabolic activity and have important functions in solute transport (e.g., phloem loading; see Chapter 5). The rhizodermal transfer cells in iron-deficient roots are most likely the sites of both H^+-efflux pumps and the release of phenolic compounds. After the supply of iron is restored, the transfer cells degenerate within 1 to 2 days. Simultaneously the enhanced rates of iron uptake, H^+ efflux, and release of phenolics fall to normal levels, the external solution pH increases, and reducing substances (reducing capacity) disappear. These changes have been discussed in Chapter 2 and are summarized in Fig. 2.21.

When the iron supply is suboptimal [e.g., when the concentrations of Fe(III) chelates are low or sparingly soluble inorganic Fe(III) compounds are supplied], rhythmic changes in root morphology and root-induced changes in the substrate pH and uptake rate of iron are observed. The growth rate of the shoots and the chlorophyll content, however, remain unaffected (Römheld and Marschner, 1981b). In response to iron deficiency, both sparingly soluble iron and manganese (MnO_2) are mobilized in the substrate. This can lead to manganese toxicity in plants growing in calcareous soils with suboptimal iron levels (Moraghan and Freeman, 1978).

In grasses the aforementioned response mechanisms are not observed. Grasses cannot be described simply as "iron inefficient," however, for they have developed a different strategy for responding to iron deficiency (Section 16.5). In general they are less susceptible to lime-induced chlorosis than most dicotyledonous species. In iron-deficient grasses nonproteinogenic amino acids accumulate in and are released from the roots (Takagi, 1976). Some of them, including avenic acid, have been identified (Fushiya *et al.*, 1982):

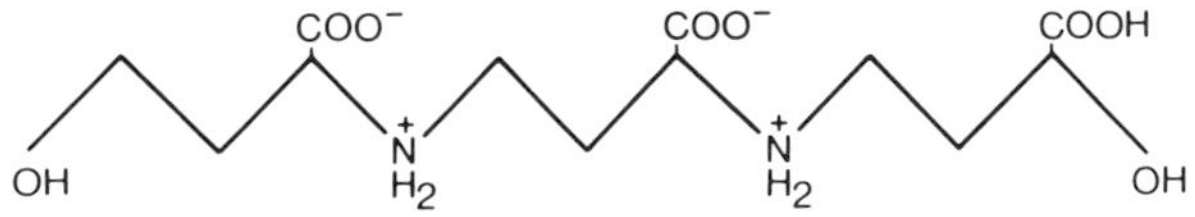

COO^- COO^- $COOH$
OH $\overset{+}{N}H_2$ $\overset{+}{N}H_2$ OH

These amino acids form highly stable complexes with Fe(III) (Mino *et al.*, 1983) but not with Fe(II) (Beneš *et al.*, 1983) and are very effective in dissolving FeOOH at high pH (Sugiura *et al.*, 1981). Chemically they are closely related to nicotianamine, which is widely distributed in higher plants and which, for example, induces regreening in a chlorophyll–defective mutant of tomato (Scholz and Böhme, 1980). Nicotianamine is an effective chelator for Fe(II) but not Fe(III) (Beneš *et al.*, 1983). In response to iron deficiency, grasses obviously transform nicotianamine into substances such as avenic acid which are released into the rhizosphere for mobilization of

Fe(III) by chelation. These Fe(III) chelates are transported into the root cells, that is, they act as phytosiderophores (Takagi *et al.*, 1984) in the same manner as the hydroxamate siderophores in microorganism (Powell *et al.*, 1982). Neither the pathway of biosynthesis of nicotianamine and the related phytosiderophores nor the mode of regulation by iron are known.

9.1.7 Iron Deficiency and Iron Toxicity

The critical deficiency level of iron in leaves is in the range of 50–150 mg/kg dry wt. Iron deficiency is a worldwide problem in crop production in calcareous soils. It is the major factor responsible for co-called lime-induced chlorosis. Iron toxicity is the second most severe yield-limiting factor in wetland rice. Both aspects are discussed in Chapter 16.

9.2 Manganese

9.2.1 General

Manganese is absorbed mainly as Mn^{2+} and is translocated predominantly as the free divalent cation in the xylem from the roots to the shoot (Graham, 1979). Of the micronutrient transition metals (manganese, iron, copper, zinc, and molybdenum), manganese has the lowest complex stability constant and thus forms the weakest bonds (Clarkson and Hanson, 1980). It can therefore replace Mg^{2+} in many reactions—for example, in its role as a bridge betwen ATP and enzyme complexes (e.g., in phosphokinases and phosphotransferases). Manganese activates a number of enzymes *in vitro*, particularly decarboxylases and dehydrogenases of the tricarboxylic acid cycle (for a comprehensive review, see Amberger, 1973). The specific requirement for manganese as a mineral nutrient, however, is presumably related to its tightly bound form in metalloproteins, where it acts as a structural constituent, as an active binding site, or, like iron, as a redox system [Mn(II)/Mn(III)].

9.2.2 Photosynthesis and Oxygen Evolution

The most well known and extensively studied function of manganese in green plants is its involvement in photosynthetic O_2 evolution. Green algae are particularly suitable for these studies because the effects of changes in manganese supply on photosynthesis can be measured immediately. In 1937 Pirson discovered that the low photosynthetic activity of manganese-deficient algae could be raised to normal levels within 1 hr by the addition of Mn^{2+} to the substrate. On the other hand, high rates of photosynthesis and growth are obtained even when the manganese supply is extremely low if H_2 rather than H_2O is used as the electron donator for CO_2 reduction (Kessler, 1955). Finally, in *Chlorella* the manganese requirement for optimal growth is ~1000 times lower under heterotrophic (darkness and an external supply of carbohydrates) than under autotrophic conditions, i.e., carbon supply via photosynthesis (Eyster *et al.*, 1958). It is now established that manganese is required in both lower and higher plants for the Hill reaction—the water-splitting and O_2-evolving system in photosynthesis (Cheniae and Martin, 1968). It is believed that photosystem II contains a manganoprotein which catalyzes the early stages of O_2 evolution (see Fig. 5.1). There is a minimum requirement of four manganese atoms for the reaction center (pigment 690) of photosystem II. According to Edwards and Walker (1983) the following reaction occurs in the water-splitting process of photosystem II:

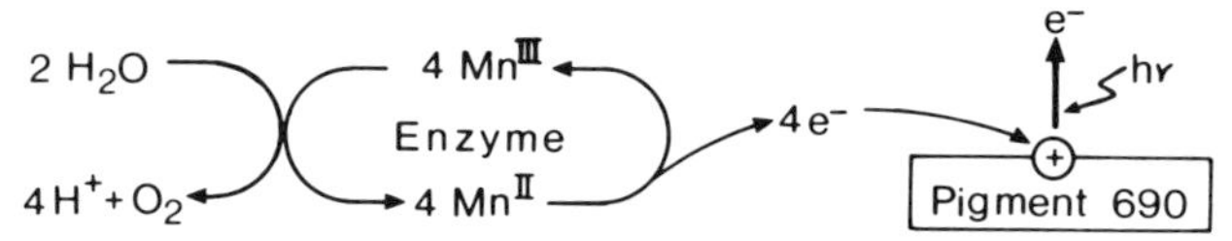

Accordingly when manganese is deficient the first step of the electron transport chain of the light reaction is impaired. The determination of the photosynthetic O_2 evolution in young leaves is therefore a sensitive and suitable test method for characterization of the manganese nutritional status of plants (Nable *et al.*, 1984). Impairment of the light reaction by manganese deficiency has corresponding negative effects on subsequent reactions such as photophosphorylation (Spencer and Possingham, 1961) and the reduction not only of CO_2 but also of nitrite and sulfate. Not only is there a decline in the rate of photosynthesis, but a progressive disorganization of the lamellar system of the chloroplasts occurs. However, other cell organelles, such as mitochondria, are not altered (Possingham *et al.*, 1964).

9.2.3 Manganese-Containing Enzymes

Only a few manganese-containing enzymes have been isolated so far. In leaf extracts of pea a superoxide dismutase (SOD) with one atom of manganese per enzyme molecule has been isolated (Sevilla *et al.*, 1980). There is increasing evidence that manganese-containing SOD is widely distributed in various families of higher plants, although it is not as widespread as the copper–zinc-containing SOD (Bridges and Salin, 1981; Sandmann and Böger, 1983). In contrast, the iron-containing SOD seems to be restricted to a few plant families (Bridges and Salin, 1981).

Superoxide dismutases are present in all aerobic organisms and play an essential role in the survival of these organisms in the presence of oxygen (Halliwell, 1978; Fridovich, 1983). They protect tissues from the deleterious effects of the oxygen free radical $O_2^{\cdot-}$ (superoxide) formed in various enzyme reactions in which a single electron is transmitted to O_2:

$$O_2 + e^- \longrightarrow O_2^{\cdot-} \text{ (superoxide)}$$

$$O_2^{\cdot-} + O_2^{\cdot-} + 2H^+ \xrightarrow[\text{dismutase (SOD)}]{\text{superoxide -}} H_2O_2 \text{ (hydrogen peroxide)} + O_2$$

$$2H_2O_2 \xrightarrow{\text{catalase}} 2H_2O + O_2$$

The inactivation of superoxide is catalyzed by SOD, and the subsequent dismutation of H_2O_2 is facilitated by catalase (Elstner, 1982). In illuminated green cells the chloroplasts are the organelles with the highest rate of oxygen turnover, including O_2 consumption. Thus the chloroplasts are also the main sites of formation of $O_2^{\cdot-}$ and H_2O_2 as intermediates (Section 5.2.4). Accordingly in leaves more than 90% of the SOD is localized in the chloroplasts and only 4–5% is in the mitchondria (Jackson *et al.*, 1978). The precise localization of the isoenzymes Cu–Zn-SOD and Mn-SOD within the chloroplasts (whether in the stroma or the thylakoid membranes) is still a matter of discussion (Jackson *et al.*, 1978; Sandmann and Böger, 1983). Nevertheless, it can be assumed that besides its role in the water-splitting system the key function of manganese in chloroplasts is its role, as a metal component of SOD, in the protection of the photosynthetic apparatus from the deleterious effects of oxygen activation. Other protective components are catalase, peroxidase, and small molecules such as glutathione and ascorbic acid. The severe disorganization of the chloroplast lamellar system (photooxidation) that occurs when manganese is deficient strongly supports the role of manganese in this protective system.

Another manganese-containing enzyme, an acid phosphatase that is violet-colored in solution, has been isolated from sweet potato (Uehara *et al.*, 1974a). It is not yet clear, however, whether this enzyme has a specific requirement for manganese or whether other divalent cations such as Zn^{2+} might be at least as effective under *in vivo* conditions (Uehara *et al.*, 1974b).

9.2.4 Modulation of Enzyme Activities

In many *in vitro* reactions manganese can replace magnesium and in some cases is even more effective in enzyme activation.

Two representative examples are the NADPH specific decarboxylating malate dehydrogenase, the malic enzyme, and the isocitrate dehydrogenase:

Malic enzyme catalyzing the reaction:
$$\text{Malate} + NADP^+ \xrightarrow[Mn^{2+}, Mg^{2+}]{} \text{pyruvate} + NADPH + H^+ + CO_2$$

Isocitrate dehydrogenase catalyzing the reaction:
$$\text{Isocitrate} + NADP^+ \xrightarrow[Mn^{2+}, Mg^{2+}]{} \text{oxalosuccinate} + NADPH$$

Whether this also reflects the situation *in vivo*, however, remains an open question (Clarkson and Hanson, 1980). In the cells, on average, the magnesium concentration is 50–100 times higher than that of manganese. In order to be of importance *in vivo*, the activation by manganese should therefore be much more effective than that by magnesium. An example of the high efficiency of manganese is shown in Fig. 9.5 for the chloroplast RNA polymerase. Activation of this enzyme requires 10 times higher concentrations of magnesium than of manganese.

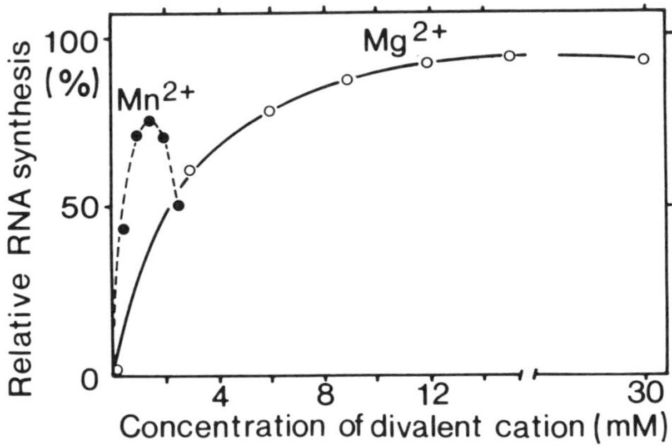

Fig. 9.5 Influence of manganese and magnesium concentrations on chloroplast RNA synthesis. (Redrawn from Ness and Woolhouse, 1980.)

An increase in peroxidase activity (R—H + R—H + $H_2O \rightarrow$ R—R + $2H_2O$) is a typical feature of manganese-deficient tissue, whereas catalase activity ($H_2O_2 + H_2O_2 \rightarrow 2H_2O + O_2$) is not affected (Vielemeyer *et al.*, 1966; Bar-Akiva and Lavon, 1967). Manganese-deficient leaves exhibit exceptionally high IAA oxidase activity (Morgan *et al.*, 1976), which might lead to enhanced auxin (IAA) degradation in the tissue. Probably the major constituent of the IAA oxidizing system is peroxidase, and cell wall–bound peroxidases are associated with the same enzyme protein (Rao *et al.*, 1982).

9.2.5 Synthesis of Proteins, Carbohydrates, and Lipids

Although manganese seems to be a structural constituent of ribosomes (Lyttleton, 1960) and also activates RNA polymerase (Fig. 9.5), protein

Table 9.5
Effect of Manganese Deficiency on the Growth and Composition of Bean Plants[a]

Parameter	Leaves		Stems		Roots	
	+Mn	−Mn	+Mn	−Mn	+Mn	−Mn
Dry wt (g/plant)	0·64	0·46	0·55	0·38	0·21	0·14
Protein nitrogen (mg/g dry wt)	52·7	51·2	13·0	14·4	27·0	25·6
Soluble nitrogen (mg/g dry wt)	6·8	11·9	10·0	16·2	17·2	21·7
Soluble carbohydrates (mg/g dry wt)	17·5	4·0	35·6	14·5	7·6	0·9

[a]From Vielemeyer *et al.* (1969).

synthesis is obviously not specifically impaired in manganese-deficient tissues. The protein content of deficient plants is either similar to (Table 9.5) or somewhat higher than that of plants adequately supplied with manganese (Lerer and Bar-Akiva, 1976).

Elevated levels of soluble nitrogen (nitrate, amino acids, etc.) in deficient plants (Table 9.5; Lerer and Bar-Akiva, 1976) are not necessarily indications of a direct involvement of manganese in nitrate assimilation. Such an involvement has not been established so far. Assimilation is much more likely to be affected indirectly by a shortage of carbohydrates for nitrate reduction in the cytosol, lower rates of nitrite reduction in manganese-deficient chloroplasts, and finally a lower demand for reduced nitrogen in the new growth of deficient plants. Manganese deficiency has the most severe effect on the level of soluble carbohydrates, which is drastically reduced, particularly in the roots (Table 9.5). Considering the role of manganese in photosynthesis, this decline in carbohydrate level is to be expected.

The role of manganese in lipid metabolism is not yet clear. In manganese-deficient cells not only the chlorophyll content but, even more so, the content of typical chloroplast membrane constituents such as glycolipids and polyunsaturated fatty acids is reduced (Constantopoulus, 1970). Besides a possible direct role in lipid synthesis (Section 8.3), manganese most likely has an indirect role in chloroplasts as a component of the pigment system of the lamella (photosystem II) and, in the form of Mn-SOD, as part of the system of protection against photooxidation of membrane constituents such as polyunsaturated fatty acids.

Distinct changes are observed in the lipid content and composition of the seeds of manganese-deficient plants (Fig. 9.6). In the deficiency range the manganese content of the leaves and both the seed yield and oil content are positively correlated. In contrast, the protein content is negatively correlated with the manganese content of the leaves, which in this case is an expression of a typical "concentration effect" due to inhibited seed growth

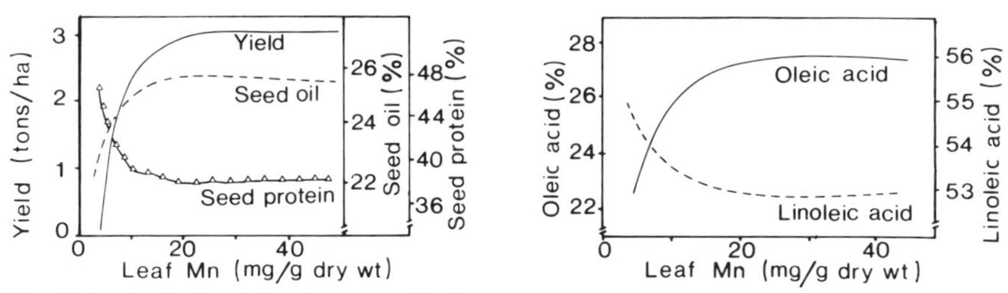

Fig. 9.6 Relationships between leaf manganese content and seed yield and seed composition of soybean. (Reproduced from Wilson *et al.*, 1982, by permission of the Crop Science Society of America.)

under conditions of manganese deficiency. The fatty acid composition of the oil is also markedly altered, the content of linoleic acid (Fig. 9.6) and certain other fatty acids increasing (Wilson *et al.*, 1982). This is counteracted by a decrease in oleic acid content. The lower oil content in the seeds of deficient plants is probably mainly the result of lower rates of photosynthesis and thus a smaller supply of carbon skeletons for fatty acids synthesis. In addition a direct involvement of manganese (as a component of the biotin enzyme) in the biosynthesis of the fatty acids might be a contributing factor. The reasons for the changes in fatty acid composition in relation to the manganese supply are obscure.

9.2.6 Cell Division and Extension

The rate of elongation seems to respond more rapidly to manganese deficiency than the rate of cell division. As shown in Fig. 9.7 with isolated tomato roots in sterile culture and an ample supply of carbohydrates, within

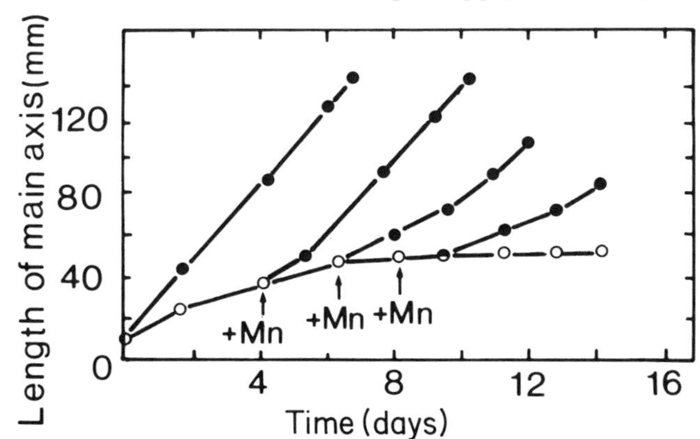

Fig. 9.7 Effect of transfer from manganese-deficient to complete medium on growth of main axis of excised tomato roots. Key: ○, manganese absent; ●, manganese present. (Based on Abbott, 1967.)

less than 2 days there is a rapid decline in extension of the main axis. After the supply of manganese is restored, the growth rate rapidly reaches normal levels if the deficiency is not too severe. In manganese-deficient plants, the formation of lateral roots ceases completely (Abbott, 1967). There is a greater abundance of small nonvacuolated cells in these roots than in control roots, which indicates that cell elongation is impaired to a greater extent by manganese deficiency than is cell division, an observation supported by tissue culture experiments (Neumann and Steward, 1968).

9.2.7 Manganese Deficiency and Toxicity

The critical deficiency levels of manganese are between 10–20 mg/g dry wt in mature leaves, and surprisingly consistent regardless of the plant species or cultivar or the prevailing environmental conditions. Below this level dry matter production (Ohki *et al.*, 1979), net photosynthesis, and chlorophyll content decline rapidly, whereas rates of respiration and transpiration remain unaffected (Ohki, 1981). Manganese-deficient plants are more susceptible to damage by freezing temperatures (Bunje, 1979).

In dicotyledons, intercostal chlorosis of the younger leaves is the most distinct symptom of manganese deficiency, whereas in cereals, greenish gray spots on the more basal leaves ("gray speck," *Dörrfleckenkrankheit*) are the major symptoms. In legumes, manganese-deficiency symptoms on the cotyledons are known as "marsh spot" in peas or as "split seed" disorder in lupins; the latter include discoloration, splitting, and deformity of seeds (Hocking *et al.*, 1977).

Under field conditions manganese deficiency is usually confined to plants growing in highly leached tropical soils or high-pH soils with a large organic matter content (Farley and Draycott, 1973). It can be readily corrected by the application of manganese (e.g., as $MnSO_4$) to the soil or leaves.

In contrast to the narrow range of critical deficiency levels in leaves, the critical toxicity levels vary widely among plant species and environmental conditions. An example of the differences among crop species is given in Table 9.6. Even within a species the critical toxicity levels can vary by a factor of 3 to 5 between cultivars (Ohki *et al.*, 1981; Edwards and Asher, 1982).

Of the environmental factors affecting critical toxicity levels, temperature and the presence of silicon are of particular importance. At high temperatures these levels in the leaves are often much higher than those at low temperatures (Heenan and Carter, 1977; Rufty *et al.*, 1979). The effect of silicon is comparable to that of high temperatures; that is, it increases tissue tolerance to manganese (Section 10.3).

Table 9.6
Critical Toxicity Levels of Manganese in the Shoots of Various Plant Species[a,b]

Species	Manganese content (mg/g dry wt)
Maize	200
Pigeon pea	300
Soybean	600
Cotton	750
Sweet potato	1380
Sunflower	5300

[a]Based on Edwards and Asher (1982).
[b]Critical toxicity levels are associated with a 10% reduction in dry matter production.

There is an extensive literature on manganese toxicity and the physiological interpretations of it (Foy, 1983). Brown spots (precipitations of MnO_2) on older leaves surrounded by chlorotic zones are a typical symptom of manganese toxicity (Bussler, 1958; Horst and Marschner, 1978b). Quite often, however, manganese-induced symptoms of deficiencies of other mineral nutrients, such as iron, magnesium, and calcium, are dominant. Induced iron deficiency is caused by both inhibited uptake of iron (Isermann, 1975) and competition (or an imbalance) between manganese and iron at the cellular level. In manganese-induced magnesium deficiency (Heenan and Campbell, 1981), competition is responsible for binding sites in the roots during uptake as well as for various metabolic reactions. It is evident from Table 9.7, however, that the inhibition of Mg^{2+} uptake by high Mn^{2+} concentrations involves more than a 1 : 1 competition for specific binding sites in the roots. Presumably, Mn^{2+} not only competes much more effectively but also in some way blocks the binding sites for Mg^{2+}. In comparison, the uptake of potassium is only slightly affected by increasing

Table 9.7
Effect of Increasing Manganese Concentrations in the Substrate on Uptake Rates of Manganese and Magnesium in Roots of Soybean Plants[a]

	Manganese supply (μM)		
Nutrient	1·8	90	275
Manganese	0·5	3·1	4·8
Magnesium	121·8	81·1	20·2

[a]Data represent micromoles of nutrient taken up per gram of root dry weight. Based on Heenan and Campbell (1981).

Mn^{2+} concentrations (Heenan and Campbell, 1981). Thus manganese toxicity can often be counteracted by a large magnesium supply (Löhnis, 1960).

Calcium deficiency is another well-known symptom (called "crinkle leaf") induced by manganese toxicity in dicotyledons such as cotton (Foy *et al.*, 1981) and bean (Horst and Marschner, 1978c). When the supply of manganese is excessive, the translocation of calcium into the shoot apex, especially, is inhibited. This might be related to the fact that high manganese levels decrease the cation-exchange capacity of the leaf tissue (Horst and Marschner, 1978c). According to results of Morgan *et al.* (1966, 1976) IAA oxidase activity is much higher not only in manganese-deficient tissues (as already described) but also in tissues with excessive manganese levels. It might be assumed, therefore, that lower auxin levels in areas of new growth are ultimately responsible for both the decline in IAA-mediated extension and the formation of new binding sites for the transport of Ca^{2+} to the apical meristems (Horst and Marschner, 1978c). Loss of apical dominance and enhanced formation of auxillary shoots ("witches' broom") constitute another symptom of manganese toxicity (Kang and Fox, 1980), further supporting the hypothesis of a relationship between auxin metabolism and manganese toxicity.

Under certain ecological conditions (e.g., temporary waterlogging), high concentrations of Mn^{2+} in the soil solution are difficult to avoid (Section 16.4). Selecting genotypes which tolerate high tissue concentrations of manganese is therefore a promising way to overcome these difficulties. Quick screening methods are based on the application of manganese to a single leaf (Horst, 1982).

9.3 Copper

9.3.1 General

The divalent copper ion (Cu^{2+}) is strongly bound in soils to humic and fulvic acids, forming copper–organic matter complexes (Stevenson and Fitch, 1981). In soil solutions up to 98% of the copper is complexed to low-molecular-weight organic compounds (Hodgson *et al.*, 1966). There are conflicting reports as to whether copper is taken up as Cu^{2+} or as copper chelate (Graham, 1981). Because of the high affinity of Cu^{2+} for various ligands (amino acids, phenolics, and synthetic chelators), even in nutrient solutions added Cu^{2+} may be rapidly complexed (Graham, 1979). If Cu^{2+} is present in equimolar concentrations, its rate of uptake is much higher than that of copper complexed to synthetic chelators such as EDTA (Coombes *et al.*, 1977) or DTPA (diethylenetriaminepentaacetic acid) (Wallace, 1980a,b). In the roots (press sap) and in the xylem sap more than 99% of the copper is present in complexed form (Graham, 1979). In the xylem and phloem sap it most likely is complexed to amino acids and related compounds (White *et al.*, 1981a,b).

Divalent copper is readily reduced to Cu^+, which is unstable. In this respect copper is similar to iron. Most of the functions of copper as a plant nutrient are based on the participation of enzymatically bound copper in redox reactions. In the redox reactions of the terminal oxidases copper enzymes react directly with molecular oxygen. Therefore, the terminal oxidation in living cells is catalyzed by copper and not by iron.

9.3.2 Copper Proteins

According to Sandmann and Böger (1983) copper is present in three different forms in proteins: (a) *blue proteins* without oxidase activity (e.g., plastocyanin), which function in one-electron transfer; (b) *non-blue proteins*, which produce peroxidases and oxidize monophenols to diphenols; and (c) *multicopper proteins* containing at least four copper atoms per molecule, which act as oxidases (e.g., ascorbate oxidase and laccase) and catalyze the reaction

$$2AH_2 + O_2 \rightarrow 2A + 2H_2O$$

Cytochrome oxidase is a mixed copper–iron protein catalyzing the terminal oxidation in mitochondria.

Under conditions of copper deficiency the activity of these copper enzymes decreases fairly rapidly. In several cases, however, it is difficult or even impossible to correlate directly the decrease in activity of a certain copper enzyme with gross metabolic changes or with inhibition of plant growth. For example, in copper-deficient cells a drastic decrease in cytochrome oxidase activity was without effect on the respiration rate, indicating that this enzyme might be present in large excess in the mitochondria (Bligny and Douce, 1977).

9.3.2.1 Plastocyanin

In general, more than 50% of the copper localized in chloroplasts is bound to plastocyanin. This compound has a molecular weight of ~10,000 and contains one copper atom per molecule. Plastocyanin is a component of the electron transport chain of photosystem I (Fig. 5.1). A proportion of 3 to 4 molecules of plastocyanin per 1000 molecules of chlorophyll appears to be

Table 9.8
Effect of Copper Deficiency in Spinach on Chloroplast Pigments and Photosynthetic Electron Transport in Photosystems II and I[a]

Treatment	Chloroplast pigment content (μg/g leaf fresh wt)			Plastocyanin (nano atoms/mg chlorophyll)	Photosystem activity (relative)	
	Chlorophyll	Carotenoids	Plasto-quinone		PS II	PS I
+Cu	1310	248	106	5·16	100	100
−Cu	980	156	57	2·08	66	22

[a]Based on Baszynski *et al.* (1978).

the rule (Sandmann and Böger, 1983). As would be expected with copper deficiency, there is a greater decrease in the plastocyanin content and the activity of photosystem I than there is in the content of other chloroplast pigments and the activity of photosystem II (Table 9.8).

In copper-deficient plants, the rate of photosynthesis can also be reduced for other reasons directly related to the role of copper in chloroplasts. Copper is a component of other chloroplast enzymes (see below) and is required for the synthesis of quinones; the decrease in plastoquinone in the chloroplasts (Table 9.8) may reflect this function of copper. In copper-deficient chloroplasts the inhibition of the electron transport is further accentuated by the lack of two polypeptides in the chloroplast membrane, which are probably necessary to maintain the appropriate membrane fluidity to ensure the mobility of plastoquinone molecules to transport electrons between the two photosystems (Droppa *et al.*, 1984). Whether copper also plays a role in the activity of fraction I protein (RuBP carboxylase/oxygenase) in the stroma of chloroplasts is still a matter of discussion (Walker and Webb, 1981).

9.3.2.2 Superoxide Dismutase

The various types of SOD isoenzymes and their requirement for the detoxification of superoxide radicals ($O_2^{\cdot-}$) have been discussed in Section 9.2. The copper–zinc SOD has a molecular weight of ~32,000, and at the active site probably one copper and one zinc atom are closely connected by a common histidine nitrogen (Sandmann and Böger, 1983). In green leaves most of this enzyme is localized in the chloroplasts (Reddy and Venkaiah, 1984), particularly in the stroma of chloroplasts (Jackson *et al.*, 1978). The copper atom is involved in the mechanism of detoxification of $O_2^{\cdot-}$ generated in photorespiration (Halliwell, 1978).

9.3.2.3 Cytochrome Oxidase

This terminal oxidase of the mitochondrial electron transport chain (Fig. 5.5) contains two copper atoms and two iron atoms in the heme configuration. The activity of the enzyme can be blocked by cyanide; the remaining respiratory O_2 consumption of cells is mediated by the quinal oxidase known as "alternative oxidase" (in the "alternative pathway", see Chapter 5). This enzyme contains copper but no heme iron. There is substantial evidence that the alternative oxidase in microsomes is involved in the desaturation of long-chain fatty acids such as oleic and linoleic acids (Wahle and Davies, 1977).

9.3.2.4 Ascorbate Oxidase

Ascorbate oxidase catalyzes the oxidation of ascorbic acid to L-dehydroascorbic acid according to the equation:

$$2 \times \text{Ascorbic acid} \xrightarrow{O_2} 2 \times \text{L-Dehydroascorbic acid} + H_2O$$

The enzyme occurs in cell walls and in the cytoplasm. Its physiological functions are still uncertain, but it may act as a terminal respiratory oxidase, as shown above, or in combination with phenolases (Section 9.3.2.5). It is certainly also linked to redox shuttle systems together with glutathione. Its activity is a very sensitive indicator of the copper nutritional status of a plant. Although in this case a decrease in enzyme activity and in plant growth are not directly related, there is a close positive correlation in the suboptimal

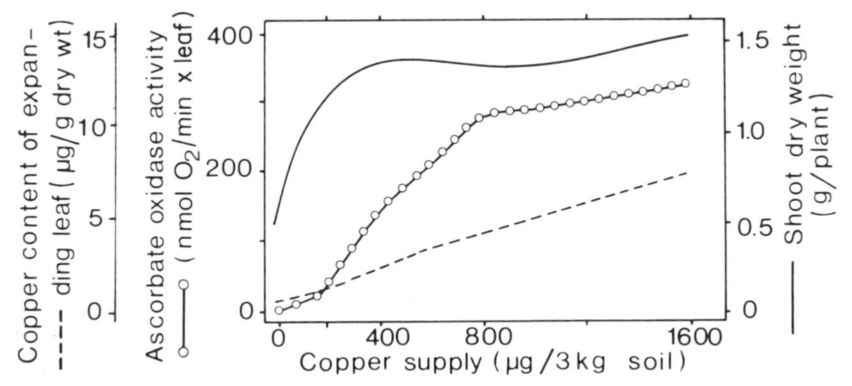

Fig. 9.8 Relationship between copper supply, shoot dry weight, ascorbate oxidase activity, and copper content of subterranean clover. (Modified from Loneragan *et al.*, 1982a.)

concentration range between the copper content of leaf tissue and its ascorbate oxidase activity (Fig. 9.8).

On the basis of this correlation a rapid and simple colorimetric field test for ascorbate oxidase activity has been developed for the diagnosis of copper deficiency. The results of this test agree closely with the diagnosis based on chemical analysis of the copper content of leaves (Delhaize *et al.*, 1982).

9.3.2.5 Phenolase and Laccase

These enzymes catalyze oxygenation reactions of plant phenols. Laccase can be found in the thylakoid membranes of chloroplasts, where it is presumably required for the synthesis of plastoquinone, a constituent of the photosynthetic e^- transport chain (Fig. 5.1). Phenolase has two distinct enzyme functions: (a) monooxygenation of monophenols, resembling tyrosinase activity, and (b) monooxygenation of *o*-diphenols such as dihydroxyphenylalanine (Dopa), resembling polyphenol oxidase activity and leading to quinones according to:

$$\text{Tyrosine} \xrightarrow{+O_2} \text{Dopa} \; +H_2O \dashrightarrow \xrightarrow{+O_2} \text{Quinone} \; +2\,H_2O \longrightarrow \text{Melanotic substances}$$

(Tyrosine: $H_2N\text{-}CH\text{-}COOH$, CH_2, OH; Dopa: $H_2N\text{-}CH\text{-}COOH$, CH_2, HO, OH; Quinone: R, O, O)

As in higher plants the two functions are associated as a phenolase complex, although the term *phenolase* is preferable (Walker and Webb, 1981). Phenolase is involved in the biosynthesis of lignin (see Section 9.3.4) and alkaloids and in the formation of brown melanotic substances, which are sometimes formed when tissues are wounded (e.g., in apple and potatoes). These substances are also active as phytoalexins, which inhibit spore germination and fungal growth. Under conditions of copper deficiency, the decrease in phenolase is quite severe (Table 9.9) and is correlated with an accumulation of phenolics (Judel, 1972) and a decrease in the formation of

Table 9.9
Effect of Copper Deficiency on Flowering and Enzyme Activities in *Chrysanthemum morifolium*[a]

Treatment	Copper content (mg/g leaf) dry wt)	No. of flowering shoots/ plant	No. of open flowers/ plant	Enzyme activity in leaves (relative)		
				Phenolase	IAA oxidase	Peroxidase
Cu sufficient	7·9	14·2	13·1	100	100	100
Cu deficient	2·4	8·3	0·5	26	52	41

[a]Based on Davies *et al.* (1978).

melanotic substances. The latter effect is reflected, for example, in the close correlation between the color of spores of *Aspergillus niger* and the copper nutritional status. With an ample copper supply the spores are black; with mild deficiency they are light brown; and with severe deficiency they are white.

The decline in phenolase activity may be at least indirectly responsible for the delay in flowering and maturation often observed in copper-deficient plants (Reuter *et al.*, 1981) and shown for the flowering of *Chrysanthemum* in Table 9.9. Copper deficiency led to a decrease in the number of flowering shoots, but mainly prevented the opening of flowers. As would be expected, the phenolase activity was much lower in copper-deficient plants, but the activities of IAA oxidase and peroxidase were also lower.

It is well documented that certain phenols are active inhibitors of IAA oxidase and that ascorbic acid also strongly inhibits peroxidase-catalyzed oxidation of IAA (Palmieri and Giovinazzi, 1982). From the changes in enzyme activities (Table 9.9), it may be assumed that in copper-deficient plants IAA accumulates and delays flowering, as is also the case with IAA application (Graves *et al.*, 1977).

9.3.2.6 Amine Oxidases

These copper proteins catalyze oxidative deamination, as follows:

$$R—CH_2—NH_2 + O_2 + 2H^+ \rightarrow R—CHO + NH_3 + 2H_2O$$

They can also use polyamines, such as putrescine and spermidine, as substrates (in which case they act as diamine oxidases) and are abundant in legumes during the seedling stage (Walker and Webb, 1981). Considering the proposed role of polyamines as secondary messengers (Chapter 5), the effects of copper deficiency on phytohormone balance and on developmental processes may also be related to polyamine metabolism.

9.3.3 Carbohydrate and Nitrogen Metabolism

In plants suffering from copper deficiency the content of soluble carbohydrates is considerably lower than normal during the vegetative stage (Brown and Clark, 1977; Mizuno *et al.*, 1982). In wheat, however, after anthesis, when the grain has developed as a dominant sink, copper-deficient plants have only a few grains (see Section 6.3.3), remain green (i.e., actively photosynthesizing) and build up excessive levels of soluble carbohydrates both in the leaves and in the roots (Fig. 9.9). At this stage, the leaves of deficient plants even release droplets of honeydew-like substances (Graham, 1980a).

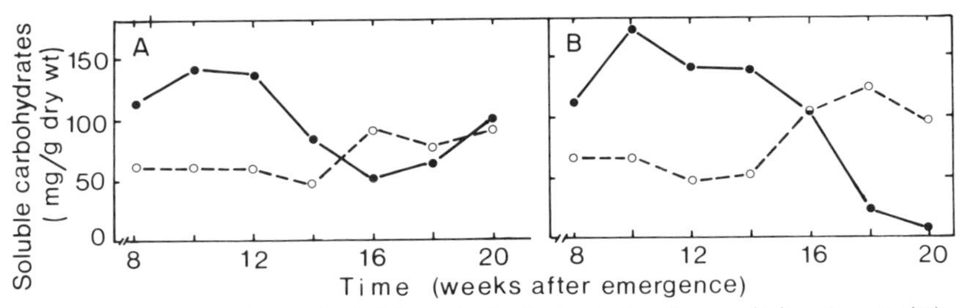

Fig. 9.9 Concentrations of soluble carbohydrates in flag leaves (A)and roots (B) of wheat plants grown at two copper levels as a function of plant age. Key: •——•, +Cu; ○ – – ○ −Cu. (Modified from Graham, 1980a.)

Given the role of copper in photosynthesis, a lower content of soluble carbohydrates would be expected during vegetative growth. The importance of lower carbohydrate levels for pollen formation and fertilization, however, is a matter of controversy. Low levels of carbohydrates in the shoot apex and a lack of starch in the pollen of copper-deficient plants (Reuter *et al.*, 1981; Bussler, 1981b) are considered by Mizuno *et al.* (1982) to be responsible for low fertility, whereas Graham (1980a) argues strongly against this relationship.

In legumes receiving a low copper supply, nodulation and N_2 fixation are depressed (Hallsworth *et al.*, 1964). Symptoms of nitrogen deficiency occur which can be overcome by the application of mineral nitrogen (Snowball *et al.*, 1980). Although this effect of copper possibly indicates a large specific copper requirement in root nodules for the N_2 fixation mechanism (Cartwright and Hallsworth, 1970), an indirect effect, involving a shortage of carbohydrate supply for nodulation and N_2 fixation in copper-deficient plants, is more likely. The close correlation between carbohydrate supply and N_2 fixation rates has been discussed in Chapter 7.

In nonlegumes as well, the effect of copper supply on nitrogen metabolism seems to be mainly indirect. In some cases free amino acids and nitrate may accumulate in copper-deficient plants (Brown and Clark, 1977). This, however, is neither typical nor the result of inhibition of protein synthesis, as shown in Table 9.10. Leaves of copper-deficient plants are often darker green (i.e., have a higher chlorophyll and protein content); however, their photosynthetic efficiency is much lower, which corresponds to their lower carbohydrate content.

It has been shown repeatedly that nitrogen application accentuates copper deficiency, and when the nitrogen supply is large, the application of copper fertilizers is required for maximum yield (Vetter and Teichmann, 1968; Thiel and Finck, 1973). In addition to unspecific interactions (e.g., growth enhancement by nitrogen), nitrogen has specific effects on copper availability and mobility, including (a) sequestration of a higher proportion

Table 9.10
Effect of Copper Deficiency on Growth, Protein and Chlorophyll Content, and Photosynthesis in Spinach[a]

Treatment	Fresh weight after 20 days (g/plant)	Protein nitrogen (mg/g fresh wt)	Chlorophyll (μg/g fresh wt)	CO_2 Fixation (mg CO_2/mg chlorophyll)
+Cu	17	2·2	546	136
−Cu	4	2·8	604	62

[a]Based on Botrill *et al.* (1970).

of copper complexed to amino acids and proteins in mature tissue and (b) a decrease in the rate of retranslocation of copper from old leaves to areas of new growth. Retranslocation of copper is closely related to leaf senescence (Chapter 3). Because a large nitrogen supply delays senescence, it also retards copper retranslocation (Hill *et al.*, 1978). In agreement with this, the critical deficiency levels of copper in the shoots required for maximum yield increase with increasing nitrogen supply (Thiel and Finck, 1973).

9.3.4 Lignification

Impaired lignification of cell walls is the most typical anatomical change induced by copper deficiency in higher plants. This gives rise to the characteristic distortion of young leaves, bending and twisting of stems and twigs ("pendula" forms in trees, Oldenkamp and Smilde, 1966), and increase in the lodging susceptibility of cereals, particularly in combination with a large nitrogen supply (Vetter and Teichmann, 1968).

As shown in Table 9.11 copper has a marked effect on the formation and chemical composition of cell walls. In deficient leaves the ratio of cell wall material to the total dry matter decreases; simultaneously the proportion of α-cellulose increases whereas the lignin content is only about half that of leaves adequately supplied with copper. This effect on lignification is even more distinct in the sclerenchyma cells of stem tissue (Fig. 9.10). In plants

Table 9.11
Effect of Copper Deficiency on the Cell Wall Composition of the Youngest Fully Emerged Leaves of Wheat[a]

Treatment	Copper content (μg/g dry wt)	Cell wall content (% dry matter)	Percentage of cell walls		
			α-Cellulose	Hemicellulose	Lignin
+Cu	7·1	46·2	46·8	46·7	6·5
−Cu	1·0	42·9	55·3	41·4	3·3

[a]From Robson *et al.* (1981b).

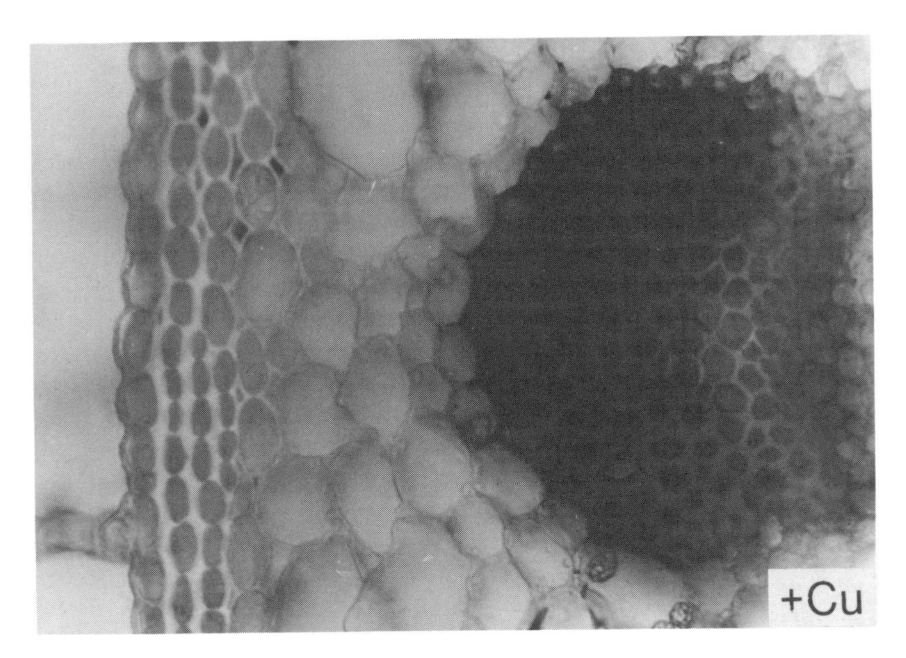

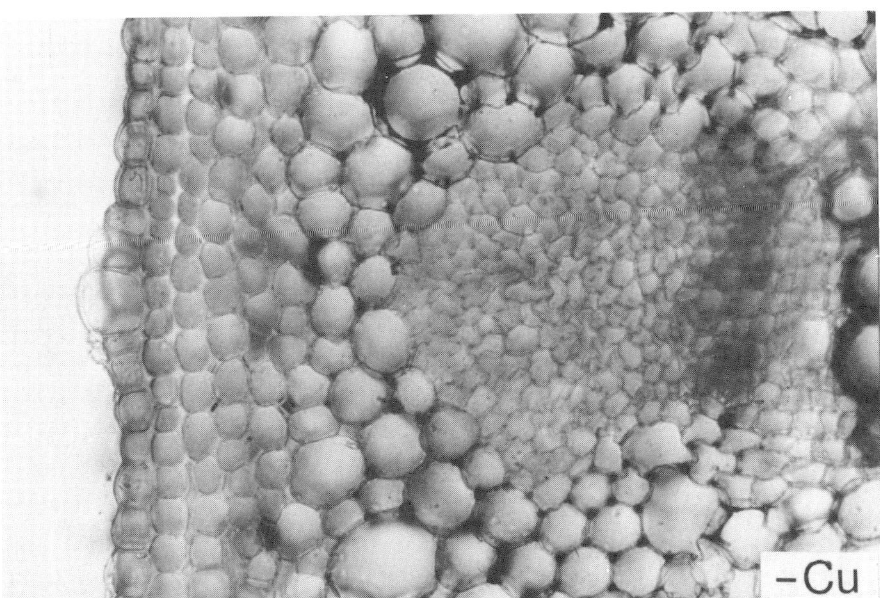

Fig. 9.10 Stem sections of sunflower plants grown with sufficient copper supply (50 μg Cu/liter) and without copper supply. (*Top*) Copper sufficient; walls of the sclerenchyma cells are thick and lignified. (*Bottom*) Copper deficient; walls of the sclerenchyma cell are thin and nonlignified. (Rahimi and Bussler, 1974.)

suffering from severe copper deficiency the xylem vessels are also insufficiently lignified. A decrease in lignification occurs even with mild copper deficiency and is thus a suitable indicator of the copper nutritional status of a plant (Rahimi and Bussler, 1974; Pissarek, 1974).

Lignification responds rapidly to copper supply; transition periods of copper deficiency during the growth period can be readily identified by variations in the degree of lignification in stem sections (Bussler, 1981b). For a histochemical test, the lignified areas can be visualized by the use of acidified phloroglucinol, which produces a red color (Rahimi and Bussler, 1974). However, this reagent also stains aromatic compounds other than lignin (Robson *et al.*, 1981b).

The inhibition of lignification in copper-deficient tissue is related to the role of the multicopper enzymes phenolase and laccase in the oxidation of phenols such as *p*-coumaric acid, one of the precursors of lignin biosynthesis:

H-C-COOH
CH
OH

$\xrightarrow[\text{Phenolase, Laccase}]{+O_2} \longrightarrow$ Lignin

In agreement with this, in copper-deficient tissue not only is the phenolase activity lower, but also phenols accumulate (Adams *et al.*, 1975; Robson *et al.*, 1981b).

9.3.5 Pollen Formation and Fertilization

Copper deficiency affects grain, seed, and fruit formation much more than vegetative growth (see also Section 6.3.3). A typical example is shown in Table 9.12. Whereas with a 0·5 μg supply of copper the maximum dry weights of roots and shoots were obtained and flower formation was nearly adequate, no fruits were formed. For fruit formation a much greater copper supply was required. The decline in root, leaf and stem dry weights with a 1·0 and 5·0 μg copper supply reflects sink competition (Chapter 5). With a 10 μg supply, toxicity occurred.

As discussed in Chapter 6, the main reason for the decrease in the formation of generative organs is the nonviability of the pollen from copper-deficient plants (Graham, 1975). When the copper supply is adequate, the anthers containing pollen and the ovaries have the highest copper content in the flowers (Knight *et al.*, 1973) and obviously also the highest copper demand. The critical stage of copper deficiency-induced pollen

Table 9.12
Relationship between Copper Supply and Growth and Dry Matter Distribution in Red Pepper[a]

Copper supply (μg/pot)	Dry weight (g/plant)			
	Roots	Leaves and stem	Buds and flowers	Fruits
0·0	0·8	1·7	0·16	None
0·5	1·6	3·3	0·28	None
1·0	1·5	3·2	0·38	0·87
5·0	1·4	3·0	0·36	1·81
10·0	1·2	2·0	0·28	1·99

[a]From Rahimi (1970).

sterility is microsporogenesis (Graham, 1975). According to Dell (1981) reduced seed set in copper-deficient plants might be the result of the inhibition of pollen release from the stamina, since lignification of the anther cell walls is required for rupture of the stamina and subsequent release of the pollen. In copper-deficient plants lignification of the anther cell walls is reduced or absent (Dell, 1981).

After grain set in wheat (Hill *et al.*, 1979c) and seed set in subterranean clover (Reuter *et al.*, 1981), further grain and seed growth, surprisingly, are not influenced by the copper nutritional status of the plants, even though at maturity the copper content of wheat grains in plants adequately supplied with copper is five to six times higher than in deficient plants. This result further emphasizes the importance of adequate copper supply during fertilization for final seed and fruit yield.

9.3.6 Copper Deficiency and Toxicity

9.3.6.1 Copper Deficiency

The critical deficiency level of copper in vegetative parts is generally in the range of 3 to 5 μg/g dry wt. Depending on the plant species, plant organ, developmental stage, and nitrogen supply, this range can be larger (Thiel and Finck, 1973; Robson and Reuter, 1981). Stunted growth, distortion of young leaves, necrosis of the apical meristem, and bleaching of young leaves ("white tip" or "reclamation disease" of cereals grown in organic soils), and/or "summer dieback" in trees are, in addition to those already discussed, typical visible symptoms of copper deficiency (Rahimi and Bussler, 1973a). Enhanced formation of tillers in cereals and of auxillary shoots in dicotyledons are secondary symptoms caused by necrosis of the apical meristem. Wilting in young leaves, also characteristic of copper-deficient plants, has been interpreted either as being the result of impairment of water transport due to insufficient lignification of the xylem vessels (Rahimi and Bussler, 1973b; Pissarek, 1974) or of structural weaknesses in the cell wall system rather than the result of a low water content per se (Graham, 1976). Inhibition of calcium transport to areas of new growth may occur in copper-deficient plants (Brown, 1979), but it is probably a secondary symptom related to the impairment of cell wall formation.

Foliar applications of copper in the form of inorganic salts, oxides, or chelates are required to correct copper deficiency rapidly in soil-grown plants. Soil applications of inorganic copper salts, oxides, or slow-release metal compounds are more appropriate for long-term effects. Selecting genotypes with highly efficient copper uptake and, particularly, highly efficient translocation of copper from the roots to the shoots and retranslocation within the shoot, is a promising long-term approach to the prevention of copper deficiency under certain ecological conditions (Section 4.3.2).

9.3.6.2 Copper Toxicity

For most crop species, the critical toxicity level of copper in the leaves is considered to be above 20 to 30 μg/g dry wt (Hodenberg and Finck, 1975; Robson and Reuter, 1981). There are, however, marked differences in copper tolerance among plant species (e.g., bean is much more tolerant than maize); these differences are directly related to the copper content of the shoots (Bachthaler and Stritesky, 1973). Copper toxicity may induce iron deficiency (Woolhouse, 1983), depending on the source of iron supply (Rahimi and Bussler, 1973a). Chlorosis can also be a direct result of the action of high copper concentrations on lipid peroxidation and thus the destruction of the thylakoid membranes (Sandmann and Böger, 1983).

A large copper supply usually inhibits root growth before shoot growth (Lexmond and Vorm, 1981). This does not mean, however, that roots are more sensitive to high copper concentrations; rather, they are the sites of preferential copper accumulation when the external copper supply is large, as shown in Table 9.13 for tomato plants. In plants receiving a large supply, the copper content of the roots rises proportionally to the concentration of copper in the external medium, whereas transport to the shoot remains strongly restricted. Without an analysis of the roots, critical toxicity levels of copper in the shoots are therefore not necessarily an appropriate indicator of the copper tolerance of plants. This is an especially important consideration when genotypes are being compared.

In nontolerant plants, inhibition of root elongation and damage to root cell membranes are an immediate response to a large copper supply (Wainwright and Woolhouse, 1977). Certain changes in root morphology,

Table 9.13
Relationship between Copper Supply (Nutrient Solution), Dry Weight, and Copper Content of Tomato Plants[a]

Copper supply (μg/liter)	Dry weight (g/plant)		Copper content (mg/kg dry wt)		
	Roots	Shoots	Roots	Stems and petioles	Leaf blades
0	0·3	2·6	4·0	2·8	3·0
2·5	2·5	9·4	3·8	2·1	3·2
5·0	3·2	11·2	6·4	2·4	4·1
50·0	3·4	12·0	64·0	4·3	14·6
250·0	1·6	9·7	360·0	6·2	20·3

[a]From Rahimi and Bussler (1974).

such as inhibited elongation and enhanced lateral root formation (Savage *et al.*, 1981), might be related to the sharp decrease in IAA oxidase activity in roots exposed to high copper concentrations (Coombes *et al.*, 1976).

For various reasons there is increasing concern about copper toxicity in agriculture (Tiller and Merry, 1981). These include the high copper levels in soils caused by the long-term use of copper-containing fungicides (e.g., in vineyards), industrial and urban activities (air pollution, city waste, and sewage sludge), and the application of pig and poultry slurries high in copper. Mechanisms of copper tolerance in plants are therefore of interest for crop production on copper-polluted soils.

9.3.6.3 Mechanisms of Copper Tolerance

Genotypical differences in tolerance to copper and other heavy metals are well-known in certain species and ecotypes of natural vegetation (Ernst, 1982; Woolhouse, 1983). It has been known for centuries that in mining areas, particularly, a special flora (*metallophytes*) that is highly tolerant to these metals develops on outcrops. In some species this tolerance is restricted to a particular heavy metal; in other species cotolerance to several heavy metals exists (Cox and Hutchinson, 1979). In some tolerant species the copper content of the leaves can be as high as 0·1% of the dry weight (Morrison *et al.*, 1981).

According to Woolhouse (1983) the mechanisms of copper tolerance in higher plants can be grouped as follows: (a) exclusion or restriction of copper uptake, (b) immobilization of copper in cell walls, (c) compartmentation of copper in insoluble complexes, (d) compartmentation of copper in soluble complexes and (e) enzyme adaptation. In general, the exclusion or restriction of copper uptake seems to be of minor importance in higher plants. Immobilization in the cell walls is considered by some to be an important mechanism (Turner, 1970) and by others to be only relatively important because of its limited binding capacity (Woolhouse, 1983). Without doubt, compartmentation of copper within the cytoplasm and in the vacuoles either as soluble or insoluble complexes are the dominant mechanisms of copper tolerance (Wu *et al.*, 1975). Binding to specific proteins, so-called metallothioneins (Section 8.3), is thereby of crucial importance. The synthesis of metallothioneins in copper-tolerant genotypes can be induced by a large supply of copper (Rauser and Curvetto, 1980; Rauser, 1983). These proteins are very high in cysteine, to which copper is bound much more firmly than is either zinc or cadmium (Rupp and Weser, 1978). From a copper-tolerant *Agrostis* species a metallothionein has been isolated which has 20·8% cysteine residues and binds 54 mg copper per gram protein (Curvetto and Rauser, 1979). High copper levels may, however, induce enhanced synthesis of low-molecular weight proteins for copper detoxification other than metallothioneins (Tukendorf *et al.*, 1984). In bean leaves, synthesis of proteins which are almost identical to plastocyanin is enhanced (Nicholson *et al.*, 1980).

Compared with this mechanism of copper detoxification, the adaptation of enzymes to high concentrations of copper is of limited importance and is presumably restricted to extracellular enzymes such as cell wall–bound phosphatases in the free space of roots (Wainwright and Woolhouse, 1975).

9.4 Zinc

9.4.1 General

Zinc is taken up predominantly as a divalent cation (Zn^{2+}); at higher pH, it is presumably also taken up as a monovalent cation ($ZnOH^{+}$). High concentrations of other divalent cations such as Ca^{2+} inhibit zinc uptake somewhat. The long-distance transport of zinc takes place mainly in the xylem, where zinc is either bound to organic acids or exists as the free divalent cation (White *et al.*, 1981a,b). In plants, zinc is not oxidized or reduced; its functions as a mineral nutrient are based primarily on its properties as a divalent cation with a strong tendency to form tetrahedral complexes (Clarkson and Hanson, 1980). Zinc acts either as a metal component of enzymes or as a functional, structural, or regulatory cofactor of a large number of enzymes. Under conditions of zinc deficiency, therefore, the changes in metabolism are quite complex. Nevertheless, some of the changes are typical and can be rather well explained by the functions of zinc in certain enzyme reactions or steps in a given metabolic pathway.

9.4.2 Zinc-Containing Enzymes

At least four plant enzymes contain bound zinc: alcohol dehydrogenase, Cu–Zn superoxide dismutase, carbonic anhydrase, and RNA polymerase.

9.4.2.1 Alcohol Dehydrogenase

This enzyme catalyzes the reduction of acetaldehyde to ethanol:

$$\text{Pyruvate} \xrightarrow{\nearrow CO_2} \text{acetaldehyde} \xrightarrow{NADH \curvearrowright NAD^+} \text{ethanol}$$

In higher plants under aerobic conditions, ethanol formation takes place mainly in meristematic zones, such as root apices. In zinc-deficient higher plants, there is a decrease in alcohol dehydrogenase activity, but there is no evidence, as there is in lower plants (e.g., yeast), that this decline in activity is responsible for gross metabolic changes.

9.4.2.2 Superoxide Dismutase

In this isoenzyme zinc is associated with copper (Cu–Zn-SOD). The localization and role of SOD have been discussed in Section 9.3. Although the functions of zinc in the enzyme are not yet clear (Walker and Webb, 1981), the activity of the Cu–Zn-SOD is much lower than normal under conditions of zinc deficiency. It can be restored *in vitro* by the addition of zinc to the assay medium (Vaughan *et al.*, 1982).

9.4.2.3 Carbonic Anhydrase

Carbonic anhydrase (CA) catalyzes the hydration of CO_2:

$$CO_2 + H_2O \rightleftharpoons HCO_3^- + H^+$$

The equilibrium depends on the pH of the solution and the catalytic action of CA, which increases the equilibration rate. Carbonic anhydrase from dicotyledons consists of six subunits and has a molecular weight of 180,000 and six zinc atoms per molecule (Sandmann and Böger, 1983). The enzyme is localized both in the cytoplasm and in the chloroplasts. The function of CA, particularly that of chloroplast CA, in photosynthetic CO_2 assimilation is still obscure. According to Edwards and Walker (1983) CA has the following effects on CO_2 equilibria in green cells:

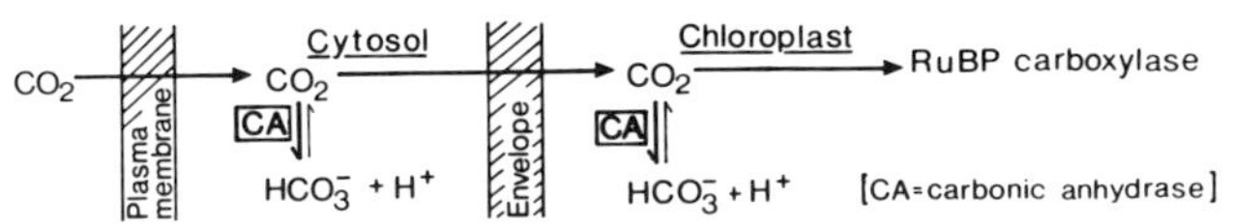

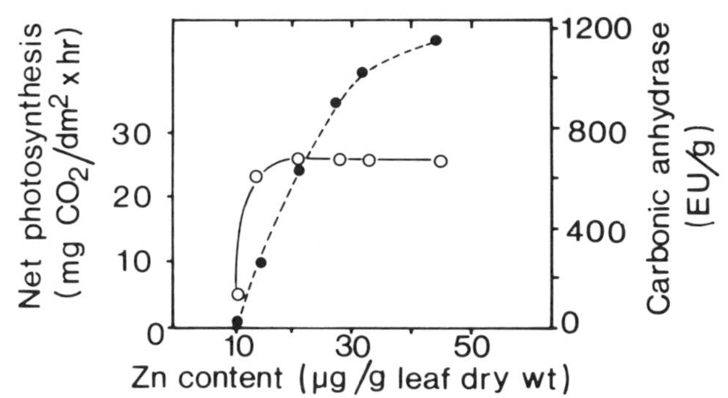

Fig. 9.11 Relationship between the zinc content of leaf blades and net photosynthesis (○——○) and carbonic anhydrase activity (● – – ●) in cotton. EU, enzyme units. (Modified from Ohki, 1976.)

In C_3 plants the CA in the cytoplasm could increase the pool size of dissolved CO_2 in the cytosol of mesophyll cells. It is not HCO_3^-, however, but CO_2 that moves freely and rapidly across the chloroplast envelope, and CA in the stroma would again increase the pool size of stored CO_2. The substrate for RuBP carboxylase, however, is CO_2 rather than HCO_3^-.

The lack of a direct relationship between CA activity and photosynthetic CO_2 assimilation is demonstrated by experiments involving variations in the zinc supply to plants. The CA activity is closely related to the zinc supply, but CO_2 assimilation per unit leaf area is affected only by very low CA activity (Randall and Bouma, 1973). Under conditions of extreme zinc deficiency CA activity is absent, but even when CA activity is low maximum net photosynthesis occurs (Fig. 9.11). Thus plants seem normally to contain a considerable excess of CA. It remains an open question whether this enzyme serves functions other than that of increasing the storage pool of CO_2.

9.4.3 Enzyme Activation

Zinc is required for the activity of various types of enzymes, including dehydrogenases, aldolases, isomerases, transphosphorylases, and RNA and DNA polymerases. It is not surprising, therefore, that zinc deficiency is associated with an impairment of carbohydrate metabolism and protein synthesis.

9.4.3.1 Carbohydrate Metabolism

In leaves suffering from increasing zinc deficiency, a sharp decline in CA activity is the most obvious of the changes that take place in the activities of

Table 9.14
Effect of Increasing Zinc Deficiency on the Activities of Enzymes in Leaves of Maize Plants Grown without Zinc Supply[a]

Enzyme	Percentage decrease in activity after days without zinc supply		
	5	10	15
Fructose-1,6-bisphosphatase	36	50	65
Carbonic anhydrase	84	76	84
PEP Carboxylase	<1	5	34
RuBP Carboxylase	9	41	38
Malic enzyme	<1	22	37

[a]Relative values; zinc-sufficient plants = 100. Based on Shrotri *et al.* (1983).

the various enzymes of the CO_2 assimilation pathway (Table 9.14). Also the activity of fructose-1,6-bisphosphatase declines, fairly rapidly; that of the other enzymes is affected to a much lesser extent, however, particularly with mild zinc deficiency.

Fructose-1,6-bisphosphatase is a key enzyme in the partitioning of C_6 sugars in the chloroplasts and the cytoplasm according to the following scheme:

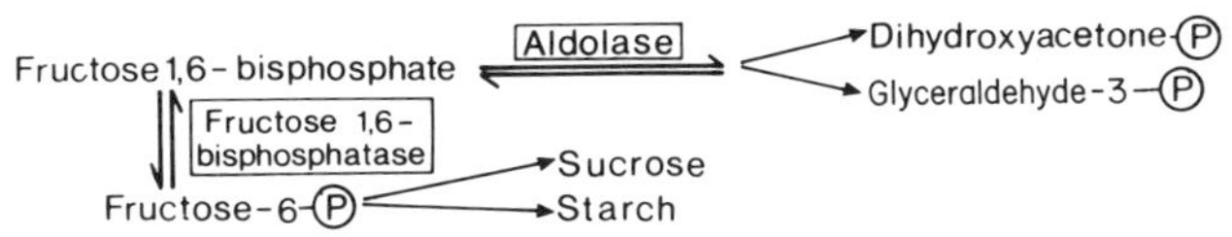

Another group of key enzymes in carbohydrate metabolism comprises the aldolase isoenzymes. In the chloroplast, they regulate the transfer of C_3 photosynthates to the cytoplasm (see diagram), and in the cytoplasm they regulate the transfer of C_6 sugars to the glycolytic pathway. According to O'Sullivan (1971) aldolase activity is drastically and specifically reduced by zinc deficiency in various plant species and may serve as an indicator of the zinc nutritional status of plants. This result has not been confirmed for citrus leaves (Bar-Akiva *et al.*, 1971), a discrepancy that might be related to a different cofactor requirement for the aldolase isoenzymes of different plant species and/or cell compartments such as chloroplasts and cytosol.

A direct connection between zinc nutritional status and starch formation has been found only in bean (Jyung *et al.*, 1975), where zinc deficiency leads to a decrease in both starch contents and starch synthase activity. As a rule, however, the carbohydrate content of leaves is either unaffected or even increased by zinc deficiency (Table 9.15) as a result of the concentration effect caused by impaired growth (i.e., lower sink activity). In extreme cases this can lead to excretion (leakage) of sugars at the leaf surface (Rahimi and Bussler, 1978).

Table 9.15
Effect of Zinc Deficiency and Resumption of Zinc Supply on Zinc and Carbohydrate Contents of Cabbage Leaves[a]

	Zinc supply (μM)		
Parameter	1·0	0·001	0·001 + 2·0[b]
Zinc content (mg/kg dry wt)	21	14	30
Sugars (mg/g fresh wt)	4·2	9·1	5·0
Starch (mg/g fresh wt)	7·5	24·6	19·2
Hill reaction activity (relative)	100	30	92

[a]Based on C. P. Sharma *et al.* (1982).
[b]Twenty-four hours after resumption of 2·0 μM zinc supply.

Despite a drastic reduction in the rate of photosynthesis (as indicated by the Hill reaction activity), sugars and starch accumulate in zinc-deficient leaves (Table 9.15). As early as 24 hr after the zinc supply is restored, the sugar levels and the Hill reaction activity are again comparable to that of the adequately supplied control plants continuously receiving 1·0 μM zinc.

In conclusion, most experimental evidence obtained with green plants supports the view that zinc deficiency-induced changes in carbohydrate metabolism are not primarily responsible for either growth retardation or the visible symptoms of zinc deficiency.

9.4.3.2 Protein Synthesis

The rate of protein synthesis and the protein content of zinc-deficient plants are drastically reduced. The accumulation of amino acids and amides in these plants demonstrates the importance of zinc for protein synthesis, as does the fact that a similar degree of deficiency of other micronutrients such as manganese and copper does not result in an accumulation of these precursors (Table 9.16).

At least three distinct mechanisms are responsible for the adverse effect of zinc deficiency on protein synthesis and protein content. From studies with

Table 9.16
Effects of Zinc, Manganese, and Copper Deficiency on Free Amino Acids and Amides in Tomato Plants[a]

		Deficiency		
Parameter	Control	Zinc	Manganese	Copper
Dry weight (mg/plant)	213·6	66·0	69·4	75·8
Amino acids (μg/mg dry wt)	16·0	31·6	18·3	21·5
Amides (μg/mg dry wt)	4·2	42·9	3·0	1·9

[a]Based on Possingham (1957).

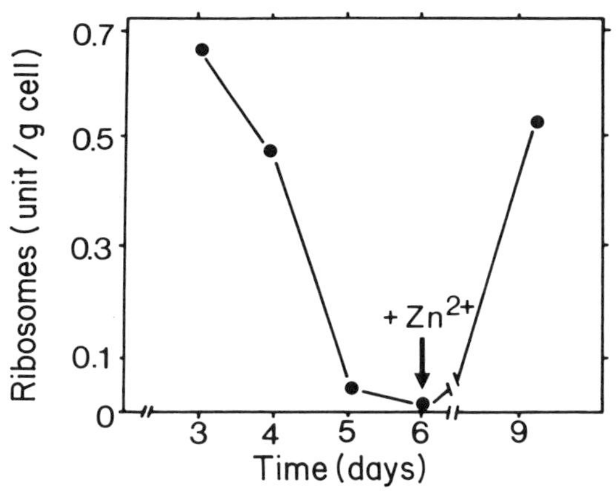

Fig. 9.12 Changes in number of ribosomes per cell (relative units) in *Euglena* in response to zinc deficiency (6 days without zinc) and resumption of zinc supply. (Data recalculated from Prask and Plocke, 1971.)

Euglena it is well established that zinc is an essential component of RNA polymerase. There are about two atoms of zinc per molecule, and if the zinc is removed, the enzyme is inactivated (Falchuk *et al.*, 1977). Furthermore, zinc is a constituent of ribosomes and is essential for their structural integrity. The zinc content of ribosomal RNA in zinc-sufficient cells of *Euglena* is in the range of 650 to 1280, whereas that in zinc-deficient cells is 300 to 380 μg per gram RNA (Prask and Plocke, 1971). In the absence of zinc, ribosomes disintegrate, but reconstitution takes place after resumption of the zinc supply (Fig. 9.12).

The decrease in the protein content of zinc-deficient plants is also the result of enhanced rates of RNA degradation. Higher rates of RNase activity are a typical feature of zinc deficiency (C. P. Sharma *et al.*, 1982). A clear inverse correlation exists between zinc supply and RNase activity and also between RNase activity and protein content (Table 9.17).

Table 9.17
Effect of Zinc Supply on Fresh Weight, RNase Activity, and Protein Nitrogen in Soybean (*Glycine wighii*)[a]

Zinc supply (mg/liter)	Fresh weight (g/plant)	RNase activity (%)[b]	Protein nitrogen (% fresh wt)
0·005	4·0	74	1·82
0·01	5·1	58	2·25
0·05	6·6	48	2·78
0·10	10·0	40	3·65

[a]Based on Johnson and Simons (1979).
[b]Percent hydrolysis of RNA substrate.

An increase in RNase activity is observed even before symptoms of zinc deficiency such as stunted growth and changes in leaf anatomy become apparent (Dwivedi and Takkar, 1974).

9.4.4 Tryptophan and Indoleacetic Acid Synthesis

There is general agreement that the most distinct zinc deficiency symptoms—stunted growth and "little leaf"—are related to disturbances in the metabolism of auxins, indoleacetic acid (IAA) in particular. The mode of action of zinc in auxin metabolism is still obscure, however. As shown by Skoog (1940), the auxin content of the shoot apices of zinc-deficient plants is extremely low. Furthermore, the auxin level decreases before the appearance of deficiency symptoms, and after the supply of zinc is restored, the level rapidly increases before the resumption of growth (Tsui, 1948). Low auxin levels in zinc-deficient plants may be the result of high IAA oxidase activity (Skoog, 1940). Most experimental evidence, however, supports the view that zinc is required for the synthesis of tryptophan, a precursor for the synthesis of IAA according to the following scheme:

Indole --(Serine)--> Tryptophan ($-CH_2-CH(NH_2)-COOH$) ---> Indoleacetic acid (IAA) ($-CH_2-COOH$)

Tryptophan levels in zinc-deficient plants are low (Tsui, 1948), and in maize mild zinc-deficiency symptoms can be corrected by supplying either zinc or tryptophan (Salami and Kenefick, 1970). In agreement with this an increasing zinc supply results in an increase in the tryptophan content in the grains of rice plants growing in zinc-deficient calcareous soil (Table 9.18).

In *Neurospora* it is well established that the enzyme tryptophan synthase has a requirement for zinc (Nason *et al.*, 1951). Experimental evidence for a similar requirement in higher plants, however, is lacking. The results of Takaki and Kushizaki (1970) obtained with zinc-deficient maize are difficult

Table 9.18
Effect of Zinc Supplied as Zn-EDTA to a Zinc-Deficient Calcareous Soil on the Tryptophan Content of Rice Grains[a]

Zinc supply (mg/kg soil)	Tryptophan content (mg/kg dry wt)
0	830
5	1476
10	2011

[a]Based on Singh (1981).

to interpret. These authors found much higher levels of tryptophan and tryptamine in zinc-deficient leaves than in normal leaves and therefore suggested that zinc is required for the synthesis of IAA from tryptophan via tryptamine. The reasons for these controversial results, even in the same plant species, are obscure.

9.4.5 Phosphorus–Zinc Interactions

Large applications of phosphorus fertilizers to soils low in available zinc may induce zinc deficiency and increase the zinc requirement of plants. As summarized by Loneragan *et al.* (1979), three different factors may be responsible for this: (a) "dilution" of zinc in plants by the increase in growth induced by phosphorus fertilizers, (b) inhibition of zinc uptake by the cations (Ca^{2+} in particular) added with phosphorus fertilizers, and (c) phosphorus-enhanced zinc adsorption in the soil to hydroxides and oxides of iron and aluminum and to $CaCO_3$. Several experimental results indicate that there are additional phosphorus–zinc interactions in plants, including inhibition of zinc translocation from the roots to the shoot (Burleson and Page, 1967; Trier and Bergmann, 1974) and "physiological inactivation " of zinc within the shoots (Boawn and Brown, 1968). The latter suggestion is based on the observation that symptoms of zinc deficiency are related to the phosphorus/zinc ratio rather than to the zinc concentration in the shoots (Millikan, 1963; Boawn and Brown, 1968).

It is not very likely, however, that the phosphorus/zinc ratio per se is of much physiological importance in this respect, as shown in Table 9.19,

Table 9.19
Effect of Phosphorus Supply on Shoot Dry Weight and Phosphorus and Zinc Content of Young Grapevine Leaves[a]

Phosphorus supply (mmol per kg soil or liter solution)	Dry weight of shoots (g)	Content of young leaves		Phosphorus/ zinc
		Phosphorus (mg/g dry wt)	Zinc (μg/g dry wt)	
Soil culture				
0·3	19·9	2·63	26·6	99
3·0	19·9	2·69	19·7	137
6·0	17·2	3·06	15·5	197[b]
Water culture				
0·1	15·7	2·72	15·7	173
1·0	15·2	8·60	13·9	618
5·0	15·5	13·47	13·8	976

[a]Average values for six cultivars. From Marschner and Schropp (1977).
[b]Zinc deficiency symptoms.

where the effects of increasing phosphorus supply to grapevines growing either in a calcareous soil (30% $CaCO_3$; pH 7·6) or in water culture are compared. In the soil-grown plants, an increase in the phosphorus supply resulted in only a slight increase in the phosphorus content of young leaves but a substantial reduction in the zinc content and induced zinc deficiency symptoms at a phosphorus/zinc ratio (molar basis) of ~200. In contrast, in the plants grown in water culture, an increase in the phosphorus supply led to a marked increase in the phosphorus content but hardly affected the zinc content of the young leaves. Zinc deficiency symptoms were not observed, even at phosphorus/zinc ratios of ~1000. Basically the same effects were found in mature leaves. These results strongly support the view that the major phosphorus–zinc interactions take place not in the plant (Ghoneim and Bussler, 1980) but in the soil, where the availability of zinc (as determined by chemical extraction procedures) and zinc diffusion rates are reduced by a large phosphorus supply (Schropp and Marschner, 1977). This decrease in zinc availability is brought about by enhanced adsorption of zinc on, for example, iron oxides (Loneragan, 1975; Loneragan *et al.*, 1979) rather than by the precipitation of zinc phosphates, which are relatively good sources of zinc for plant growth (Jurinak and Inouye, 1962).

Phosphorus–zinc interactions in soil are complicated by the infection of roots with vesicular–arbuscular mycorrhiza. Infected roots take up more zinc than noninfected roots (Pairunan *et al.*, 1980). As discussed later (Section 15.7) mycorrhizal infection of roots is strongly depressed by an increase in phosphorus supply. Lambert *et al.* (1979) found that an increasing phosphorus supply to soybean plants reduced the zinc content of mycorrhizal plants but hardly affected the zinc content of noninfected plants.

New insight into phosphorus–zinc interactions in plants has been provided by Loneragan and his group from Western Australia. Under conditions of zinc deficiency and ample supply of phosphorus, the phosphorus uptake increases and toxic concentrations of phosphorus may accumulate (Loneragan *et al.*, 1979). Under these experimental conditions, the symptoms of phosphorus toxicity (e.g., necrosis on mature leaves) in zinc-deficient plants may be mistaken for evidence of an accentuation of the zinc deficiency because of the large phosphorus/zinc ratio. In agreement with the results obtained with grapevine, the values for ochra (Table 9.20) show that zinc uptake is not affected by increasing phosphorus concentrations in the external solution. In the absence of zinc or with low external concentrations, however, the levels of phosphorus in the shoots of ochra are very high, leading to toxicity symptoms. In general, a phosphorus content of greater than ~2% leaf dry wt can be considered toxic.

Although the connection between zinc deficiency and phosphorus toxicity is not yet fully understood, there is substantial evidence that zinc affects

Table 9.20
Effects of Zinc and Phosphorus Concentrations in the Nutrient Solution on the Growth and Zinc and Phosphorus Content of the Shoots of Ochra (*Abelmoschus exculentum* L.)[a]

Zn supply (μM)	Dry weight (g/plant)[b]		Zinc content (μg/g dry wt)[b]		Phosphorus content (mg/g dry wt)[b]	
	P_1	P_2	P_1	P_2	P_1	P_2
0	8·3	9·5	15	15	11·0	24·1
0·25	9·6	9·9	27	27	9·6	20·2
1·0	9·8	11·6	54	57	8·7	11·8

[a]Based on Loneragan *et al.* (1982b).
[b]P_1, 0·25 m*M* phosphate; P_2, 2·0 m*M* phosphorus.

phosphorus metabolism in the roots (Loughman *et al.*, 1982) and increases the permeability of the plasma membranes of root cells to phosphorus, and also to chloride (Welch *et al.*, 1982). It is well documented that zinc stabilizes biomembranes, presumably by reacting with the sulfhydryl groups (R—SH) of membrane proteins (Chvapil, 1973). Zinc may therefore have specific functions in the structural orientation of macromolecules within membranes and thus in membrane integrity (Welch *et al.*, 1982), an aspect which deserves more attention. These effects of zinc are presumably related also to the protective role of SOD enzymes against lipid peroxidation of membranes, the latter process being closely related to the senescence of tissues (Pauls and Thompson, 1984).

9.4.6 Zinc Deficiency and Toxicity

9.4.6.1 Zinc Deficiency

Zinc deficiency is widespread among plants grown in highly weathered acid soils and in calcareous soils. In the latter case zinc deficiency is often associated with iron deficiency ("lime chlorosis"). The low availability of zinc in calcareous soils of high pH results mainly from the adsorption of zinc to clay or $CaCO_3$ rather than from the formation of sparingly soluble $Zn(OH)_2$ or $ZnCO_3$ (Trehan and Sekhon, 1977). In addition zinc uptake and translocation to the shoot are strongly inhibited by high concentrations of bicarbonate, HCO_3^- (Forno *et al.*, 1975; Dogar and van Hai, 1980). This effect has striking similarities to the effect of HCO_3^- on iron (Section 16.5), in neither case is it well understood. In contrast to iron deficiency, zinc deficiency in plants grown in calcareous soils can be corrected fairly readily by the soil application of inorganic zinc salts such as $ZnSO_4$ (Nayyar and Takkar, 1980).

Fig. 9.13 Zinc deficiency in apple with typical inhibition of internode elongation ("rosetting") and reduction in leaf size ("little leaf").

The most characteristic visible symptoms of zinc deficiency in dicotyledons are stunted growth due to shortening of internodes ("rosetting") and a drastic decrease in leaf size ("little leaf"), as shown in Fig. 9.13. Quite often these symptoms are combined with chlorosis, which is either highly contrasting or diffuse. In cereals such as maize, chlorotic bands along the midrib and a red, spotlike discoloration (caused by anthocyanins) on the leaves often occur (Rahimi and Bussler, 1979). Symptoms of chlorosis and necrosis on older leaves of zinc-deficient plants (Rahimi and Bussler, 1979) are most likely the result of phosphorus toxicity. In dicotyledons zinc deficiency symptoms may be similar in appearance to virus infections and may be mistaken as such, as happened in the case of "crinkle disease" in hop (*Humulus scandens*) (Schmidt *et al.*, 1972).

The critical deficiency levels are below ~15–20 mg zinc per kilogram dry weight of leaves. Grain and seed yield are depressed to a relatively greater extent by zinc deficiency than the total dry matter production, perhaps because zinc seems to play a specific role in fertilization. Pollen grains have a very high zinc content, and during fertilization most of the zinc is incorporated into the developing seed (Polar, 1975).

9.4.6.2 Zinc Toxicity

When the zinc supply is large, zinc toxicity can readily be induced in nontolerant plants, inhibition of root elongation being a very sensitive parameter (Godbold *et al.*, 1983). Quite often zinc toxicity leads to chlorosis of young leaves, which may be, at least in part, an induced deficiency since hydrated Zn^{2+} and Fe^{2+} have similar ionic radii (Woolhouse, 1983). In crop plants zinc toxicity occurs mainly when sewage sludge with a high zinc content has been applied. The critical toxicity levels of zinc in leaves of crop plants are more than ~400–500 mg/kg dry wt. Increasing soil pH by liming is the most effective procedure for decreasing both zinc content and zinc toxicity in plants (White *et al.*, 1979). In comparison with the genotypical differences among plants of the natural vegetation (see below), genotypical differences in zinc tolerance among crop plants are small, but nevertheless marked, even within the same species. In soybean genotypes zinc tolerance is positively correlated with the zinc content of the leaves. In other words, in this example mechanism of tolerance is not exclusion of uptake but tolerance of the tissue for high levels of zinc (White *et al.*, 1979).

As is the case for copper tolerance (Section 9.3.5) there are impressive genotypical differences in zinc tolerance among certain species and ecotypes of the natural vegetation (Ernst, 1982; Woolhouse, 1983). The mechanisms of zinc tolerance are in principle the same as those of copper tolerance (Section 9.3.6.3). The relative importance of these mechanisms may be somewhat different, however. Compartmentation of zinc as soluble complexes in the vacuoles seems to be of particular importance for zinc tolerance, at least in *Deschampsia*, as shown in Table 9.21. Whereas in the

Table 9.21
Effect of Zinc Supply on Concentrations of Zinc in the Cytoplasm and Vacuoles of Roots of a Zinc Tolerant and Nontolerant Clone of *Deschampsia caespitosa*[a]

External concentration (m*M* Zn^{2+})	Bound zinc in the cytoplasm (m*M*)		Soluble zinc in the vacuoles (m*M*)	
	Nontolerant	Tolerant	Nontolerant	Tolerant
0·10	7·1	10·6	3·7	5·3
0·75	33·4	6·2	2·1	33·4

[a]Based on Brookes *et al.* (1981).

nontolerant clone receiving an abundant supply of zinc there is a preferential accumulation of zinc in the cytoplasm, in the tolerant clone the zinc concentration in the cytoplasm remains low; instead, zinc is sequestered in the vacuoles, most likely in the form of soluble zinc complexes of malate and oxalate (Mathys, 1977).

The formation of insoluble complexes such as metallothioneins (Section 4.3) in the cytoplasm has not yet been studied in relation to zinc tolerance in plants. The stability of such zinc complexes is low, however, in comparison with that of the corresponding copper complexes (Rupp and Weser, 1978). Enzyme adaptation to high concentrations of ionic zinc is probably not involved in zinc tolerance (Woolhouse, 1983). Calcium increases the tolerance of plants to zinc (Baker, 1978). Inhibition of zinc uptake by calcium is involved, but there is also an increase in tissue tolerance to high zinc levels, presumably because of the role of calcium in membrane stabilization, which is required for sequestering zinc in the vacuoles.

9.5 Molybdenum

9.5.1 General

Although molybdenum is a metal, it occurs in aqueous solution mainly as molybdate oxyanion, MoO_4^{2-}, in its highest oxidized form [Mo(VI)]. Its properties resemble those of nonmetals and other divalent inorganic anions. In mineral soils of low pH, phosphate and molybdate are similar with respect to their strong adsorption to iron oxide hydrates, and in uptake by roots sulfate and molybdate are competing anions. Molybdate is a weak acid; with decreasing pH from 6·5 to 4·5 and below the dissociation decreases ($MoO_4^{2-} \rightarrow HMoO_4^{-} \rightarrow H_2MoO_4$) and the formation of polyanions is favored (molybdate → tri- → hexamolybdate). The rate of molybdate uptake by roots is closely related to metabolic activity (Kannan and Ramani, 1978). In long-distance transport molybdenum seems to be reasonably mobile. Its mobility in the phloem is indicated by the retranslocation of leaf-applied molybdenum (Kannan and Ramani, 1978). The form in which molybdenum is translocated is unknown, but its chemical properties indicate that it is most likely transported as MoO_4^{2-} rather than in complexed form.

The requirement of plants for molybdenum is lower than that for any of the other mineral nutrients. The functions of molybdenum as a plant nutrient are related to the valency changes it undergoes as a metal component of enzymes. In its oxidized stage it exists as Mo(VI); it is reduced to Mo(V) and eventually to Mo(IV).

9.5.2 Molybdenum-Containing Enzymes

In plants only a few enzymes have been found to contain molybdenums as cofactor. These include xanthine oxidase/dehydrogenase, aldehyde oxidase, sulfite oxidase, nitrate reductase, and nitrogenase. Molybdenum appears to have similar catalytic functions in all of these enzymes, and even the protein components of the enzymes are similar (Nicholas *et al.*, 1962). In higher plants, however, the existence of only two molybdenum-containing enzymes, nitrogenase and nitrate reductase, is well established. The molybdenum requirement of higher plants therefore depends on the mode of nitrogen supply.

9.5.2.1 *Nitrogenase*

All biological systems fixing N_2 require nitrogenase (Chapter 7). Each nitrogenase molecule contains two molybdenum atoms, which are associated with iron (24–36 iron atoms per molecule), some of it as a 4Fe–4S cluster, as in ferredoxin. It is likely that molybdenum is directly involved in the reduction of N_2, whereas iron acts as a transmitter of electrons:

$$\xrightarrow{e^-} [\text{Fe-S cluster}]_n \xrightarrow{e^-} \text{Mo} \xrightarrow{\boxed{N_2}} \text{Mo-N}\equiv\text{N} \xrightarrow[e^-]{H^+} \text{Mo N}\equiv\text{NH} \xrightarrow[e^-]{H^+} \text{Mo}\equiv\text{N-NH}_2 \longrightarrow$$

$$\xrightarrow[e^-]{H^+} \text{Mo}=\text{NH-NH}_2 \xrightarrow[e^-]{H^+} \text{Mo}\equiv\text{NH} + \boxed{NH_3} \xrightarrow[2e^-]{2H^+} \text{Mo} + \boxed{NH_3}$$

According to this scheme (Chatt, 1979), N_2 is bound to molybdenum, and electrons are fed stepwise to it, gradually degrading the strong N≡N bond by protonation until two molecules of NH_3 are released.

The molybdenum requirement of root nodules in legumes and nonlegumes (e.g., alder) is therefore relatively high. When the external supply is low, the molybdenum content of root nodules is usually higher than that of the leaves, whereas when the external supply is high, the content in the leaves rises much more than in the nodules (Becking, 1961; Franco and Munns, 1981). The preferential accumulation in root nodules may lead to a considerably lower molybdenum content in the shoots and seeds of nodulated legumes (Ishizuka, 1982).

As would be expected, the growth of plants relying on N_2 fixation is particularly stimulated by the application of molybdenum to deficient soils (Table 9.22). The response of nodule dry weight to molybdenum is spectacular and indirectly reflects the increase in the capacity for N_2 fixation brought about by molybdenum.

In soils with low molybdenum availability, the effect of application of molybdenum to legumes depends on the form of nitrogen supply. As shown in Table 9.23 molybdenum applied to both nodulating and nonnodulating

Table 9.22
Effect of Molybdenum on the Growth and Nitrogen Content of Alder Plants (*Alnus glutinosa*) Grown in a Molybdenum-Deficient Soil[a]

Parameter	Molybdenum application (μg/pot)	Dry weight (g/pot)			
		Leaves	Stems	Roots	Nodules
Dry weight (g)	0	1·79	0·59	0·38	0·007
	150	5·38	2·20	1·24	0·132
Nitrogen content (%)	0	2·29	0·92	1·79	2·77
	150	3·58	1·17	1·83	3·26

[a]From Becking (1961).

Table 9.23
Influence of Nitrogen and Molybdenum Supply on Leaf Nitrogen Content and Seed Yield of Nonnodulating and Nodulating Soybean Plants[a]

Treatment (g Mo/ha)	Nonnodulating (kg N/ha)				Nodulating (kg N/ha)			
	0	67	134	201	0	67	134	201
	Nitrogen (% leaf dry wt)							
0	3·1	4·6	5·3	5·6	4·3	5·1	5·4	5·6
34	3·6	4·7	5·3	5·6	5·7	5·5	5·6	5·6
	Seed yield (tons/ha)							
0	1·71	2·66	3·00	3·15	2·51	2·76	3·08	3·11
34	1·62	2·67	2·94	3·16	3·05	3·11	3·23	3·13

[a]Plants were grown in a soil of pH 5·6. Based on Parker and Harris (1977).

soybean plants increased the nitrogen content and seed yield only in the nodulated plants without or with insufficient supply of nitrogen fertilizer. This demonstrates the greater requirement for molybdenum in N_2 fixation than in nitrate reduction. It also indicates that on soils with low molybdenum availability it is possible to replace the application of nitrogen fertilizer to legumes by the application of molybdenum fertilizer combined with proper rhizobial infection.

9.5.2.2 Nitrate Reductase

Nitrate reductase (NR) contain both heme iron and two atoms of molybdenum and catalyzes the reduction of nitrate to nitrite by a reversible valency change:

$$\xrightarrow{2e^-} \text{2Cyt Fe(II/II)} \xrightarrow{2e^-} \text{2Mo(V/VI) [or Mo(IV/VI)]} \xrightarrow{2e^-} NO_3^-/NO_2^-$$

The details of this reduction are described in Section 8.2. Nitrate reductase activity (NRA) is low in leaves of molybdenum-deficient plants but can be

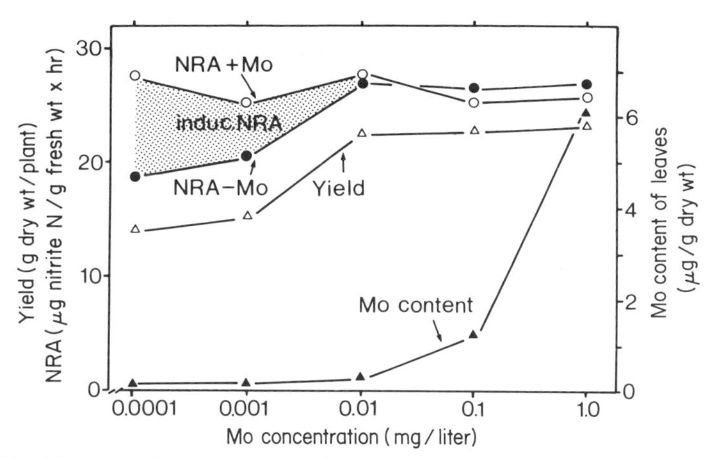

Fig. 9.14 Nitrate reductase activity (NRA) of spinach leaves from plants grown with different levels of molybdenum. Leaf segments were incubated with (NRA + Mo) or without (NRA − Mo) molybdenum for 2 hr. Stippled area represents "inducible NRA." (Redrawn from Witt and Jungk, 1977.)

readily induced within a few hours by infiltration of the leaf segments with molybdenum (Randall, 1969). As shown in Fig. 9.14 there is a close connection between molybdenum supply, the NRA of the leaves (NRA − Mo), and the yield of spinach. Among leaf segments incubated for 2 hr with molybdate (NRA + Mo) there was an increase in the NRA of the leaf tissue of deficient plants only. "Inducible NRA" can therefore be used as a test for the molybdenum nutritional status of plants, as recommended for citrus by Shaked and Bar-Akiva (1967).

The critical deficiency level of molybdenum in the leaves of nitrate-fed plants is usually far below 1 μg per gram leaf dry weight (Fig. 9.14; see also the review by Gupta and Lipsett, 1981). There are conflicting reports as to whether there is any molybdenum requirement when plants are supplied with reduced nitrogen such as ammonium or urea. Also with these sources of nitrogen the growth rate is lower and molybdenum deficiency symptoms occur, particularly when plant species that are very sensitive to molybdenum deficiency are grown on deficient soils (Trobisch and Germar, 1959) or in nutrient solutions lacking molybdenum (Table 9.24). When nitrate is supplied in the absence of molybdenum, plants grow poorly, have a low chlorophyll content (chlorosis), a low ascorbic acid content, but high nitrate levels, and show typical deficiency symptoms on the leaves ("whiptail"; see Fig. 9.16). When ammonium is supplied, the response to molybdenum is much less marked but still present in terms of both dry weight and ascorbic acid content. In the absence of molybdenum ammonium-fed plants also develop whiptail symptoms.

Table 9.24
Effects of Molybdenum and Source of Nitrogen on Growth and Chlorophyll, Nitrate, and Ascorbic Acid Content of Tomato[a]

Treatment ($CaCO_3$ + nitrogen form)	Dry weight (g/plant)		Chlorophyll (mg/100 g fresh wt)		Nitrate (mg/g dry wt)		Ascorbic acid (mg/100 g fresh wt)	
	−Mo	+Mo	−Mo	+Mo	−Mo	+Mo	−Mo	+Mo
Nitrate	9·6	25·0	8·9	15·8	72·9	8·7	99	195
Ammonium	15·9	19·4	21·6	17·4	10·4	8·7	126	184

[a]Based on Hewitt and McCready (1956).

Under nonsterile conditions nitrification of ammonium occurs in the substrate, and thus the uptake and accumulation of nitrate cannot be prevented (Table 9.24). Among cauliflower plants growing under sterile conditions (Hewitt and Gundry, 1970) those supplied with ammonium but not molybdenum did not develop any deficiency symptoms and seemed to have no molybdenum requirement, a result which confirms the corresponding results in green algae (Ichioka and Arnon, 1955). It is supposed (Hewitt and Gundry, 1970) that even low nitrate levels induce the synthesis of nitrate reductase and that this enzyme without the appropriate molybdenum cofactor may have other activities leading to metabolic disturbances. This view is supported by experiments in which tungsten was applied to molybdenum-deficient plants (Fido *et al.*, 1977). The tungsten was incorporated into nitate reductase and prevented the development of typical molybdenum deficiency symptoms, but it did not restore nitrate reductase activity. It is well known that certain metalloenzymes even within the same plant species are not abolutely metal specific. Similar metals can be incorporated and may restore the original catalytic reaction (Sandmann and Böger, 1983), or develop a modified type of catalytic reaction.

9.5.3 Gross Changes in Metabolism

In molybdenum-deficient plants there is a whole range of changes both in organic constituents and in the activities of certain enzymes. Some of these changes are difficult to reconcile with our current knowledge of metabolic functions of molybdenum. For example, the proportion of inorganic phosphorus increases at the expense of organically bound phosphorus (Possingham, 1954), and organic acids accumulate (Höfner and Grieb, 1979). High levels of low-molecular-weight organic nitrogen compounds such as amides (Gruhn, 1961), high ribonuclease activity, and low aminotransferase activity (Agarwala *et al.*, 1978) may indicate the involvement of molybdenum in

Table 9.25
Effect of Molybdenum Supply to Maize Plants on Pollen Production and Viability[a]

Molybdenum supply (mg/kg)	Molybdenum concentration in pollen grains (μg/g)	Pollen-producing capacity (no. of pollen grains/anther)	Pollen diameter (μm)	Pollen viability (% germination)
20	92	2437	94	86
0·1	61	1937	85	51
0·01	17	1300	68	27

[a]From Agarwala *et al.* (1979).

protein synthesis. The presence of xanthine dehydrogenase (a molybdenum-containing enzyme) in pea leaves (Nguyen and Feierabend, 1978) lends support to the idea that molybdenum has some other discrete functions in higher plants.

Molybdenum deficiency also has striking effects on pollen formation in maize (Table 9.25). In deficient plants not only was tasseling delayed, but a large proportion of the flowers failed to open and the capacity of the anthers for pollen production was reduced. Furthermore, the pollen grains were smaller, were free of starch, had much lower invertase activity, and showed poor germination. Impaired pollen formation may also explain the failure of fruit formation in molybdenum-deficient melon plants growing in acid soil (Gubler *et al.*, 1982).

As shown in Fig. 9.15 the risk of premature sprouting of maize grains on

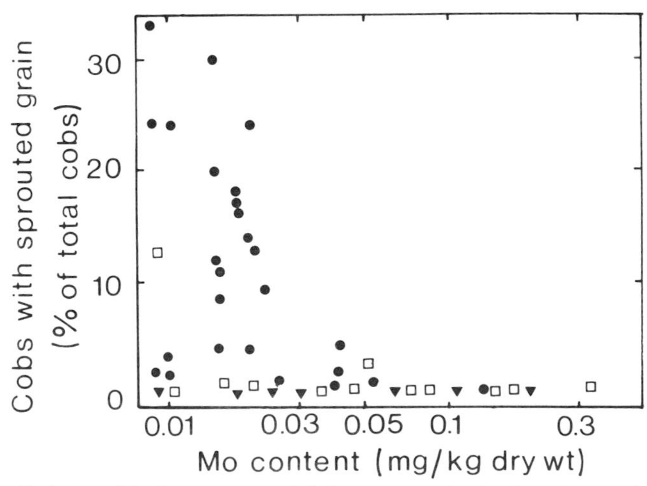

Fig. 9.15 Relationship between molybdenum content of maize grains, time of nitrogen top dressing, and percentage of sprouted cobs of maize. Top dressing with nitrogen at (▾) 30 days; (□) 40–55 days; (•) 70–85 days. (Based on Tanner, 1978.)

the cob greatly increases when the molybdenum content of the grains falls below 0·03 mg/kg dry wt. The extent of sprouting also appears to be related to the time of nitrogen application. Little sprouting occurred when top dressing with NH_4NO_3 took place before 60 days after germination. Sprouting of grains low in molybdenum, however, was strongly enhanced by very late nitrogen application. The causal relationships are not known. Direct stimulatory effects of high nitrate tissue levels on sprouting (Tanner, 1978) or indirect effects via other metabolic changes (e.g., enzyme activities) in deficient grains may be involved.

9.5.4 Molybdenum Deficiency and Toxicity

Depending on the plant species, the critical deficiency levels of molybdenum vary between 0·1 and 1·0 μg per gram leaf dry weight (Gupta and Lipsett, 1981; Bergmann, 1983). In legumes relying on N_2 fixation, symptoms of nitrogen deficiency dominate in molybdenum-deficient plants. When combined nitrogen is supplied, in many plant species the most typical symptom of molybdenum deficiency is a drastic reduction and irregularities in leaf blade formation known as whiptail (Fig. 9.16), caused by local necrosis in the tissue and insufficient differentiation of vascular bundles at early stages of leaf development (Bussler, 1970a).

Local chlorosis and necrosis along the main veins of mature leaves (e.g., "yellow spot" in citrus) and whiptail in young leaves may reflect the same type of local metabolic disturbances, occurring however, at different stages of leaf development (Bussler, 1970b). Marginal chlorosis and necrosis on older leaves with a high nitrate level also occur when there is severe deficiency.

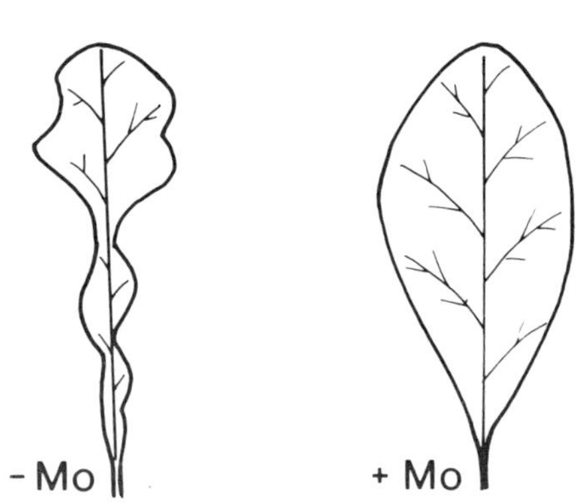

Fig. 9.16 Schematic representation of changes in leaf morphology in molybdenum-deficient cauliflower ("whiptail" symptom).

Table 9.26
Relationship between Soil pH, Molybdenum Supply, and Dry Weight and Molybdenum Content of Soybean[a]

Parameter	Molybdenum supply (mg/pot)	Soil pH		
		5·0	6·0	7·0
Dry weight (g/pot)	0	14·9	18·9	22·5
	5	19·6	19·5	20·4
Molybdenum content of shoots (μg/g dry wt)	0	0·09	0·82	0·90
	5	1·96	6·29	18·50

[a]Based on Mortvedt (1981).

Molybdenum deficiency is widespread in legumes and certain other plant species (particularly of the Cruciferae and in maize) grown in acid mineral soils with a large content of reactive iron oxide hydrates and thus a high capacity for adsorbing MoO_4^{2-}. The application of foliar sprays is the most appropriate procedure for correcting acute deficiency; soil application or seed treatment is preferable for preventing deficiency (Gupta and Lipsett, 1981). Although metabolically coupled (Section 9.5.1), the uptake rate of molybdenum by roots seems not to be controlled by the internal concentration and is merely proportional to the external concentration (Franco and Munns, 1981). This aspect should be taken into account when molybdenum is applied, particularly in combination with liming of acid soils (Table 9.26).

As shown in Table 9.26, both with and without molybdenum supply the molybdenum content of the shoots of soybean increases by a factor of ~10 when the soil pH is increased from 5·0 to 7·0 by liming. The effect of liming only on the plant dry weight is similar to the application of molybdenum to the unlimed soil. Thus, to a certain extent liming and molybdenum application can be seen as alternatives for legume growth on acid mineral soils. A combination of both procedures often leads to distinct luxury uptake and a very high content of molybdenum in the shoots.

A unique feature of molybdenum nutrition is the wide variation between the critical deficiency and toxicity levels. These levels may differ by a factor of up to 10^4 (e.g., 0·1–1000 μg molybdenum per gram dry weight) as compared to a factor of 10 or less for boron or manganese. Under conditons of molybdenum toxicity, malformation of the leaves and a golden yellow discoloration of the shoot tissue occur, most likely owing to the formation of molybdocatechol complexes in the vacuoles (Hecht-Buchholz, 1973). Genotypical differences in molybdenum toxicity are closely related to differences in the translocation of molybdenum from the roots to the shoots (Chapter 3).

Table 9.27
Relationship between the Molybdenum Content of Soybean Seeds and the Subsequent Seed Yield of Plants Growing in a Molybdenum-Deficient Soil[a]

Molybdenum content of seeds (mg/kg dry wt)	Seed yield of the crop (kg/ha)
0·05	1505
19·0	2332
48·4	2755

[a]Based on Gurley and Giddens (1969).

High but nontoxic levels of molybdenum in plants are advantageous for seed production, but such levels in forage plants are dangerous for ruminants. A high molybdenum content in seeds ensures proper seedling growth and higher final grain yields in plants growing in soils low in available molybdenum (Table 9.27). Correspondingly, the effect of molybdenum application to a deficient soil on plant growth is inversely related to the seed content (Tanner, 1982) and amount of molybdenum applied to the seed crop (Weir and Hudson, 1966).

In contrast to the uptake rates of other micronutrients, the rate of molybdenum uptake by soybean plants during the first 4 weeks after germination is extremely low; thus the molybdenum requirement for growth has to be met by retranslocation from the seed (Ishizuka, 1982). Large-seeded cultivars in combination with high molybdenum availability during the seed-filling period are therefore very effective in the production of seeds suitable for soils low in available molybdenum (Franco and Munns, 1981).

Seed pelleting with molybdenum is another procedure for preventing deficiency during early growth and establishing a vigorous root system for subsequent uptake from soils low in available molybdenum (Tanner, 1982). As shown in Table 9.28 seed pelleting with the relatively insoluble molyb-

Table 9.28
Effect of Molybdenum Trioxide Application on Dry Matter Production and Nitrogen Content of the Subtropical Pasture Legume *Desmodium intartum* Grown in a Soil of pH 4·7[a]

Molybdenum-application (g/ha)	Dry weight (kg/ha)	Nitrogen content (% dry wt)
0	70	1·9
100 (soil application)	1220	3·2
100 (seed pelleting)	1380	3·4

[a]From Kerridge *et al*. (1973).

Table 9.29
Suppression of Molybdenum Uptake in Tomato and Pea Plants by Soil Application of Gypsum, $CaSO_4$[a]

Rates of gypsum application (mg/kg soil)	Molybdenum content (μg/g dry wt)	
	Tomato	Pea
0	5·25	10·80
100	3·25	8·05
400	2·45	5·07

[a]Based on Stout *et al.* (1951).

denum trioxide at a rate of 100 g molybdenum per hectare is even somewhat more effective than soil application.

In contrast to plants, animals—particularly ruminants—are very sensitive to excessive concentrations of molybdenum. Molybdenum levels above 5 to 10 mg/kg dry wt of forage are high enough to induce toxicity known as molybdenose (or "teart"). This occurs, for example, in western parts of the United States, in Australia, and in New Zealand, often in soils with poor drainage and high in organic matter (Gupta and Lipsett, 1981). Molybdenose is actually caused by an imbalance of molybdenum and copper in the ruminant diet. Sulfur-containing fertilizers or gypsum can be used to depress molybdenum uptake (Table 9.29). The sulfate in gypsum is more effective in this respect than equivalent amounts of sulfate in superphosphate (Pasricha *et al.*, 1977).

Molybdenum nutrition of plants growing in mixed pastures of legumes, herbs and grasses therefore requires special consideration. On the one hand, the relatively large requirement of legumes for N_2 fixation and for molybdenum in the seeds must be met, but at the same time toxic levels must not be allowed to accumulate in the forage of grazing animals.

9.6 Boron

9.6.1 General

In aqueous solution boron occurs mainly as boric acid, H_3BO_3 [or $B(OH)_3$], a very weak acid that accepts OH^- rather donates H^+:

$$B(OH)_3 + 2H_2O \rightleftharpoons B(OH)_4^- + H_3O^+$$

At physiological pH (<8) mainly the undissociated boric acid is present in the soil or nutrient solution. This is also the preferred form taken up by the roots (Chapter 2).

There is considerable disagreement concerning the relative importance of active and passive boron uptake by plant roots (Raven, 1980). Boron is strongly complexed to cell wall constituents even in the roots (Thellier *et al.*, 1979), and the size of this fraction roughly reflects genotypical differences in the boron requirement of plants (see Sections 9.6.2 and 9.6.8). Long-distance transport of boron from the roots to the shoot is confined to the xylem, and uptake and translocation are closely related not only to the mass flow of water to the root surface but also to the xylem water flow (Chapter 3). The regulation of boron uptake and translocation by plants is rather limited in comparison to that of other mineral nutrients. Nevertheless, distinct genotypical differences in the root-to-xylem transfer of boron were demonstrated by Brown and Jones (1971) in tomato; in roots with similar levels of accumulated boron, the rate of boron translocation into the shoot differed among genotypes by a factor of up to 5.

There is no indication that boron is an enzyme component, and there is only little evidence (Birnbaum *et al.*, 1977) that the activity of any enzyme is enhanced or inhibited by boron. With respect to its role in plant nutrition, boron is still the least understood of all the mineral nutrients, and what is known of the boron requirement arises mainly from studies of what happens when boron is withheld or resupplied after deficiency. This lack of information is surprising, because on a molar basis the requirement for boron, at least among dicotyledons, is higher than that for any other micronutrient. It is very easy in certain plant species (e.g., sunflower) to induce very rapidly a range of distinct metabolic changes and visible deficiency symptoms by withholding boron.

Boron is an essential mineral element for all vascular plants. Neither fungi nor freshwater algae (e.g. *Chlorella*), however, seem to have a boron requirement (Lewis, 1980a; Dugger, 1983). According to McClendon's (1976) classification of the origins of mineral nutrient requirements, the requirement for boron is evolutionarily derived. According to Lewis (1980a), the functions of boron are primarily extracellular and related to lignification (see Section 9.6.2) and xylem differentiation (see Section 9.6.5).

9.6.2 Boron Complexes with Organic Structures

Boric acid forms stable mono- [Eq. (1)] and diesters (Eq. (2)] with *cis*-diols:

$$(1)\quad \begin{matrix}=C-OH\\ | \\ =C-OH\end{matrix} + \begin{matrix}HO\\HO\end{matrix}{>}B-OH \rightleftharpoons \left[\begin{matrix}=C-O\\ | \\ =C-O\end{matrix}{>}B{<}\begin{matrix}OH\\OH\end{matrix}\right]^- + H_3O^+$$

$$(2)\quad \left[\begin{matrix}=C-O\\ | \\ =C-O\end{matrix}{>}B{<}\begin{matrix}OH\\OH\end{matrix}\right]^- + \begin{matrix}HO-C=\\ | \\ HO-C=\end{matrix} \rightleftharpoons \left[\begin{matrix}=C-O\\ | \\ =C-O\end{matrix}{>}B{<}\begin{matrix}O-C=\\ | \\ O-C=\end{matrix}\right]^- + 2H_2O$$

Polyhydroxil compounds with an adjacent *cis*-diol configuration are required for the formation of such complexes; the compounds include a number of sugars and their derivatives (e.g., sugar alcohols and uronic acids), in particular mannitol, mannan, and polymannuronic acid. They serve, for example, as constituents of the hemicellulose fraction of cell walls. In contrast, glucose, fructose, and galactose and their derivatives (e.g., sucrose) do not have this *cis*-diol configuration and thus do not form stable borate complexes. Some *o*-diphenolics, such as caffeic acid and hydroxyferulic acid, which are important precursors of lignin biosynthesis in dicotyledons (McClure, 1976), possess the *cis*-diol configuration and hence form stable borate complexes.

A substantial proportion of the total boron content of higher plants seems to be complexed as stable *cis*-borate esters in the cell walls (Thellier *et al.*, 1979). The fact that the boron requirement of dicotyledons is greater than that of monocotyledons is presumably related to higher proportions of compounds with this *cis*-diol configuration in the cell walls, mainly in the hemicellulose fraction and in lignin precursors (Lewis, 1980a). It has been shown by Tanaka (1967) that the content of strongly complexed boron in the root cell walls is 3–5 μg/g dry wt in monocotyledons (e.g., wheat) and up to 30 μg/g dry wt in certain dicotyledons (e.g., sunflower). These differences roughly reflect the differences between the species in the boron requirement for optimal growth (see Section 9.6.8). It might be assumed that the functions of this apoplastic boron are somewhat similar to those of calcium in both regulating synthesis and stabilizing certain cell wall constituents, including the plasma membrane (Fig. 8.19).

Attempts have been made to compile the many, often contradictory results on boron deficiency-induced metabolic changes in higher plants and to develop a more unified model of the action of boron. All these models are based on the capacity of boron to form reversible or irreversible diol–borate complexes of different stabilities with substrates, enzymes, and/or membranes and in this way affect enzyme activities and metabolic pathways. A primary function, proposed by Lee and Aronoff (1967), is based on the capacity of boron to form stable 6-Ⓟ-gluconate–borate complexes and thus restrict both the flux of substrate into the pentosephosphate pathway and the synthesis of phenols. As a result, glycolysis and the synthesis, for example of hemicellulose and related cell wall material increase:

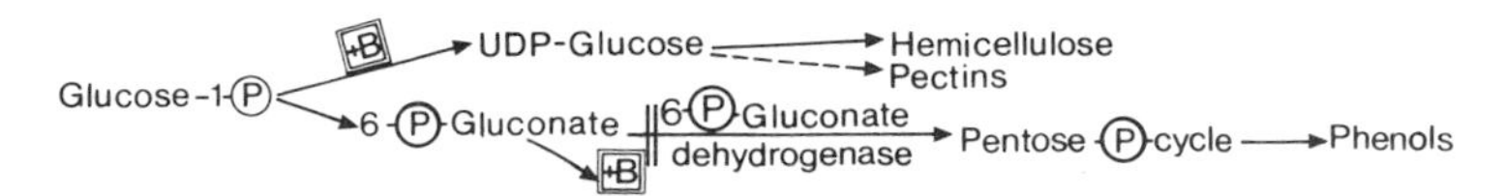

Sound evidence exists for a shift toward an increase in the substrate flux into the pentose-phosphate pathway under boron deficiency (Eichhorn and

Augsten, 1974; Birnbaum *et al.*, 1977), which is also reflected in an accumulation of phenolic substances in boron-deficient plants (Perkins and Aronoff, 1956).

Recent interpretations of the role of boron are also based on the formation of stable *cis*-diol borate complexes. However, views differ considerably as to what are considered as the primary effects of boron; lignin biosynthesis and xylem differentiation (Lewis, 1980a), membrane stabilization (Pilbeam and Kirkby, 1983), and altered enzyme reactions (Dugger, 1983). In the following section examples are given of some of the most distinct and typical effects of boron on metabolism and growth of higher plants.

9.6.3 Cell Elongation, Cell Division, and Nucleic Acid Metabolism

One of the most rapid responses to boron deficiency is inhibition or cessation of elongation growth of both primary and lateral roots, giving the roots a stubby and bushy appearance. As shown in Fig. 9.17A, inhibition occurs as soon as 3 hr after the boron supply is interrupted, it is severe after 6 hr, and root elongation growth stops after 24 hr. Twelve hours after the boron supply has been restored, however, elongation growth is again rapid.

Between 6 and 12 hr after the boron supply is cut off, there is a dramatic increase in the activity of IAA oxidase in the roots (Fig. 9.17B). When boron is resupplied, there is an immediate reduction in IAA oxidase activity. The similarities in the responses of elongation growth and IAA oxidase activity to boron deficiency and resupply are striking. There is, however, a distinct difference in the time of response to deficiency: Elongation growth is

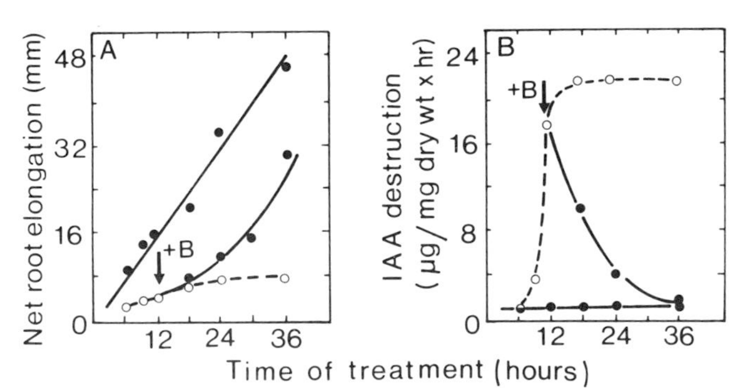

Fig. 9.17 Effect of boron deficiency on elongation growth (A) and IAA oxidase activity (B) in apical 5-mm root sections of squash (*Cucurbita pepo* L.). Resumption of boron supply after 12 hr (arrow) of boron deficiency. Key: ●——●, +B; ○ – –○, –B. (Redrawn from Bohnsack and Albert, 1977.)

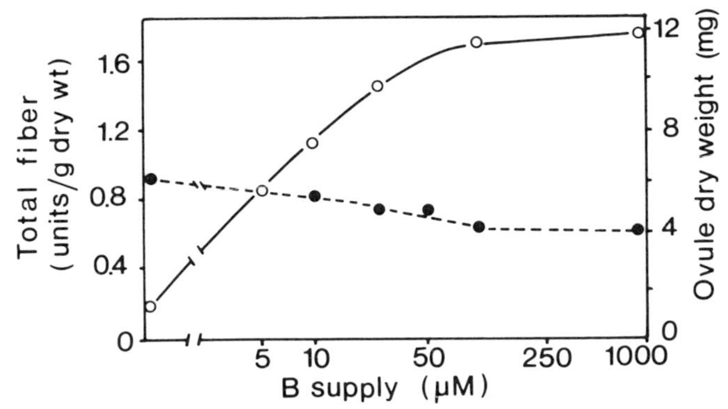

Fig. 9.18 Effect of boron supply on dry weight (● – – ●) and fiber development (○——○) of unfertilized cotton ovules cultured in the presence of IAA, gibberellic acid, and cytokinin. Total fiber units per gram dry wt represent the ratio of fiber length to gram of dry weight. (Redrawn from Birnbaum *et al.*, 1974.)

inhibited ~3 hr before IAA oxidase activity increases. It is assumed by Bohnsack and Albert (1977) that an accumulation of supraoptimal levels of IAA in the boron-deficient root tips is responsible for both the decrease in root elongation and the subsequent induction of IAA oxidase synthesis (but see Section 9.6.5).

Similar responses to boron can be demonstrated in cotton ovules cultured *in vitro*. Epidermal cells of cotton ovules that form lint fibers begin to elongate on the day of anthesis. The degree of extension is closely related to the external boron concentration, as shown in Fig. 9.18. It is evident that boron is necessary for fiber elongation and the prevention of callusing of the epidermal cells, as indicated indirectly by the decline in ovule dry weight. From additional observations it has been concluded that boron is required primarily for cell elongation rather than for cell division (Birnbaum *et al.*, 1974). However, most of the recent experimental evidence indicates that boron is also required for cell division, although the views are conflicting on what is primarily affected by boron deficiency: cell division (Cohen and Lepper, 1977; Kouchi, 1977) or cell elongation (Lovatt *et al.*, 1981).

The decrease in nucleic acid content that takes place when boron is withheld for several days is a well-documented phenomenon (Hundt *et al.*, 1970b; Shkol'nik, 1974). In the case of DNA this is presumably a secondary effect since DNA synthesis continues for several hours after the cessation of elongation growth due to boron deficiency (Cohen and Albert, 1974; Moore and Hirsch, 1983). On the other hand, RNA, which has a much higher turnover rate than DNA (Section 8.4), is strongly and rapidly affected by boron deficiency (Albert, 1965). The synthesis of uracil seems to be particularly impaired. This nucleotide is a major constituent of RNA and also a precursor of energy-rich phosphates:

$$\text{Uracil} \xrightarrow{+\text{ribose}} \text{uridine} \begin{cases} \nearrow \text{UTP/UDP} \longrightarrow \text{carbohydrates} \\ \searrow \text{RNA} \end{cases}$$

Lower levels of RNA in boron-deficient tissues may also be the result of enhanced rates of degradation, as indicated by much higher RNase activities (Dave and Kannan, 1980). Contradictory results on the effect of boron deficiency on RNA metabolism are in part caused by differences in the duration of experiments: A few hours after boron is withheld, the turnover rate of RNA increases (Cory and Finch, 1967) without there being much change in net content, and only in severely deficient tissues does degradation dominate and the net content of RNA decrease.

9.6.4 Carbohydrate and Protein Metabolism

General interest in the role of boron in carbohydrate metabolism is focused on two aspects: the synthesis of cell wall material and the transport of sugars. In agreement with the proposed role of boron in stimulating the utilization of glucose 1-phosphate, in boron-deficient roots the incorporation of labeled phosphorus (^{32}P) into the nucleotide fraction is inhibited and hexosephosphates accumulate (Table 9.30). A 1-hr pretreatment of deficient roots with boron was sufficient to increase the incorporation of ^{32}P into the nucleotide fraction to equivalent levels to those of boron-sufficient roots.

Table 9.30
Effect of Boron on the Incorporation of Labeled Phosphorus (^{32}P) after 1 Hour of Absorption by the Terminal 0·5 cm of Roots of Faba Bean[a]

	^{32}P in each fraction (total = 100)		
Phosphorus fraction	+B+[b]	−B−[c]	−B+[d]
Insoluble	12·3	11·1	7·9
Nucleotides	13·2	4·0	14·4
Hexosephosphates	11·2	25·3	19·4
Triosephosphates	6·4	6·4	7·8
Inorganic phosphate	57·0	53·4	50·6

[a]From Robertson and Loughman (1974a).
[b]Boron-sufficient plants plus 1 hr pretreatment with 10 μM boron.
[c]Boron-deficient plants without pretreatment.
[d]Boron-deficient plants plus 1 hr pretreatment with 10 μM boron.

Accordingly, the utilization of carbohydrates for the synthesis of cellulose or RNA would also be impaired at the step in which nucleotides are formed from hexosephosphate. In agreement with this, the utilization of externally supplied glucose for cell wall synthesis in cotton fibers is impaired by boron deficiency (Dugger and Palmer, 1980). Somewhat surprisingly, the cell wall diameter and the proportion of cell wall material to the total dry weight are both higher in boron-deficient tissues (Rajaratnam and Lowry, 1974; Hirsch and Torrey, 1980). The role of boron, however, is reflected in the chemical composition and the ultrastructure of the cell walls: There is a higher concentration of pectic substances (Rajaratnam and Lowry, 1974), and a larger proportion of glucose is incorporated into β-1,3-glucan, the main component of callose (Dugger and Palmer, 1980), which also accumulates in the sieve tubes of boron-deficient plants (Venter and Currier, 1977). The primary cell walls of boron-deficient cells are not smooth but are characterized by an irregular deposition of vesicular aggregations intermixed with membranous material (Hirsch and Torrey, 1980). These results indicate that boron not only complexes strongly with cell wall constituents (Section 9.6.2) but is required for structural integrity and acts together with calcium as an "intercellular cement" (Ginzburg, 1961).

In 1954 Gauch and Dugger (see Dugger, 1983) proposed that boron plays a key role in higher plants by facilitating the short- and long-distance transport of sugars via the formation of borate–sugar complexes. Although some results might be interpreted as supporting this hypothesis (Dugger, 1983), it is doubtful that boron fulfills this function, at least for sucrose (Section 9.6.2), the dominant sugar in phloem transport. While boron may facilitate sugar uptake by leaves, the export of photosynthates from the leaves either is unaffected (Weiser *et al.*, 1964) or is impaired by callose formation in the sieve tubes (Venter and Currier, 1977) or by the lack of sink activity of roots and shoot apices in plants suffering from severe boron deficiency. It has also been shown by Yih and Clark (1965) that in boron-deficient plants the rate of root elongation decreases irrespectively of the sugar content of the root tips.

Under conditions of severe boron deficiency the protein content of young leaves decreases and soluble nitrogen compounds, particularly nitrate, accumulate (Hundt *et al.*, 1970a). The decrease in protein content is confined to the cytoplasm, whereas the chloroplast protein content is not affected, a result in agreement with the fact that chlorosis is not a common symptom of boron deficiency. Leaf metabolism and composition might be affected by boron deficiency indirectly via its effect on cytokinin synthesis in the root tips: When the supply of boron is withheld, both the production and export of cytokinins into the shoots decrease (Wagner and Michael, 1971). In tobacco the same is true of nicotine alkaloids (Scholz, 1958).

9.6.5 Tissue Differentiation, Auxin and Phenol Metabolism

Striking interactions are observed between auxins (IAA) and boron (Lewis, 1980a). Among green plants auxin and its most prominent and abundant representative, IAA, are synthesized only in vascular species and are involved in the differentiation of xylem vessels. A boron requirement is, with a few exceptions, also confined to vascular plants, and the effects of boron deficiency on tissue differentiation are typical. In root tips, for example, boron deficiency results in a reduction in elongation growth associated with changes in the direction of cell division from normal longitudinal to radial (Robertson and Loughman, 1974b). Enhanced cell division in a radial direction with a distinct proliferation of cambial cells and impaired xylem differentiation are also typical in the subapical shoot tissue of boron-deficient plants (Bussler, 1964; see Fig. 9.19).

Enhancement of cell division in cambial stem tissue and impaired xylem differentiation are not, however, direct effects of boron deficiency. Similar morphological changes can be obtained in boron-sufficient plants by mechanical destruction of the apical meristem of the shoot (Krosing, 1978). It can be concluded, therefore, that inhibition or even a lack of xylem differentiation is only indirectly related to boron nutrition.

The relationships between boron nutrition, auxin level, differentiation, and lignification are still not well understood. In boron-deficient plants, auxin levels are often much higher than normal (Coke and Whittington, 1968), and an exogenous supply of IAA induces anatomical changes in the root tips similar to those caused by boron deficiency. This has led to the interpretation that boron-deficiency symptoms are a reflection of increased auxin levels (Robertson and Loughman, 1974b). However, the ultrastructural changes brought about by boron deficiency and excessive IAA levels are quite different (Hirsch and Torrey, 1980). Furthermore, typical symptoms of boron deficiency may occur without any increase in the IAA level of the same tissue (Hirsch *et al.*, 1982). In early stages of deficiency there is even a tendency to lower IAA levels in apical tissues (Fackler *et al.*, 1985). Also no significant correlations between IAA level and boron deficiency symptoms were found by Smirnov *et al.* (1977) when they compared different plant species or plant organs. High IAA levels may occur only in those plant species that, in response to boron deficiency, accumulate certain phenolics such as caffeic acid, which is an effective inhibitor of IAA oxidase activity (Birnbaum *et al.*, 1977). Certain phenolic compounds not only are effective inhibitors of root elongation growth but also simultaneously enhance radial cell division (Svensson, 1971); that is, they induce anatomical changes that are similar to those caused by IAA.

It seems more likely, therefore, that the accumulation of phenolics in

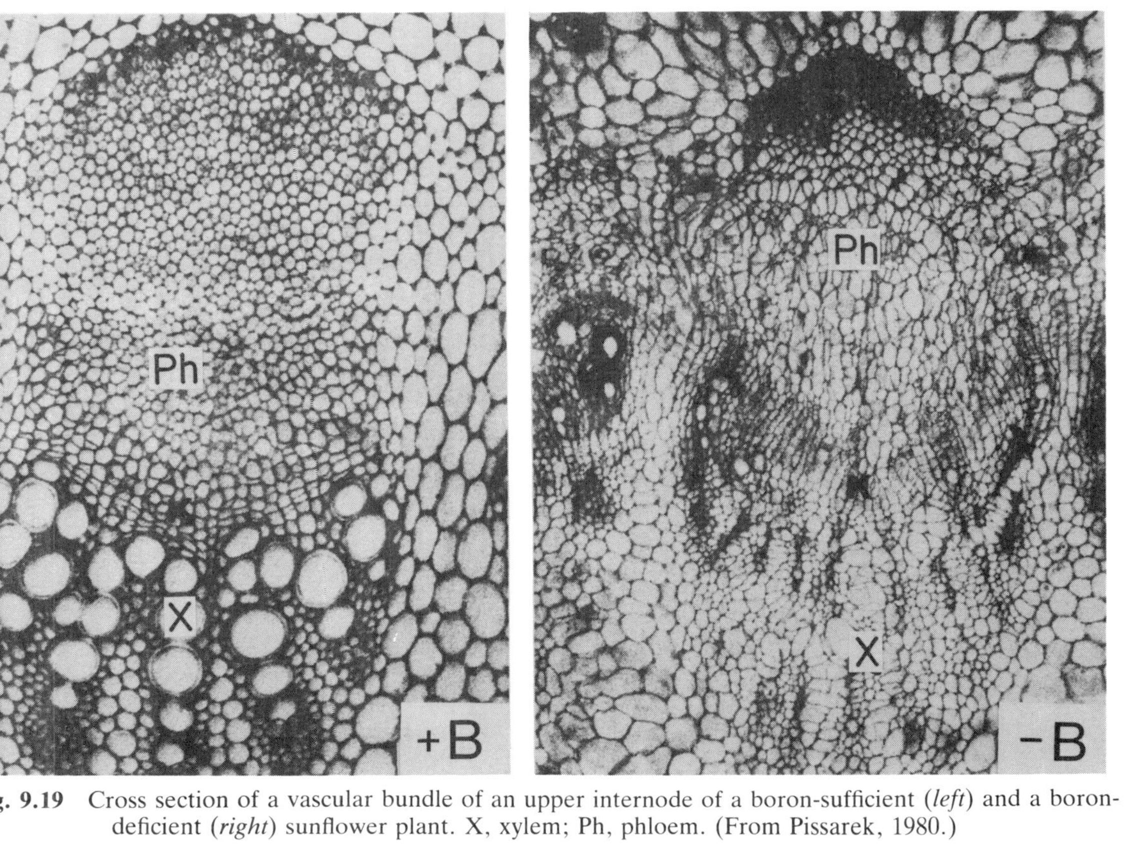

Fig. 9.19 Cross section of a vascular bundle of an upper internode of a boron-sufficient (*left*) and a boron-deficient (*right*) sunflower plant. X, xylem; Ph, phloem. (From Pissarek, 1980.)

boron-deficient tissues per se (Perkins and Aronoff, 1956) is responsible for the metabolic changes and cell damage in boron-deficient tissue (Lewis, 1980a, Pilbeam and Kirkby, 1983). The proposed role of boron in controlling phenol metabolism and thus lignification can be summarized as follows:

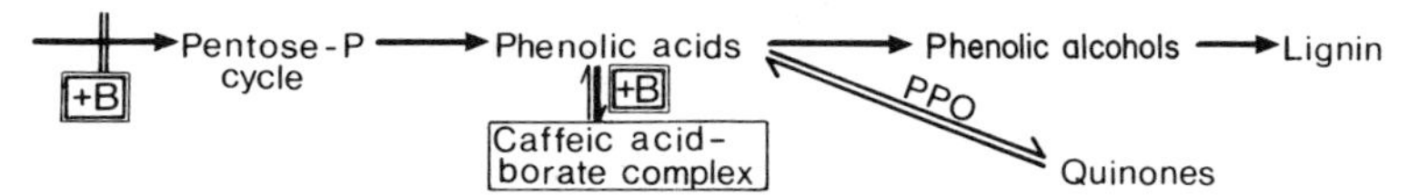

Boron seems to regulate not only the flux of substrate into the pentose-phosphate cycle, but also lignin biosynthesis via the formation of stable phenolic acid–borate complexes, particularly with caffeic acid. With boron deficiency, therefore, phenolics accumulate. This may enhance the activity of polyphenol oxidase (PPO), leading to elevated concentrations of highly reactive intermediates such as caffeic quinone in the cell walls (Shkol'nik *et al.*, 1981). The injurious effects of certain phenolics on the permeability of the plasma membrane and on membrane-bound enzymes are well documented, and corresponding changes in cell wall synthesis may therefore be expected when boron is withheld (Hirsch *et al.*, 1982).

9.6.6 Membrane Permeability

There is increasing evidence that boron has a direct effect on membranes, probably by the formation of *cis*-diol borate complexes with appropriate membrane constituents. It has been shown by Tanada (1978) that the formation and maintenance of bioelectrical potentials across membranes induced by infrared irradiation or by gravity require the presence of boron. The interaction of boron with membrane constituents and its stabilizing effects on membranes are reflected in its influence on the turgor-regulated nyctinastic movements of the leaflets of *Albizzia* (Tanada, 1982) and its enhancement of both ^{86}Rb influx and stomatal opening in *Commelina communis* (Roth-Bejerano and Itai, 1981). Boron pretreatment of the root tips of faba bean and maize for only 1 hr markedly enhanced phosphate uptake in both boron-sufficient and boron-deficient roots and nearly restored the uptake rate of the originally boron-deficient maize roots (Table 9.31), another indication of the effect of boron on membranes.

The effect of boron pretreatment on the uptake rates of chloride and rubidium was similar to that on phosphate uptake rates, and membrane-bound ATPase activity, which was low in boron-deficient maize roots, was restored to the same level as that in boron-sufficient roots within 1 hr (Pollard *et al.*, 1977). That boron affects other membrane-bound enzymes was shown by Dave and Kannan (1980), who observed a severalfold increase

Table 9.31
Effect of Boron Pretreatment on the Subsequent Uptake of Inorganic Phosphate by Root Tip Zones of Faba Bean and Maize[a]

Pretreatment of root tips for 1 hr	P_i Uptake (nmol/g × hr)			
	Faba bean grown with or without B		Maize grown with or without B	
	+B	−B	+B	−B
No boron	112	52	116	66
10^{-5} *M* H_3BO_3	152	108	190	171

[a]Root tip zones were 0–2 cm from the apex. From Pollard *et al.* (1977).

in membrane-bound RNase activity in boron-deficient plants. Water-soaked areas in the tissue of boron-deficient plants are also indirect indicators of the effects of boron on membrane stabilization.

9.6.7 Pollen Germination and Pollen Tube Growth

The supply of boron required for seed and grain production is usually higher than that needed for vegetative growth only. This has been shown to be the case, for example, for maize (Vaughan, 1977; see also Chapter 6) and white clover (Johnson and Wear, 1967). Boron has both indirect and direct effects on fertilization. Indirect effects are probably related to the increase in amount and change in sugar composition of the nectar, whereby the flowers of species that rely on pollinating insects become more attractive to insects (Smith and Johnson, 1969; Erikson, 1979). Direct effects of boron are reflected by the close relationship between boron supply and the pollen-producing capacity of the anthers, as well as the viability of the pollen grains (Agarwala *et al.*, 1981). Moreover, boron stimulates germination, particularly pollen tube growth (Fig. 9.20). Finally, leakage of sugars from the pollen decreases with increasing external boron concentration.

In maize a minimum boron content of 3 μg/g dry wt in the silk is required for pollen germination and fertilization (Vaughan, 1977). The critical deficiency levels in the stigma may, however, vary considerably among cultivars and species. In grapevine (*Vitis vinifera*) which is known for its high boron requirement, with sufficient boron supply the boron content of the stigma is 50–60 μg/g dry wt and even at contents of 8–20 μg/g dry wt fertilization is impaired (Gärtel, 1974). According to Lewis (1980b) high boron levels in the stigma and style are required for physiological inactivation of callose from the pollen tube walls by the formation of borate–

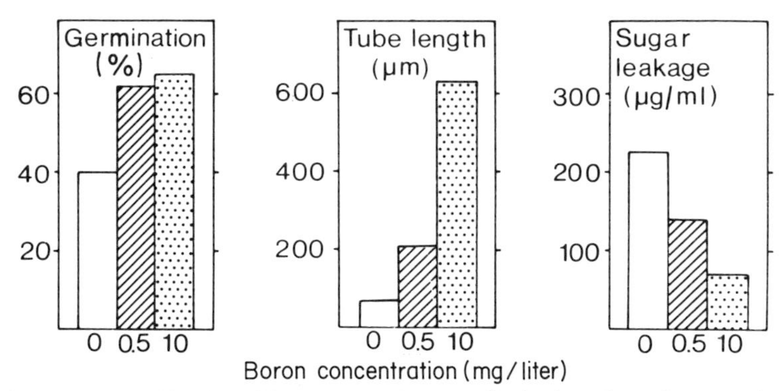

Fig. 9.20 Effect of boric acid concentrations on lily (*Lilium longiflorum* L.) pollen germination, tube growth, and leakage of sugar to the medium. (Redrawn from Dickinson, 1978.)

callose complexes. When boron levels are low, callose levels increase and induce the synthesis of phytoalexins (including phenols), a defense mechanism similar to that in response to microbial infections.

9.6.8 Boron Supply, Plant Growth and Quality, and Boron Toxicity

Boron deficiency is a widespread nutritional disorder. Under high rainfall conditions boron is readily leached from soils as $B(OH)_3$. Boron availability to plants decreases with increasing soil pH, particularly in calcareous soils and soils with a high clay content, presumably as a result of the formation of $B(OH)_4^-$ and anion adsorption. Availability also decreases sharply under drought conditions, probably because of both a decrease in boron mobility by mass flow to the roots (Kluge, 1971) and polymerization of boric acid:

$$3B(OH)_3 \rightleftharpoons B_3O_3(OH)_4^- + H_3O^+ + H_2O$$

Interactions between boron and other mineral nutrients during uptake and in the plant itself seem to be of minor importance. Reports on the physiological significance of the calcium/boron ratio in plants are inconclusive (Gupta, 1979; Bergmann, 1983). Both elements are restricted in their mobility and have extracellular functions. Certain similarities in the visible symptoms of deficiency of both mineral nutrients might thus be expected and have been reported, for example, in peanut kernels (Cox and Reid, 1964) and lettuce (Crisp *et al.*, 1976).

Plant species differ characteristically in their capacity for boron uptake when grown in the same soil (Table 9.32), which generally reflects typical species differences in the boron requirement for growth. For example, the

Table 9.32
Boron Content of the Leaf Tissue of Plant Species from the Same Location[a]

Plant species	Boron content (mg/kg dry wt)
Wheat	6·0
Maize	8·7
Timothy	14·8
Tobacco	29·4
Red clover	32·2
Alfalfa	37·0
Brussel sprouts	50·2
Carrots	75·4
Sugar beet	102·3

[a]Based on Gupta (1979).

critical deficiency range, expressed as milligrams boron per kilogram dry weight, is about 5–10 in monocotyledons (e.g., wheat), 25–60 in red clover, 30–80 in carrots, and 40–100 in sugar beet (Bergmann, 1983).

These differences in boron requirement are most likely related mainly to differences in cell wall composition (Section 9.6.2). Accordingly, sensitivity to boron deficiency varies considerably, dicotyledons being much more sensitive than monocotyledons. Differences among cultivars may also exist. High light intensities seem to increase sensitivity to boron deficiency (MacInnes and Albert, 1969) by raising the requirement for boron in the tissue (Tanaka, 1966). This might be related to the elevated contents of phenolics often observed in plants exposed to high light intensities.

Symptoms of boron deficiency in the shoots are noticeable at the terminal buds or youngest leaves, which become discolored and may die. Internodes are shorter, giving the plants a bushy or rosette appearance. Interveinal chlorosis on mature leaves may ocur, as might misshaped leaf blades. An increase in the diameter of petioles and stems is particularly common and may lead to symptoms such as "stem crack" in celery. Drop of buds, flowers, and developing fruits is also a typical symptom of boron deficiency. In the heads of vegetable crops (e.g. lettuce), water-soaked areas, tipburn, and brown- or blackheart occur. In storage roots of celery or sugar beet, necrosis of the growing areas leads to heart rot (Fig. 9.21). With severe deficiency the young leaves also turn brown and die, subsequent rotting and microbial infections of the damaged tissue being common.

Boron deficiency-induced reduction or even failure of seed and fruit set are well known. In boron-deficient fleshy fruits, not only is the growth rate lower, but the quality may also be severely affected by malformation (e.g., "internal cork" in apple) or, in citrus, by a decrease in the pulp/peel ratio

Fig. 9.21 Boron deficiency in sugar beet. (*Left*) Severe boron deficiency (heart and crown rot). (*Middle*) Mild boron deficiency (heart rot). (*Right*) Boron sufficient. (By courtesy of W. Bussler.)

(Foroughi *et al.*, 1973). Boron-deficiency symptoms have been described in detail by Gupta (1979) and illustrated with color plates by Bergmann (1983).

For the application of boron either to the soil or as a foliar spray, different sodium borates, including borax or sodium tetraborate, can be used. Boric acid can also be used for foliar sprays. The amount of boron applied varies from 0·3 to 3·0 kg/ha depending on the requirement and sensitivity of the crop to boron toxicity. The narrow concentration range between boron deficiency and toxicity requires special care in the application of boron fertilizers.

Boron toxicity may occur when large amounts of municipal compost are applied (Purves and Mackenzie, 1974), and it is of much concern in semiarid and arid areas where irrigation water contains high levels of boron. The criticial boron concentration in irrigation water varies between 1 and 10 mg/liter for sensitive and tolerant crops, respectively. For example, it is between 3 and 4 mg/liter for wheat and between 4 and 6 mg/liter for peas (Chauhan and Powar, 1978). According to El-Sheikh *et al.* (1971) the critical toxicity levels, expressed as milligrams boron per kilogram dry weight of leaves, is 100 for maize, 400 for cucumber, and 1000 for squash. For more information on critical toxicity levels, see the reviews by Gupta (1979) and Bergmann (1983). Typical boron toxicity symptoms on mature leaves are marginal or tip chlorosis or both and necrosis. They reflect the distribution of boron in shoots, following the transpiration stream (Chapter 3).

9.7 Chlorine

9.7.1 General

Chlorine is a strange mineral nutrient. Its normal concentration in plants is between 70 and 700 mmol/kg dry wt (~2000 and 20,000 mg/kg dry wt), which is typical of the level of macronutrients. On the other hand, the chlorine requirement for optimal growth is between 10 and 30 mmol/kg dry wt (~340 and 1200 mg/kg dry wt), which is in the range of micronutrient levels. Chlorine is ubiquitous in nature, and it occurs in aqueous solution as chloride (Cl^-). It is readily taken up by plants, and the uptake is closely coupled to metabolic activity. The mobility of chloride is high, both in short- and long-distance transport; it is also reasonably well retranslocated from mature leaves. Chlorinated organic compounds are synthesized by microorganisms (e.g., chloroamphenicol) and have also been found in higher plants. However, all evidence indicates that chlorine exerts its functions in higher plants mainly as the highly mobile inorganic anion that is as Cl^-, in processes related to charge compensation and osmoregulation.

9.7.2 Chloride Supply and Plant Growth

Because chloride is usually supplied to plants from various sources (soil reserves, rain, fertilizers, and air pollution) there is much more concern about toxic levels in plants than about deficiency. In fact, it is difficult to induce chlorine deficiency under normal experimental conditions.

The essentiality of chlorine as a mineral nutrient was first demonstrated unequivocally by Broyer *et al.* (1954) in experiments with tomato plants in which contamination with chlorine (water, chemicals, and air) was kept to a minimum. At ~250 μg chloride per gram of leaf dry weight, severe deficiency symptoms occurred and the plants finally died. Severely deficient plants recovered within a few days when the supply of chloride was restored. The growth of various plant species was affected to different degrees by an interruption in the chloride supply (Fig. 9.22). Growth reduction was most severe in lettuce, whereas growth of squash was not affected. The chloride content of the deficient plants was still quite high because of contamination by seeds, chemicals, water, and air. Environmental factors which enhance growth rate and transpiration rate (e.g., high temperatures plus high irradiation) increase the susceptibility of plants to chlorine deficiency (Ozanne *et al.*, 1957).

Wilting of leaves, especially leaf margins, is a typical symptom of chlorine deficiency, even in water culture, when plants are exposed to full sunlight

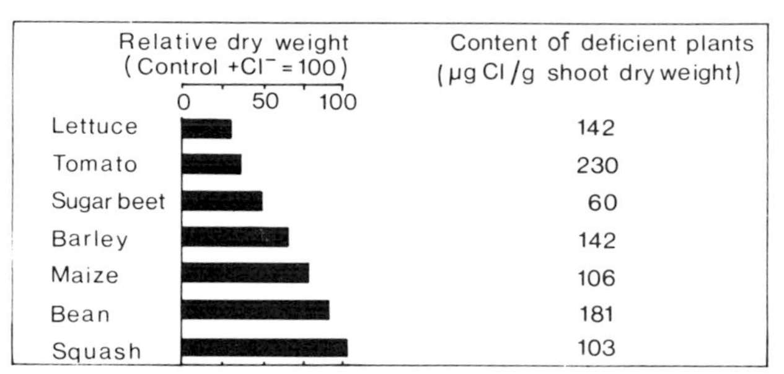

Fig. 9.22 Relative shoot dry weight and chlorine content of chlorine-deficient plants. (Redrawn from Johnson *et al.*, 1957.)

(Broyer *et al.*, 1954). Root elongation growth is severely inhibited, but the formation of short lateral roots is enhanced, giving the root system a stubbly appearance (Johnson *et al.*, 1957). Wilting might therefore be related to inhibited water uptake or insufficient regulation of the transpiration rate by the stomata.

Growth reduction and deficiency symptoms could be restored to 90% of the levels in plants adequately supplied with chloride by supplying bromide (Broyer, 1966). Chloride and bromide have similar physicochemical properties; for example, their hydrated ionic radii are nearly the same: 0·332 nm (Cl^-) and 0·330 nm (Br^-). Substitution of chloride by bromide is of no practical significance, however, because of the difference in their natural abundances. In the earth's crust, the sea, and the air, as well as in plants, chlorine is ~1000 times more abundant than bromine (McClendon, 1976).

9.7.3 Photosynthesis and Stomatal Regulation

In 1944 Warburg discovered that the Hill reaction in isolated chloroplasts requires chloride, and since that time the involvement of chloride in the splitting of water at the oxidizing site of photosystem II (Chapter 5) has been confirmed by various authors. It is assumed (Kelley and Izawa, 1978) that chloride acts as a cofactor of the manganese-containing O_2-evolving system:

$$H_2O \xrightarrow[Mn + Cl^-]{O_2 \nearrow \quad e^-} \text{PS II} \xrightarrow{e^-} \text{PS I} \xrightarrow{e^-}$$

The enhancing effect of chloride can be quite dramatic and is correlated with a corresponding increase in photophosphorylation (Table 9.33). Bromide can replace chloride completely in this function, but iodide is without effect. The dependence of photosynthetic O_2 evolution has also been demonstrated with isolated thylakoid membranes (Ball *et al.*, 1984).

In spinach and sugar beet the chloride concentration in the chloroplasts has been calculated to be close to 100 m*M*, compared with less than 10 m*M* in

Table 9.33
Effect of Halides on O_2 Evolution and ATP Formation in Isolated Spinach Chloroplasts[a]

Halide added (~4 m*M*)	O_2 Evolved (μmol)	ATP Formed (μmol)
None	0	0·3
Cl^-	4·0	3·7
Br^-	4·0	4·0
I^-	0	0·7

[a]From Bové *et al.* (1963).

the leaf tissue, implying preferential accumulation in the chloroplasts (Robinson and Downton, 1984). Nevertheless, the role of chloride in photosynthesis in intact plants is still not clear. In sugar beet, chlorine deficiency can lead to a 60% reduction in growth rate without affecting the net photosynthesis per unit of chlorophyll (Fig. 9.23). The principal effect of chlorine deficiency is a reduction in cell multiplication rates in leaf blades, which in turn leads to a decrease in leaf surface area and thereby plant growth (Terry, 1977). However, the results on the preferential accumulation of chloride in chloroplasts and on proton–chloride cotransport in thylakoid membranes (Theg and Homann, 1982) support the assumption of a chloride requirement in photosystem II reactions, at least in isolated systems. The discrepancy between these results and the lack of a chloride effect on photosynthesis in intact plants (Terry, 1977) requires further clarification as to whether the relatively high chloride concentrations necessary for maximum photosynthetic O_2 evolution in isolated systems (e.g., >25 m*M* chloride in thylakoid membranes) is an artefact, caused by the loss of certain polypeptides during the isolation procedure (Andersson *et al.*, 1984).

According to Fig. 9.23 the critical deficiency level of chlorine in the leaf blades of sugar beet is ~20 μmol/dry wt (~0·7 mg/g). Somewhat higher critical levels of 50 and 200 μmol/g dry wt were found for leaves and petioles,

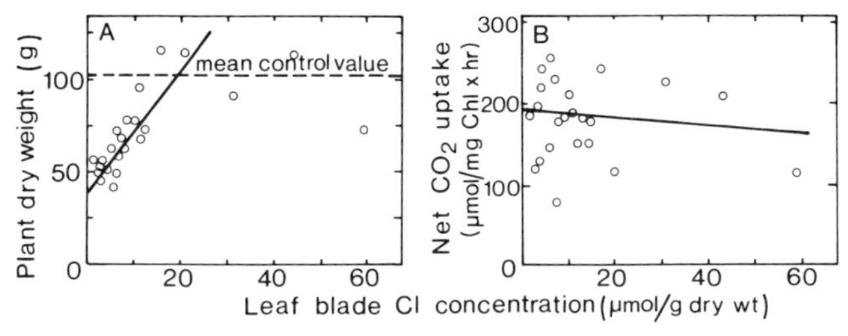

Fig. 9.23 Effect of chlorine deficiency on growth (A) and photosynthesis (B) of sugar beet. (From Terry, 1977.)

respectively, in the same crop by Ulrich and Ohki (1956). This discrepancy in critical levels can be accounted for by differences in cultivars and experimental conditions.

Chloride can affect photosynthesis and plant growth indirectly via stomatal regulation. In plant species such as *Vicia faba* with well-developed chloroplasts in the guard cells, K^+ influx during stomatal opening is compensated for by the synthesis of equivalent concentrations of malate at the expense of starch in the guard cells (Section 8.7). In other plant species (e.g., onion) the guard cells possess poorly developed chloroplasts and thus lack starch for malate synthesis. The K^+ influx, therefore, has to be connected with an equivalent anion influx, mainly chloride (Section 8.7). Accordingly, stomatal opening in onion is inhibited in the absence of chloride (Schnabl and Ziegler, 1977; Schnabl, 1980). Stomatal closure is correlated with a corresponding efflux of K^+ and accompanying anions from the guard cells. In plant species with guard cells which lack starch, chlorine deficiency can therefore impair the control functions of the stomata for water loss, which is especially important when the water supply is limited and short-term changes in the fine regulation of stomata are required.

9.7.4 Membrane-Bound Proton-Pumping ATPase

It is generally agreed that monovalent cations, K^+ in particular, stimulate an Mg·ATPase at the plasma membrane and induce H^+ efflux from the cytosol (Chapter 2). There is increasing evidence for the existence of another type of H^+-pumping ATPase located in membranes of cytoplasm vesicles or in the tonoplast or both (Hager and Helmle, 1981). This second type of pump, (Mg, Mn)·ATPase, is not affected by monovalent cations but rather is stimulated specifically by chloride (Table 9.34). Bromide is somewhat less

Table 9.34
Effect of Salts on the Membrane-Bound H^+-Pumping ATPase[a]

Salt (10 m*M* monovalent ion)	ATPase stimulation (% of control)
No monovalent ion	10
KCl (control)	100
NaCl	102
NaBr	87
KNO_3	21
K_2SO_4	3

[a]Based on Mettler *et al.* (1982).

effective; sulfate has an inhibitory effect. Nitrate either stimulates the pump only slightly (Table 9.34) or even inhibits its activity (O'Neill *et al.*, 1983).

The chloride-stimulated H^+-ATPase is probably of particular importance, since it operates as an electrogenic H^+ pump at the tonoplast that transports H^+ from the cytoplasm into the vacuole (Churchill and Sze, 1983), thereby creating the typical pH gradient between the two compartments (cytosol, pH > 7; vacuole, pH ≪ 6). The close relationship between KCl supply and ATPase activity in roots (Section 8.7) is therefore a reflection of two different mechanisms located at different membranes:

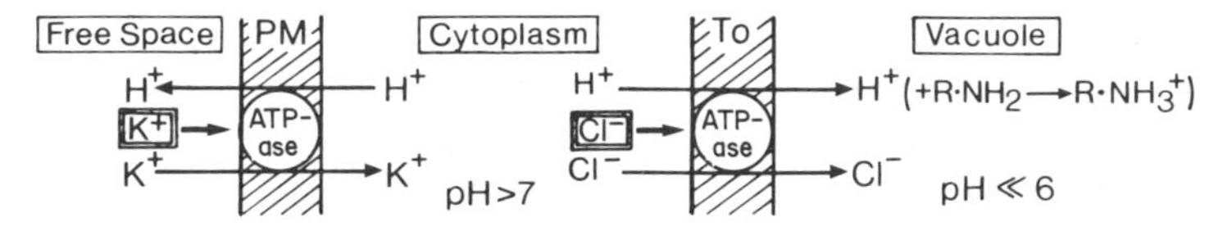

There are also striking similarities between the chloride-stimulated H^+-ATPase and the mechanisms regulating elongation growth in coleoptiles (Hager and Helmle, 1981). Severe inhibition of root elongation growth in chlorine-deficient plants might also be related to this function of chloride. The lack of stimulation of H^+-ATPase by nitrate is probably an important regulatory mechanism for ensuring rapid nitrate transport into the shoot—the main site of nitrate reduction and simultaneously preferentially transporting chloride into the vacuoles of root cells where it can function as an osmotically active solute.

9.7.5 Other Effects

Not much is known about the functions of chloride in other metabolic processes. The levels of certain amino acids and amides are exceptionally high in chlorine-deficient plants (Freney *et al.*, 1959) as a result of either inhibition of synthesis or degradation of proteins. A specific role of chloride in nitrogen metabolism is indicated by its stimulating effect on asparagine synthetase, which uses glutamine as a substrate:

$$\text{Glutamine} \xrightarrow[\text{asparagine synthetase}]{(NH_3)} \text{asparagine} + \text{glutamic acid}$$

Either chloride or bromide enhances this transfer by a factor of ~7, whereas sulfate has an inhibitory effect. Furthermore, chloride increases the affinity of the enzyme for the substrate by a factor of 50 (Rognes, 1980). In plant species in which asparagine is the major compound in the long-distance transport of soluble nitrogen (Section 8.2), chloride might therefore also play an important role in nitrogen metabolism.

When supplied in large quantities, chloride has a whole range of less specific effects on metabolism and growth, such as competition with nitrate uptake and modulation of cation–anion balance and organic acid metabolism (Chapter 2). Chloride is one of the major osmotically active solutes in the vacuoles and thus affects, for example, the turgor potential of leaves or of specialized tissues such as the pulvini in *Mimosa* (Section 8.7.5.3). In principle, the role of chloride in stomatal movement is also a type of osmoregulation, restricted to a small proportion of the leaf cells. Therefore, on a whole-leaf basis the requirement for chloride in stomatal regulation is in the range of the requirement for micronutrients, whereas the concentration of chloride required for a substantial contribution to osmoregulation in the vacuoles of the remainder of the leaf tissue is about 10 to 100 m*M*. In halophytes the chloride concentrations in the vacuoles of leaf cells can be even much higher.

9.7.6 Chloride Requirement and Toxicity

A comparison between the chloride requirement for growth and supply by various sources reveals that chlorine deficiency seldom occurs under field conditions. Assuming a minimal requirement for optimal growth of 1 g/kg dry wt, one would expect, on average, a crop requirement of between 4 and 8 kg chloride per hectare. This is the amount of chloride supplied by rain. In highly leached soils, however, the chloride levels are very low and may even be insufficient to cover the requirement of plants, as has been shown by Ozanne (1958). For maximal growth of white clover, for example, concentrations of 8 μM phosphorus and 100 μM chloride in the nutrient solution are required; at concentrations of only 10 μM chloride, the dry matter yield drops to ~50% (Chisholm and Blair, 1981).

On the other hand, concentrations of chloride in the external solution of more than 20 m*M* can lead to chloride toxicity in sensitive plant species. In tolerant species the external concentration can be four to five times higher without reducing growth. Differences in chloride toxicity are related mainly to differences in the sensitivity of leaf tissue to excessive chloride levels. More than 3·5 g chloride per kilogram of leaf dry weight (~10 m*M* Cl^- in the leaf water) are toxic to sensitive species such as most fruit trees, as well as to bean and cotton. In contrast, 20–30 g chloride per gram of leaf dry weight (~60–90 m*M* Cl^- in the leaf water) do not have a harmful effect on tolerant species such as barley, spinach, lettuce, and sugar beet. Genotypical differences in chloride tolerance are closely related to salt tolerance mechanisms, which are discussed in Section 16.6.

10

Beneficial Mineral Elements

10.1 Definition

Mineral elements which either stimulate growth but are not essential (for a definition of essentiality see Chapter 1) or which are essential only for certain plant species, or under given conditions, are usually defined as *beneficial elements*. This definition applies in particular to sodium, silicon, and cobalt. Making the distinction between beneficial and essential is especially difficult in the case of some trace elements, as will become obvious in the following sections. Developments in analytical chemistry and in methods of minimizing contamination during growth experiments may well lead in the future to an increase in the list of micronutrient elements and a corresponding decrease in the list of beneficial mineral elements.

10.2 Sodium

10.2.1 General

The sodium content of the earth's crust is ~2·8%, compared with 2·6% for potassium. In temperate regions the Na^+ concentration in the soil solution is on average 0·1–1 m*M*—similar to or higher than the K^+ concentration. In semiarid and arid regions, particularly under irrigation, concentrations of 50 to 100 m*M* Na^+ (mostly as NaCl in the soil solution) are typical and have a rather detrimental effect on the growth of most crop plants (Section 16.6). The hydrated Na^+ ion has a radius of 0·358 nm, whereas that of K^+ is 0·331 nm. Most higher plants have developed high selectivity in the uptake of K^+ as compared with that of Na^+, and this is particularly obvious in transport to the shoot (Chapter 3). Plant species are characterized as *natrophilic* or *natrophobic* depending on their growth response to sodium and their capacity for long-distance transport of sodium to the shoots. The role of Na^+ in the mineral nutrition of higher plants has to be considered from two main viewpoints: its essentiality and/or the extent to which it can replace K^+ functions in plants.

10.2.2 Essentiality

The essentiality of Na^+ as a micronutrient has been established for the halophyte *Atriplex vesicaria* (Brownell, 1965). When contamination with Na^+ in the basic nutrient solution was kept to a minimum (below 0·1 μM Na^+), plants became chlorotic and necrotic and no further growth occurred, despite high K^+ levels in the plants (Table 10.1). The growth response to Na^+ in the low concentration range (0·02 mM) was quite dramatic, although the sodium content (~0·1% in the dry wt) was still in the range typical for a micronutrient. At higher supply, however, the sodium content was in the range of a macronutrient. The growth responses to Na^+ in the latter case, therefore, are presumably more related to the functions of particularly K^+, such as osmoregulation and enzyme activation.

In further studies on various halophilic plant species, responses to Na^+ that were similar to those shown in Table 10.1 were found in species characterized by the C_4 photosynthetic pathway (Brownell and Crossland, 1972) and the CAM pathway (Brownell and Crossland, 1974). According to Brownell (1979) Na^+ is a micronutrient in the strict sense for C_4 plants but not for C_3 plants. The mechanism by which Na^+ functions as a micronutrient in such C_4 species as *Atriplex vesicaria, Amaranthus tricolor* or *Kochia childsii* is not well understood, since neither the enzymes nor the overall mechanism of the C_4 pathway seem to be affected by Na^+ (Boag and Brownell, 1979). More recent evidence (Nable and Brownell, 1984) has shown that in sodium-deficient *Amaranthus tricolor* high levels of alanine accumulate. According to the proposed role of amino acids in the chloroplasts of C_4 plants (Section 5.2.3), an involvement of Na^+ in the shuttle of metabolites between the mesophyll and bundle sheath chloroplasts is assumed. This assumption is supported by the effect of varied ambient CO_2

Table 10.1
Effect of Sodium Sulfate Concentrations on the Growth and Sodium and Potassium Content of Leaves of *Atriplex vesicaria* L.[a]

Treatment (mM Na^+)	Dry wt (mg/ four plants)	Content of leaves (mmol/kg dry wt) Na	K
None	86	10	2834
0·02	398	48	4450
0·04	581	78	2504
0·20	771	296	2225
1·20	1101	1129	1688

[a]The basic nutrient solution contained 6 mM potassium. From Brownell (1965).

concentration on the growth of a C_4 species, *Amaranthus*, as compared to a C_3 species, tomato (Johnston *et al.*, 1984). The much higher efficiency of C_4 plants compared to C_3 plants in utilizing low ambient CO_2 concentrations—which is mainly based on the shuttle system—was confined to the sodium-sufficient plants, containing about 0·02% Na in the dry weight. Despite these interesting results, a generalization of results on the function of Na^+ from these natrophylic C_4 species to C_4 plants in general remains questionable until typical natrophobic C_4 species such as maize and sorghum are also included in these studies.

Whether C_4 or C_3 species, many halophytes show a distinct growth response to high Na^+ concentrations in the substrate (10–100 m*M* Na^+). Sodium is not a macronutrient, however, even for extreme halophytes (Flowers *et al.*, 1977). Growth responses of halophytes to Na^+ are merely reflections of a high salt requirement for osmotic adjustment (Flowers and Läuchli, 1983), for which, however, Na^+ can be much more suitable than K^+ (Eshel, 1985).

10.2.3 Substitution of Potassium by Sodium

The beneficial effects of Na^+ on the growth of nonhalophytes (glycophytes) are well known in agriculture and horticulture (for reviews, see, e.g., Lehr, 1953; Marschner, 1971). In general, plant species can be classified into four groups according to the differences in their growth response to Na^+ (Fig. 10.1).

In group A not only can a high proportion of K^+ be replaced by Na^+ without an effect on growth; additional growth stimulation occurs which

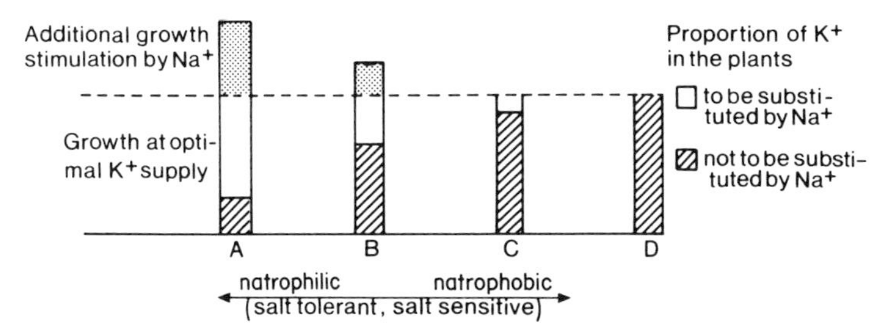

Fig. 10.1 Tentative scheme for the classification of crop plants according to both the extent to which sodium can replace potassium in plants and additional growth stimulation by sodium. Group A: mainly members of Chenopodiaceae (e.g., sugar beet, table beet, turnip, swiss chard) and many C_4 grasses (e.g., Rhodes grass). Group B: cabbage, radish, cotton, pea, flax, wheat, and spinach. Group C: barley, millet, rice, oat, tomato, potato, and ryegrass. Group D: maize, rye, soybean *Phaseolus* bean, lettuce, and timothy.

cannot be achieved by increasing the potassium content of the plants. In group B specific growth responses to Na^+ are observed, but they are much less distinct. Also, a much smaller proportion of K^+ can be replaced without a decline in growth. In group C only minor substitution is possible and Na^+ has no specific effect on growth. In group D no substitution of K^+ is possible. This classification cannot be used in a strict sense, of course, because it does not take into account, for example, differences among cultivars within a species in the substitution of K^+ by Na^+. These differences can be substantial, as has been shown in tomato by Makmur *et al.*, (1978).

The differences in the growth responses of natrophilic and natrophobic species to Na^+ are related to differences in uptake, particularly in the translocation of Na^+ to the shoots (Chapter 3.) An example of this is shown in Fig. 10.2 for sugar beet and bean. In sugar beet Na^+ is readily translocated to the shoots, where it replaces most of the K^+. This substitution increases the dry weight of the plants to values above those of deficient plants (0.05 m*M* K^+) as well as above those of plants receiving a large K^+ supply (5·0 m*M* K^+). In contrast, the growth of potassium-deficient bean plants (0·5 m*M* K^+) is further depressed by Na^+. The reasons for the lack of growth response in bean are quite obvious, at least for the shoots; in bean roots an effective mechanism (*exclusion mechanism*) exists for blocking Na^+ transport to the shoots (Chapter 3). The potential for replacement of K^+ by Na^+ is therefore very limited or absent in group D species such as bean.

Similar differences between plants species have been found in forage grasses, such as the natrophilic ryegrass and the natrophobic timothy, in experiments under controlled conditions (Smith *et al.*, 1980; Jarvis, 1982). The different strategies for regulating Na^+ transport to the shoots have important consequences in pasture plants for animal nutrition and in crop

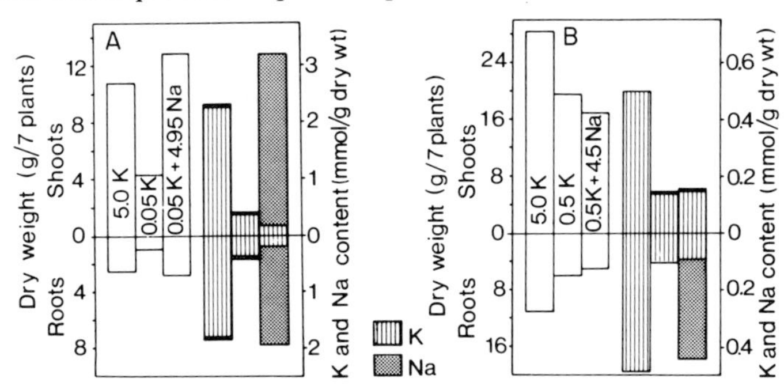

Fig. 10.2 Dry weight and potassium and sodium content of (A) sugar beet (cv. Sharpes Klein E type) and (B) bean (cv. Windsor Long Pod) grown in nutrient solutions with different concentrations of potassium and sodium. Total concentration of potassium and potassium + sodium, respectively, 5 m*M*. (Based on Hawker *et al.*, 1974.)

plants in general for salt tolerance (Greenway and Munns, 1980). The majority of agriculturally important crops are characterized by a more or less distinct natrophobic behavior (Groups C and D, Fig. 10.1) with correspondingly low salt tolerance. When exposed to high NaCl concentrations the exclusion mechanism in natrophobic species (referred to as *excluders*) cannot prevent massive transport of Na^+ to the shoot, where it is detrimental to the fine structure of chloroplasts (Marschner and Mix, 1974). In soybean a close positive correlation therefore exists between the capacity to prevent Na^+ transport to the shoots and growth depression by high NaCl concentrations in the substrate (Läuchli and Wieneke, 1979). In contrast, natrophilic species, especially those of group A, have a moderate to high salt tolerance and behave as *includers*. Under saline conditions they accumulate a large amount of Na^+ in the shoots, where it is utilized either in the vacuoles of leaf cells for osmotic adjustment, much as it is in typical halophytes (Harvey *et al.*, 1981), or it may also fulfil specific functions in the cytoplasm and its organelles. Genotypical differences among plant species in the strategies adopted are well documented in tomato (Rush and Epstein, 1981) and lupins (Van Steveninck *et al.*, 1982) but they may also occur within a species among genotypes (Marschner *et al.*, 1981a).

In natrophilic species also, the replacement of K^+ by Na^+ in the shoots is limited. The extent of substitution differs among individual organs and even among cell compartments. Average values for substitution for the whole shoot are therefore misleading and underestimate the essentiality of K^+ for growth and metabolism. In tomato, for example, replacement of K^+ by Na^+ takes place mainly in the petioles of expanded leaves (Besford, 1978a). In sugar beet the substitution can be very high in mature leaves (including blades), but it is much lower in expanding leaves (El-Sheikh and Ulrich, 1970), leading to a steep and converse gradient in the K^+/Na^+ ratios of leaves of different age (Table 10.2).

In old leaves nearly all the K^+ can be replaced by Na^+ and made available for specific functions in meristematic and expanding tissues. This substitution is not limited to vacuolar K^+ but also takes place in the chloroplasts of sugar beet leaves (Marschner and Döring, 1977), as is also known for halophytes (Larkum 1968). In contrast, in young expanding leaves there is a threshold level of substitution of ~0·5 mmol K^+ per gram dry weight (Table 10.2), which corresponds to a concentration of ~50 m*M* K^+ per kilogram fresh weight. At this concentration, K^+ is obviously essential for cell division and differentiation and cannot be replaced by Na^+ even in natrophilic species. Evidence exists for the essentiality of K^+ in expanding leaf tissue of sugar beet for chlorophyll formation (Marschner and Possingham, 1975), for the induction of nitrate reductase in spinach leaves (Pflüger and Wiedemann, 1977), and for the translation of m-RNA on wheat germ ribosomes

Table 10.2
Effect of the Replacement of Potassium by Sodium in the Nutrient Solution on the Potassium and Sodium Content of Sugar Beet (cv. Fia)[a]

Age and position of leaves	Supply of K^+ and Na^+ (m*M*)					
	5·0 K^+		0·25 K^+ + 4·75 Na^+		0·10 K^+ + 4·90 Na^+	
	K	Na	K	Na	K	Na
Whole shoot	3·0	<0·03	0·24	2·72	0·10	3·29
Old leaves (nos. 1–7)	3·43	<0·03	0·18	3·05	0·05	4·20
Middles leaves (nos. 8–15)	2·36	<0·03	0·34	2·01	0·14	2·97
Young leaves (nos. 16–22)	1·87	<0·03	0·52	1·75	0·48	1·82

[a]Sodium and potassium contents expressed as millimoles per gram dry weight. From Marschner *et al.*, 1981b).

(Gibson *et al.*, 1984). In the latter case, the specific K^+ requirement is 100 to 120 m*M* and independent of the salt tolerance of the wheat genotypes.

In natrophobic species such as maize and bean there is an absolute requirement for K^+ in most of its metabolic functions (Section 8.7). Replacement of K^+ by Na^+ may occur to some extent in the root vacuoles, whereas such substitution in the cytoplasm causes dramatic changes in the fine structure of the cytoplasm and its organelles (Hecht-Buchholz *et al.*, 1971).

10.2.4 Growth Stimulation by Sodium

In addition to K^+ substitution (the *sparing effect* of Na^+), growth stimulation by Na^+ is of great practical and scientific interest. It raises the possibility of applying inexpensive, low-grade potash fertilizers with a high proportion of sodium, and it increases the potential of successfully selecting and breeding for adaptation of crop plants to saline soils.

Responses to Na^+ differ not only among plant species but also among genotypes of a species, as shown in Table 10.3. Compared with the effect of a K^+ supply only, substitution of half the K^+ in the substrate by Na^+ led to an increase in the dry weight of the plants and the sucrose content of the storage root in all three genotypes. When 95% of the K^+ in the substrate (and ~90% within the plants) was replaced by Na^+, the dry weight of the plants was not affected further; instead, sucrose production per plant was severely reduced in the two genotypes Monohill and Ada. The decrease in Monohill resulted from a lower sucrose concentration (Table 10.3) and that in Ada from a shift

Table 10.3
Genotypical Differences in the Response of Sugar Beet Plants to the Replacement of Potassium by Sodium in the Nutrient Solution[a]

Genotype	Treatment (m*M*)		Dry wt	Sucrose in storage root	
	K^+	Na^+	(g/plant)	(% fresh wt)	(g/storage root)
Monohill	5·0	—	115	9·2	54·4
	2·5	2·5	133	11·9	49·6
	0·25	4·75	126	7·6	34·2
Ada	5·0	—	86	4·9	19·0
	2·5	2·5	131	7·1	43·3
	0·25	4·75	132	7·7	20·9
Fia	5·0	—	44	10·0	13·7
	2·5	2·5	65	10·4	20·3
	0·25	4·75	84	11·2	27·9

[a]Duration of the experiment 9 weeks. From Marschner *et al.* (1981b).

in shoot growth at the expense of storage root growth (Marschner *et al.*, 1981b). In Fia, however, sucrose concentration and production were enhanced when more K^+ was replaced by Na^+. Salt tolerance differed among the three genotypes, in agreement with the general pattern of classification (Fig. 10.1). Genotype Fia tolerated up to 150 m*M* NaCl in the external medium without significant growth reduction, whereas growth was severely depressed at that concentration in the other two genotypes (Marschner *et al.*, 1981a).

Growth stimulation by Na^+ is caused mainly by its effect on cell expansion and on the water balance of plants. Not only can Na^+ replace K^+ in its contribution to the solute potential in the vacuoles and consequently in the generation of turgor and cell expansion (Section 8.7); it may surpass K^+ in this respect since it accumulates preferentially in the vacuoles (Jeschke, 1977). The superiority of Na^+ can be demonstrated by the expansion of sugar beet leaf segments *in vivo* (Marschner and Possingham, 1975) as well as in intact sugar beet plants, where leaf area, thickness, and succulence are distinctly greater when a high proportion of K^+ is replaced by Na^+ (Milford *et al.*, 1977). An example of this effect is shown in Table 10.4. The leaves are thicker and store more water per unit leaf area; that is, they are more succulent. Succulence is a morphological adaptation which is usually observed in salt-tolerant species growing in saline substrates (Jennings, 1976) and is considered an important buffer mechanism against deleterious changes in leaf water potential under conditions of moderate water stress.

Sodium increases not only the leaf area but also the number of stomata per unit leaf area (Table 10.5). The chlorophyll content, however, is lower in these plants, and this may be responsible for the lower rate of net photo-

Table 10.4
Effect of Replacement of Potassium by Sodium in the Nutrient Solution on Sugar Beet Leaves (cv. Monohill)[a]

Supply (m*M*)	Dry wt (g leaves/plant)	Content of leaf blades (mmol/g dry wt)		Leaf area (cm^2/leaf)	Leaf thickness (μm)	Succulence (g H_2O/dm^2)
		K	Na			
5 K^+	7·6	2·67	0·03	233	274	3·07
0·25 K^+ + 4.75 Na^+	9·7	0·43	2·45	302	319	3·71

[a]Based on Hampe and Marschner (1982).

synthesis per unit leaf area. The higher growth rates of sugar beet plants with a high sodium but low potassium content are therefore the result, not of increased photosynthetic efficiency, but of a larger leaf area (Lawlor and Milford, 1973).

When the availability of water in the substrate is high, Na^+ increases the water consumption per unit fresh weight increment (Table 10.5); that is, it decreases the water use efficiency. If, however, the availability of water in the substrate is lowered to an osmotic potential of -4 bar by the addition of mannitol, water consumption decreases slightly in plants supplied with Na^+, but increases sharply in plants receiving a K^+ supply only. In the latter case the growth rate is much more depressed than the transpiration rate by the increase in osmotic potential to -4 bar.

Sodium improves the water balance of plants when the water supply is limited. This obviously occurs via stomatal regulation (Fig. 10.3). With a sudden decrease in the availability of water in the substrate (*drought stress*)

Table 10.5
Effect of the Replacement of Potassium by Sodium in the Nutrient Solution on Properties of Sugar Beet Leaves[a]

Supply (m*M*)	Stomata lower surface (no./cm^2)	Chlorophyll (mg/g dry wt)	Net photo-synthesis (mg $CO_2/dm^2 \times$ hr)	Water consumption (g H_2O/g fresh wt increment)	
				$-0{\cdot}2$ bar[b]	$-4{\cdot}0$ bar[b]
5·0 K^+	11,807	12·1	15·2	17·7	28·2
0·25 K^+ + 4·75 Na^+	15,127	9·2	14·4	26·5	24·6

[a]Based on Hampe and Marschner (1982).
[b]Osmotic potential of the nutrient solution.

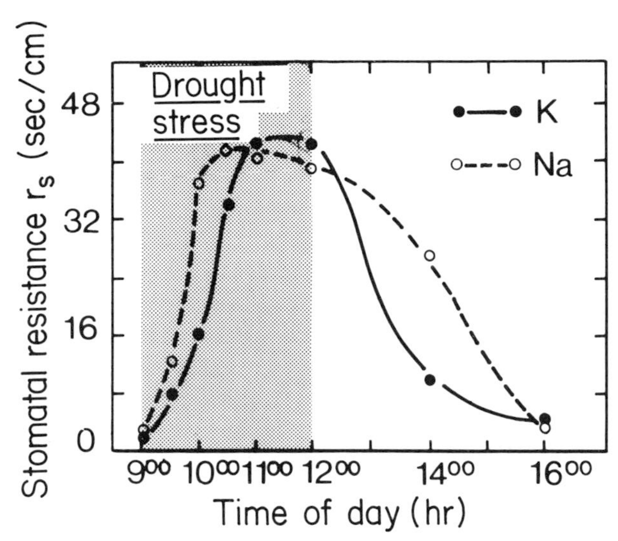

Fig. 10.3 Effect of transient drought stress (decrease in solution water potential to −7·5 bar by the addition of mannitol) on stomatal resistance to water vapor exchange in leaves of sugar beet (cv. Monohill). Plants were grown in nutrient solutions with either 5 m*M* K^+ (●——●) or 0·25 m*M* K^+ + 4·75 m*M* Na^+ (○ – – ○). (Based on Hampe and Marschner, 1982.)

the stomata of plants supplied with Na^+ close more rapidly than plants supplied with K^+ only and, after stress release, exhibit a substantial delay in opening. As a consequence, in plants supplied with Na^+ the relative leaf water content is maintained at a higher level even with low levels of water available in the substrate (drought periods, saline soils). Replacement of K^+ by Na^+ in its role in stomatal opening has been shown in epidermal strips of *Commelina* species (Willmer and Mansfield, 1970; Raghavendra *et al.*, 1976) and might therefore be a common feature of natrophilic species with high levels of Na^+ in the leaves.

Presumably, the effects of Na^+ on cell expansion and water balance in plants are also mainly responsible for the fact that, under field conditions, the yields of sugar beet obtained by the application of sodium fertilizers are often higher than those produced by potassium fertilizers. The application of sodium fertilizers results in an increase in the leaf area index early in the growing season and a corresponding increase in light interception, and it improves the water use efficiency of leaves under conditions of moderate water stress during the growing season (Durrant *et al.*, 1978).

Replacement at the cellular level of high proportion of K^+ by Na^+ also affects the activation of enzymes which are particularly sensitive to K^+ (Section 8.7). For example, the activation of starch synthase by K^+ is three to four times higher that its activation by Na^+ (Hawker *et al.*, 1974). This enzyme catalyzes the reaction

$$\text{Glucose-1}\cdot\text{Ⓟ} \xrightarrow[\text{ADP-glucose pyrophosphorylase}]{} \text{ADP-glucose} \xrightarrow[K^+ >> Na^+]{\text{starch synthase}} \text{starch}$$

In agreement with this, in the leaves of plants in which a high proportion of K^+ is replaced by Na^+ the starch content is much lower but the content of soluble carbohydrates, particularly sucrose, is much higher (Hawker *et al.*, 1974). This shift may favor both cell expansion in the leaf tissue and phloem transport to sinks such as the storage roots of sugar beet. Furthermore, Na^+ is more effective than K^+ in stimulating sucrose accumulation in the storage tissue of sugar beet (Chapter 5). This effect of Na^+ on sucrose storage seems to be related to an ATPase located at the tonoplast of beet storage cells (Willenbrink, 1983). The existence of K^+, Na^+-ATPases which require the presence of both K^+ and Na^+ for maximal activity is well documented in roots of natrophilic species (Kylin and Hansson, 1971).

10.2.5 Application of Sodium Fertilizers

Given the genotypical differences in growth response to Na^+ and the abundance of sodium in the biosphere, one can expect the application of sodium to have beneficial effects (a) in natrophilic plant species, (b) when soil levels of available potassium and/or sodium are low, and (c) in areas with irregular rainfall and/or transient drought during the growing season.

The potential replacement of K^+ by Na^+ must be taken into account when the application of fertilizers to natrophilic species is being considered. When sodium levels in the leaves are high, the potassium levels required for optimal growth decrease from 3·5 to 0·8% of the leaf dry weight in Italian ryegrass (Hylton *et al.*, 1967) and from 2·7 to 0·5% in Rhodes grass (Smith, 1974). Corresponding differences in optimal levels in leaves exist in tomato (Besford, 1978a) and sugar beet.

The sodium content of forage and pasture plants is an important factor in animal nutrition. The sodium requirement for lactating dairy cows is ~0·20% dry wt (Smith *et al.*, 1978; Zehler, 1981), which is distinctly higher than the average sodium content of natrophobic pasture species (Smith *et al.*, 1980). In contrast, the potassium content in these species is usually at least adequate but often in excess of animal needs, which is in the range of 2 to 2·5% dry wt. The use of sodium fertilizer to increase the sodium content of forage and pasture plants is thus an important procedure in large areas of the world. A high sodium content increases the acceptability of forage to animals and enhances daily food intake (Zehler, 1981). However, sodium fertilizers are effective only when applied to grassland or mixed pastures with a reasonably high proportion of natrophilic species.

10.3 Silicon

10.3.1 Uptake, Content and Distribution

Silicon is the second most abundant element in the earth's crust. In soil solutions the prevailing form is monosilicic acid, $Si(OH)_4$, with a solubility in water (at 25°C) of ~2 m*M* (equivalent to 120 mg SiO_2 per liter). On average, the concentrations in soil solutions are 30 to 40 mg SiO_2 per liter (range between about 7 and 80 mg) with a tendency to lower concentrations at high pH (>7) and when large amounts of sesquioxides are present in soils and anion adsorption is dominant (Jones and Handreck, 1967). Concentrations of SiO_2 in aqueous solutions higher than 120 mg/liter indicate either supersaturation of $Si(OH)_4$ or partial polymerization of monosilicic acid.

Higher plants differ characteristically in their capacity to take up silicon. Depending on their SiO_2 content (expressed as a percentage of shoot dry weight), they can be divided into three major groups: wetland Gramineae, such as wetland rice or horsetails (*Equisetum*), 10–15%; dryland Gramineae, such as sugarcane and most of the cereal species, and a few dicotyledons, 1–3%; and most dicotyledons, especially legumes, <0·5%. In a survey of 175 plant species grown in the same soil, Takahashi and Miyake (1977) distinguished between *silicon accumulators*—plants in which silicon uptake largely exceeds water uptake—and *nonaccumulators*—plants in which silicon uptake is similar to or less than water uptake. The ratio of silicon uptake to water uptake in a given plant species, however, is also a function of the external concentration (Jones and Handreck, 1969). Table 10.6 illustrates the relationship between plant species, silicon concentration, and silicon uptake. In this experiment, the silicon content of the shoots was both measured and calculated. In the latter case, the uptake of silicon and water were assumed to occur in the same ratio as that in the external solution. It is evident that, at low external concentrations, rice, and to a lesser extent wheat, took up more silicon than calculated, indicating an active transport mechanism between root and xylem. In rice but not wheat, even at the highest external concentration an active component could still be observed. In contrast, in soybean, exclusion mechanisms restricted the mass flow-driven (passive) transport of silicon across the roots to the xylem vessels. High levels of silicon in the shoots of rice and wheat corresponded to distinctly lower transpiration coefficients (Table 10.6), a side effect of silicon which can be considered beneficial (see Section 10.3.3).

Table 10.6
Measured and Calculated Silicon Content of Shoots of Plant Species Grown in Nutrient Solutions with Different Silicon Concentrations[a]

Plant species	SiO_2 concentration in nutrient solution (mg/liter)	Transpiration coefficient (liters H_2O/kg dry wt)	SiO_2 content (mg/kg dry wt)		Ratio measured/ calculated
			Measured	Calculated[b]	
Rice	0·75	286	10·9	0·2	54·5
	30	248	94·5	7·4	12·7
	162	248	124	40·2	3·1
Wheat	0·75	295	1·2	0·22	5·5
	30	295	18·4	8·9	2·1
	162	267	41·0	43·3	0·9
Soybean	0·75	197	0·2	0·15	1·3
	30	197	1·7	5·9	0·3
	162	197	4·0	31·9	0·1

[a]From Vorm (1980).
[b]Assuming "nonselective" silicon uptake by mass flow.

Since under field conditions the average SiO_2 concentration of the soil solution is 30 mg/liter or higher, in long-term experiments (with uncontrolled fluctuations in the silicon concentration) the silicon content of dryland cereals such as wheat can be quantitatively explained by nonselective uptake of silicon via mass flow in the transpiration stream from the soil solution to the shoots (Hutton and Norrish, 1974). Similar results have been obtained with oat in pot experiments with soils (Jones and Handreck, 1965). In contrast, in wetland rice and other species of the accumulator type, silicon uptake is closely related to root metabolism and not greatly affected by the transpiration rate (Okuda and Takahashi, 1965).

The long-distance transport of silicon in plants is confined to the xylem. Its distribution within the shoots and shoot organs is therefore determined by the transpiration rate in the organs (Jones and Handreck, 1967). Most silicon remains in the apoplast and is deposited after water evaporation at the termini of the transpiration stream, mainly the outer walls of the epidermal cells on both surfaces of the leaves. Silicon is deposited either as amorphous silica ($SiO_2 \cdot nH_2O$, "opal") or as so-called opal phytoliths with distinct three-dimensional shapes (Parry and Smithson, 1964). The epidermal cell walls are impregnated with a firm layer of silicon and become effective barriers against water loss by cuticular transpiration and fungal infections (Chapter 11). In grasses a considerable portion of silicon in the epidermis on both leaf surfaces is located intracellularly in so-called *silica cells* (Sangster, 1970) or "bulliform" cells (Takeoka *et al.*, 1984). Silicon is also deposited in the cell walls of xylem vessels, preventing compression of the vessels under conditions of high transpiration (Raven, 1983), and in the endodermis cells of roots, where it acts as a barrier against invasion of the stele by parasites and pathogens (Bennett, 1982).

The deposition of silicon in plant hairs on leaves, culms, inflorescence bracts and brush hairs of cereal grains such as wheat (Bennett and Parry, 1981) is suspected of posing a potential threat to human health. There is substantial evidence that the inflorescence bracts of grasses of the genus *Phalaris* and foxtail millet (*Setaria italica*) contain sharp, elongated siliceous fibers which fall into the critical size range of fibers that have been classified as carcinogenic (Sangster *et al.*, 1983). There are striking correlations between esophageal cancer and the consumption of either foxtail millet in, for example, North China (Parry and Hodson, 1982), or of wheat contaminated with *Phalaris* in the Middle East (Sangster *et al.*, 1983).

10.3.2 Metabolic Functions

The essentiality of silicon in unicellular organisms such as diatoms is well documented, and many details of its metabolic functions in these organisms are known (Werner and Roth, 1983). In higher plants the essentiality of silicon has been established for several species of silicon accumulators (*silicophile* species) such as *Equisetum arvense* (Chen and Lewin, 1969) and certain wetland grass species (Takahashi and Miyake, 1977). In wetland rice lacking silicon, vegetative growth and grain production are severely reduced and deficiency symptoms, such as necrosis on mature leaves and wilting of plants, occur (Lewin and Reimann, 1969). Nevertheless, the actual requirement for silicon as a mineral nutrient for vegetative growth seems to be extremely low even for wetland rice; and large silicon requirement appears to be confined to the reproductive stage (Table 10.7).

Table 10·7
Effect of Silicon Supplied at Different Growth Stages on Growth and Grain Yield of Wetland Rice[a,b]

Treatment at vegetative stage:	−Si	+Si	−Si	+Si
Treatment at reproductive stage:[c]	−Si	−Si	+Si	+Si
% SiO_2 (shoot dry wt)	0·05	2·2	6·9	10·4
Dry wt (g/pot)				
Roots	4·0	4·3	4·2	4·7
Shoots	23·5	26·5	31·0	33·6
Grain	5·3	6·6	10·3	10·8

[a]Based on Okuda and Takahashi (1965) and Takahashi and Miyake (1977).
[b]+Si, 100 mg SiO_2 per liter; −Si, no silicon supply.
[c]Ear emergence.

Sugarcane is another silicon accumulator plant. Under field conditions an SiO_2 content in leaf blades of below ~3% (about 1·4% silicon) dry wt is associated with a drastic reduction in growth and typical visible deficiency symptoms ("leaf freckling") on leaf blades directly exposed to full sunlight (Elawad *et al.*, 1982a,b). In contrast, under greenhouse conditions the requirement of sugarcane for silicon seems to be extremely low (Gascho, 1977).

The beneficial effects of silicon on the growth of accumulator plants are well documented, but according to Takahashi and Miyake (1977) and Miyake and Takahashi (1978, 1983) silicon is also an essential mineral element for tomato and cucumber. Tomato plants grow normally without silicon until the onset of flowering; then the newly developed leaves are malformed, pollination is impaired, and in severe cases no fruits are formed (Takahashi and Miyake, 1977). The leaves wilt and the plants wither, symptoms which might be related to the role of silicon in the compression resistance of xylem vessels when transpiration rates are high. Similar results have been obtained with cucumber (Table 10.8). Again, silicon affects vegetative growth to a much lesser extent than it does reproductive growth.

The reasons for the silicon requirement for reproductive growth in tomato and cucumber are not known, but the effects are similar in wetland rice (Table 10.7). Nevertheless, the results obtained with tomato and cucumber should be critically reexamined since in the absence of silicon, or at low levels of silicon supply, the phosphorus content of the leaves was excessively high (Table 10.8). Furthermore chloride was supplied only occasionally—if at all—as HCl for pH stabilization.

Table 10.8
Effect of Silicon Supply as Silicic Acid on Cucumber Grown in Solution Culture[a]

	Silicon supply (silicic acid) (mg SiO_2/liter)			
Property	0	5	20	100
Shoot dry weight (g)	61	71	79	95
No. of fruits	0·3	1·7	2·5	4·5
Fruit fresh weight (g)	8	67	142	261
Mineral element content (% leaf dry wt)				
SiO_2	0·07	0·44	1·37	5·54
P	2.10	1·74	1·70	0·87
Mildew disease[b]	>1	>1	0·2–0·5	<0·1

[a]From Miyake and Takahashi (1983).
[b]Number of colonies/cm^2 leaf, older leaves.

Because of its abundance in the biosphere the essentiality of silicon as a micronutrient for higher plants is very difficult to prove. Even highly purified water contains 2×10^{-5} m*M* silicon (Werner and Roth, 1983), and correspondingly the leaves of silicon accumulator plants that were subjected to a so-called no-silicon treatment contained at least 700 mg SiO_2 per kilogram dry weight (Table 10.8).

There have been only a few in-depth studies on metabolic changes in higher plants when silicon is omitted from the external solution or when a specific inhibitor of silicon metabolism, germanic acid, is supplied (Werner, 1967). In the absence of silicon a considerable decrease in the incorporation of inorganic phosphate into ATP, ADP, and sugar phosphates is observed in sugarcane (Wong You Cheong and Chan, 1973); in wheat root cell walls the proportion of lignin declines and that of phenolic compounds increases (Jones *et al.*, 1978). This latter aspect deserves particular attention for various reasons. Some of the cell wall–bound silicon is presumably present as an ester-like derivative of silicic acid (R^1—O—Si—O—R^2) acting as a bridge in the structural organization of polyuronides (Jones, 1978). Furthermore, evidence exists for specific interactions between silicon and the content and metabolism of polyphenoles in xylem cell walls (Parry and Kelso, 1975). As shown by Weiss and Herzog (1978) silicic acid, like boric acid, has a high affinity for *o*-diphenols such as caffeic acid and corresponding esters, forming mono-, di- and polymeric silicon complexes of high stability and low solubility:

Silicon may therefore affect the stability of higher plants not only as an inert deposition in lignified cell walls but also by modulating lignin biosynthesis. As stressed by Raven (1983), silicon as a structural material requires much less energy than lignin. About 2 g of glucose are necessary for the synthesis of 1 g of lignin; the ratio of the energy requirement for lignin to that of silicon is 20 : 1.

10.3.3 Beneficial Effects

Besides the lower energy costs of silicon as a structural material, which are rather speculative, silicon has a number of other, well documented and readily visible and/or measurable beneficial effects. Under field conditions, particularly in dense stands of cereals, silicon can stimulate growth and yield by several indirect actions. These include decreasing mutual shading by improving leaf erectness, decreasing susceptibility to lodging, decreasing the incidence of fungal infections, and preventing manganese and/or iron toxicity.

Leaf erectness is an important factor affecting light interception in dense plant populations. For a given cultivar, leaf erectness decreases with increasing nitrogen supply (Table 10.9), for reasons discussed in Section 8.2. Silicon increases leaf erectness and thus to a large extent counteracts the negative effects of a large nitrogen supply on light interception. In a similar manner, silicon counteracts the negative effects of an increasing nitrogen supply on haulm stability and lodging susceptibility (Idris *et al.*, 1975).

Table 10.9
Relationship between Silicon[a] and Nitrogen Supply and Leaf Openness[b] in Rice Plants (cv. IR8) at Flowering[c]

Nitrogen supply (mg/liter)	SiO_2 supply (mg/liter)		
	0	40	200
5	23°	16°	11°
20	53°	40°	19°
200	77°	69°	22°

[a]As sodium silicate.
[b]Angle between the culm and the tip of the leaves.
[c]Based on Yoshida *et al.* (1969).

The deposition of silicon in the epidermal layer of leaves is particularly effective in increasing the mechanical resistance of tissue against attacks by fungi such as powdery mildew (Table 10.8), blast infection in rice (Volk *et al.*, 1958), and insect pests. This aspect deserves increasing attention in intensive agriculture (Chapter 11).

Stimulation of growth by silicon can also be caused by the prevention or depression of manganese and iron toxicity (Vlamis and Williams, 1967). As shown in Fig. 10.4, silicon has no effect on the growth of bean plants at low manganese levels. At high levels, however, it either prevents (5·0 μM) or at least reduces (10·0 μM) the severe growth depression induced by manganese toxicity. Although the effect of silicon on growth is quite impressive at high manganese levels, it nevertheless has to be classified as a beneficial effect since it is restricted to conditions of excessive manganese supply.

The increase in manganese tolerance in bean plants is not due to lower manganese uptake but to an increase in the tolerance of the leaf tissue to high manganese levels (Horst and Marschner, 1978a). In plant species such as barley and bean with low tissue tolerance to high manganese levels,

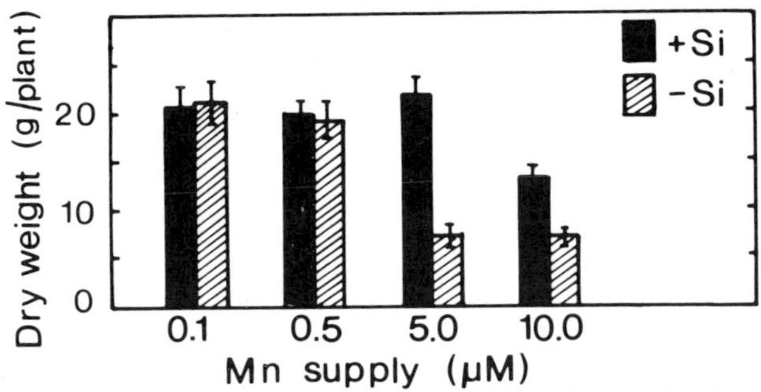

Fig. 10.4 Effect of manganese on the dry weight of bean in the absence and presence of silicon (1·55 mg SiO_2/liter). Vertical lines represent standard deviation. (Modified from Horst and Marschner, 1978.)

silicon alters the microdistribution of manganese within the leaf tissue (Williams and Vlamis, 1957). In the absence of silicon the distribution of manganese is nonhomogeneous and characterized by a spot-like accumulation (Fig. 10.5). These spots resemble the typical visible symptoms of manganese toxicity, namely, brown spots of manganese oxides surrounded by chlorotic or/and necrotic zones. Silicon prevents this accumulation by bringing about homogeneous distribution (Fig. 10.5) and presumably thereby increases tissue tolerance to high manganese levels. The critical toxicity levels of manganese in the leaf tissue of a given plant species can differ considerably, therefore, depending on the silicon content of the leaves (Horst and Marschner, 1978a). The mechanism of this silicon effect is not fully understood.

In wetland species, such as wetland rice, silicon increases tolerance to excessive levels of manganese and iron in the rooting medium in another way, namely, by depressing the rates of iron uptake and particularly manganese uptake (Vorm and Diest, 1979). In these plant species silicon increases the "oxidizing power" of the roots (Okuda and Takahashi, 1965) by increasing the volume and rigidity of the aerenchyma (air-filled spaces in shoots and roots), thereby enhancing O_2 transport from the shoots to the submerged root system exposed to toxic concentrations of reduced manganese and iron. The application of silicon fertilizers in various forms (e.g., electric furnace slag or rice husks) is therefore an effective procedure, especially in sugarcane (Elawad *et al.*, 1982a) and wetland rice production. In rice, yield responses to silicon application can be expected at SiO_2 levels below 11% of the leaf dry weight (Okuda and Takahashi, 1965).

Other beneficial effects of silicon application, such as the "mobilization" of soil phosphorus (Scheffer *et al.*, 1982), reduced water loss by cuticular transpiration, and increased resistance against lodging and pests, deserve

Fig. 10.5 Autoradiograph showing the effect of silicon (0·75 mg SiO_2/liter) on ^{54}Mn distribution in bean leaves supplied with 0·1 m*M* ^{54}Mn for 6 days. Manganese content of the primary leaves: 22·0 μg/g dry wt at –Si and 16·7 μg/g dry wt at +Si. (From Horst and Marschner, 1978a.)

more attention in the future in crops other than rice and sugarcane (Munk, 1982).

Silicon is an essential mineral element for animals, where it is a constituent of certain mucopolysaccharides in connective tissues (Jones, 1978). In grazing animals the uptake of a large amount of phytoliths may lead to excessive abrasion of the rumen wall, and dissolved silicon may form secondary depositions in the kidney, thereby causing serious economic losses (Jones and Handreck, 1969).

10.4 Cobalt

The role of cobalt as an essential mineral element for ruminants was discovered in 1935 in field investigations of livestock production in Australia. The requirement of cobalt for N_2 fixation in legumes and in root nodules of nonlegumes (e.g., alder) was established 25 years later (Ahmed and Evans, 1960; Delwiche *et al.*, 1961). In 1963 Kliewer and Evans (1963a) isolated from the root nodules of legumes and nonlegumes the cobalamin coenzyme B_{12}, and they demonstrated the close correlation between cobalt supply, the B_{12} coenzyme content of *Rhizobium*, the formation of leghemoglobin, and N_2 fixation (Kliewer and Evans, 1963b). On the basis of these studies and later reports by other authors, it has been established that *Rhizobium* and other N_2-fixing microorganisms have an absolute cobalt requirement regardless of whether or not they are growing within nodules or regardless whether they rely on N_2 fixation or are being supplied with mineral nitrogen. However, the demand for cobalt is much higher for N_2 fixation than for ammonium nutrition (Kliewer and Evans, 1963a).

In the coenzyme cobalamin (vitamin B_{12} and its derivatives) Co (III) is the metal component, which is chelated to four nitrogen atoms at the center of a porphyrin structure similar to that of iron in hemin. As summarized by Dilworth *et al.* (1979), in *Rhizobium* three specific cobalamin-dependent enzyme systems are known, and cobalt-induced changes in their activities are presumably responsible for the relationship between cobalt supply, nodulation, and N_2 fixation in legumes. These symptoms are the following:

1. Methionine synthase. Under conditions of cobalt deficiency, impaired protein synthesis may occur in *Rhizobium*.
2. Ribonucleotide reductase. This enzyme is involved in the reduction of ribonucleotides to deoxyribonucleotides and therefore in DNA synthesis. In agreement with this, there are fewer and longer bacteroids in the root nodules of cobalt-deficient plants than in normal plants, indicating depressed rhizobial cell division (Chatel *et al.*, 1978).

3. Methylmalonyl-coenzyme A mutase. This enzyme is involved in the synthesis of heme (iron porphyrins) in the bacteroids and thus—in cooperation with the host nodule cells—in the synthesis of leghemoglobin. Under conditions of cobalt deficiency the synthesis of leghemoglobin is therefore reduced, which affects the N_2 fixation rate adversely.

A cobalt-induced increase in the growth and nitrogen content of the shoots of nodulated legumes under field conditions was reported shortly after the discovery of the role of cobalt in N_2 fixation (Ozanne *et al.*, 1963; Powrie, 1964); cobalt application was without effect, however, in legumes supplied with mineral nitrogen.

In legumes grown in cobalt-deficient soils, *Rhizobium* infection is much less extensive than in plants supplied with cobalt, and the onset of N_2 fixation, as indicated by the nitrogen accumulation in the plants, is delayed for several weeks (Fig. 10.6). Inoculation with *Rhizobium* makes little difference when cobalt has been supplied but is quite effective in deficient plants at later growth stages.

The effect of cobalt on root nodules in 6-week old plants (day 42 in Fig. 10.6) is shown in Table 10.10. The weight and cobalt content of the nodules increased, as did the number of bacteroids and amount of cobalamin and leghemoglobin per unit nodule fresh weight. In experiments with purified sand and minimum contamination, the effects of cobalt on nodule growth and leghemoglobin content were much more distinct (Chapter 7) than in cobalt-deficient soil (Table 10.10). Somewhat surprisingly but in agreement with the results of other authors, only ~12% of the total nodule cobalt was

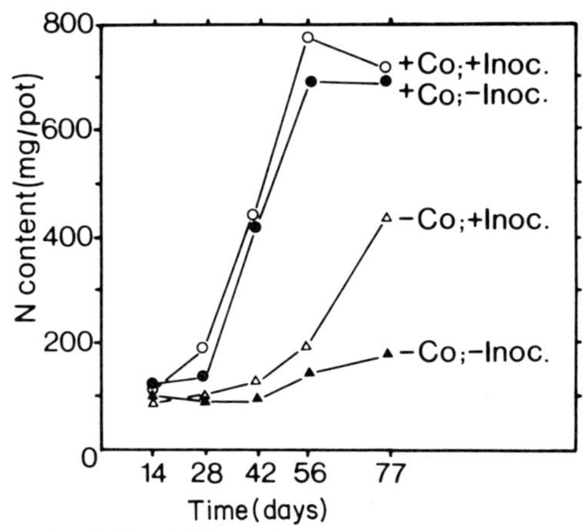

Fig. 10.6 Effect of cobalt and inoculation with *Rhizobium* on the time course of nitrogen accumulation in *Lupinus angustifolius* L. grown in a cobalt-deficient soil (eight plants/pot). (From Dilworth *et al.*, 1979.)

Table 10.10
Effect of Cobalt on Nodule Growth and Composition in *Lupinus angustifolius* Grown in a Cobalt-Deficient Soil and Inoculated with *Rhizobium lupini*[a,b]

Cobalt treatment	Crown nodule fresh wt (g/plant)	Cobalt content (ng/g nodule dry wt)	No. of bacteroids $\times 10^{-9}$ per g nodule fresh wt	Cobalamin (ng/g nodule fresh wt)	Leghemoglobin (mg/g nodule fresh wt)
−	0·1	45	15	5·9	0·71
+	0·6	105	27	28·3	1·91

[a]Based on Dilworth *et al.* (1979).
[b]0·19 mg cobalt as the sulfate salt was supplied per pot. Harvest after 6 weeks.

bound to cobalamin in the deficient plants (Table 10.10). Nothing is known about the location, chemical binding, and possible functions of the cobalt in nodules that is not available for cobalamin formation.

With cobalt deficiency, there is a preferential accumulation of cobalt in the nodules. On a whole-plant basis, however, the roots have the highest cobalt content. The proportion of cobalt in the shoots, nodules, and roots is 1 : 6 : 15 and 1 : 3 : 25 in cobalt-deficient and cobalt-sufficient plants, respectively (Robson *et al.*, 1979).

The critical deficiency levels of cobalt in root nodules are not known; levels in deficient plants vary between 0·02 and 0·17 mg/g nodule fresh wt, depending on the plant species (Robson *et al.*, 1979). The cobalt content of the shoots can be used as an indicator of cobalt deficiency in legumes, where the critical deficiency levels are between 0·04 (Ozanne *et al.*, 1963) and 0·02 mg per kilogram of shoot dry weight (Robson *et al.*, 1979).

The cobalt content of seeds of the same species varies widely among plants grown in different localities. In *Lupinus angustifolius* between 0·006 and 0·73 mg cobalt per kg seed weight have been found (Robson and Mead, 1980). Accordingly, plants grown from seeds with less than 0·03 mg/kg respond strongly to cobalt supply, whereas those grown from seeds with 0·73 mg/kg do not. In large-seeded lupins ~0·1 mg cobalt per kilogram seed weight is enough to prevent cobalt deficiency in plants grown in cobalt-deficient soils (Gladstones *et al.*, 1977). Treating seeds with cobalt is therefore an effective procedure for supporting N_2 fixation and the growth of legumes in cobalt-deficient soils, as shown in a field experiment with peanut (Table 10.11). Foliar sprays alone were also quite effective but were less efficient than the combination of seed treatment and foliar sprays. The effect of foliar sprays indicates a reasonable retranslocation of cobalt from leaves, as has also been shown after the application of labeled cobalt to clover and alfalfa leaves (Handreck and Riceman, 1969). In the phloem, cobalt seems to be translocated largely as a negatively charged complex (Wiersma and Goor, 1979).

Table 10.11
Effect of Cobalt on Peanut[a]

Cobalt treatment	No. of nodules per plant	Total nitrogen at maturity (% dry wt)	Pod yield (kg/ha)
Control (−Co)	91	2·38	1232
Seed treatment	150	2·62	1687
Foliar sprays (2×)	123	3·14	1752
Seed treatment + foliar sprays (2×)	166	3·38	1844

[a]Based on Reddy and Raj (1975).

There are considerable differences in the sensitivities of legume species to cobalt deficiency. *Lupinus angustifolius* is particularly sensitive in comparison with *Trifolium subterraneum* (Gladstones *et al.*, 1977). In *L. angustifolius* cobalt increased the yield and total amount of nitrogen per plant but, unexpectedly, decreased significantly the nitrogen content as a percentage of the dry weight. Nevertheless, the cobalt-treated plants looked healthy and were dark green, whereas the cobalt-deficient plants looked unhealthy and had yellowish leaves. One may therefore speculate that growing root nodules support plant growth not only by N_2 fixation but also by other factors such as cytokinins (Gladstones *et al.*, 1977).

It is still not clear whether cobalt also has direct functions in higher plants. A distinct growth response to cobalt was reported by Wilson and Hallsworth (1965) in clover supplied with mineral nitrogen or inoculated with ineffective *Rhizobium* strains. A similar response was noted in wheat (Wilson and Nicholas, 1967) and *Hevea* (Bolle-Jones and Mallikarjuneswara, 1957). These responses to cobalt in non-N_2-fixing higher plants are always small, and there is no evidence that under these conditions cobalt fulfils the requirements of an essential mineral element, despite the occurrence of a cobalamin-dependent enzyme, the leucine-2,3 aminomutase, for example, in potato (Poston, 1978). In contrast, in photosynthetic lower plants such as *Euglena gracilis*, cobalamin is essential for growth and is localized in various subcellular fractions as well as in the thylakoids of chloroplasts (Isegawa *et al.*, 1984).

Cobalt stimulates extension growth when added to excised plant tissue or organs such as coleoptiles and hypocotyls (Bollard, 1983). Inhibition of endogenous ethylene formation by cobalt seems to be involved in this stimulation (Samimy, 1978), an effect which is also responsible for the extension of life of cut roses by cobalt application (Venkatarayappa *et al.*, 1980).

In deficient soils (mainly sandy soils) cobalt application is important not only for the N_2 fixation of legumes, but also for increasing the nutritional quality of forage plants. The critical cobalt level for ruminants is about 0·07 mg per kilogram dry weight of forage, that is, higher than the critical level for N_2 fixation in legumes. On average the cobalt content of plants varies between 0·05 and 0·30 mg/kg and is usually higher in legumes than in grasses (Kubota and Allaway, 1972).

There are contradictory reports on critical toxicity levels of cobalt, values varying from 0·4 mg in clover (Ozanne *et al.*, 1963) up to a few milligrams per kilogram dry weight in bean and cabbage (Bollard, 1983). In crop and pasture species there are also distinct genotypical differences in tolerance to excessive levels of cobalt in the shoots. Some plant species adapted to metalliferous soils contain more than 4000 mg/kg dry wt (Brooks *et al.*, 1978).

10.5 Nickel

The nickel (Ni) content of crop plants normally ranges from about 0·1 to 1·0 mg/kg dry wt. Critical toxicity levels for sensitive and moderately tolerant crop species are >10 and >50 mg, respectively (Welch, 1981; Bollard, 1983). In plants growing in serpentine soils the nickel content might be 100 or 1000 times higher. Such levels are highly toxic for most plant species. Nevertheless, there are a number of adapted species (*hyperaccumulators*) in serpentine soils that may contain more than 3% Ni in the shoot dry weight (Bollard, 1983). A high proportion of nickel is complexed to organic acids in these species, which may contribute to their tolerance, but other mechanisms must also be involved (Woolhouse, 1983).

Nickel is readily taken up by most crop species. As a divalent cation (Ni^{2+}) it competes with other cations, including Ca^{2+}, Mg^{2+}, Fe^{2+} and Zn^{2+}. Therefore, high levels of nickel in the substrate and in plants may induce zinc or iron deficiency and lead to characteristic symptoms of chlorosis (Anderson *et al.*, 1973). There is concern that the application of sewage sludge to crop plants results in elevated levels of nickel in these species (Marschner, 1983). There were a number of earlier reports on the stimulation of germination and growth of various crop species by low concentrations of nickel in the substrate (for a review, see Mishra and Kar, 1974). As a rule these reports were not very conclusive with respect to the role of nickel in higher plants. Evidence for such a role came from biochemical studies which demonstrated that nickel is the metal component of urease, the enzyme that catalyzes the reaction of $CO(NH_2)_2 + H_2O \rightarrow 2NH_3 + CO_2$. Highly purified preparations of urease from seeds of jack bean (*Canavalia ensiformis* L.) contain nickel (Dixon *et al.*, 1975). The enzyme consists of six subunits with two nickel atoms in each subunit (Dixon *et al.*, 1980c; Alagna *et al.*, 1984). Nickel is not required for the synthesis of the enzyme protein (Winkler *et al.*, 1983) but, as the metal component, it is essential for the structure and functioning of the enzyme (Dixon *et al.*, 1980b; Klucas *et al.*, 1983).

The function of nickel in urease is supported by results of Polacco (1977) with cell cultures of soybean. In the absence of added nickel, urease activity was low and growth was poor when urea was the source of nitrogen. The addition of nickel increased both growth and urease activity more than fivefold. When ammonium was the source of nitrogen, added nickel did not stimulate cell growth. Apparently, nickel is essential for plants supplied with urea and for plants in which ureides (e.g., allantoin) play a significant role in nitrogen metabolism. This holds true in particular for several legume species in which ureides are important forms of soluble nitrogen during transport from root nodules to the shoots as well as via the phloem to the seeds. The stimulation by nickel of the nodule weight and seed yield of soybean further supports the role of nickel in legumes (Bertrand and De Wolf, 1973). Ureides are also important metabolites of nitrogen metabolism during seed germination. According to Welch (1981) the stimulation of germination by nickel (Mishra and Kar, 1974) may be based on the function of nickel as the metal component of urease.

As is to be expected, particularly in growth experiments the response to nickel supply depends on the use of highly purified solutions and of seeds with low nickel content (Dixon *et al.*, 1980a). Under such conditions severe leaflet tip necrosis occurred in soybean and cowpea in the absence of added nickel, irrespective of the form of nitrogen nutrition (urea; NO_3-N; NH_4-N; N_2 fixation) (Eskew *et al.*, 1984). These results clearly demonstrate the beneficial role played by nickel for legumes with their particular type of nitrogen metabolisms, that is, the preferential formation of ureides (Chapter 5). The same might be true for other plant families with this type of nitrogen metabolism (Chapter 5), or plants supplied with urea, for example, as a foliar spray. In order to classify nickel as an essential mineral element, however, additional criteria have to be fulfilled (Chapter 1) and thus, further studies are necessary, particularly those that include other plant families. It might well be that the classification will be as difficult as is the case with cobalt or sodium—in other words, nickel would be considered essential only for certain types of nitrogen nutrition and certain plant families.

10.6 Selenium

The average selenium content of crop plants varies between 0·01 and 1·0 mg/kg dry wt. The presence of selenium in plants first attracted attention in the 1930s when it was realized that selenium toxicity is responsible for certain livestock disorders ("alkali disease" and "blind staggers") in animals grazing on soils with a high selenium content, so-called seleniferous soils (Brown and Shrift, 1982). For ruminants the toxic levels of selenium are above 5 mg/kg dry wt (Kubata and Allaway, 1972). However, certain species of the genus *Astraglus* (milk vetch) and other species such as *Neptunia ambexicaulis* may contain more than 4000 mg/kg dry wt (Peterson and Butler, 1967; Brown and Shrift, 1982). When accumulator and nonaccumulator species are growing in the same soil, the selenium content of the accumulator species can be about 4000 mg/kg dry wt without affecting the growth of the plants negatively whereas a selenium content in a nonaccumulator species (e.g., sunflower) of only ~2 mg/kg dry wt is sufficient to retard growth (Shrift, 1969).

Reports of a selenium requirement for high growth rates in accumulator species could not be confirmed by Broyer *et al.* (1972). These authors demonstrated that in plants grown in solution culture without selenium, toxic levels of phosphate accumulated in the leaves. The addition of selenium prevented this excessive phosphate uptake and thereby stimulated growth. At nontoxic phosphorus levels, selenium was without beneficial effect on the growth of accumulator plants of the genus *Astragalus*. This is another instructive example of the necessity of critically evaluating the so-called beneficial mineral elements.

Selenium is taken up as selenate (SeO_4^{2-}) or selenite (SeO_3^{2-}). Selenate and sulfate (SO_4^{2-}) compete for the same binding sites at the plasma membrane of root cells (Chapter 2). The uptake and transport as well as the assimilation of selenate follow the same pathway as sulfate (Section 8.3), leading to the formation of selenium analogues of cysteine and methionine, namely, selenocysteine and selenomethionine. The pathway of assimilation is the same in accumulator and nonaccumulator plants (Burnell, 1981). In nonaccumulator plants, selenoamino acids are incorporated into proteins which are either nonfunctioning or at least much less capable of functioning as enzyme proteins than the corresponding sulfur-containing proteins (Brown and Shrift, 1982). In contrast, in accumulator plants, the selenoamino acids are transformed into nonprotein amino acids such as selenomethylcysteine:

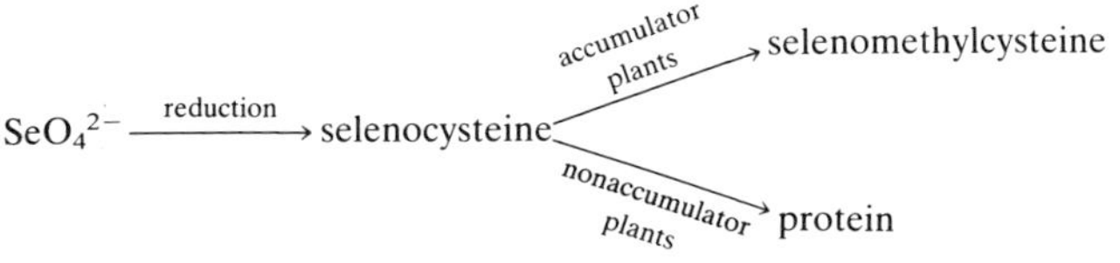

This exclusion of selenoamino acids from incorporation into proteins resembles a special case of "physiological inactivation" and can be considered one of the most important mechanisms of selenium tolerance in accumulator plants (Brown and Shrift, 1982).

Whereas selenium toxicity in animals is usually restricted to regions characterized by seleniferous soils and the typical flora of accumulator species, latent selenium deficiency in animals may be more abundant. In large areas in the United States, for example, the selenium content of forage and grains is below 0·05 mg/kg dry wt; to prevent selenium deficiency in ruminants ("white muscle disease"), however, the selenium content of the diet must be between 0·05 and 0·10 mg/kg dry wt (Kubota and Allaway, 1972).

10.7 Aluminum

Aluminum concentrations in mineral soil solutions are below 1 mg/liter at pH values higher than 5·5 but rise sharply at lower pH. Most crop plants are sensitive to high aluminum concentrations, and aluminum toxicity in acid soils is a serious problem that requires much effort in the selection and breeding of plants for higher aluminum tolerance (Section 16.3). There have been several reports that low aluminum concentrations in the soil or nutrient solution stimulate growth [for a review see Bollard (1983) and Foy (1983)]. These concentrations vary between 0·2 and 5·0 mg/liter in sugar beet, maize, and some tropical legumes. In the tea plant, which is one of the most aluminum-tolerant crop species, growth stimulation is observed at aluminum concentrations as high as 27 mg/liter. A possible explanation for the stimulation of growth by low aluminum concentrations is the prevention of copper, manganese, or phosphorus toxicity by aluminum. The last possibility is of particular importance in solution culture experiments, in which excessive phosphorus levels are quite common.

A general problem in most studies on the effect of low levels of aluminum on plant growth is the contamination of the nutrient solution with aluminum. Normally, the roots of plants growing in nutrient solutions with supposedly zero levels of aluminum contain 50–100 mg aluminum per kilogram dry weight (Bollard, 1983). In only a few experiments has special care been taken to keep the contamination as low as possible. Under these conditions, the growth rate of maize was stimulated fourfold by an aluminum concentration of 0·2 mg/liter, compared to the control which contained 0·06 mg/liter as contamination (Bertrand and De Wolf, 1968). This result is astonishing and requires critical examination to rule out the possibility of the side effects mentioned above.

In conclusion, under certain conditions and in species and genotypes with high aluminum tolerance, low levels of aluminum may have beneficial effects on the growth of higher plants. Moreover, there have been reports that aluminum may serve as a fungicide against certain types of root rot. These effects are the exception, however, and negative effects of aluminum on plant growth in soils of low pH are the rule (Section 16.3).

10.8 Other Mineral Elements

The requirement for such mineral elements as iodine and vanadium is fairly well established for certain lower plant species, including marine algae (iodine) and fungi and freshwater algae (vanadium). The reports on the stimulation of growth of higher plants by iodine and vanadium are rare and vague. An example of this is the effect of vanadium on the growth of tomato (Basiouny, 1984), or the effect of titanium on the growth of various crop species (Pais, 1983). For further information see Bollard (1983).

In recent years there have been a vast number of reports on the presence of heavy metals, such as cadmium, chromium, lead, and mercury, in higher plants. Most of these reports are concerned mainly with environmental pollution, the presence of heavy metals in the food chain, and genotypical differences in the critical toxicity levels of heavy metals in plants. For further studies the reader is referred to Ernst and Joosse-van Damme (1983) and Marschner (1983).

11

Relationship between Mineral Nutrition and Plant Diseases and Pests

11.1 General

The effects of mineral nutrients on plant growth and yield are usually explained in terms of the functions of these elements in plant metabolism. However, mineral nutrition may also exert secondary, often unpredicted influences on the growth and yield of crop plants: By effecting changes in growth pattern, plant morphology and anatomy, and particularly chemical composition, mineral nutrients may either increase or decrease the resistance of plants to pathogens and pests. Impressive progress has been made in breeding and selection for increased resistance to diseases and pests. Resistance can be increased by changes in anatomy (e.g., thicker epidermal cells and a higher degree of lignification and/or silification), and in physiological and biochemical properties (e.g., higher production of inhibitory or repelling substances). Resistance can be particularly increased by altering plant responses to parasitic attacks through enhanced formation of mechanical barriers (lignification) and the synthesis of toxins (phytoalexins). Apparent resistance can be achieved when the most susceptible growth stages of the host plant are not synchronized with the period of highest activity of parasites and pests (known as "escape from attack").

Although resistance is genetically controlled, it is considerably influenced by environmental factors. As a rule the effects are relatively small in highly susceptible or highly resistant cultivars but very substantial in moderately susceptible or partially resistant cultivars. Mineral nutrition is an environmental factor that can be manipulated relatively easily. In order to complement disease and pest control methods (e.g., the application of fungicides and pesticides) by the manipulation of mineral nutrition (via fertilizer application), one must have a detailed knowledge of the means by which mineral nutrients increase or decrease plant resistance.

The range of nitrogen fertilizer effects on a plant disease is illustrated in Table 11.1. As the nitrogen supply is increased, the incidence of leaf blotch rises in all three cultivars. However, the absolute levels of infection, as

Table 11.1
Nitrogen Fertilizer Supply and the Incidence of Leaf Blotch (*Rhynchosporium scalis*) in Spring Barley Cultivars[a]

Nitrogen supply (kg/ha)	Flag leaf area infected by leaf blotch (%)		
	Proctor	Cambrinus	Deba Abed
0	0·4	15·4	3·6
66	1·3	21·3	20·5
132	4·5	30·5	57·3

[a]Based on Jenkyn (1976).

expressed as a percentage of flag leaf affected, are different. With Proctor the increase has no physiological or economic importance, whereas with the other two cultivars detrimental effects on photosynthesis and on grain yield would be expected.

The close correlation between nitrogen supply and leaf blotch shown in Table 11.1 cannot be generalized, however, to all fungal and parasitic diseases. Usually, a "balanced" nutrient supply which ensures optimal plant growth is also considered optimal for plant resistance. Such an ideal situation is shown in Fig. 11.1 for pelargonium plants. An inverse relationship exists between nutrient supply and plant growth, on the one hand, and severity of bacterial infection, on the other. From this finding, one can conclude that plants with an optimal nutritional status have the highest resistance to diseases and that susceptibility increases as nutrient concentrations deviate from this optimum. This ideal situation, however, is not the rule, as explained later in this chapter.

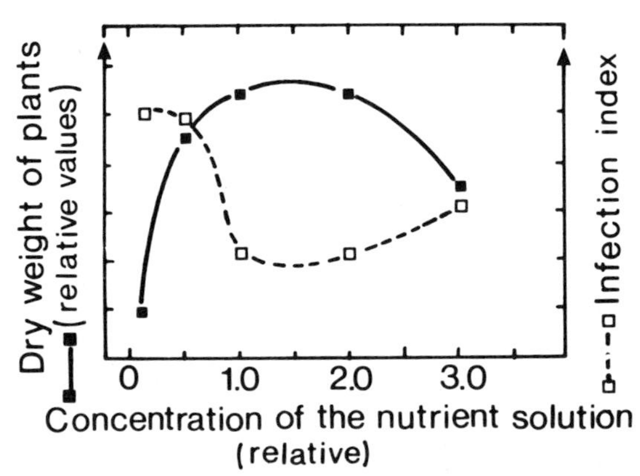

Fig. 11.1 Effect of the nutrient solution concentration on growth (noninfected plants) and on degree of infection (inoculation) with bacterial stem rot (*Xanthomonas pelargonii*) in *Pelargonium*. Relative values; water only = 0; basic nutrient solution = 1; twofold concentration of nutrient solution = 2; threefold concentration of nutrient solution = 3. (Modified from Kivilaan and Scheffer, 1958.)

The interactions between higher plants and parasites and pests are very complex, and to give a short outline of the role of mineral nutrients in these interactions requires considerable simplification. Nevertheless, there are some principal areas of host–parasite interactions where the roles of mineral nutrients not only are well established, but are predictable and can readily be demonstrated. It is the aim of this chapter to highlight these interactions with a few representative examples in order to demonstrate both the potential possibilities and the limitations of disease and pest control by mineral nutrition and fertilizer application. Comprehensive reviews on this subject have been presented by Fuchs and Grossmann (1972) on a general level and by Perrenoud (1977) with special emphasis on potassium.

11.2 Fungal Diseases

11.2.1 Principles of Infection

The germination of spores on leaf and root surfaces is greatly stimulated by the presence of plant exudates. The flow of exudates contributes to the success or failure of infection in most fungal diseases by soil- and airborne pathogens. The rate of flow and composition of exudates depend on the cellular concentration and the corresponding diffusion gradient (Fig. 11.2). The concentrations of sugars and amino acids are high in leaves, for example, when potassium is deficient. Amino acid and amide concentrations are high when the nitrogen supply is excessive (Section 8.2). The concentration of assimilates in the apoplast and at the leaf surface depends on the permeability of the plasma membrane. On average, the concentra-

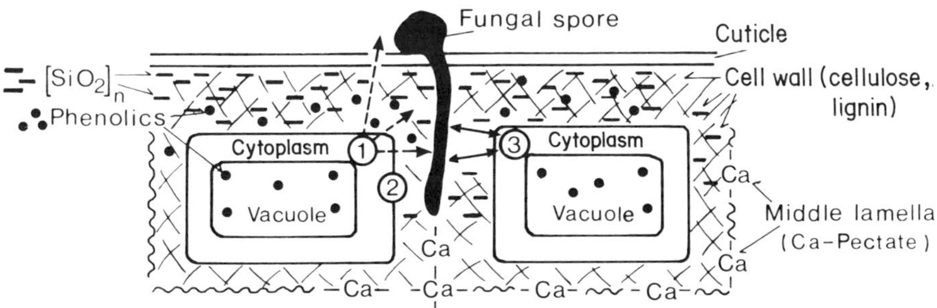

① Diffusion of low-molecular-weight assimilates (sugars, amino acids)
② Plasma membrane permeability
③ Interactions between fungus/epidermal cell (formation of toxins, phenolics)

Fig. 11.2 Schematic representation of the penetration of a fungal hypha on the leaf surface into the epidermal cell layer (apoplast), and some factors which affect the penetration and growth rate of the hypha and are closely related to mineral nutrition.

tions of amino acids and sugars in the apoplast of both leaf and stem tissue is in the range of 1 to 5 m*M* (Hancock and Huisman, 1981) but may rise considerably with calcium or boron deficiency (which causes increased membrane permeability) and potassium deficiency (which impairs polymer synthesis).

The concentrations of soluble assimilates in the apoplast of the host determine the growth of the parasite during the penetration and postinfection phases. Only a few groups of plant parasites are truly intracellular with direct access to assimilates in the symplast (Hancock and Huisman, 1981). Some parasites, such as powdery mildew of barley, have access only to epidermal cells. In these cases, the physical and chemical properties of the epidermal cells are of much more importance with respect to susceptibility and resistance than those of the bulk leaf tissue (Hwang *et al.*, 1983). Epidermal cell layers of leaves (Kojima and Conn, 1982) and stems and roots (Barz, 1977) are characterized by much higher levels of phenolic compounds and flavanoids (i.e., substances with distinct fungistatic properties). The role of mineral nutrients in phenol metabolism is well documented, and examples of phenol accumulation have been discussed in relation to boron and copper deficiency (Chapter 9).

Most parasitic fungi invade the apoplast by releasing pectolytic enzymes, which dissolve the middle lamella (Fig. 11.2). The activity of these enzymes is strongly inhibited by Ca^{2+}, which explains the close correlation between the calcium content of tissues and their resistance to fungal diseases (see Section 11.2.4).

During infection a whole range of interactions occur between the hyphae and the host cells (Fig. 11.2). Inducible resistance mechanisms are associated mainly with the epidermis, the effectiveness of these mechanisms depending on the type of pathogen and the resistance of the host. Mineral nutrients and the mineral nutritional status of plants are involved in these mechanisms. Lewis (1980b) has put forward the hypothesis that boron is involved in the enhanced formation of phenolic phytoalexins in response to microbial infections. Borate-complexing compounds (e.g., callose) either released from invading hyphae or produced by the host cells in response to infection would trigger enhanced formation of phenolics at the sites of infection. It is well established that the phenolic content of nitrogen-deficient plants is high and that both the level of these substances (Kiraly, 1964) and their fungistatic effects (Kirkham, 1954) decrease when the supply of nitrogen is large.

As tissues (particularly leaves) age, lignification and/or the accumulation and deposition of silicon in the epidermal cell layer may form an effective physical barrier to hyphal penetration (Fig. 11.2). These processes provide the main structural resistance of plants to diseases, especially in the leaves of grasses (Sherwood and Vance, 1980). Both lignification and silicon deposition are affected by mineral nutrition in various ways.

11.2.2 Role of Silicon

Grasses in general and wetland rice in particular are typical silicon accumulator plants (Section 10.3). As the silicon supply increases the silicon content of the leaves also rises, inducing a corresponding decline in susceptibility to fungal diseases such as rice blast (Fig. 11.3). The increase in resistance (which manifests as a decrease in the number of eyespots) appears to be related directly to the silicon concentration in the external solution and in the leaves.

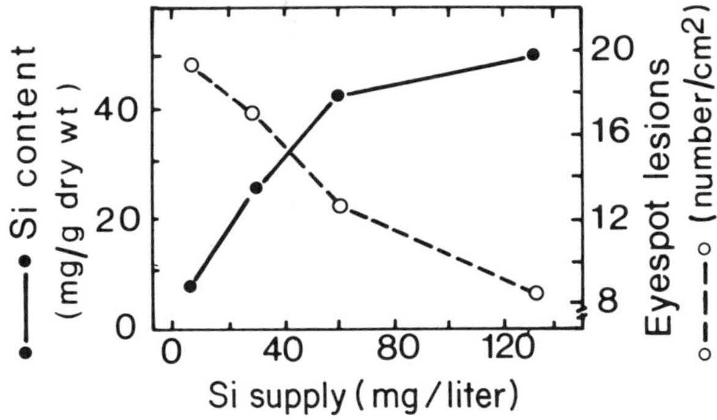

Fig. 11.3 Silicon content and susceptibility to blast fungus (*Piricularia oryzea* Cav.) of fully expanded rice leaves. (Modified from Volk *et al.*, 1958.)

The limitations of silicon in the control of fungal disease are also evident from Fig. 11.3. The solubility of monosilicic acid in water is ~120 mg SiO^2/liter, and silicon is translocated in the xylem preferentially to mature leaves (Chapter 2). Rice blast infection, however, occurs mainly in young leaves. As shown in Fig. 11.4, with maturation (full expansion at about day 8) and ageing of the leaves, resistance increases rapidly and becomes virtually complete whether the silicon supply is high or low. Nevertheless, the effect of silicon on the resistance of young leaves is substantial.

Although there is no doubt about the role of silicon in the epidermal cells as a physical barrier to hyphal penetration, other mechanisms are also probably involved in the higher resistance of young leaves supplied with silicon. The formation of organosilico compounds (Section 10.3) may increase the stability of the cell walls to enzymatic degradation (Volk *et al.*, 1958). Silicon also seems to take part in the defense of the host cells against hyphal penetration. After infection of the leaves, an enhanced deposition of

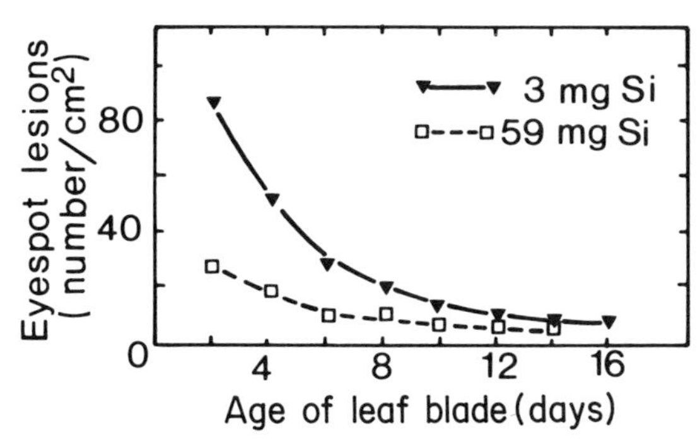

Fig. 11.4 Decline in the number of eyespot lesions (which indicates increased resistance to blast fungus) with the age of rice leaves and silicon concentration in the nutrient solution of 3 mg Si/liter and 59 mg Si/liter. (Modified from Volk *et al.*, 1958.)

silicon is observed around the infection hypha. The degree of deposition is metabolically controlled by the surrounding host cells; that is, it is not a purely passive transport by mass flow in the apoplast (Heath, 1979). Thus the function of silicon deposition in these defense reactions might be similar to that of enhanced synthesis of polyphenols and lignin at the sites of infection (Vance *et al.*, 1980).

11.2.3 Role of Nitrogen and Potassium

There is an extensive literature on the effect of both nitrogen and potassium on parasitic diseases because their role in modulating disease resistance is quite readily demonstrated and, furthermore, is of particular importance for fertilizer application. However, the results are often inconsistent and in some cases controversial, mainly for two reasons. (a) It is not clearly stated whether the levels of these mineral nutrients represent a deficiency, an optimal supply, or an excessive supply (see Fig. 11.1). (b) The distinction between obligate and facultative parasites in the infection pattern is not clearly made. As shown in Table 11.2 a high nitrogen concentration increases the severity of infection by obligate parasites but has the opposite effect on diseases caused by facultative parasites, such as *Alternaria* and *Fusarium*, and most bacterial diseases, as represented in the table by *Xanthomonas* spp. In contrast to nitrogen, potassium elicits uniform responses: High concentrations increase the resistance of host plants to both obligate and facultative parasites.

The principal differences in the response of obligate and facultative parasites to nitrogen are shown in Fig. 11.5. The susceptibility of wheat plants to stem rust, caused by an obligate parasite, increases with increasing

Table 11.2
Tentative Summary of the Effects of Nitrogen and Potassium Levels on the Severity of Diseases (+ → ++++) Caused by Parasites[a]

Pathogen and disease	Nitrogen level		Potassium level	
	Low	High	Low	High
Obligate parasites				
Puccinia spp. (rust diseases)	+	+++	++++	+
Erysiphe graminis (powdery mildew)	+	+++	++++	+
Facultative parasites				
Alternaria spp. (leaf spot diseases)	+++	+	++++	+
Fusarium oxisporum (wilt and rot disease)	+++	+	++++	+
Xanthomonas spp. (bacterial spots and wilt)	+++	+	++++	+

[a]Based on Kiraly (1976) and Perrenoud (1977).

nitrogen supply, the nitrogen-deficient plants being the most resistant. In contrast, the susceptibility of tomato plants to bacterial leaf spot, caused by a facultative parasite, decreases with increasing nitrogen supply even at levels required for optimal growth of the host plant.

These differences in response are based on the nutritional requirements of the two types of parasite. Obligate parasites rely on assimilates supplied by living cells. On the other hand, facultative parasites are semisaprophytes which prefer senescing tissue or which release toxins in order to damage or kill the host plant cells. As a rule, all factors which support the metabolic and synthetic activities of host cells and which delay senescence of the host plant also increase resistance to facultative parasites.

The role of nitrogen in increasing the susceptibility of host plants to obligate fungal parasites (see Table 11.1) is a matter of concern in both

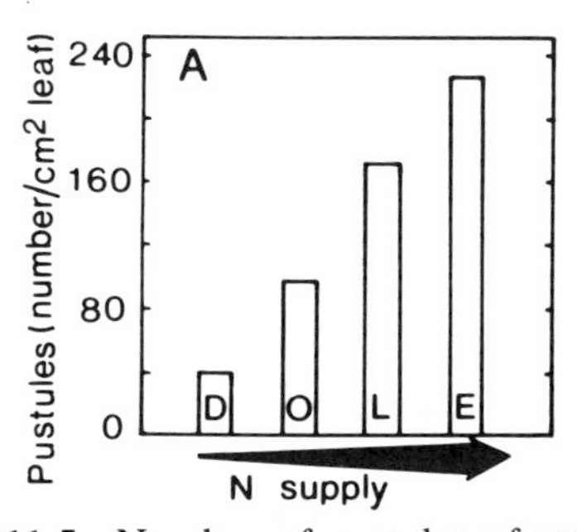

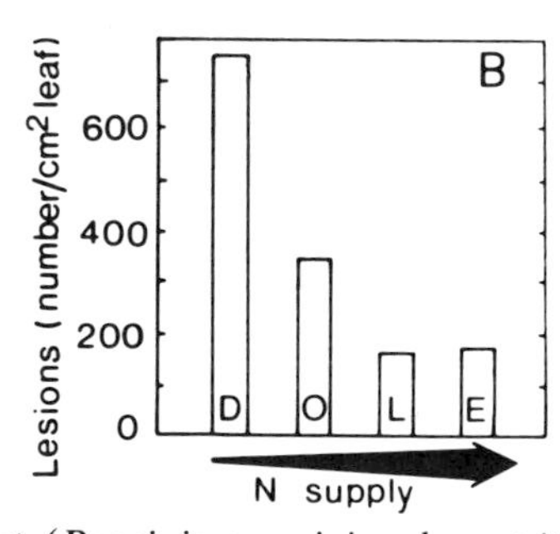

Fig. 11.5 Number of pustules of stem rust (*Puccinia graminis* subsp. *tritici*) in wheat (A) and number of necrotic lesions caused by bacterial spot (*Xanthomonas vesicatoria*) in tomato (B) grown in nutrient solutions with increasing nitrogen concentration. D, Deficient; O, optimal; L, luxurious; E, excessive. (Based on Kiraly, 1976.)

agricultural and horticultural practice. This effect is related to both the nutritional requirements of the parasite and changes in the anatomy and physiology of the host plant in response to nitrogen. As discussed in Section 8.2, nitrogen in particular enhances growth rate, and during the vegetative stage of growth the proportion of young to mature tissue shifts in favor of the young tissue, which is more susceptible. In addition an increase in amino acid and amide concentration in the apoplast and at the leaf surface seems to have a greater influence than sugars on the germination and growth of conidia (Robinson and Hodges, 1981). Specific metabolic changes also occur which are of importance in relation to resistance. When the nitrogen supply is large, the activities of some key enzymes of phenol metabolism are depressed (Matsuyama and Dimond, 1973), as is the content of phenolics (Kiraly, 1964). In rice leaves the lignin content of low-nitrogen plants is ~1100 μg per 100 g dry weight, whereas that of high-nitrogen plants is ~500 μg (Matsuyama, 1975).

A decrease in silicon content is another change usually observed in plants in response to increasing nitrogen supply (Grosse-Brauckmann, 1957; Volk *et al.*, 1958). This, however, is an unspecific response referred to as *dilution by growth*. The various anatomical and biochemical changes together with the increase in the content of low-molecular-weight organic nitrogen compounds as substrates for parasites are the main factors responsible for the close correlation between nitrogen supply and susceptibility to obligate parasites.

With potassium the situation is less complex. Potassium decreases the suceptibility of host plants to both types of parasites. This effect can be quite dramatic, as shown in Fig. 11.6 for stem rot in rice. In this situation the disease could be controlled simply by the application of potassium fertilizer. In most cases, however, the effect of potassium is confined to the deficiency range; that is, potassium-deficient plants are more susceptible than potassium-sufficient plants to parasitic diseases of both groups. As a rule,

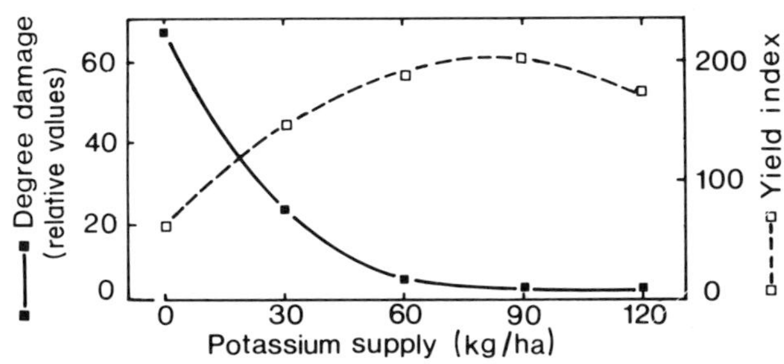

Fig. 11.6 Effect of potassium supply on grain yield of wetland rice and incidence of stem rot (*Helminthosporium sigmoideum*). Basal dressing of nitrogen and phosphorus constant at 120 and 60 kg/ha, respectively. (Based on Ismunadji, 1976.)

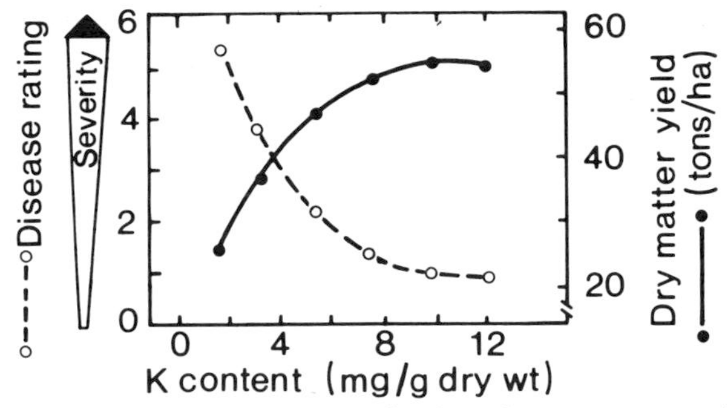

Fig. 11.7 Severity of leaf spot disease (*Helminthosporium cynodontis*) and dry matter yield in coastal bermudagrass (*Cynodon dactylon* L. Pers.) versus leaf potassium content. (Reproduced from Matocha and Smith, 1980, by permission of the American Society of Agronomy.)

susceptibility decreases (or resistance increases) in response to potassium in the same way plant growth responds to increasing potassium supply (Fig. 11.7). Beyond the optimal potassium supply for growth, no further increase in resistance can be achieved by increasing the supply of potassium and its level in plants.

Results similar to those shown in Fig. 11.7 have been obtained with oil palms infected with *Fusarium* (Ollagnier and Renard, 1976) and wheat infected with stripe rust (Kovanci and Colakoglu, 1976). In both cases, the increase in resistance was confined to the deficiency range.

The high susceptibility of potassium-deficient plants to parasitic diseases is related to the metabolic functions of potassium (Section 8.7). In deficient plants the synthesis of high-molecular-weight compounds (proteins, starch, and cellulose) is impaired and low-molecular-weight organic compounds accumulate. In the deficiency range an increase in the potassium supply therefore leads to an increase in growth and a decrease in the content of low-molecular-weight organic compounds until growth is maximal. A further increase in supply and plant level of potassium is without substantial effect on the organic constituents of plants and thus usually on resistance. The characteristic pattern of these changes is given in Fig. 11.8.

In the deficiency range, potassium-induced growth enhancement also causes nonspecific decreases in the content of other mineral elements (dilution by growth). Beyond maximal growth there may continue to be a slight decrease in the levels of other cations such as Ca^{2+} and Mg^{2+} due to competition at uptake sites in the roots (Chapter 2). In plants receiving a suboptimal calcium supply the risk of both calcium deficiency-related physiological disorders and disease susceptibility may therefore increase (see Section 11.2.4).

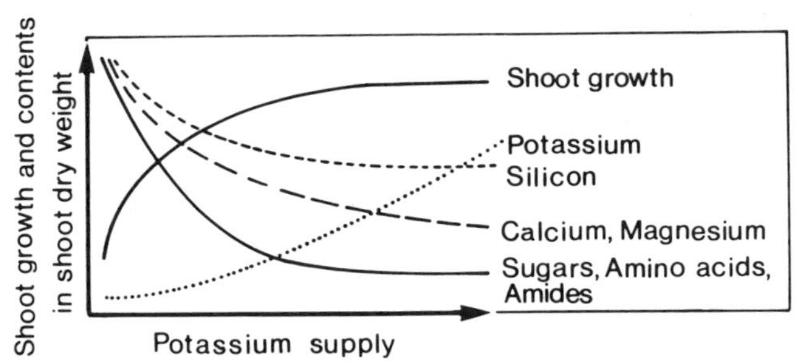

Fig. 11.8 Tentative scheme of growth response and major changes in plant composition with increasing potassium supply.

The relationship between potassium and resistance is more complex in seeds and fruits that are supplied with potassium primarily by retranslocation from vegetative organs. In certain soybean varieties the incidence of pod blight rapidly increases late in the season in the upper pods, and this is correlated with a sharp drop in the potassium content of the pods. With an exceptionally high soil application of potassium the disease is markedly suppressed (Table 11.3). Although the effect on seed yield is only marginal, seed quality is very much improved by a large potassium supply. Heavily infected seeds are unsuitable for animal feed and other purposes.

Table 11.3
Effect of Potassium Supplied to Soybean on Seed Yield and Percentage of Seeds Infected with *Diaporthe sojae*[a]

Potassium supply (kg/ha)	Seed yield (g/plant)	Infected seeds (% of total)
0	30·9	75·2
82	32·2	61·6
410	33·8	27·2
1640	34·5	13·3

[a]Based on Crittenden and Svec (1974).

11.2.4 Role of Calcium and Other Mineral Nutrients

The calcium content of plant tissues affects the incidence of parasitic diseases in two ways. First, calcium is essential for the stability of biomembranes; when calcium levels are low, therefore, the efflux of low-molecular-weight compounds (e.g., sugars) from the cytoplasm into the apoplast is enhanced. Second, calcium polygalacturonates are required in the middle lamella for cell wall stability (Section 8.6). Many parasitic fungi and bacteria

Table 11.4
Relationship between Cation Content and Severity of Infection with *Botrytis cinera* Pars. in Lettuce[a]

Cation content (mg/g dry wt)			Infection with *Botrytis*[b]
K	Ca	Mg	
14·4	10·6	3·2	4
23·8	5·4	4·1	7
34·2	2·2	4·7	13
48·9	1·8	4·2	15

[a]Based on Krauss (1971).
[b]Infection index: 0–5 slight infection; 6–10 moderate infection; 11–15 severe infection.

invade plant tissue by producing extracellular pectolytic enzymes such as polygalacturonase, which dissolves the middle lamella. The activity of this enzyme is drastically inhibited by calcium (Bateman and Lumsden, 1965). The susceptibility of plants to infection with such parasites is therefore inversely related to the calcium content of the tissue, as shown in Table 11.4. In this experiment the total concentration in the nutrient solution of three cations, K^+, Ca^{2+}, and Mg^{2+}, was kept constant and only the K^+/Ca^{2+} ratio was altered. Thus a decrease in the calcium content of the plants was ultimately correlated with an increase in the potassium content. Additional experiments showed that an increase in potassium content does not necessarily lead to an increase in infection as long as the calcium content of plants is kept at a high level.

In soybean, "twin stem" abnormality is endemic on many acid tropical soils. Necrosis of the apical meristem occurs, and the plants are simultaneously heavily infected with *Sclerotium* spp. (Muchovej and Muchovej, 1982). Increasing the calcium supply suppresses both fungal infection and twin stem. It seems likely that the latter abnormality is the direct result of calcium deficiency (symptoms of which are apical meristem necrosis and loss of apical dominance) and that the fungal infection is a secondary event.

Various parasitic fungi preferentially invade the xylem and dissolve the cell walls of conduction vessels. This leads to plugging of the vessels and subsequent wilting symptoms (e.g., *Fusarium* wilt). The growth and activity of these fungi are closely related to the calcium concentration in the xylem sap (Table 11.5). Bean plants were grown before infection at high and uniform calcium level. Thereafter, the plants were inoculated with *Fusarium oxysporum* and transferred to solutions with different calcium levels. It is evident from Table 11.5 that the calcium concentration in the xylem exudate is the controlling factor in resistance to *Fusarium* wilt. In contrast, there was

Table 11.5
Relationship between the Severity of *Fusarium* Wilt and the Calcium Concentration in the Xylem Exudate of Tomato Plants Supplied with Calcium in the Substrate[a]

Calcium supply in substrate (mg/liter)	Calcium concentration in xylem exudate (mg/liter)	Disease index[b]
0	73	1·00
50	219	0·92
200	380	0·80
1000	1081	0·09

[a]From Corden (1965).
[b]0 = healthy plants; 1 = severely infected plants.

no correlation between the calcium content of the tomato stem tissue and the disease index or the growth of the parasite in the host tissue (Corden, 1965).

Plant tissues low in calcium are also much more susceptible than tissues with normal calcium levels to parasitic diseases during storage. This is of particular concern in the case of fleshy fruits with their typically low calcium content. During storage the fruits are more susceptible not only to so-called physiological disorders (Section 8.6) but also to fungal diseases that cause fruit rotting (Fig. 11.9). Calcium treatment of fruits before storage is

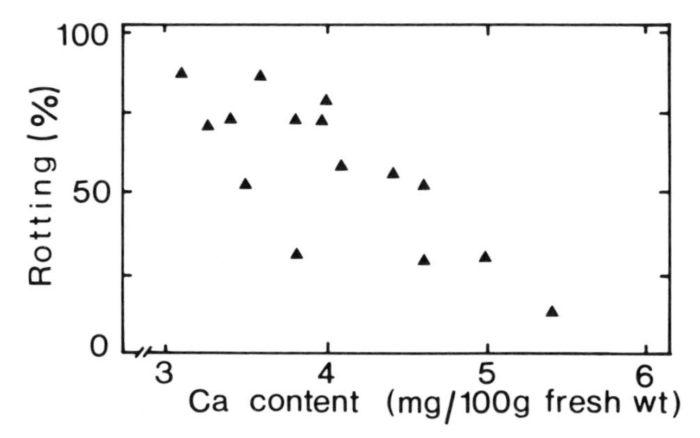

Fig. 11.9 Relationship between the calcium content of apples (cv. Cox orange) and the incidence of rotting due to *Gloesporium perennans* infection after the apples were stored for 3 months at 3°C. (Modified from Sharpless and Johnson, 1977.)

therefore an effective procedure for preventing losses both from physiological disorders and from fruit rotting.

There is an extensive literature on the effects of micronutrients on parasitic diseases. However, the results are often contradictory and are based more on observation than on systematic studies (for a review, see Fuchs and Grossmann, 1972). In general, similar principles govern the effect of both micronutrients and macronutrients on resistance: Deficiencies of nutrients which lead to an accumulation of low-molecular-weight organic substances lower plant resistance. It seems most likely that, in deficient plants with impaired metabolic activity, defense mechanisms are also less effective against both obligate and facultative parasites. With zinc deficiency, a leakage of sugars onto the leaf surface of *Hevea brasilensis* increases the severity of infection with *Oidium* (Bolle-Jones and Hilton, 1956). In boron-deficient wheat plants the rate of infection with powdery mildew is severalfold higher than that in boron-sufficient plants, and the fungus also spreads more rapidly over the plant (Schütte, 1967).

Copper has been widely used as a fungicide. The amount required, however, is very much higher than the requirement of plants. The action of copper as a fungicide relies on direct application to the plant (and fungal) surface. At least in wheat the copper nutritional status does not significantly affect the rate of infection by powdery mildew; with severe deficiency, however, the development of "adult plant resistance" to powdery mildew is inhibited (Graham, 1980b). Inhibition of lignification, impaired phenol metabolism, accumulation of soluble carbohydrates, and delay in leaf senescence (Section 9.3) are probably the main reasons for the higher susceptibility of mature copper-deficient plants.

11.3 Bacterial and Viral Diseases

11.3.1 Bacterial Diseases

Bacterial diseases, which are caused by various facultative parasites, can be divided into three main types: leaf spot diseases, soft rots, and vascular diseases (Grossmann, 1976). In leaf spot diseases (e.g., bacterial leaf blight, *Xanthomonas oryzae*), pathogens usually enter the host plant through the stomata. Thus the epidermal layer is a rather ineffective barrier to infection. Having entered the plant, the bacteria spread and multiply in the intercellular spaces. The effect of the mineral nutritional status of the host plant on this process is similar to its effect on facultative fungal parasites: The multiplication and severity of the disease are enhanced when potassium and calcium levels are deficient and often (Kiraly, 1976), but not always (Fuchs and Grossmann, 1972), when the nitrogen level is deficient. Multiplication and spreading are facilitated by bacterial pectolytic enzymes and toxins that attack and finally kill the host cells. There is very little detailed information on the effect of the mineral nutritional status of plants on their mechanisms of defense against bacterial invasion.

Soft rot diseases are caused, for example, by *Erwinia* spp. and *Xanthomonas* spp. The parasites usually enter the host plant tissue through wounds. The rate of wound cork formation is therefore important for the resistance of the host plant. Rates of wound cork formation are higher in potassium-sufficient than in potassium-deficient plants (Leuchs, 1959). Growth-induced stem and petiole cracking (i.e. wound formation) in boron-deficient plants can be considered as a factor in increasing susceptibility to soft rot diseases.

Table 11.6
Relationship between the Calcium Content of Bean, the Activity of Pectolytic Enzymes in the Plant Tissue, and the Severity of Soft Rot Disease Caused by *Erwinia carotovora*[a]

Calcium content (mg/g dry wt)	Pectolytic activity (relative units)[b]				Severity of symptoms[c]
	Polygalacturonase		Pectate transeliminase		
	−	+	−	+	
6·8	0	62	0	7·2	4
16	0	48	0	4·5	4
34	0	21	0	0	0

[a]From Platero and Tejerina (1976).
[b]+, Bacterial inoculation; −, no inoculation.
[c]4 = Complete decay of plants within 6 days; 0 = no symptoms.

Spreading of bacteria within the host tissue, as in many fungal diseases, is facilitated by polygalacturonases and related pectolytic enzymes. Accordingly, the resistance of plants is closely related to their calcium content, as shown in Table 11.6 for bean plants. In infected tissue the activity of pectolytic enzymes is very high but inversely related to the calcium content of the tissue. The severity of the disease symptoms also reflects the role of calcium in resistance.

Bacterial vascular diseases caused, for example, by *Pseudomonas solanacearum* spread within plants through the xylem; they lead to slime formation and finally plugging of the vessels ("bacterial wilt"). The relationship between these diseases and mineral nutrition is obscure. There are a few reports of a slight increase in severity of the disease in tomato suffering from potassium deficiency (Perrenoud, 1977).

11.3.2 Viral Diseases

The multiplication of viruses is confined to living cells, and their nutritional requirements are restricted to amino acids and nucleotides. As a rule, nutritional factors which favor the growth of the host plant also favor viral multiplication. This holds true particularly for nitrogen and phosphorus (Fuchs and Grossmann, 1972), but the same tendency is observed for potassium (Perrenoud, 1977). Despite the stimulation of multiplication, expression of the visible symptoms of viral diseases does not necessarily correspond to an increase in mineral nutrient supply to the host plant. Sugar beet yellow or potato leaf roll virus symptoms may disappear when the supply of nitrogen is large even though the plants are totally infected, whereas symptoms of yellow dwarf in onion or of tobacco mosaic virus in *Nicotiana tobaccum* are exaggerated by high nitrogen levels. The stimulation or repression of visible symptoms and the degree of growth depression depend mainly on the competition for nitrogen between virus and host cells. This competition differs in various viral diseases but is also influenced by environmental factors such as temperature. In reports on the effect of micronutrients on viral diseases no consistent pattern has been observed (Fuchs and Grossmann, 1972; Martin, 1976).

11.4 Soilborne Fungal and Bacterial Diseases

The population density of microorganisms at the root surface is several times higher than that in the bulk soil. Microorganisms at the root surface include various pathogens. Competition between and repression of microorganisms, as well as chemical barriers (e.g., high concentrations of polyphenols in the rhizodermis; Barz, 1977) and physical barriers (e.g., silicon deposition at the endodermis; see Section 10.3) ensure that microbial invasion of both roots and shoots via the roots is restricted. Invasion and infection by certain microorganisms, however, is beneficial for higher plants (e.g., *Rhizobium* or endomycorrhiza).

Mineral nutrition affects soilborne fungal and bacterial diseases in various ways. For example, in manganese-deficient Norway spruce, fungistatic activity against *Fomes amosus* (Fr.) Cooks is much lower in the inner bark of roots, leading to heart root disease (Wenzel and Kreutzer, 1971); when the plant is supplied with high manganese and low nitrogen levels, both the content of these nutrients and the fungistatic activity of the inner bark increase (Alcubilla *et al.*, 1971). The incidence of infection of growing potato tubers with *Streptomyces scabies*, giving rise to common potato scab, is reduced either by lowering of the soil pH or by application of manganese

(Mortvedt *et al.*, 1961; McGregor and Wilson, 1964). Manganese exerts its effect not by increasing the resistance of the tuber tissue to the fungus but by directly inhibiting the vegetative growth of *S. scabies* before the onset of infection (Mortvedt *et al.*, 1963).

In peanut, preharvest pod rot is caused by severe infection with *Phytium myriotylum* and *Rhicoctonia solani*. The occurrence of this disease is closely related to the calcium content of the pod tissue and can be kept at a low level by the soil application of calcium (e.g., as gypsum) (Hallock and Garren, 1968). Evidently the same causal relationship exists between calcium content and fungal infection of the pods as those described for aerial parts of other plant species.

There are a large number of reports on the effects of soil pH and the form of fertilizer nitrogen (ammonium nitrogen versus nitrate nitrogen) on soilborne pathogens and disease severity. "Take-all" is a root rot disease in wheat and barley that is caused by *Gaeumannomyces graminis* and seriously limits grain production in many regions of the world. The fungus has a growth optimum at pH 7 and is very sensitive to low pH (Smiley and Cook, 1973). A decrease in the severity of take-all is observed even at a pH below ~6·8. Liming of acid soils therefore increases the risk of root infections and yield losses by take-all. Figure 11.10 shows that, in a soil of pH 3·8, inoculation with *G. graminis* was without significant effect on growth or yield. Liming enhanced growth and increased yield in noninfected plants but had a severely depressing effect on infected plants. The incidence of take-all became more severe as the soil pH increases.

The severity of take-all is thus mainly dependent on soil pH. For various reasons control of the disease by acidification of bulk soil is either impossible or inadvisable (Section 16.3). However, the same inhibitory effect on the growth of *G. graminis* can be achieved by acidification of the rhizosphere

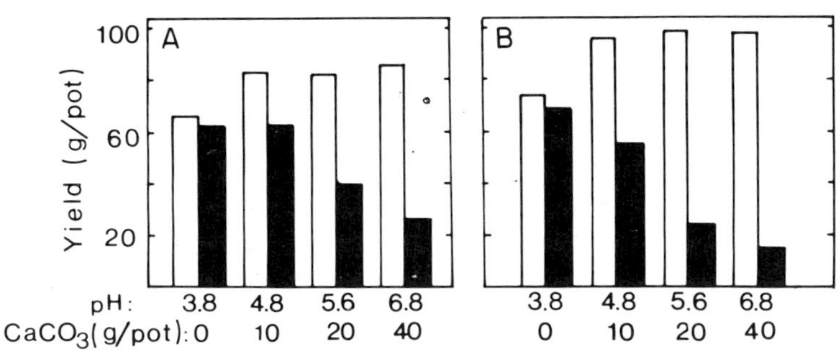

Fig. 11.10 Effect of liming and inoculation with *Gaeumannomyces graminis* var. *tritici* (take-all) on (A) straw yield and (B) grain yield of spring wheat (*Triticum sativum*). Open bars, noninoculated; solid bars, inoculated. (Modified from Trolldenier, 1981.)

with ammonium-based fertilizers such as $(NH_4)_2SO_4$. Nitrate-based fertilizers have the opposite effect, increasing both the rhizosphere pH and the incidence of take-all (Smiley and Cook, 1973). Hyphal extension along wheat roots varied between 5·8 mm (ammonium sulfate + N-Serve) and 15·8 mm (calcium nitrate) within a period of 23 days (Smiley, 1978). In agreement with this, inoculation of wheat with *G. graminis* had no effect on grain yield when ammonium sulfate + N-Serve was the nitrogen source (rhizosphere pH 4·1–4·4) but reduced grain yield to ~15% of the noninoculated control plants when sodium nitrate (rhizosphere pH 6·1–6·4) was applied (Trolldenier, 1981). Although biological control of take-all via selective ammonium fertilization cannot solve the problem entirely, under field conditions it can make an important contribution to an integrated system of disease control. Ammonium fertilizer may also affect soil pathogens directly. For example, ammonia is highly toxic to certain *Fusarium* species, and nitrite ($NH_4^+ \rightarrow NO_2^- \rightarrow NO_3^-$) is toxic to *Pythium* and *Phytophthora* (Henis, 1976).

11.5 Pests

Pests are animals such as insects, mites, and nematodes which are harmful to cultivated plants. In contrast to fungal and bacterial pathogens, they have digestive and excretory systems and their dietary requirements are often less specific. Furthermore, visual factors such as color of leaves are important for "recognition" or "orientation." For example, many aphid species tend to settle on yellow-reflecting surfaces (Beck, 1965). The main types of resistance of host plants to pests are (a) physical (e.g., color, surface properties, hairs), (b) mechanical (e.g., fibers, silicon), (c) chemical/biochemical (e.g., content of stimulants, toxins, repellents). Mineral nutrition can affect all three factors to various degrees.

Generally, young or rapidly growing plants are more likely to suffer attack by pests than are old and slow-growing plants. Therefore, there is often a positive correlation between nitrogen application and pest attack. On the other hand, with potassium no clear-cut relationship seems to exist (Table 11.7), although in ~60% of the 231 cases cited in an extensive literature review dealing with potassium effects only, a large potassium supply had resulted in a decreased attack by pests (Perrenoud, 1977).

Although sugar can act as a feeding stimulant (Beck, 1965) for most sucking insects such as the rice brown planthopper (Sogawa, 1982), amino acids and amides are more important in this respect. This is illustrated in Table 11.8 for squash bugs. Nitrogen-deficient plants showed the lowest feeding response. The number of squash bugs per plant was clearly related

Table 11.7
Number of Citations in the Literature of Effects of Mineral Nutrient Supply on the Susceptibility of Plants to Insect, Mite, and Nematode Pests[a]

Nitrogen[b]			Phosphorus[b]			Potassium[b]		
+	0	−	+	0	−	+	0	−
40	11	8	6	4	3	7	7	7

[a]Based on Jones (1976).
[b]+, Susceptibility increased; 0, unaffected; −, decreased.

Table 11.8
Relationship between Mineral Nutrient Deficiencies, Number of Squash Bugs (*Anasa tristis*) per Plant, and Soluble Nitrogen Content of Squash[a]

Nutrient supply	Squash bugs (no./plant)	Soluble nitrogen (μg/g fresh wt)
Complete	1·70	32·1
−N	0·66	4·5
−P	2·11	93·7
−K	2·45	98·9
−S	3·42	143·7

[a]From Benepal and Hall (1967).

to the content of total soluble nitrogen in the leaves. In contrast, the protein content of the leaves did not appear to be associated with feeding preference (Benepal and Hall, 1967).

From this result one can conclude that whether or not mineral nutrients and fertilizers can ward off an attack by sucking parasites depends on whether these substances increase or decrease the content of soluble organic nitrogen in plants. The striking effects of various forms of fertilizer on the attack of oak trees by cup-shield lice (Table 11.9) and of citrus trees by

Table 11.9
Effects of Fertilizers Applied on a Soil Low in Available Potassium on Infestation of Oak Trees (*Quercus pendula*) by Cup-shield Lice (*Eulecanium refulum* Ckll.)[a]

	Fertilizer			
	K + Mg	N + P + K + Mg	Mg	N + P + Mg
No. of lice/10-cm stem section	0·72	0·82	4·32	8·78

[a]Based on Brüning (1967).

purple and black scale (Chaboussou, 1976) is probably related to such changes in plant composition. Magnesium applied alone or (especially) in combination with nitrogen and phosphorus (Table 11.9) increased the nutrient imbalance (giving rise to potassium deficiency), which resulted in a corresponding increase in soluble organic nitrogen content (i.e., a more favorable diet for the lice). Changes in the mechanical properties of the plant tissue might have been a contributing factor.

The behavior of nitrogen with respect to pest attack varies strikingly between field crops and forest trees. In trees supplied with high levels of nitrogen there is a spectacular decline in the number of biting insects—sawfly and caterpillars—that feed on needles or leaves (Merker, 1961). The reasons for this decline are not clear; enhanced formation of toxic substances, an excessive nitrate content, or an unfavorable carbon/nitrogen ratio have been discussed as possible factors (Bogenschütz and König, 1976).

Whether micronutrients, like macronutrients, affect the resistance of plants to pests depends, in principle, on whether or not a deficiency of a micronutrient results, for example, in the accumulation of low-molecular-weight organic compounds, or a decline in compounds which act as repellents. With a few exceptions, information about micronutrient–pest relationships is sketchy and not very specific. Rajaratnam and Hock (1975) put forward the hypothesis that the decrease in attack by red spider mites of oil palm seedlings observed as the boron content of the leaves is increased (from the deficiency to the sufficiency range) is causally related to inhibition of the synthesis of flavanoids in boron-deficient plants. These polyphenols are involved in resistance not only to fungi but also to insects.

Epidermal cell walls containing silicon deposits act as a mechanical barrier to the stylet and particularly the mandibles of sucking and biting insects. It has been demonstrated that the mandibles of larvae of the rice stem borer are damaged when the silicon content of rice plants is high (Jones and Handreck, 1967). For the rice brown planthopper, however, soluble silicic acid rather than the deposited silicon in the leaves is the effective sucking inhibitor. Silicon concentrations as low as 10 mg/liter appear to be effective (Sogawa, 1982).

The physical properties of leaf surfaces are of great importance for the severity of attack by sucking insects. Labial exploration of the surface takes place before insertion of the stylet into the tissue (Sogawa, 1982). Changes in the surface properties of leaves were presumably the main reason for the striking decrease in the attack of wheat plants by aphids when several foliar sprays containing sodium silicate were applied (Fig. 11.11). As the level of nitrogen nutrition increased, so also did the number of aphids of the species *Sitbion avenae*. However, foliar sprays with silicon reduced the number to

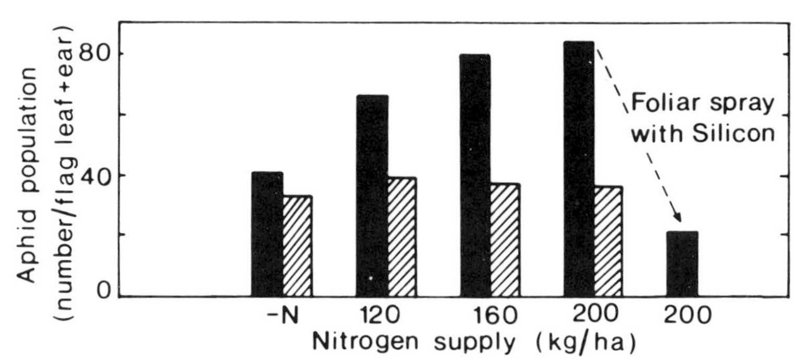

Fig. 11.11 Effect of nitrogen supply and foliar sprays containing silicon (1% Na_2SiO_2) on population density of two aphid species in winter wheat. Striped bars, *Metopolophium dirhodum*; shaded bars, *Sitobion avenae*; −N denotes nitrogen-deficient control plants. (Based on Hanisch, 1980.)

below that in nitrogen-deficient control plants. These beneficial effects of silicon-containing foliar sprays deserve particular attention in intensive crop production.

The results of this experiment also illustrate the difficulties of making generalizations about the relationship between increasing nitrogen supply and attack by sucking insects: In contrast to *S. avenae*, which is a typical ear feeder, the other aphid species, *M. dirhodum*, did not respond to increasing nitrogen supply. Differences in feeding habits and in preferences for plant organs (*M. dirhodum* prefers leaf blades) are possible reasons for the differences in response to nitrogen.

In wetland rice several species of leaf hoppers pose a more serious threat as vectors of viruses than as juice-sucking pests (Beck, 1965), which is another important reason for controlling sucking insects.

11.6 Direct and Indirect Effects of Fertilizer Application on Diseases and Pests

Under field conditions fertilizer application affects diseases and pests indirectly by producing denser stands and higher humidity within a crop. Many pathogens require free water for infection. This aspect is of particular importance for nitrogen fertilizer application, because high levels of nitrogen applied early in the growing season favor abundant tillering and dense, tall crops stands. As a consequence, chemical disease control is not only most effective but usually essential with high levels of nitrogen. These relationships are demonstrated in Fig. 11.12 for wheat plants naturally infected with yellow rust.

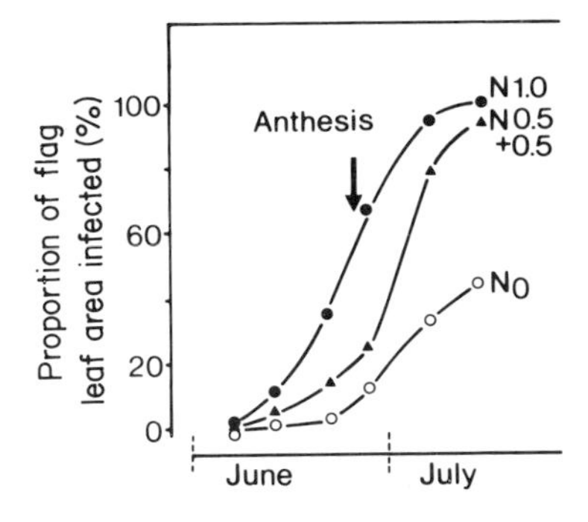

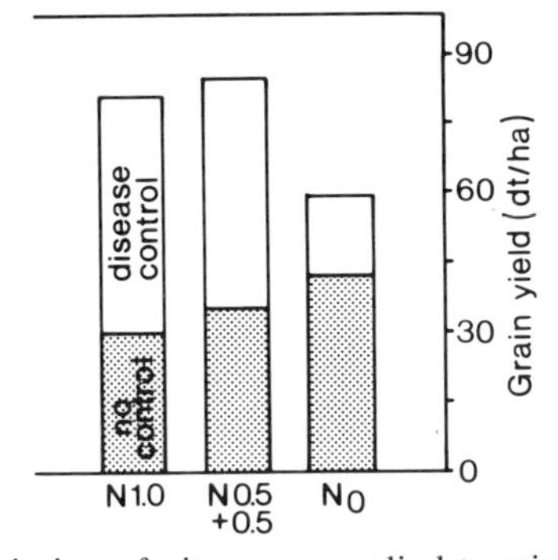

Fig. 11.12 Relationship between level and timing of nitrogen supplied to winter wheat and yellow rust infections (*Puccinia striiformis* Westend) and grain yield with and without chemical disease control. N 1·0 denotes 160 kg N/ha as early dressing; N 0·5 + 0·5, split application: 80 kg N early and 80 kg N at anthesis; N_0, no nitrogen fertilizer. (Based on Darwinkel, 1980a.)

The effects of both level and timing of nitrogen supply on the rust infection are striking, the most severe infection occurring with the large early single dressing (N 1·0). A split application of nitrogen decreased infection to a large extent in the early growth stages, but after the second application (at anthesis) fungal growth increased rapidly. Nevertheless, the epidemic was considerably postponed by the split application. In the plants not receiving nitrogen (N_0) the infection remained low. Similar results have been obtained in wheat infected with powdery mildew (Darwinkel, 1980b).

Without chemical disease control, the infection reduced the grain yield in all treatments (Fig. 11.12). The reduction, however, differed among the treatments, being drastic with nitrogen treatment but relatively small with N_0 treatment. Thus, without chemical disease control, the highest grain yield was obtained in plants not receiving nitrogen, and with disease control the highest yield was obtained in plants receiving the split application of nitrogen.

In order to ensure a high grain yield per unit area, a large application of nitrogen fertilizer is required. However, a large application simultaneously increases the risk and usually the incidence of disease and pest attack and thus the necessity of control. In programs of integrated disease and pest management, the application of fungicides and pesticides should be kept to a minimum and the partial replacement (at least) of these compounds by other, nontoxic substances should be attempted. In this respect, the application of silicon either as a foliar spray or in the soil deserves more attention. As shown in Table 11.10 the soil application of silicon alone is quite effective, although less so than a foliar spray with fungicide. Future studies should concentrate more on the extent to which fungicide application can be

Table 11.10
Response of Spring Barley Yield to Control of Powdery Mildew (*Erysiphe graminis*) with a Fungicide (Calixin) or with Silicon Supplied to the Soil[a]

SiO_2 Supply (g per pot or 10 kg soil)	Dry matter (g/pot)				SiO_2 Content in straw (% dry wt)	
	Grain		Straw			
	−Fungicide	+Fungicide	−Fungicide	+Fungicide	−Fungicide	+Fungicide
0	43	68	62	76	1·33	1·15
5	58	70	70	80	3·80	3·09
10	61	71	73	80	5·21	4·41
20	61	71	77	84	6·81	6·40

[a]Based on Grosse-Brauckmann (1979).

reduced by simultaneous silicon application both to the soil and as a foliar spray. It is also necessary to intensify studies on the effects of micronutrient application, especially in intensive crop production, in relation not only to yield but also to the resistance of plants to diseases and pests.

A final aspect of disease and pest control and resistance must be mentioned. Severe disease and pest attacks may decrease root surface area and root activity either directly (e.g., root rot, nematodes) or indirectly by decreasing the green leaf area which acts as a source of assimilates for the roots. Water and mineral nutrient uptake is then impaired, and latent nutrient deficiencies may be induced, or become acute, contributing to the reduction in growth of infected plants. In cotton, for example, root infection with nematodes had no effect on the shoot growth when the level of potassium was high, but shoot growth decreased in plants when the level of potassium was inadequate, although the infection rate of the roots was even higher in plants receiving a large potassium supply (Oteifa and Elgindi, 1976).

12

Diagnosis of Deficiency and Toxicity of Mineral Nutrients

12.1 Nutrient Supply and Growth Response

The usual growth (dry matter production) versus nutrient supply curve (growth response curve) has three well-defined regions (Fig. 12.1). In the first, the growth rate increases with increasing nutrient supply (deficient range). In the second, the growth rate reaches a maximum and remains unaffected by nutrient supply (adequate range). Finally, in the third region, the growth rate falls with increasing nutrient supply (toxic range).

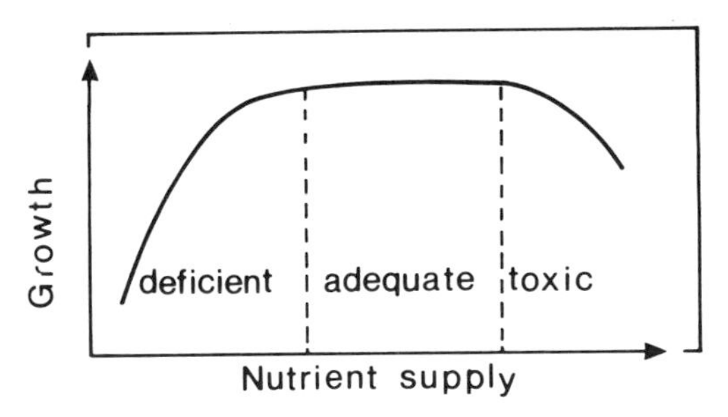

Fig. 12.1 Relationship between nutrient supply and growth.

In crop production, optimal nutrient supply is usually achieved by the application of fertilizer. Rational fertilizer application requires information on the nutrients that are available in the soil, on the one hand, and the nutritional status of the plants, on the other. The possibilities and limitations of using visual diagnosis and plant analysis as a basis for recommending whether or not to use fertilizer, and what type and quantity, are discussed in this chapter.

12.2 Diagnosis of Nutritional Disorders by Visible Symptoms

As a rule, nutritional disorders that inhibit growth and yield only slightly are not characterized by specific visible symptoms. Symptoms become clearly visible when a deficiency is acute and the growth rate and yield are severely depressed. Some exceptions exist. For example, transient visible symptoms of magnesium deficiency in cereals are observed fairly often under field conditions during stem extension but this is without detrimental effect on the final grain yield (Pissarek, 1979).

Diagnosis based on visible symptoms requires a systematic approach, as summarized in Table 12.1. Symptoms appear preferentially on either older or younger leaves, depending on whether the mineral nutrient in question is readily retranslocated (see Chapter 3). Chlorosis or necrosis and the pattern of both are important criteria for diagnosis. As a rule, visible symptoms of nutrient deficiency are much more specific than those of nutrient toxicity, unless the toxicity of one mineral nutrient induces a deficiency of another. Visible deficiency symptoms of individual nutrients were briefly described in Chapters 8 and 9. For details (including color pictures) of symptoms of nutritional disorders the reader is referred to Wallace (1961), Bergmann and Neubert (1976) and Bergmann (1983).

Diagnosis may be especially complicated in field-grown plants when more than one mineral nutrient is deficient and/or there is a deficiency of one mineral nutrient and simultaneously toxicity of another—for example, in

Table 12.1
Some Principles of Visual Diagnosis of Nutritional Disorders[a]

Plant part	Prevailing symptom		Disorder
			Deficiency
Old and mature leaf blades	Chlorosis	Uniform	N (S)
		Interveinal or blotched	Mg (Mn)
	Necrosis	Tip and marginal scorch	K
		Interveinal	Mg (Mn)
Young leaf blades and apex	Chlorosis	Uniform	Fe (S)
		Interveinal or blotched	Zn (Mn)
	Necrosis (chlorosis)		Ca, B, Cu
	Deformations		Mo (Zn, B)
			Toxicity
Old and mature leaf blades	Necrosis	Spots	Mn (B)
		Tip and marginal scorch	B, salt (spray injury)
	Chlorosis, necrosis		Nonspecific toxicity

[a]Letters in parentheses indicate that symptoms are variable.

waterlogged acid soils, both manganese toxicity and magnesium deficiency (complex symptoms). Diagnosis may be further complicated by the presence of diseases, pests, and other symptoms caused, for example, by mechanical injuries including spray damage. In order to differentiate the symptoms of nutritional disorders from these other symptoms, it is important to bear in mind that the former always have a typical symmetric pattern: Leaves of the same or similar position (physiological age) on a plant show nearly identical patterns of symptoms, and there is a marked gradation in the severity of the symptoms from old to young leaves (Table 12.1).

In order to make a more precise visual diagnosis, it is helpful to acquire additional information, including soil pH, results of soil testing for mineral nutrients, soil water status (dry/waterlogged), weather conditions (low temperature or frost) and application of fertilizers, fungicides, or pesticides. In some instances the type and amount of fertilizer to be used can be recommended on the basis of visual diagnosis immediately. This is true of foliar sprays containing micronutrients (iron, zinc, or manganese) or magnesium. In most cases, however, visual diagnosis is an insufficient basis for making fertilizer recommendations. Nevertheless, it offers the possibility of focusing further chemical and biochemical analysis of leaves and other plant parts (plant analysis) on selected mineral nutrients. This is of particular importance for annual crops, because the results are required immediately and seasonal fluctuations in the nutrient content of the plants often do not justify the high cost of performing a complete mineral nutrient analysis.

12.3 Plant Analysis

12.3.1 Relationship between Growth Rate and Mineral Nutrient Content

Typical relationships for a given mineral nutrient are shown in Fig. 12.2. There is an ascending portion of the curve where growth either increases sharply without changes in nutrient content (I and II) or where increases in growth and mineral nutrient content are closely related (III). This is followed by a more or less level portion where growth is not limited by the nutrient in question (IV and V) and, finally, by a portion where excessive nutrient content causes toxicity and a corresponding decline in growth (VI).

Occasionally, with an extreme deficiency of copper (Reuter *et al.*, 1981) or zinc (Howeler *et al.*, 1982b), for example, a C-shaped response curve is obtained (Fig. 12.2, region I, dashed line) in which a nutrient-induced increase in growth rate is accompanied by a decrease in its content in the dry weight, which is often referred to as the "Piper–Steenbjerg" effect (Bates,

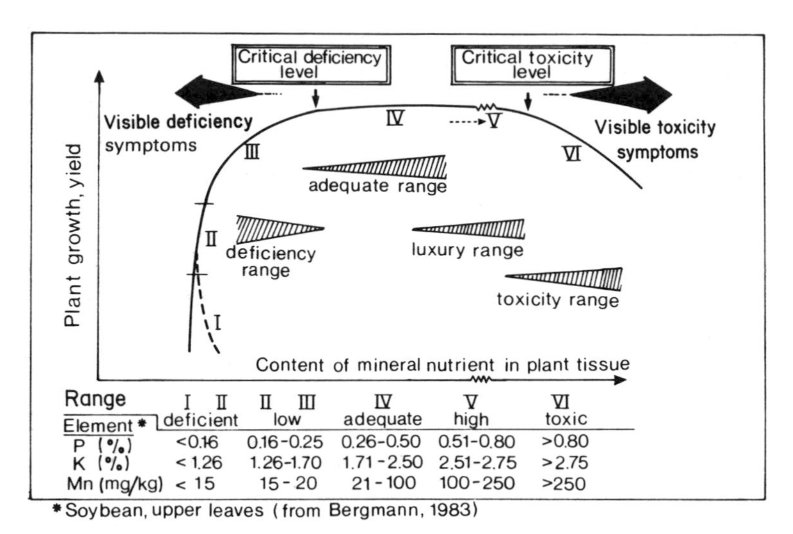

Range Element*	I II deficient	II III low	IV adequate	V high	VI toxic
P (%)	<0.16	0.16-0.25	0.26-0.50	0.51-0.80	>0.80
K (%)	< 1.26	1.26-1.70	1.71-2.50	2.51-2.75	>2.75
Mn (mg/kg)	< 15	15-20	21-100	100-250	>250

*Soybean, upper leaves (from Bergmann, 1983)

Fig. 12.2 Relationship between nutrient content and growth or yield (upper) and examples of the nutrient content of soybean leaves (percent dry weight) in the various nutrient supply regions (lower).

1971). Possible explanations are a lack of remobilization from old leaves and stem (Reuter *et al.*, 1981) or necrosis of the apical meristems with a corresponding cessation of growth despite further uptake of small amounts of the mineral nutrient in extremely deficient plants.

Concentration and dilution effects of mineral nutrients in plants are common phenomena which must be carefully considered in interpretations of contents in terms of ion antagonism and/or synergism during uptake. This holds true in particular when the levels of mineral nutrients are in the deficiency or toxicity range (Jarrell and Beverly, 1981). If, for example, the concentrations of two mineral nutrients are in the deficiency range and only one of them is supplied, growth enhancement causes a "dilution" of the other mineral nutrient (decrease in the content) and induces severe deficiency without any competition occurring in uptake or within the plant.

Central to the diagnosis of nutritional disorders by plant analysis are the critical deficiency and toxicity levels of each mineral nutrient in the plant tissue (Fig. 12.2). Growth is maximal between the critical deficiency and toxicity levels. In practice, for various reasons (see below), critical levels are not fixed points (contents) but a range of values. Usually, the critical levels are defined as those levels at which growth or yield is 5 to 10% below maximum (Bouma, 1983). One can define critical deficiency or toxicity levels quite precisely in growth experiments under controlled environmental conditions by varying the supply of a mineral nutrient over a wide concentration range. For practical purposes, the interpretation of plant analysis data must be less ambitious. Therefore, mineral nutrient levels are grouped into ranges, as shown in the lower portion Fig. 12.2 for soybean. If nutrient levels are in the adequate range there is a high statistical probability that these nutrients are not growth-limiting factors. Certainly, levels in the luxury range further decrease the risk that these nutrients will become deficient under conditions unfavorable for root uptake (e.g., dry topsoil) or when the demand is very high (e.g., retranslocation to fruits). However, there is a greater risk of growth reduction by direct toxicity of these nutrients or by their induction of a deficiency of other nutrients.

In general the nutritional status of a plant is better reflected in the mineral element content of the leaves than in that of other plant organs. Thus leaves are usually used for plant analysis. With some species and with certain mineral nutrients, levels may differ between leaf blades and petioles, and sometimes the petioles are a more suitable indicator of nutritional status (Bouma, 1983). In fruit trees the analysis of the fruits themselves is a better indicator, especially for calcium and boron in relation to fruit quality and storage properties (Bould, 1966).

12.3.2 Developmental Stage of Plants and Age of Leaves

Next to the mineral nutrient supply, the physiological age of a plant or plant part is the most important factor affecting the mineral nutrient content of the plant. There is a fairly distinct decline with age of most mineral nutrient levels (except that of calcium) in plants and organs. This decline is caused mainly by a relative increase in the proportion of structural material (cell walls and lignin) and of storage compounds (e.g., starch) in the dry matter. Mineral nutrient levels corresponding to the adequate or the critical deficiency range are therefore lower in old than in young plants. This dilution effect has been shown for potassium in field-grown barley plants throughout the growing season (Leigh *et al.*, 1982). In young plants the potassium content was 5–6% of the shoot dry weight but decreased to ~1% toward maturation, although the plants were well supplied with potassium. This was reflected in the potassium concentration of the tissue water (mainly representative of vacuolar sap), which remained fairly constant at ~100 m*M* throughout the season.

Complications arising from changes in the critical deficiency content with age can be avoided by sampling tissues at specific physiological ages. For example, as shown in Table 12.2, the critical deficiency level of copper in the whole plant tops decreases drastically in clover with age but remains fairly constant at ~3 μg in the youngest leaf blades throughout the season.

Table 12.2
Critical Level of Copper (at Maximum Yield) in Subterranean Clover and Plant Organs and Plant Age[a]

	Age of plant (days after sowing)				
Plant part	26	40	55	98	F[b]
Whole plant tops	3·9	3·0	2·5	1·6	1·0
Youngest open leaf blade	3·2	~3	~3	~3	~3

[a]Critical copper levels expressed as micrograms per gram dry weight. Based on Reuter *et al.* (1981).
[b]Early flowering.

However, this method of using only the youngest leaves is suitable only for those mineral nutrients which either are not retranslocated or are retranslocated to only a very limited extent from the mature leaves to areas of new growth, that is, when deficiency occurs first in young leaves and at the shoot apex (Table 12.1). The situation is different for potassium, nitrogen, and magnesium; since the levels of these mineral nutrients are maintained fairly constant in the youngest leaves, the mature leaves are a much better indicator of the nutritional status of a plant, as shown for potassium in Fig. 12.3. Here, the youngest leaf is an unsuitable indicator because the potassium levels indicating deficiency and toxicity vary only between 3·0 and 3·5%, respectively, compared with 1·5 and 5·5% in mature leaves. This illustrates the necessity of using mature leaves to assess the nutritional status of mineral nutrients which are readily retranslocated in plants.

If young and old leaves of the same plant are analyzed separately, additional information can be obtained on the nutritional status of those

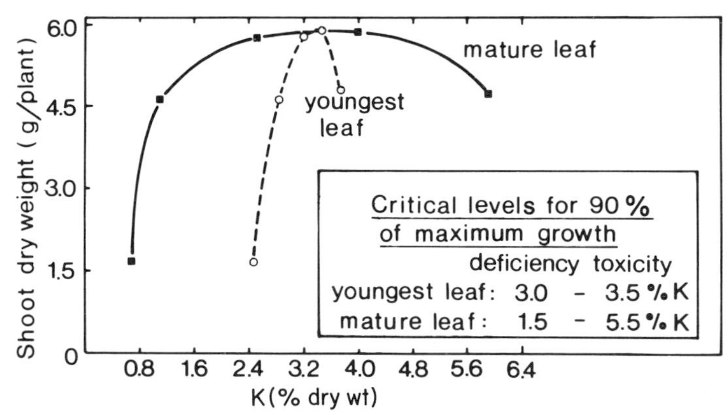

Fig. 12.3 Relationship between shoot dry weight and potassium content of mature and youngest leaves of tomato plants grown in nutrient solutions with various potassium concentrations. Inset: calculated critical levels.

mineral nutrients which are readily retranslocated. A much higher level of, say, potassium in the mature leaves indicates luxury consumption or even toxicity. The reserve gradient, on the other hand, is an indicator of the transition stage between the adequate and deficient ranges; if this gradient is steep, latent or even acute deficiency may exist. The use of gradients is particularly helpful under conditions in which relevant reference data on critical levels are lacking (e.g., for a species or cultivar) or under certain ecological conditions. If toxicity is suspected, the old leaves are the most suitable organs for plant analysis.

Compared with the changes in the mineral nutrient content of annual species, the short-term fluctuations throughout the growing season of the mineral nutrient content of leaves and needles of trees are relatively small because of the nutrient buffering capacity of twigs and trunk. In evergreen trees the simultaneous analysis of leaves and needles differing in age offers the possibility of minimizing the effects of short-term fluctuations (Table 12.3). With increasing age of the needles, the content of all macronutrients decreases except that of calcium. This reduction may be indicative of remobilization. Presumably, however, it is mainly an expression of a dilution effect resulting from increased lignification of the old needles. Only with calcium is dilution overcompensated for by a continuation of high influx into the old needles. With the exception of magnesium, the data of Table 12.3 are indicative of trees well supplied with macronutrients.

Table 12.3
Mineral Nutrient Content of Spruce (*Picea abies* Karst) Needles of Different Age[a]

Mineral nutrient	Age of needles (years)			
	1	2	3	4
Nitrogen	1·79	1·76	1·46	1·22
Phosphorus	0·20	0·17	0·14	0·13
Potassium	0·63	0·56	0·47	0·44
Magnesium	0·04	0·04	0·03	0·03
Calcium	0·28	0·40	0·50	0·59

[a]Levels of mineral nutrients expressed as a percentage of the dry weight. Based on Bosch (1983).

12.3.3 Plant Species

Critical deficiency levels differ among plant species even when comparisons are made between the same organs at the same physiological age. This is also true for the adequate range. These variations are based on differences in plant metabolism and plant constitution (e.g., differences in the role of calcium and boron in cell walls) and have been discussed in Chapters 8 and 9.

Table 12.4
Mineral Nutrient Levels in the Adequate Range of Some Representative Annual and Perennial Species[a]

	Level (% dry wt)					Level (mg/kg dry wt)				
Species (organ)	N	P	K	Ca	Mg	B	Mo	Mn	Zn	Cu
Spring wheat (whole shoot, booting stage)	3·0–4·5	0·3–0·5	2·0–3·8	0·4–1·0	0·15–0·3	5–10	0·1–0·3	35–150	20–70	5–10
Ryegrass (whole shoot)	3·0–4·2	0·35–0·5	2·5–3·5	0·6–1·2	0·2–0·5	6–12	0·15–0·5	40–150	20–50	6–12
Sugar beet (mature leaf)	4·0–6·0	0·35–0·6	3·5–6·0[b]	0·7–2·0	0·3–0·7	40–100	0·25–1·5	35–150	20–80	7–15
Cotton (mature leaf)	3·6–4·7	0·3–0·5	1·7–3·5	0·6–1·5	0·35–0·8	20–80	0·6–2·0	35–150	25–80	8–20
Tomato (mature leaf)	4·0–5·5	0·4–0·65	3·0–6·0	3·0–4·0	0·35–0·8	40–80	0·3–1·0	40–150	30–80	6–12
Orange (*Citrus* spp.) (mature leaf)	2·4–3·5	0·15–0·3	1·2–2·0	3·0–7·0	0·25–0·7	30–70	0·2–0·5	25–125	25–60	6–15

[a]Based on Bergmann (1983).
[b]Sodium content below 1·5%.

Representative data for the adequate range in selected species are given in Table 12.4. More extensive and detailed data, including toxicity levels, can be found in Chapman (1966), Jones (1967), Bergmann and Neubert (1976) and Bergmann (1983).

As shown in Table 12.4 the levels of macronutrients in the adequate range are of a similar order of magnitude in the various plant species; an exception is calcium, the content of which is substantially lower in the two monocotyledons. In all species the adequate range is relatively narrow for nitrogen, because luxury levels of nitrogen have unfavorable effects on growth and plant composition (Chapter 8). In apple leaves, for example, a nitrogen level of more than 2·4% often affects fruit color and storage adversely (Bould, 1966). On the other hand, the adequate range for magnesium is usually broader, due mainly to competing effects of potassium; at higher potassium levels, higher magnesium levels are required to ensure an adequate magnesium nutritional status.

The levels of micronutrients in the adequate range vary by a factor of 2 or more (Table 12.4). Manganese levels show the greatest variation, which may indicate that for manganese in particular, leaf tissue is capable of buffering fluctuations in the root uptake of manganese. In plants growing in soil, manganese exhibits more rapid and distinct fluctuations in uptake than any other mineral nutrient, the rate depending on fluctuations in soil redox potential and thus on the concentrations of Mn^{2+} (Section 16.4).

The data given in Table 12.4 are average values and offer no more than a guide as to whether a mineral nutrient is in the deficient, adequate, or toxic range. This should be borne in mind when only one or a few mineral nutrients have been analyzed and the information on possible nutrient interactions is therefore insufficient.

The critical toxicity levels of sodium and chloride are in general closely related to genotypical differences in salt tolerance. The interpretation of these levels is complicated because in saline substrates a decline in growth is often caused in the first instance by effects on the water balance of plants and not necessarily by direct toxicity of sodium and/or chloride in the leaf tissue (Section 16.6).

12.3.4 Nutrient Interactions

There are a whole range of nonspecific as well as specific interactions between mineral nutrients in plants (Robson and Pitman, 1982) which affect the critical levels. A typical example of a nonspecific interaction is shown in Table 12.5 for nitrogen and phosphorus. The critical deficiency level of nitrogen increases as the phosphorus content increases and vice versa.

Table 12.5
Effect of Foliage Phosphorus Content on the Critical Deficiency Level of Nitrogen and Vice Versa in *Araucaria cunninghamii*[a]

Foliage phosphorus content (% dry wt)	Critical level of nitrogen (% dry wt)	Foliage nitrogen content (% dry wt)	Critical level of phosphorus (% dry wt)
0·06	1·07	0·60	0·07
0·09	1·18	1·05	0·08
0·12	1·24	1·35	0·10
0·16	1·31	1·65	0·11
0·21	1·35	1·80	0·12

[a]Based on Richards and Bevege (1969).

Interactions between two mineral nutrients are important when the levels of both are near the deficiency range. Increasing the supply of only one mineral nutrient stimulates growth, which in turn can induce a deficiency of the other by the dilution effect. In principle, these unspecific interactions hold true for any mineral nutrients with levels at or near the critical deficiency level. Optimal ratios between nutrients in plants are therefore often as important as absolute levels. For example, a ratio of nitrogen to sulfur of ~17 is considered to be adequate for the sulfur nutrition of wheat (Rasmussen *et al.*, 1977) and soybean (Bansal *et al.*, 1983). It should be borne in mind, however, that optimal ratios considered alone are insufficient criteria because they can also be obtained when both mineral nutrients are in the deficiency range (Jarrell and Beverly, 1981), as well as in the toxicity range.

Specific interactions which affect critical deficiency levels were discussed in Chapters 8 and 9; therefore, only two examples are reiterated here: (a) competition between K^+ and and Mg^{2+} on the cellular level, which usually involves the risk of K^+-induced Mg^{2+} deficiency; and (b) replacement of K^+ by Na^+ in natrophilic species, which has to be considered in the evaluation of potassium content (see Table 12.4).

Specific interactions are also important in evaluating critical toxicity levels. The critical level of manganese, for example, differs not only among species and cultivars of a species (Section 9.3), but within the same cultivar, the difference depending on the silicon supply. In bean leaves the critical toxicity level of manganese can increase from 100 mg/kg dry wt in the absence of silicon to ~1000 mg (i.e., by a factor of 10) in the presence of silicon (Horst and Marschner, 1978a).

12.3.5 Environmental Factors

Fluctuations in environmental factors such as temperature and soil moisture can affect the mineral nutrient content of leaves considerably. These factors influence both the availability and uptake of nutrients by the roots and the shoot growth rate. Their effects are more distinct in shallow-rooted annual species than in deep-rooted perennial species, which have a higher nutrient buffer capacity within the shoot. This aspect must be considered in interpretations of both critical deficiency and toxicity levels in leaf analysis. If fluctuations in soil moisture are high, then as a rule for a given plant species the critical deficiency levels of nutrients such as potassium and phosphorus are also somewhat higher in order to ensure a higher capacity for retranslocation during periods of limited root supply. The effects of irradiation and temperature on the nutrient content of leaves are described in detail by Bates (1971).

Rain and dust are other environmental factors to be considered in leaf analysis. Substantial leaching of certain mineral nutrients from the leaves may occur with high rainfall (Chapter 4). Dust on leaf surfaces, especially of mature leaves, has to be removed carefully in order to prevent severe contamination. This is particularly important for micronutrients such as iron (Jones, 1972). The treatment of plants with fungicides and pesticides containing large amounts of mineral elements (e.g., copper and zinc) may create additional problems in leaf analysis.

12.3.6 Nutrient Efficiency

Genotypical differences in the critical deficiency levels of a nutrient can also be brought about by differences in the utilization of a nutrient. In a physiological sense, this may be expressed in terms of unit dry matter produced per unit nutrient in the dry weight (e.g., mg P/kg dry wt). As an example, the difference in nitrogen efficiency between C_3 and C_4 grasses is shown in Table 12.6. Much more dry matter is produced in C_4 grasses than in

Table 12.6
Relationship between Dry Matter Production and Nitrogen Content of C_3 and C_4 Grasses[a,b]

Nitrogen supply (equivalent to kg/ha)	Dry matter (g/pot)		Nitrogen content (% dry wt)	
	C_3	C_4	C_3	C_4
0	11	22	1·82	0·91
67	20	35	2·63	1·18
134	27	35	2·77	1·61
269	35	48	2·78	2·00

[a]Based on Colman and Lazemby (1970).
[b]C_3 Grasses: *Lolium perenne* and *Phalaris tuberosa*; C_4 grasses: *Digitaria macroglossa* and *Paspalum dilatatum*.

C_3 grasses per unit leaf nitrogen. This is a general phenomenon which is observed in comparisons of other C_3 and C_4 species (R. H. Brown, 1978). The higher nitrogen efficiency of C_4 species is presumably related to the lower investment of nitrogen in enzyme proteins used in chloroplasts for CO_2 fixation. In C_4 species only ~10% of the soluble leaf protein is found in ribulosebisphosphate carboxylase, compared with ~50% in C_3 species (R. H. Brown, 1978). For CO_2 fixation via the PEP carboxylase pathway in C_4 species (Chapter 6), less enzyme protein is required.

Differences in the utilization of mineral nutrients are also found among cultivars, strains, and lines of a species. These differences are a component of the nutrient efficiency in general as will be discussed in detail in Section 16.2.3. In an agronomical sense, nutrient efficiency is usually expressed by the yield differences of genotypes growing in a soil with insufficient amounts of nutrients. In most instances, high nutrient efficiency is related primarily to root growth and activity, and in some instances also to the transport from the roots to the shoots (Läuchli, 1976b). Only relatively few data indicate a higher nutrient efficiency in terms of utilization within the shoots—for example, utilization of phosphorus in bean (Whiteaker *et al.*, 1976) and maize genotypes (Elliott and Läuchli, 1985), and potassium in bean and tomato (Shea *et al.*, 1967; Gerloff and Gabelman, 1983).

In principle, higher nutrient efficiency, as reflected by lower critical deficiency levels, in one genotype compared with those in another genotype of the same species can be based on various mechanisms:

1. Higher rates of retranslocation during either vegetative or reproductive growth [e.g., zinc in maize (Massey and Loeffel, 1967)].
2. Higher nitrate reductase activity in the leaves and thus more efficient utilization of nitrogen for protein storage [e.g., in wheat grains (Dalling *et al.*, 1975) and potato tubers (Kapoor and Li, 1982))].
3. Higher proportion of replacement of potassium by sodium and thus lower critical deficiency levels of potassium [e.g., in tomato (Gerloff and Gabelman, 1983)].
4. Lower proportion of nutrients which are not—or only poorly—available for metabolic processes, either due to compartmentation or chemical binding, e.g., calcium in tomato (Greenleaf and Adams, 1969) and tobacco (next section), or phosphorus in maize genotypes (Elliott and Läuchli, 1985).
5. Differences in the ratio of vegetative shoot growth (source) to the growth of reproductive and/or storage organs (sink). This aspect (Chapter 6) is probably in part responsible for the general pattern in so-called modern cultivars of many crop species with a high harvest index in which the critical deficiency levels of mineral nutrients in the leaves are usually higher than those of traditional cultivars.

12.3.7 Total Analysis versus Fractionated Extraction

Most frequently it is the total dry matter content of a nutrient that is determined in plant analysis (e.g., after ashing). The determination of only a fraction of the content—for example, that is soluble in water or in dilute acids or chelators—sometimes provides a better indication of nutritional status. This has been shown for zinc where the water-soluble fraction reflects the nutritional status much better than the total zinc or carbonic anhydrase activity (Rahimi and Schropp, 1984). In species or tissues with preferential nitrate accumulation, the nitrate content is often a better indicator of nitrogen nutritional status than is the total nitrogen level (Hylton *et al.*, 1965). The nitrate concentration in the stem base of wheat measured either semiquantitatively (Beringer and Hess, 1979) or quantitatively (Papastylianou *et al.*, 1982), as well as the nitrate concentration in the petioles of fully expanded leaves of sugar beet (Gilbert *et al.*, 1983) and cotton (Tabor *et al.*, 1984) is a reliable indicator of nitrogen nutritional status. For a few years this rapid test has been successfully used as a basis for recommendations as to whether, and if so, what level of nitrogen fertilizer should be applied during the growing season of wheat (Wollring and Wehrmann, 1981; Wehrmann *et al.*, 1982). In principle, this method is suitable for all those crop species, such as sugar beet (Gilbert *et al.*, 1983), in which nitrate is the prevailing form of nitrogen taken up by the roots and translocated to the shoots. In species which preferentialy reduce nitrate in the roots (e.g., members of the Rosaceae) or when ammonium nitrogen is supplied and taken up, rapid tests for certain amino acids or amides may provide alternatives to the rapid nitrate test.

For assessing the sulfur nutritional status of plants, the content of sulfate—the main storage form of sulfur—is a better indicator than the total sulfur content only (Ulrich and Hylton, 1968). The best indicator of sulfur nutritional status of wheat (Freney *et al.*, 1978) and wetland rice (Islam and Ponnamperuma, 1982) seems to be the ratio of sulfate to total sulfur.

The importance of determining only a defined nutrient fraction is illustrated in Table 12.7. Differences in the susceptibility of tobacco cultivars to calcium deficiency were not related to the total calcium content but to the soluble fraction in the buds. These differences were caused by differences in the rate of oxalic acid synthesis and thus in the precipitation of sparingly soluble calcium oxalate. Accordingly, the critical deficiency level of total calcium is higher in B 21 than in Ky 10. Determination of only the soluble fraction would be a more appropriate method for assessing the calcium nutritional status of the two cultivars.

For various reasons, total iron content is not a reliable indicator of iron nutritional status (Section 9.1). Thus data on critical deficiency levels or on

Table 12.7
Calcium and Oxalic Acid Content of Two Cultivars of Burley Tobacco Differing in Susceptibility to Calcium Deficiency[a]

Cultivar	Plants with Ca-deficiency symptoms (% of total)	Content of buds (meq/g dry wt)			Content of upper leaves (meq/g dry wt)		
		Ca	Oxalic acid	Ca^- oxalic acid (soluble Ca)	Ca	Oxalic acid	Ca^- oxalic acid (soluble Ca)
Ky 10	0	0·25	0·08	0·17	0·28	0·11	0·17
B 21	50	0·23	0·16	0·07	0·30	0·15	0·15

[a]Based on Brumagen and Hiatt (1966).

levels in the adequate range are given, if at all, only with reservation (Chapman, 1966). Promising methods for measuring "physiologically available" or "active" iron involve extracting iron with dilute acids or Fe(II) with chelators (Section 9.1).

12.4 Histochemical and Biochemical Methods

Nutritional disorders are generally related to typical changes in the fine structure of cells and their organelles (Vesk *et al.*, 1966; Hecht-Buchholz, 1972) and of tissue. Light microscopic studies on changes in anatomy and morphology of leaf and stem tissue can be helpful in the diagnosis of deficiencies of copper, boron, and molybdenum (Pissarek, 1980; Bussler, 1981a). A combination of histological and histochemical methods is useful in the diagnosis of copper (Section 9.3.4) and phosphorus deficiencies (Besford and Syred, 1979).

Enzymatic methods involving marker enzymes offer another approach to assessing the mineral nutritional status of plants. These methods are based on the fact that the activity of certain enzymes is lower or higher (depending on the nutrient) in deficient than in normal tissue. Examples were given in Chapter 9 for copper and ascorbic acid oxidase; zinc and aldolase or carbonic anhydrase; and molybdenum and nitrate reductase. Either the actual enzyme activity is determined in the tissue after extraction or the leaves are incubated with the mineral nutrient in question for 1 to 2 days to determine the inducible enzyme activities of, for example, nitrate reductase by molybdenum (Section 9.5.2) and peroxidase activity by iron (Bar-Akiva *et al.*, 1970).

In principle these enzymatic methods could be very valuable if the total content or the soluble fraction of a mineral nutrient were poorly correlated with its physiological availability. Whether enzymatic methods can replace chemical analysis as a basis for making fertilizer recommendations depends on their selectivity, accuracy, and particularly whether they are sufficiently simple for the spot test. In the case of iron and peroxidase (Bar-Akiva *et al.*, 1978; Bar-Akiva, 1984) and copper and ascorbic acid oxidase (Delhaize *et al.*, 1982), enzymatic methods seem to fulfil these requirements. Nevertheless, calibration of the methods remains a problem when a suitable standard (nondeficient plants) is not available and there are no visible deficiency symptoms.

Biochemical methods can also be used for macronutrients. The accumulation of putrescine in potassium-deficient plants (Section 8.7) is a biochemical indicator of the potassium requirement of lucerne (Smith *et al.*, 1982). Inducible nitate reductase activity can be used as an indicator of nitrogen nutritional status (Bar-Akiva *et al.*, 1970; Witt and Jungk, 1974). Pyruvate kinase activity in leaf extracts depends on the potassium and magnesium content of the leaf tissue (Besford, 1978b). In phosphorus-deficient tissue, phosphatase activity is much higher, especially the activity of a certain fraction (fraction B; isoenzyme) of the enzyme (Table 12.8). The increase in

Table 12.8
Growth, Phosphorus Content, and Phosphatase Activity of Young Wheat Plants[a]

Phosphorus supply	Shoot dry wt (mg/plant)	Phosphorus content of shoot (%)	Phosphatase activity (μmol NPP[b]/g fresh wt × min)		
			Total	Fraction A	Fraction B
High	223	0·8	5·6	4·4	0·5
Low	135	0·3	11·1	6·7	2·9

[a]From Barrett-Lennard and Greenway (1982).
[b]NPP, p-nitrophenylphosphate.

phosphatase activity in deficient tissue is an interesting physiological and biochemical phenomenon which might be related to enhanced turnover rates and/or remobilization of phosphorus (Smyth and Chevalier, 1984). It is an open question, however, as to whether enzymatic test methods, especially for macronutrients in general, will become as important as supplementary tools for assessing the nutritional status of plants as is the rapid nitrate test for nitrogen fertilizer recommendations.

12.5 Plant Analysis versus Soil Analysis

There is a long history of controversy over whether soil or plant analysis provides a more suitable basis for making fertilizer recommendations. Both methods rely in a similar manner on calibration, that is, the determination of

o

the relationship between levels in soils or plants and the corresponding growth and yield response curves, usually obtained in pot or field experiments using different levels of fertilizers. Both methods have advantages and disadvantages, and they also give qualitatively different results (Schlichting, 1976). Chemical soil analysis indicates the potential availability of nutrients that roots may take up under conditions favorable for root growth and root activity (Chapters 13 and 14). Plant analysis in the strict sense reflects only the actual nutritional status of plants. Therefore, in principle a combination of both methods provides a better basis for recommending fertilizer applications than one method alone. The relative importance of each method for making recommendations differs, however, depending on such conditions as plant species, soil properties, and the mineral nutrient in question.

In fruit or forest trees soil analysis alone is not a satisfactory guide for fertilizer recommendations, mainly because of the difficulty of determining with sufficient accuracy the root zones in which deep-rooting plants take up most of their nutrients. On the other hand, in these perennial plants seasonal fluctuations in the mineral nutrient content of leaves and needles are relatively small compared with those in annual species. The nutrient content of mature leaves and needles is therefore also an accurate reflection of the long-term nutritional status of a plant. Furthermore, calibrations of critical deficiency level and adequate range can be made very precisely and refined for a special location, plant species, and even cultivar. Therefore, in perennial species foliar and needle analysis is in most instances the method of choice. In this instance, however, chemical soil analysis, performed once, is helpful for characterizing the overall level of potentially available nutrients.

In annual crops the short-term fluctuations of mineral nutrient levels place a severe limitation on plant analysis as a basis for making fertilizer recommendations. Chemical soil analysis is required for predicting the variation in plant nutrient content throughout the growing season. In annual crops a large proportion of the mineral nutrients are taken up from the topsoil, which makes soil analysis easier and increases its importance as tool for making fertilizer recommendations (but see also Chapter 13). There is no doubt, however, that for various reasons plant analysis will also become more important for annual crops in the future. In intensive agriculture, nutrient imbalances in plants, especially latent micronutrient deficiencies, are becoming increasingly serious (Franck and Finck, 1980). For economic and ecological reasons, fertilizer application should be kept at a level which ensures that the nutrient content of plants will not far exceed the critical deficiency level, and this must be checked by plant analysis. The rapid nitrate test illustrates the development of this field.

In extensive agriculture with very low fertilizer application, plant analysis is particularly important for identifying those mineral nutrients which have the most limiting effect on growth and yield. Plant analysis might also become increasingly important in connection with Beaufils's Diagnosis and Recommendation Integrated System (DRIS), which is based on the collection of as many data as possible on soil properties and plant composition and the development of computer models for predicting fertilizer requirements according to the relative adequacy of mineral nutrient levels in plants (Sumner, 1977). On the one hand, this system requires more plant analysis data; on the other, it may also permit a distinct refinement in the interpretation of these data in terms of fertilizer recommendations, as has been demonstrated for sugar cane (Elwali and Gascho, 1984).

In pastures, plant analysis is used more frequently than soil analysis, not only because of the peculiarities of the rooting pattern in mixed pastures (deep- and shallow-rooting species) but also because of the importance of the mineral composition of pasture and forage plants for animal nutrition.

Part II

Soil–Plant Relationships

13

Nutrient Availability in Soils

13.1 Chemical Soil Analysis

The most direct way of determining nutrient availability in soils is to measure the growth responses of plants by means of field plot fertilizer trials. This is a time-consuming procedure, however, and the results are not easily extrapolated from one location to another. In contrast, chemical soil analysis—soil testing—is a comparatively rapid and inexpensive procedure for obtaining information on nutrient availability in soils as a basis for recommending fertilizer application. Soil testing has been practised in agriculture and horticulture for many years with relative success. The effectiveness of the procedure is closely related to the extent to which the data can be calibrated with field fertilizer trials, as well as to the interpretation of the analysis. Quite often, expectations by far exceed the capability of soil testing. The reasons for this discrepancy are discussed in this chapter in detail, with special reference to phosphorus and potassium.

Soil testing makes use of a whole range of conventional extraction methods involving different forms of dilute acids, salts, or complexing agents. Depending on the method used, quite different amounts of plant nutrient are extracted, as shown for phosphorus in Table 13.1. As a guide, 1 mg phosphorus per 100 g soil may be taken to be ~30 kg phosphorus per hectare at a 20-cm profile depth (specific soil weight 1·5 = 3 million kg soil per hectare). Which method is the most suitable for characterizing the availability of a given mineral nutrient and thus for predicting fertilizer response must be evaluated by means of growth experiments (see Section 12.5).

Quite often several methods are equally suitable for soil testing of the same mineral nutrient (Vetter *et al.*, 1978). For phosphorus, for example, water extraction might be as suitable for determining availability as extraction with diluted acids, despite the different amounts of phosphorus extracted by these methods (Schachtschabel and Beyme, 1980). Predicting fertilizer response on the basis of chemical soil analysis alone is not satisfactory, however, for various reasons: It indicates only the potential

Table 13.1
Mean Content of Readily Soluble Phosphorus in 40 Soils Extracted with Various Solutions[a]

Extraction solution	Readily soluble phosphorus (mg/100 g of air-dried soil)
Neutral NH_4F (pH 7·0)	14·8
Acidic NH_4F (pH < 2)	7·4
Truog, $H_2SO_4 + (NH_4)_2SO_4$ (pH 3·0)	3·6
Acetic acid (pH 2·6)	2·5
Bicarbonate, $NaHCO_3$ (pH 8·5)	2·4
Calcium lactate (pH 3·8)	1·2

[a]Based on Williams and Knight (1963).

capacity of a soil to supply nutrients to plants; it does not sufficiently characterize the mobility of the nutrients in the soil; it does not provide any information on the plant factors, such as root growth and root-induced changes in the rhizosphere, that are of decisive importance for nutrient uptake under field conditions. In the following three sections these factors are briefly discussed, beginning with nutrient availability in relation to mobility and root growth. For a comprehensive treatment of the bioavailability of nutrients, the reader is referred to Barber (1984).

13.2 Movement of Nutrients to the Root Surface

13.2.1 Principles of Calculations

The importance of the mobility of nutrients in soils for nutrient availability to plants was first stressed by Barber (1962). His concept was based on three components: root interception, mass flow, and diffusion (Fig. 13.1). As roots proliferate through the soil they move into spaces previously occupied by soil with available nutrients, adsorbed, for example, to clay surfaces. Root surfaces come into close contact (that is, they intercept) with nutrients during this displacement process. On the basis of the idea of displacement, Jenny and Overstreet (1939) proposed a contact exchange mechanism for cation uptake without transport through the soil solution (see Chapter 15). Barber (1966) interpreted this displacement process in a more general manner as a factor which might contribute to meeting the total mineral nutrient requirement of plants. Calculations of root interception are based on (a) the amount of available nutrients in the soil volume occupied by the roots; (b) root volume as a percentage of the total soil volume—on average

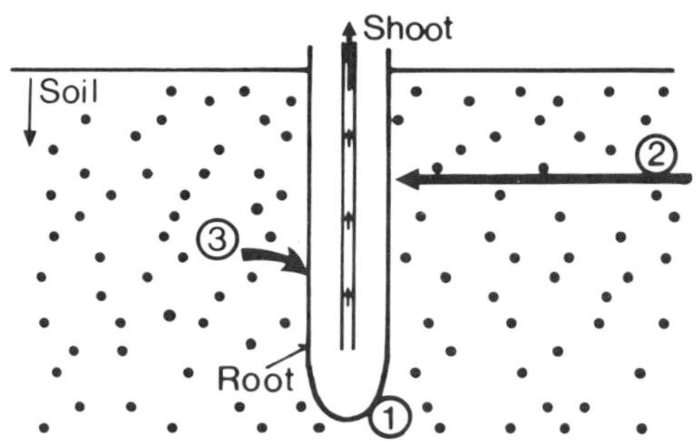

Fig. 13.1 Schematic representation of the movement of mineral elements to the root surface of soil-grown plants. (1) Interception: Soil volume displaced by root volume ("uptake without transport"). (2) Mass flow: Transport of bulk soil solution along the water potential gradient (driven by transpiration). (3) Diffusion: Nutrient transport along the concentration gradient. ● = Available nutrients (as determined, e.g., by soil testing).

1% of the topsoil volume; and (c) the proportion of the total soil volume occupied by pores (on average 50%).

The second component is the mass flow of water and dissolved nutrients to the root surface, which is driven by the transpiration of plants. Estimates of the quantity of nutrients supplied to plants by mass flow are based on the nutrient concentration in the soil solution and the amount of water transpired either per unit weight of shoot tissue (transpiration coefficient, e.g., 300–600 liters H_2O/kg shoot dry wt) or per hectare of a crop. The contribution of diffusion to the delivery of nutrients to the root surface cannot be measured directly but must be calculated from the difference between total uptake by plants and the contribution by root interception plus mass flow. An example of such a calculation is shown in Table 13.2. It is apparent that, in this soil, mass flow makes the main contribution to calcium and magnesium supply whereas the supply of potassium and phosphorus depends mainly on diffusion. Furthermore, the supply of calcium and magnesium by mass flow is greater than uptake, and thus an accumulation of these nutrients at the root surface would be expected. Results similar to those shown in Table 13.2 have been obtained by other authors (e.g., Mengel *et al.*, 1969).

The term *root interception* has been criticized by Brewster and Tinker (1970) on the grounds that in a dynamic model the effects of root growth are subsumed under the concept of diffusion. Current theories of solute movement in the soil–root system therefore concentrate on the mass flow and diffusion components only (Nye and Tinker, 1977; Barber, 1984). As will be shown in Chapter 15, however, conditions at the soil–root interface are sometimes considerably different in various aspects from those at a

Table 13.2
Estimated Amounts of Mineral Nutrients Supplied to Maize Roots in a Fertile Silt Loam Soil by Root Interception, Mass Flow, and Diffusion[a]

Nutrient	Amount "available" in the topsoil[b] (kg/ha)	Total uptake by crops (kg/ha)	Supply (kg/ha) by		
			Interception	Mass flow	Diffusion
Calcium	4000	45	40	90	—
Magnesium	800	35	8	75	—
Potassium	300	110	3	12	95
Phosphorus	100	30	1	0·12	28·9

[a]Estimated root volume equal to 1% of the soil volume. From Barber (1974).
[b]According to soil testing.

distance from the roots. These conditions are not sufficiently described by a mechanistic model treating roots only as a sink for mineral nutrients delivered by mass flow and/or diffusion.

13.2.2 Concentration of Nutrients in the Soil Solution

The concentration of mineral nutrients in the soil solution varies over a wide range, depending on such factors as soil moisture, pH, cation-exchange capacity, redox potential, quantity of soil organic matter and microbial activity, and fertilizer application. Figure 13.2 presents data for the equilibrium concentration of macronutrients in soil solutions from agricultural and horticultural areas.

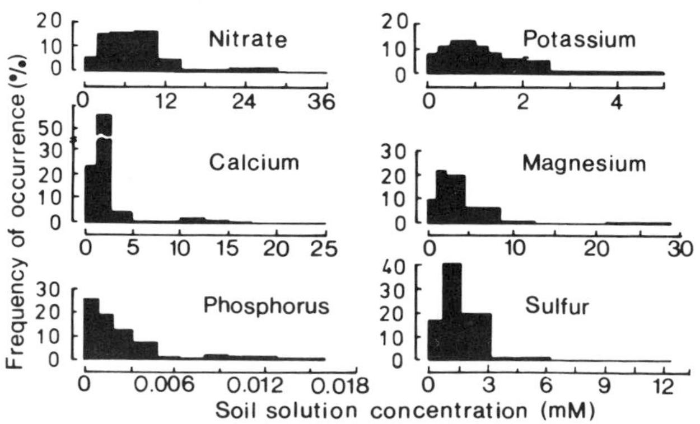

Fig. 13.2 Frequency distribution of the equilibrium concentrations of nutrients in soil solutions from agricultural and horticultural areas. (Redrawn from Asher, 1978; with permission from CRC Press, Inc.)

The concentration of mineral nutrients in the soil solution is an indicator of the mobility of nutrients both toward the root surface and in the vertical direction (i.e., in humid climates it indicates the potential for leaching). Compared with the concentration of other nutrients, that of phosphorus is extremely low (Fig. 13.2), leaching or transport by mass flow to root surfaces thus being generally of minor importance in mineral soils. In contrast to other anions such as nitrate and sulfate, phosphate strongly interacts with surface-active sesquioxides and oxidhydrates of clay minerals; between 20 and 70% of the phosphate in soil solution is present in organically bound form (Welp *et al.*, 1983).

Not only the concentration of nutrients in the soil solution (the so-called *intensity*), but also the buffer power (the so-called *capacity*) of a soil is extremely important for mineral nutrition. The buffer power is described in terms of the relation between the amount of adsorbed nutrients on the solid phase (and/or bound in labile organic structures) and those in the soil solution. The buffer power determines especially the degree, but also the rate of replenishment of nutrients from the solid phase into the nutrient solution (Section 13.6). Since nitrate (and chloride) does not form insoluble compounds with soil constituents, there is a great deal of fluctuation in its concentration in the soil solution.

The concentration in the soil solution, particularly of the cations, is strongly influenced by the pH and, depending on the cation, also by the redox potential. With a fall in the pH, or in redox potential, the concentration of the micronutrients manganese, iron, zinc and copper increases to various degrees (Sims and Patrick, 1978; Herms and Brümmer, 1980; Sanders, 1983).

Chelation with low-molecular-weight organic substances is another factor which exerts a dominant influence on the concentration of micronutrient cations in the soil solution and their transport to root surfaces by means of mass flow and diffusion. In soil solutions of calcareous soils, between 40 and 75% of the zinc and 98–99% of the copper have been found in organic complexes (Hodgson *et al.*, 1966; Sanders, 1983). The proportion of complexed manganese varies between 84 and 99% (Geering *et al.*, 1969), and it may drop at high soil pH to 40% (Sanders, 1983). For the plants, the importance of complexed micronutrients in the soil solution is particularly evident in calcareous soils (Section 16.5), and also is indicated by the fact that soil extractions with synthetic chelators are suitable soil tests for available micronutrients (Lindsay and Norvell, 1978).

There is, however, often a poor correlation between the concentration of chelated micronutrients in the soil solution, on the one hand, and the availability of these micronutrients as indicated by plant uptake, on the other (Gilmour, 1977). This is because the metal–organic complexes in the soil solution differ both in electrical charge [they are negatively charged, uncharged, or positively charged (Sims and Patrick, 1978)] and in size. In general, the rate of uptake of metal cations from metal–organic complexes is lower than that of free cations (Chapter 2) and decreases with the size of the organic ligand, as has been demonstrated for copper (Jarvis, 1981).

Table 13.3
Effect of a Metal Chelator Added to the Nutrient Solution or Soil on the Trace Element Content of Bean Leaves[a,b]

	Content in leaves (μg/g dry wt)				
	Zn	Cu	Fe	Mn	Ni
Nutrient solution					
Control	34·0	37·3	125	132	32·7
+ 10^{-4} M DTPA	19·2	3·8	149	118	0·0
Soil culture					
Control	23·4	7·6	124	108	2·0
+ 10^{-3} M DTPA	26·8	18·6	230	136	12·8

[a]Based on Wallace (1980a,b).
[b]Chelator: diethylethylenetriamine pentaacetate (DTPA).

Table 13.3 indicates that the chelation of micronutrient cations and particularly of nickel produces opposite effects on the availability of these elements for bean plants, depending on whether the chelator is added to the nutrient solution or to the soil. In aerated or stirred nutrient solution, chelation (except that of iron) does not increase the concentration and/or mobility of those cations; rather it exerts a negative influence on uptake. In contrast, in soils, where the most limiting factors are low concentration and thus mobility (i.e., transport to root surfaces), chelation substantially increases the availability of micronutrient cations, but also nickel to plants.

13.2.3 Role of Mass Flow

Calculations of the contribution of mass flow to the nutrient supply of field-grown plants rely on detailed data both on the concentration of nutrients in the soil solution throughout the season and on the water consumption of the plants in question. The results of such experiments are shown in Table 13.4. Expressed as average values for the whole growing season, the contribution of mass flow differs not only among mineral nutrients but also among plant species (owing to differences in transpiration rate and/or uptake rate of a given mineral nutrient). Mass flow is more than sufficient for supply of all nutrients only in podzol, a sandy soil with high nutrient mobility. In contrast, mass flow is unimportant for potassium supply in the luvisol, a loamy silt soil with low potassium equilibrium concentrations in the soil solution. Moreover, the supply of nitrate by mass flow to spring wheat in this soil is only 40% of the total uptake. The high supply of nitrate to sugar beet by mass flow on this soil is more the exception than the rule (Section 13.5). On the basis of the data in Table 13.4 one would predict a substantial accumulation of calcium and magnesium at the root surface (Chapter 15).

Table 13.4
Contribution of Nutrient Transport by Mass Flow to the Roots Compared with Nutrient Uptake during the Growing Season[a]

		Contribution of mass flow (% of uptake)[b]				
Plant species	Soil	K	Nitrate N	Mg	Na	Ca
Sugar beet	Luvisol[c]	7	100	60	n.d.[d]	640
Spring wheat	Luvisol[c]	4	40	150	2500	1700
Spring barley	Podzol	130	110	180	610	700

[a]Field experiments, Renger *et al.* (1981).
[b]Total uptake = 100%.
[c]Derived from loess.
[d]n.d., Not determined.

No data are given in Table 13.4 for phosphorus, but a rough calculation can be made. The amount of water transpired by a crop varies in the range of 2 to 4 million liters/ha (Barber, 1974; Strebel *et al.*, 1983). Assuming a phosphorus concentration in the soil solution of 5 μM (0·15 mg/liter) and a total water consumption of the crop by transpiration of 3 million liters, a phosphorus supply by mass flow of ~0·45 kg is calculated, which corresponds to about 2 to 3% of the total demand of a crop.

The contribution of mass flow depends on the plant species (Table 13.4) and is, for example, higher for potassium in onion than in maize since onion roots have a higher water uptake rate per unit length (Baligar and Barber, 1978). The relative contribution of mass flow also varies with the age of plants (Brewster and Tinker, 1970) and relative humidity, that is, transpiration rate (Elgawhary *et al.*, 1972).

When the soil water potential is high (e.g., at field capacity), mass flow is unrestricted and maintains a similar water potential at the root surface. As the soil water potential falls (i.e. becomes more negative), the rate of water uptake by the roots may exceed the supply by mass flow and the soil may then dry out. This is observed around the roots, particularly when the transpiration rate is high (Nye and Tinker, 1977), and often occurs in the topsoil during the growing season. Under field conditions the rainfall pattern (or irrigation cycle) therefore has an important influence on the contribution of mass flow to the total nutrient supply.

Since mass flow and diffusion to the root surface usually occur simultaneously, it is not possible to strictly separate these processes. In order to define the amount of solutes transported to the root by mass flow, Brewster and Tinker (1970) have recommended the term *apparent mass flow* (Nye and Tinker, 1977).

13.2.4 Role of Diffusion

Diffusion is the main driving force for the movement at least of phosphorus and potassium to the root surface. In contrast to mass flow, diffusion is an important factor of ion mobility only in the immediate vicinity of the root surface and thus is closely related not only to soil conditions but also to plant factors such as root growth and root surface area.

The important parameter for effective diffusion D_e can be summarized by a simplified formula:

$$D_e = D_l \theta F_l \, dC_l/dC_x \qquad (1)$$

where D_l is the diffusion coefficient of the nutrient in free solution; θ is the fraction of soil volume occupied by solution (i.e., the cross section of diffusion); F_l is the impedance factor, which takes account of the tortuous pathway of ions and other solutes through pores, increasing path length and thus decreasing concentration gradients, and also includes an increase in water viscosity and anion adsorption or repulsion; and dC_l/dC_x the concentration gradient between solution concentration in the bulk soil (C_l) and at the root surface (C_x), respectively.

13.2.4.1 Soil Factors

As a rule, the concentration of potassium and phosphorus is much lower at the root surface than in the bulk soil, creating a typical concentration profile around the roots. As shown in Fig. 13.3, with increasing potassium concentrations in the soil the diffusion coefficient increases. The nutrient-depleted area surrounding the plant roots, the depletion zone, also increases from ~4 mm in depleted soil (depleted by intensive cropping) to ~6 and ~10 mm in unfertilized and fertilized soil, respectively. Hence, raising the level of exchangeable potassium by fertilizer application increases the potassium supply via diffusion more than would be expected from the increase in the amount of exchangeable potassium per unit soil weight only. This was shown by Kuchenbuch and Jungk (1984) in a fertilizer experiment with rape, where

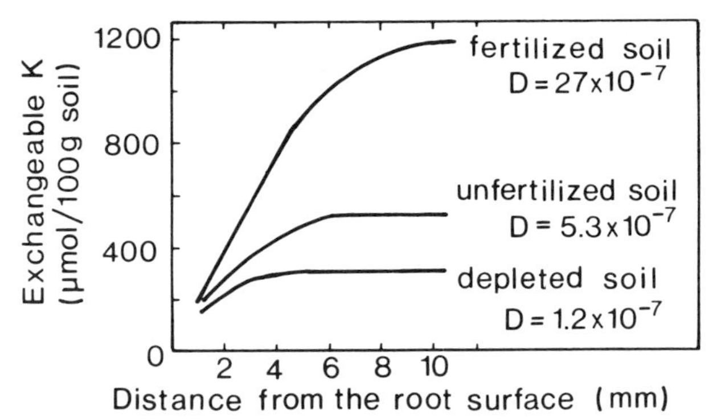

Fig. 13.3 Concentration profile of potassium around a maize root grown in a silt loam soil with various levels of exchangeable potassium (D = diffusion coefficient expressed as square centimeters per second). (Modified from Jungk *et al.*, 1982.)

the amount of exchangeable potassium per unit soil weight was increased by a factor of ~4, whereas the amount of exchangeable potassium delivered by diffusion to the root surface was increased by a factor of ~20. Application of NaCl or $MgCl_2$ also increased the extension of the depletion zone and thus the delivery of potassium to the root surface.

The shape of the depletion zone also differs typically among soils, depending on their clay content (Fig. 13.4). In soil A, with 21% clay and a correspondingly higher cation-exchange capacity, the equilibrium concentration of K^+ in the soil solution was much lower than in soil B, with only 4% clay. In both soils, roots depleted the soil solution to about 2 to 3 μM K^+. The depletion zone in soil B was much wider than that in soil A, however, reflecting the much lower capacity of soil B to replenish the K^+ in the soil solution.

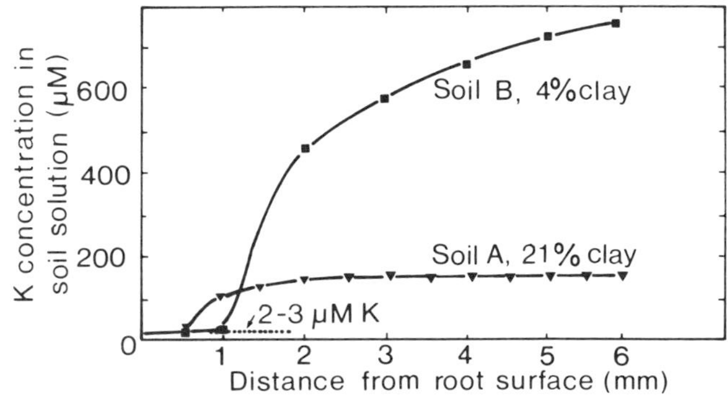

Fig. 13.4 Concentration profile of potassium in the soil solution around maize roots growing in soils with different clay contents. Potassium concentration at the root surface, 2–3 μM. (Modified from Claassen and Jungk, 1982.)

Under otherwise constant conditions, soil moisture affects nutrient supply via diffusion to the root surface. As the soil water potential falls, the diffusion coefficient decreases [decrease in θ and increase in F_1, see Eq. (1)]. This effect is shown in Table 13.5 for calcium, magnesium, and potassium in a model experiment with an H^+-loaded cation exchanger resin simulating the root as a physiological sink for mineral nutrients. With a fall in soil moisture content, the diffusion rate of all three cations declined (Table 13.5) and correspondingly so did the extension of the depletion zones of potassium (Mengel *et al.*, 1969) and magnesium (Grimme, 1973).

Table 13.5
Effect of Soil Moisture Content on the Diffusion of Calcium, Magnesium, and Potassium to a Cation Exchanger Resin within 96 Hours at 15°C[a]

	Content in cation exchanger (meq $\times 10^{-2}$)		
Soil moisture content (%)	Ca^{2+}	Mg^{2+}	K^+
28, w/v (~field capacity)	38	32	5·9
17	27	19	3·8
10 (approx. −15 bar)	19	12	2·1

[a]Based on Schaff and Skogley (1982).

It is well known that at low soil moisture levels the reduction of potassium and phosphorus uptake is greater than that of calcium or magnesium, which may even be increased (Talha *et al.*, 1979). As has been demonstrated by Zur *et al.* (1982) at low soil moisture levels the soil water potential at the root surface is much lower (i.e. more negative) than that of the bulk soil. Contact may thus be lost between root surface and soil. It has been postulated that long root hairs are of particular importance in preventing this loss of contact (Nye and Tinker, 1977). When the soil water potential is low, the mechanical impedance of the soil increases and root elongation is also inhibited, which further limits nutrient supply to the root surface by diffusion. The supply of boron to the roots is depressed to a greater degree by low soil water potential than by that of any other micronutrient (Kluge, 1971). It is not known whether mass flow or diffusion is mainly responsible for boron supply.

13.2.4.2 Plant Factors

The extension of the depletion zones of potassium and phosphorus is more or less a reflection of root hair length. This has been convincingly demonstrated by Lewis and Quirk (1967) and Bhat and Nye (1974) for labeled phosphorus (^{32}P). In autoradiographs the depletion zone is highly visible (Fig. 13.5).

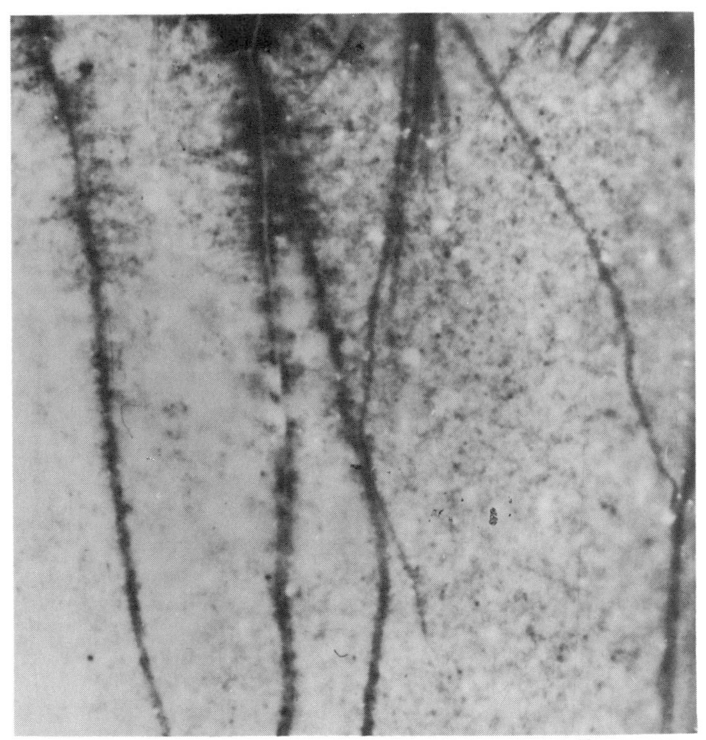

Fig. 13.5 Autoradiograph of maize roots in a soil labeled with ^{32}P showing zones of phosphorus depletion around the roots (removal of ^{32}P indicated by black zones).

Root hair formation is modified by environmental factors (Chapter 14) and also differs typically among plant species. Genotypical differences in root hair length are particularly important for the concentration profile of phosphorus and potassium around roots. An example is given in Fig. 13.6. The maximum depletion resembles the average root hair length (0·7 and 1·3 mm in maize and rape, respectively), and the extension of the depletion zone is nearly identical with the maximal root hair length.

When the uptake rates of phosphorus or potassium per unit root length of different plant species are compared, a close positive correlation can be demonstrated between the uptake rate per unit root length and volume of the root hair cylinder. This is shown for potassium in Fig. 13.7. In onion, root hairs were virtually absent, whereas of the plant species tested rape had the longest root hairs. Per unit length of onion root, the potassium of only 2 to 3 mm^3 soil was available, as compared with ~60 mm^3 for rape. Results similar to those shown in Fig. 13.7 were obtained for phosphorus in comparative studies with plant species of different root hair lengths (Itoh

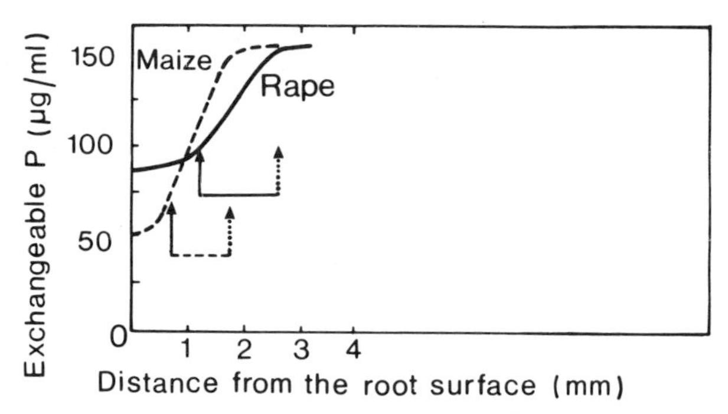

Fig. 13.6 Concentration profile of phosphate around individual roots of maize and rape growing in a sandy soil. Solid arrows denote average and dotted arrows maximal root hair length. (Modified from Hendriks *et al.*, 1981.)

and Barber, 1983a). In agreement with this, even within a given plant species (white clover) selection of genotypes results in more efficient phosphorus uptake in genotypes with long root hairs than in those with short root hairs (Caradus, 1982). The fact that grasses compete more effectively than legumes in a mixed stand for potassium (Steffens and Mengel, 1980) is probably also related to some extent to the wider root hair cylinder of grasses, but certainly to the total root length (see below).

The extension of the depletion zone of a given species is also affected by other plant factors such as the nutritional status (which may affect the root hair length; Section 14.2.1) and particularly by root activity (e.g., supply of

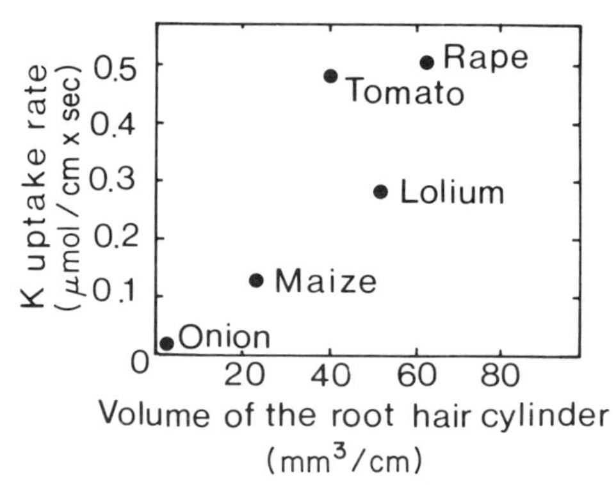

Fig. 13.7 Rate of potassium uptake per unit length of root in relation to the volume of the root hair cylinder. The plant species were grown in a silt loam with 21% clay. (Modified from Jungk *et al.*, 1982.)

O_2 and carbohydrates), considered in terms of C_{min} (the minimum concentration at which roots can deplete the external solution, see Section 2.4.1). Average values for soil-grown plants are 2–3 μM potassium (Claassen and Jungk, 1982), 1 μM phosphorus (Hendriks *et al.*, 1981), and 6 μM nitrate (Cumbus and Nye, 1982). These C_{min} values are considerably higher than those found in nutrient solutions, indicating that the soil values reflect an equilibrium between removal by the roots and replenishment by the solid phase. The C_{min} values in solution culture also differ considerably among plant species (Asher, 1978)—for phosphorus, for example, between 0·12 μM in tomato (Itoh and Barber, 1983a), 0·04 μM in soybean (Silberbush and Barber, 1984) and 0·01 μM in ryegrass (Breeze *et al.*, 1984). Within a given species the C_{min} values of genotypes may (Schenk and Barber, 1979b) or may not (Silberbush and Barber, 1984) differ.

Among the soil factors which affect the extension of the depletion zone, the buffer power of the soil is of particular importance (Fig. 13.4). On average, the potassium depletion zone for maize extends to 5 mm in sandy soils and to 2 mm in loamy or clay soils (Claassen *et al.*, 1981). This has important consequences for the rate of replenishment of the soil solution in the root hair cylinder, as is demonstrated by the following calculation for phosphorus: concentration in the soil solution, 5 $\mu M \cong$ 0·15 mg phosphorus per liter; amount of soil solution in the topsoil (0–30 cm), ~500,000 liters= 75 g phosphorus per hectare; requirement during phase of rapid growth (e.g. in cereals between tillering and heading), ~300–500 g phosphorus per hectare × day. Since only ~25% of the topsoil is explored by roots during one growing season (Jungk, 1984) the rate of replenishment within the root hair cylinder has to be at least 10–20 times per day in order to meet the requirement of plants. For potassium, too, the rate of replenishment in the root hair cylinder has to be high. Within 2·5 days more than half of the potassium taken up by maize is derived from the so-called nonexchangeable fraction of the soil in the root hair cylinder (Claassen and Jungk, 1982), and in rape, within 7 days the contribution by the nonexchangeable fraction in an unfertilized and fertilized soil was 71% and 20%, respectively, of the total uptake (Kuchenbuch and Jungk, 1984). From these data it can be concluded that field-grown plants do not uniformly deplete even the densely rooted topsoil; near the root surface a high proportion of the nonexchangeable potassium contributes to the total uptake, whereas at a distance from the root (that is, in the bulk soil) even the readily exchangeable potassium is not utilized.

For phosphorus the situation is even more complicated due to the high dynamics of organically bound phosphorus in the rhizosphere (Helal and Sauerbeck, 1984a,b), and the root-induced changes in the rhizosphere, especially in response to phosphorus deficiency (e.g., increase in root hair length, Section 14.2.1; and in root exudation, Section 15.4). Current mechanistic models of phosphorus uptake which do not consider these roots responses therefore considerably underestimate the predicted uptake of phosphate (Silberbush and Barber, 1984).

The spatial availability of nutrients in the soil is not only determined by the volume of the root hair cylinder but also by the total root length, the second component of the root surface area. The importance of the total root length is readily demonstrated in studies comparing different plant species. The higher capacity of ryegrass compared to red clover in acquiring potassium (Steffens and Mengel, 1980, 1981) and phosphorus (Steffens, 1984) is mainly due to the fact that the total root length of ryegrass is five times larger. As shown by Claassen and Jungk (1984), the volume of the root hair cylinder of different plant species was closely correlated with the uptake rate of potassium per unit root length (Fig. 13.7) but not with the potassium content of the shoots. A close correlation with the contents in the shoots existed, however, when the average root age and the ratio of root length (cm) per unit shoot dry wt (mg) was included in the calculation. The importance of the total root length for nutrient acquisition can also be demonstrated within a species such as cowpea, when the phosphorus uptake from a deficient soil of different cultivars is compared (Adeptu and Akapa, 1967).

13.3 Role of Rooting Density

Although high rooting density and long root hairs are important factors in the uptake of nutrients supplied by diffusion, the relationship between rooting density and uptake rate is not linear, as shown in Fig. 13.8. When the

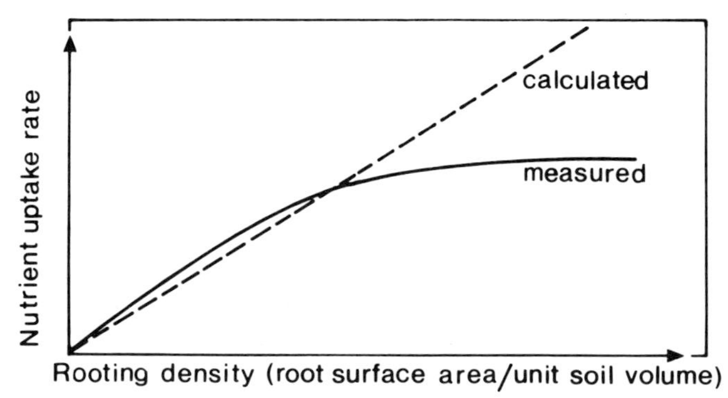

Fig. 13.8 Relationship between rooting density and uptake rate of nutrients delivered by diffusion.

root density is high, the uptake rate levels off. This is caused by overlapping of the depletion zones of individual roots (Nye and Tinker, 1977) and reflects competition for nutrients.

In principle the same curvilinear relationship should exist between the rate of phosphorus uptake and root hair density, because of competition among individual root hairs (Itoh and Barber, 1983b). This competition has to be considered when attempts are made to correlate rooting density, for example, in different soil layers or horizons, and their contribution to the nutrient supply.

13.4 Nutrient Availability and Distribution of Water in Soils

Under field conditions the level of chemically available nutrients is usually much higher in the topsoil than in the subsoil. Generally, rooting density follows a similar pattern, and the logarithm of the rooting density declines linearly with increasing depth (Greenwood *et al.*, 1982). Variations in the water supply, however, modify both root distribution and nutrient uptake from various soil depths. The effect of water supply on the root distribution of spring barley was clearly demonstrated in two successive years (Scott-Russell, 1977). In the first year, with high rainfall (82 mm) occurring a month after planting, more than 70% of the total root mass was found in the topsoil (2·5–12·5 cm) 2 months after planting and only ~10% of the roots had penetrated deeper than 22·5 cm; in contrast, in the following year, with inadequate rainfall (24 mm) during the first month after planting, the corresponding values for the distribution of the root mass were about 40 and 30%, respectively. This type of shift in root distribution has important consequences for nutrient uptake from various soil horizons. In spring wheat growing on loess soil, on average ~50% of the total potassium taken up later in the growing season is derived from the subsoil (Grimme *et al.*, 1981). However, depending on the rainfall during the growing season (i.e. the amount of water available in the topsoil), the percentage varies substantially from year to year, being ~60% during a dry year and ~30% during a wet year (Fleige *et al.*, 1983).

It was shown by Fox and Lipps (1960) that ~3% of the total root mass of alfalfa takes up more than 60% of the total nutrients from the subsoil during drought periods. But even when the rainfall distribution is reasonably uniform during the growing season, in semihumid regions of Central Europe the soil moisture content in the topsoil is often a limiting factor, for example, for phosphorus availability (Table 13.6). Despite the much higher level of available phosphorus in the topsoil (as indicated by soil testing) at later growing stages between 40 and 30% of the total phosphorus uptake comes from the subsoil.

Table 13.6
Phosphorus Uptake by Spring Wheat and Phosphorus Delivery as a Function of Soil Depth and Time[a]

Available P[b] (mg/100 g soil)	Soil depth (cm)	Delivery of P from soil (%) Booting stage: 0·345 P uptake[c]	Anthesis: 0·265 P uptake[c]	Milky stage: 0·145 P uptake[c]
11·5	0–30	83·3	58·8	67·4
4·5	31–50	8·1	17·8	15·5
2·5	51–75	5·9	16·3	12·0
2·0	76–90	2·7	7·1	5·1

[a]Luvisol derived from loess. Based on data from Fleige *et al.* (1981).
[b]Determined by extraction with calcium ammonium lactate.
[c]Phosphorus uptake expressed as kilograms per hectare per day.

13.5 Shift in Mass Flow–Diffusion Supply during Ontogenesis

The relative proportions of nutrient supplied by mass flow and by diffusion shift most noticeably for nitrate, the buffer power of which is low in the soil. The concentration is usually high in the topsoil early in the growing season but rapidly declines thereafter as a result of plant uptake. In spring wheat the change in nitrate concentration in the soil solution is correlated with a decline in the amount of nitrate supplied by mass flow and an increase in supply via diffusion, which then supplies more than 50% of the total nitrate (Strebel *et al.*, 1980).

For sugar beet the supply by mass flow during the entire growing season is even less, an average of 32 kg nitrate nitrogen, compared with 181 kg supplied by diffusion (Strebel *et al.*, 1983). A time course study (Fig. 13.9) demonstrated that the supply by mass flow is restricted to the early growing period; during this time, nitrate is taken up fom the topsoil, which has a large nitrate concentration in the soil solution. Thereafter, the nitrate in the topsoil is depleted, and upon root proliferation into the subsoil nitrate is

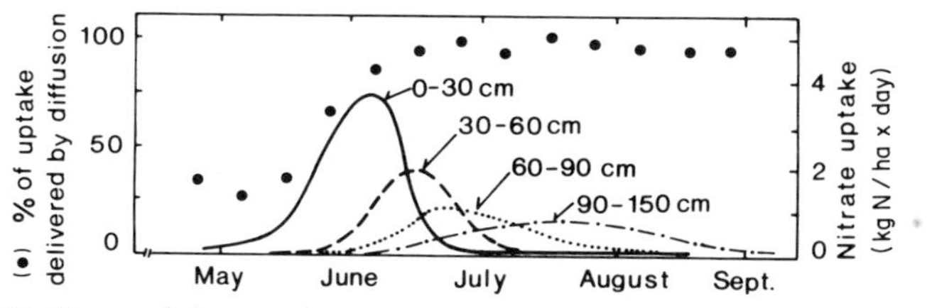

Fig. 13.9 Delivery of nitrate nitrogen to sugar beet plants as a function of soil depth (centimeters) and time. Soil: luvisol derived from loess. (Based on Strebel *et al.*, 1983).

supplied exclusively by diffusion. This example convincingly illustrates that average data on the contribution of mass flow and diffusion (as well as of different soil horizons) to the total supply have to be interpreted with care, as such data insufficiently reflect the high variability and dynamics of these processes.

13.6 Intensity Quantity Ratio and Its Consequences for Soil Testing

The most important soil factors affecting nutrient supply to the plant roots are the nutrient concentration in the soil solution, the rate of nutrient replenishment, and the amount of readily soluble nutrients in the soil profile. As Schofield (1955) proposed, the nutrient concentration in the soil solution can be characterized by an intensity factor, and the fraction adsorbed to the solid phase or bound to labile organic compounds by a quantity factor. This concept has been extended by various authors and is summarized in Fig. 13.10. Within the root hair zone the rate of replenishment from the labile pool has to be extremely high for phosphorus and potassium (see Section 13.5). The rate of replenishment of nitrate depends on nitrification of ammonium formerly adsorbed to the solid phase or supplied by mineralization of organically bound nitrogen. As these rates are often too low, in order to cover the demand of fast growing plants the supply must be maintained either by root exploration of the subsoil or by split application of nitrogen fertilizer.

These factors must be considered when chemical soil analysis is used as a basis for recommending fertilizer applications. The method of electroultrafiltration (EUF), which involves the use of different electrical field strengths and temperatures in an aqueous soil suspension (Nemeth, 1982), provides more information on the rate of replenishment and the intensity/quantity ratio than do conventional chemical extraction methods. This method also seems to be a suitable indicator of the size of the readily soluble, organically bound nitrogen pool, which may contribute substantially to the nitrogen

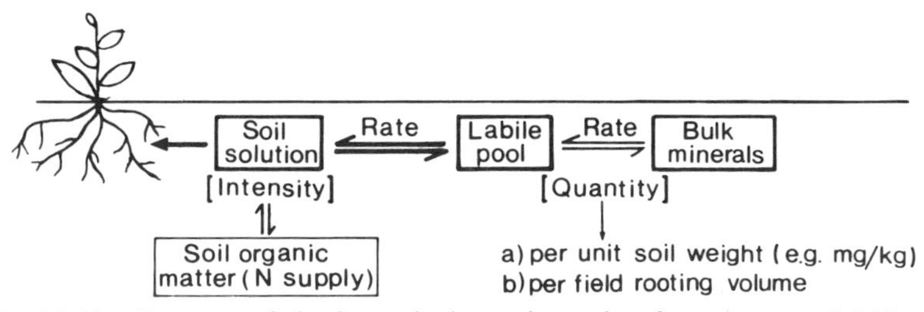

Fig. 13.10 Concept of the intensity/quantity ratio of nutrient availability.

supply of a crop in the following growing season (Nemeth *et al.*, 1979; Nemeth, 1985).

A soil analysis as a basis for recommending nitrogen fertilizer application has been improved considerably by the N_{min} method (Wehrmann and Scharpf, 1979; Wehrmann *et al.*, 1982). With this method the amount of mineralized nitrogen, mainly nitrate, in the soil profile (usually in 0–90 cm depth) is measured at the beginning of the growing season, thus taking into account the high mobility of nitrate in the soil profile and nitrogen uptake from the subsoil. In principle, the contribution of the subsoil to the supply of other mineral nutrients also deserves more attention (Vetter and Früchtenicht, 1979) as has been demonstrated in this chapter. This holds true unless subsoil penetration by roots is impaired for various reasons, which are discussed in Chapters 14 and 16.

Improving the reliability of fertilizer recommendations based on chemical soil analysis does not depend primarily on the extraction method used but rather on the systematic consideration of the roots as decisive factors in the "spatial" availability of nutrients. Models for predicting nutrient availability and nutrient uptake under field conditions must therefore be based on both soil and plant factors in which root parameters are the key element (Claassen and Barber, 1976; Schenk and Barber, 1979a; Greenwood, 1983). Studies on root growth and nutrient dynamics in the rhizosphere are therefore a topic of great interest for both scientific and practical reasons.

14

Effect of External and Internal Factors on Root Growth and Development

14.1 Hormonal Control

The development of root systems is characterized by a very high adaptability and involves complex interactions between roots and shoots and between roots and their environment. Root growth and development are under hormonal control. Environmental factors such as the supply of mineral nutrients as well as soil aeration and soil mechanical factors influence root growth, at least in part, by modulating the phytohormone balance. Extension of the main axes and initiation of lateral root formation are regulated primarily by auxin (IAA) derived from the shoot and cytokinins (CYT) produced in the apical meristem (Fig. 14.1). Auxin is a strong promoter whereas CYT are strong inhibitors of lateral root formation (Wightman *et al.*, 1980). Elongation is severely inhibited by high concentrations of CYT (Stenlid, 1982) or IAA (Butcher and Pilet, 1983). In agreement with this, inhibition of extension of the main axes by mechanical impedance is correlated with a threefold increase in the IAA content of the apical zone and enhanced initiation of lateral root formation (Lachno *et al.*, 1982).

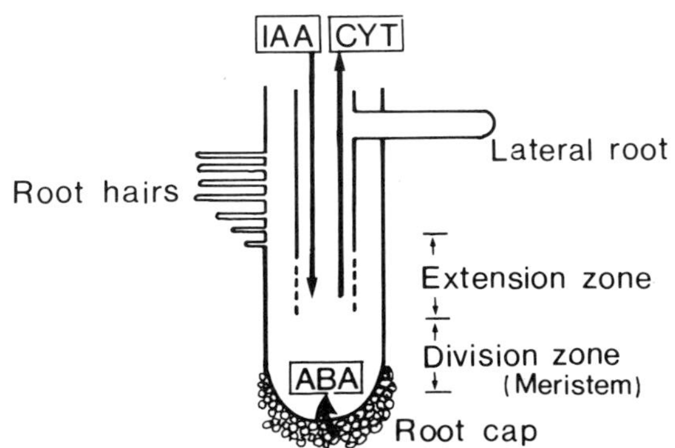

Fig. 14.1 Schematic representation of some aspects of phytohormone balance in a growing root (IAA, indoleacetic acid auxin; CYT, cytokinins; ABA, abscisic acid).

Supraoptimal IAA supply from the shoots to the apical zone may be in part responsible for the inhibition of elongation of the main axes and the increase in lateral root formation when the nitrogen supply is high (Brouwer, 1981).

There is increasing evidence that root cap cells (Fig. 14.1) are the main sites of abscisic acid (ABA) synthesis (Pilet, 1981). ABA has primarily an inhibitory effect on root extension (Barlow and Pilet, 1984) but it enhances the initiation of lateral root formation, that of root hairs in particular (Biddington and Dearman, 1982). The root cap not only protects the root meristem cells and facilitates penetration through pores (acting as a "lubricant") but is also the site of perceiving light and gravity signal (Feldman, 1984) and the synthesis and release of growth regulators, especially ABA. Root cap cells, furthermore, are the sensors which after making contact with a mechanical impedance, send a message (presumably a hormone) to the elongation zone and thereby depress extension growth, for example in barley, within 10 min (Goss and Scott-Russell, 1980).

Ethylene, which can be considered a secondary messenger (Chapter 5), has effects on roots that are similar to those of supraoptimal IAA concentrations. It inhibits extension of the main axes and strongly enhances the formation of lateral roots (Crossett and Campbell, 1975). Whether both IAA and CYT exert their inhibitory influence on extension at least in part through an enhancement of ethylene biosynthesis (Stenlid, 1982) is still an open question (Butcher and Pilet, 1983; Feldman, 1984). Ethylene also promotes the formation of lysigenous cavities (aerenchyma) in the root cortex (Section 16.4). It should be kept in mind that causal interpretations of the effects of phytohormones (either externally supplied or the endogenous levels) on root growth and development have to take into account the necessity of receptor sites for the phytohormone action (Section 5.6).

Root hairs originate from the rhizodermal cell layer and usually develop 5–10 mm behind the apex (Fig. 14.1). They differ very much in density, length, and longevity; frequently they persist only a few days. The formation and characteristics of root hairs are highly influenced by phytohormones, IAA in particular (Tien *et al.*, 1979) and by environmental factors. In general root hair formation is stimulated by a low nutrient supply and/or all factors which inhibit root extension.

14.2 Soil Chemical Factors

14.2.1 Mineral Nutrient Supply

The distribution of roots in soils can be modified by the placement of fertilizers. Rooting density increases severalfold in zones where the concen-

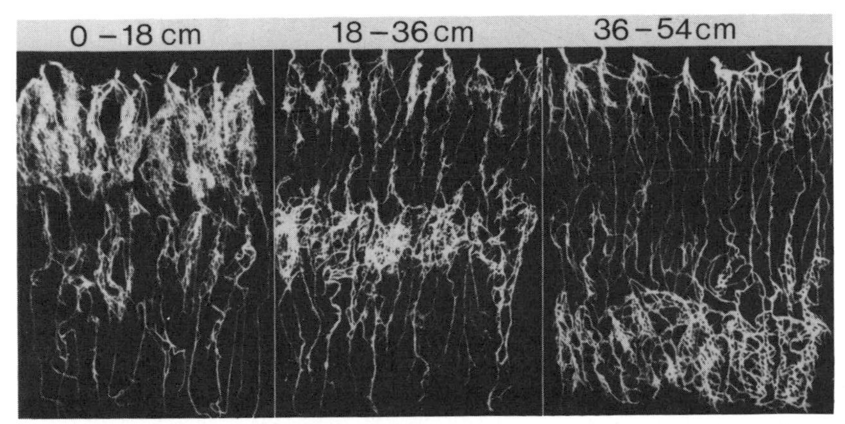

Fig. 14.2 Effect of nitrogen fertilizer placement in different soil depths on the root distribution of barley growing in a sandy soil. (From Gliemeroth, 1953.)

tration of nutrients, especially nitrogen, is high (Fig. 14.2). Deep placement of fertilizer enhances plant growth and yield, particularly under drought conditions when the water potential of the surface soil decreases but ample water is available in the subsoil (Garwood and Williams, 1967).

Placement of phosphorus fertilizers is a common and effective practice in soils low in readily soluble phosphorus and/or in order to ensure a sufficient supply to the roots, especially in the early growing stages when soil temperatures are low. Rooting density increases in zones of phosphorus fertilizer placement, although the effect is relatively small compared with that of combined nitrogen–phosphorus fertilizers (Böhm, 1974).

The effects of localized nutrient supply on root morphology can be demonstrated in more detail if the nutrients are supplied in higher concentrations to a certain root zone only. As shown in Fig. 14.3, in the root zone supplied with a large amount of nitrate, a spectacular enhancement of lateral root formation took place.

Although the effect of a localized supply of phosphorus under field conditions is usually less dramatic than that of nitrate, in principle similar effects are obtained if phosphorus is supplied to one root zone only and the remainder of the root system is kept in a solution without phosphorus (Table 14.1). The total length of the lateral roots increased 15-fold over that of controls in 21 days. The corresponding increase in dry weight (by a factor of 10) took place partially at the expense of the remaining root zones, A and C.

The question arises as to how mineral nutrients exert this morphogenetic influence on root systems. It can be readily demonstrated that the removal of part of the root system is compensated for by enhanced growth of the remaining parts. Obviously, localized nutrient supply has similar effects on

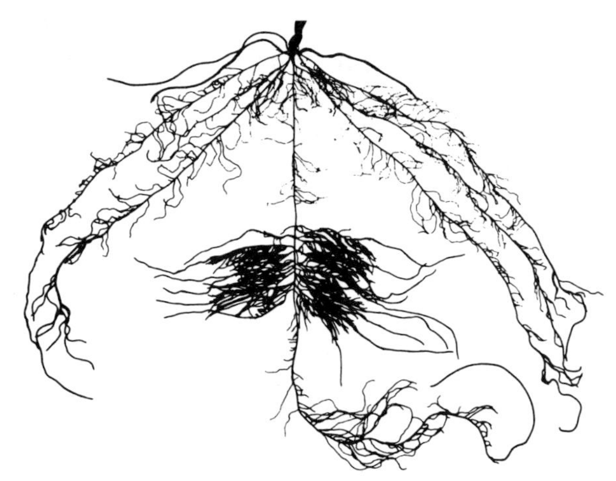

Fig. 14.3 Modification of the root system of barley by providing 1 m*M* nitrate to the midpart of one root axis for 15 days; the remainder of the root system received only 0·01 m*M* nitrate. (From Drew and Saker, 1975.)

the root growth pattern. It can be assumed that competition for photosynthates occurs between the parts of a root system and that the sink activity of a particular zone for photosynthates (phloem unloading) is increased when a growth-limiting mineral nutrient is supplied to this zone only. However, since phytohormones, such as IAA, are translocated in the phloem basipetaly and released at the sites of phloem unloading, it is likely that the enhancement effect of localized mineral nutrient supply on lateral root initiation is also under hormonal control (Brouwer, 1981). Weakening the apical dominance of the apical meristem at the root axes (e.g., by ammonium supply) would have similar effects on lateral root initiation.

Table 14.1
Effect of Localizing the Supply of Phosphorus to a Root Segment on the Length of Lateral Roots and the Dry Weight of Barley Plants[a,b]

	Uniform supply		Localized supply	
Root zone	Length of lateral roots (cm)	Dry weight (mg)	Length of lateral roots (cm)	Dry weight (mg)
A (basal)	40·0	8·9	14·3	3·5
B (middle)	27·2	3·7	332·0	37·8
C (apical)	17·5	10·2	11·1	4·9

[a]Based on Drew and Saker (1978).
[b]Phosphorus was supplied to a 4-cm segment (middle, or B, zone) of a single seminal root axis. Duration of experiment: 21 days.

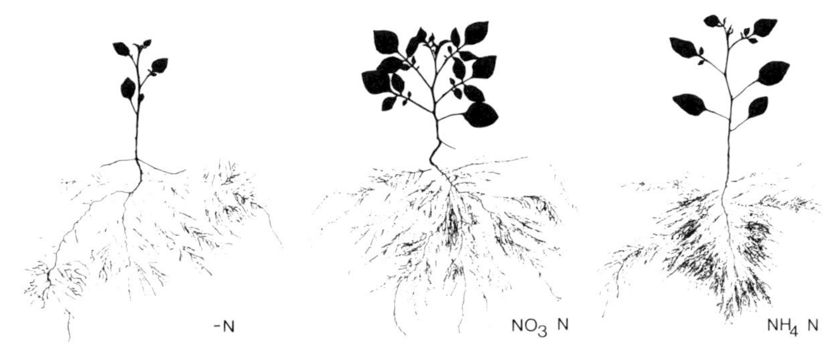

Fig. 14.4 Potato seedlings grown in a sandy soil. (*Left*) No nitrogen supplied. (*Middle*) Nitrate nitrogen supplied. (*Right*) Ammonium nitrogen supplied. (By courtesy of B. Sattelmacher.)

Enhancement of extension of the main axes and retarded lateral root formation are often observed when there is a suboptimal supply or deficiency of mineral nutrients—for example, potassium (Jensen, 1982) or nitrogen (Bergmann, 1954). Enhanced extension of both main and lateral roots may also occur at suboptimal nitrogen levels (Fußeder, 1984). The variability in the results of nitrogen effects is presumably caused by the modulating effect of the temperature; there is a tendency for the inhibition of root growth by high nitrogen levels to be accentuated by high root temperatures (Geisler and Krützfeldt, 1983). In any case, a suboptimal supply of nutrients, and of nitrogen in particular, increases the ratio of roots to shoots (Fig. 14.4). A low nitrogen supply during early growth stages therefore leads to an increase in the soil area explored by the roots, the subsoil in particular. The effects of ammonium and nitrate on root morphogenesis differ markedly (Fig. 14.4). With ammonium there is a greater reduction in elongation, the roots are short and thick (Bhat, 1983), and lateral root formation is stimulated (Klemm, 1966). These morphogenetic effects are most obvious in nutrient solutions but can also be observed in roots of soil-grown plants (Fig. 14.4).The reasons for the different effects of nitrogen forms on root morphogenesis are unknown; differences in both pathways of assimilation in the roots and in plant hormonal balance (Chapter 6) are probably involved. Hormonal effects are probably also responsible for the increase in the formation of aerenchyma in the cortex of maize roots even in well-aerated solutions when the nitrogen supply is low (Konings and Verschuren, 1980). An increase in aerenchyma formation is a typical root response to elevated levels of ethylene (Section 16.4).

Phosphorus deficiency, like nitrogen deficiency, leads to an increase in the ratio of roots to shoots (Table 14.2). Increasing the duration of phosphorus starvation results in an increase in root dry weight, root length in particular. The roots also become finer.

Table 14.4
Effect of Liming an Acid Subsoil (pH 4·6) on the Elongation of Cotton Root[a]

Percentage of subsoil mass limed[b]	Distance between limed layers (cm)	Relative root length
Unlimed	—	32
10	4·5	38
20	4·0	57
40	3·0	57
60	2·0	70
100	—	100

[a]Based on Pearson *et al.* (1973).
[b]Application of the same dose, but differently distributed.

tive species such as cotton the root elongation rate may be severely inhibited at aluminum/calcium molar activity ratios of greater than 0·02 (Lund, 1970). In soils this critical ratio may vary widely, however, presumably depending on the proportion of chelated aluminum (Adams and Moore, 1983). Chelation of aluminum most likely also contributes to the beneficial effects of mulching on plant growth, especially in highly permeable acid mineral soils of the humid tropics. Under these conditions yield responses to relatively small amounts of calcium fertilizer (applied in order to change the Al/Ca ratio) are often similar to or even higher than responses to heavy liming.

The calcium/total cation ratio is also of importance for root growth under saline conditions (Section 16.6) and in relation to the application of ammonium phosphate fertilizer. In acid soils with a low cation-exchange capacity, ammonium phosphate can severely inhibit root growth by inducing calcium deficiency (Bennett and Adams, 1970).

At high pH, root growth can be inhibited either indirectly or directly. An example of an indirect effect is the change in root morphology induced by iron deficiency (Section 9.1). The most well known direct effect at high pH is ammonia toxicity. Root elongation is severely inhibited by ammonia concentrations of as low as 0·05 m*M* (Schenk and Wehrmann, 1979). Ammonia toxicity is probably also the reason for the inhibition of root growth in neutral or alkaline soils after the application of ammonium phosphate (Bennett and Adams, 1970) or the band application of urea (Creamer and Fox, 1980).

14.2.3 Aeration

Soil aeration is essential for meeting the respiratory requirements of both roots and soil microorganisms. In a soil with a dense crop stand, O_2

consumption and CO_2 evolution may be as high as 17 liters/m^2 $\times$ day (Cannell, 1977). The transfer of gases between soil and atmosphere occurs mainly in air-filled pores since gas diffusion in air is about 10^4 times more rapid than in water. Although in many species adapted to waterlogging (e.g., wetland rice) sufficient internal diffusion of O_2 from leaves to roots takes place in the aerenchyma (Section 16.4), in mesophytic (nonwetland) species this internal transfer is either unimportant or at least insufficient to meet the requirement of large root systems.

In most mesophytic species O_2 concentrations can be lowered to about 15 to 10% in the gas phase without severe effects on root growth (Geisler, 1967). With lowering of the O_2 concentration from 21 to about 10%, the root respiration in maize remains almost unaffected whereas root extension is severely impaired, indicating that at least in this concentration range processes other than respiration are responsible for the inhibitory effect of poor aeration on root growth (Saglio *et al.*, 1984). Definitive conclusions on the critical O_2 concentrations in soils are difficult to reach, however, for various reasons. For example, poor soil aeration and a decrease in O_2 are correlated with a simultaneous increase in CO_2. Depending on its concentration, CO_2 itself has either stimulatory (1–2% CO_2) or inhibitory effects (>5% CO_2) on root growth (Geisler, 1968). An accumulation of CO_2 at the root surface seems to favor root hair formation (Bergmann, 1958).

Lack of aeration in the subsoil restricts root growth in soils with a high water table. Cessation of root growth occurs even at some distance above the water table (Wiersum, 1967). The soil volume supplying mineral nutrients to the roots is correspondingly restricted by high water tables.

14.2.4 Low-Molecular-Weight Organic Solutes and Ethylene

Root growth is affected in various ways by the water-soluble fraction of soil organic matter. Whereas the high-molecular-weight organic acid fraction (especially fulvic acids) when present in low concentrations may enhance root initiation and root elongation (Mylonas and McCants, 1980), at higher concentrations the low-molecular-weight fractions are associated with inhibition of root growth and are usually of much concern. This is particularly true of phenolic and short-chain fatty acids, which often accumulate in poorly aerated or waterlogged soils during decomposition of organic material (e.g., straw or green manure). In well-structured soils, an uneven distribution of organic matter (clumps) may cause local anaerobiosis (the formation of anaerobic microhabitats). Water extracts from those soils, especially 3–4 weeks after incorporation of the organic matter, are highly detrimental to root respiration, root growth, and root hair formation (Patrick, 1971).

P

During the decomposition of organic substances with a large lignin content (e.g., straw), phytotoxic substances, including phenolic acids such as *p*-coumaric and *p*-hydroxybenzoic acid, may accumulate and severely inhibit root elongation in sensitive species such as rye and wheat even at concentrations of between 1 and 10 mg/liter (Börner, 1957) and in tolerant species such as sugarcane at 100 mg/liter (Wang *et al.*, 1967). These and other phenolic acids are known for their effects on IAA metabolism. They either stimulate or inhibit IAA oxidase activity (Baranov, 1979). For example, *p*-coumaric acid has morphogenetic effects on roots that are comparable to those of supraoptimal IAA supply (Svensson, 1972). In agreement with this low concentrations of these phenolic acids enhance the initiation of lateral root formation but severely inhibit further extension (Pingel, 1976).

Under waterlogging, other phytotoxic substances, primarily acetic acid and other volatile (short-chain) fatty acids may accumulate in phytotoxic concentrations (Harper and Lynch, 1982). These acids are detrimental to root elongation (Lynch, 1978), and inhibit root and shoot growth, even in plant species adapted to waterlogging (Table 14.5).

Table 14.5
Effect of Volatile Fatty Acids on Growth of Rice Plants 28 Days after Transplanting[a]

	Dry weight (g/plant)	
Volatile fatty acid	Roots	Shoots
None	1·70	4·56
Acetate (C_2)	1·40	3·70
Propionate (C_3)	1·00	3·54
Butyrate (C_4)	0·90	2·99
Valerate (C_5)	0·29	1·20

[a]Amount of each acid supplied: 1 mmol per 100 g soil.
From Chandrasekaran and Yoshida (1973).

As a rule phytotoxicity increases with chain length (Table 14.5) and with decreasing substrate pH (Jackson and St. John, 1980). This pH effect is related to the high rates of permeation of the plasma membrane by the undissociated species of acids (Chapter 2). The inhibitory effect of decomposition products (e.g., of straw) on root growth therefore depends not only on the concentration of phenolics and volatile fatty acids but also on the pH of the rooting medium, as demonstrated in Fig. 14.6. Between pH 4 and 8 the root elongation of the control plants was not much different. With the addition of the extracts, however, root elongation clearly became pH dependent. The phytotoxicity of organic acids produced by *Fusarium*

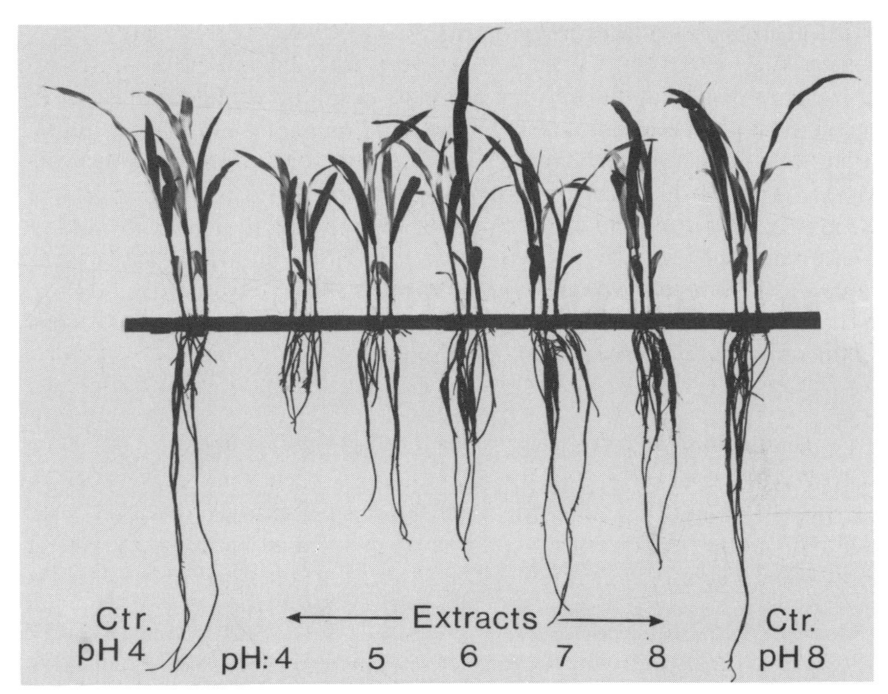

Fig. 14.6 Relationship between the pH of the nutrient solution and root growth of maize. Control plants (Ctr.) in nutrient solution only. (*Center*) Addition of 25 ml/liter each of extracts from anaerobically fermented oat straw. (Courtesy of A. Wolf.)

culmarum on root extension and root hair formation in barley is also inversely related to the substrate pH (Katouli and Marchant, 1981).

In poorly aerated soils high in organic matter and at high temperatures, an abnormally large amount of ethylene may accumulate, especially in the rhizosphere. Ethylene is produced by microbial activity as well as by the roots themselves. Although low concentrations of ethylene may stimulate root growth, higher concentrations (1–10 μl/per liter of soil atmosphere*) in most plant species inhibit root extension, although they stimulate the initiation of lateral root (Jackson *et al.*, 1981) and aerenchyma formation in the root cortex (Section 16.4). Higher concentrations of ethylene are found in compacted soils of vineyards with chlorotic grapevines (Perret and Koblet, 1981). It has been postulated by Perret and Koblet (1981) that inhibition of root extension by ethylene is mainly responsible for insufficient iron uptake and thus for chlorosis in grapevines growing in compacted soils.

**Soil atmosphere* refers to the gaseous phase of the soil.

14.2.5 Rhizosphere Microorganisms

Rhizosphere microorganisms may stimulate or inhibit root growth depending on the type of microorganism and on environmental conditions. Inoculation with a complex rhizosphere microflora, or growth in nonsterile compared to sterile culture, usually has inhibitory effects on root growth (Schönwitz and Ziegler, 1982; Fußeder, 1984) and root hair formation (Bowen and Rovira, 1961; Rovira *et al.*, 1983). As shown in Table 14.6, however, these results cannot be generalized for the rhizosphere microflora in total. For example, some soil fungi that are not regarded as pathogens retard root growth, whereas others stimulate it.

Table 14.6
Effect of Some Naturally Occurring Soil Fungi on Root Growth[a]

Fungus	Wheat	Pea	Rape
Emericellopsis minima	58	76	64
Phoma exigua	75	64	84
Prichocladium opacum	92	105	104
Verticillium lateritium	104	121	102

[a]Values represent the root weight of inoculated plants as a percentage of noninoculated controls. Based on Domsch (1969).

Inhibition may be caused by the production of phytotoxins; stimulation by the mobilization of mineral nutrients (Chapter 15), N_2 fixation (Chapter 7), and/or the production of phytohormones. The morphogenetic effects of N_2-fixing microorganisms on the root system, such as the stimulation of root extension by *Azotobacter* (Harper and Lynch, 1980) or of root hair formation by *Azospirillum* (Martin and Glatzle, 1982), are probably hormonal (Section 7.5.2). These effects are advantageous for the uptake of mineral nutrients with low mobility in soils. The changes in root morphology and in nutrient uptake brought about by the infection of roots with mycorrhizae are discussed in Chapter 15.

14.3 Soil Physical Factors

14.3.1 Mechanical Impedance

When soils are compacted, the bulk density increases and the number of larger pores is reduced. Roots cannot decrease in diameter in order to enter

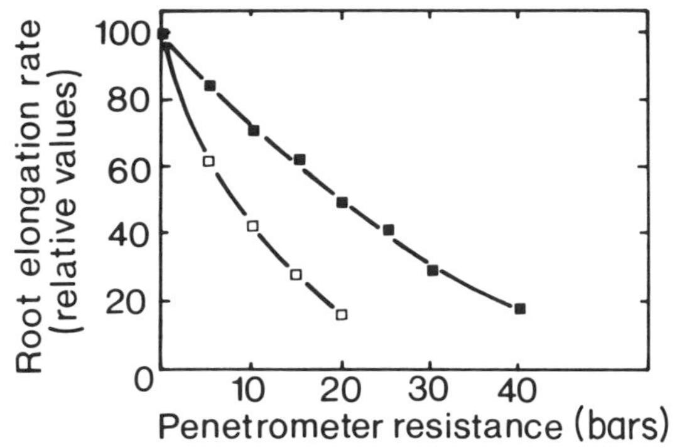

Fig. 14.7 Root elongation rates as a function of soil strength (resistance to root penetration). Key: ■, peanut; □, cotton. (Based on Taylor and Ratliff, 1969.)

pores smaller than the diameter of the root tip (Wiersum, 1957). Thus roots growing through compacted soil have to displace soil particles. The forces necessary for this displacement readily become limiting, and root elongation rates fall (Fig. 14.7). There may be various reasons for differences in root responses to soil strength among plant species, the difference in the average diameter of the roots being one of them.

Depending on the soil type and the soil moisture level, in bulk densities of up to about 1·3–1·4 (expressed as megagrams of dry soil per cubic meter) this inhibition in root elongation is not necessarily correlated with inhibited uptake of mineral nutrients—their uptake might even be increased, as had been shown for potassium (Talha *et al.*, 1979). This increase is probably in part due to the increase in the buffer power for nutrients at higher bulk densities (Silberbush *et al.*, 1983). Furthermore, inhibition of the extension of the main axes of roots by mechanical impedance is usually correlated with enhanced development of lateral roots, which have a smaller diameter than the main axes, leading to a dense superficial root system (Fig. 14.8). This modification of root morphology does not necessarily decrease the root surface area and may thus be without effect on nutrient and water uptake, provided that sufficient amounts of these are available in the restricted rooting zone (Goss, 1977; Shierlaw and Alston, 1984). During the growing season in temperate and semiarid climates, the latter is more likely to be the exception than the rule. In any case plants growing in compacted soils are more susceptible to drought (Scott-Russell and Goss, 1974); moreover, the requirements for readily soluble mineral nutrients are higher in the topsoil. For example, a maximum yield of barley was obtained in a soil of low bulk density at levels of "available" phosphorus of 18 mg phosphorus per 100 g soil, compared with 30 mg per 100 g soil with high bulk density (Prummel, 1975).

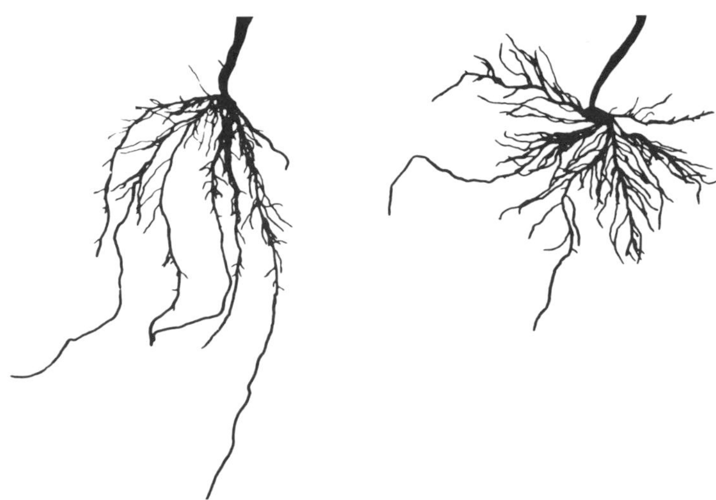

Fig. 14.8 Root system of young barley plants grown in the field in a soil with different bulk density. (*Left*) $1{\cdot}35\ g/cm^3$; (*right*) $1{\cdot}50\ g/cm^3$. (From Scott-Russell and Goss, 1974.)

Inhibition of root extension in compacted soils is not necessarily caused by mechanical impedance only; other factors may change simultaneously, which can affect extension growth:

Dry soils
(a) Increase in mechanical impedance ↘
(b) Decrease in soil water potential → Decrease in extension growth

Wet soils
(a) Decrease in aeration
(b) Accumulation of phytotoxins ↗ Decrease in extension growth

A fall in soil water potential may affect elongation via osmotic water uptake. Roots, however, are quite capable of increasing the internal osmotic potential and compensating for a decrease in soil water potential, at least down to −15 bar (Greacen and Oh, 1972). The rate of root elongation is usually reduced by much lower external pressure (Fig. 14.7). Using a solid substrate (glass ballotini) with an unlimited supply of mineral nutrients, water, and O_2, Scott-Russell and his group at the Letcombe Laboratory (Great Britain) demonstrated that a pressure of only 0·5 bar (50 kPa) required to enlarge the pores of the substrate was sufficient to inhibit root extension in barley by 80% (Scott-Russell and Goss, 1974). An example of such an effect on root morphology is shown in Table 14.7. Extension of the main axes was drastically reduced; lateral root growth was initiated a few

Table 14.7
Effects of Mechanical Impedance on the Development and Extension of Seminal Roots and Primary Lateral Roots of Barley[a]

	Applied pressure (kPa)	
	0	50
Seminal roots		
Mean length (cm)	8·6	2·0
Distance of youngest lateral root from the apex (mm)	30	4
Lateral roots		
Mean number per root	19	10
Number per centimeter of root	3·5	6·7

[a]Based on Goss (1977).

millimeters behind the apex, and the density of these roots was higher than normal per centimeter of root length.

That physical phenomena do not exert the only influence on elongation is evidenced by (*a*) the low external pressure required to inhibit elongation, (*b*) the role of the root cap as a sensor (see Section 14.1) and (*c*) the fact that after stress is relieved it usually take 3 days for extension to increase again (Goss, 1977). The stress factor of mechanical impedance seems to be transformed on the root cap cells into a hormonal signal which leads to a decrease in cell expansion of the main axes and depression of lateral root initiation.

14.3.2 Moisture

The effect of soil water content (water potential) on root growth is expressed by a typical optimum curve for which chemical and physical factors in the soil are the responsible parameters. In dry soils mechanical impedance is the dominant stress factor, although chemical factors, such as a loss of soil–root contact (roots and soil particles shrink as they dry; Faiz and Weatherley, 1982), may also be involved, which may lower the supply of mineral nutrients (Chapter 13). In saline soils at the root surface, salt concentrations can be much higher and thus water potential much lower than in the bulk soil (Section 15.2). In addition, ion imbalance in the rhizosphere soil solution may have direct negative effects on root metabolism and growth.

Increasing soil moisture relieves these stress factors; there is either minimal or no impairment of extension growth by mechanical impedance, and nutrient supply via diffusion and mass flow is maximal. Soil aeration and the accumulation of phytotoxins then become the dominant stress factors in root growth unless they are counterbalanced by morphological and physiological adaptation by the roots (Section 16.2).

14.3.3 Temperature

Root growth is often limited by low or high soil temperatures. The temperature optimum varies among species and tends to be lower for root growth than for shoot growth (Brouwer, 1981). Optimum temperatures are usually in the range of 20 to 25°C; for cotton the value is ~30°C (Pearson *et al.*, 1970) and that for cassava ~35°C (B. Sattelmacher, personal communication). Minimum temperatures for the root growth of species native to warm climates is usually between 8 and 15°C. A typical temperature response curve is shown in Fig. 14.9. The detrimental effects of high temperatures on root growth are probably related to an insufficient carbohydrate supply to the root meristems (Cumbus and Nye, 1982), although the morphology of heat-stressed roots is similar to that of roots growing in compacted soils, which would indicate hormonal effects.

Establishing a stand of vigorous plants in early spring is often a major problem in temperate zones. Low shoot growth rates at low soil temperature

Fig. 14.9 Effect of root zone temperature on root morphology and shoot growth of potato seedlings. (By courtesy of B. Sattelmacher.)

seem to be mainly a matter of insufficient supply of mineral nutrients and water to the shoot. At low soil temperatures, therefore, higher concentrations of chemically available nutrients are required per unit soil volume (Chapter 13) in order to supply the shoot with the same amount of nutrients as at higher temperatures. The effect of low temperatures on the nutrient supply to the shoot is primarily caused by inhibited root growth and to a much lesser degree is a direct effect on the uptake kinetics of nutrients such as phosphorus (Mackay and Barber, 1984). At low soil temperature (e.g., 10 compared to 25°C) soil compaction also inhibits root growth of several cereal species more severely (Al-Ani and Hay, 1983). For a review of the extended classical work on the effects of soil temperature on root and shoot growth the reader is referred to Richards *et al.* (1952). Interpretations of the effects of root temperatures on the shoot growth however, also have to take into account the involvement of root-borne phytohormones (Chapter 6). At low root temperatures the root export of CYT to the shoot is substantially depressed (Atkin *et al.*, 1972). In grapevine roots at a temperature of 12°C the concentration of CYT in the xylem sap drops to nearly one-half the value at 25°C, and in addition qualitative changes occur in the CYT spectrum (Zelleke and Kliewer, 1981).

14.4 Shoot/Root Ratio

The ratio of shoot to root growth is modified by external factors and also changes during ontogenesis (Brouwer, 1967). Competition for photosynthates between shoots and roots (including reproductive and storage organs) plays an important role. For example, with a decrease in light intensity to 30% in grapevines, the dry weight (grams per plant) of the shoots and roots decreased from 20·0 to 17·8g and from 9·6 to 4·0g, respectively; that is, the shoot/root ratio shifted from about 2·1 to 4·5. On the other hand, when the supply of mineral nutrients is insufficient, as a rule the roots become the dominant sink for photosynthates and root growth is favored over shoot growth, as shown in Table 14.8 for nitrogen. With increasing nitrogen supply, shoot dry weight increases faster than root dry weight; that is, ultimately, a relatively small root system has to supply a large shoot with mineral nutrients and water.

The involvement of phytohormones in the regulation of shoot growth by rootborne phytohormones has to be considered when other unfavourable soil conditions, besides low soil temperatures, exist for root growth and activity. Waterlogging, for example, drastically decreases the export of CYT (Burrows and Carr, 1969) and gibberellins (Reid *et al.*, 1969) within 1 or 2 days. Simultaneously with waterlogging, shoot elongation is drastically

Table 14.8
Effect of Nitrogen Level on Dry Weight, Shoot/Root Ratio, and Total Root Length per Plant[a]

Nitrogen supply (mg/liter)	Dry weight (g/plant)		Shoot/root ratio	Root length (m)
	Shoot	Root		
0	0·24	0·38	0·63	4·7
21	0·75	0·84	0·89	6·2
42	1·34	1·30	1·03	6·8
105	2·40	1·97	1·25	8·1
210	4·49	2·89	1·55	10·2

[a]Experiment was performed on 17-day-old maize plants. Based on Maizlich *et al.* (1980).

depressed and leaf senescence enhanced. Foliar sprays containing CYT can counteract at least some of the negative effects of waterlogging on shoot growth (Reid and Railton, 1974).

A large proportion of the photosynthates of plants are required for root growth and root metabolism. In annual crop species this figure is ~30% (Chapter 15), and in perennials the proportion can even be higher. In a scotch pine stand in Sweden, more than 50% of the photosynthates produced were used for the development of fine roots (Persson, 1979). This seems to be a high cost in terms of the carbon economy of plants. At least during vegetative growth, however, photosynthesis usually is not carried out at full capacity (Chapter 5). Furthermore, a large root system and a high rate of root replacement, ensuring a high proportion of young roots, are advantageous for water and mineral nutrient uptake, particularly in soils with low fertility. On the other hand, when there is a large and continuous supply of water and nutrients only 10–20% of the root system may be required (Greenwood, 1983) as shown by plant production in water culture (e.g., the "Nutrient Film Technique") in commercial horticulture.

15

The Soil–Root Interface (Rhizosphere) in Relation to Mineral Nutrition

15.1 General

The conditions in the rhizosphere differ in many respects from those in the soil some distance from the root (bulk soil). Roots not only act as a sink for mineral nutrients transported to the root surface by mass flow and diffusion. In addition, they take up either ions or water preferentially, which may lead to the depletion or accumulation of ions (Chapter 13), they release H^+ or HCO_3^- which changes the pH, and they consume or release O_2, which may cause alterations in the redox potential. Root exudation may mobilize mineral nutrients directly, but it primarily provides the energy for microbial activity in the rhizosphere. These root-induced modifications are of crucial importance for the mineral nutrition of plants. Although the chemical properties of the bulk soil (e.g., the pH) are very important for root growth and mineral nutrient availability, the conditions in the rhizosphere and the extent to which roots can modify them are decisive for mineral nutrient uptake.

15.2 Ion Concentration in the Rhizosphere

In the rhizosphere there is usually a depletion of phosphate and potassium (Chapter 13). In soils low in available potassium, this can lead to disaggregation of polymineralytic shale particles and the accumulation of amorphous iron and aluminum oxyhydrates, indicating enhanced "weathering" of soil material at the soil–root interface (Sarkar *et al.*, 1979).

On the other hand, a greater uptake of water than of ions leads to ion accumulation in the rhizosphere. This can be predicted from calculations based on the model of solute transport by diffusion and mass flow to the root surface of those ions which are present in high concentrations in the soil solution (Section 13.2). This accumulation can also be demonstrated in the soil that one collects by carefully picking the roots from the soil and shaking

Table 15.1
Estimated Supply of Calcium to the Roots Compared with Calcium Uptake and Level of Soluble Calcium in the Rhizosphere[a]

Species	Water consumption (ml/2 plants)	Supply (mg Ca/ 2 plants)	Uptake (mg Ca/ 2 plants)	Difference	Soluble Ca[b] (μg/cm root length)
Capeweed (*Artotheca calendula*)	6	0·6	0·4	+0·2	+0·20
Wimmera ryegrass (*Lolium rigidum*)	28	2·8	0·6	+2·2	+1·41
Blue lupin (*Lupinus digitatus*)	80	8·0	9·0	−1·0	−0·56

[a]Reproduced from Barber and Ozanne (1970) by permission of the Soil Science Society of America.
[b]Compared with calcium in the bulk soil.

them. This soil is defined as *rhizosphere soil* or loosely adhering soil (~1–4 mm from the surface of the root axis). The soil remaining on the root is called *rhizoplane soil* or closely adhering soil (~0–2 mm from the surface of the root axis). The mechanical separation into these two fractions can only be tentative—usually the term rhizosphere soil includes both fractions. Whether or not ions accumulate in the rhizosphere depends on both transport by mass flow (transpiration rate) and rate of uptake by the roots. Plant species differ in both respects, as shown for calcium in Table 15.1. The three plant species were grown in the same soil with a calcium concentration in the soil solution of $2{\cdot}5 \times 10^{-3}\,M$. The calculated difference between supply to the root surface (water consumption) and rate of calcium uptake (measured) predicted either on accumulation or depletion of calcium in the rhizosphere, depending on the species. There was good agreement between the calculated data and the actual content of soluble calcium in the rhizosphere soil.

The accumulation of ions in the rhizosphere is closely related to the transpiration rate. This has been directly measured for sulfate in onion (Nye and Tinker, 1977). At sufficiently high Ca^{2+} and SO_4^{2-} concentrations in the soil solution, the precipitation of $CaSO_4$ at the root surface can be demonstrated (Malzer and Barber, 1975). Over a long period in soil-grown plants, these precipitations occasionally may form a solid mantle around the roots (*pedotubules*) a few millimeters or even more than 1 cm in diameter (S. A. Barber, 1974).

Ion accumulation in the rhizosphere is of particular importance in saline soils with high concentrations of water-soluble salts such as sodium chloride. As shown in Table 15.2 there is a concentration gradient for both chloride and sodium from the bulk soil to the root surface, and this gradient becomes

Table 15.2
Relationship between Water Uptake per Unit Length of Root and Sodium and Chloride Accumulation around Maize Roots[a]

Water uptake (transpiration, ml/cm × 10)	Chloride (mg/100 g soil)			Sodium (mg/100 g soil)			Electro-conductivity, close[d] (mmhos/cm)
	Bulk[b]	Loose[c]	Close[d]	Bulk[b]	Loose[c]	Close[d]	
0·38	31	41	58	22	34	41	1·38
0·46	36	43	65	28	33	45	2·28
0·82	43	66	97	36	49	68	3·79
0·95	44	64	128	38	57	90	5·02

[a]Based on Sinha and Singh (1974).
[b]Bulk soil.
[c]Loosely adhering soil.
[d]Closely adhering soil.

steeper as the transpiration rate increases. Accordingly, the electrical conductivity of the soil increases near the root surface, especially at high transpiration rates, and one can predict a decrease in the availability of water.

This accumulation of salt at the root surface is important for plant growth and irrigation in saline soils. Estimations of expected growth reduction of plants growing in saline soils are usually based on calculations of salt concentrations in soil-saturated extracts (Mass and Hoffman, 1977). The salt concentration in the soil solution under field conditions is estimated to be about two to four times higher than that in the saturation extract (soil paste). This, however, does not necessarily reflect the actual condition in the rhizosphere. In field experiments with onion (Schleiff, 1980–1981) the rhizosphere soil immediately after irrigation contained ~3·0–4·5 meq Cl^- per 100 g soil, which was three to five times higher than the concentration in the bulk soil. Four days later the concentrations were about 10 and 1 meq Cl^- per 100 g in the rhizosphere and bulk soil, respectively. Within these 4 days the osmotic potential in the rhizosphere soil solution increased from about −0·6 to −2·5 MPa. Thus water may be unavailable to plants long before the critical conductivity levels (see Section 16.6.3) are obtained in the bulk soil.

15.3 Rhizosphere pH and Redox Potentials

15.3.1 Cation–Anion Uptake and Rhizosphere pH

Root-induced changes in the rhizosphere pH are brought about by the excretion of H^+ or HCO_3^-, by the evolution of CO_2 by respiration, and by

Table 15.3
Effect of Nitrogen Form (Ammonium Sulfate or Calcium Nitrate) on the pH of the Bulk Soil and Rhizosphere Soil and on Phosphorus Uptake by Soybean Plants[a,b]

	Soil pH after 3 weeks					
	Bulk soil with		Rhizosphere soil with			
Bulk soil pH	NH_4^+	NO_3^-	NH_4^+	NO_3^-	Rhizosphere soil pH[c]	Phosphorus in shoots (% dry wt)[c]
5·2	5·0	5·4	4·7	6·6	4·7	0·21
6·3	5·9	7·0	5·6	7·1	6·2	0·16
6·7	6·6	7·0	6·3	7·2	7·1	0·13

[a]Based on Riley and Barber (1971).
[b]Bulk soil: unfertilized soil without plants; pH adjusted with $Ca(OH)_2$. Rhizosphere soil: strongly adhering soil.
[c]Recalculated data, compiled from NH_4^+ and NO_3^- supply.

the presence of root exudates (organic acids and amino acids). Major changes in pH induced by root exudates are more the exception than the rule. In aerated soils, CO_2 is also of minor importance for the rhizosphere pH since it rapidly diffuses away from the root through air-filled pores. Moreover, HCO_3^- is confined to the soil solution, where its mobility is very low. Changes in rhizosphere pH are therefore brought about predominantly by the net excretion of H^+ or HCO_3^- from the roots. This excretion is related to the cation/anion uptake ratio (Section 2.9)—and is an indication of the need to approach electrochemical balance both in root cells and in the external solution. The form of nitrogen supply has the most prominent influence on the cation/anion uptake ratio (Section 2.9) and thus on rhizosphere pH. Nitrate supply is correlated with a higher rate of HCO_3^- net excretion than of H^+ net excretion, and ammonium supply with the reverse.

The effects of different forms of nitrogen supply to plants on the soil pH are shown in Table 15.3. The effects on bulk soil pH are small compared with those on the rhizosphere soil, where the pH differences were as high as 1·9 units. As would be expected from the relationship between solution pH and phosphorus uptake (Section 2.6), there is an inverse relationship between the rhizosphere pH and the phosphorus content in the shoot dry weight. Thus, ammonium-fed plants usually have higher phosphorus contents in the shoots than nitrate-fed plants (Soon and Miller, 1977). Differences in rhizosphere pH are probably responsible for the fact that iron deficiency in plants grown in calcareous soils ("lime chlorosis") can be alleviated by ammonium fertilizer application but becomes more severe with nitrate fertilizer application (Farrahi-Aschtiani, 1972).

Average values of rhizosphere pH can be somewhat misleading and may lead to inaccurate conclusions about the nutrient dynamics in the rhizosphere. For example, even within the root system of an individual plant, distinct pH gradients exceeding 2 pH units are sometimes observed along the root axis (Marschner and Römheld, 1983; Römheld, 1984).

Striking differences in the rhizosphere pH exist among plant species growing in the same soil and supplied with nitrate nitrogen. Buckwheat (Raij and Diest, 1979) and chickpea (Marschner and Römheld, 1983) have a very low rhizosphere pH compared, for example, with that of wheat or maize. These genotypical differences reflect differences in cation/anion uptake ratios (Raij and Diest, 1979; Bekele *et al.*, 1983) and/or in the dominant sites of nitrate reduction (roots or shoots). As the cation-exchange capacity of plants increases (e.g., dicots > monocots), the ratio of cation to anion uptake usually increases, with a corresponding lowering of rhizosphere pH (Smiley, 1974). The utilization of sparingly soluble phosphate fertilizer such as rock phosphate therefore differs among plant species according to their cation/anion uptake ratio (Raij and Diest, 1979; Bekele *et al.*, 1983). A decrease in rhizosphere pH also occurs in response to iron deficiency (Marschner *et al.*, 1982) or phosphorus deficiency (Grinstead *et al.*, 1982). In both cases an increase in the ratio of cation to anion uptake is the responsible factor.

When rape plants were grown in a phosphorus-deficient soil (Table 15.4), a decrease in the phosphorus concentration in rhizosphere soil solution was associated during the first 2 weeks with an increase in rhizosphere pH. Thereafter, this relationship was reversed. The changes in rhizosphere pH were closely related to the cation/anion uptake ratio, which increased in the older plants. This shift in response to phosphorus deficiency has striking similarities to the response to iron deficiency in dicotyledonous species (Chapter 2).

Table 15.4
Time Course of Dry Matter Production, Phosphorus Concentration, and pH of the Rhizosphere and Ion Uptake of Rape Plants Grown in a Low-Phosphorus Soil[a]

Age of plants (days)	Dry weight (g/vessel)	Phosphorus concentration in rhizosphere soil solution (μM)	Rhizosphere pH	Uptake of cations and anions[b]
0	—	5·17	6·1	—
7	0·16	2·56	6·3	Cat < An
14	0·89	0·82	6·5	Cat < An
20	1·89	1·40	5·3	Cat > An
28	3·69	2·47	4·3	Cat > An

[a]Based on Grinsted *et al.* (1982) and Hedley *et al.* (1982).
[b]Nitrogen supplied as $Ca(NO_3)_2$.

15.3.2 Nitrogen Fixation and Rhizosphere pH

Plants that meet their nitrogen requirement by N_2 fixation rather than nitrate uptake take up more cations than anions since uncharged N_2 enters the root. Thus, the cation-anion uptake ratio of plants with N_2 fixation is comparable to that of plants with ammonium nutrition. The consequences of the different cation-anion uptake ratio in alfalfa are reflected in differences in the acidity (net release of H^+) and alkalinity (net release of HCO_3^- or OH^-) and in the rhizosphere pH (Table 15.5). The capacity of plants to utilize phosphorus from rock phosphate is therefore higher in N_2-fixing plants than in nitrate-fed plants. In soybean, N_2-fixing plants also had higher iron and manganese contents than did nitrate-fed plants and did not show iron-deficiency symptoms (Wallace, 1982).

Table 15.5
Effects of Nitrogen Sources on Acidity and Alkalinity Generated by Roots of Alfalfa Plants, on the Soil pH, and on the Utilization of Rock Phosphate[a]

Treatment		Acidity	Alkalinity		Phosphorus	Yield
Nitrogen source	Rock phosphate	(meq/g dry wt)	(meq/g dry wt)	Soil pH (H_2O)	uptake (mg/pot)	(g dry wt/pot)
Nitrate	−	—	1·1	6·3	1	2·5
Nitrate	+	—	0·8	7·3	23	18·8
N_2	−	0·5	—	6·2	4	4·7
N_2	+	1·4	—	5·3	49	26·9

[a]From Aguilar S. and Diest (1981).

Differences in rhizosphere pH similar to those shown in Table 15.5 were observed in long-term experiments comparing N_2-fixing legumes (red clover) with nonlegumes (ryegrass) supplied with ammonium nitrate (Mengel and Steffens, 1982). In ryegrass the rhizosphere pH remained constant at ~7·0 for more than 1 year, whereas in red clover the pH gradually fell to ~4·4. At the final harvest, in red clover the concentration (milliequivalent per 100 g shoot dry weight) of cations was 1402, whereas the anion concentration was 425. In pea, the growth on an acid soil of pH 4·0 was much more inhibited in the N_2-fixing plant than in the nitrate-fed plants, most likely a reflection of the different rhizosphere pH values (Beusichem and Longelaan, 1984). Nye (1981) has developed a model for predicting changes in rhizosphere pH depending on the source of nitrogen as well as the pH buffering capacity and moisture content of the soil. In the long run, symbiotic nitrogen nutrition affects the acidification also of the bulk soil and thus the lime requirement. In alfalfa, N_2 fixation required for an annual shoot dry matter production of 10 tons/ha would produce soil acidity equivalent to 600 kg $CaCO_3$ per hectare (Nyatsanaga and Pierre, 1973). As shown in a review paper by Haynes (1983) without liming a distinct negative correlation exists between age of legume pastures and the soil pH. In humid climates the loss of symbiotically fixed nitrogen from the system through leaching of nitrate and an equivalent amount of cations such as magnesium and calcium may contribute to soil acidification under leguminous pastures. A similar impact on the long-term soil acidification by N_2 fixation can be observed in forest ecosystems when the pH under red alder is compared with douglas fir (Van Miegroet and Cole, 1984).

15.3.3 Redox Potentials

Reliable data on redox potentials in the rhizosphere are lacking; it is difficult to make precise measurements even in the bulk soil. Also, in well-aerated arable soils there is a mosaic of anaerobic microsites varying in location and size. Most likely, these are more abundant in the rhizosphere because of O_2 consumption by both microbial and root respiration. Higher rates of N_2 fixation and of denitrification in the rhizosphere than in the bulk soil indicate an abundance of such microsites. Although in upland soils root exudates themselves may directly affect redox processes in the rhizosphere (see Section 15.4), the main root-induced changes are brought about indirectly via the supply of organic substrates for microbial activity (Section 15.6.2).

As the soil water content increases, redox potentials tend to decrease until in submerged soils negative values are obtained. The fall in redox potential is correlated with a range of changes in the solubility of mineral nutrients (e.g., manganese and iron, occasionally phosphorus) and also the accumulation of phytotoxic organic solutes (Chapter 14). The interrelationships among nutrients in submerged soils have been reviewed thoroughly by Ponnamperuma (1972). Plants adapted to waterlogging and to submerged soils maintain high redox potentials in the rhizosphere via O_2 transport from the shoots to the roots and the release of O_2 into the rhizosphere (Fig. 15.1).

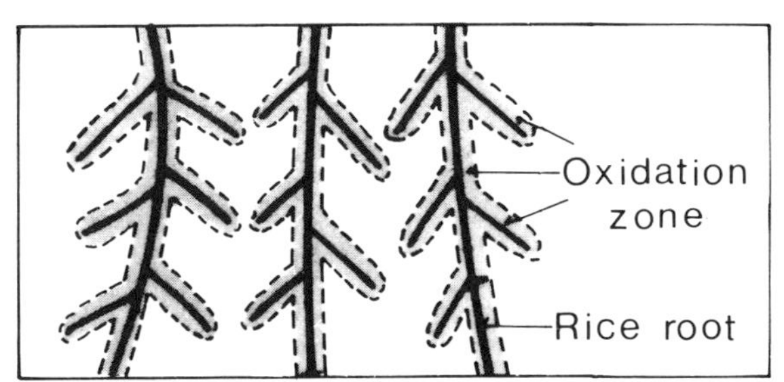

Fig. 15.1 Schematic representation of wetland rice roots in submerged soil.

The oxidation power* of rice roots relies on the release of free O_2 into the rhizosphere and on an enzymatic reaction at the root surface (Ando *et al.*, 1983). The latter process is reflected by a heavy precipitation of FeOOH at the external surface of rhizodermal cells (Chen *et al.*, 1980). The oxidation capacity is a function of both the rate of O_2 transport into the roots and the rate of O_2 consumption in the roots and the rhizosphere, the latter process being strongly affected by mineral nutrition (Section 15.6.2).

15.4 Root Exudates

15.4.1 Characterization and Amounts

Growing roots release an appreciable amount of organic carbon into the rhizosphere. Three major components are involved: low-molecular-weight organic compounds (free exudates), high-molecular-weight gelatinous material (mucilage), and sloughed-off cells and tissues and their lysates (Fig 15.2). In contrast to lysates, free exudates and mucilage are released from healthy roots either as a secretion (i.e., by metabolically coupled transport) or as leakage. It is neither possible nor important, however, strictly to differentiate the roles of compounds secreted or leaked into the rhizosphere.

In most studies no clear distinction is made between the three major components, and also the steady supply of decomposed root cap cells, dead root hairs, and other cells included in the category of root exudates; in these cases the term *rhizodeposition* would be more appropriate (Section 15.5.1). In a nonsterile environment the situation is complicated by the continuous turnover of organic material by rhizosphere microorganisms.

The amount of organic carbon released into the rhizosphere expressed as a fraction of the total dry matter production of young plants varies over a wide range, from less than 1% to more than 30%. The proportion decreases with plant age, in maize from ~7 to ~5% in 4 and 12-week old plants, respectively (Haller and Stolp, 1985). Various forms of stress, such as nutrient deficiency (Trolldenier and Rheinbaben, 1981b), water deficiency, anaerobiosis, or mechanical impedance increase the amount released. According to Barber and Gunn (1974), under sterile conditions the release of soluble organic compounds from roots increased to 9% of the root dry weight in solid substrates, where stress was induced by mechanical impedance, compared with ~5% in nutrient solution. Corresponding differences

*The term *oxidation power* refers to the capacity of roots to maintain a high redox potential at the root surface, despite a very low (negative) redox potential in the bulk soil.

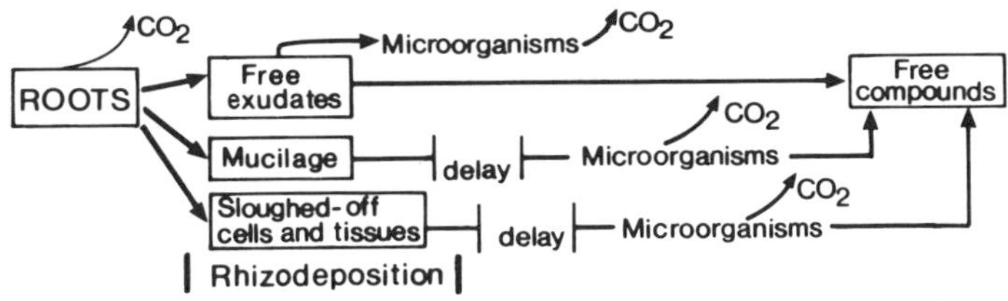

Fig. 15.2 Model of carbon flux in the rhizosphere. (Modified from Warembourg and Billes, 1979.)

were found in root exudation between the two substrates by a factor of 3 for sugars and vitamins (Schönwitz and Ziegler, 1982), and a factor of 10 for phenolics (Lameta D'Arcy, 1982). The presence of microorganisms in the rhizosphere increases the release of organic carbon considerably (Schönwitz and Ziegler, 1982). Over a 3-week period, roots of wheat seedlings released 7–13% of the carbon assimilated when grown in the absence and 18–25% when grown in the presence of microorganisms (Barber and Martin, 1976).

15.4.2 Low-Molecular-Weight Organic Solutes

The main constituents of root exudates (free exudates; see Fig. 15.2) are sugars, organic acids, amino acids, and phenolic compounds. Quantitatively, sugars and organic acids are generally the predominant compounds. These proportions vary considerably, however, depending, for example, on plant species and nutritional status. With potassium deficiency, the proportions of sugars and organic acids in the exudates of maize are shifted in favor of the latter (Kraffczyk *et al.*, 1984), whereas under iron deficiency in grasses specific amino acids dominate in the exudates (Section 9.1.6). It should not be overlooked, however, that in studies under nonsterile conditions the chemical composition of root exudates may have been altered by rhizosphere microorganisms (Fig. 15.2).

Sugars have only minor effects on the mobilization of mineral nutrients. There are several possible ways in which root exudates affect nutrient mobilization and uptake (Fig. 15.3). The increase in the solubilization of MnO_2 by root exudates (Bromfield, 1958; Uren, 1982) seems to be due mainly to organic acids (Godo and Reisenauer, 1980). Mobilization cannot be explained simply by a pH effect but is brought about by reduction and perhaps also by subsequent chelation. At a given pH, root exudates from wheat dissolved 10–50 times more manganese from MnO_2 than did a buffer solution only (Godo and Reisenauer, 1980). Malic acid is an important component of these root exudates. During the oxidation of 1 mol of malic acid to CO_2 at the surface of MnO_2, 6 mol of Mn^{2+} are released (Jauregui

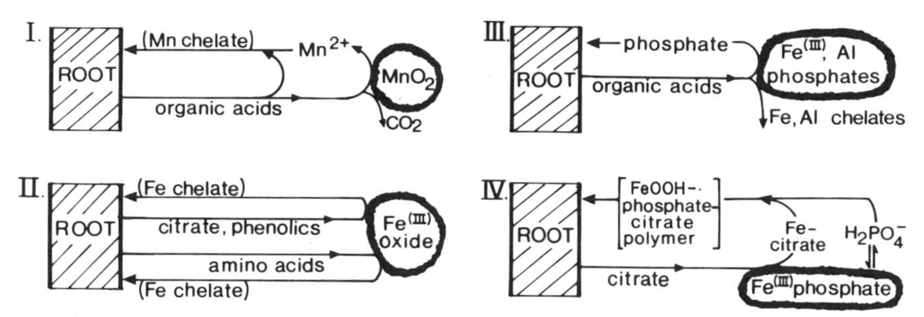

Fig. 15.3 Schematic representation of various mechanisms in the rhizosphere for the solubilization of sparingly soluble inorganic compounds by root exudates in relation to the mineral nutrition of plants (see text).

and Reisenauer, 1982); chelation of Mn^{2+} prevents its reoxidization and increases the mobility of reduced manganese in the rhizosphere (Fig. 15.3, mechanism I). The reduction of Fe(III) oxide by malate is not as great as that of MnO_2 and is enhanced by acidity and low levels of MnO_2. Phenolics contribute to the promotion of manganese reduction (Bromfield, 1958; Marschner *et al.*, 1982) and particularly iron reduction by root exudates (Section 2.8). In many dicotyledonous species, iron deficiency leads to an increase in the release of phenolics, which enhance the mobilization of both iron and manganese in the rhizosphere. This effect is probably followed by chelation, which facilitates transport to the uptake sites at the plasma membrane (Fig. 15.3, mechanism II). Chelation of inorganic Fe(III) is the dominant mechanism by which certain nonproteinogenic amino acids in the root exudates of grasses enhance solubilization and uptake of Fe(III) under iron deficiency (Sections 2.8 and 9.1).

The organic acids in root exudates are also considered important for the solubilization of sparingly soluble inorganic phosphates. Root exudates of wheat contain relatively high concentrations of 2-ketogluconic acid; this is the main component responsible for the acidification of the rhizosphere by the root exudates and for the increase in phosphate uptake from rock phosphate (Moghimi *et al.*, 1978). The means by which organic acids mobilize phosphate are not confined to lowering the rhizosphere pH. Citrate, for example, is well known for its capacity to desorb phosphate from sesquioxide surfaces by anion exchange (Parfitt, 1979). As a rule, however, a combination of both desorption and chelation is responsible for phosphate mobilization from iron and/or aluminum phosphates (Fig. 15.3, mechanisms III and IV). Citric and malic acids form relatively stable chelates with Fe(III) and aluminum, thereby increasing the solubility and rate of phosphorus uptake. The significance of this effect for phosphorus nutrition should not be overestimated since only a small fraction of the total root exudates is usually made up of organic acids. However, in certain plant species adapted to acid mineral soils with extremely low phosphorus availability, such as *Eucalyptus* spp. (Mulette *et al.*, 1974) and tea plants (Jayman and Sivasubramanian, 1975), this mechanism is of major importance for phosphorus nutrition. As a side-effect, chelation of aluminum by organic acids alleviates the harmful effects on plants of high aluminum concentrations in acid mineral soils (Section 16.3). The annual species *Lupinus albus* also releases large quantities of citrate, representing between 1·5 and 12% of the root dry weight (Gardner *et al.*, 1983a). It has been postulated that citrate reacts in soil, forming ferric hydroxyphosphate citrate polymers which diffuse to the root surface, where Fe(III) is reduced and phosphate is taken up by the roots (Fig. 15.3, mechanism IV). This mechanism would enable lupins to use citrate as a "shuttle" in the rhizosphere for the mobilization of iron phosphates and the acquisition of phosphorus (Gardner *et al.*, 1983a).

The capacity of root exudates for the mobilization of sparingly soluble inorganic compounds also depends on their concentration per unit volume of soil. Thus, it can be postulated that with increasing density of root hairs or of lateral roots (Section 15.5) the chances are increased of root exudates mobilizing these compounds (Gardner *et al.*, 1983c). This spatial effect of root exudates is opposite to that described for the acquisition of nutrients such as potassium and (inorganic) phosphorus which are transported by diffusion along the concentration gradient to the root surface, and where the overlapping of depletion zones decreases efficiency (Section 13.3).

15.4.3 Mucilage and Mucigel

Root surfaces, particularly in apical zones, are covered by high-molecular-weight gelatinous material (*mucilage*), which consists mainly of polysaccharides and polygalacturonic acids. This material is secreted by the root cap cells and is also released by epidermal cells (Vermeer and McCully, 1981). The production of the mucilage is positively correlated with the root growth rate (Trolldenier and Hecht-Buchholz, 1984). In nonsterile media it also includes substances produced by bacterial degradation of the cell walls (Rovira *et al.*, 1983). In soil-grown plants the mucilage is usually invaded by microorganisms, and both organic and inorganic soil particles are embedded in it. This mixture of gelatinous material, microorganisms, and soil particles is termed *mucigel* (Jenny and Grossenbacher, 1963).

Several functions have been proposed for the mucigel of soil-grown plants (Oades, 1978), including protecting roots from desiccation, acting as a lubricant, and improving root–soil contact, especially in dry soils. A contact between soil and the root surface containing high concentrations of polygalacturonic acids in the mucilage can be of major importance for mineral nutrition under certain circumstances. A close contact with sufficient soil material can lead to solubilization of enough micronutrients to meet the needs of plants. Finely divided MnO_2 (Passioura and Leeper, 1963) or iron oxides (Jenny, 1965) are utilized by plants in this way. In the ill-defined transition zone at the soil-root interface effects take place which are different from those occurring in the free solution. These effects were formerly termed as *contact exchange* (Jenny and Overstreet, 1939), and later as *two-phase-effects* (Matar *et al.*, 1967), meaning that in this transition zone of the two phases the soil and the root surfaces interact in a particular manner. Today these effects are usually subsummized as rhizosphere effects. It was demonstrated by Matar *et al.* (1967) that in phosphorus-deficient soil, plants take up phosphorus which is not in equilibrium with the soil solution but is mobilized at the root–soil interface presumably via phosphate desorption from clay surfaces by the polygalacturonic acid component of mucilage (Nagarajah *et al.*, 1970). Two-phase effects supply only a minor fraction of the total demand for macronutrients such as phosphorus. The situation is different with micronutrients, however (Jenny, 1965). As shown in Table 15.6 maize plants grown in quartz sand with FeOOH take up sufficient iron for normal growth and chlorophyll formation. This iron was mobilized at the sand–root interface and was not in equilibrium with the free solution. This is indicated by the extremely low iron content of the plants grown in a nutrient solution in which iron was supplied only at the equilibrium concentration in the quartz sand. Most probably, in the quartz sand iron was mobilized at the sand-root interface by localized high concentrations of phytosiderophores in root exudates (Section 15.4.2).

Mucigel may also be of importance for micronutrient uptake from dry soils. Nambiar (1976a) presented evidence that plant roots growing through a layer of soil drier than the wilting point can take up a significant amount of zinc, provided that the roots have access to water elsewhere (e.g., in the subsoil). In dry soils more mucilage is released in response to mechanical

Table 15.6
Rhizosphere Effect on the Utilization of Sparingly Soluble FeOOH by Maize[a,b]

Treatment	Dry weight (g/6 plants)	Chlorophyll (mg/g dry wt)	^{59}Fe Content (μg/g dry wt)
Sand + ^{59}FeOOH	2·85	13·3	26
Nutrient solution (−Fe)	1·45	1·7[c]	0·3

[a]Based on Azarabadi and Marschner (1979).
[b]The plants were grown in a sand and water culture system connected by circulating an iron-free nutrient solution.
[c]Severe chlorosis.

Table 15.7
Effect of Mucilage on the Growth and Aluminum Content of Roots of Cowpea (cv. Tvu 354) Grown in Nutrient Solutions with or without Aluminum[a]

Treatment	Mucilage	Root growth (cm/day)	Al Content of root tips (0–5 mm): Roots (μg Al/25 tips)	Mucilage (μg Al/25 tips)	Roots (mg Al/g dry wt)	Mucilage (mg Al/g dry wt)
−Al	+	6·3	—	—	—	—
	−[b]	5·9	—	—	—	—
+Al[c]	+	4·8	12·4	16·6	2·1	16·6
	−[b]	2·1	20·6	3·6	3·2	14·5

[a]From Horst *et al.* (1982).
[b]Mucilage removed mechanically three times per day.
[c]±5 mg Al/liter.

impedance, and this probably facilitates zinc transport from the soil particles within the mucigel to the plasma membrane of root cells (Nambiar, 1976b).

Mucigel probably has an additional function in acid mineral soils, namely, protection of the root meristems from aluminum. All solutes entering the roots must pass through the mucigel–mucilage. The thickness of this layer increases from basal to apical root zones, where it is continuously produced. In the apical 5 mm of cowpea roots it may represent more than 10% of the root dry weight (Horst *et al.*, 1982). In roots exposed to aluminum a high proportion of the aluminum is bound specifically to the mucilage. On a dry weight basis the mucilage contains about eight times more aluminum than the root tissue (Table 15.7). Accordingly, removal of the mucilage leads to an increase in the aluminum content of the root tissue and severe inhibition of root extension.

Because of the stimulation of mucilage production by mechanical impedance, it can be assumed that in acid mineral soils the protective effect of the mucilage against aluminum penetration into the apical meristems is at least as high as that shown above for nutrient solution cultures.

15.5 Cluster-Rooted Plants

The efficiency with which a root system mobilizes mineral nutrients in the rhizosphere via exudates and acidification is partly a function of the local rooting density. In many plant species that are specially adapted to infertile, acid mineral soils, the root system is characterized by cluster of finely divided, highly branched sections of roots. The best known of these are the proteoid roots found in many genera of the Proteaceae. In tree species (e.g., *Banksiaa* spp.) the clusters form a dense mat near the soil surface beneath the

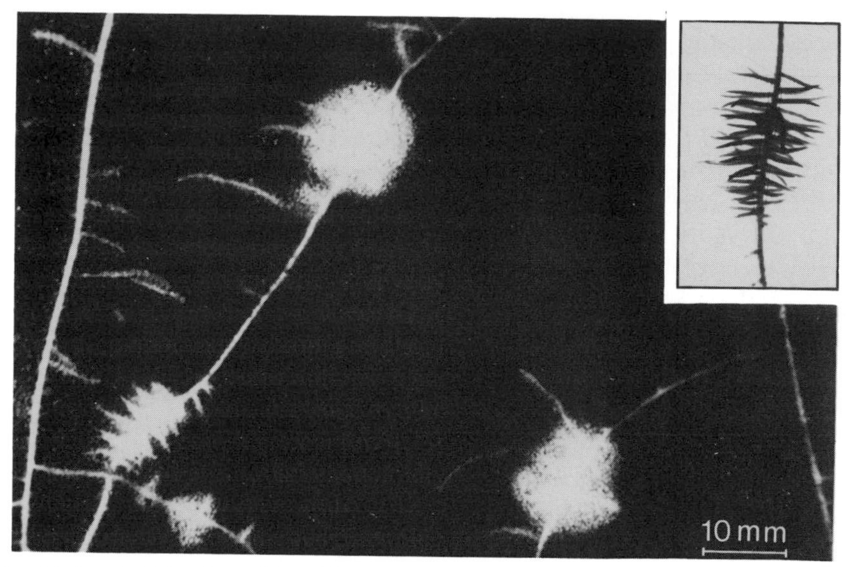

Fig. 15.4 Acidification of the rhizosphere in *Lupinus albus* L. (bromocresol purple as pH indicator); pH of the bulk soil and of the zones with proteoid roots (white areas): 6·0 and 4·5, respectively. Inset: Section of a lateral root with the typical dense cluster of rootlets. (By courtesy of V. Römheld.)

canopy after the onset of the rainy season (Lamont, 1982). Within the clusters, rhizosphere effects (see Section 15.4.3) are particularly important and intensive extraction of a limited soil volume is made possible by root exudates which would otherwise diffuse into a larger soil volume with corresponding dilution (Gardner *et al.*, 1982a, 1983c). There are an increasing number of reports of proteoid roots in legume species such as *Lupinus albus*, showing the formation of dense bottle-brush-like clusters of rootlets of limited growth (Fig. 15.4, inset) covered with a dense mat of root hairs (Gardner *et al.*, 1981). Rhizosphere soil from proteoid roots is more acidic and contains more reductants and chelating compounds than rhizosphere soil from other root zones (Gardner *et al.*, 1982b). An example of the differences in the acidification of different sections of rhizosphee soil is shown in Fig. 15.4.

The rhizosphere effect in proteoid root zones is particularly evident when *Lupinus albus* is compared, for example, with rape or buckwheat (Gardner *et al.*, 1982b). The displaced rhizosphere soil solution of rape and lupin mobilized 2·0 and 49 mg manganese (reduction of MnO_2), respectively, and 0·3 and 48 mg iron [chelation of Fe(III)], respectively. Root exudation of citrate is considered to be responsible for iron mobilization. The acquisition of phosphorus from iron phosphates by *Lupinus albus* (Fig. 15.3, mechanism IV) is presumably based on this mechanism of iron chelation by citrate.

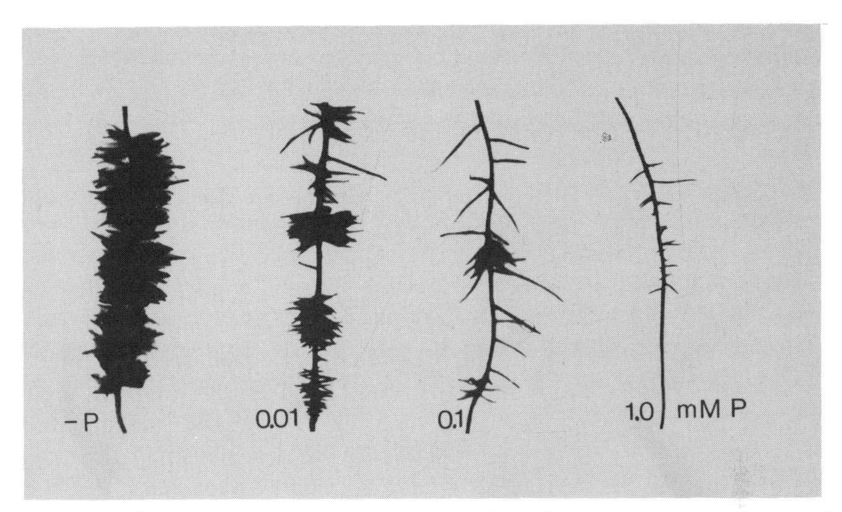

Fig. 15.5 Effect of the phosphorus concentration of the nutrient solution on the root morphology of *Lupinus albus* L. (By courtesy of V. Römheld.)

The proportion of the root system which develops proteoid roots is inversely related to the phosphorus supply (Fig. 15.5). In soils the corresponding decrease in efficiency of mobilization of sparingly soluble soil phosphates is obviously compensated for by higher rates of uptake of readily soluble fertilizer phosphorus (Table 15.8). The decrease in the proportion of proteoid roots is, however, closely correlated with a decline in the concentrations of reduced manganese and chelated iron in the rhizosphere, again indicating the role of proteoid roots in the mobilization of these two micronutrients.

Table 15.8
Effect of Phosphorus Supply on the Growth and Mineral Composition of *Lupinus albus* L. Grown in a Calcareous Soil, pH 8·6[a]

	Superphosphate added (mg/kg soil)		
	0	334	667
Shoot dry wt (g)	1·93	2·01	2·02
Proteoid roots (%)	46	28	16
Phosphorus in shoots (%)	0·17	0·20	0·23
Water extractable[b] (in rhizosphere soil)			
Manganese (mg/liter)	18·7	16·7	5·3
Iron (mg/liter)	1·5	0·9	0

[a]Based on Gardner *et al.* (1982c).
[b]Extracted by leaching soil with water.

15.6 Organic Carbon Supply and Microbial Activity in the Rhizosphere

15.6.1 Rhizodeposition and Population Density

Roots in nonsterile media support a large population of microorganisms on their external surface (the rhizoplane) and in the rhizosphere soil. The number of bacteria (but not necessarily fungi) is much larger at the rhizoplane and in the rhizosphere than in the bulk soil. The relative increase in the number of microorganisms is expressed as an *R/S* ratio, *R* and *S* being, respectively, the numbers per gram of soil in the rhizosphere and in the bulk soil. The *R/S* ratio varies greatly depending, for example, on plant species and plant age, being in the range of 5 to 50. Usually the number of microorganisms per unit surface area of roots increase basipedally, and on average they cover less than 10% of the rhizoplane (Rovira *et al.*, 1983). For some higher plants preferential invasion by certain species is advantageous (e.g., in the case of associative N_2 fixation), whereas for others it is a disadvantage (e.g., invasion by pathogens).

The population density of bacteria in the rhizosphere depends on the amount of exudate, mucilage, and sloughed-off cells (Fig. 15.2). In soil-grown plants between 75% (Haller and Stolp, 1985) and more than 85% (Barber and Martin, 1976) of the total organic supply for microbial activity in the rhizosphere is represented by sloughed-off cells and tissues. In soil-grown wheat plants under nonsterile conditions during a period of from 3 to 8 weeks, between 20 and 40% of the carbon translocated from the shoots to the roots was lost as organic carbon into the rhizosphere (Martin, 1977). Most of this carbon is utilized fairly rapidly by rhizosphere microorganisms. Assuming a microbial maintenance coefficient of 0·03 g/g × hr (turnover rate of microbial biomass), it has been calculated that 1 g of wheat root will probably have an attendant microflora in the rhizosphere weighing between 40 and 100 mg (Gardner *et al.*, 1983b). These exceptionally high values of microbial biomass deserve critical reexamination.

Despite the substantial release of organically bound carbon into the rhizosphere, there is even a net decrease in organic soil carbon in the rhizosphere, probably as a result of a destruction of (organic matter)-Fe/Al-(mineral particle) linkages by chelating agents released by the roots (Reid *et al.*, 1982) in combination with an enhancement effect of the low-molecular-weight organic root exudates on the microbial activity (Helal and Sauerbeck, 1984a). Such a priming effect on soil organic matter is well known following the incorporation of readily decomposable organic material (e.g., green manure) into the soil.

The large turnover of organic carbon by microbial activity in the rhizosphere has important implications for both the carbon balance of plants and mineral nutrient relationships in the rhizosphere. At maturity only a small fraction of the total root-derived organic carbon is retained in the root system. Most of it is lost by root respiration, exudation, and root decomposition. In a comprehensive study using sealed systems over an entire growing period, Sauerbeck and Johnen (1976) found the highest release of organic carbon by the roots of soil-grown wheat during the period of rapid vegetative growth. At harvest the following amounts of carbon (in grams per pot) were measured: root dry weight, 3·0; root respiration, 1·9; root exudation and decomposition, 7·6. In other words, more than twice as much organic carbon was released into the rhizosphere as was retained in the root system at harvest. Similar data have been obtained with other annual species (Sauerbeck *et al.*, 1981) and probably also apply to forest trees (Section 14.4).

15.6.2 The Role of Noninfecting Rhizosphere Microorganisms in Mineral Nutrition

Noninfecting microorganisms affect the mineral nutrition of plants through their influence on (*a*) the growth and morphology of roots, (*b*) the physiology and development of plants, (*c*) the availability of nutrients, and (*d*) nutrient uptake processes (Rovira *et al.*, 1983). Points (*a*) and (*b*) were discussed in Chapter 14. In this section some examples are given of the effects of noninfecting microorganisms on the availability and uptake of mineral nutrients. The role of N_2-fixing rhizosphere bacteria and associations is omitted since this was discussed in Section 7.5. The subject of the role of rhizosphere microorganisms on mineral nutrient uptake in general has been reviewed by Tinker (1984).

Examples of direct effects on mineral nutrient availability and uptake are relatively rare but well documented (Barber, 1978). In most instances it is difficult to predict these effects. For example, if root exudates consist mainly of organic acids with a large capacity for mobilizing manganese, iron, or phosphorus, utilization of the acids by rhizosphere microorganisms will have a detrimental effect on the uptake of these mineral nutrients. On the other hand, minor effects are to be expected if exudated sugars are utilized by rhizosphere microorganisms. Because the main carbon source for the rhizosphere microorganisms is sloughed-off cells and tissues, at least some of the end products of microbial activity (e.g., organic acids) may have effects on mineral nutrient mobilization that are similar to those of root exudates (Fig. 15.2). These interactions may well be one of the reasons for contradictory reports on the effects of rhizosphere microorganisms on manganese uptake by plants (Barber, 1978). Mineralization of the rhizodeposition also provides a possibility for "recycling" organically bound mineral nutrients such as phosphorus and nitrogen. The beneficial effects of rhizosphere microorganisms on the mineralization of organically bound soil phosphorus deserve particular attention (Helal and Sauerbeck, 1981), because this form of phosphorus represents between 10 and 60% of the total soil phosphorus in the topsoil and is usually considered to be poorly available to plant roots.

Although compared to the bulk soil, the total amounts of both inorganic and organically bound phosphorus decrease in the rhizosphere soil, nevertheless, there is an increase in some fractions such as phytate phosphorus, phospholipids, and bicarbonate-soluble organic phosphorus; simultaneously, the phosphatase activity increases by a factor of 2 to 3 (Helal and Sauerbeck, 1984a,b). Although these studies in nonsterile culture do not allow differentiation between the proportion of microbial phosphatase activity and that of root cells (particularly root hairs), a fairly high phosphatase activity is always associated with root cell walls, and this activity increases with plant age and in particular in response to phosphorus deficiency (Dracup *et al.*, 1984). In any case, the high turnover rate of organically bound phosphorus in the rhizosphere demonstrates the incompleteness of studies on adsorption and desorption of inorganic phosphorus in the bulk soil for characterization of the phosphorus dynamics in the rhizosphere.

There is an extensive literature on so-called phosphate-dissolving bacteria, and there has been considerable speculation as to whether such bacteria might allow increased utilization of soil and fertilizer phosphorus. Although these bacteria are capable of dissolving sparingly soluble inorganic phosphates (e.g., rock phosphates), it is unlikely that this mechanism operates to any greater extent in the rhizosphere (Tinker, 1980). These bacteria have to compete with other rhizosphere microorganisms for organic carbon as an energy substrate. They are therefore difficult to establish and maintain in high quantity in the rhizosphere. The situation is different, however, for microorganisms such as mycorrhizas which receive photosynthates directly from the root cells (see Section 15.7).

When a large organic carbon supply from the roots is combined with low O_2 partial pressure in the rhizosphere, competition between roots and microorganisms for O_2 may increase the loss of nitrate nitrogen through denitrification (Rheinbaben and Trolldenier, 1983, 1984). This is an important factor, especially in submerged soils; it may also decrease the oxidation power of roots of wetland rice (Fig. 15.1) and lead to an excessive uptake of manganese and iron. A deficiency of phosphorus or potassium increases root exudation (Rheinbaben and Trolldenier, 1983), the number of

Table 15.9
Potassium Nutritional Status of Wetland Rice Plants, Number of Bacteria, and Oxygen and Iron Concentration in the Nutrient Solution[a]

Supply of K^+	Number of bacteria ($\times 10^6$)	O_2 Concentration (mg/liter)	FeII Concentration (mg/liter)
High K^+	1244	17·0	1·0
Low K^+	1686	8·6	2·4
High K^+/no K^+[b]	2036	0·5	10·6

[a]Based on Trolldenier (1973).
[b]55 days high K^+, then 20 days without K^+.

bacteria, and O_2 consumption in the rhizosphere, as can be concluded from results on the potassium nutrition of rice plants (Table 15.9). An inverse relationship exists between bacterial number and O_2 concentration. The concentration of reduced iron therefore increases with an increasing number of bacteria. Since FeII is rapidly taken up by roots, the risk of iron toxicity in wetland rice ("bronzing") is increased by all factors which increase root exudation and microbial activity in the rhizosphere, such as a deficiency of potassium, phosphorus, or calcium (Ottow *et al.*, 1982).

Enhanced root exudation in phosphorus-deficient plants growing in aerated soils can have important consequences not only for microbial activity in the rhizosphere but also for the subsequent enhancement of phosphorus uptake if conditions in the rhizosphere permit the establishment of an effective population of mycorrhizas.

15.7 Mycorrhizas

15.7.1 Ecto- and Endomycorrhizas

Among the most widespread associations between microorganisms and higher plants are mycorrhizas. The roots of most soil-grown plants are usually mycorrhizal. As a rule the fungus is strongly or wholly dependent on the higher plant, whereas the plant may or may not benefit. In some instances mycorrhizas are essential.

Two major groups are involved: ectomycorrhizas and endomycorrhizas. Ectomycorrhizas predominate in tree species of the temperate zone but can also be found in tree and shrub species of semiarid zones (Högberg and Nylund, 1981). Formation results from penetration of the fungal hyphae into the intercellular spaces of the cortex (free space), where they form a network of fungal mycelium (Hartig net) and an interwoven mantle around the root (sheath-forming mycorrhizas). From the extension of the sheath,

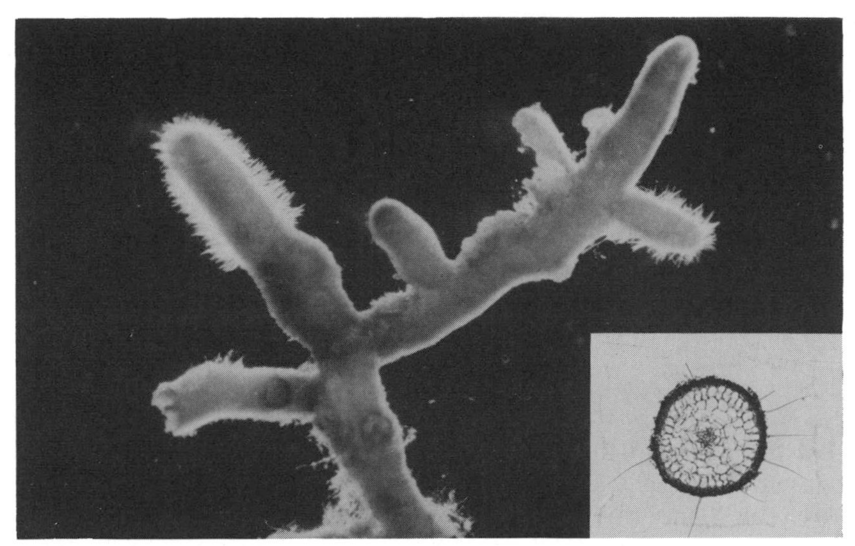

Fig. 15.6 Ectomycorrhizal short roots of oak tree. Inset: Root cross section with hypha mantle and strands of external mycelium. (From Egli, 1983.)

mycelium strands penetrate the surrounding soil (Fig. 15.6). Frequently the infected roots are fully mantled, short, and highly branched.

Endomycorrhizas grow both inter- and intracellularly in the root cortex. The fungal mantle extends well into the soil (Fig. 15.7). Although there are several groups, the most common represent the vesicular–arbuscular mycorrhiza (VAM). The fungus grows in the cortex and develops lipid-rich, ovoid bodies (vesicles) and highly branched, haustorium-like structures within the

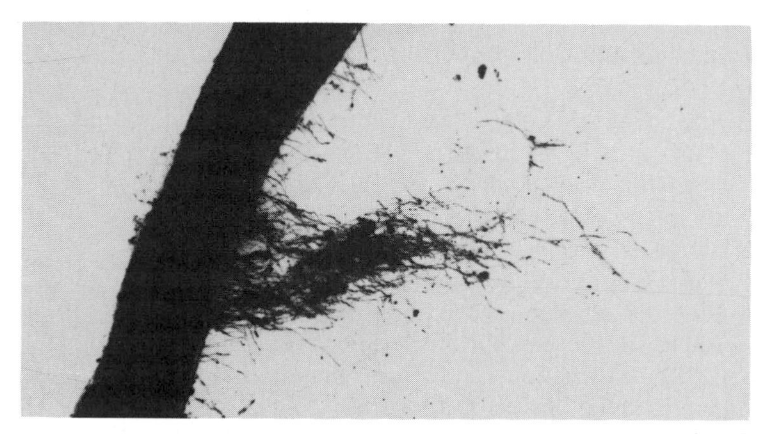

Fig. 15.7 Photomicrograph of an onion root with external mycelium of VAM. (By Courtesy of A. Gnekow.)

host cells (arbuscules), which appear to be the sites of nutrient transfer. The fungus is a member of the family *Endogonaceae*; the major genus is *Glomus*, considered to be the most abundant of all soil fungi (Lamont, 1982). It is an obligate symbiotic fungus and is not very host specific; VAMs are found on the majority of the world's vegetation. Some host species may carry both ecto- and endomycorrhizas (Mejstrik and Cudlin, 1983). Mycorrhizas are, however, rarely found or are even absent in certain families such as Cruciferae and Chenopodiaceae.

15.7.2 Infection and Energetics

The rate of infection is affected by various soil factors, such as pH and temperature, showing typical optimum values at slightly acid soil pH and temperatures in the range of 20 to 25°C. The infection rate is very low in nitrogen-deficient plants and increases with increasing nitrate supply and nitrogen content of the roots (Hepper, 1983). Ammonium supply has the opposite effect, possibly a response to the elevated phosphorus levels in ammonium-fed plants (Johnson *et al.*, 1984). For a comprehensive review of the effects of these environmental factors see Gianinazzi-Pearson and Diem (1982). Within a given plant species the infection rate is closely positively correlated with the root content of soluble carbohydrates (Same *et al.*, 1983) and the root exudation of sugars (Azcón and Ocampo, 1981). No correlation exists, however, between sugar exudation and infection rate when different plant species are compared (Azcón and Ocampo, 1984). Since the host plant also supplies photosynthates for subsequent fungal growth, a positive correlation exists between the efficiency of VAM to stimulate the host plant growth and the light intensity (Diedrichs, 1983a) or the day length (Diedrichs, 1983b). Unfavorable environmental conditions such as shading and defoliation depress mycorrhizal growth (Same *et al.*, 1983), but to a lesser degree than that of the host root growth and, in nodulated legumes, the nodule weight (Bayne *et al.*, 1984). The most prominent influence is exerted by phosphorus. With phosphorus deficiency the rate of exudation of reducing sugars and amino acids increases, and this is often correlated with enhanced infection rate (Table 15.10). This may explain, at least in part, the general pattern of an inverse relationship between the application of readily soluble phosphorus (e.g., superphosphate) and root infection with VAM fungi. In contrast to superphosphate, rock phosphate has very little effect on infection (Powell, 1980).

The extent of infection is controlled not by the phosphorus concentration in the soil solution but by the phosphorus content of the plant. This was convincingly demonstrated by the foliar application of phosphorus, which

Table 15.10
Relationship between Phosphorus Supply, Root Exudate Content, and Root Infection with Vesicular–Arbuscular Mycorrhizas in Sudangrass (*Sorghum vulgare* Pers.)[a]

Phosphorus supply (mg/kg)	Root exudate per gram root dry wt: Reducing sugars (mg)	Amino acids (μg)	Root infection with VAM (%)
0	3·89	271	89
28	1·22	60	24
56	1·08	50	10
228	1·19	50	5

[a]Based on J. H. Graham *et al.* (1981).

depressed both the infection rate and dry weight of the mycelium (Fig. 15.8). It is an open question as to whether the effect of high internal phosphorus concentration can be explained only by lower root contents of sugars and free amino acids. Jasper *et al.* (1979) made the interesting observation that VAM infection of ryegrass growing in unfertilized virgin soil of low phosphorus content was more sensitive to increasing phosphorus supply than that of ryegrass growing in an adjacent fertilized agricultural soil. This result probably arises from different indigenous VAM species in these soils. The effect of increasing phosphorus supply on the infection rate varies markedly between VAM species. At extremely low soil phosphorus levels the infection rate may be impaired by phosphorus limitation on the growth of the fungi itself (Bolan *et al.*, 1984a). With increasing supply, root growth and infection rate increase until an optimum supply of phosphorus is approached which differs remarkably between the VAM species; beyond this level, however, the infection rate is depressed, to different degrees in all VAM species (Bolan *et al.*, 1984a) and also dependent on the host species (Davis *et*

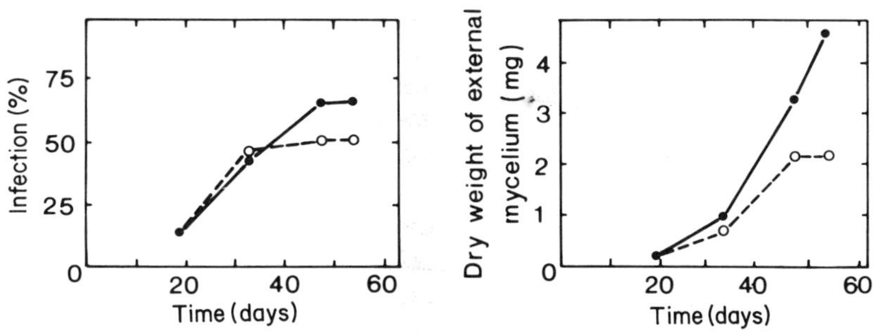

Fig. 15.8 Effect of phosphorus on percentage of infection and dry weight of external mycelium of VAM in onion roots with time. Phosphorus supplied to roots only (●——●); additional foliar application of phosphorus (○ – – – ○). (Redrawn from Sanders, 1975.)

al., 1984). These complicated interactions explain the often contradictory results obtained from studies of relationships between phosphorus supply and infection rates.

With increasing degrees of infection, there is a progressively greater carbon drain into the fungus. In VAM the external hyphae are extremely fragile and a high proportion might be lost during preparation. Thus, only rough calculations of the fungal biomass can be made. In bean with a 62% root infection the fungal biomass may represent more than 5% of the total root biomass (Kucey and Paul, 1982). Mycorrhizal roots also experience a considerably higher carbon loss through respiration than do nonmycorrhizal roots; in pine the difference per unit root weight was 3·6-fold (Reid *et al.*, 1983). Accordingly, a higher proportion of the total photosynthates is translocated to mycorrhizal roots (Snellgrove *et al.*, 1982). This proportion is in the range of 6–10% of the net photosynthesis (Koch and Johnson, 1984). Mycorrhizal plants may compensate for this, at least in part, by enhanced photosynthesis (Gianinazzi-Pearson and Gianinazzi, 1983), provided that there is still sufficient reserve in increasing the source capacity (Section 5.7). Under conditions of low phosphorus availability in soils, plants benefit more in terms of growth from carbon allocation to VAM fungal growth than from carbon release from the roots to support the growth of rhizosphere microorganisms. In young plants, this carbon drain into VAM may become critical, however, during the establishment of the infection before phosphorus is supplied to the host plants. The situation is comparable to that of legumes during the early stages of nodulation (Chapter 7).

15.7.3 Phosphorus Uptake

In the majority of cases enhanced phosphorus uptake and improved phosphorus nutrition are the primary causes of growth and yield increases in mycorrhizal plants. Thus growth responses to infection decrease when readily soluble phosphorus sources are supplied to the host plant (Menge *et al.*, 1978). This improvement in growth may itself lead to a more rapid uptake of other mineral nutrients.

Mycorrhizal roots can take up several times more phosphorus per unit root length than nonmycorrhizal roots. This is primarily because of the greater surface area resulting from the growth of hyphae, which may reach distances of several centimeters from the root surface. This extension permits phosphorus uptake outside of the depletion zone in the rhizocylinder. Phosphorus transport in VAM hyphae of up to 8 cm from the root surface to the host root cells has been demonstrated (Rhodes and Gerdemann, 1975).

Table 15.11
Effect of Fungal or Bacterial Root Infection on the Dry Weight and Phosphorus Content of Shoots of Lavender Plants (*Lavendula spica*) Grown in Calcareous Soil[a]

	Dry weight (mg/3 plants)		Phosphorus content (mg/g dry wt)	
Treatment[b]	Shoots	Roots	Shoots	Roots
Control	180	110	0·60	0·72
+ PDB	200	200	1·20	1·28
+ VAM	1550	450	1·48	1·06
+ VAM + PDB	1670	500	1·88	1·48

[a]From Azcón *et al.* (1976).
[b]PDB, Phosphate-dissolving bacteria; VAM, vesicular–arbuscular mycorrhiza.

This "long-distance transport" of phosphorus in the hyphae and the larger energy supply from the host directly to the fungi are the main reasons for the much higher contribution of mycorrhizas than of rhizosphere bacteria to the phosphorus nutrition of plants (Table 15.11).

It seems that, as a rule, mycorrhizas absorb phosphorus from the same source as noninfected roots: the soil solution (Tinker, 1980). According to Bolan *et al.* (1984b) it is difficult to disprove that mycorrhizal roots can utilize more phosphorus sources than can nonmycorrhizal roots. In ectomycorrhizas, high phosphatase activity at the hyphal surface might be an additional factor for enhancement of solubilization and utilization of organically bound phosphorus (Dighton, 1983). It is not clear whether differences in rhizosphere pH between mycorrhizal and nonmycorrhizal roots (Section 15.7.4) contribute to the higher phosphorus uptake of the mycorrhizal roots.

The main effects of mycorrhizas on phosphorus nutrition are brought about by accelerated rates of phosphorus uptake from low-solubility sources. For example, the low phosphorus requirement of field-grown cassava (1·3 μM phosphorus in soil solution) for maximal growth rates is related to the association with VAM (Kang *et al.*, 1980). The fungal membrane transport system seems to have a higher affinity for phosphorus (lower K_m value) than the host roots (Cress *et al.*, 1979). This may contribute to the higher efficiency of mycorrhizal roots. The main factors, however, are the lower C_{min} values and the larger surface area per unit root length. Higher rates of removal of phosphorus from the soil solution by hyphae promote the dissolution of adjacent particles of sparingly soluble inorganic phosphorus, either from the soil or from fertilizer phosphorus, as shown for red pepper in Table 15.12. In contrast to red pepper, sunflower utilized sparingly soluble phosphorus sources quite well and showed no growth responses to VAM infection, in spite of an intensive development of mycorrhizas and a significantly higher phosphorus uptake (Graw *et al.*, 1979). Various factors

Table 15.12
Effect of Inoculation with VAM Fungi and Application of Phosphate Fertilizer on the Shoot Dry Weight of Plants Grown in a Phosphorus-Deficient Soil[a]

		Form of phosphate in fertilizer[b]			
Species	Inoculation	No P	MCP	HA	Al–P
Red pepper	−	0·03	0·54	0·03	0·11
	+	0·41	0·85	0·54	0·87
Sunflower	−	2·69	6·83	5·76	6·45
	+	2·90	6·47	5·22	6·01

[a]Shoot dry weight expressed as grams per plant. Based on Graw *et al.* (1979).
[b]MCA, Monocalcium phosphate; HA, hydroxylapatite; Al–P, aluminum phosphate.

may be responsible for the genotypical difference in mycorrhizal dependency (see Section 15.7.5).

In red pepper and other species that are highly dependent on mycorrhizas for the utilization of sparingly soluble phosphates, the formation of polyphosphates in the external VAM hyphae (Martin *et al.*, 1983) and the transport of phosphorus in the hyphae as polyphosphates (Cox *et al.*, 1980) presumably play a key role in both the high efficiency of uptake and the large capacity for phosphorus transport in the hyphae. In the internal hyphae, transfer to the host cell probably occurs in the fine branches of the arbuscules as a metabolically controlled process involving hydrolysis to orthophosphate before release into the host cell (Marx *et al.*, 1982).

High infection rates with VAM fungi are not necessarily an indication of a large contribution to the phosphorus nutrition of the host plant. Considerable differences exist not only among species of higher plants but also among species or strains of VAM fungi (Gram *et al.*, 1979). As yet these interactions are not well understood; they are of particular importance, however, for the potential utilization of VAM in agriculture.

15.7.4 Other Mycorrhizal Effects

Mycorrhizal infection can increase the uptake of mineral nutrients other than phosphorus. In forest trees, particularly in the seedling stage, root infection with ectomycorrhizas enhances the uptake rate of potassium (Rygiewicz and Bledsoe, 1984) and ammonium (Rygiewicz *et al.*, 1984). In annual species, root infection with VAMs may increase the uptake rates and the shoot contents (per unit dry weight) of micronutrient cations, for example, copper (Gildon and Tinker, 1983b) or both copper and zinc (Lambert *et al.*, 1979). It has been shown that VAM hyphae can readily transport zinc as well as sulfate to the host roots (Cooper and Tinker, 1978). VAMs may also enhance the uptake of other heavy metals such as nickel (Killham and Firestone, 1983). On the other hand, at elevated heavy metal concentrations in soils, the high binding capacity, particularly of the fungal sheath in ectomycorrhizas, is important for the heavy metal tolerance of the host plant (Bradley *et al.*, 1981). A similar protecting effect has been observed in plants infected with VAM (Gildon and Tinker, 1983a). Increase or decrease in solubilization and uptake of heavy metal cations in mycorrhizal roots are perhaps related to the ability of most fungi, including mycorrhizas, to produce hydroxamate siderophores, particularly in response to iron deficiency (Section 9.1.6). In agreement with this, in the rhizosphere soil of ectomycorrhizal pine seedlings the concentrations of hydroxamate siderophores are 11 to 56 times higher than in the bulk soil (Powell *et al.*, 1982). The particular ability of some mycorrhizas to accumulate cadmium and other heavy metals is also of concern for human nutrition (Marschner, 1983).

As a rule, mycorrhizas affect only those nutrients which have a very low mobility in soils and which are present in the soil solution in very low concentrations relative to the requirement of plants. An increase in the levels of other mineral elements such as nitrogen or potassium may often be an indirect effect brought about, for example, by the stimulation of N_2 fixation in legumes (see Section 15.7.5) and an enhancement of growth by the alleviation of phosphorus deficiency, especially in seedlings of fruit trees such as avocado (Menge *et al.*, 1980) and hardwood species (Schultz *et al.*, 1981). These species are very dependent on mycorrhizal infection and show extremely poor growth in phosphorus-deficient soils.

It well may be that several of the effects of mycorrhizas on growth, morphology, photosynthesis, and water balance are hormonal. Infection with VAM can increase levels of CYT and increase or decrease the levels of abscisic acid and gibberellin-like substances in the host plant. Ectomycorrhizas produce substantial amounts of auxins (Strzelczyk and Pokojska-Burdziej, 1984) which is in agreement with observations of Slankis (1973) that cell-free extracts from ectomycorrhizas induce changes in root morphology similar to those induced by auxin treatment. In general, however, it is often difficult to determine the extent to which the effects of mycorrhizas on the photosynthesis or water balance of the host plant are direct results of fungal hormonal production, or indirect effects that occur via the enhancement of growth of the host. To give an example—increase in the hydraulic conductance of mycorrhizal roots (Graham and Syvertsen, 1984) is probably the result of the improved phosphorus nutrition of the host (Sections 5.6.4 and 8.4.6). On the other hand, the enhanced shoot growth of mycorrhizal plants increases the water consumption, and thus also the susceptibility to water stress (Sweatt and Davies, 1984; Sieverding, 1984). Infection with VAM may influence the host growth via a qualitative shift in the rhizosphere bacterial populations (Ames *et al.*, 1984). Furthermore, VAM infections may influence the colonization of roots by other symbiotic microorganisms, reduce the susceptibility, or increase the tolerance, of roots to soil–borne pathogens such as nematodes, or phytopathogenic fungi such as *Fusarium oxisporum*. These various interactions have been reviewed by Gianinazzi-Pearson and Gianinazzi (1983).

Another effect of VAM infection in wheat and barley which cannot have been caused by improved phosphorus nutrition was reported by Buwalda *et al.* (1983). Infection increased the content not only of phosphate, but also of other anions such as chloride and sulfate, whereas the cation content remained basically unchanged. As a consequence, the cation–anion balance was altered in the plants in favor of anions. For electrochemical reasons the rhizosphere pH of mycorrhizal roots is expected to become more neutral or alkaline, and this is indeed the case, as demonstrated by Bledsoe and Zasoski (1983) in the rhizosphere of mycorrhizal Douglas fir seedlings. Although the reasons for the shift in cation–anion uptake in mycorrhizal plants are unknown (Buwalda *et al.*, 1983), the corresponding changes in rhizosphere pH affect the availability and uptake of other mineral elements.

15.7.5 Mycorrhizal Dependency

In most soils low in available phosphorus, roots are heavily infected with indigenous VAM fungi (Table 15.13). Well-known exceptions are species of the Chenopodiaceae and Cruciferae, cabbage being an example of the latter.

Table 15.13
Effect of Suppressing and Reintroducing VAM Fungi on the Growth of Plants[a]

	Soil treatment[b]		
Plant species	Nonfumigated	Fumigated	Fumigated–reinoculated
Carrot	8·5 (61)	0·4 (0)	7·4 (60)
Leek	4·4 (50)	0·4 (0)	4·0 (67)
Tomato	4·1 (61)	2·5 (0)	5·1 (90)
Wheat	2·0 (63)	1·7 (0)	2·1 (79)
Cabbage	11·9 (0)	14·2 (0)	13·6 (0)

[a]Plant growth expressed on a dry weight basis as grams per pot. From Plenchette *et al.* (1983).
[b]Values in parentheses indicate root colonization (percentage of total root) with VAM fungi.

The results shown in Table 15.13 represent the range of mycorrhizal dependency of crop species grown in a soil low in available phosphorus. Destruction of VAM by soil fumigation elicited three kinds of growth responses: (*a*) Carrot and leek grew very poorly; growth was restored to about the level of growth in nonfumigated soil after reinoculation with VAM fungi. (*b*) Tomato and wheat exhibited small or negligible growth reduction, despite high infection rates. (*c*) Cabbage was not infected by VAM and fumigation increased growth, presumably due to the destruction of soil-borne pathogens. The results in Table 15.13 demonstrate that one should not expect a large stimulation of growth by the inoculation of field-grown plants unless indigenous VAM fungi have been damaged—for example, by fumigation or heavy application of fungicides such as Benomyl (Hale and Sanders, 1982) or Oxadiazon (Sieverding and Leihner, 1984). This is an important consideration in nurseries for seedling growth of citrus (Menge *et al.*, 1978), avocado (Menge *et al.*, 1980), and hardwood species (Schultz *et al.*, 1981). In the case of avocado and citrus, seedling growth of nonmycorrhizal plants requires extremely high levels of readily soluble phosphorus fertilizers. In hardwood species and grapevines (Eibach, 1982), mycorrhizas seem to stimulate growth other than, or in addition to, that induced by phosphorus uptake.

It is generally agreed that VAMs stimulate growth and phosphorus uptake either not at all or only to a limited degree in such plant species as grasses and cereals (e.g., wheat) which have extensive, highly branched root systems and long root hairs. In contrast the responses are high in species with coarse root systems that are not highly branched, such as many woody species and crop species such as cassava (Fig. 15.9). In nonmycorrhizal plants the critical deficiency level of available soil phosphorus is 190 mg, compared with only 15 mg in mycorrhizal plants.

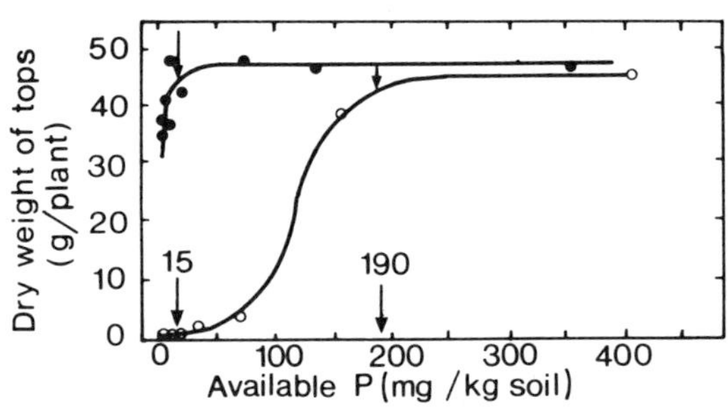

Fig. 15.9 Relationship between the dry weight of tops of inoculated (with VAM) (●——●) and noninoculated (○——○) cassava and level of available phosphorus (soil testing method, Bray II) in sterilized soil. Arrows indicate the critical phosphorus levels corresponding to 95% of maximal yield. (Redrawn from Howeler *et al.*, 1982a.)

The promotion of phosphorus uptake by VAMs in many crop species has implications for soil testing. The relationship between chemical soil analysis and growth response to applied phosphorus is often poor, and it has been demonstrated that this is partly due to differences in mycorrhizal infection (Stribley *et al.*, 1981).

In legumes, growth stimulation by VAM fungi can improve phosphorus nutrition and, as a consequence, increase nodulation and N_2 fixation. In subterranean clover, inoculation stimulated nodulation and growth only when insufficient phosphorus was applied for maximum growth of the nonmycorrhizal plants (Robson *et al.*, 1981a), whereas in other species such as *Stylosanthes guyanensis* specific interactions between VAMs and *Rhizobium* cannot be discounted (Table 15.14).

The different responses of *Stylosanthes* and *Lotus* to VAM infection is another instructive example of both the role of root morphology in general and root hair length in particular in the acquisition of phosphorus from deficient soils. The results also show that mycorrhizal hyphae can counterbalance a small surface area of host roots. Mycorrhizas are therefore considered in many situations to comprise an alternate root system, especially for phosphorus uptake. In addition, the dual system of nutrient uptake in mycorrhizal plants offers a higher degree of flexibility in response to unfavorable soil factors such as pH, moisture level, and temperature (Rovira *et al.*, 1983).

Table 15.14
Interaction between VAM and Rhizobial Infection and Phosphorus Nutritional Status of Plants Grown in a Phosphorus-Deficient Soil[a]

Species	Treatment	Fresh wt (g)	Nodules[b]	Shoot (% P)	VAM Infection (%)
Stylosanthes guyanensis	−VAM	0·47	0	0·20	0
	+VAM	1·63	5	0·44	74
	−VAM + P	0·91	5	0·58	0
Lotus pedunculatus	−VAM	2·54	4	0·29	0
	+VAM	2·01	4	0·27	54
	−VAM + P	3·86	5	0·53	0

[a]Based on Crush (1974).
[b]0 = none; 5 = numerous.

	Root diameter (average) (mm)	Roots with hairs (average) (%)	Length of root hairs (average) (μm)
Stylosanthes	285	6	108
Lotus	229	99	809

15.7.6 Outlook

Many ectomycorrhizal fungi can be multiplied on synthetic culture media, and production of these fungi on these media is therefore a common procedure for inoculating tree species. To establish seedling growth of fruit trees such as citrus, methods for inoculation with VAM in the nursery and during transplanting have been developed (Menge, 1983). The potential of using VAM fungi as "biofertilizers" on a large scale and in agricultural production is fascinating. Nonetheless, expectations should be tempered by a realistic appraisal, for various reasons:

1. Lack of inoculum. At present it is not possible to obtain sufficient quantities of VAM inoculum because the host plant is necessary for multiplication.
2. Difficulties in the production of pathogen-free inoculum in sufficient quantities.
3. Competition with indigenous VAM fungi. In phosphorus-deficient soils the infection rate with indigenous VAM fungi is already high, and the introduction of new VAM fungi requires a very high inoculum density because of competition (Sanders and Sheikh, 1983). On the other hand, soil sterilization on a large scale is neither possible nor advisable for obvious reasons.

So far, field results showing a clear yield response to inoculation with VAM fungi are still scanty (Kang *et al.*, 1980; Islam and Ayanaba, 1981). Unless new massive propagation techniques are developed, the introduction of new and more efficient VAM fungi on a large scale in agriculture will remain wishful thinking. Moreover, in most situations it is an open question as to whether any potential benefit of VAM on the host is not already being obtained by natural infection (Tinker, 1980).

16

Adaptation of Plants to Adverse Chemical Soil Conditions

16.1 Natural Vegetation

Soil chemical factors such as pH, salinity, and nutrient availability determine the distribution of natural vegetation. Species and ecotypes can be classified in ecophysiological terms according to their distribution in soils. Some examples of species grouped in this way are acidophobes and acidophiles; calcifuges and calcicoles; halophytes and glycophytes; and metallophytes (adapted to metalliferous soils). Reviews have been published on ecophysiological aspects of plant responses, especially those of natural vegetation, to soil pH (Kinzel, 1983), salinity (Munns *et al.*, 1983), and heavy metals (Woolhouse, 1983). In this chapter the main emphasis is on crop plants, where differences among species in yield response to soil pH and salinity, for example, have been well known in agriculture for many years but where extensive research on the mechanisms involved has begun only relatively recently.

The mechanisms by which wild plants adapt to soils of low nutrient availability are of particular interest in relation to the mineral nutrition of crop plants. In phosphorus-deficient soils, for example, crop plants usually grow poorly, have low phosphorus levels in their shoots, and develop visible phosphorus deficiency symptoms. In contrast, wild plants adapted to such soils have a relatively high phosphorus content, and deficiency symptoms are absent (Chapin and Bieleski, 1982). As a rule, however, the higher efficiency of wild plants is related not to a higher capacity of the roots to extract soil phosphorus or to a more efficient use of phosphorus within the plants but rather to a slow growth rate, whereby the dilution effect is avoided (Chapin, 1983). This is just the opposite of what is considered to reflect high nutrient efficiency in crop plants, namely, a high ratio of tissue dry weight to unit weight of nutrient in the tissue (Bieleski and Läuchli, 1983). This strategy of wild plants is retained in certain tropical root crop species, where in spite of large differences in yield brought about by soil acidity (see Fig.

16.6) the mineral element content of the leaves remains about the same (Abruna-Rodriguez *et al.*, 1982).

16.2 High-input versus Low-input Approach

16.2.1 The Problem

In the past, the approach to soil fertility problems in crop production emphasized changing the soils to fit the plants. Soil fertility factors, such as pH and nutrient availability, were adjusted to optimum levels for a given plant species. This high-input approach, coupled with the heavy use of chemical fertilizers, was very successful in the temperate zones in increasing yields of crops grown in soils that do not, as a rule, have extreme chemical properties. However, the high-input genotypes of crop plants usually have a limited adaptability to the adverse chemical soil conditions that usually prevail in the tropics and subtropics. These conditions cannot easily be ameliorated because of their extent, the cost of improving the soils, or both (Vose, 1983). In tropical America, for example, ~70% of the soils are acid and infertile (Sanchez and Salinas, 1981). In subtropical and semiarid regions, soil salinity and alkalinity and related nutritional problems such as iron and zinc deficiency are widespread. About 25% of the world's area of cultivable soil has acute chemical problems (Vose, 1983).

The realization of the difficulties or even failure of the high-input approach in most tropical and subtropical soils led two decades ago to a shift in approach toward fitting the plants to the soils. This requires genotypes better adapted to given ecological conditions, as well as selection and breeding programs for high nutrient efficiency and high tolerance to such constraints as aluminum and manganese toxicity, waterlogging, and salinity. For some crop species such as soybean, the existing genetic potential for fitting plants nutritionally to soils is quite high (Brown and Jones, 1977a).

This low-input approach using adapted genotypes with a more efficient use of nutrients from soil reserves and fertilizers leads to yields that are only 80–90% of the maximum. This low-input approach should address itself not only to extreme soil chemical conditions (e.g., salinity) but also to the selection and breeding of genotypes that utilize soil and fertilizer nutrients, especially phosphorus and nitrogen, with high efficiency.

16.2.2 Genetic Basis for Mineral Nutrition

That the mineral nutrition of crop plants is under genetic control is indicated by the numerous nutritional differences among cultivars and strains. More specific evidence comes from inheritance studies involving cultivars and strains differing in nutritional requirements. In some cases major nutritional features are under the control of a single gene pair; in most situations more complex genetic systems are involved (Gerloff and Gabelman, 1983; Graham, 1984).

Since the early 1960s there has been an enormous increase in interest in and research on genetically based mineral nutrition of plants—for example, by Brown and his group and Foy and his group, at the Plant Stress Laboratory in Beltsville, Maryland, and by Epstein and Läuchli at the University of California, Davis. There has been impressive progress in both breeding programs for improving the adaptation of crop species to problem soils and in research on physiological mechanisms which are responsible for, or at least involved in, adaptation. Although practical goals can be achieved in breeding even when physiological explanations are lacking (Vose, 1983), progress will be more rapid when the mechanisms controlling genotypical variations in nutrient efficiency and tolerance to adverse chemical soil conditions are better understood.

16.2.3 Nutrient Efficiency

From the agronomical point of view and in an operational sense genotypical differences in the nutrient efficiency of crop plants are usually defined by the differences in the relative growth or in the yield when grown in a deficient soil. For a given genotype, nutrient efficiency is reflected by the ability to produce a high yield in a soil that is limiting in one or more mineral nutrients for a standard genotype (Graham, 1984). This definition can be applied to comparisons of both genotypes (cultivars or lines) within a species or to plant species. This is an important topic for the selection and breeding of genotypes adapted to low nutrient availability in soils, and with a high efficiency in the utilization of soil and fertilizer nutrients (Barrow, 1978). There have been a large number of reports in recent years on nutrient efficiency, with the main emphasis on these agronomical aspects, comparing the yield, or the percentage of yield reduction, in genotypes supplied with insufficient amounts of mineral nutrients (Blair and Cordero, 1978; Graham *et al.*, 1982; Gabelman and Gerloff, 1983). A typical example of the phosphorus efficiency of three pasture species is given in Fig. 16.1. Despite a similar final dry weight at the highest phosphorus supply, the growth response of the three species to a given supply increased from *Trifolium cherleri* to *T. subterraneum* and *Lolium rigidum*. The minimum levels of applied phosphorus required in the three species to give 90% of maximum yield were 302, 87, and 26 mg per kilogram of soil, respectively. As one

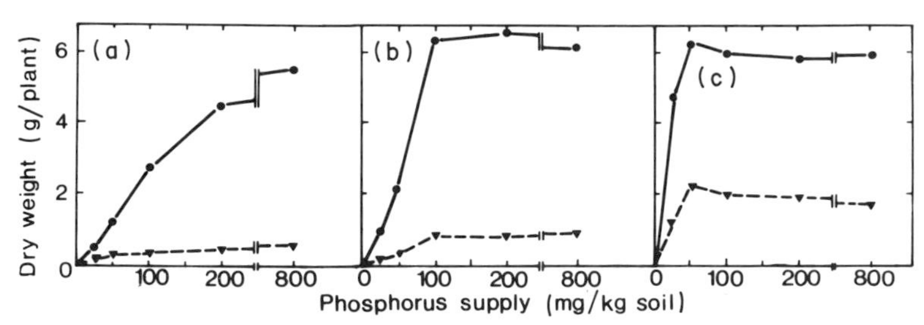

Fig. 16.1 Growth response of three pasture species to phosphorus fertilizer applied to a phosphorus-deficient soil. (a) *Trifolium cherleri*; (b) *Trifolium subterraneum*; (c) *Lolium rigidum* (●, total plants; ▾, roots). (Based on Ozanne *et al.*, 1969.)

would expect from the role of root growth and root surface area in phosphorus acquisition (Section 13.3) there is a close correlation between the efficiency of fertilizer phosphorus utilization and root dry weight (Fig. 16.1).

In mixed pastures on phosphorus-deficient soils, with low levels of fertilizer phosphorus a shift in botanical composition can be expected, favoring the growth of grass, as also occurs in a mixed stand of legumes and grasses growing in potassium-deficient soils (Section 13.3).

The differences in yield response among crop species and cultivars within a species when grown in a copper-deficient soil provide another impressive example of genetically controlled mineral nutrition and nutrient efficiency (Table 16.1). Wheat and oat are generally sensitive to a low copper supply, whereas rye is relatively insensitive. Important differences also exist in the copper efficiency of wheat cultivars. When the supply is suboptimal, Gabo

Table 16.1
Responses of Grain Yield of Various Genotypes to Copper[a,b]

	Copper supply (mg/11 kg soil)			
Species, cultivars	0	0·1	0·4	4·0
Wheat				
cv. Gabo	0	0	9·5	100
cv. Halberd	1·6	7·1	52·0	100
cv. Chinese spring	0	25·5	44·0	100
Rye				
cv. Imperial	100	114	114	100
Triticale				
cv. Beagle	98·6	95·2	98·6	100

[a]Data compiled from Nambiar (1967c) and Graham and Pearce (1979).
[b]Data represent grain yields expressed as relative values. Plants were grown in a copper-deficient soil and supplied with different levels of copper.

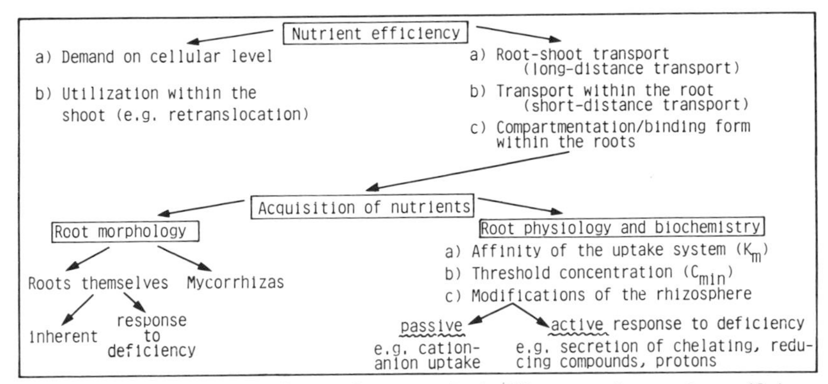

Fig. 16.2 Possible mechanisms of genotypical differences in nutrient efficiency.

either fails totally or has only a very low grain yield compared with the relatively copper efficient Chinese spring. Triticale, a hybrid of wheat and rye, also has a high copper efficiency similar to its parent rye, indicating that the specific mechanisms for copper uptake in rye (R. D. Graham *et al.*, 1981) are genetically controlled and transferable to triticale.

Genotypical differences in nutrient efficiency occur for a number of reasons, these being related to uptake, transport, and utilization within plants (Fig. 16.2). Typical differences in nitrogen requirement at the cellular level exist, for example, between C_3 and C_4 species (Section 12.3.6) or in calcium and boron requirements between monocots and dicots (Chapters 8 and 9). Differences in calcium efficiency in tomato strains are presumably related to the binding stage in the plants (English and Maynard, 1981). There are many examples of genotypical differences among cultivars or mutants in the short-distance transport of mineral nutrients within the roots or in the long-distance transport from the roots to the shoots (Läuchli, 1976b). The "iron-inefficient" PI soybean cultivar represents a genotype in which a slow rate of root to shoot transport, in this case of iron, is responsible for the low nutrient efficiency (Brown *et al.*, 1967). A combination of differences in the rates of both magnesium uptake and transport to the shoot has been reported by Clark (1975) in maize; in the efficient inbred line B 57 maximal dry weight was obtained with a magnesium concentration in the substrate of only 0·12 m*M* compared with 4·1 m*M* in the inefficient inbred line Oh 40 B. High magnesium efficiency is important for the adaptation of crop plants to acid soils.

It is generally agreed that in most instances the acquisition of nutrients by roots plays the most important role in nutrient efficiency. Genotypes within a species can differ very much in both the affinity of the uptake system (K_m) and the threshold concentration (C_{min}), as has been demonstrated, for example, for phosphorus in corn inbred lines (Schenk and Barber, 1979b).

Table 16.2
Root and Shoot Growth of Two Strains of Bean Receiving Adequate and Inadequate Phosphorus Supplies[a]

	Adequate P supply			Inadequate P supply		
Strain no.	Root dry wt (mg/plant)	Shoot dry wt (mg/plant)	Root/shoot ratio	Root dry wt (mg/plant)	Shoot dry wt (mg/plant)	Root/shoot ratio
6	242	1465	0·17	124	777	0·16
11	181	1233	0·15	365	1141	0·31

[a]Based on Whiteaker *et al.* (1976).

Root-induced modifications of the rhizosphere (pH, release of organic acids, etc.) are another important component of nutrient efficiency (Chapter 15).

In the case of phosphorus efficiency, root morphology is the most important factor. Of particular interest are changes in the roots in response to phosphorus deficiency. Although an increase in the relative rate of root growth under these conditions is a general phenomenon, genotypes can behave quite differently, as shown in Table 16.2. Growth of strain 6 was greatly reduced, but the root/shoot ratio was approximately the same. In contrast, root growth and root/shoot ratio of strain 11 nearly doubled. The capacity to distribute a higher proportion of photosynthates to the roots is obviously under genetic control and is an important aspect of phosphorus efficiency for plants grown in deficient soils.

In the basis of physiological (K_m, C_{min}, etc.) and morphological root characteristics (e.g., the number and length of root hairs) of various species, Barber (1982) demonstrated a close correlation between observed uptake of phosphorus from a deficient soil and uptake predicted by a model, indicating that other factors such as VAM were of minor importance for phosphorus efficiency. This result cannot be generalized, however, as had been discussed in principle in Chapter 15, and as is shown in the example in Table 16.3 for alfalfa. Without VAMs there was virtually no growth, and all three cultivars responded similarly to the increase in phosphorus supply. With VAMs only, growth increased nearly 100-fold in all three cultivars, but there were obvious differences in the efficiency of the VAM association with the host plants. In Du Puits, VAMs can totally replace phosphorus application but in Buffalo this replacement is less than 50%.

These results on VAM effects may raise the question as to the relevance of screening for phosphorus efficiency in species such as alfalfa in nutrient solutions and/or in the absence of VAMs. In agreement with this, no correlation was found in maize inbred lines between phosphorus efficiency

Table 16.3
Shoot Dry Weight of Three Alfalfa Cultivars with or without Vesicular–Arbuscular Mycorrhizae (VAM) at Different Levels of Phosphorus Supply[a]

Soil treatment		Shoot dry weight (mg/plant)		
mg P/kg	±VAM	Buffalo	Cherokee	Du Puits
0	–	11	18	32
20	–	114	235	375
80	–	2389	2058	2115
0	+	1113	1740	2177

[a]Plants were grown in a phosphorus-deficient soil, pH 7·2. From Lambert *et al.* (1980).

in nutrient solutions and the field responses of lines grown in a phosphorus-deficient soil (Fox, 1978).

16.2.4 Tolerance to Excessive Supply of Mineral Elements

In many instances adaptation to adverse chemical soil conditions requires tolerance to excessive levels of mineral elements such as aluminum and manganese in acid mineral soils, manganese in waterlogged soils, and sodium chloride in saline soils. Thus, *multiple stress tolerance* is often necessary for adaptation. Despite the complexity, impressive progress has been made in selection and breeding for this adaptation, as shown in Table 16.4 for wetland rice.

In a large systematic program, genotypes have been selected and bred and released to farmers. These have had increased yields by ~2 tons/ha in unamended soils with adverse chemical properties. This has allowed

Table 16.4
Grain Production of Wetland Rice Genotypes with Different Degrees of Adaptation to Adverse Chemical Soil Conditions[a]

Soil condition	No. of cultivars or lines (tested)	Mean grain yield (tons/ha)		
		Minimum	Maximum	Advantage
Phosphorus deficiency	118	1·9	4·3	2·4
Zinc deficiency	107	0·8	2·9	2·1
Salinity	64	1·5	3·6	2·1
Alkalinity	50	0·8	3·7	2·9
Aluminum–manganese toxicity	36	1·8	3·6	1·8
Iron toxicity	58	2·1	4·7	2·6

[a]From Ponnamperuma (1982).

marginal land to be brought into production without the need to take costly reclamation measures.

16.3 Acid Mineral Soils

16.3.1 Major Constraints

In acid mineral soils a variety of individual chemical constraints and interactions betwen them limit plant growth. At pH > 4 it is not the low pH per se (i.e., the high H^+ concentration) that limits growth but toxicity and/or deficiency of mineral elements. The main factors are excessive levels of free and exchangeable aluminum. In some instances an excessive level of manganese and deficiencies of phosphorus, calcium, and magnesium are also involved. Less frequently, levels of sulfur, potassium, and the micronutrients molybdenum, zinc, and copper are important. The nitrogen level is generally very low. Fertility in most of these acid soils is low. In more than 70% of the acid soils of tropical America, aluminum toxicity and calcium and magnesium deficiences exist, and nearly 100% of the soils are phosphorus-deficient or have a high phosphorus-fixing capacity (Sanchez and Salinas, 1981). Besides the economic problems of liming these soils, it is generally difficult to incorporate lime deeper than 30 cm. Subsoil acidity is a potential growth-limiting factor throughout many areas of the United States (Foy *et al.*, 1974); it restricts root penetration and thus water and nutrient uptake from the subsoil.

In order for plants to adapt to acid mineral soils, both high tolerance to aluminum (and manganese) and highly efficient utilization of the mineral nutrients mentioned above are required. It is generally agreed, that, as a rule, high aluminum tolerance is the key factor in the adaptation of crop plants to soils of pH < 5.

16.3.2 Solubility of Aluminum and Manganese

Whereas below pH 4 soluble aluminum exists as Al^{3+}, above this pH the concentration of Al^{3+} decreases drastically and becomes negligible above pH ~5·5 (Fig. 16.3). In acid soils below pH 5·5 an increasing proportion of the cation-exchange sites of clay minerals is occupied by aluminum, where it especially replaces other polyvalent cations (Ca^{2+}, Mg^{2+}) and simultaneously acts as a strong adsorber of phosphate (phosphate fixation). A close correlation therefore exists between the proportion of exchangeable aluminum in soils, pH, and inhibition of growth of most plant species, provided

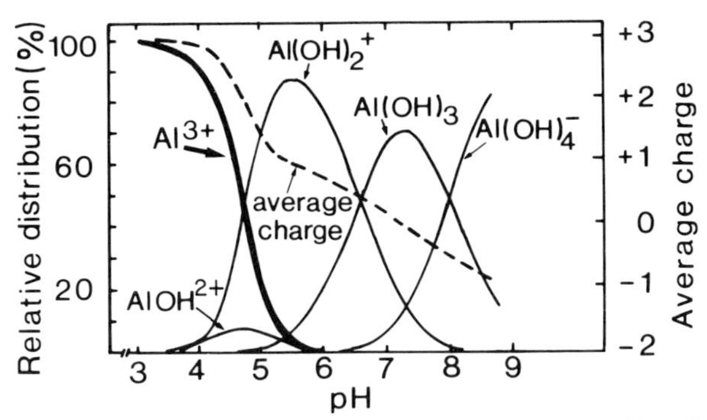

Fig. 16.3 Relative distribution and average charge of soluble aluminum as a function of pH. (Modified from Marion *et al.*, 1976.)

the soils or soil horizons compared are similar in organic matter content (Adams and Moore, 1983). Above pH 4 not only does the total concentration of aluminum decrease markedly, but there is also a shift in the aluminum species: Hydroxyaluminum species, such as a $Al(OH)^{2+}$ and $Al(OH)_2{}^+$, are formed (Fig. 16.3). This shift is correlated with a decline in average charge until $Al(OH)_3$ is formed and, at higher pH, partially dissolves as the anion $Al(OH)_4{}^-$.

There is increasing evidence that in addition to these monomeric ions polymeric ions such as $Al_2(OH)_2{}^{4+}$ and $Al_7(OH)_{17}{}^{4+}$ also exist. Their concentration depends on both the pH and the aluminum concentration (Nair and Prenzel, 1978). This complex situation makes uptake studies and physiological interpretations of aluminum effects on root growth quite difficult. A further complication is the interaction with phosphate and the formation of soluble polymeric aluminophosphate complexes, which is maximal at pH 5 (White, 1977). This may explain why the roots of alfalfa plants take up and translocate much more aluminum and phosphate to the shoot at pH 5·0 than at pH 4·5 but experience less aluminum injury (White, 1977). The majority of the results on aluminum toxicity indicate that Al^{3+} is the most toxic species, but nothing is definitely known about the toxicity of the polymeric species. In soils, aluminum toxicity is not only a function of the pH and of the total aluminum concentration but also of the soil solution ratio of inorganic aluminum/organically bound aluminum (e.g., aluminum complexes of organic acids and phenolics). As demonstrated by Adams and Moore (1983), root growth of soybean was depressed in the topsoil (high in organic matter) at concentrations of between 9 and 134 μM Al (depending on the soil type) in the soil solution, compared to 4 μM Al in the soil solution of the subsoils (low in organic matter).

The solubility of manganese and its equilibrium concentration in the soil solution are less complicated; the pH and redox potential are the dominant factors ($MnO_2 + 4H^+ + 2e^- \rightleftharpoons Mn^{2+} + 2H_2O$). High concentrations of Mn^{2+} in the soil solution are therefore to be expected only in acid soils with high levels of readily reducible manganese in combination with a large content of organic matter, high microbial activity, and anaerobiosis, either temporarily (e.g., short-term flooding) or permanently (Section 16.4). Many acid soils of the tropics are highly weathered, and their total manganese content is often low. Thus there is a smaller risk of manganese toxicity than of aluminum toxicity in these soils. Manganese toxicity is a much more severe problem in poorly drained soils even at a pH as high as 6·0 (Section 16.4).

16.3.3 Physiological Effects of Excessive Aluminum Levels

Signs of aluminum toxicity first appear in the root system, which becomes stubby as a result of the inhibition of elongation of the main axis and lateral roots. The severity of inhibition of root growth is a suitable indicator of genotypical differences in aluminum tolerance (Fig. 16.4). That aluminum toxicity was the cause of the acid soil injury of the root system of the cultivar Kearney was confirmed by water culture experiments (Table 16.5).

The physiological mechanisms of the toxic effects of aluminum on root growth are not yet fully understood. Clarkson (1966) and Morimura *et al.* (1978) described inhibition of cell division in root apical meristems as the primary effect of aluminum. Cell division ceases within a few hours after the exposure of roots to aluminum (Fig. 16.5). Although thereafter cell division resumes, it remains at a lower level than in controls not exposed to aluminum. Genotypical differences in aluminum tolerance are reflected in the rate of recovery from primary aluminum shock. Inhibition of cell division may be due to binding of aluminum to DNA (Morimura and Matsumoto, 1978), which is in accord with the localization of aluminum in the nuclei after short-term treatment with aluminum (Matsumoto *et al.*, 1976a). Aluminum accumulation is especially high in the nuclei of root cap cells (Naidoo *et al.*, 1978); inhibition of root elongation might therefore be the result, at least in part, of injury to root cap cells, which act as sensors for environmental stress (Section 14.1).

Aluminum may affect the uptake of phosphorus directly through the precipitation of aluminum phosphate at the root surface and/or in the free space, although complex formation can interfere with or even stimulate phosphorus uptake. Inhibition of magnesium and calcium uptake is often observed and is mainly the result of cation competition for or blocking of binding sites. A close relationship exists, for example, between the aluminum concentration in the soil solution and the magnesium concentration

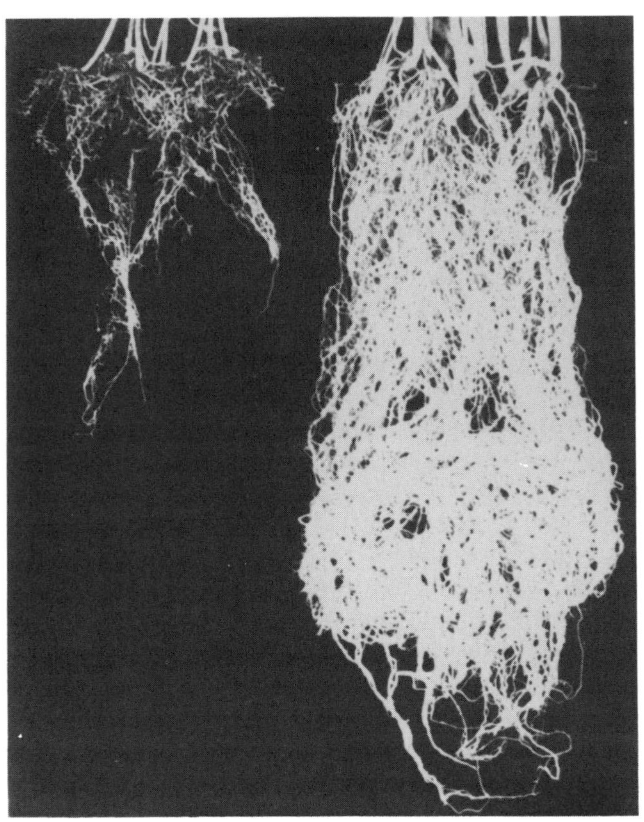

Fig. 16.4 Roots of Kearney (left) and Dayton (right) barley cultivars grown in acid soil of pH 4·6. (From Foy, 1974.)

Table 16.5
Difference in Aluminum Tolerance of Barley Cultivars in Relation to pH Changes and Uptake of Mineral Elements from Nutrient Solution[a]

					Mineral element composition (meq/100 g dry wt)				
			Dry wt (g/pot)		Roots		Shoots		
Cultivar	Al Added (3 mg/liter)	Final pH[b]	Roots	Shoots	Al	P[c]	Al	P[c]	Ca
Dayton	–	6·8	0·7	2·4	2	19	5	20	28
	+	6·7	0·8	2·2	21	14	5	10	27
Kearney	–	7·2	1·1	3·2	4	20	4	21	49
	+	4·7	0·2	2·4	35	22	5	16	12

[a]Based on Foy *et al.* (1967).
[b]Initial pH 4·8.
[c]Calculated as $H_2PO_4^-$.

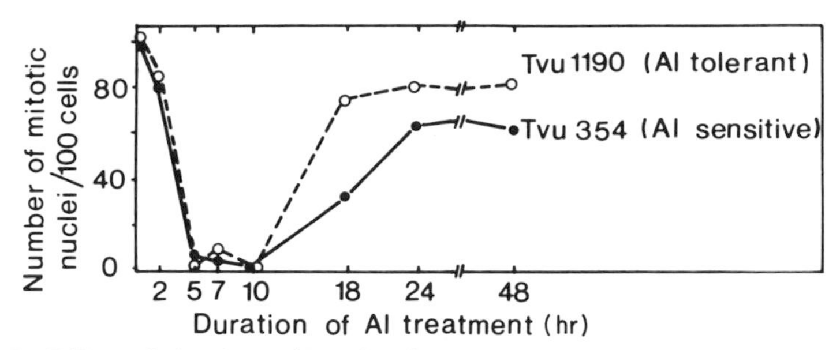

Fig. 16.5 Effect of aluminum (5 mg/liter) on the rate of cell division in root tips of two cowpea genotypes. Relative values; control (no aluminum) = 100. (Based on Horst *et al.*, 1982.)

required in the soil solution for optimal growth of oat (Grimme, 1982). In oat, aluminum drastically depresses the shoot content of magnesium, whereas levels of calcium are only slightly lowered and those of potassium not affected (Grimme, 1983).

Calcium deficiency in the apical meristems is a well-documented manifestation of aluminum toxicity in soybean (Foy *et al.*, 1969), snapbean (Foy *et al.*, 1972) and maize; with an increase in the external calcium concentration, the aluminum-induced inhibition of root elongation is reduced in highly sensitive and completely prevented in less sensitive cultivars (Rhue and Grogan, 1977). In wheat, however, the results on aluminum-induced calcium deficiency are inconsistent (Foy *et al.*, 1974), and in cowpea calcium deficiency is not a primary effect of aluminum toxicity (Horst *et al.*, 1983). Long-term studies have revealed a range of changes in mineral composition in both roots and shoots. Many of the effects, however, are probably secondary and are due to impairment of root metabolism and root growth. This impairment might be in part responsible for the inability of sensitive genotypes to increase the solution pH in the presence of aluminum (Table 16.5).

It is characteristic of crop species as nonaccumulators of aluminum (Section 16.3.5.2) that, regardless of their tolerance, the transport of aluminum into the shoots is severely restricted and accumulation is confined to the roots (Table 16.5). The difficulty of interpreting the effects of aluminum on phosphorus uptake is reflected in the negative correlation between inhibition of phosphorus uptake and aluminum tolerance, which is probably a secondary effect related to the role of pH in the formation of aluminophosphate complexes. In soils the major influence of aluminum on phosphorus uptake is presumably exerted by strong phosphorus adsorption to clay minerals and by a reduction of the root surface area, which is the main prerequisite for a high rate of phosphorus uptake.

The inhibition of shoot growth of legumes in acid soils may be related to

Table 16.6
Effect of Liming a Highly Aluminum Saturated Soil on the Growth and Nodulation of Soybean[a]

Soil pH	Al Saturation[b] (%)	Dry wt (g/plant)		Nodulation		
		Shoot	Root	No./plant	Dry wt (mg/nodule)	N Content (mg N/shoot)
4·55	81	2·4	1·07	21	79	65
5·20	28	3·2	1·08	65	95	86
5·90	4	3·6	1·08	77	99	93

[a]Reproduced from Sartain and Kamprath (1975) by permission of the American Society of Agronomy.
[b]Percentage of cation-exchange sites (CEC) saturated by aluminum.

impaired nodulation (Table 16.6). Liming decreases aluminum saturation and increases nodulation and N_2 fixation. Contrary to what might be expected, the shoot dry weight but not the root dry weight is affected by the soil pH. The root morphology changes, however, and with an increase in soil pH the proportion of small-diameter lateral roots increases (Sartain and Kamprath, 1975). The correlation between soil acidity, nodulation, and N_2 fixation is a general phenomenon in legume species (Munns *et al.*, 1977) and is presumably caused by the interaction of several factors such as changes in root morphology and thus in infection sites, calcium concentrations that are suboptimal for nodulation (Chapter 7), and aluminum-inhibited calcium uptake. There is substantial evidence that changes in the root morphology of *Stylosanthes* is responsible for the inhibition of nodulation by high aluminum concentrations (Carvalho *et al.*, 1982).

Although less common, inhibition of nodulation in legumes can be caused by limited persistence of *Rhizobium* strains in acid soils, as shown for *R. meliloti* and the nodulation of annual *Medicago* species (Robson and Loneragan, 1970b).

16.3.4 Physiological Effects of Excessive Manganese Levels

Manganese, unlike aluminum, does not injure the roots directly but affects the shoot, regardless of the manganese tolerance of a species or cultivar (Table 16.7). Genotypical differences in manganese tolerance are related not to differences in uptake or transport to the shoot but to the manganese tolerance of the shoot tissue.

The physiological effects of an excessive manganese supply on the uptake of other mineral nutrients, on metabolism, and on phytohormone balance were discussed in Section 9.2. Of particular importance for plant growth in acid mineral soils is the inhibition of calcium and magnesium uptake by high manganese concentrations. Crinkle leaf in young leaves and chlorotic

Table 16.7
Effect of the Manganese Concentration in the Nutrient Solution on the Dry Weight and Manganese Content of Soybean Cultivars[a]

Cultivar	Mn Supply (mg/liter)	Dry weight (g/plant)		Mn Content of shoots (μg/g dry wt)
		Shoots	Roots	
T 203	1·5	5·4	0·61	208
	4·5	6·6	0·55	403
	6·5	7·0	0·55	527
Bragg	1·5	5·7	0·59	297
	4·5	5·3	0·64	438
	6·5	4·5	0·68	532

[a]Based on Brown and Devine (1980).

speckling in mature leaves are well-known symptoms of manganese toxicity in dicotyledonous species grown in acid soils and are probably expressions of induced deficiency of calcium and magnesium, respectively. Under these conditions visible symptoms of manganese toxicity are observed even at levels which may decrease growth only slightly, in contrast to aluminum toxicity, which severely inhibits growth without producing clearly identifiable symptoms in the shoot. Hence in acid mineral soils with high exchangeable levels of both aluminum and manganese the growth depression observed may be erroneously attributed to manganese toxicity when in fact aluminum toxicity is the more important of the two factors (Foy *et al.*, 1978). In legumes manganese toxicity also depends on the mode of nitrogen nutrition. When the supply of manganese to bean is large, the nitrogen content of the shoots decreases to a much greater extent in plants relying on N_2 fixation than in plants fed bound nitrogen (Döbereiner, 1966). Similar negative effects of high manganese levels on nodulation have been observed in other legume species (Foy *et al.*, 1978), although, at least in isolated culture, most *Rhizobium* strains are more sensitive to aluminum than to manganese (Keyser and Munns, 1979). In conclusion, nodulation is a very critical step for legumes in acid mineral soils; it is adversely affected by a combination of high aluminum and/or manganese and low calcium concentrations. Low phosphorus availability also has a negative effect (Chapter 7).

16.3.5 Mechanism of Adaptation to Acid Mineral Soils

16.3.5.1 General

Plants adapted to acid mineral soils have a variety of mechanisms to cope with adverse chemical soil factors. These mechanisms are regulated sepa-

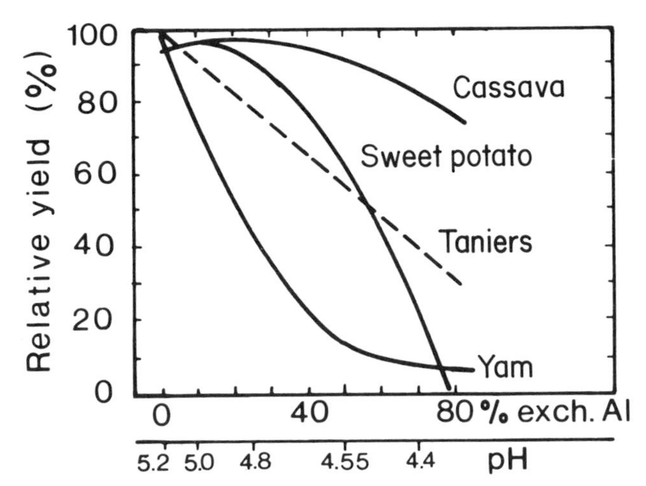

Fig. 16.6 Relationship between exchangeable aluminum (percentage of total cation-exchange capacity), soil pH and yield of four tropical root crops. (Redrawn from Abruna-Rodriguez *et al.*, 1982, by permission of the Soil Science Society of America.)

rately (e.g., those of aluminum and manganese tolerance), although some are interrelated (e.g., those of aluminum tolerance and phosphorus efficiency). From the agronomical point of view, only the sum of the individual mechanisms is of importance because it determines the requirement for fertilizers in general and for lime in particular.

Amongst the annual root crop species, cassava is known for its high tolerance to acid soils, compared, for example, with sweet potatoes, taniers, and yams (Fig. 16.6). Other acid soil–tolerant crops are cowpea, peanut, and potato, whereas maize, soybean, and wheat are nontolerant species (Sanchez and Salinas, 1981). A large input of nutrients is necessary to adjust the soil chemical properties, mainly by liming, to the requirements of nontolerant plants (Fig. 16.6). It is interesting that despite the large yield differences brought about by an alteration in soil pH by liming in three of the four root crops, the macronutrient and manganese content of the leaves was hardly affected, except that of calcium (Abruna-Rodriguez *et al.*, 1982). Here, leaf analysis would be of limited value for determining both the mechanisms of adaptation and nutritional status of plants.

Differences in acid soil tolerance among cultivars of a given species can be quite large. For example, in an unlimed soil of pH 4·5 and 80% aluminum saturation, a traditional, adapted dryland rice cultivar produced ~2·3 tons of grain per hectare, compared with a modern "high-yielding" cultivar, which produced only 1 ton; the latter required ~6 tons/ha lime and a corresponding decrease in aluminum saturation to 15% to achieve the grain yield of the traditional, adapted cultivar in the unlimed soil (Spain *et al.*, 1975).

16.3.5.2 Aluminum Tolerance

Especially in the tropics, aluminum toxicity is the most important individual limitation on crop production in acid soils. Some of the existing genetic variability in aluminum tolerance of a certain crop species appears to have been introduced unintentionally by breeding the same crop species in different regions with high or low soil pH, as in the case of wheat (Foy *et al.*, 1974) or soybean (Lafever *et al.*, 1977).

Three major mechanisms are involved in aluminum tolerance:

1. Exclusion from uptake (excluder plants)
2. Inactivation in the roots (excluder/includer plants)
3. Accumulation in the shoots (includer plants).

Mechanism 3 exists mainly in highly aluminum tolerant species of natural vegetation, and only a few cultivated species are aluminum includers, such as tea plants, the old leaves of which may contain up to 30 mg aluminum per gram dry weight (Matsumoto *et al.*, 1976b). In crop species, mechanisms 1 and 2 predominate, and it is often difficult to differentiate between the two (e.g., to differentiate between precipitation or polymerization of aluminum at the root surface and/or in the free space and accumulation within the root cells). Speculations by Aniol (1984) on the induction of aluminum tolerance in wheat by formation of particular proteins for the endogenous inactivation of aluminum—similar to the metallothioneins (Chapter 8)—deserve further attention. The following discussion mainly considers factors which may be involved in the exclusion mechanism.

Rhizosphere pH. A slight increase in pH above 4 drastically reduces the concentration of Al^{3+} (Fig. 16.3). There is evidence that, in the presence of aluminum, tolerant cultivars of certain crop species tend to raise the external pH faster than sensitive cultivars both in the nutrient solution (Table 16.5) and in the rhizosphere of soil culture (Mugwira and Patel, 1977). A slight pH increase at the root surface or in the free space is probably sufficient to lower not only the aluminum solubility but also its charge and to lead to the formation of aluminum polymer species. This polymerization may facilitate phosphorus uptake and may be one of the explanations for the well-documented stimulatory effect of low aluminum concentrations on root and shoot growth of tolerant crop species and cultivars in general (Foy, 1983) and of tea plants in particular (Matsumoto *et al.*, 1976b).

An increase in the rhizosphere pH is brought about primarily by a shift in the cation/anion uptake ratio in favor of the anions. In aluminum-tolerant cultivars of wheat, a relatively higher rate of anion uptake (and corresponding OH^- or HCO_3^- release) is maintained even in the presence of aluminum (Dodge and Hiatt, 1972), probably due to higher rates of NO_3^- uptake and utilization by plants even in the presence of aluminum (Foy and Fleming, 1982). This mechanism of aluminum exclusion is not, however, observed in bean (Foy *et al.*, 1972). The ecological importance of water culture studies on root-induced pH changes as a mechanism of aluminum tolerance should not be overestimated, as only average values are obtained, whereas in soils the rhizosphere pH along the roots can differ remarkably (Section 15.3). In an acid mineral soil of pH 4·75 (bulk soil) the rhizosphere pH of an acid tolerant species (Norway spruce) supplied with ammonium nitrogen increased to 5·2 in the apical root zones and decreased to 4·2 in basal root zones (Häußling *et al.*, 1985). A high rhizosphere pH at apical zones will be particularly effective in preventing aluminum injury to root growth.

Mucilage–Mucigel. The primary toxic effect of aluminum on root growth is the inhibition of cell division in the apical meristem. The penetration of aluminum into the meristem is restricted by mucilage secreted in the apical root zones (Section 15.4). It well might be that genotypical differences in the rate of secretion, especially in response to root exposure to aluminum, are important factors in the capacity of plants to restrict penetration of aluminum into apical meristems (Horst *et al.*, 1982). However, experimental evidence for this assumption is lacking.

Cation-Exchange Capacity. As the cation-exchange capacity (CEC) of root tissue increases, there is a greater exchange adsorption of polyvalent cations in the apoplast. In acid soils this could lead to a higher aluminum accumulation as indicated, for example, by a close positive correlation between the CEC of different plant species and the aluminum content in their roots (Wagatsuma, 1983). There is indeed also some evidence for a negative correlation between CEC and aluminum tolerance in wheat and barley cultivars (Foy *et al.*, 1967) and in populations of certain species of wild plants (Büscher and Koedam, 1983). A general role of the CEC in this respect is unlikely, however; for example, dicotyledons, which have a high CEC, generally are not less aluminum tolerant than monocotyledons, which have a low CEC.

Complexation of Aluminum. Aluminum forms stable complexes with various synthetic and naturally occurring low-molecular-weight organic solutes such as polyphenols and organic acids. Citric acid is very effective in this

Table 16.8
Effect of Aluminum Treatments on the Dry Weight and Mineral Element Content of Maize[a]

Treatment[b]	Dry weight (g/plant)	Mineral element content (μmol/g dry wt)		
		Al	P	Ca
Shoots				
Control (−Al)	1·96	0·5	58	55
$Al(OH)_2Cl$	1·05	1·9	24	28
Al Citrate	2·09	0·9	54	55
Al–Soil organic matter extract	1·85	0·5	68	63
Roots				
Control (−Al)	1·14	0·2	96	103
$Al(OH)_2Cl$	0·51	276·0	91	37
Al Citrate	1·17	62·0	99	89
Al–Soil organic matter extract	1·03	14·0	77	71

[a]From Barlett and Riego (1972).
[b]The concentration of aluminum in aluminum-containing treatments was 0·33 m*M*.

respect; between pH 3·8 and 4·9 $[Al_3(OH)_4Citrate_2]^-$ is the predominant species (Wiese and Veith, 1975). Such complexes protect the roots from the harmful effects of high concentrations of free aluminum (Table 16.8). With aluminum citrate, aluminum accumulation in the roots is much lower than with inorganic aluminum, and there is only a minor reduction in the phosphorus and calcium content of the roots and shoots. Root and shoot growth is comparable to that of controls (−Al). Soil organic matter extracts containing various soil-borne complexing agents are about as effective as citrate.

The release of relatively large amounts of organic acids is a typical feature of plant species adapted to highly acid mineral soils, for example, certain lupin species (Chapter 15) and tea plants (Jayman and Sivasubramanian, 1975). Complexation of aluminum by organic acids not only provides protection against the harmful effects of free aluminum on root growth; it is also important for the acquisition of phosphorus (Section 15.4). In highly tolerant species such as tea plants, aluminum complexes formed in the rhizosphere are readily taken up and translocated to the leaves, where they are preferentially deposited in the epidermal layer (Matsumoto *et al.*, 1976b).

A similar mechanism is assumed to operate in certain *Eucalyptus* species adapted to extremely acid, phosphorus-deficient soils (Mulette *et al.*, 1974). Complexation of aluminum by polyphenols or organic acids leached from leaves or litter might offer an indirect way for certain *Eucalyptus* species to achieve both high aluminum tolerance and acquisition of phosphorus from extremely deficient soils (Ellis, 1971).

Screening for Aluminum Tolerance. Field screening in acid soils is labor intensive, requires several months for completion, and is often influenced by secondary factors such as differential disease resistance of cultivars. To avoid these problems, rapid screening methods have been developed, mainly in water culture systems. In most crop species the relative root length of plants exposed versus those not exposed to aluminum is the most appropriate parameter. Because aluminum toxicity to roots is affected by a number of factors (e.g., concentration of aluminum, calcium, and magnesium, solution pH) the biggest problem in developing rapid screening techniques is finding an appropriate combination of these factors to use (Rhue and Grogan, 1977). In some instances the classification of genotypes (cultivars) based on their aluminum tolerance in the rapid screening methods correlates well with the growth response of these genotypes in acid soils, for example, with wheat and barley (Foy *et al.*, 1967, 1974) and dryland rice (Howeler and Cadavid, 1976). Quite often, however, the correlations are poor, for example, with dryland rice (Nelson, 1983), or the results differ among the various acid soils, for example, with wheat and triticale (Mugwira *et al.*, 1981). These discrepancies are not surprising and indicate that factors such as rhizosphere pH have been insufficiently considered, and/or that factors other than excessive aluminum levels were involved and may have had an even more harmful effect on growth.

16.3.5.3 Manganese Tolerance

Mechanisms of manganese tolerance are located in the shoots, as indicated, for example, by reciprocal root stock–scion grafts of tolerant and nontolerant genotypes (Heenan *et al.*, 1981). It is the genotype of the scion that determines the tolerance of the plant to high manganese concentrations in the substrate and within the leaf tissue.

The critical toxicity concentrations of manganese in leaf tissue differ to a great extent among plant species (Section 9.2), as well as cultivars of a species such as soybean (Table 16.7) and cowpea (Horst, 1983). In mature leaves of tolerant cultivars manganese is uniformly distributed, whereas in nontolerant cultivars with the same manganese content, a local spotlike accumulation of manganese occurs, which is correlated with chlorosis and necrosis at the sites of accumulation (Horst, 1983). The manganese distribution in leaf blades of nontolerant and tolerant cultivars resembles that obtained in a given cultivar supplied with an excessive amount of manganese in the absence and presence of silicon, respectively (Section 10.3). Restricted translocation of manganese to young leaves might also be involved in the high manganese tolerance of certain lettuce cultivars (Blatt and Diest,

1981). It well might be that the formation of manganese-oxalate complexes with high stability found in tolerant species (Memon and Yatazawa, 1984) is also of importance for genotypical differences in manganese tolerance.

Tolerance to excessive levels of manganese is not necessarily correlated with tolerance to excessive levels of aluminum (Nelson, 1983) and vice versa, as has been shown for wheat cultivars (Foy *et al.*, 1973). Separate screening for manganese tolerance and aluminum tolerance is therefore necessary; a combination is required only under certain soil conditions. The application of manganese to the petioles of leaves provides a simple, rapid, and nondestructive method for screening cowpea for manganese tolerance during vegetative growth (Horst, 1982).

Manganese tolerance during vegetative growth is not necessarily correlated with tolerance during reproductive growth. At least in legumes (soybean, cowpea) manganese toxicity may reduce grain yield more than vegetative growth. For example, under field conditions some cowpea genotypes growing in a soil with an excessive manganese level produce little or no grain despite vigorous vegetative growth (Kang and Fox, 1980). The application of manganese to the peduncle seems promising as a technique for screening for manganese tolerance during reproductive growth (Horst, 1982).

16.3.5.4 Nutrient Efficiency

Adaptation to acid soils requires highly efficient uptake and/or utilization of nutrients, especially phosphorus, calcium, and magnesium. Many of the plant species considered to be adapted to acid mineral soils are usually heavily infected with mycorrhizas (Sanchez and Salinas, 1981). A typical example is cassava, the high phosphorus efficiency of which relies entirely on mycorrhizal infection (Section 15.7). Tolerance to aluminum and high phosphorus efficiency also coexist in certain cultivars of wheat and dryland rice (Sanchez and Salinas, 1981), as would be expected from the effect of aluminum on root growth, on the one hand, and the importance of root surface area for the acquisition of phosphorus from soils, on the other.

In contrast to phosphorus efficiency, a high calcium efficiency is usually related to utilization within plants. In this respect genotypical differences can be quite distinct, as shown for cowpea in Table 16.9. With a large calcium supply, the growth and calcium content of both cultivars are similar. With a small supply, however, the growth of Solojo is drastically reduced despite the relatively higher calcium content of the tissue compared with that of Tvu 354. Solojo would probably fail in an acid soil due to calcium deficiency unless a relatively large amount of calcium fertilizer were supplied.

Table 16.9
Effect of the Calcium Concentration in the Nutrient Solution on the Dry Weight and Calcium Content of Cowpea Cultivars[a]

	1250 μM Ca		5·0 μM Ca	
	Tvu 354	Solojo	Tvu 354	Solojo
Dry weight (g/plant)	3·38	3·80	2·86	0·94
Calcium content (mg/g dry wt)				
Leaves	21·6	20·0	1·15	1·67
Stem	10·7	10·5	0·81	1·01
Roots	2·4	2·4	0·65	0·83

[a]From W. J. Horst (unpublished data).

In acid mineral soils molybdenum availability is very low (Section 9.5). Thus molybdenum efficiency is sometimes involved in adaptation to acid mineral soils. This was demonstrated by Brown and Clark (1974) in a comparison of two maize inbred lines, Pa 36 and WH, grown in an acid soil (pH 4·3). The poorer growth of Pa 36 compared with that of WH was caused by insufficient molybdenum uptake. Although Pa 36 was much more efficient than WH in taking up phosphorus from a soil with a very low phosphorus content, even in the presence of aluminum, low molybdenum efficiency may limit the overall adaptation of Pa 36 to certain acid mineral soils.

16.3.5.5 Source of Nitrogen

In strongly acid soils nitrification is inhibited and NH_4^+ is the most important source of nitrogen for plant species other than legumes. Usually at low pH, plants adapted to acid soils also tolerate NH_4^+ levels which are toxic to nonadapted plants (Gigon and Rorison, 1972). On the other hand, acid-tolerant plants supplied only with nitrate often grow poorly and develop chlorosis. This is due either to induced iron deficiency or to inherently low levels of nitrate reductase activity in some plant species (Kinzel, 1983).

With only a few exceptions, acid mineral soils have very low levels of both total and available nitrogen. Under tropical conditions, associative N_2 fixation in the rhizosphere may meet a substantial proportion of the nitrogen requirement of nonlegume C_4 species (Section 7.5). In legumes the large phosphorus, calcium, and molybdenum requirement for nodulation and excessive aluminum and/or manganese levels may become limiting factors for N_2 fixation. Low rates of fertilizer application (e.g., molybdenized rock phosphate) are usually essential to cover the demand for mineral nutrients, but high aluminum concentrations remain a problem for nodulation

(Carvalho *et al.*, 1982). The aluminum tolerance of legumes therefore plays a key role in the low-input approach to tropical crop production.

16.4 Waterlogged and Flooded Soils

16.4.1 Soil Chemical Factors

In waterlogged* soils, air is displaced from the pore spaces either to different depths of subsoil (which results in a high water table) or in the topsoil. This often occurs in temperate climates during the winter and spring and also, temporarily, during the summer as a result of heavy rainfall or irrigation. Flooded or submerged soils are permanently below the water table, or at least under water for several months every year. Paddy soils are the most prominent agricultural example of the latter (Ponnamperuma, 1972). Since oxygen (and other gases) diffuse in air about 10^3 to 10^4 times more rapidly than in water or water-saturated soils (Armstrong, 1979), oxygen is depleted more or less rapidly by the respiration of soil microorganisms and plant roots in waterlogged soils. Various degrees of anaerobiosis, including anoxia (the absence of molecular oxygen), occur. The reduction of mineral nutrients in these soils takes place roughly in the sequence shown in Table 16.10.

Table 16.10
Redox Reactions in Relation to the Soil Redox Potential

Redox reaction	Redox potential *E* (mV) at pH 7
Onset of nitrate reduction (denitrification)	450–550
Onset of Mn^{2+} formation	350–450
Absence of free O_2	350
Absence of nitrate	250
Onset of Fe^{2+} formation	150
Onset of sulfate reduction (H_2S formation)	−50
Absence of sulfate	−180

Based on Brümmer (1974).

After the disappearance of molecular oxygen, other compounds are used by soil microorganisms as alternative electron acceptors for respiration. The first is nitrate, which is reduced to nitrite (NO_2^-), nitrous oxide (N_2O), and molecular nitrogen (N_2). Especially in soils that are alternately wet and dry, nitrite can accumulate temporarily during denitrification and, to a certain degree, also during nitrification ($NH_4^+ \rightarrow NO_3^-$). Manganese oxides

*The terms *waterlogged* and *flooded* are, as a rule, used synonymously to describe soils with excessive water levels.

[mainly Mn(IV)] are the next electron acceptors. In acid soils high in manganese oxides and organic matter but low in nitrate, very high levels of water-soluble and exchangeable Mn^{2+} can build up within a few days. After prolonged waterlogging the reduction of Fe(III) occurs. For plant species adapted to flooded soils (wetland species, e.g., rice) enhanced iron reduction increases the iron availability, and quite often plants may suffer from excessive iron levels. Iron reduction increases phosphate solubility and availability to wetland species if $Fe(III)PO_4$ is present in the soil in sufficient amounts. The reduction of sulfate to H_2S in submerged soils decreases the solubility of iron, zinc, copper, and cadmium by the formation of sparingly soluble sulfides (Ponnamperuma, 1972; Herms and Brümmer, 1979). For wetland species such as rice, the formation of heavy-metal sulfides is both an advantage and a disadvantage: It may prevent iron toxicity but can lead to zinc deficiency.

Various products of microbial carbon metabolism, such as ethylene, accumulate in waterlogged soils. During prolonged waterlogging, volatile fatty acids and phenolics accumulate in soils high in readily decomposable organic matter (e.g., green manure, straw), which has a detrimental effect on root metabolism and growth (Chapter 14).

16.4.2 Plant Responses

16.4.2.1 Waterlogging Injury

Most plant species not adapted to waterlogging (nonwetland species) develop injury symptoms sequentially over a period of several days if waterlogging continues. Wilting and, in herbaceous species, epinasty (downward bending of leaves) are likely to be the first symptoms. A decrease in the water permeability of the roots and an accumulation of ethylene in the shoots are responsible for wilting and epinasty, respectively (Bradford *et al.*, 1982). A decline in or cessation of shoot extension is another typical symptom, followed after several days of waterlogging by enhanced senescence of the lower leaves, indicating nitrogen deficiency. Waterlogging can impair the nitrogen nutrition of legumes by interfering with nodulation (Tran Dang Hong *et al.*, 1977).

The severity of the effects of waterlogging on growth and yield depends on the plant species, developmental stage of the plants, soil properties (e.g., pH, organic matter content), and soil temperature in particular. Peas are especially sensitive to waterlogging. After only one day or several days, enhanced senescence occurs, and shortly after flowering 24 hr of waterlogging can depress yields appreciably (Cannell *et al.*, 1979). Because low soil temperatures decrease the oxygen requirement for respiration, the drop in redox potential, and thus injury from waterlogging is less severe at low than at high soil temperatures (Trought and Drew, 1982). Flooding of wheat for 30 days during grain filling at soil temperatures of 15 and 25°C reduced grain yield by about 20 and 70%, respectively (Luxmoore *et al.*, 1973).

Whether the principal cause of waterlogging injury to the shoots is related directly to oxygen deficiency in the roots or more indirectly to the production of toxic substances in the soil depends on the circumstances, which are discussed in the following section, with emphasis on plant metabolism and mineral nutrition. Comprehensive reviews of the subject are those of Cannell (1977) and Drew (1979).

16.4.2.2 Growth and Nutrient Uptake

Short-term responses of plants to anaerobic soil conditions can readily be demonstrated by waterlogging of a previously well-aerated soil. The growth of existing roots ceases immediately (Fig. 16.7), and they may die within a few days. In contrast, shoot growth continues for several days at a similar or even somewhat higher rate, although visible symptoms of waterlogging injury (transient wilting, inhibition of leaf extension, and chlorosis) are observed within a few days (Trought and Drew, 1980a).

Cessation of root growth and root respiration leads to a drastic drop in the uptake and transport of mineral nutrients to the shoot within a few days of waterlogging (Table 16.11). Because the shoot dry weight continues to increase, the nutrient concentration in the shoot declines by "dilution." There is evidence that inhibited nutrient uptake and thus nutrient deficiency are directly responsible for enhanced leaf senescence and cessation of shoot growth in plants subjected to waterlogging (Trought and Drew, 1980a). For example, symptoms of enhanced leaf senescence induced by waterlogging

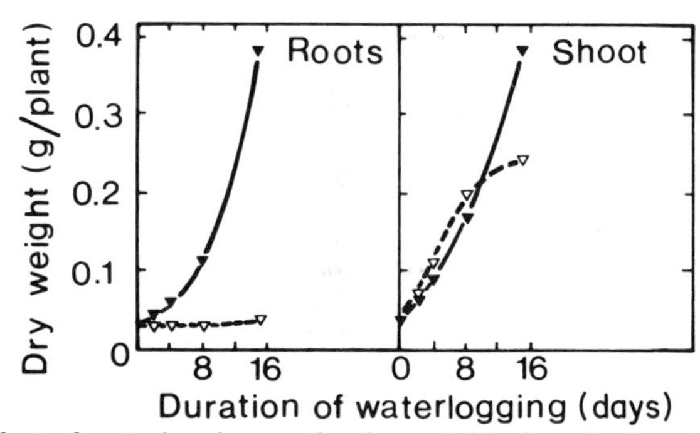

Fig. 16.7 Effect of waterlogging on the dry weight of seminal roots and shoots of winter wheat seedlings (▼, control; ▽, waterlogged). (Based on Trought and Drew, 1980a.)

Table 16.11
Effect of Waterlogging on the Growth and Mineral Nutrient Content of the Shoots of Barley Seedlings[a]

	2 Days of waterlogging		6 Days of waterlogging		Net uptake, 2–6 days
	Control	Waterlogged	Control	Waterlogged	(% of control)
Leaf extension[b] (cm)	6·4	4·2	12·3	5·2	—
Shoot dry weight (mg/plant)	170	170	380	360	—
Content of the shoots (μmol/g dry wt)					
Nitrate	390	139	470	14·3	9·9
Phosphorus	217	149	210	71	2·9
Potassium	1540	1190	1420	615	9·6

[a]From Drew and Sisworo (1979).
[b]Leaf no. 3 (youngest).

can be prevented by the daily application of nitrogen (nitrate or ammonium) to the surface of a waterlogged soil where new roots are developing (Trought and Drew, 1980b). Also, in long-term experiments, heavy nitrogen fertilizer dressing alleviated waterlogging injury to cereal crops (Watson *et al.*, 1976) due both to compensation for losses by denitrification and to impaired uptake from poorly aerated soils.

The beneficial effect of additional nitrogen application on shoot growth should not be overestimated and generalized, however, because the uptake of other mineral nutrients is also impaired (Table 16.11). Furthermore, nutrient deficiency is only one aspect of waterlogging injury. In soils high in organic matter and nitrate, sudden waterlogging might lead to an accumulation of nitrite in the soil solution to concentrations which are toxic to the roots of sensitive plant species. Tobacco, for example, is injured by nitrite concentrations as low as 5 mg/liter, but values 10 times higher than this are often found in waterlogged soils (Hamilton and Lowe, 1981). Waterlogging injury caused primarily by manganese toxicity occurs in plant species with low inherent manganese tolerance such as alfalfa (Table 16.12), especially in

Table 16.12
Effect of a 3-Day Period of Waterlogging on the Shoots of Alfalfa Grown in a Soil with a Large Organic Matter Content[a]

Soil application (g lime/kg soil)	Water-logging	Soil pH	Shoot dry weight (g/pot)	Mn Content of shoots (mg/kg dry wt)
0	–	4·8	3·1	426
0	+	5·2	1·2	6067
2·5	–	6·4	5·7	99
2·5	+	6·7	3·0	957

[a]Based on Graven *et al.* (1965).

R

acid soils with high levels of manganese oxides. At pH ~5, a lack of a substantial amount of nitrate as an alternative electron acceptor for even a 3-day period of waterlogging leads to a manganese content in the leaves which is extremely toxic to alfalfa. Although liming cannot prevent manganese toxicity induced by short-term waterlogging it can at least lower considerably the manganese content and thus the detrimental effects on growth.

A particular type of waterlogging injury may occur in semiarid and arid regions after heavy irrigation with saline water. In most crop species salt tolerance is based on mechanisms which prevent or at least restrict salt accumulation in the shoots (exclusion mechanisms, Section 16.6). These mechanisms rely on a high metabolic activity in the roots and thus, in nonwetland species, on soil aeration. At a given salinity level the levels of sodium and chloride in the leaves are therefore increased by both an increase in temperature and waterlogging (Table 16.13). Because of the higher oxygen requirement at 20°C than at 10°C, waterlogging is much more effective at 20°C in increasing the levels of sodium and chloride in the leaves. Salt injury is therefore more likely to take place at 20°C than at 10°C. Similar results have been found in apple trees (West, 1978). In sunflower, waterlogging increased the sodium content in the leaves much more than the chloride content, indicating the impairment of an exclusion mechanism (Na^+ efflux pump, Section 16.6.5) in the roots not adapted to O_2 deficiency in the root environment (Kriedemann and Sands, 1984). This interaction between salinity and soil aeration should be considered in the practice of irrigation when saline water has to be used, especially on poorly structured soils.

Table 16.13
Sodium and Chloride Content of Tomato Leaves after 15 Days of Exposure to Saline Solution[a]

		Leaf content (% dry wt)	
Temperature	Treatment	Sodium	Chloride
10°C	Drained	1·53	2·61
	Waterlogged	1·84	3·04
20°C	Drained	2·96	3·25
	Waterlogged	5·82	9·49

[a]Saline solution: 90 m*M* NaCl. From West and Taylor (1980).

16.4.2.3 Phytohormones and Ethylene

Waterlogging inhibits the synthesis and shoot export of cytokinins and gibberellins (Chapter 14). Inhibition of leaf extension and enhanced leaf senescence in waterlogged plants are thus in part a reflection of a deficiency of these phytohormones in the shoots. Correspondingly, foliar application of cytokinins and gibberellins may counteract at least temporarily the inhibition of shoot elongation and enhanced leaf senescence imposed by waterlogging (Jackson and Campbell, 1979).

The hormonal effects of ethylene are attracting a great deal of interest. In waterlogged soils as well as in the roots of plants growing in them, exceptionally high levels of ethylene build up as a result of slow diffusion rates of the gas in water as compared to air. Water acts as a barrier to the escape of ethylene produced in roots and other submerged tissue ("waterjacket effect"). Accumulation of ethylene in soils is important even at oxygen concentrations in the soil atmosphere below 9% (Hunt *et al.*, 1981). In waterlogged soils, ethylene is produced both by microorganisms and by roots. A substantial proportion is translocated from the roots to the shoots (Table 16.14), where it is responsible for the epinasty of leaves. Strict anaerobiosis (anoxia) stimulates the synthesis in the roots of the immediate precursor of ethylene (1-aminocyclopropane-1-carboxylic acid), which is translocated to the leaves and there converted to ethylene (Bradford *et al.*, 1982).

Table 16.14
Effect of Flooding Pea Plants on the Concentration of Ethylene in the Soil Solution and Roots and Shoots[a,b]

	Ethylene (μl/liter)		
	Soil solution	Roots	Shoots
Control (aerated)	0·01	0·33	0·25
Flooded	0·97	1·01	0·93

[a]Based on Jackson and Kowalewska (1983).
[b]Pea plants were flooded for 4 days. Ethylene was extracted under partial vacuum.

Although elevated levels of ethylene in the rooting medium inhibit root elongation (Chapter 14), they simultaneously trigger anatomical and morphological adaptations to waterlogging and submerged conditions. Ethylene either supplied externally or produced endogenously by a lack of aeration promotes the development of prominent air spaces (aerenchyma) in the root cortex (Fig. 16.8). This morphogenetic response is not restricted to the roots but is also observed in the basal part of the stems (Kuznetsova *et al.*, 1981) and is essential for the internal ventilation of roots growing in anaerobic environments (see Section 16.4.3.2).

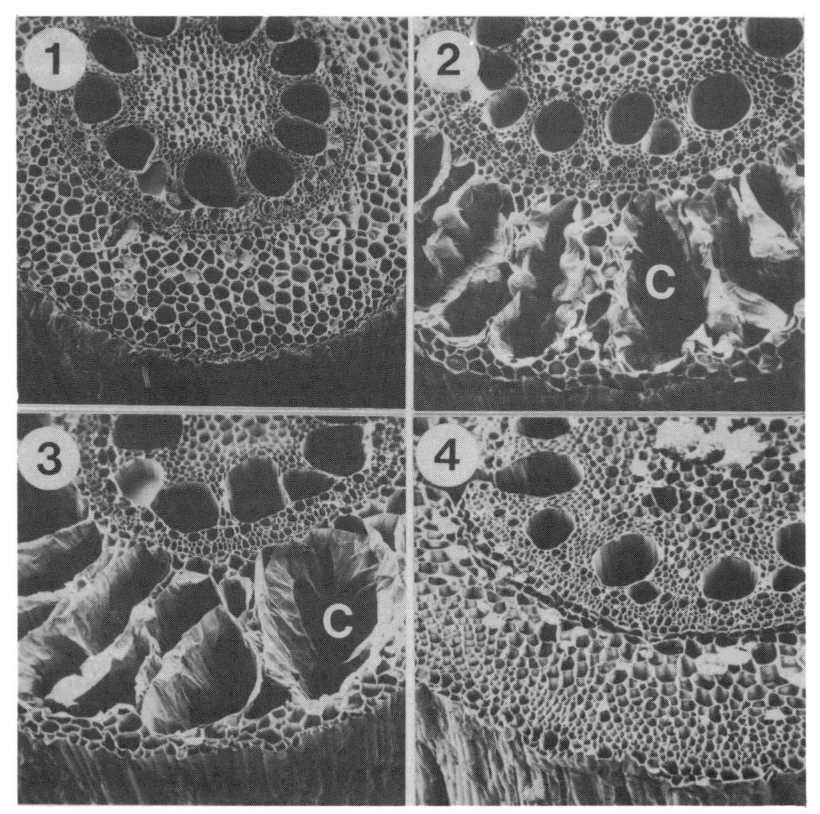

Fig. 16.8 Transverse sections of maize roots under a scanning electron microscope. (1) Control grown in well-aerated solution; (2) root receiving 5 μl/liter ethylene in air; (3) root from nonaerated solution; (4) root receiving nitrogen gas (anoxic treatment). C, Cortical air space. (From Drew *et al.*, 1979.)

16.4.2.4 Ethanol and Other Organic Phytotoxins

Anaerobic root respiration leads to the formation of acetaldehyde and, particularly, ethanol. In tomato even a 24-hr period of waterlogging was sufficient to increase the ethanol concentration in the xylem exudate to 5 to 7 m*M* which is toxic to tomato shoots (Fulton and Erickson, 1964). Similarly, after 3 to 4 days of flooding, the ethanol concentration increased sharply in the roots of pea plants, and the release of ethanol both into the rhizosphere and into the xylem also increased, the concentration rising to 2·1 m*M*, compared with 0·07 m*M* in the nonflooded control (Jackson *et al.*, 1982). It has been suggested by McManmon and Crawford (1971) and Crawford and Zochowski (1984) that the accumulation of ethanol is mainly responsible for flooding injury to nonwetland species. There are substantial doubts, however, as to the generalizability of this concept (see Section 16.4.3.3).

In soils high in organic matter and simultaneously high in microbial activity prolonged periods of waterlogging lead to the accumulation of volatile fatty acids and phenolics in the soil, which are additional stress factors affecting root metabolism and growth, especially at low soil pH (Chapter 14).

The relative importance of toxic substances that accumulate in the soil, and of root-borne toxins from waterlogging depends on particular circumstances. For example, in nonwetland species sudden waterlogging at a high soil temperature primarily affects root metabolism via anaerobiosis, whereas after prolonged periods of waterlogging of soils high in organic matter accumulation of soil-borne toxins may become increasingly injurious (Chapter 14). However, both the severity and type of injury depend mainly on the plant species and even the cultivar within a species.

16.4.3 Mechanisms of Adaptation

16.4.3.1 Avoidance versus Tolerance

Plants differ widely in their capacity to adapt to waterlogging, as is apparent from the differences between nonwetland and wetland species. Extreme examples of crop plants are barley and wetland rice. Other well-known genotypical differences exist among forage species (Cannell, 1977). The principal stress factor for plants in waterlogged soils is an oxygen deficit. According to the general stress concept of Levitt (1980) adaptation can be achieved by avoidance of the stress factor and/or stress tolerance:

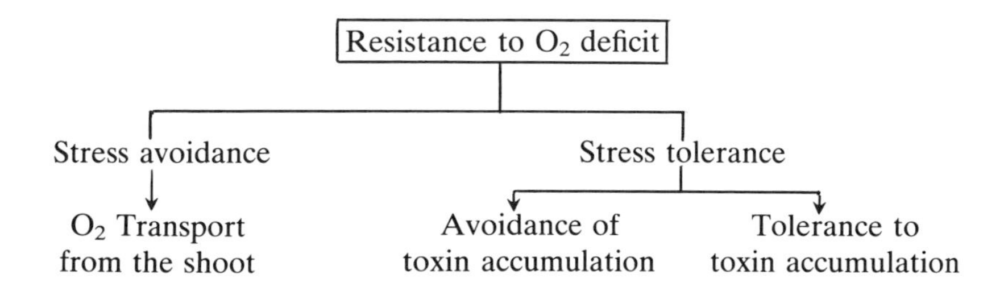

There is increasing evidence that, in general, avoidance is the principal mechanism and that tolerance plays an additional role in adaptation to an oxygen deficit in the soil.

16.4.3.2 Anatomical and Morphological Adaptations

Transport of oxygen from the shoots to the roots and the rhizosphere is readily demonstrated in both wetland and nonwetland species (Greenwood, 1967). Oxygen transport takes place to a limited extent in intercellular spaces; the main pathway, however, is the aerenchyma in the root cortex (Fig. 16.8). The proportion of air-filled intercellular spaces and of aerenchyma is an expression of root porosity (air-filled spaces as a proportion of the total root volume). Root porosity differs among plant species and is also adaptive (Fig. 16.8). For wetland rice, maize, and barley grown in an aerated nutrient solution, the relative values for root porosity are 1·0, 0·25, and 0·10, respectively (Jensen *et al.*, 1967). In a given range, the root system of nonwetland species has the capacity to adapt to waterlogging conditions (Table 16.15). Plants were grown in well-drained soil for 2 weeks, and thereafter some of them were flooded. With the exception of barley, the root porosity of all plants tested was higher under conditions of flooding than of nonflooding. Maize showed the greatest degree of adaptation; the wheat cultivar Pato behaved similarly. The differences in the root porosity of wheat cultivars corresponded well with the higher tolerance of Pato than of Inia to waterlogging under field conditions (Yu *et al.*, 1969).

Table 16.15
Rooting Depth and Root Porosity of Plants Grown under Drained and Flooded Conditions in a Loam Soil[a]

	Root porosity (%)		Rooting depth (cm)	
Plant species, cultivars	Drained	Flooded	Drained	Flooded
Maize	6·5	15·5	47	17
Sunflower	5	11	33	15
Wheat				
cv. Pato	5·5	14·5	10	5
cv. Inia	3·0	7·5	23	10
Barley	3·5	2·0	32	15

[a]Data recalculated from Yu *et al.* (1969).

Changes in root anatomy are closely correlated with changes in root morphology. After waterlogging many old roots die, but numerous adventitious roots with well-developed aerenchyma emerge from the base of the stem and grow to a limited extent into the anaerobic soil. Whether existing roots die on sudden waterlogging (anaerobiosis) and new (adapted) roots have to be formed, or whether the development of aerenchyma is enhanced in the existing roots, depends mainly on the plant species. The principal differences among species are shown schematically in Fig. 16.9.

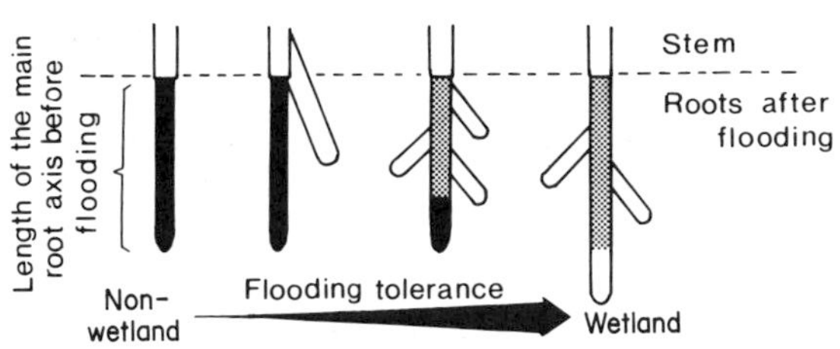

Fig. 16.9 Suggested relationship between the responses of roots of nonwetland and wetland species to a limited period of soil flooding. Black areas, dead tissue; stippled areas, surviving tissue; white areas, regrowth. (Based on Armstrong, 1979.)

In nonwetland species the internal ventilation (Armstrong, 1979) is insufficient to prevent an oxygen deficit from occurring in the roots, the apical meristems in particular, when the external supply of oxygen is cut off by flooding. New roots have to be formed in the basal zones or at the stem. Roots of wetland species, such as rice, with an inherently higher ventilation are capable of adapting to anaerobiosis even in aerated substates by enlargement of the aerenchyma. Air spaces are formed behind the apical meristem, and the aerenchyma in the basal zones may occupy 20 and up to 50% of the total root volume in well-drained and flooded soils, respectively (Armstrong, 1979). The aerenchyma fulfils a dual role; it provides a low diffusion resistance pathway for oxygen transport from the shoots, and it simultaneously diminishes the respiratory demand of the basal root zones. Stimulation of aerenchyma formation by flooding also takes place in the basal part of the stems, even in nonwetland species such as maize (Kuznetsova *et al.*, 1981). In deep-water rice, where a major part of the leaves is submerged in the early growing stages, the main pathway of oxygen diffusion seems to be not in the leaf aerenchyma but along a continuous air layer between the hydrophobic leaf surface and the surrounding water (Raskin and Kende, 1983).

Although the molecular mechanism is still not fully understood, it is well established that elevated levels of ethylene are partly responsible for aerenchyma formation in roots and stems (Kawase, 1981; see also Fig. 16.8). There is also some evidence that genotypical differences in tolerance to flooding are positively correlated with differences in the rate of ethylene formation in roots in response to flooding (Blake and Reid, 1981).

In wetland species oxygenation of the rhizosphere through the release of molecular oxygen by the roots is vital for plants growing in anaerobic soils (Yoshida and Tadano, 1978). Differences in the capacity of cultivars of wetland rice to release oxygen from the roots and thus oxidize Fe^{2+} in the rhizosphere are correlated with a resistance to bronzing disease, or Akagare disease (Armstrong, 1969), which is caused by iron toxicity (Howeler, 1973). These differences in the "oxidation power" of roots are readily apparent from differences in the amount of FeOOH precipitated on the roots when grown under field conditions in flooded soil (Table 16.16). The largest amount of oxygen was released from the roots of the cultivar Brazos. This cultivar also had the highest grain yield capacity when grown in different locations. It has been calculated that at maturity ~500 kg FeOOH per hectare is present as root coating each season (Chen *et al.*, 1980).

Table 16.16
Relationship between the Rate of Oxygen Release from the Roots and FeOOH Precipitation on the Roots of Wetland Rice Cultivars[a]

	Brazos	Bluebelle	Lebonnet	Labelle
Oxygen release[b]	0·947	0·692	0·528	0·359
FeOOH (% of total root dry wt)	14·10	8·45	7·15	5·09

[a]Cultivars were grown in Katy fine sandy loam soil. Based on Chen *et al.* (1980).
[b]Increase in O_2 saturation in 10 ml H_2O per 5 min; laboratory studies.

In rice and probably other wetland species, the formation and stability of the aerenchyma are dependent on the silicon supply. When the silicon content of rice plants increased from 0·2 to 5·7% dry wt, the iron uptake from flooded soil decreased to about one-third (Okuda and Takahashi, 1965). Mineral nutrition also affects "oxidation power" indirectly: Nutrient deficiencies that increase the exudation of photosynthates from the roots simultaneously enhance microbial activity and oxygen consumption in the rhizosphere (Chapter 15) and thus increase the risk of iron toxicity (Yoshida and Tadano, 1978; Ottow *et al.*, 1982).

16.4.3.3 Metabolic Adaptations

Besides anatomical and morphological adaptations, metabolic adaptations are also involved in tolerance to waterlogging. On the basis of studies with plant species of different flooding tolerance a scheme has been proposed (McManmon and Crawford, 1971; Crawford and Baines, 1977) in which intolerant species suffer from accelerated glycolysis and ethanol production. In contrast, tolerant species avoid this acceleration and also undergo a metabolic switch from ethanol to malate production. The principles of this concept can be summarized as follows:

PEP → Pyruvate → Acetaldehyde → Ethanol
ADH
PEP Carboxylase
Malic enzyme
Oxaloacetate → Malate
– – → = flooding intolerant
——→ = flooding tolerant

Whether this concept is of general applicability, however, has been questioned for various reasons (Jackson *et al.*, 1982). For example, in flooding-tolerant species or cultivars exposed to O_2 deficiency (hypoxia) the level of alcohol dehydrogenase (ADH) activity or of ethanol may even be higher than that in nontolerant species (Wignarajah *et al.*, 1976; Tripepi and Mitchell, 1984). Lower levels of ethanol in roots of tolerant than in roots of intolerant species may therefore reflect a more extensive or rapid aerenchyma formation in response to flooding (i.e., stress avoidance rather than stress tolerance). Another argument against the metabolic flooding theory has been presented by Webb and Armstrong (1983), who showed that under conditions of anoxia, root tip viability is lower in wetland than in nonwetland species. Interestingly, the detrimental effect of anoxia on root tip viability could be postponed for several hours by an exogenous supply of glucose, indicating that a shortage of carbohydrates (perhaps due to impaired phloem unloading) in the root meristem is one of the main causes of injury.

Table 16.17
Effect of Manganese Supply on the Growth and Manganese Content of Mature Leaves of Barley and Wetland Rice[a]

Mn Supply (mg/liter)	Shoot dry weight (g/plant)		Mn Content (mg/kg dry wt)	
	Barley	Rice	Barley	Rice
0·2	14·0	14·5	70	100
0·5	12·1	15·5	190	400
2·0	6·5	15·0	310	2200
5·0	6·0	12·1	960	5300

[a]Data recalculated from Vlamis and Williams (1964).

Genotypical differences in the manganese tolerance of shoot tissue are closely related to tolerance to waterlogging (Finn *et al.*, 1961); barley and wetland rice represent extremes in this respect (Table 16.17). Whereas even less than 200 mg manganese per kilogram leaf dry weight are toxic to barley, more than 10 times this amount is tolerated by the leaf tissue of rice without any grown depression. In barley grown in soils high in manganese, sensitivity to waterlogging may well be related in part to the low manganese tolerance of the shoots. In any approach to improving the adaptation of crop plants to waterlogging, manganese tolerance should not be overlooked. In view of the impressive differences in the manganese tolerance of the shoot tissue of cultivars of species such as lettuce (Sonneveld and Voogt, 1975) or soybean (Section 9.2.7) the prospects of selecting for higher manganese tolerance in other crop species is promising (Section 16.3).

16.5 Alkaline Soils

16.5.1 Soil Characteristics

Alkaline soils (pH > 7) are very common in semiarid and arid climates. Calcareous soils cover more than 25% of the earth's surface. Their content of free $CaCO_3$ in the upper horizon varies from a few percent up to 95%. The predominant soil associations* are the rendzinas, which are shallow soils with high organic matter content overlying calcareous material. Less common are the chernozems and xerosols. The pH of calcareous soils is determined by the presence of $CaCO_3$, which buffers the soils in the pH range

Table 16.18
Relative Abundance of Alkaline Soils and Major Constraints on Plant Growth in These Soils

	Soil pH	
	7	8 9
Dominant soil associations	Rendzinas (chernozems, xerosols)	Solonetz (sodic soils) (solonchaks, saline soils)
Relative abundance		
Major nutritional constraints	Deficiency: Fe, Zn, P (Mn)[a]	Toxicity: Na, B Deficiency: Zn, Fe, P (Ca, K, Mg)
Other constraints	Excess of HCO_3^- Water deficit Mechanical impedance	Poor aeration Excess of HCO_3^- Water deficit Mechanical impedance

[a]Parentheses indicate less frequently, or only in certain situations.

7·5–8·5 (Table 16.18). At pH > 8, solonetz is the predominant soil association. Solonetz or sodic (alkali) soils are characterized by a sodium adsorption ratio (SAR)† of the clay greater than 15, and they often contain free sodium carbonate. Sodic soils usually occur in association with saline soils (solonchaks), and saline–sodic soils are more abundant than purely sodic soils. In the context of constraints on plant growth it is necessary, however, to make a clear distinction between salinity and sodicity. Saline soils are not necessarily alkaline, and plant growth is affected mainly by high levels of sodium chloride and impairment of water balance (Section 16.6).

*The term *soil association* refers to a unit composed of two or more kinds of soils and miscellaneous land areas, used in soil surveys.

†*The sodium adsorption ratio* or *SAR* describes a relationship between soluble sodium and soluble divalent cations which can be used to predict the exchangeable sodium percentage of soil equilibrated with a given solution:

$$\frac{\text{sodium}}{(\text{calcium} + \text{magnesium})^{1/2}}$$

The major nutritional constraints in calcareous soils differ from those in sodic soils (Table 16.18). The differences are directly related either to other soil chemical factors such as bicarbonate concentration or to soil physical factors. In sodic soils, poor physical conditions and correspondingly poor soil aeration are the major constraints and are often correlated with sodium and boron toxicity.

Nitrogen is the growth-limiting factor for most crop species (other than legumes) growing in alkaline soils. More than 90% of the soil nitrogen is organical bound nitrogen (humus) which becomes available to plants after mineralization by soil microorganisms. Both the total amount of soil nitrogen and its availabilty to plants are therefore closely related to the soil organic matter content and conditions of mineralization (soil moisture, temperature, aeration). The soil pH is only of minor importance for the level and turnover of nitrogen in alkaline soils.

16.5.1.1 Iron

Mineral soils have, on average, a total iron content of ~2%. Most crop species remove only between 1 and 2 kg iron per hectare annually. In well-aerated soils with a high pH, however, the concentration of Fe^{2+} and Fe^{3+} in the soil solution becomes negligible, and the total concentration of inorganic iron species [between pH 7 and 9 mainly $Fe(OH)_2{}^+$, $Fe(OH)_3$, and $Fe(OH)_4{}^-$)] in the soil solution is only around $10^{-10}M$ (Fig. 16.10). Although definite conclusions about the forms or concentrations of inorganic iron at

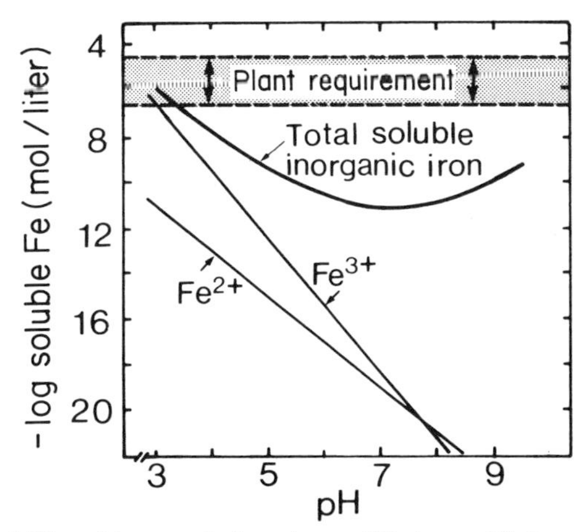

Fig. 16.10 Solubility of inorganic iron in equilibrium with iron oxide and 0·2 atm oxygen in comparison with the plant requirement for soluble iron at the root surface. (Modified from Lindsay, 1974.)

the root surface cannot be made, the concentration of chelated iron required for optimal growth is of the order of 10^{-6} to $10^{-5}M$ (Römheld and Marschner, 1981b; Lindsay and Schwab, 1982). These values have to be interpreted with reservation, however, as they are based on supply of synthetic iron chelates (e.g., Fe EDTA) which are utilized relatively poorly, at least in grasses (Section 16.5.3).

In aerated soils the solubility of iron is governed not by inorganic iron species but by the formation of chelated iron. The chelators are derived from soil organic matter, microorganisms, and from root exudates. For example, in a soil of pH 7·9 more than 35,000 times the concentration of soluble iron was found than predicted from inorganic equilibrium constants (O'Connor *et al.*, 1971). In alkaline soils with high organic matter content the concentration of organic iron chelates in the soil solution can reach concentrations of 10^{-4} to 10^{-3} M (Mashhady and Rowell, 1978). The application of farmyard manure to calcareous soils low in organic matter may therefore be an effective procedure for increasing iron solubility and iron uptake by crop species with low iron efficiency such as sorghum (Mathers *et al.*, 1980).

Besides humic acids, organic acids and phenolics, hydroxamate siderophores have been identified as major iron-chelating compounds. These are produced by soil microorganisms (mainly fungi) and have a very high affinity for Fe^{3+}, particularly at high pH. It has been shown by Powell *et al.* (1982) and Reid *et al.* (1984b) that the level of hydroxamate siderophores in the rhizosphere soil is about 11 to 56-fold higher than that in the bulk soil. In nutrient solutions the iron from these siderophores can be utilized at least to some extent by higher plant species such as oat, sorghum and sunflower (Reid *et al.*, 1984a); Cline *et al.*, 1984). In oat, the rate of shoot transport of iron supplied as hydroxamate siderophore was even higher than that from the synthetic iron chelate Fe EDDHA (Reid *et al.*, 1984a). The ecological importance of the hydroxamate siderophores for the iron nutrition of higher plants, however, should not be overestimated for the following reasons: (*a*) these siderophores are strongly adsorbed to the soil matrix at higher pH (Cline *et al.*, 1983), and (*b*) the higher plants have efficient mechanisms to mobilize iron in the rhizosphere (Section 16.5.3).

In sodic soils ($pH > 8{\cdot}5$), sodium carbonate disperses organic matter, and low-molecular-weight organic substances (mainly sodium humates) form soluble complexes with iron (and manganese). Increasing the $NaHCO_3$ concentration from 12 to 75 mM (pH $8{\cdot}0 \rightarrow 8{\cdot}8$) increased the concentration of iron and manganese in the soil solution by a factor of 18 and 2·3, respectively (Mashhady and Rowell, 1978). This humate effect was also demonstrated in solution culture, where the addition of sodium humates prevented iron deficiency in tomato plants grown at high pH in the presence of high bicarbonate concentrations (Badurowa *et al.*, 1967).

16.5.1.2 Zinc

The solubility of inorganic zinc, like that of inorganic iron, decreases with increasing pH. Again, the decrease is much less than predicted from the equilibrium-constants of inorganic zinc compounds, such as zinc carbonate and zinc phosphate. There is growing evidence (Brümmer, 1981) that the concentration of zinc (and phosphate and boron) in the soil solution is determined by adsorption and desorption processes. The relationship between soil pH and zinc concentration in the soil solution differs therefore among soils, depending on other solute components (e.g., Ca^{2+}, organic acids) and the characteristics of the adsorbers (e.g., iron oxides, $CaCO_3$). The application of farmyard manure to alkaline soils low in organic matter therefore increases the solubility and uptake of zinc (Srivastava and Sethi, 1981).

Besides salinity and iron toxicity, the next most important nutritional limitation on the yield of wetland rice is zinc deficiency (Ikehashi and Ponnamperuma, 1978). In the early growing stages this deficiency is widespread, especially in rice growing in alkaline paddy soils with more than 1% organic matter (Forno *et al.*, 1975).

16.5.1.3 Phosphorus

In alkaline soils (except chernozems) the availability of phosphorus is generally low. As with zinc the concentration of phosphorus in the soil solution is determined primarily not by dissolution or precipitation of definite inorganic compounds such as tricalcium phosphate or apatite but by the desorption and adsorption of phosphorus, particularly in soils with more than 1% organic matter. In these soils, at least in the pH range of 6 to 8, the phosphorus concentration in the soil solution may not decline but rather may increase (Welp *et al.*, 1983). In alkaline soils of increasing pH and decreasing soil organic matter content, the equilibrium constants of inorganic phosphates become increasingly important for the concentration of phosphorus in the soil solution. In general, however, phosphorus deficiency in crop plants growing in alkaline soils is caused primarily by very low levels of total phosphorus and low soil moisture levels, that is, when the mobility of phosphorus is limited and root growth is restricted.

16.5.1.4 Manganese

In well-aerated calcareous soils (rendzinas), the solubility of manganese decreases with increasing levels of both $CaCO_3$ and MnO_2 due to the adsorption of manganese at $CaCO_3$ and its oxidation on MnO_2 surfaces and probably to the precipitation of manganese calcite (Jauregui and Reisenauer, 1982). The effect of soil pH and $CaCO_3$ on the availability of manganese is well known in temperate zones as a result of the manganese deficiency induced by overliming of acid soils low in manganese. On calcareous soils, however, manganese deficiency in crop plants is not a major problem for reasons related to the high microbial activity in the topsoil, anaerobic microsites, high microbial activity in the rhizosphere, and root-induced mobilization of manganese in the rhizosphere (see Section 16.5.3.3).

In poorly structured soils, manganese solubility is usually relatively high. In general, in alkaline soils leaching is not important, and the total manganese content is higher than that of soils in humid climates.

16.5.1.5 Boron

Boron adsorption to clay minerals increases sharply above pH 6·5 and is maximal at pH ~9 (Kluge and Beer, 1979; Keren *et al.*, 1981). Heavy liming of acid soils low in boron therefore induces boron deficiency in crop plants (Scott *et al.*, 1976). In alkaline soils, the low boron solubility dictated by boron adsorption to clay minerals is usually compensated for, however, by a lack of leaching or by boron supplied by irrigation water. Boron toxicity is thus much more likely, particularly in sodic soils, than is boron deficiency.

16.5.2 Major Chemical Constraints on Plant Growth

16.5.2.1 Lime-Induced Chlorosis

The most prominent nutritional disorder of crop plants grown in soils with more than 20% of $CaCO_3$ is lime-induced chlorosis (Schinas and Rowell, 1977). Plant species that are mainly affected include apple, citrus, grapevine, peanut, soybean, sorghum, and dryland rice. It is the major problem in sorghum and soybean production in the Great Plains of the United States (Clark, 1982a). In most instances lime-induced chlorosis is related directly or indirectly to iron deficiency, as indicated, for example, by the alleviation of chlorosis by foliar sprays containing iron salts or iron chelates. For reviews on this topic the reader is referred to Chen and Barak (1982) and Vose (1982).

Leaves suffering from lime-induced chlorosis often have a content of total iron similar to or even higher than that of green leaves, indicating a "physiological inactivation" of iron (Section 9.1). Although elevated phosphorus levels in soils or plants might sometimes be involved, a high

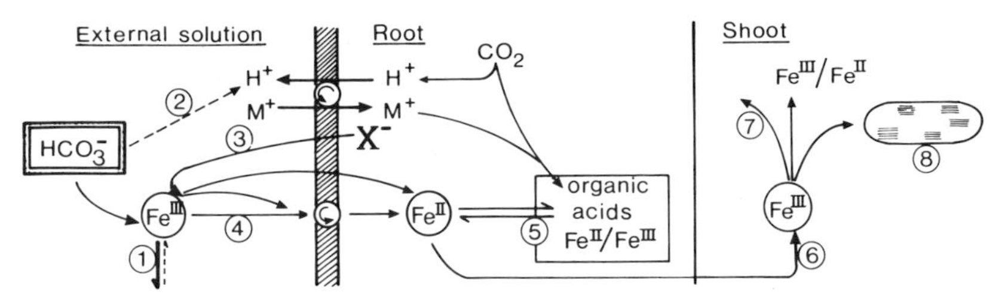

Fig. 16.11 Eight possible effects of a high bicarbonate concentration in the substrate on the uptake, transport, and availability of iron for chlorophyll formation in leaves. For a description of mechanisms (1)–(8), see text.

bicarbonate (HCO_3^-) concentration in the soil solution is mainly responsible for lime-induced chlorosis (Boxma, 1972; Chen and Barak, 1982; Coulombe *et al.*, 1984; Mengel *et al.*, 1984). A high soil moisture level (poor aeration) therefore tends to increase the incidence of this nutritional disorder. Despite the vast number of reports on this topic, the mechanism by which high HCO_3^- concentrations induce chlorosis is still not well understood. Quite contradictory results are often obtained, although this is not surprising considering the complexity of the problem. Figure 16.11 summarizes some of the major means by which high HCO_3^- concentration may affect the uptake, translocation, and utilization of iron in plants. A high HCO_3^- concentration in the soil solution both raises and buffers the pH and thus further lowers the concentration of soluble inorganic iron [mechanism (1)]. Simultaneously in dicotyledonous species, root responses to iron deficiency are severely inhibited by high pH; these responses include impairment of the H^+-efflux pump by neutralization of H^+ [mechanism (2)], lowering the release of phenolics [mechanism (3)] and FeIII reduction at the plasma membrane [mechanism (4)] (Römheld and Marschner, 1983; see also Chapter 2). In agreement with this a high HCO_3^- concentration leads to a decrease in the uptake and transport of iron to the shoot (Boxma, 1972; Venkat-Raju and Marschner, 1981; Kolesch *et al.*, 1984).

In roots supplied with a high HCO_3^- concentration, CO_2 fixation and organic acid synthesis increase, particularly in so-called calcifuge species (Bedri *et al.*, 1960; Lee and Woolhouse, 1969b). It is not clear to what extent sequestering of iron in vacuoles by certain organic acids [mechanism (5)] contributes to the inhibition of iron transport to the shoot [mechanism (6)]. A high HCO_3^- concentration inhibits root extension (Lee and Woolhouse, 1969a) and decreases the root pressure–driven solute flux in the xylem from the roots to the shoot (Wallace *et al.*, 1971). The transport of iron to expanding leaves is especially impaired (Rutland and Bukovac, 1971), and the distribution of iron within the leaf tissue is uneven [mechanism (7)] (Rutland, 1971). These effects are discussed in relation to the alkalinization of the tissue (Mengel and Malissiovas, 1981) and of the cytoplasm in particular (Kolesch *et al.*, 1984).

It should be kept in mind that severe symptoms of lime-induced chlorosis are often correlated not only with higher levels of total iron in the leaves on a dry weight basis but also with severe inhibition of leaf growth and chloroplast development [mechanism (8)]. Inhibition of root growth by a high HCO_3^- concentration also decreases the rate of cytokinin export to the shoot. Cytokinins are necessary for protein synthesis and chloroplast development (Parthier, 1979). An accumulation of Fe(III) in chlorotic leaves may therefore also be the consequence of a limitation on other factors required for chlorophyll formation.

The role of phosphorus in lime-induced chlorosis is not clear. It may impair iron nutrition at various levels—for example, precipitation as Fe(III) phosphates in the soil and at the root surface, interference with Fe(III) reduction from iron chelates (Brown and Olsen, 1980), mobilization of iron oxides at the root surface (Chapter 2), or immobilization of iron within plants (Cumbus *et al.*, 1977). Phosphorus-efficient genotypes of sorghum are more susceptible to lime-induced chlorosis than are phosphorus-inefficient genotypes (Brown *et al.*, 1977). A large phosphorus/iron ratio in leaves suffering from lime-induced chlorosis was found by Atkas and Egmond (1979) and others. A similar ratio may also exist, however, when chlorosis is induced by other means (DeKock and Hall, 1955). Elevated phosphorus levels in chlorotic leaves are therefore probably the result of growth inhibition and are thus the consequence (concentration effect), not the cause of iron chlorosis (Mengel *et al.*, 1984). Speculations about an "inactivator" of iron formed in plants in the presence of both HCO_3^- and phosphorus (Kinzel, 1982) require support by experimental data. Although iron deficiency can be induced in crop plants growing in calcareous soils supplied with very high levels of fertilizer phosphorus, there is substantial doubt that phosphorus is responsible for the occurrence of lime-induced chlorosis under field conditions (Kovanci *et al.*, 1978; Mengel *et al.*, 1979). Many laboratory and greenhouse studies on phosphorus–iron interactions have been conducted with phosphorus concentrations that are orders of magnitude higher than those typical of soil solutions in calcareous soils.

Lime-induced chlorosis is of minor importance in sodic soils, mainly for two reasons: the increase in iron solubility by low-molecular-weight organic compounds and growth inhibition by soil constraints other than iron deficiency (Mashhady and Rowell, 1978).

16.5.2.2 Other Constraints

Low soil moisture levels and a corresponding water deficit in plants is a typical constraint on plant growth in alkaline soils of low rainfall areas. At low soil moisture levels both the diffusion of nutrients to the root surface and root extension are impaired. The uptake of phosphorus and potassium is closely related to soil moisture content and root extension (Chapter 14). In alkaline soils with low levels of labile phosphorus (Chapter 13), phosphorus deficiency is therefore more closely related to decreasing soil moisture levels than to increasing pH. Soil physical factors (low moisture or poor aeration) are also in part responsible for excessive sodium uptake in sodic soils.

There are various reasons for the widespread occurrence of zinc deficiency in crop plants growing in alkaline soils, including the low solubility of zinc, restricted root growth, and high HCO_3^- concentrations. In wetland rice zinc deficiency is a serious yield-limiting factor, especially in modern cultivars growing in sodic soils (B. D. Sharma *et al.*, 1982). High HCO_3^- concentrations drastically inhibit the transport of zinc to the shoot of wetland rice (Forno *et al.*, 1975). In paddy soils 3–6 weeks after planting, HCO_3^- concentrations in excess of 10 m*M* are common; at these levels, HCO_3^- not only inhibits zinc uptake but also seems to injure rice roots directly (Dogar and Hai, 1980). This response of wetland rice to high HCO_3^- concentrations seems to resemble closely that typical of certain calcifuge species.

16.5.3 Mechanisms of Adaptation

16.5.3.1 Calcicoles versus Calcifuges

Plant species and populations within species of the natural vegetation that preferentially grow in calcareous soils (calcicoles) possess adaptive mechanisms for coping with constraints on growth such as low iron availability and high HCO_3^- concentrations. For example, calcicoles have a higher iron efficiency than calcifuges (which have adapted to acid soils), and high HCO_3^- concentrations have only a negligible inhibitory effect on root extension (Lee and Woolhouse, 1969a). Calcicoles are often highly efficient in phosphorus uptake (Musick, 1978), presumably in many instances because of heavy root infection with mycorrhizas (Kianmehr, 1978).

The role of high calcium concentrations in the adaptation of plants to calcareous soils seems to be rather complex. In certain calcicole species the uptake of calcium is more restricted than in calcifuge species (Bousquet *et al.*, 1981), presumably due to a lower affinity of root plasma membranes for Ca^{2+} (Monestiez *et al.*, 1982). A higher level of calcium–binding proteins in the cytoplasm of calcicole species (Le Gales *et al.*, 1980) could be another mechanism involved (Section 8.6.4).

16.5.3.2 Iron Deficiency

Crop species or cultivars within species which grow in alkaline soils without developing symptoms of lime-induced chlorosis are called iron efficient; those that become chlorotic are called iron inefficient (Brown and Jones, 1976). This is basically an ecological classification. For detailed studies on the mechanisms and also screening methods of iron efficiency, nutrient solutions are often used, containing varied levels of iron, phosphate, and $CaCO_3$, different nitrogen sources, and different pH levels (Brown and Jones, 1976). There are impressive differences in the iron efficiency of different species as well as cultivars within a species. Although the response of genotypes in nutrient solution and in calcareous soils are often comparable (J. C. Brown, 1978), this is not always the case (Cianzio and Fehr, 1982). Rapid screening methods involving nutrient solutions may merely offer the possibility of reducing the number of genotypes to be tested under field conditions (Clark *et al.*, 1982). Addition of excess $CaCO_3$ to the nutrient solution is considered essential for diminishing the gap between iron efficiency reflected in the nutrient solution and resistance of plants to lime-induced chlorosis in calcareous soils (Brown and Jones, 1976).

In future, the basic qualitative differences between plant species in responding to iron deficiency (Table 16.19) will require particular attention, using nutrient solution-based screening methods.

Mobilization of iron in the rhizosphere can be brought about by both nonspecific and specific mechanisms:

Nonspecific mechanisms:
These are not related to the iron nutritional status of the plant, and have been discussed in detail in Chapter 15.

1. Root-induced decrease in pH as a consequence of preferential cation uptake (e.g., induced by ammonium sulfate or N_2 fixation in legumes)
2. Release of organic acids by the roots
3. Release by the roots of photosynthates as substrates for rhizosphere microorganisms, which in turn affect pH, redox potential and chelator concentration (e.g., siderophores)

Specific mechanisms (strategies):
These are the mechanisms of root response to iron deficiency which differ among plant species (Table 16.19). The majority of detailed studies on specific mechanisms have focussed on dicotyledonous species, and thus Strategy I (see Table 16.19) is much better understood than Strategy II. The principal differences between the two strategies can be demonstrated in solution culture by measuring the rates of Fe(III) reduction at the root

Table 16.19
Summary of the Major Strategies of Roots in Response to Iron Deficiency. A: Individual Components and their Relative Importance for Mobilization and Uptake of Iron; B: Major Ecological Asvantages/Disadvantages for Regulation of Iron Nutrition

	Strategy I (e.g., in soybean, peanut, sunflower, and most other dicots)	Strategy II (e.g., in barley, oats, rice, and probably most other grasses)
A: Type of response		
Enhancement of H^+ release	Important[a]	Unimportant or absent[b]
Enhancement of Fe(III) reduction at the plasma membrane	Important[c]	Of limited importance or absent
Enhancement of release of phenolics	Important[d]	No experimental evidence
Enhancement of release of nonproteinogenic amino acids which chelate iron	No experimental evidence	Highly important[e]
Formation of rhizodermal transfer cells	Important[f]	Absent[f]
B: Advantages/disadvantages		
Fine regulation of iron uptake rates	Important[g]	Important[h]
Specificity of mobilization processes	Low[i]	Not known
Sensitivity to high phosphate levels	Relatively low[a]	High[j]
Impairment of response by high pH and high HCO_3^-	Relatively high[k]	Relatively low[k]

[a]See Chapter 2. [b]Römheld and Kramer (1983). [c]Chaney *et al.* (1972); Bienfait *et al.* (1982). [d]Olsen *et al.* (1981). [e]See Section 9.1. [f]Römheld and Kramer (1983). [g]Römheld and Marschner (1981a). [h]Takagi *et al.*, 1984; V. Römheld (unpublished data). [i]Moraghan (1979). [j]Azarabadi and Marschner (1979). [k]Römheld *et al.* (1982); V. Römheld (unpublished data).

surface (reducing capacity of the roots) and iron uptake rates in iron-adequate and iron-deficient plants (Table 16.20). Whereas in iron-deficient peanut the reducing capacity of the roots and iron uptake increase sharply, no response is observed in barley. This would indicate that Strategy II offers no advantage to barley when grown in alkaline soils. Based on such studies in water culture with synthetic iron chelates, barley and other grass species had been erroneously defined as iron-inefficient. When the same cultivars of the two species shown in Table 16.19 are grown together in calcareous soil, however, barley remains green whereas peanut becomes chlorotic. Several factors are responsible for the discrepancy in the results. For example, high concentrations of HCO_3^- impair Strategy II to a much lesser extent than Strategy I. The most important factor, however, is that the key component of Strategy II, the mobilization of inorganic Fe(III) by phytosiderophores released by the roots under iron deficiency (Section 9.1.6), is not appreciated in nutrient solution cultures with supply of synthetic iron chelates. In

Table 16.20
Relationship between the Iron Nutritonal Status of Peanut (cv. A 124B) and Barley (cv. Karina), the Reducing Capacity of the Roots, and Uptake of ^{59}Fe from ^{59}Fe-EDDHA[a]

Pretreatment	Chlorophyll content (mg/g dry wt)	Reducing capacity of the roots (nmol Fe^{2+}/g fresh wt × hr)	Iron uptake (nmol ^{59}Fe/g dry wt × hr)	
			Roots	Shoots
		Peanut		
+Fe	8·6	40	22	0·4
−Fe	2·8	2570	1042	181·0
		Barley		
+Fe	9·0	60	14	0·2
−Fe	2·1	68	13	0·4

[a] ^{59}Fe-EDDHA (ferric ethylenediaminedi [o-hydroxyphenyl acetate]) was supplied for 10 hr. From V. Römheld (unpublished data).

plant species with Strategy II these synthetic iron chelates are furthermore relatively poor sources of iron compared to inorganic Fe(III) compounds (Mino *et al.*, 1983; Takagi *et al.*, 1984; Reid *et al.*, 1984a). Studies on plants grown in nutrient solutions and supplied with synthetic iron chelates are therefore not suitable for evaluating Strategy II.

16.5.3.3 Manganese Efficiency

Relatively few reports exist on genotypical differences in manganese efficiency (for oat, see Brown *et al.*, 1977). Manganese efficiency is of minor importance for adaptation to alkaline soils because manganese is readily reduced by soil microorganisms and by roots of dicots in response to iron deficiency (Section 9.1.6). The reduction and uptake of manganese in iron-deficient plants increase in nutrient solutions (Römheld *et al.*, 1982) as well as in soils and can lead to manganese toxicity in plants grown in calcareous soils (Table 16.21).

Table 16.21
Iron Deficiency-Induced Manganese Toxicity in Flax (*Linum usitatissimum* L.) Grown in a Calcareous Soil of pH 8·0[a]

Treatment	Shoot dry wt (g/pot)	Contents of shoot (dry wt)		
		Manganese (mg/kg)	Iron (mg/kg)	Phosphorus (%)
Control (−Fe)	3·60	881	83	0·32
2 mg Fe/pot (Fe-EDDHA)	5·55	64	174	0·32

[a] From Moraghan (1979).

The manganese content of flax plants (controls) was poorly related to the amount of manganese extracted from the soils but, significantly, was inversely related to the amount of extractable iron (Moraghan, 1978). In agreement with this, manganese toxicity could be eliminated by the application of Fe-EDDHA, which dramatically decreased the manganese content of the plants (Table 16.21). Therefore, high iron efficiency may in most instances prevent manganese deficiency in plants with Strategy I growing in calcareous soils; it may even increase the risk of manganese toxicity, as shown for flax (Table 16.21) and for an iron-efficient genotype of soybean (Brown and Jones, 1977b).

16.5.3.4 Differences among Cultivars

Within the framework of the two major strategies (Table 16.19) there are large differences among cultivars and strains of a given crop species in iron efficiency and in resistance to lime-induced chlorosis. In fact, differences in tolerance to lime-induced chlorosis among soybean cultivars are classical examples of genetically controlled mineral nutrition in general and iron nutrition in particular (Weiss, 1943). Under conditions of iron deficiency soybean and tomato cultivars with high iron efficiency release more protons and/or have a higher root reducing capacity than cultivars with low iron efficiency (Olsen and Brown, 1980). Genotypical differences in resistance to lime-induced chlorosis also exist in crop species characterized by strategy II, such as oat (McDaniel and Dunphy, 1978), bermudagrass (McCaslin *et al.*, 1981), and lovegrass (Voigt *et al.*, 1982). These differences are important considerations in selection and breeding programs designed to improve the adaptation of crop species to alkaline soils. Wutscher *et al.* (1970) demonstrated the possibility of increasing the adaptability of citrus by the selection of rootstocks with high resistance to lime-induced chlorosis—an achievement of some significance for *Citrus trifoliata*, which is extremely sensitive to chlorosis (Hamze and Nimah, 1982).

Among annual crop species, cultivars of chickpea (Saxena and Sheldrake, 1980) have been selected for resistance to lime-induced chlorosis. Moreover, yield reduction in calcareous soils varied between 6·4 and 81·9% for adapted and nonadapted soybean cultivars, respectively (Froehlich and Fehr, 1981).

The genetic potential within a plant species of improving adaptation to calcareous soils is demonstrated in Table 16.22 for peanut. The nonadapted cultivar Congo Red, originating from acid soils, became severely chlorotic when grown in a calcareous soil; iron chelates had to be applied to overcome chlorosis and to obtain a reasonable yield. In contrast, in the adapted cultivar 71–238 chlorosis was absent, the yield was much higher, and iron

Table 16.22
Effect of Iron Chelate Application (10 kg Fe/ha as Fe-EDDHA) on the Pod Yield of Peanut Grown in a Calcareous Soil (23% $CaCO_3$) of pH 7·8 [a,b]

Cultivar	Iron chelate application	Pod yield (kg/ha)	Yield increase (%)
Congo red	–	833	—
	+	2583	210
Shulamit	–	3305	—
	+	4749	44
71–238	–	4388	—
	+	4777	9

[a]Based on Hartzook *et al.* (1974).
[b]Ten kilograms of iron (as Fe-EDDHA) were applied per hectare.

application had only a slightly beneficial effect. Resistance to lime-induced chlorosis is based on several factors and thus genetic control is complex. In soybean, for example, resistance cannot be adequately described by a single-gene model but is influenced by genes with additive effects (Prohaska and Fehr, 1981).

16.5.3.5 Zinc Efficiency

Differences in the zinc-efficiency of crop species are well documented. Zinc deficiency symptoms are severe in bean, maize, soybean, and tomato grown in alkaline soil (pH 8·2), whereas in wheat, barley, sugar beet, and potato grown in the same soil, symptoms are lacking (Viets *et al.*, 1954). Differences in the zinc efficiency of bean cultivars can be quite distinct and are related to higher rates of zinc uptake by the efficient cultivar (Judy *et al.*, 1965). The same situation exists in pigeon pea; the leaves of an efficient cultivar had much higher zinc levels, and only a small application of zinc fertilizer was necessary to obtain maximal yield (Table 16.23).

Table 16.23
Zinc Content of Leaves at Maturity and Grain Yield of Pigeon Pea Cultivars Grown in a Zinc-Deficient soil (pH 7·8)[a]

	Zn Content of leaves (mg/kg dry wt)			Grain yield (g/pot)		
Cultivar	0	5 mg Zn	50 mg Zn	0	5 mg Zn	50 mg Zn
T 21	15·0	19·7	37·1	3·8	8·5	10·4
Plant A-3	21·2	30·8	90·8	6·7	10·1	10·0

[a]The cultivars were supplied with 5 and 50 mg zinc (as $ZnSO_4$) per kilogram soil. From Shukla and Raj (1980).

These and other reports on differences in zinc efficiency demonstrate that sufficient genetic variability exists to develop systematic selection and breeding programs as part of an overall program to improve the adaptation of crop species to alkaline soils. In the case of wetland rice, in which zinc deficiency is a severe problem, progress has already been made, and most of the new breeding lines of the International Rice Research Institute (IRRI) are at least moderately zinc efficient (Ikehashi and Ponnamperuma, 1978). The benefit accruing to farmers who use zinc efficient cultivars is substantial when wetland rice is produced in alkaline soils with low zinc availability (Table 16.4).

16.5.3.6 Other Mineral Nutrients

Although iron and zinc efficiency are the two major mineral nutrition requirements for adaptation to alkaline soils, other aspects must also be considered. High phosphorus efficiency must always be achieved, although it seems to be negatively correlated in some instances with zinc (Shukla and Raj, 1980) and iron efficiency (Brown *et al.*, 1977).

In soils with low levels of total phosphorus, high phosphorus efficiency is necessary for better utilization of fertilizer phosphorus and not for mining of soil phosphorus.

Compared with the large number of genetic studies on calcareous soils, there are very few studies on the adaptation of crop species to sodic soils. Nevertheless, the possibilities for improving the adaptation of wheat and barley (Chandra, 1979) and wetland rice to sodic soils have been demonstrated (Table 16.4). The extent to which individual factors, such as zinc and phosphorus efficiency or tolerance to an excessive supply of sodium and boron, contribute to this progress is not clear.

16.6 Saline Soils

16.6.1 The Problem

The saline areas of the world consist of salt marshes of the temperate zones, the mangrove swamps of the subtropics and tropics, and the interior salt marshes adjacent to salt lakes. Saline soils are abundant in semiarid and arid regions, where the amount of rainfall is insufficient for substantial leaching. Salinity problems occur in nonirrigated croplands and rangelands either as a result of evaporation and transpiration of saline underground water or due to salt input from rainfall. They are particularly critical in irrigated areas. Salinity has been an important historical factor and has influenced the life spans of agricultural systems. It frequently destroyed ancient agrarian societies, and more recently large areas of the Indian subcontinent have been rendered unproductive by salt accumulation and poor water management. In Pakistan, for example, about 10 million of 15 million hectares of canal-irrigated land are becoming saline (Wyn Jones, 1981). It has been suggested that worldwide there is more land going out of irrigation due to salinity than there is new land coming into irrigation (Vose, 1983). Salinity is the major nutritional constraint on the growth of wetland rice.

Even water of good quality may contain from 100 to 1000 g of salt per cubic meter. With an annual application of 10,000 m^3/ha, between 1 and 10 tons of salt are added to the soil. As a result of transpiration and evaporation of water, soluble salts further accumulate in the soil and must be removed periodically by leaching and drainage. But even when proper technology is applied to these soils, they contain salt concentrations which often impair the growth of crop plants with low salt tolerance.

Although salt tolerance is relatively low in most crop species and cultured woody species, it is encouraging that genetic variability exists not only among species but also among cultivars within a species. Therefore, not only are selection and breeding for salt tolerance an important issue for traditional agricultural production in semiarid and arid regions; they may also offer the potential for utilizing the unlimited resource of seawater for irrigation. Examples of this have been highlighted by Epstein *et al.* (1980), who showed that, with certain barley strains, grain yields of up to 1 ton/ha can be obtained even when undiluted seawater, supplemented with nitrogen and phosphorus, is used for irrigation.

16.6.2 Soil Characteristics and Classification

Soils are considered saline if they contain soluble salts in quantities sufficient to interfere with the growth of most crop species. This, however, is not a fixed amount of salt but depends on the plant species, the texture and water capacity of the soil, and the composition of the salts. Thus the criterion for distinguishing saline from nonsaline soils is arbitrary. According to the definition of the U.S. Salinity Laboratory the saturation extract (the solution extracted from a soil at its saturation water content) of soil has an electrical conductivity (EC_e) greater than 4 mmhos/cm (equivalent to $\sim$40 m*M* NaCl per liter) and an exchangeable sodium percentage (ESP) of less than 15. Due to the excess of salt, the soils are usually flocculated and aerated. Although the pH of saline soils can vary over a wide range, it is usually around neutrality, with a tendency toward slight alkalinity. The dominant soil association consists of solonchaks (Table 16.18). Saline soils

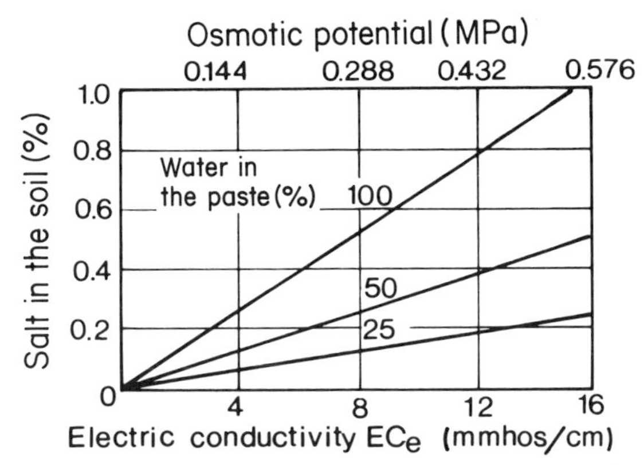

Fig. 16.12 Relationship between the conductivity of the saturation extract at 25°C and the salt content of the soil. Water in the paste (%); relative values; saturation extract = 100; 50 and 25 = calculated changes in EC_e and MPa in the soil solution when the water content drops to 50 and 25% respectively of that at saturation (50% is about field capacity of soils). (Based on U.S. Salinity Laboratory Staff, 1954.)

with an ESP of greater than 15 are termed saline–alkali soils (or saline–sodic soils), have high pH values, and tend to become rather impermeable to both water and aeration when the soluble salts are removed by leaching.

In evaluations of the suitability of saline soils for crop production, the measurement of EC_e offers a rapid and simple method for characterizing the salt content (Fig. 16.12). From the EC_e the osmotic potential of the saturation extract can also be calculated: $EC_e \times -0{\cdot}36$ bar, or $-0{\cdot}036$ MPa/mmhos $\times$ cm.

Since the EC_e is measured in the saturation extract (the value 100 in Fig. 16.12), the salt concentration in the soil solution at field capacity will be about twice that in the saturation extract (i.e., 50 in Fig. 16.12) and correspondingly higher when the soil moisture level declines below field capacity. For comparison, the EC_e of seawater is in the range of 44 to 55 mmhos/cm, and as a rule irrigation water of good quality must have an EC_e below 2 mmhos/cm. For plant growth in saline soils, however, the EC_e of the saturation extract only is an insufficient indicator, mainly for two reasons: (*a*) The actual salt concentration at the root surface can be much higher than in the bulk soil (Chapter 15) and (*b*) the EC_e characterizes only the total salt content, not its composition. Although NaCl is usually the dominant salt, others may be abundant in various combinations, depending on the origin of the saline water and the solubility of the salts (Table 16.24).

Furthermore, in both saline soils and irrigation waters high boron concentrations might become more critical for plant growth than salt concentra-

Table 16.24
Solubility of Salts at 25°C

Salt	Solubility (mEq/ liter H_2O)
Calcium chloride ($CaCl_2 \cdot 6H_2O$)	25,470
Magnesium chloride ($MgCl_2 \cdot 6H_2O$)	14,955
Sodium chloride (NaCl)	6,108
Magnesium sulfate ($MgSO_4 \cdot 7H_2O$)	5,760
Sodium bicarbonate ($NaHCO_3$)	1,642
Sodium sulfate ($NaSO_4 \cdot 10H_2O$)	683
Calcium sulfate ($CaSO_4 \cdot 2H_2O$)	30

tions *per se*: Irrigation water with more than 0·5 mg boron per liter might injure sensitive species such as citrus and walnut, and more than 2·0 mg/liter is harmful to most crop species.

16.6.3 Salinity and Plant Growth

16.6.3.1 Water Deficit versus Ion Excess

Plants growing in saline soils face two problems: high salt concentrations in the soil solution (i.e., high osmotic pressure and correspondingly low soil water potential) and high concentrations of potentially toxic ions such as Cl^- and Na^+ or unfavorable combinations of salt ions (e.g., a high Na^+/Ca^{2+} ratio). Salt exclusion minimizes ion toxicity but accelerates water deficit in plants, whereas salt absorption facilitates osmotic adjustment but can lead to ion toxicity and nutritional imbalance. It is often impossible to assess the relative contribution of ion excess and water deficit to growth inhibition at high soil salinity levels. In most instances, however, growth inhibition in salt-sensitive species even at low salinity levels is caused primarily by ion toxicity.

16.6.3.2 Genotypical Differences in Growth Response to Salinity

Plant species differ greatly in their growth response to salinity, as shown schematically in Fig. 16.13. The growth of halophytes is optimal at relatively high levels of NaCl, a response which can be explained only in part by the role of sodium as a mineral nutrient in these species (Section 10.2). Only a few crop species are slightly stimulated by low salinity levels (group II). Most are nonhalophytes (glycophytes), and either their salt tolerance is relatively low (group III) or their growth is severely inhibited even at low substrate salinity levels (group IV).

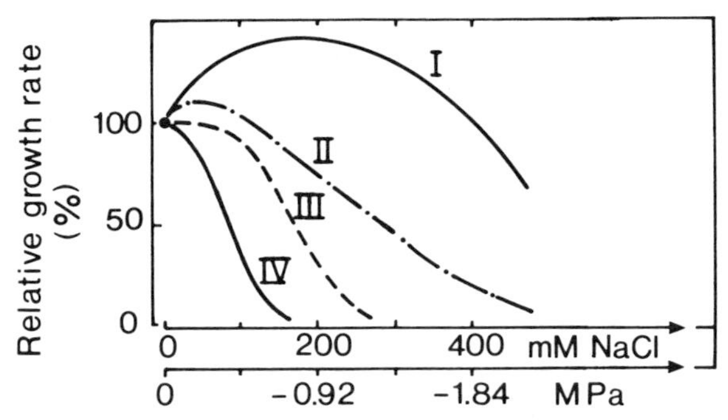

Fig. 16.13 Growth response of various plant species to increasing substrate salinity. I, Halophytes; II, halophylic crop species related to I (e.g., sugar beet); III, salt-tolerant crop species (e.g., barley); IV, salt-sensitive crop species (e.g., bean). (Modified from Greenway and Munns, 1980, with permission from the Annual Review of Plant Physiology. Copyright 1980 by Annual Review Inc.)

Classification of the salt tolerance (or sensitivity) of a crop species can be based on the EC_e of the soil saturation extract and the expected yield reduction related to a given EC_e value (Bernstein, 1964). More recently, evaluation of the relative salt tolerance of crop species as well as forage species and fruit trees has been based on two parameters: the threshold EC_e (i.e., maximum salinity without yield reduction) and the percent yield decrease per unit of salinity increase (i.e., 1 EC_e) beyond the threshold. Examples taken from an extensive study are given in Table 16.25. It is evident that barley tolerates relatively high salinity levels in comparison, for example, with maize and especially bean.

It has been known for many years that large differences in salt tolerance exist among cultivars within a crop species. Some of the literature data have

Table 16.25
Salt Tolerance of Agricultural Crops: Threshold EC_e (25°C) at the Point of Initial Yield Decline and Yield Decrease per Unit EC_e beyond Threshold[a]

Crop species	Threshold EC_e (mmhos/cm)	Yield decrease beyond threshold EC_e (%)
Barley	8·0	5·0
Sugar beet	7·0	5·9
Wheat	6·0	7·1
Soybean	5·0	20·0
Tomato	2·5	9·9
Maize	1·7	12·9
Bean	1·0	19·0

[a]Based on Maas and Hoffman (1977).

Table 16.26
Effect of Salinity on the Growth Depressions of Cultivars within Crop Species[a]

Species	Treatment (m*M* NaCl)	Growth parameter	Percentage of growth depression of cultivars	Reference
Barley	125	Grain yield	45 to <5	Greenway (1962)
Wheat	~50	Grain yield	90–50	Bernal *et al.* (1974)
Sugar beet	150	Total dry weight	93–49	Marschner *et al.* (1981a)
Soybean	50	Total dry weight	75–44	Läuchli and Wieneke (1979)
Tobacco	500	Surviving plants	100–15	Nabors *et al.* (1980)
Bean	Saline–sodic soil	Surviving plants	79–1	Ayoub (1974)

[a]Differences in the salt tolerance of the cultivars are indicated by the range of relative values; control (without salt) = 100.

been summarized by Duvick *et al.* (1981) and Vose (1983). A few examples of the differences are given in Table 16.26. The genetic variability within a species is not only a valuable tool for studying mechanisms of salt tolerance but also an important basis for screening and breeding for higher salt tolerance. Progress in this field has been particularly impressive with wetland rice (Table 16.4) and barley (Epstein *et al.*, 1980).

The sensitivity to salinity of a given species or cultivar may change during ontogeny. It may decrease or increase, depending on the plant species, cultivar, or environmental factors. Sugar beet, for example, is highly tolerant during most of its life cycle but sensitive during germination. In contrast, the salt sensitivity of rice, tomato, wheat, and barley usually increases after germination (Maas and Hoffman, 1977). In maize, salt sensitivity is particularly high and low at tasseling and grain filling, respectively (Maas *et al.*, 1983). Reports are often contradictory in respect to changes in salt sensitivity during ontogenesis. The difficulties of generalizing even for a given species have been demonstrated by Lynch *et al.* (1982), who found that the most sensitive cultivar in seedling stage was rather tolerant at maturity, whereas another cultivar showed the opposite pattern.

16.6.3.3 Environmental Factors Affecting Growth Response to Salinity

Climatic factors may significantly modulate plant responses to salinity. At a given salinity level, the salt tolerance of such crop species as barley and maize (but not wheat) is increased by high relative humidity (Hoffman and

Jobes, 1978). It is not clear whether this increase is a reflection of improved water balance in the shoots or a lower rate of transport of Na^+ and Cl^- to the shoots of plants with a lower transpiration rate. Lowering the transpiration rate by increasing the relative humidity is difficult to realize under field conditions, but application of antitranspirants seems to be more feasible. Antitranspirants may considerably increase the salt tolerance of wheat by both improving the water balance and increasing the K^+/Na^+ ratio in the shoots (Malash and Flowers, 1984).

At high irradiation, the salt tolerance of muskmelon (Meiri *et al.*, 1982) and faba bean (Helal and Mengel, 1981) increases significantly. Under these conditions the roots of faba bean have a higher capacity to restrict the transport of Cl^- and Na^+ to the shoots; that is, the exclusion mechanism is supported by high irradiation, most likely due to a better supply of carbohydrates to the roots. Under field conditions, however, the beneficial effect of high irradiation on salt tolerance might be counteracted by negative effects of low relative humidity and correspondingly high transpiration rates, leading to salt accumulation at the root surface (Section 15.2).

High salinity levels hardly affect germination at low soil temperatures but become increasingly inhibitory above 25°C in sugar beet (Mahmoud and Hill, 1981) and senna (*Cassia acutifolia*) (Ayoub, 1977). The reasons for this temperature effect are obscure; a shortage of carbohydrates due to enhanced respiration rates might be involved.

The detrimental effects of soil salinity are amplified by poor soil aeration (Table 16.13). This is particularly true for plant species relying on salt extrusion when growing in saline–sodic soils. Under these conditions the application of gypsum can considerably increase the salt tolerance. In potato, for example, addition of gypsum remarkably increases the tuber yield (Table 16.27). Even at high salinity levels, the yield is only slightly lower than that of the control, although 1·2% salt corresponds to an EC_e of

Table 16.27
Effect of Salinization and of Gypsum on the Growth of Potato (cv. Red Lasoda) in a Sandy Loam Soil[a]

	Tuber yield (g fresh wt/plant)	
Treatment	Without gypsum	With gypsum (2%)
Control (no salt)	221	226
0·6% salt	183	280
1·2% salt	149	207

[a]Salination was achieved with a mixture of sodium, magnesium, and calcium salts ($Cl^- > SO_4^{2-} > HCO_3^-$). Based on Abdullah and Ahmad (1982).

~20 mmhos/cm and the average threshold EC_e for the species potato is ~2 mmhos/cm (Maas and Hoffman, 1977). As has been shown in soybean growing in a saline–sodic soil (Coale *et al.*, 1984) addition of gypsum has a dual effect: (*a*) It improves soil structure and thus soil aeration and (*b*) increases the Ca^{2+}/Na^+ ratio and thus supports the capacity of roots to restrict Na^+ influx.

This example illustrates that for crop production in saline soils the salt tolerance of plants is a relative parameter; other agronomic procedures, such as irrigation cycling, are at least as important as the proper selection of species and cultivars for crop performance and yield. For a comprehensive review of these agronomic procedures, the reader is referred to Hoffman (1981).

16.6.4 Physiological and Biochemical Effects of Salinity

16.6.4.1 Salt Exclusion versus Salt Inclusion

Frequently, at low or moderate salinity levels, growth reduction of glycophytes is not correlated with specific symptoms such as leaf scorch or chlorosis. Plants are stunted and may have darker green leaves than normal control plants. Very often, the leaves of dicots become succulent, that is, the water content per unit leaf area increases. Generally, there is a greater reduction in shoot growth than in root growth. Without performing a chemical analysis, one cannot draw definite conclusions as to whether a water deficit or ion toxicity/imbalance, or both, is the predominant constraint on shoot growth (Fig. 16.14). The effects of salinity on the growth of plant species which restrict salt uptake (excluders) differ from those on the growth of those species with relatively high rates of uptake and translocation of salt ions (mainly Cl^- and/or Na^+) to the shoots (includers). Classifying species as excluders or includers (Fig. 16.14) is helpful in demonstrating the principles of adverse effects of (or adaption to) salinity but in reality very few glycophytes are strict excluders or includers; most are intermediate types. In various herbaceous and most woody species, marginal chlorosis and necrosis of leaves are widespread under conditions of NaCl salinity; the leaf contents indicate that Cl^- toxicity is the major constraint.

Whether a water deficit or ion toxicity is the main constraint on plant growth also depends on the type of salinity (e.g., whether the predominant anion is Cl^- or SO_4^{2-}, and the size of the cation ratio Ca^{2+}/Na^+), the duration of exposure, and the salinity level. In plants exposed to high salinity levels for short periods, water deficit is the principal constraint. In plants exposed for long periods, which is more typical of field-grown plants, in addition to

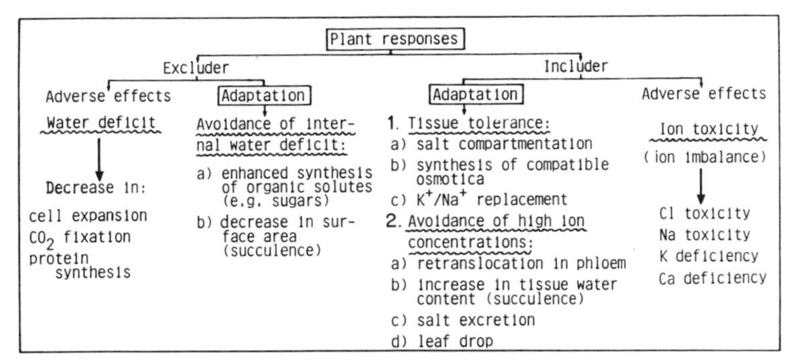

Fig. 16.14 Adverse effects of salinity and possible mechanisms of adaptation. (Reproduced from Greenway and Munns, 1980, with permission from the Annual Review of Plant Physiology. Copyright 1980 by Annual Review Inc.)

water deficit, especially in expanding leaves (Munns *et al.*, 1982), ion toxicity and imbalance are increasingly important.

16.6.4.2 Photosynthesis and Respiration

Salinity level and leaf area are usually inversely related. This could be explained by a water deficit in expanding tissues. However, not only the total leaf area but also net CO_2 fixation per unit leaf area may decline, whereas respiration (dark respiration) increases, leading to a drastic reduction in net CO_2 assimilation per unit leaf area per day (Table 16.28). Lower rates of net CO_2 fixation during the light period are not necessarily caused by water deficit and partial stomatal closure but can also be the result of direct adverse effects of Cl^-. In grapevine, for example, rates of CO_2 fixation are inversely related to thc Cl^- levels of the leaves and to an increase in mesophyll resistance and not to stomatal resistance, as would be expected from water deficit (Downton, 1977a). After long-term exposure to high salinity levels, lower rates of net photosynthesis may also be caused by a drop in chlorophyll content per unit leaf area, but not per unit weight of

Table 16.28
Effect of NaCl Salinity on the CO_2 Balance of Cotton (cv. Acala SJ-1)[a]

Salinity (MPa)	Leaf area (dm^2/plant)	CO_2 Balance (mg $CO_2/dm^2 \times 24$ hr) Net fixation light period	Evolution dark period	Net assimilation
−0·04	30	57	11	46
−0·64	24	44	16	29
−1·24	18	41	19	23

[a]Based on Hoffman and Phene (1971).

chlorophyll (Robinson *et al.*, 1983) indicating that the remaining chlorophyll is fully photosynthetic active.

Salinity may also increase the respiration rate of the roots, which have a higher carbohydrate requirement for maintenance respiration in saline substrates (Schwarz and Gale, 1981). This higher requirement presumably results from the compartmentation of ions, ion secretion (e.g., Na^+-efflux pump), or the repair of cellular damage. However, the "salt-induced" enhancement of root respiration probably does not account for more than 25% of the growth reduction caused by salinity. The remainder has been ascribed by Schwarz and Gale (1981) to reduced photosynthesis. In certain salt-tolerant species of the natural vegetation, root respiration even declines in saline substrates, and carbohydrates formerly used for "alternative respiration" (Chapter 5) are then preferentially utilized for the synthesis of sorbitol, a compatible organic solute for osmotic adaptation in the cytoplasm (Lambers *et al.*, 1981a).

16.6.4.3 Protein Synthesis

Protein synthesis in the leaves of plants growing in saline substrates may decline in response either to a water deficit or to a specific ion excess. Where there is a low substrate water potential imposed by Carbowax (a high-molecular-weight organic solute) or NaCl, protein synthesis in the leaves of bean is inhibited by both substrates, but inhibition is more severe with salt stress than with water stress only (Frota and Tucker, 1978). The effects of NaCl salinity on protein synthesis might be due to Cl^- toxicity in sensitive species (e.g., soybean), whereas in the more salt tolerant barley, Na^+/K^+ imbalance in the leaves is probably the responsible factor (Table 16.29). The

Table 16.29
Effect of Substrate Salinization on Growth, Mineral Element Content, and Protein Synthesis in Barley (cv. Miura)[a]

Treatment	Shoot dry weight (mg/plant)	Content (mmol/100 g fresh wt)		^{15}N Content (% of total ^{15}N)[b]	
		K	Na	Protein N	Inorganic N
Control	371	126	14	44	3
+80 m*M* NaCl	286	80	208	29	20
+80 m*M* NaCl +10 m*M* KCl	323	136	160	49	1

[a]Based on Helal and Mengel (1979).
[b]After supply of $^{15}NH_4\,^{15}NO_3$ for 24 hr.

adverse effect of high NaCl concentrations on both potassium content and protein synthesis in barley can be counterbalanced by KCl, despite the further decrease in the osmotic potential and increase in the Cl^- concentration of the substrate (Table 16.29). In the barley cultivar, replacement of K^+ by Na^+ may allow osmotic adjustment in expanded leaves but not the maintenance of protein synthesis. Except in a few halophytes, a K^+ concentration in the cytoplasm of between 100 and 150 m*M* is required for protein synthesis (Section 9.7). Sodium cannot replace K^+ in this function, irrespective of the salt tolerance of cultivars within a given species, as has been shown in wheat (Gibson *et al.*, 1984).

16.6.4.4 Ion Toxicity and Ion Imbalance

In many herbaceous crop species, grapevine, and many fruit trees, growth inhibition occurs even at very low levels of Cl^- salinization, where water deficit is not a constraint (Greenway and Munns, 1980). In these species Cl^- toxicity is responsible for growth inhibition. Many leguminous species belong to this group. Isoosmotic concentrations of NaCl are therefore much more inhibitory than Na_2SO_4 on the growth of peanut (Chavan and Karadge, 1980) and bean (Hajrasuliha, 1980). Outstanding examples in this respect are certain soybean cultivars: In poorly drained coastal soils the application of potash fertilizer containing KCl can raise the Cl^- levels of some cultivars to ~1% of the leaf dry weight, causing leaf scorch and a severe reduction in grain yield (Parker *et al.*, 1983). In sensitive woody species such as *Picea omoika* Cl^- levels as low as 0·2 and 0·3% of the needle dry weight are toxic and lead to chlorosis and necrosis (Alt *et al.*, 1982).

Sodium toxicity seems not to be as widespread as Cl^- toxicity and is mainly related to low absolute Ca^{2+} levels in saline substrates or high Ca^{2+}/Na^+ ratios in combination with poor soil aeration (saline–sodic soils). Many crop species with relatively low salt tolerance are typical Na^+ excluders. They are capable at low and moderate salinity levels of restricting the transport of Na^+ to the leaves, where it is highly toxic. This exclusion mechanism relies on relatively high external Ca^{2+} concentrations and Ca^{2+}/Na^+ ratios, respectively. As shown by LaHaye and Epstein (1971) in bean exposed to 50 m*M* NaCl, increasing the supply of $CaSO_4$ from 0·1 to 10·0 m*M* stimulated shoot dry weight from 0·46 to 0·74 g/plant, which was associated with a drastic decline in the sodium content of the leaves from ~1·4 to less than 0·1 mmol/g dry weight.

Soil salinity increases the incidence of calcium-related physiological disorders such as tipburn in lettuce and blossom end rot in tomato (Sonneveld and Ende, 1975) either by competition betwen Na^+ and Ca^{2+} during uptake, or by decreasing the soil water potential and thus root pressure

(Section 3.7). When salinity is correlated with alkalinization, decrease in calcium solubility by precipitation of $CaCO_3$ can be an additional factor.

Although high levels of Cl^- may inhibit NO_3^- uptake (Chapter 2), induced nitrogen deficiency is not likely to be an important factor in the growth depression caused by soil salinity. In contrast, under field conditions salt tolerance is apparently reduced in various crop species when a large amount of nitrogen fertilizer is applied (Maas and Hoffman, 1977). This decrease is probably related mainly to a change in water balance imposed by nitrogen, namely, a change in rooting pattern and in phytohormone level (decrease in the abscisic acid/cytokinin ratio; see Chapter 6). In legumes relying on N_2 fixation the situation might be different because salinity impairs nodulation either directly by interfering with *Rhizobium* colonization or indirectly by inhibiting root hair development (Tu, 1981).

In substrates with high phosphorus availability, salinity imposed by Cl^- salts stimulates phosphorus uptake and can lead to toxic phosphorus levels in the leaves (Cerda *et al.*, 1977). In soybean there exist distinct differences between cultivars in the salinity-induced phosphorus toxicity (Grattan and Maas, 1984). The toxic phosphorus levels in the leaves are not a result of a concentration effect due to growth depression, but of enhanced rates of phosphorus uptake by the roots and of translocation to the shoots (Roberts *et al.*, 1984). In most soils, however, it is unlikely that the phosphorus availability is so high that salt-induced phosphorus toxicity will play a major role in growth depression in saline soils. On the contrary, the salt-stimualted phosphorus uptake may improve phosphorus nutrition of plants growing in saline soils.

16.6.4.5 Phytohormones

Changes in phytohormone levels are involved in the response of plants to salinity. Elevated levels of abscisic acid (ABA) are a typical response to water deficit induced by drought or salinity. Elevated ABA levels in the leaves are important for rapid osmotic adjustment via stomatal closure. They also favor the accumulation of amino acids in general and proline in particular, and improve adaptation to salinity (Eder and Huber, 1977). Salinity may enhance leaf senescence, indicated by inhibition of protein synthesis and more rapid chlorophyll degradation. Kinetin partially counteracts these effects (Katz *et al.*, 1978). Although there have been reports on the reduction or even prevention of salt-induced growth depression by cytokinins (CYT), gibberellic acid (GA), or indoleacetic acid (IAA) (Starck and Karwowska, 1978; Bejaoui, 1985), the current picture of the interactions between salinity, levels of phytohormones and growth depression is still confused. At least the endogenous levels of ABA and CYT are not related to differences in salt tolerance among soybean cultivars (Roeb *et al.*, 1982). There is a need for more systematic studies on these interactions in which genotypical differences in plant responses to salinity are considered.

16.6.5 Mechanisms of Adaptation to Saline Substrates

16.6.5.1 Avoidance versus Tolerance

In principle, salt tolerance can be achieved in salt excluders by the avoidance of an internal water deficit, and in salt includers by high tissue tolerance to or avoidance of high tissue concentrations of salts (Fig. 16.14). The situation is usually more complex, however, and salt tolerance is actually produced by various integrated strategies with anatomical, morphological, physiological, and biochemical features. For comprehensive reviews the reader is referred to Greenway and Munns (1980), Wyn Jones (1981) and Yeo (1983).

In terrestrial halophytes of the Chenopodiaceae, high salt tolerance is based mainly on the inclusion of salts and their utilization for turgor maintenance or for the replacement of K^+ in various metabolic functions by Na^+. The crop species sugar beet is included in this group. Among the monocotyledons the highly salt tolerant kallar grass (*Diplachne fusca*) is also a salt includer (Sandhu *et al.*, 1981), although components of excluders, such as intensive retranslocation from the shoot and root release of Na^+ and Cl^- can also be observed (Bhatti and Wieneke, 1984). In the halophylic monocotyledons *Puccinellia peisonis* (Stelzer and Läuchli, 1977) and *Festuca rubra* (Khan and Marshall, 1981) exclusion is also at least an important contributory factor to high salt tolerance.

In glycophytes, which comprise most crop species, there is generally an inverse relationship between salt uptake and salt tolerance; that is, exclusion is the predominant strategy (Greenway and Munns, 1980). However, the classification of glycophytes as excluders is an oversimplification. Exclusion is only a relative term, that is, it means a much lower salt uptake in comparison to includers. It usually applies only to the salt transport from the roots to the leaves in general, and expanding leaves and the shoot apex in particular.

Typical differences among crop species in terms of growth and the mineral element content of the shoots are shown in Table 16.30. Sugar beet has the typical features of a salt-tolerant halophytic includer. Growth is enhanced by NaCl salinity and the levels of chloride and especially sodium in the shoot increase with increasing external supply. On the other hand, the potassium and calcium levels decline due to cation competition. Maize is much less salt tolerant than sugar beet; growth is inhibited, although the levels of chloride and especially sodium in the shoot remain relatively low. Induced potassium or calcium deficiency is an unlikely cause of growth depression; impaired osmoregulation is probably the main factor. Of the three species shown in Table 16.30 bean has the lowest salt tolerance, and chloride toxicity is the main reason for growth depression at the low salinity level. In contrast to Cl^-, the shoot transport of Na^+ is effectively restricted in bean. Thus bean, like many other salt-sensitive crop species, is an effective excluder of Na^+ but not of Cl^-.

Table 16.30
Effect of Increasing NaCl Concentration in the Substrate on Three Crop Species[a]

Species	Concentration of NaCl (m*M*)	Dry weight (relative)	Content (meq/g dry wt)			
			Na	Cl	K	Ca
Sugar beet (cv. Monohill)	0	100	0·1	0·05	3·3	1·6
	25	108	1·7	1·0	2·2	0·5
	50	115	2·1	1·2	2·0	0·4
	100	101	2·6	1·5	1·9	0·3
Maize (cv. DC 790)	0	100	0·02	0·01	1·6	0·5
	25	90	0·2	0·5	1·8	0·3
	50	70	0·2	0·6	2·0	0·3
	100	62	0·3	0·8	2·0	0·3
Bean (cv. Contender)	0	100	0·02	0·01	1·7	2·9
	25	64	0·04	1·0	2·2	3·7
	50	47	0·2	1·4	1·9	3·4
	100	37	0·4	1·5	2·2	3·6

[a]From Lessani and Marschner (1978) and H. Marschner (unpublished data).

Differences in the capacity for Na^+ and Cl^- exclusion also exist among cultivars of species. For example, the higher salt tolerance of certain cultivars of wheat (Bernal *et al.*, 1974), barley (Greenway and Munns, 1980), and citrus (Maas and Hoffman, 1977) is related to a more effective restriction of shoot transport of both Na^+ and Cl^-; that of soybean cultivars (Abel, 1969; Läuchli and Wieneke, 1979) and grapevine rootstocks is related to the restrictions of Cl^- transport only (Downton, 1977b; Antcliff *et al.*, 1983). The differences between soybean cultivars in the restriction of Cl^- transport to the leaves are particularly impressive. When grown in a saline soil the salt-sensitive includer Jackson contained ~0·9% chloride on a leaf dry weight basis, compared with the salt-tolerant excluder Lee, which contained only 0·05% chloride. The inheritance of the capacity for Cl^- exclusion in these soybean cultivars is controlled by a single gene pair (Abel, 1969).

Restrictions on excessive Na^+ and Cl^- accumulation in the shoots of plants grown in saline substrates begin with root permeability and selectivity. Halophytes growing in saline media also require barriers in the roots to prevent passive salt influx. The roots of these halophytes have special anatomical features; for example, the width of the Casparian strip is two or three times greater than that of glycophytes (Poljakoff-Mayber, 1975), and the inner cortex cell layer may be differentiated into a second endodermis (Stelzer and Läuchli, 1977).

In glycophytes differences in passive membrane permeability to Na^+ and Cl^- or efflux pumps are the main mechanisms at the root level for restricting uptake and root-to-shoot transfer. An example of the relationship between the lipid composition of root membranes and Cl^- transport to the shoots is

Table 16.31
Relationship between the Lipid Composition of the Roots of Grapevine Rootstocks and Chloride Translocation to the Leaves[a]

	Rootstock		
	Sc	1613-3	TS
Distribution factor (Cl in leaves/Cl in the substrate)	0·06	0·11	0·39
Monogalactose diglyceride[b]	9·5	22·4	39·9
Phosphatidylcholine[b]	20·8	14·6	3·7
Palmitic acid[b]	26·5	23·4	22·9

[a]Based on Kuiper (1968).
[b]Data represent percentage of total lipids.

presented for different rootstocks of grapevine in Table 16.31. The monogalactose diglyceride content of the roots (Chapter 2) was directly related to the Cl^- accumulation in the leaves, and a striking negative correlation was observed between the phosphatidylcholine and Cl^- content. On the other hand, in root stocks of citrus there is some evidence that genotypical differences in Cl^- exclusion are related to the levels of free steroles in the membranes (Douglas and Walker, 1984).

No such obvious correlations for Na^+ uptake and transport in the roots have been found so far. Nevertheless, exclusion of Na^+ from leaves is much more common (Section 10.2) and is based on several mechanisms, including efflux pumps in the roots, preferential accumulation in root vacuoles, resorption in the xylem parenchyma, retranslocation out of the leaves, and subsequent extrusion into the rooting medium (Chapter 3). The role of those mechanisms in the salt tolerance of bersim (*Trifolium alexandrinum* L.) was demonstrated by Winter (1982). Transfer cells play a key role in this internal cycling of Na^+ (D. Kramer, 1983).

16.6.5.2 Osmotic Adjustment

With a sudden increase in salinity, osmotic adjustment is achieved first by a decrease in tissue water content (partial dehydration). Salt tolerance and further growth in a saline substrate, however, require a net increase in the quantity of osmotically active solutes in the tissue (Yeo, 1983). In genotypes in which salt exclusion is the principal mechanism of salt tolerance, either the synthesis of organic solutes such as sugars and amino acids or the uptake rate of, for example, K^+, Ca^{2+}, or NO_3^- must be increased.

In genotypes in which salt inclusion is the predominant strategy, osmotic adjustment is achieved by the accumulation of salts (mainly NaCl) in the vacuoles of leaf cells. In natrophilic species Na^+ can replace K^+ not only in its functions as an osmotically active solute in the vacuoles but to some extent also in specific functions in cell metabolism. In these species Na^+ can even be more effective than K^+ in preventing water-deficiency stress (Section 10.2). However, in saline substrates, osmotic adjustment of the cytoplasm (cytosol) and its organelles (e.g., the chloroplast would require unphysiologically high concentrations of Na^+ and Cl^-. Thus organic solutes must take over the function of osmotically active substances in the cytosol.

16.6.5.3 Compartmentation and Compatible Osmotica

Certain enzymes, such as membrane-bound ATPase, in the roots are either activated or inhibited *in vitro* by high salt concentrations, depending on the salt tolerance of the intact plants; that is, membrane-bound ATPases of halophytes are less sensitive to salt than those of glycophytes (Lerner *et al.*, 1983). As a rule, however, the sensitivity of enzymes of halophytes to high salt concentrations is similar to that of corresponding enzymes of glycophytes (Fig. 16.15).

Thus halophytic species which accumulate large quantities of NaCl or other soluble salts in their leaves for osmotic adjustment must protect their enzymes in the cytoplasm and in the chloroplasts (Robinson *et al.*, 1983) from high concentrations of Na^+ and Cl^-. Such a strict compartmentation of Na^+ and Cl^- in the vacuoles has been demonstrated, for example, in spinach growing in saline substrates and containing high NaCl levels in the leaf tissue (Coughlan and Wyn Jones, 1980). Even typical halophytes such as *Atriplex spongiosa* maintain an Na^+/K^+ ratio of 0·29 in the cytoplasm of expanding cells as compared to 12·4 in the vacuoles of fully differentiated cells (Storey *et al.*, 1983). For osmotic adjustment of the cytoplasm, however, K^+ is not a suitable cytosolute as, for example, concentrations of higher than 125 m*M* K^+ inhibit protein synthesis (Brady *et al.*, 1984). In leaves with high vacuolar concentrations of Na^+ and Cl^-, osmotic adjustment of the cytoplasm and its organelles, including the chloroplasts, therefore requires an accumulation

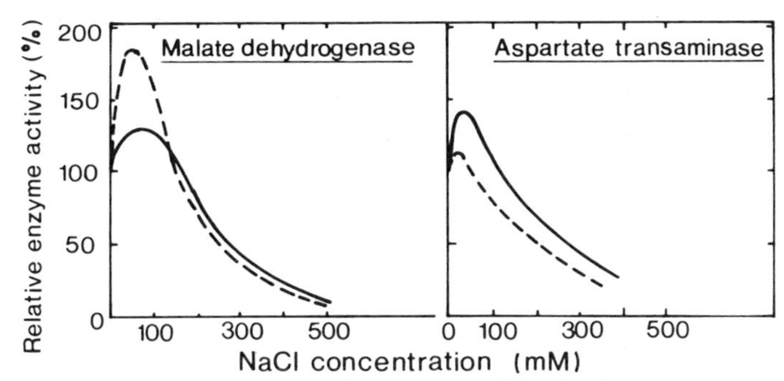

Fig. 16.15 Effect of increasing NaCl concentrations in the assay on the *in vitro* activity of malate dehydrogenase and aspartate transaminase from extracts of *Atriplex spongiosa* (– – –) and bean, *Phaseolus vulgaris* L. (——). (Based on Greenway and Osmond, 1972.)

of compatible organic solutes (cytosolutes), the quantity depending on the osmotic pressure of the vacuolar sap (Coughlan and Wyn Jones, 1980). The type of compatible solutes involved has a strong taxonomic basis: glycine betaine (Section 8.2) in Chenopodiaceae, sorbitol in Plantaginaceae and possibly proline in Gramineae and other families (Wyn Jones, 1981). Glycine betaine (synonym for betaine) is a very effective cytosolute because it is highly water soluble and does not carry a net charge, which could affect the charge balance of the cytoplasm. As shown in Fig. 16.16 enzyme activities are not inhibited even at extremely high concentrations of glycine betaine, whereas even much lower concentrations of NaCl severely inhibit these enzymes.

Glycine betaine is almost exclusively sequestered in the cytoplasm, that is, in ~5% of the total volume of fully differentiated leaf cells, where it may reach concentrations of more than 200 mM in tissues with a large NaCl content (Leigh *et al.*, 1981).

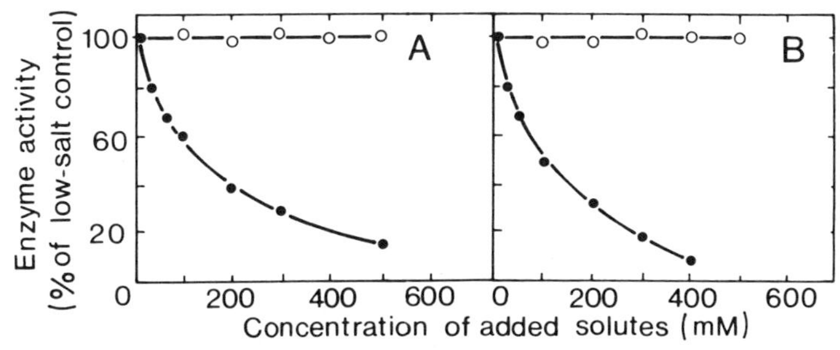

Fig. 16.16 Comparative effects of NaCl (●) and glycine betaine (○) on soluble enzymes of barley. A, Malate dehydrogenase; B, pyruvate kinase. (Based on Pollard and Wyn Jones, 1979.)

Proline accumulation is a well-known response to water deficit; it also occurs with salt stress and has a protective effect on seed germination in saline media (Bar-Nun and Poljakoff-Mayber, 1977). The major function of proline seems to be the protection of the cytoplasm against short-term (e.g., diurnal) fluctuations in leaf water potential. However, other cytosolutes, including sugar and sugar alcohols, probably provide greater protection against long-term effects imposed by high salt concentrations in the vacuoles.

Osmotic adjustment in plants via salt inclusion or exclusion has important implications for energy balance. Since NaCl and other soluble salts are abundant in saline substrates, they can be regarded as potentially "cheap," although dangerous, osmotica. According to Wyn Jones (1981) the approximate energy cost of acumulating 1 osmol of solute for osmoregulation is as follows:

	Energy requirement (mol ATP)
NaCl uptake	0·54
Synthesis of K^+-malate (CO_2 fixation)	13
Accumulation of C_6 sugars	54

A concentration of 300 m*M* C_6 sugars in the cell sap would account for 20 to 30% of the tissue dry weight. Thus, species with effective salt exclusion from the shoot tissue do not have much chance of achieving substantial growth rates in highly saline substrates.

In most species with moderate or high salt tolerance, average values for the levels of salts and other osmotically active solutes in the shoot do not represent sufficient information for understanding the mechanism of salt tolerance as it has been shown, for example, in two wheat genotypes differing in salt tolerance (Kingsbury *et al.*, 1984). An instructive example of the diversity of osmotic adjustment in various organs of the same plant is shown in Table 16.32 for *Aster tipolium* growing in a saline substrate. Osmotic adjustment in the leaves is mediated by Na^+, and Cl^-, whereas that in the flowers is mediated by K^+, glycine betaine, and sugars. But even within a given leaf the role of cytosolutes may vary; in young leaves of sorghum, glycine betaine is important for osmotic adjustment only in the leaf blades, not in the leaf sheaths (Grieve and Maas, 1984). Also the accumulation of Na^+ and Cl^- in the free space (the apoplast) may deserve more attention in evaluations of compartmentation of the various solutes for osmoregulation of various tissues (Leigh and Tomos, 1983).

In crop species with moderate salt tolerance there is a general pattern between mature and expanding leaves in the mode of osmoregulation. For example, in barley growing in a saline substrate, sugars play an insignificant

Table 16.32
Chemical Composition of *Aster tipolium* Leaves and Florets[a]

Component	Leaves (m*M*)[b]	Florets (m*M*)[b]
Sodium	360	56
Chloride	320	51
Potassium	72	133
Glycine betaine	18	82
Total soluble sugars	53	493

[a]Based on Gorham *et al.* (1980).
[b]Plant water basis.

role, compared with Na^+ and Cl^-, in osmotic adjustment in mature leaves, whereas they contribute more than 20% to the osmotic adjustment in expanding leaves (Delane *et al.*, 1982). Growth retardation of barley in saline substrates is therefore attributable to a water deficit in the expanding tissue due to insufficient phloem import of osmotically active solutes (Munns *et al.*, 1982).

Succulence is a typical morphological adaptation to high substrate salinity in most dicotyledonous species of both salt excluders (Longstreth and Nobel, 1979) and salt includers (Longstreth *et al.*, 1984). It is another mechanism for avoiding excessive salt accumulation in the leaf tissue (Jennings, 1968). Succulence is caused either by preferential transport of Na^+ to the vacuole (Section 10.1) or is induced by Cl^- (Smith and Struckmeyer, 1977). In salt includers not only the volume of the vacuoles increases but also the proportion of the intercellular spaces (Hajibagheri *et al.*, 1984). Succulence is not a specific response to salinity, however, because it is also induced by water deficit, application of phytohormones, decapitation of shoot apices, or in certain species under short-day conditions.

16.6.5.4 Salt Excretion and Leaf Drop

In many halophytic species additional mechanisms control the salt content of aerial tissue. For example, salt glands secrete large quantities of salt to the leaf surface, where it can be washed off by rain or dew. The importance of salt excretion to the salt tolerance of these halophytes is indicated by the fact that the salt tolerance of intact plants (*Sueda* or *Atriplex*) cannot be reproduced in callus cultures; in contrast, the salt tolerance of callus cultures of sugar beet is more related to that of intact plants (Smith and McComb, 1981). Salt excretion from the leaves is also important for the high salt tolerance of kallar grass (*Diplachne fusca* L.), a forage grass of high agronomical potential for growth in highly saline as well as alkaline soils (Sandhu *et al.*, 1981).

In certain moderately salt tolerant species such as tomato, the deposition of salts in leaf hairs is another mechanism for preventing excessive salt accumulation in photosynthetically active leaf cells and providing the opportunity for salt loss by leaching or mechanical means (Chapter 4).

Certain desert shrubs shed their lower leaves when salt accumulation reaches a critical level. This method of removing excess salt is an important strategy for the survival of natural vegetation. However, it has little potential as a type of adaptation to be introduced into crop plants.

16.6.5.5 Outlook

Selection and breeding programs designed to improve the adaptation of crop plants to saline soils have to consider the various mechanisms responsible for salt tolerance and sensitivity. In Cl^--sensitive species such as soybean and grapevine, efforts should be focused on the exclusion of Cl^- from leaf tissue. However, effective excluders of both Na^+ and Cl^- will probably not become very productive crops in highly saline soils because water deficit and large photosynthate requirements for osmotic adjustment seriously restrict their growth. Crop plants of the includer type have a much greater potential for better adaptation coupled with sufficient productivity when grown in highly saline soils. Includers of both Na^+ and Cl^- rely on a strict compartmentation of salts within individual leaf cells and on their capacity to maintain a high K^+/Na^+ ratio in growing tissue. Salt deposition in nonphotosynthetic tissue and/or excretion to the leaf surface are also important for adaptation.

Tissue cultures can be used to select lines with increased salt tolerance. With this procedure, however, only the cellular components of salt tolerance can be deducted and evaluated, such as compartmentation within the cells, or K^+/Na^+ replacement (Orton, 1980; Nabors *et al.*, 1980; Smith and McComb, 1981; McHughen and Swartz, 1984). However, the morphological, physiological, and biochemical complexities of salt tolerance of intact plants (e.g., root to shoot transport of Na^+ and Cl^-, role of compatible organic solutes in different organs) must be tested in solution culture under controlled conditions and by field screening of plants growing in highly saline soils. In addition to using existing genetic variability within crop species for breeding (Epstein *et al.*, 1980) it is possible to introduce important components of salt tolerance into crop species from their wild relatives, as successfully demonstrated through interspecific hybridization in tomato by Rush and Epstein (1976) and Tal and Shannon (1983). The prospects for improvements in the salt tolerance of crop species by conventional selection and breeding as well as by genetic engineering have been critically reviewed by Downton (1984).

References

Abbot, A. J. (1967). Physiological effects of micronutrient deficiencies in isolated roots of *Lycopersicon esculentum*. *New Phytol.* **66**, 419–437.

Abdullah, Z. and Ahmad, R. (1982). Salt tolerance of *Solanum tuberosum* L. growing on saline soils amended with gypsum. *Z. Acker- Pflanzenbau* **151**, 409–416.

Abel, G. H. (1969). Inheritance of the capacity for chloride inclusion and chloride exclusion by soybeans. *Crop Sci.* **9**, 697–698.

Abou, A. A. and Volk, O. H. (1971). Nachweis von Cytokinin-Aktivität in rost-infizierten Pelargonium-Blättern. *Z. Pflanzenphysiol.* **65**, 240–247.

Abruna-Rodriguez, F., Vicente-Chandler, J., Rivera, E. and Rodriguez, J. (1982). Effect of soil acidity factors on yield and foliar composition of tropical root crops. *Soil Sci. Soc. Am. J.* **46**, 1004–1007.

Adams, F. (1966). Calcium deficiency as a causal agent of ammonium phosphate injury to cotton seedlings. *Soil Sci. Soc. Am. Proc.* **30**, 485–488.

Adams, F. and Moore, B. L. (1983). Chemical factors affecting root growth in subsoil horizons of coastal plain soils. *Soil Sci. Soc. Am. J.* **47**, 99–102.

Adams, P., Graves, C. J. and Winsor, G. W. (1975). Some effects of copper and boron deficiencies on the growth and flowering of *Chrysanthemum moriflorum* cv. Hurricane. *J. Sci. Food Agric.* **26**, 1899–1901.

Adeptu, J. A. and Akapa, L. K. (1967). Root growth and nutrient uptake characteristics of some cowpea varieties. *Agron. J.* **69**, 940–943.

Agarwala, S. C., Sharma, C. P., Farooq, S. and Chatterjee, C. (1978). Effect of molybdenum deficiency on the growth and metabolism of corn plants raised in sand culture. *Can. J. Bot.* **56**, 1905–1908.

Agarwala, S. C., Chatterjee, C., Sharma, P. N., Sharma, C. P. and Nautiyal, N. (1979). Pollen development in maize plants subjected to molybdenum deficiency. *Can. J. Bot.* **57**, 1946–1950.

Agarwala, S. C., Sharma, P. N., Chatterjee, C. and Sharma, C. P. (1981). Development and enzymatic changes during pollen development in boron deficient maize plants. *J. Plant Nutr.* **3**, 329–336.

Aguilar S. A. and van Diest, A. (1981). Rock-phosphate mobilization induced by the alkaline uptake pattern of legumes utilizing symbiotically fixed nitrogen. *Plant Soil* **61**, 27–42.

Ahmed, C. M. S. and Sagar, G. R. (1981). Effects of a mixture of NAA + BA on numbers and growth rates of tubers of *Solanum tuberosum* L. *Potato Res.* **24**, 267–278.

Ahmed, S. and Evans, H. J. (1960). Cobalt: A micronutrient element for the growth of soybean plants under symbiotic conditions. *Soil Sci.* **90**, 205–210.

Alagna, L., Hasnain, S. S., Piggott, B. and Williams, D. J. (1984). The nickel ion environment in jack bean urease. *Biochem. J.* **220**, 591–595.

Al-Ani, M. K. A. and Hay, R. K. M. (1983). The influence of growing temperature on the growth and morphology of cereal seedling root systems. *J. Exp. Bot.* **34**, 1720–1730.

Albert, L. S. (1965). Ribonucleic acid content, boron deficiency systems, and elongation of tomato root tips. *Plant Physiol.* **40**, 649–652.

Alcubilla, M., Diaz-Palcio, M. P., Kreutzer, K., Laatsch, W., Rehfuess, K. E. and Wenzel, G. (1971). Beziehungen zwischen dem Ernährungszustand der Fichte (*Picea abies* Karst.), ihrem Kernfäulebefall und der Pilzhemmwirkung ihres Basts. *Eur. J. For. Pathol.* **1**, 100–114.

Al-Khatib, K. and Paulsen, G. M. (1984). Mode of high temperature injury to wheat during grain development. *Physiol. Plant.* **61**, 363–368.

Allen, M. (1960). The uptake of metallic ions by leaves of apple trees. II. The influence of certain anions on uptake from magnesium salts. *J. Hortic. Sci.* **35**, 127–135.

Allen R. D. (1969). Mechanism of the seismonastic reaction in *Mimosa pudica*. *Plant Physiol.* **44**, 1101–1107.

Alleweldt, G., Düring, H. and Waitz, G. (1975). Untersuchungen zum Mechanisms der Zuckereinlagerung in wachsende Weinbeeren. *Angew. Bot.* **49**, 65–73.

Allinger, P., Michael, G. and Martin, P. (1969). Einfluss von Cytokininen und anderen Wuchsstoffen auf die Stoffverlagerung in abgeschnittenen Blättern. *Flora (Jena), Abt. A* **160**, 538–551.

Alt, D. and Stüwe, S. (1982). Decline of the nitrate content in lettuce (*Lactuca sativa* var. Capitata L.) by means of monitoring the nitrogen content of the nutrient solution in hydroponic systems. *In* "Proceedings of the Ninth International Plant Nutrition Colloquium, Warwick, England" (A. Scaife, ed.), pp. 17–21. Common. Agric. Bur., Farnham Royal, Bucks.

Alt, D., Zimmer, R., Stock, M., Peters, I. and Krupp, J. (1982). Erhebungsuntersuchungen zur Nährstoffversorgung von *Picea omorika* im Zusammenhang mit dem Omorikasterben. *Z. Pflanzenernaehr. Bodenkd.* **145**, 117–127.

Altman, A. (1982). Retardation of radish leaf senescence by polyamines. *Physiol. Plant.* **54**, 189–193.

Amberger, A. (1954). Einfluss von Kalium und Stickstoff auf Ferment- und Kohlehydrathaushalt von Grünlandpflanzen. *Z. Pflanzenernaehr., Dueng., Bodenkd.* **66**, 211–221.

Amberger, A. (1973). Die Rolle des Mangans im Stoffwechsel der Pflanzen. *Agrochimica* **17**, 165–183.

Amberger, A. and Gutser, R. (1978). Umsatz und Wirkung von Harnstoff-Dicyandiamid sowie Ammonsulfat-Dicyandiamid-Produkten zu Weidelgras und Reis. *Z. Pflanzenernaehr. Bodenkd.* **141**, 553–566.

Ames, R. N., Reid, C. P. P. and Ingham, E. R. (1984). Rhizosphere bacterial population responses to root colonization by a vesicular-arbuscular mycorrhizal fungus. *New Phytol.* **96**, 555–563.

Anderson, A. J. and Spencer, D. (1950). Sulphur in nitrogen metabolism of legumes and non-legumes. *Aust. J. Sci. Res., Ser. B* **3**, 431–449.

Anderson, A. J., Meyer, D. R. and Mayer, F. K. (1973). Heavy metal toxicities: Levels of nickel, cobalt and chromium in the soil and plants associated with visual symptoms and variation in growth of an oat crop. *Aust. J. Agric. Res.* **24**, 557–571.

Anderson, W. P. and Collins, J. C. (1969). The exudation from excised maize roots bathed in sulphate media. *J. Exp. Bot.* **20**, 72–80.

Andersson, B., Critchley, C., Ryrie, I. J., Jansson, C., Larsson, C. and Anderson, J. M. (1984). Modification of the chloride requirement for photosynthetic O_2 evolution. The role of the 23 KDa polypeptide. *FEBS Lett.* **168**, 113–117.

Ando, T., Yoshida, S. and Nishiyama, I. (1983). Nature of oxidizing power of rice roots. *Plant Soil* **72**, 57–71.

Anghinoni, I. and Barber, S. A. (1980). Phosphorus influx and growth characteristics of corn roots as influenced by phosphorus supply. *Agron. J.* **72**, 685–688.

Aniol, A. (1984). Induction of aluminium tolerance in wheat seedlings by low doses of aluminium in the nutrient solution. *Plant Physiol.* **75**, 551–555.

Antcliff, A. J., Newman, H. P. and Barrett, H. C. (1983). Variation in chloride accumulation in some American species of grapevine. *Vitis* **22**, 357–362.

App, A., Santiago, T., Daez, C., Menguito, C., Ventura, W., Tirol, A., Po, J., Watanabe, I., De Datta, S. K. and Roger, P. (1984). Estimation of the nitrogen balance for irrigated rice and the contribution of phototropic nitrogen fixation. *Field Crops Res.* **9**, 17–27.

Ariovich, D. and Cresswell, C. F. (1983). The effect of nitrogen and phosphorus on starch accumulation and net photosynthesis in two variants of *Panicum maximum* Jecq. *Plant, Cell Environ.* **6**, 657–664.

Armstrong, M. J. and Kirkby, E. A. (1979a). Estimation of potassium recirculation in tomato plants by comparison of the rates of potassium and calcium accumulation in the tops with their fluxes in the xylem stream. *Plant Physiol.* **63**, 1143–1148.

Armstrong, M. J. and Kirkby, E. A. (1979b). The influence of humidity on the mineral composition of tomato plants with special reference to calcium distribution. *Plant Soil* **52**, 427–435.

Armstrong, W. (1969). Rhizosphere oxidation in rice: An analysis of intervarietal differences in oxygen flux from the roots. *Physiol. Plant.* **22**, 296–303.

Armstrong, W. (1979). Aeration in higher plants. *Adv. Bot. Res.* **7**, 225–332.

Arneke, W. W. (1980). Der Einfluss des Kaliums auf die Komponenten des Wasserpotentials und auf die Wachstumsrate von *Phaseolus vulgaris*. Ph.D. Thesis, Universität Giessen.

Arnon, D. I. and Stout, P. R. (1939). The essentiality of certain elements in minute quantity for plants with special reference to copper. *Plant Physiol.* **14**, 371–375.

Arora, S. K. and Luchra, Y. P. (1970). Metabolism of sulphur containing amino acids in *Phaseolus aureus* Linn. *Z. Pflanzenernaehr. Bodenkd.* **126**, 151–158.

Arzet, H. R. (1972). Änderung des Kalium- und Magnesiumgehaltes einiger wirtschaftseigener Futtermittel in den letzten 100 Jahren. *Landwirtsch. Forsch.* **25**, 226–271.

Asher, C. J. (1978). Natural and synthetic culture media for Spermatophytes. *CRC Handb. Ser. Nutr. Food, Sect. G* **3**, 575–609.

Asher, C. J. and Edwards, D. G. (1983). Modern solution culture techniques. *In* "Encyclopedia of Plant Physiology, New Series" Vol. 15A (A. Läuchli and R. L. Bieleski, eds.), pp. 94–119. Springer-Verlag, Berlin and New York.

Ashworth, E. N., Christiansen, M. N., John, J. B. St. and Patterson, G. W. (1981). Effect of temperature and BASF 13 338 on the lipid composition and respiration of wheat roots. *Plant Physiol.* **67**, 711–715.

Aslam, M. and Huffaker, R. C. (1982). *In vivo* nitrate reduction in roots and shoots of barley (*Hordeum vulgare* L.) seedlings in light and darkness. *Plant Physiol.* **70**, 1009–1013.

Aslam, M. and Huffaker, R. C. (1984). Dependency of nitrate reduction on soluble carbohydrates in primary leaves of barley under aerobic conditions. *Plant Physiol.* **75**, 623–628.

Atkas , M. and Egmond, F. van (1979). Effect of nitrate nutrition on iron utilization by an Fe-efficient and an Fe-inefficient soybean cultivar. *Plant Soil* **51**, 257–274.

Atkin, R. K., Barton, G. E. and Robinson, D. K. (1972). Effect of root growing temperature on growth substances in xylem exudate of *Zea mays*. *J. Exp. Bot.* **24**, 475–487.

Atkins, C. A., Kuo, J., Pate, J. S., Flynn, A. M. and Steele, T. W. (1977). Photosynthetic pod wall of pea (*Pisum sativum* L.). Distribution of carbon dioxide-fixing enzymes in relation to pod structure. *Plant Physiol.* **60**, 779–786.

Atkinson, D. and Wilson, S. A. (1979). The root-soil interface and its significance for fruit tree roots of different ages. *In* "The Soil-Root Interface" (J. L. Harley and R. Scott-Russell, eds.), pp. 259–270. Academic Press, London.

Avivi, Y. and Feldman, M. (1982). The response of wheat to bacteria of the genus Azospirillum. *Isr. J. Bot.* **31**, 237–245.

Ayoub, A. T. (1974). Causes of inter-varietal differences in susceptibility to sodium toxicity injury in *Phaseolus vulgaris*. *Agric. Sci.* **83**, 539–543.

Ayoub, A. T. (1977). Some primary features of salt tolerance in senna (*Cassia acutifolia*). *Exp. Bot* **28**, 484–492.

Azarabadi, S. and Marschner, H. (1979). Role of the rhizosphere in utilization of inorganic iron III compounds by corn plants. *Z. Pflanzenernaehr. Bodenkd.* **142**, 751–764.

Azcón, R. and Ocampo, J. A. (1981). Factors affecting the vesicular–arbuscular infection and mycorrhizal dependency of thirteen wheat cultivars. *New Phytol.* **87**, 677–685.

Azcón, R. and Ocampo, J. A. (1984). Effect of root exudation on VA mycorrhizal infection at early stages of plant growth. *Plant Soil* **82**, 133–138.

Azcón, R., Barea, J. M. and Hayman. D. S. (1976). Utilization of rock phosphate in alkaline soils by plants inoculated with mycorrhizal fungi and phosphate solubilizing bacteria. *Soil Biol. Biochem.* **8**, 135–138.

Bachthaler, G. and Stritesky, A. (1973). Wachstumsuntersuchungen an Kulturpflanzen auf einem mit Kupfer überversorgten Mineralboden. *Bayer. Landwirtsch. Jahr.* **50**, 73–81.

Badurowa, M. Guminski, S. and Suder-Morav, A. (1967). Die Wirkung steigender Konzentrationen von Natriumhydrogenkarbonat in Wasserkulturen und die Gegenwirkung des Na-Humats. *Biol. Plant* **9**, 92–101.

Baier D. and Latzko, E. (1976). Properties and regulations of C-1-fructose 1,6-diphosphatase from spinach chloroplasts. *Biochim. Biophys. Acta* **396**, 141–148.

Baker, A. J. M. (1978). The uptake of zinc and calcium from solution culture by zinc-tolerant and non-tolerant *Silene maritima* With. in relation to calcium supply. *New Phytol.* **81**, 321–330.

Baker, D. A., Malek, F. and Dehvar, F. D. (1980). Phloem loading of amino acids from the petioles of ricinus leaves. *Ber. Dtsch. Bot. Ges.* **93**, 203–209.

Balandreau, J., Rinaudo, G., Fares-Hamad, I. and Dommergues, Y. (1975). Nitrogen fixation in the rhizosphere of rice plants. *In* "Nitrogen Fixation by Free-living Micro-organisms" (W. D. P. Stewart, ed.), pp. 57–70. Cambridge University Press, Cambridge.

Baligar, V. C. and Barber, S. A. (1978). Potassium uptake by onion roots characterized by potassium/rubidium ratio. *Soil Sci. Soc. Am. J.* **42**, 618–622.

Balke, N. E. and Hodges, T. K. (1975). Plasma membrane adenosine triphosphatase of oat roots. *Plant Physiol.* **55**, 83–86.

Ball, M. C., Taylor, S. E. and Terry, N. (1984). Properties of thylakoid membranes of the mangrove *Avicennia germinans* and *Avicennia marina*, and the sugar beet, *Beta vulgaris*, grown under different salinity conditions. *Plant Physiol.* **76**, 531–535.

Ballio, A., De Michelis, M. I., Lado, P. and Randazzo, G. (1981). Fusicoccin structure-activity relationships: Stimulation of growth by cell enlargement and promotion of seed germination. *Physiol. Plant.* **52**, 471–475.

Bamji, M. S. and Jagendorf, A. T. (1966). Amino acid incorporation by wheat chloroplasts. *Plant Physiol.* **41**, 764–770.

Bangerth, F. (1976). A role of auxin and auxin transport inhibitors on the Ca content of artificially induced parthenocarpic fruits. *Physiol. Plant.* **37**, 191–194.

Bangerth, F. (1979). Calcium-related physiological disorders of plants. *Annu. Rev. Phytopathol.* **17**, 97–122.

Bangerth, F., Dilley, D. R. and Dewey, D. H. (1972). Effect of calcium infusion on internal break-down and respiration of apple fruits. *J. Am. Soc. Hortic. Sci.* **97**, 679–682.

Bansal, K. N., Motiramani, D. P. and Pal, A. R. (1983). Studies on sulphur in vertisols, I. Soil and plant tests for diagnosing sulphur deficiency in soybean (*Glycine max* (L.) Merr.). *Plant Soil* **70**, 133–140.

Bar-Akiva, A. (1984). Substitutes for benzidine as H-donor in the peroxidase assay, for rapid diagnosis of iron deficiency in plants. *Commun. Soil Sci. Plant Anal.* **15**, 929–934.

Bar-Akiva, A. and Lavon, R. (1967). Visible symptoms and metabolic patterns in micronutrient-deficient Eureka lemon leaves. *Isr. J. Agric. Res.* **17**, 7–16.

Bar-Akiva, A., Sagiv, J. and Leshem, J. (1970). Nitrate reductase activity as an indicator for assessing the nitrogen requirement of grass crops. *J. Sci. Food Agric.* **21**, 405–407.

Bar-Akiva, A., Sagiv, J. and Hasdai, D. (1971). Effect of mineral deficiencies and other co-factors on the aldolase enzyme activity of citrus leaves. *Physiol. Plant.* **25**, 386–390.

Bar-Akiva, A., Maynard, D. N. and English, J. E. (1978). A rapid tissue test for diagnosing iron deficiencies in vegetable crops. *HortScience* **13**, 284–285.

Baranov, V. I. (1979). Biological activity of oxidized phenolic compounds and their role in break of indolyl-3-acetic acid. *Sov. Plant Physiol. (Engl. Transl.)* **26**, 688–695.

Barber, D. A. (1972). "Dual-isotherms" for the absorption of ions by plant tissues. *New Phytol.* **71**, 255–262.

Barber, D. A. (1978). Nutrient uptake. *In* "Interactions between Non-pathogenic Soil Microorganisms and Plants" (Y. R. Dommergues and S. V. Krupa, eds.), pp. 131–162. Elsevier, Amsterdam.

Barber, D. A. and Gunn, K. B. (1974). The effect of mechanical forces on the exudation of organic substances by the roots of cereal plants grown under sterile conditions. *New Phytol.* **73**, 30–45.

Barber, D. A. and Lee, R. B. (1974). The effect of micro-organisms on the absorption of manganese by plants. *New Phytol.* **73**, 97–106.

Barber, D. A. and Martin, J. K. (1976). The release of organic substances by cereal roots into soil. *New Phytol.* **76**, 69–80.

Barber S. A. (1962). A diffusion and massflow concept of soil nutrient availability. *Soil Sci.* **93**, 39–49.

Barber, S. A. (1966). The role of root interception, mass flow and diffusion in regulating the uptake of ion by plants from soil. *Tech. Rep. Ser.—I.A.E.A.* **65**, 39–45.

Barber, S. A. (1974). Influence of the plant root and ion movement in soil. *In* "The Plant Root and its Environment" (E. W. Carson, ed.), pp. 525–564. Univ. Press of Virginia, Charlottesville.

Barber, S. A. (1979). Growth requirement for nutrients in relation to demand at the root surface. *In* "The Soil-Root Interface" (J. L. Harley and R. Scott-Russell, eds.), pp. 5–20. Academic Press, London and Orlando.

Barber, S. A. (1982). Soil-plant root relationships determining phosphorus uptake. *In* "Proceedings of the Ninth International Plant Nutrition Colloquium, Warwick, England" (A. Scaife, ed.), pp. 39–44. Commonw. Agric. Bur., Farnham Royal, Bucks.

Barber, S. A (1984). "Soil Nutrient Bioavailability. A Mechanistic Approach". John Wiley, New York.

Barber, S. A. and Ozanne, P. G. (1970). Autoradiographic evidence for the differential effect of four plant species in altering the calcium content of the rhizosphere soil. *Soil Sci. Soc. Am. Proc.* **34**, 635–637.

Barel, D. and Black. C. A. (1979). Effect of neutralization and addition of urea, sucrose and various glycols on phosphorus absorption and leaf damage from foliar-applied phosphate. *Plant Soil* **52**, 515–525.

Barker, A. V. (1979). Nutritional factors in photosynthesis of higher plants. *J. Plant Nutr.* **1**, 309–342.

Barlett, R. J. and Riego, D. C. (1972). Effect of chelation on the toxicity of aluminium. *Plant Soil* **37**, 419–423.

Barlow, P. W. and Pilet, P. E. (1984). The effect of abscisic acid on cell growth, cell division and DNA synthesis in the maize root meristem. *Physiol. Plant.* **62**, 125–132.

Barneix, A. J., Breteler, H. and Geijn, S. C. van de (1984). Gas and ion exchanges in wheat roots after nitrogen supply. *Physiol. Plant.* **61**, 357–362.

Barnett, K. H. and Pearce, R. B. (1983). Source-sink ratio alteration and its effect on physiological parameters in maize. *Crop Sci.* **23**, 294–299.

Bar-Nun, N. and Poljakoff-Mayber, A. (1977). Salinity stress and the content of proline in roots of *Pisum sativum* and *Tamarix tetragyna*. *Ann. Bot. (London) [N.S.]* **41**, 173–179.

Barrett-Lennard, E. G. and Greenway, H. (1982). Partial separation and characterization of soluble phosphatases from leaves of wheat grown under phosphorus deficiency and water deficit. *J. Exp. Bot.* **33**, 694–704.

Barrett-Lennard, E. G., Marschner, H. and Römheld, V. (1983). Mechanism of short term Fe(III) reduction by roots. Evidence against the role of secreted reductants. *Plant Physiol.* **73**, 893–898.

Barrow, N. J. (1978). Problems of efficient fertilizer use. *Inf. Ser.—N.Z. Dep. Sci. Ind. Res.* **134**, 37–52.

Barz, W. (1977). Degradation of polyphenols in plants and cell suspension cultures. *Physiol. Veg.* **15**, 261–277.

Basiouny, F. M. (1984). Distribution of vanadium and its influence on chlorophyll formation and iron metabolism in tomato plants. *J. Plant Nutr.* **7**, 1059–1073.

Basyzynski, T., Ruszkowska, M., Krol, M., Tukendorf, A. and Wolinska, D. (1978). The effect of copper deficiency on the photosynthetic apparatus of higher plants. *Z. Pflanzenphysiol.* **89**, 207–216.

Baszynski, T., Warcholowa, M., Krupa, Z., Tukendorf, A., Krol, M. and Wolinska, D. (1980). The effect of magnesium deficiency on photochemical activities of rape and buckwheat chloroplasts. *Z. Pflanzenphysiol.* **99**, 295–303.

Bateman, D. F. and Lumsden, R. D. (1965). Relation between calcium content and nature of the pectic substances in bean hypocotyles of different ages to susceptibility to an isolate of *Rhizoctonia solani. Phytopathology* **55**, 734–738.

Bates, T. E. (1971). Factors affecting critical nutrient concentrations in plants and their evaluation: A review. *Soil Sci.* **112**, 116–126.

Bayne, H. G., Brown, M. S. and Bethlenfalway, G. J. (1984). Defoliation effects on mycorrhizal colonization, nitrogen fixation and photosynthesis in the *Glycine-Glomus-Rhizobium* symbiosis. *Physiol. Plant.* **62**, 576–580.

Beçana, M., Aparicio-Tejo, P. M. and Sánchez-Diaz, M. (1985). Nitrate and nitrite reduction by alfalfa root nodules: Accumulation of nitrite in *Rhizobium meliloti* bacteroids and senescence of nodules. *Physiol. Plant.* **64**, 353–358.

Beck, S. T. (1965). Resistance of plants to insects. *Annu. Rev. Entomol.* **10**, 207–232.

Becking, J. H. (1961). A requirement of molybdenum for the symbiotic nitrogen fixation in alder. *Plant Soil* **15**, 217–227.

Bedri, A. A., Wallace, A. and Rhoads, W. A. (1960). Assimilation of bicarbonate by roots of different plant species. *Soil Sci.* **89**, 257–263.

Beevers, L. and Hageman, R. H. (1983). Uptake and reduction of nitrate: Bacteria and higher plants. *In* "Encyclopedia of Plant Physiology, New Series" (A. Läuchli and R. L. Bieleski, eds.), Vol. 15A, pp. 351–375. Springer-Verlag, Berlin and New York.

Behl, R. and Hartung, W. (1984). Transport and compartmentation of abscisic acid in roots of *Hordeum distichon* under osmotic stress. *J. Exp. Bot.* **35**, 1433–1440.

Behl, R. and Jeschke, W. D. (1982). Potassium fluxes in excised barley roots. *J. Exp. Bot.* **33**, 584–600.

Bejaoui, M. (1985). Interactions entre NaCl et quelques phytohormones sur la croissance du soja. *J. Plant Physiol.* **120**, 95–110.

Bekele, T., Cino, B. J., Ehlert, P. A. I., Mass, A. A. van der and Diest, A. van (1983). An evaluation of plant-borne factors promoting the solubilization of alkaline rock phosphates. *Plant Soil* **75**, 361–378.

Bel, A. J. E. and Patrick, J. W. van (1985). Proton extrusion in seed coats of *Phaseolus vulgaris* L. *Plant, Cell Environ.* **8**, 1–6.

Bell, C. W. and Biddulph, O. (1963). Translocation of calcium. Exchange versus mass flow. *Plant Physiol.'* **38**, 610–614.

Belucci, S., Keller, E. R. and Schwendimann, F. (1982). Einfluss von Wachstumsregulatoren auf die Entwicklung und den Ertragsaufbau der Ackerbohne (*Vicia faba* L.). I. Wirkung von Gibberellinsäure (GA_3) auf die Ertragskomponenten und die Versorgung der jungen Früchte mit ^{14}C. *Angew. Bot.* **56**, 35–53.

Benepal, P. S. and Hall, C. V. (1967). The influence of mineral nutrition of varieties of *Cucurbita pepo* L. on the feeding response of Squash bug *Anasa tristis* De Geer. *Proc. Am. Soc. Hortic. Sci.* **90**, 304–312.

Beneš, I., Schreiber, K., Ripperger, H. and Kircheiss, A. (1983). Metal complex formation by nicotianamine, a possible phytosiderophore. *Experientia* **39**, 261–262.

Bennett, A. B. and Spanswick, R. M. (1984). H^+-ATPase activity from storage tissue of *Beta vulgaris*. II. H^+/ATP stoichiometry of an anion-sensitive H^+-ATPase. *Plant Physiol.* **74**, 545–548.

Bennett, A. B., O'Neill, S. D. and Spanswick, R. M. (1984). H^+-ATPase activity from storage tissue of *Beta vulgaris*. I. Identification and characterization of an anion-sensitive H^+-ATPase. *Plant Physiol.* **74**, 538–544.

Bennett, A. C. and Adams, F. (1970). Calcium deficiency and ammonia toxicity as separate causal factors of $(NH_4)_2HPO_4$ injury to seedlings. *Soil Sci. Soc. Am. Proc.* **34**, 255–259.

Bennett, D. M. (1982). Silicon deposition in the roots of *Hordeum sativum* Jess, *Avena sativa* L. and *Triticum aestivum* L. *Ann. Bot. (London)* [N.S.] **50**, 239–245.

Bennett, D. M. and Parry, D. W. (1981). Electronprobe microanalysis studies of silicon in the epicarp hairs of the caryopses of *Hordeum sativum* Jess., *Avena sativa* L., *Secale cereale* L. and *Triticum aestivum* L. *Ann. Bot. (London)* [N.S.] **48**, 645–654.

Bennett, J. H., Lee, E. H., Krizek, D. T., Olsen, R. A. and Brown, J. C. (1982). Photochemical reduction of iron. II. Plant related factors. *J. Plant Nutr.* **5**, 335–344.

Ben-Zioni, A., Vaadia, Y. and Lips, S. H. (1971). Nitrate uptake by roots as regulated by nitrate reduction products of the shoot. *Physiol. Plant.* **24**, 288–290.

Bergmann, W. (1954). Wurzelwachstum und Ernteertrag. *Z. Acker- Pflanzenbau* **97**, 337–368.

Bergmann, W. (1958). Über die Beeinflussung der Wurzelbehaarung von Roggenkeimpflanzen durch verschiedene Aussenfaktoren. *Z. Pflanzenernaehr., Dueng., Bodenkd.* **80**, 218–224.

Bergmann, W. (1983). "Ernährungsstörungen bei Kulturpflanzen, Entstehung und Diagnose." Fischer, Jena.

Bergmann, W. and Neubert, P. (1976). "Pflanzendiagnose und Pflanzenanalyse." Fischer, Jena.

Beringer, H. (1966). Einfluss von Reifegrad und N-Düngung auf Fettbildung und Fettsäurezusammensetzung in Haferkörnern. *Z. Pflanzenernaehr. Bodenkd.* **114**, 117–127.

Beringer, H. and Forster, H. (1981). Einfluss variierter Mg-Ernährung auf Tausendkorngewicht und P-Fraktionen des Gerstenkorns. *Z. Pflanzenernaehr. Bodenkd.* **144**, 8–15.

Beringer, H. and Hess, G. (1979). Brauchbarkeit der Pflanzenanalyse zur Bemessung später N-Gaben zu Winterweizen. *Landwirtsch. Forsch.* **32**, 384–394.

Berkum, P. van and Sloger, C. (1982). Physiology of root-associated nitrogenase activity in *Oryza sativa*. *Plant Physiol.* **69**, 1161–1164.

Bernal, C. T., Bingham, F. T. and Oertli, J. (1974). Salt tolerance of Mexican wheat. II. Relation to variable sodium chloride and length of growing season. *Soil Sci. Soc. Am. Proc.* **38**, 777–780.

Bernhard-Reversat, F. (1975). Nutrients in through fall and their quantitative importance in rain forest mineral cycle. *Ecol. Stud.* **11**, 153–159.

Bernstein, L. (1964). Salt tolerance of plants. *Agric. Inf. Bull. (U.S. Dep. Agric.)* **283**.

Bernstein, L. and François, L. E. (1975). Effects of frequency of sprinkling with saline waters compared with daily drip irrigation. *Agron. J.* **67**, 185–190.

Bertl, A., Felle, H. and Bentrup, F.-W. (1984). Amine transport in *Riccia fluitans*. Cytoplasmic and vacuolar pH recorded by a pH-sensitive microelectrode. *Plant Physiol.* **76**, 75–78.

Bertrand, D. and De Wolf, A. (1968). L'aluminium oligio-élément nécessaire au mais. *C.R. Hebd. Seances Acad. Sci. Ser. D* **267**, 2325–2327.

Bertrand, D. and De Wolf, A. (1973). Importance du nickel, comme oligo élément, pour les Rhizobium des nodosites des legumineuses. *C.R. Hebd. Seances. Acad. Sci.* **276**, 1855–1858.
Besford, R. T. (1978a). Effect of replacing nutrient potassium by sodium on uptake and distribution of sodium in tomato plants. *Plant Soil* **50**, 399–409.
Besford, R. T. (1978b). Use of pyruvate kinase activity of leaf extracts for the quantitative assessment of potassium and magnesium status of tomato plants. *Ann. Bot. (London)* [N.S.] **42**, 317–324.
Besford, R. T. and Syred, A. D. (1979). Effect of phosphorus nutrition on the cellular distribution of acid phosphatase in the leaves of *Lycopersicon esculentum* L. *Ann. Bot. (London)* [N.S.] **43**, 431–435.
Bethlenfalvay, G. J., Abu-Shakra, S. S., Fishbeck, K. and Phillips, D. A. (1978). The effect of source-sink manipulations on nitrogen fixation in peas. *Physiol. Plant.* **43**, 31–34.
Beusichem, M. L. and Langelaan, J. G. van (1984). Nitrogen accumulation in nodulated and non-nodulated pea plants, grown in a sandy soil at different acidities. *Commun. Soil Sci. Plant Anal.* **15**, 493–506.
Bhat, K. K. S. (1983). Nutrient inflows into apple roots. *Plant Soil* **71**, 371–380.
Bhat, K. K. S. and Nye, P. H. (1974). Diffusion of phosphate to plant roots in soil. II. Uptake along the roots at different times and the effect of different levels of phosphorus. *Plant Soil* **41**, 365–382.
Bhat, K. K. S., Nye, P. H. and Brereton, A. J. (1979). The possibility of predicting solute uptake and plant growth response from independently measured soil and plant characteristics. VI. The growth and uptake of rape in solutions of constant nitrate concentration. *Plant Soil* **53**, 137–167.
Bhatti, A. S. and Wieneke, J. (1984). Na^+ and Cl^--leaf extrusion, retranslocation and root efflux in *Diplachne fusca* (Kallar grass) grown in NaCl. *J. Plant Nutr.* **7**, 1233–1250.
Biddington, N. L. and Dearman, A. S. (1982). The effect of abscisic acid on root and shoot growth of cauliflower plants. *Plant Growth Regul.* **1**, 15–24.
Bieleski, R. L. (1968). Effect of phosphorus deficiency on levels of phosphorus compounds in *Spirodela*. *Plant Physiol.* **43**, 1309–1316.
Bieleski, R. L. and Ferguson, I. B. (1983). Physiology and metabolism of phosphate and its compounds. *In* "Encyclopedia of Plant Physiology, New Series" (A. Läuchli and R. L. Bieleski, eds.), Vol. 15A, pp. 422–449. Springer-Verlag, Berlin and New York.
Bieleski, R. L. and Läuchli, A. (1983). Synthesis and outlook. *In* "Encyclopedia of Plant Physiology, New Series" (A. Läuchli and R. L. Bieleski, eds.), Vol. 15B, pp. 745–755. Springer-Verlag, Berlin and New York.
Bienfait, H. F. (1985). Regulated redox processes at the plasmalemma of plant root cells and their function in iron uptake. *J. Bioenerg. Biomembr.* **17**, 73–83.
Bienfait, H. F., Duivenvoorden, J. and Verkerke, W. (1982). Ferric reduction by roots of chlorotic bean plants: Indications for an enzymatic process. *J. Plant Nutr.* **5**, 451–456.
Birnbaum, E. H., Beasley, C. A. and Dugger, W. M. (1974). Boron deficiency in unfertilized cotton (*Gossypium hirsutum*) ovules grown in vitro. *Plant Physiol.* **54**, 931–935.
Birnbaum, E. H., Dugger, W. M. and Beasley, B. C. A. (1977). Interaction of boron with components of nucleic acid metabolism in cotton ovules cultured in vitro. *Plant Physiol.* **59**, 1034–1038.

Blair, G. J. and Cordero, S. (1978). The phosphorus efficiency of three annual legumes. *Plant Soil* **50**, 387–398.
Blake, T. J. and Reid, D. M. (1981). Ethylene, water relations and tolerance to water logging of three Eucalyptus species. *Aust. J. Plant Physiol.* **8**, 497–505.
Blatt, C. R. and Diest, A. van (1981). Evaluation of a screening technique for manganese toxicity in relation to leaf manganese distribution and interaction with silicon. *Neth. J. Agric. Sci.* **29**, 297–304.
Bledsoe, C. S. and Zasoski, R. J. (1983). Effects of ammonium and nitrate on growth and nitrogen uptake by mycorrhizal Douglas-fir seedlings. *Plant Soil* **71**, 445–454.
Blevins, D. G., Barnett, N. M. and Frost, W. B. (1978). Role of potassium and malate in nitrate uptake and translocation by wheat seedlings. *Plant Physiol.* **62**, 784–788.
Bligny, R. and Douce, R. (1977). Mitochondria of isolated plant cells (*Acer pseudoplatanus* L.). II. Copper deficiency effects on cytochrome *c* oxidase and oxygen uptake. *Plant Physiol.* **60**, 675–679.
Blom-Zandstra, G. and Lampe, J. E. M. (1983). The effect of chloride and sulphate salts on the nitrate content in lettuce plants. (*Lactuca sativa* L.). *J. Plant Nutr.* **6**, 611–628.
Boag, T. S. and Brownell, P. F. (1979). C_4 photosynthesis in sodium deficient plants. *Aust. J. Plant Physiol.* **6**, 431–434.
Boawn, L. C. and Brown, J. C. (1968). Further evidence for P-Zn imbalance in plants. *Soil Sci. Soc. Am. Proc.* **32**, 94–97.
Bogenschütz, H. and König, E. (1976). Relationships between fertilization and tree resistance to forest insect pests. *Proc. 12th Colloq. Int. Potash Inst. Bern,* pp. 281–289.
Boguslawski, E. von (1958). Das Ertragsgesetz. *In* "Encyclopedia of Plant Physiology" (W. Ruhland, ed.), Vol. 4, pp. 943–976. Springer-Verlag, Berlin and New York.
Böhm, W. (1974). Phosphatdüngung und Wurzelwachstum. *Phosphorsaeure* **30**, 141–157.
Bohnsack, C. W. and Albert, L. S. (1977). Early effects of boron deficiency on indoleacetic acid oxidase levels of squash root tips. *Plant Physiol.* **59**, 1047–1050.
Bolan, N. S., Robson, A. D. and Barrow, N. J. (1984a). Increasing phosphorus supply can increase the infection of plant roots by vesicular–arbuscular mycorrhizal fungi. *Soil Biol. Biochem.* **16**, 419–420.
Bolan, N. S., Robson, A. D., Barrow, N. J. and Aylmore, L. A. G. (1984b). Specific activity of phosphorus in mycorrhizal and non-mycorrhizal plants in relation to the availability of phosphorus to plants. *Soil Biol. Biochem.* **16**, 299–304.
Bollard. E. G. (1983). Involvement of unusual elements in plant growth and nutrition. *In* "Encyclopedia of Plant Physiology, New Series" (A. Läuchli and R. L. Bieleski, eds.), Vol. 15B, pp. 695–755. Springer-Verlag, Berlin and New York.
Bolle-Jones, E. W. and Hilton, R. N. (1956). Zinc-deficiency of *Hevea braziliensis* as a predisposing factor to *Oidium* infection. *Nature (London)* **177**, 619–620.
Bolle-Jones, E. W. and Mallikarjuneswara, V. R. A. (1957). A beneficial effect of cobalt on the growth of *Hevea braziliensis. Nature (London)* **179**, 738–739.
Börner, H. (1957). Die Abgabe organischer Verbindungen aus Karyopsen, Wurzeln und Ernterückständen von Roggen, Weizen und Gerste und ihre Bedeutung bei der gegenseitigen Beeinflussung der höheren Pflanzen. *Beitr. Biol. Pflanz.* **33**, 33–83.

Borstlap, A. C. (1983). The use of model-fitting in the interpretation of "dual" uptake isotherms. *Plant, Cell Environ.* **6**, 407–416.
Bosch, C. (1983). Ernährungskundliche Untersuchung über die Erkrankung der Fichte (*Picea abies* Karst.) in den Hochlagen des Bayrischen Waldes. Diplomarbeit, Universität München.
Bothe, H., Yates, M. G. and Cannon, F. C. (1983). Physiology, biochemistry and genetic dinitrogen fixation. *In* "Encyclopedia of Plant Physiology, New Series" (A. Läuchli and R. L. Bieleski, eds.), Vol. 15A, pp. 241–285. Springer-Verlag, Berlin and New York.
Bottrill, D. E., Possingham, J. V. and Kriedemann, P. E. (1970). The effect of nutrient deficiencies on photosynthesis and respiration in spinach. *Plant Soil* **32**, 424–438.
Bould, C. (1966). Leaf analysis of deciduous fruits. *In* "Temperate to Tropical Fruit Nutrition" (N. F. Childers, ed.), pp. 651–684. Horticultural Publications, Rutgers University, New Brunswick, New Jersey.
Bould, C. and Parfitt, R. I. (1973). Leaf analysis as a guide to the nutrition of fruit crops. X. Magnesium and phosphorus sand culture experiments with apple. *J. Sci. Food Agric.* **24**, 175–185.
Bouma, D. (1983). Diagnosis of mineral deficiencies using plant tests. *In* "Encyclopedia of Plant Physiology, New Series" (A. Läuchli and R. L. Bieleski, eds.), Vol. 15A, pp. 120–146. Springer-Verlag, Berlin and New York.
Bouma, D., Dowling, E.J. and Wahjoedi, H. (1979). Some effects of potassium and magnesium on the growth of subterranean clover *(Trifolium subterraneum). Ann. Bot. (London)* [N.S.] **43**, 529–538.
Bousquet, U., Scheidecker, D. and Heller, R. (1981). Effet des conditions de culture sur la nutrition calcique de plantules calcifuge ou calcicoles. *Physiol. Veg.* **19**, 253–262.
Bové, J. M., Bové, C., Whatley, F. R. and Arnon, D. J. (1963). Chloride requirement for oxygen evolution in photosynthesis. *Z. Naturforsch. B: Anorg. Chem., Org. Chem., Biochem., Biophys., Biol.* **18B**, 683–688.
Bowen, G. D. and Rovira, A. D. (1961). The effect of micro-organisms on plant growth. I. Development of roots and root hairs in sand and agar. *Plant Soil* **15**, 166–188.
Bowes, G. and Ogren, W. L. (1972). Oxygen inhibition and other properties of soybean ribulose 1,5-diphosphate carboxylase. *J. Biol. Chem.* **247**, 2171–2176.
Bowling, D. J. F. (1981). Release of ions to the xylem in roots. *Physiol. Plant* **53**, 392–397.
Boxma, R. (1972). Bicarbonate as the most important soil factor in lime-induced chlorosis in the Netherlands. *Plant Soil* **37**, 233–243.
Bradfield. E. G. (1976). Calcium complexes in the xylem sap of apples shoots. *Plant Soil* **44**, 495–499.
Bradfield, E. G. and Guttridge, C. G. (1984). Effects of night-time humidity and nutrient solution concentration on the calcium-content of tomato fruit. *Sci. Hortic. (Amsterdam)* **22**, 207–217.
Bradford, K. J., Hsiao, T. C. and Yang, S. F. (1982). Inhibition of ethylene synthesis in tomato plants subjected to anaerobic root stress. *Plant Physiol.* **70**, 1503–1507.
Bradley, R., Burt, A. J. and Read, D. J. (1981). Mycorrhizal infection and resistance to heavy metal toxicity in *Calluna vulgaris*. *Nature (London)* **292**, 335–337.

Brady, C. J., Gibson, T. S., Barlow, E. W. R., Speirs, J. and Wyn Jones, R. G. (1984). Salt-tolerance in plants. I. Ions, compatible organic solutes and the stability of plant ribosomes. *Plant, Cell Environ.* **7**, 517–578.

Branton, D. (1969). Membrane structure. *Annu. Rev. Plant Physiol.* **20**, 209–238.

Bravo-F., P. and Uribe, E. G. (1981). Temperature dependence of the concentration kinetics of absorption of phosphate and potassium in corn roots. *Plant Physiol.* **67**, 815–819.

Breeze, V. G., Wild, A., Hopper, M. J. and Jones, L. H. P. (1984). The uptake of phosphate by plants from flowing nutrient solution. II. Growth of *Lolium perenne* L. at constant phosphate concentrations. *J. Exp. Bot.* **35**, 1210–1221.

Breimer, T. (1982). Environmental factors and cultural measures affecting the nitrate content of spinach. *Fert. Res.* **3**, 191–292.

Breteler. H. and Nissen, P. (1982). Effect of exogenous and endogenous nitrate concentration on nitrate utilization by dwarf bean. *Plant Physiol.* **70**, 754–759.

Breteler, H. and Smit. A. L. (1974). Effect of ammonium nutrition on uptake and metabolism of nitrate in wheat. *Neth. J. Agric. Sci.* **22**, 73–81.

Brevedan, R. E., Egli, D. B. and Leggett, J. E. (1977). Influence on N nutrition on total N, nitrate and carbohydrate levels in soybeans. *Agron. J.* **69**, 965–969.

Brevedan, R. E., Egli, D. B. and Leggett, J. E. (1978). Influence of N nutrition on flower and pod abortion and yield of soybeans. *Agron. J.* **70**, 81–84.

Brewster, J. L. and Tinker, P. B. (1970). Nutrient cation flows in soil around plant roots. *Soil Sci. Soc. Am. Proc.* **34**, 421–426.

Bridges, S. M. and Salin, M. L. (1981). Distribution of iron-containing superoxide dismutase in vascular plants. *Plant Physiol.* **68**, 275–278.

Briskin, D. P. and Poole, R. J. (1983). Characterization of a K^+-stimulated adenosine triphosphatase associated with the plasma membrane of red beet. *Plant Physiol.* **71**, 350–355.

Bromfield, S. M. (1958). The solution of γ-MnO_2 by substances released from soil and from the roots of oats and vetch in relation to manganese availability. *Plant Soil* **10**, 147–160.

Brookes, A., Collins, J. C. and Thurman, D. A. (1981). The mechanism of zinc tolerance in grasses. *J. Plant Nutr.* **3**, 695–705.

Brookes, R. R., Morrison, R. S., Reeves, R. D. and Malaisse, F. (1978). Copper and cobalt in African species of *Aeolanthis* Mert. *(Plectranthirae, Labiatae)*. *Plant Soil* **50**, 503–507.

Brouwer, R. (1967). Beziehungen zwischen Spross- und Wurzelwachstum. *Angew. Bot.* **41**, 244–254.

Brouwer, R. (1981). Co-ordination of growth phenomena within a root system of intact maize plants. *Plant Soil* **63**, 65–72.

Brown, J. C. (1978). Plant tolerance to alkaline soils. *ASA Spec. Publ.* **32**, 257–267.

Brown, J. C. (1979). Role of calcium in micronutrient stresses of plants. *Commun. Soil Sci. Plant Anal.* **10**, 459–472.

Brown, J. C. and Clark, R. B. (1974). Differential response of two maize inbreds to molybdenum stress. *Soil Sci. Soc. Am. Proc.* **38**, 331–333.

Brown, J. C. and Clark, R. B. (1977). Copper as essential to wheat reproduction. *Plant Soil* **48**, 509–523.

Brown, J. C. and Devine, T. E. (1980). Inheritance of tolerance or resistance to manganese toxicity in soybeans. *Agron. J.* **72**, 898–904.

Brown, J. C. and Jones, W. E. (1971). Differential transport of boron in tomato (*Lycopersicon esculentum* Mill.). *Physiol. Plant.* **25**, 279–282.

Brown, J. C. and Jones, W. E. (1976). A technique to determine iron efficiency in plants. *Soil Sci. Soc. Am. J.* **40**, 398–405.
Brown, J. C. and Jones, W. E. (1977a). Fitting plants nutritionally to soils. I. Soybeans. *Agron. J.* **69**, 399–404.
Brown, J. C. and Jones, W. E. (1977b). Manganese and iron toxicities dependent on soybean variety. *Commun. Soil Sci. Plant Anal.* **8**, 1–15.
Brown, J. C. and Olsen, R. A. (1980). Factors related to iron uptake by dicotyledonous and monocotyledonous plants. III. Competition between root and external factors for Fe. *J. Plant Nutr.* **2**, 661–682.
Brown, J. C., Weber, C. R. and Caldwell, B. E. (1967). Efficient and inefficient use of iron by two soybean genotypes and their isolines. *Agron. J.* **59**, 459–462.
Brown, J. C., Chaney, R. L. and Ambler, J. E. (1971). A new tomato mutant inefficient in the transport of iron. *Physiol. Plant* **25**, 48–53.
Brown, J. C., Clark, R. B. and Jones, W. E. (1977). Efficient and inefficient use of phosphorus by sorghum. *Soil Sci. Soc. Am. J.* **41**, 747–750.
Brown, J. C., Foy, C. D., Bennett, J. H. and Christiansen, M. N. (1979). Two light sources differentially affected ferric iron reduction and growth of cotton. *Plant Physiol.* **63**, 692–695.
Brown, R. H. (1978). A difference in N use efficiency in C_3 and C_4 plants and its implication in adaptation and evolution. *Crop. Sci.* **18**, 93–98.
Brown, T. A. and Shrift, A. (1982). Selenium: Toxicity and tolerance in higher plants. *Biol. Rev. Cambridge Philos. Soc.* **57**, 59–84.
Brownell, P. F. (1965). Sodium as an essential micronutrient element for a higher plant *(Atriplex vesicaria). Plant Physiol.* **40**, 460–468.
Brownell, P. F. (1979). Sodium as an essential micronutrient element for plants and its possible role in metabolism. *Adv. Bot. Res.* **7**, 117–224.
Brownell, P. F. and Crossland, C. J. (1972). The requirement for sodium as a micronutrient by species having the C_4 dicarboxylic photosynthetic pathway. *Plant Physiol.* **49**, 794–797.
Brownell, P. F. and Crossland, C. J. (1974). Growth responses to sodium by *Bryophyllum tibuflorum* under conditions inducing crassulacean acid metabolism. *Plant Physiol.* **54**, 416–417.
Broyer, T. C. (1966). Chlorine nutrition of tomato: Observations on inadvertent accretion and loss and their implications. *Physiol. Plant.* **19**, 925–936.
Broyer, T. C., Carlton, A. B., Johnson, C. M. and Stout, P. R. (1954). Chlorine—a micronutrient element for higher plants. *Plant Physiol.* **29**, 526–532.
Broyer, T. C., Johnson, C. M. and Huston, R. P. (1972). Selenium and nutrition of *Astragalus*. I. Effect of selenite or selenate supply on growth and selenium content. *Plant Soil* **36**, 635–649.
Brümmer, G. (1974). Redoxpotentiale und Redoxprozesse von Mangan-, Eisen- und Schwefelverbindungen in hydromorphen Böden und Sedimenten. *Geoderma* **12**, 207–222.
Brümmer, G. (1981). Ad- und Desorption oder Ausfällung und Auflösung als Lösungskonzentration bestimmende Faktoren in Böden. *Mitt. Dtsch. Bodenkd. Ges.* **30**, 7–18.
Bruinsma, J. (1977). Rolle der Cytokinine bei Blüten- und Fruchtentwicklung. *Z. Pflanzenernaehr. Bodenkd.* **140**, 15–23.
Brumagen, D. M. and Hiatt, A. J. (1966). The relationship of oxalic acid to the translocation and utilization of calcium in *Nicotiana tabacum. Plant Soil* **24**, 239–249.

Brüning, D. (1967). Befall mit *Eulecanium corni* Bché. f. *robinarium* Dgl. und *Eulecanium rufulum* Ckll. in Düngungsversuchen zu Laubgehölzen. *Arch. Pflanzenschutz* **3**, 193–200.
Brunold, C. (1981). Regulation of adenosine 5′-phosphosulfate sulfotransferase in higher plants. *In* "Biology of Inorganic Nitrogen and Sulfur" (H. Bothe and A. Trebst, eds.), pp. 352–358. Springer-Verlag, Berlin and New York.
Brunold, C. and Schmidt, A. (1978). Regulation of sulfate assimilation in plants. 7. Cysteine inactivation of adenosine 5-phosphosulfate sulfotransferase in *Lemna minor* L. *Plant Physiol.* **61**, 342–347.
Brunold, C. and Suter, M. (1984). Regulation of sulfate assimilation by nitrogen nutrition in the duckweed *Lemna minor* L. *Plant Physiol.* **76**, 579–583.
Buban, T., Varga, A., Tromp, J., Knegt, E. and Bruinsma, J. (1978). Effects of ammonium and nitrate nutrition on the level of zeatin and amino nitrogen in xylem sap of apple rootstocks. *Z. Pflanzenphysiol.* **89**, 289–295.
Bukovac, M. J. and Wittwer, S. H. (1957). Absorption and mobility of foliar applied nutrients. *Plant Physiol.* **32**, 428–435.
Bunje, G. (1979). Untersuchungen zum Einfluss der Mangan- und Kupferversorgung auf die Kälteresistenz von Winterweizen, Hafer und Mais anhand von Gefässversuchen. Dissertation, Universität Kiel.
Burleson, C. A. and Page, N. R. (1967). Phosphorus and zinc interactions in flax. *Soil Sci. Soc. Am. Proc.* **31**, 510–513.
Burnell, J. N. (1981). Selenium metabolism in *Neptunia amplexicauli. Plant Physiol.* **67**, 316–324.
Burrows, W. J. and Carr, D. J. (1969). Effects of flooding the root system of sunflower plants on the cytokinin content in the xylem sap. *Physiol. Plant* **22**, 1105–1112.
Burström, H. (1968). Calcium and plant growth. *Biol. Rev. Cambridge Philos. Soc.* **43**, 287–316.
Büscher, P. and Koedam, N. (1983). Soil preference of populations of genotypes of *Asplenium trichomanes* L. and *Polypodium vulgare* L. in Belgium as related to cation exchange capacity. *Plant Soil* **72**, 275–282.
Bussler, W. (1958). Manganvergiftung bei höheren Pflanzen. *Z. Pflanzenernaehr., Dueng., Bodenkd.* **81**, 256–265.
Bussler, W. (1963). Die Entwicklung von Calcium-Mangelsymptomen. *Z. Pflanzenernaehr., Dueng., Bodenkd.* **100**, 53–58.
Bussler, W. (1964). Die Bormangelsymptome und ihre Entwicklung. *Z. Pflanzenernaehr., Dueng., Bodenkd.* **105**, 113–136.
Bussler, W. (1970a). Die Entwicklung der Mo-Mangelsymptome an Blumenkohl. *Z. Pflanzenernaehr., Bodenkd.* **125**, 36–50.
Bussler, W. (1970b). Die Molybdän-Mangelsymptome une ihre Entwicklung. *Z. Pflanzenernaehr., Bodenkd.* **125**, 50–64.
Bussler, W. (1981a). Microscopic possibilities for the diagnosis of trace element stress in plants. *J. Plant Nutr.* **3**, 115–128.
Bussler, W. (1981b). Physiological functions and utilization of copper. *In* "Copper in Soils and Plants" (J. F. Loneragan, A. D. Robson and R. D. Graham, eds.), pp. 213–234. Academic Press, London and Orlando.
Butcher, D. and Pilet, P. E. (1983). Auxin effects on root growth and ethylene production. *Experientia* **39**, 493–494.
Buwalda, J. G., Stribley, D. P. and Tinker, T. B. (1983). Increased uptake of anions by plants with vesicular-arbusclar mycorrhizas. *Plant Soil* **71**, 463–467.

Caldwell, C. R. and Haug, A. (1981). Temperature dependence of the barley root plasma membrane-bound Ca^{2+} and Mg^{2+}-dependent ATPase. *Physiol. Plant* **53**, 117–124.

Cammarano, P., Felsani, A., Gentile, M., Gualerzi, C., Romeo, C. and Wolf, G. (1972). Formation of active hybrid 80-S particles from subunits of pea seedlings and mammalian liver ribosomes. *Biochim. Biophys. Acta* **281**, 625–642.

Campbell, L. C., Miller, M. H. and Loneragan, J. F. (1975). Translocation of boron to plant fruits. *Aust. J. Plant Physiol.* **2**, 481–487.

Campbell, N. A. and Thomson, W. W. (1977). Effects of lanthanum and ethylenediaminetetraacetate on leaf movements of *Mimosa. Plant Physiol.* **60**, 635–639.

Campbell, N. A., Stika, K. M. and Morrison, G. H. (1979). Calcium and potassium in the motor organ of sensitive plant: Localization by ion microscopy. *Science* **204**, 185–187.

Cannell, R. Q. (1977). Soil aeration and compaction in relation to root growth and soil management. *Appl. Biol.* **2**, 1–86.

Cannell, R. Q., Gales, K., Snaydon, R. W. and Suhail, B. A. (1979). Effects of short-term water logging on the growth and yield of peas *(Pisum sativum). Ann. Appl. Biol.* **93**, 327–335.

Caradus, J. R. (1982). Genetic differences in the length of root hairs in white clover and their effect on phosphorus uptake. *In* "Proceedings of the Ninth International Plant Nutrition Colloquium, Warwick, England" (A. Scaife, ed.), pp. 84–88. Commonw. Agric. Bur., Farnham Royal, Bucks.

Carmi, A. and Koller, D. (1979). Regulation of photosynthetic activity in the primary leaves of bean (*Phaseolus vulgaris* L.) by materials moving in the water-conducting system. *Plant Physiol.* **64**, 285–288.

Carmi, A. and Van Staden, J. (1983). Role of roots in regulating the growth rate and cytokinin content in leaves. *Plant Physiol.* **73**, 76–78.

Carpita, N., Sabularse, D., Montezinos, D. and Delmer, D. P. (1979). Determination of the pore size of cell walls of living plant cells. *Science* **205**, 1144–1147.

Carroll, B. J. and Gresshoff, P. M. (1983). Nitrate inhibition of nodulation and nitrogen fixation in white clover. *Z. Pflanzenphysiol.* **110**, 77–88.

Cartwright, B. and Hallsworth. E. G. (1970). Effects of copper deficiency on root nodules of subterranean clover. *Plant Soil* **33**, 685–698.

Carvalho, M. M. de, Edwards, D. G. and Asher, C. J. (1982). Effects of aluminium on nodulation of two stylosanthes species grown in nutrient solution. *Plant Soil* **64**, 141–152.

Cassells, A. L. and Barlass, M. (1976). Environmentally induced changes in the cell walls of tomato leaves in relation to cell and protoplast release. *Physiol. Plant.* **37**, 239–246.

Cassman, K. G., Whitney, A. S. and Stockinger, K. R. (1980). Root growth and dry matter distribution of soybean as affected by phosphorus stress, nodulation, and nitrogen source. *Crop. Sci.* **20**, 239–244.

Cerda, A., Bingham, F. T. and Hoffman, G. (1977). Interactive effect of salinity and phosphorus on sesame. *Soil. Sci. Soc. Am. J.* **41**, 915–918.

Chaboussou, F. (1976). Cultural factors and the resistance of citrus plants to scale insects and mites. *Proc. 12th Colloq. Int. Potash Inst. Bern,* pp. 259–280.

Chamberlain, I. S. and Spanner, D. C. (1978). The effect of law temperatures on the phloem transport of radioactive assimilates in the stolon of *Saxifraga sarmentosa* L. *Plant, Cell Environ.* **1**, 285–290.

Chandra, S. (1979). Genetics and plant breeding. *In* "A Decade of Research," pp. 80–98. Central Soil Salinity Research Institute, Karnal, India.
Chandrasekaran, S. and Yoshida, T. (1973). Effects of organic acid transformation in submerged soils on growth of the rice plant. *Soil Sci. Plant Nutr.* **19**, 39–45.
Chaney, R. L., Brown, J. C. and Tiffin, L. O. (1972). Obligatory reduction of ferric chelates in iron uptake by soybeans. *Plant Physiol.* **50**, 208–213.
Chapin, F. S., III (1983). Adaptation of selected trees and grasses to low availability of phosphorus. *Plant Soil* **72**, 283–297.
Chapin, F. S., III and Bieleski, R. L. (1982). Mild phosphorus stress in barley and a related low-phosphorus-adapted barley grass: Phosphorus fractions and phosphate absorption in relation to growth. *Physiol. Plant.* **54**, 309–317.
Chapman, H. D. (1966). "Diagnostic Criteria for Plants and Soils." Riverside Div. Agric. Sci., University of California.
Charyuly, P. B. B. N., Nayak, D. N. and Rao, V. R. (1981). $^{15}N_2$ incorporation by rhizosphere soil. Influence of rice variety, organic matter and combined nitrogen. *Plant Soil* **59**, 399–405.
Chatel, D. L., Robson, A. D., Gartrell, J. W. and Dilworth, M. J. (1978). The effect of inoculation and cobalt application on the growth of and nitrogen fixation by sweet lupinus. *Aust. J. Agric. Res.* **29**, 1191–1202.
Chatt, J. (1979). Problems of dinitrogen reduction and its prospects. *In* "Nitrogen Assimilation in Plants" (E. J. Hewitt and C. V. Cutting, eds.), pp. 17–26. Academic Press, New York.
Chauhan, R. P. S. and Powar, S. L. (1978). Tolerance of wheat and pea to boron in irrigation water. *Plant Soil* **50**, 145–149.
Chavan, P. D. and Karadge, B. A. (1980). Influence of sodium chloride and sodium sulfate salinization on photosynthetic carbon assimilation in peanut. *Plant Soil* **56**, 201–207.
Cheeseman, J. M. and Hanson, J. B. (1979). Energy-linked potassium influx as related to cell potential in corn roots. *Plant Physiol.* **64**, 842–845.
Chen, C. C., Dixon, J. B. and Turner, F. T. (1980). Iron coatings on rice roots: Mineralogy and quantity influencing factors. *Soil Sci. Soc. Am. J.* **44**, 635–639.
Chen, C. H. and Lewin, J. (1969). Silicon as a nutrient element for *Equisetum arvense*. *Can. J. Bot.* **47**, 125–131.
Chen, Y. and Barak, P. (1982). Iron nutrition of plants in calcareous soils. *Adv. Agron.* **35**, 217–240.
Cheniae, G. M. and Martin, I. F. (1968). Sites of manganese function in photosynthesis. *Biochim. Biophys. Acta* **153**, 819–837.
Cheniae, G. M. and Martin, I. F. (1969). Photoreactivation of manganese catalyst in photosynthetic oxygen evolution. *Plant Physiol.* **44**, 351–360.
Chereskin, B. M. and Castelfranco, P. A. (1982). Effects of iron and oxygen on chlorophyll biosynthesis. II. Observations on the biosynthetic pathway in isolated etio-chloroplasts. *Plant Physiol.* **68**, 112–116.
Chhabra, R., Ringoet, A., Lamberts, D. and Scheys, I. (1977). Chloride losses from tomato plants. (*Lycopersicon esculentum* Mill.). *Z. Pflanzenphysiol.* **81**, 89–94.
Chino, M. (1979). Calcium localization with plant roots by electron probe X-ray microanalysis. *Commun. Soil. Sci. Plant Anal.* **10**, 443–457.
Chino, M., Fukumorita, T., Kawabe, S. and Ando, Y. (1982). Chemical composition of rice phloem sap collected by "insect technique." *In* "Proceedings of the Ninth International Plant Nutrition Colloquium, Warwick, England" (A. Scaife, ed.), pp. 105–110. Commonw. Agric. Bur., Farnham Royal, Bucks.

Chisholm, R. H. and Blair, G. J. (1981). Phosphorus uptake and dry weight of stylo and white clover as affected by chlorine. *Agron. J.* **73**, 767–771.

Cho, B.-H. and Komor, E. (1980). The role of potassium in charge compensation for sucrose-proton-symport by cotyledons of *Ricinus communis. Plant Sci. Lett.* **17**, 425–435.

Christiansen, M. N., Carns, H. R. and Slyter, D. J. (1970). Stimulation of solute loss from radicles of *Gossypium hirsitum* L. by chilling, anaerobiosis, and low pH. *Plant Physiol.* **46**, 53–56.

Churchill, K. A. and Sze, H. (1983). Anion-sensitive, H^+-pumping ATPase in membrane vesicles from oat roots. *Plant Physiol.* **71**, 610–617.

Chvapil, M. (1973). New aspects in the biological role of zinc: A stabilizer of macromolecules and biological membranes. *Life Sci.* **13**, 1041–1049.

Cianzio, S. R. and Fehr, W. R. (1982). Variation in the inheritance of resistance to iron deficiency chlorosis in soybeans. *Crop Sci.* **22**, 433–434.

Claassen, N. and Barber, S. A. (1976). Simulation model for nutrient uptake from soil by a growing root system. *Agron. J.* **68**, 961–964.

Claassen, N. and Jungk, A. (1982). Kaliumdynamik im wurzelnahen Boden in Beziehung zur Kaliumaufnahme von Maispflanzen. *Z. Pflanzenernaehr. Bodenkd.* **145**, 513–525.

Claassen, N. and Jungk A. (1984). Bedeutung von Kaliumaufnahmerate, Wurzelwachstum und Wurzelhaaren für das Kaliumaneignungsvermögen verschiedener Pflanzenarten. *Z. Pflanzenernaehr. Bodenkd.* **147**, 276–289.

Claassen, N., Hendriks, L. and Jungk, A. (1981). Rubidium-Verarmung des wurzelnahen Bodens durch Maispflanzen. *Z. Pflanzenernaehr. Bodenkd.* **144**, 533–545.

Clark, R. B. (1975). Differential magnesium efficiency in corn inbreds. I. Dry-matter yields and mineral element composition. *Soil Sci. Soc. Am. Proc.* **39**, 488–491.

Clark, R. B. (1982a). Iron deficiency in plants grown in the great plains of the U.S. *J. Plant Nutr.* **5**, 251–268.

Clark, R. B. (1982b). Nutrient solution growth of sorghum and corn in mineral nutrition studies. *J. Plant Nutr.* **5**, 1039–1057.

Clark, R. B., Tiffin, L. O. and Brown, J. C. (1973). Organic acids and iron translocation in maize genotypes. *Plant Physiol.* **52**, 147–150.

Clark, R. B. Yusuf, Y., Ross, W. M. and Maranville, J. W. (1982). Screening for sorghum genotypic differences to iron deficiency. *J. Plant Nutr.* **5**, 587–604.

Clarkson, D. T. (1966). Effect of aluminium on the uptake and metabolism of phosphorus by barley seedlings. *Plant Physiol.* **41**, 165–172.

Clarkson, D. T. (1977). Membrane structure and transport. *In* "The Molecular Biology of Plant Cells" (H. Smith, ed.), pp. 24–63. Blackwell, Oxford.

Clarkson, D. T. and Hanson, J. B. (1980). The mineral nutrition of higher plants. *Annu. Rev. Plant Physiol.* **31**, 239–298.

Clarkson, D. T. and Sanderson, J. (1978). Sites of absorption and translocation of iron in barley roots. *Plant Physiol.* **61**, 731–736.

Clarkson, D. T. and Scattergood, C. B. (1982). Growth and phosphate transport in barley and tomato plants during the development of, and recovery from, phosphate-stress. *J. Exp. Bot.* **33**, 865–875.

Clarkson, D. T. and Warner, A. J. (1979). Relationships between root temperature and the transport of ammonium and nitrate ions by Italian and perennial ryegrass. (*Lolium multiflorum* and *Lolium perenne*). *Plant Physiol.* **64**, 557–561.

Clarkson, D. T., Robards, A. W. and Sanderson, J. (1971). The tertiary endodermis in barley roots: Fine structure in relation to radial transport of ions and water. *Planta* **96**, 292–305.

Clarkson, D. T., Sanderson, J. and Scattergood, C. B. (1978a). Influence of phosphate-stress and phosphate absorption and translocation by various parts of the root system of *Hordeum vulgare* L. (Barley). *Planta* **139**, 47–53.

Clarkson, D. T., Robards, A. W., Sanderson, J. and Peterson, C. A. (1978b). Permeability studies on epidermal hypodermal sleeves isolated from roots of *Allium cepa* (onion). *Can. J. Bot.* **56**, 1526–1532.

Claussen, W. and Biller, E. (1976). Einfluss der Frucht auf den Saccharose- und Stärkestoffwechsel in den Wurzeln der Aubergine (*Solanum melongena* L.). *Angew. Bot.* **50**, 217–232.

Claussen, W. and Biller, E. (1977). Die Bedeutung der Saccharose und Stärkegehalte der Blätter für die Regulierung der Nettophotosyntheserate. *Z. Pflanzenphysiol.* **81**, 189–198.

Claussen, W. and Lenz, F. (1979). Die Bedeutung des Assimilatstaus in den Blättern für Regulierung der Nettophotosyntheseraten bei Auberginen (*Solanum melongena* L.). *Angew. Bot.* **53**, 41–52.

Cleland, R. E. (1982). The mechanism of auxin-induced proton efflux. *In* "Plant Growth Substances" (P. F. Wareing, ed.), pp. 23–31. Academic Press, London.

Cleland, R. E. and Rayle, D. L. (1977). Reevaluation of the effect of calcium ions on auxin-induced elongation. *Plant Physiol.* **60**, 709–712.

Clement, C. R., Hopper, M. J. and Jones, L. H. P. (1978a). The uptake of nitrate by *Lolium perenne* from flowing nutrient solution. I. Effect of NO_3-concentration. *J. Exp. Bot.* **29**, 453–464.

Clement, C. R., Hopper, M. J., Jones, L. H. P. and Leafe, E. L. (1978b). The uptake of nitrate by *Lolium perenne* from flowing nutrient solution. II. Effect of light, defoliation, and relationship to CO_2 flux. *J. Exp. Bot.* **29**, 1173–1183.

Clement, C. R., Jones, L. H. P. and Hopper, M. J. (1979). Uptake of nitrogen from flowing nutrient solution: Effect of terminated and intermittent nitrate supplies. *In* "Nitrogen Assimilation in Plants" (E. J. Hewitt and C. V. Cutting, eds.), pp. 123–133. Academic Press, London and Orlando.

Cline, G. R., Powell, P. E., Szaniszlo, P. J. and Reid, C. P. P. (1983). Comparison of the abilities of hydroxamic and other organic acids to chelate iron and other ions in soils. *Soil Sci.* **136**, 145–157.

Cline, G. R., Reid, C. P. P., Powell, P. E. and Szaniszlo, P. J. (1984). Effects of a hydroxamate siderophore on iron absorption by sunflower and sorghum. *Plant Physiol.* **76**, 36–39.

Clutterbuck, B. J. and Simpson, K. (1978). The interactions of water and fertilizer nitrogen in effects on growth pattern and yield of potatoes. *J. Agric. Sci.* **91**, 161–172.

Coale, F. J., Evangelou, V. P. and Grove, J. H. (1984). Effects of saline-sodic soil chemistry on soybean mineral composition and stomatal resistance. *J. Environ. Qual.* **13**, 635–639.

Cohen, D. J. and Nadler, K. D. (1976). Calcium requirement for indoleacetic acid-induced acidification by Avena coleoptiles. *Plant Physiol.* **57**, 347–350.

Cohen, E., Okon, Y., Kigel, J., Nur, I. and Henis, Y. (1980). Increase in dry weight and total nitrogen content in *Zea mays* and *Setaria italica* associated with nitrogen-fixing *Azospirillum* ssp. *Plant Physiol.* **66**, 746–749.

Cohen, M. S. and Albert, L. S. (1974). Autoradiographic examination of meristems of intact boron-deficient squash roots treated with tritiated thymidine. *Plant Physiol.* **54**, 766–768.
Cohen, M. S. and Lepper, R., Jr. (1977). Effect of boron on cell elongation and division in squash roots. *Plant Physiol.* **59**, 884–887.
Coke, L. and Whittington, W. J. (1968). The role of boron in plant growth, IV. Interrelationships between boron and indol-3-yl acetic acid in the metabolism of bean radicles. *J. Exp. Bot.* **19**, 295–308.
Collier, G. F. and Tibbitts, T. W. (1984). Effects of relative humidity and root temperature on calcium concentration and tipburn development in lettuce. *J. Am. Soc. Hortic. Sci.* **109**, 128–131.
Collins, M. and Duke, S. H. (1981). Influence of potassium-fertilization rate and form on photosynthesis and N_2 fixation of alfalfa. *Crop Sci.* **21**, 481–485.
Colman, R. L. and Lazemby, A. (1970). Factors affecting the response of tropical and temperate grasses to fertilizer nitrogen. *Proc. 11th, Int. Grassl. Conf. Surf. Paradise,* pp. 393–397.
Constantopoulus, G. (1970). Lipid metabolism of manganese-deficient algae. I. Effect of manganese deficiency on the greening and the lipid composition of *Euglena gracilis* Z. *Plant Physiol.* **45**, 76–80.
Conway, B. E. (1981). "Ionic Hydration in Chemistry and Biophysics." Elsevier, Amsterdam.
Cooil, B. J. (1974). Accumulation and radial transport of ions from potassium salts by cucumber roots. *Plant Physiol.* **53**, 158–163.
Coombes, A. J., Lepp, N. W. and Phipps, D. A. (1976). The effect of copper on IAA-oxidase activity in root tissue of barley (*Hordeum vulgare,* cv. Zephyr). *Z. Pflanzenphysiol.* **80**, 236–242.
Coombes, A. J., Phipps, D. A. and Lepps, N. W. (1977). Uptake pattern of free and complexed copper ions in excised roots of barley (*Hordeum vulgare* L. cv. Zephyr). *Z. Pflanzenphysiol.* **82**, 435–439.
Cooper, K. M. and Tinker, P. B. (1978). Translocation and transfer of nutrients in vesicular–arbuscular mycorrhizas. II. Uptake and translocation of phosphorus, zinc and sulphur. *New Phytol.* **81**, 43–52.
Cooper, T. and Bangerth, F. (1976). The effect of Ca and Mg treatment on the physiology, chemical composition and bitter-pit development of "Cox orange" apples. *Sci. Hortic. (Amsterdam)* **5**, 49–57.
Corden, M. E. (1965). Influence of calcium nutrition on Fusarium wilt of tomato and polygalacturonase activity. *Phytopathology* **55**, 222–224.
Cory, S. and Finch. L. R. (1967). Further studies on the incorporation of ^{32}P-phosphate into nucleic acids of normal and boron-deficient tissue. *Phytochemistry* **6**, 211–215.
Coughlan, S. J. and Wyn Jones, R. G. (1980). Some responses of *Spinacea oleracea* to salt stress. *J. Exp. Bot.* **31**, 883–893.
Coulombe, B. A., Chaney, R. L. and Wiebold, W. J. (1984). Bicarbonate directly induces iron chlorosis in susceptible soybean cultivars. *Soil Sci. Soc. Am. J.* **48**, 1297–1301.
Cowan, I. R., Raven, J. A., Hartung, W. and Farquhar, G. D. (1982). A possible role for abscisic acid in coupling stomatal conductance and photosynthetic carbon metabolism in leaves. *Aust. J. Plant Physiol.* **9**, 489–498.
Cowling, D. W. and Lockyer, D. R. (1981). Increased growth of ryegrass exposed to ammonia. *Nature (London)* **292**, 337–338.

Cox, F. R. and Reid, P. H. (1964). Calcium-boron nutrition as related to concealed damage in peanuts. *Agron. J.* **56**, 173–176.
Cox, G., Moran, K. J., Sanders, F., Nockolds, C. and Tinker, P. B. (1980). Translocation and transfer of nutrients in vesicular–arbuscular mycorrhizas. III. Polyphosphate granules and phosphorus translocation. *New Phytol.* **84**, 649–654.
Cox, R. and Hutchinson, T. C. (1979). Metal co-tolerances in the grass *Deschampsia cespitosa. Nature (London)* **279**, 231–233.
Crafts, A. S. and Broyer, T. C. (1938). Migration of salts and water into xylem of roots of higher plants. *Am. J. Bot.* **24**, 415–431.
Craig, T. A. and Crane, F. L. (1981). A transmembrane electron transport system in plant cells. *Plant Physiol.* **67**, 599.
Cram, W. J. (1973). Internal factors regulating nitrate and chloride influx in plant cells. *J. Exp. Bot.* **24**, 328–341.
Cram, W. J. (1980). Chloride accumulation as a homeostatic system: Negative feedback signals for concentrations and turgor maintenance in a glycophyte and a halophyte. *Aust. J. Plant Physiol.* **7**, 237–249.
Cram, W. J. (1983). Characteristics of sulfate transport across plasmalemma and tonoplast of carrot root cells. *Plant Physiol.* **72**, 204–211.
Crawford, R. M. M. and Baines, M. A. (1977). Tolerance of anoxia and the metabolism of ethanol in tree roots. *New Phytol.* **79**, 519–526.
Crawford, R. M. M. and Zochowski, Z. M. (1984). Tolerance of anoxia and ethanol toxicity in chickpea seedlings (*Cicer arietium* L.) *J. Exp. Bot.* **35**, 1472–1480.
Creamer, F. L. and Fox, R. H. (1980). The toxicity of banded urea or diammonium phosphate to corn as influenced by soil temperature, moisture and pH. *Soil Sci. Soc. Am. J.* **44**, 296–300.
Cress, W. A., Throneberry, G. O. and Lindsey, D. L. (1979). Kinetics of phosphorus absorption by mycorrhizal and nonmycorrhizal tomato roots. *Plant Physiol.* **64**, 484–487.
Crisp, P., Collier, G. F. and Thomas, T. H. (1976). The effect of boron on tipburn and auxin activity in lettuce. *Sci. Hortic. (Amsterdam)* **5**, 215–226.
Crittenden, H. W. and Svec, C. V. (1974). Effect of potassium on the incidence of *Diaporthe sojae* in soybean. *Agron. J.* **66**, 696–698.
Crooke, W. M. and Knight, A. H. (1962). An evaluation of published data on the mineral composition of plants in the light of cation exchange capacities of their roots. *Soil Sci.* **93**, 365–373.
Crooke, W. M., Knight, A. H. and MacDonald, I. R. (1960). Cation exchange capacity and pectin gradients in leek root segments. *Plant Soil* **13**, 123–127.
Crossett, R. N. (1968). Effect of light upon the translocation of phosphorus by seedlings of *Hordeum vulgare* (L.) *Aust. J. Biol. Sci.* **21**, 225–233.
Crossett, R. N. and Campbell, D. J. (1975). The effects of ethylene in the root environment upon the development of barley. *Plant Soil* **42**, 453–464.
Crush, J. R. (1974). Plant growth responses to vesicular–arbuscular mycorrhiza. VII. Growth and nodulation of some herbage legumes. *New Phytol.* **73**, 743–752.
Cumbus, I. P. and Nye, P. H. (1982). Root zone temperature effects on growth and nitrate absorption in rape (*Brassica napus* cv. Emerald). *J. Exp. Bot.* **33**, 1138–1146.
Cumbus, I. P., Hornsey, D. J. and Robinson, L. W. (1977). The influence of phosphorus, zinc and manganese on absorption and translocation of iron in water cress. *Plant Soil* **48**, 651–660.

Curvetto, N. R. and Rauser, W. E. (1979). Isolation and characterization of copper binding proteins from roots of *Agrostis gigantea* tolerant to excess copper. *Plant Physiol.* **63**, Suppl., 59.

Cutsem P. van and Gillet, C. (1982). Activity coefficient and selectivity values of Cu^{2+}, Zn^{2+} and Ca^{2+} ions adsorbed in the *Nitella flexilis* L. cell wall during triangular ion exchanges. *J. Exp. Bot.* **33**, 847–853.

Dadson, R. B. and Acquaah, G. (1984). *Rhizobium japonicum*, nitrogen and phosphorus effects on nodulation, symbiotic nitrogen fixation and yield of soybean (*Glycine max* (L.) Merill) in the southern savanna of Ghana. *Field Crops Res.* **9**, 101–108.

Dai, Y.-R., Kaur-Sawhney, R. and Galston, A. W. (1982). Promotion by gibberellic acid of polyamine biosynthesis in internodes of light-grown dwarf peas. *Plant Physiol.* **69**, 103–105.

Dalling, M. J., Halloran, G. M. and Wilson, J. H. (1975). The relationship between nitrate reductase activity and grain nitrogen productivity in wheat. *Aust. J. Agric. Res.* **26**, 1–10.

Danielli, J. F. and Davson, H. A. (1935). A contribution to the theory of the permeability of thin films. *J. Cell. Comp. Physiol.* **5**, 495–508.

Darwinkel, A. (1980a). Grain production of winter wheat in relation to nitrogen and diseases. I. Relationship between nitrogen dressing and yellow rust infection. *Z. Acker- Pflanzenbau* **149**, 299–308.

Darwinkel, A. (1980b). Grain production of winter wheat in relation to nitrogen and diseases. II. Relationship between nitrogen dressing and mildew infection. *Z. Acker- Pflanzenbau* **149**, 309–317.

Dave, I. C. and Kannan, S. (1980). Boron deficiency and its associated enhancement of RNAase activity in bean plants. *Z. Pflanzenphysiol.* **97**, 261–264.

Davies, D. D. (1973). Control of and by pH. *Symp. Soc. Exp. Biol.* **27**, 513–529.

Davies, J. N., Adams, P. and Winsor, G. W. (1978). Bud development and flowering of *Chrysanthemum morifolium* in relation to some enzyme activities and to the copper, iron and manganese status. *Commun. Soil Sci. Plant Anal.* **9**, 249–264.

Davis, E. A., Young, J. L. and Rosc, S. L. (1984). Detection of high-phosphorus tolerant VAM-fungi colonizing hops and peppermint. *Plant Soil* **81**, 29–36.

Deane-Drummond, C. E. (1984). The apparent induction of nitrate uptake by *Chara corallina* cells following pretreatment with or without nitrate and chlorate. *J. Exp. Bot.* **35**, 1182–1193.

Deane-Drummond, C. E., Clarkson, D. T. and Johnson, C. B. (1979). Effects of shoot removal and malate on the activity of nitrate reductase assayed in vivo in barley roots (*Hordeum vulgare* cv. Midas). *Plant Physiol.* **64**, 660–662.

DeBoer, D. L. and Duke, S. H. (1982). Effects of sulphur nutrition on nitrogen and carbon metabolism in lucerne (*Medicago sativa* L.). *Physiol. Plant.* **54**, 343–350.

DeKock, P. C. and Hall, A. (1955). The phosphorus-iron relationship in genetical chlorosis. *Plant Physiol.* **30**, 293–295.

DeKock, P. C., Hall, A. and Inkson, R. H. E. (1979). Active iron in plant leaves. *Ann. Bot. (London)* [N.S.] **43**, 737–740.

DeLane, R., Greenway, H., Munns, R. and Gibbs, J. (1982). Ion concentration and carbohydrate status of the elongating leaf tissue of *Hordeum vulgare* growing at high external NaCl. I. Relationship between solute concentration and growth. *J. Exp. Bot.* **33**, 557–573.

Delhaize, E., Loneragan, J. F. and Webb, J. (1982). Enzymic diagnosis of copper deficiency in subterranean clover. II. A simple field test. *Aust. J. Agric. Res.* **33**, 981–987.

Dell, B. (1981). Male sterility and outer wall structure in copper-deficient plants. *Ann. Bot. (London)* [N.S.] **48**, 599–608.

Deloch, H. W. (1960). Über die analytische Bestimmung des Schwefels in biochemischen Substanzen und die Schwefelaufnahme durch landwirtschaftliche Kulturpflanzen in Abhängigkeit von der Düngung. Dissertation, Universität Giessen.

Delrot, S. and Bonnemain, J.-L. (1981). Involvement of protons as a substrate for the sucrose carrier during phloem loading in *Vicia faba* leaves. *Plant Physiol.* **67**, 560–564.

Delwiche, C. C., Johnson, C. M. and Reisenauer, H. M. (1961). Influence of cobalt on nitrogen fixation by Medicago. *Plant Physiol.* **36**, 73–78.

Demmig, B. and Gimmler, H. (1983). Properties of the isolated intact chloroplast at cytoplasmic K^+ concentrations. I. Light-induced cation uptake into intact chloroplasts is driven by an electrical potential difference. *Plant Physiol.* **73**, 169–174.

De-Polli, H., Boyer, C. D. and Neyra, C. A. (1982). Nitrogenase activity associated with roots and stems of field-grown corn (*Zea mays* L.) plants. *Plant Physiol.* **70**, 1609–1613.

Deroche, M.-E., Carrayol, E. and Jovivet, E. (1983). Phosphoenolpyruvate carboxylase in legume nodules. *Physiol. Veg.* **21**, 1075–1081.

Desai, N. and Chism, G. W. (1978). Changes in cytokinin activity in the ripening tomato fruit. *J. Food Sci.* **43**, 1324–1326.

Dhillon, S. S. (1978). Influence of varied phosphorus supply on growth and xylem sap cytokinin level of sycamore (*Platanus occidentalis* L.) seedlings. *Plant Physiol.* **61**, 521–524.

Diedrichs, C. (1983a). Influence of light on the efficacy of vesicular–arbuscular mycorrhiza in tropical and subtropical plants. II. Effect of light intensity under growth chamber conditions. *Angew. Bot.* **57**, 45–53.

Diedrichs, C. (1983b). Influence of light on the efficacy of vesicular–arbuscular mycorrhiza in tropical and subtropical plants. III. Influence of daylength. *Angew. Bot.* **57**, 55–67.

Dieter, P. (1984). Calmodulin and calmodulin-mediated processes in plants. *Plant, Cell Environ.* **7**, 371–380.

Dickinson, D. B. (1978). Influence of borate and pentaerythriol concentrations on germination and tube growth of *Lilium longiflorum* pollen. *J. Am. Soc. Hortic. Sci.* **103**, 413–416.

Dighton (1983). Phosphatase production by mycorrhizal fungi. *Plant Soil* **71**, 455–462.

Dijkshoorn, W. and Wijk, A. L. van (1967). The sulphur requirement of plants as evidenced by the sulphur-nitrogen ratio in the organic matter. A review of published data. *Plant Soil* **26**, 129–157.

Dilworth, M. J., Robson, A. D. and Chatel, D. L. (1979). Cobalt and nitrogen fixation in *Lupinus angustifolius* L. II. Nodule formation and functions. *New Phytol.* **83**, 63–79.

Dixon, N. E., Gazola, C., Blakeley, R. L. and Zerner, B. (1975). Jack bean urease (EC 3.5.1.5), a metalloenzyme. A simple biological role for nickel? *J. Am. Chem. Soc.* **97**, 4131–4133.

Dixon, N. E., Gazola, C., Asher, C. J., Lee, D. W., Blakeley, R. L. and Zerner, B. (1980a). Jack bean urease (EC 3.5.1.5). II. The relationship between nickel, enzymatic activity, and the "abnormal" ultraviolet spectrum. The nickel content of Jack bean. *Can. J. Biochem.* **58**, 474–480.

Dixon, N. E., Blakeley, R. L. and Zerner, B. (1980b). Jack bean urease (EC 3.5.1.5). III. The involvement of active-site nickel ion in inhibition by ß-mercaptoethanol, phosphoramidate, and fluoride. *Can. J. Biochem.* **58**, 481–488.

Dixon, N. E., Hinds, J. A., Fihelly, A. K., Gazola, C., Winzor, D. J., Blakeley, R. L. and Zerner, B. (1980c). Jack bean urease (EC. 3.5.1.5) IV. The molecular size and the mechanism of inhibition by hydroxamic acids. Spectrophotometric titration of enzymes with reversible inhibitors. *Can. J. Biochem.* **58**, 1323–1334.

Döbereiner, J. (1966). Manganese toxicity effects on nodulation and nitrogen fixation of beans (*Phaseolus vulgaris* L.) in acid soils. *Plant Soil* **24**, 153–166.

Döbereiner, J. (1983). Dinitrogen fixation in rhizosphere and phyllosphere associations. *In* "Encyclopedia of Plant Physiology, New Series" (A. Läuchli and R. L. Bieleski, eds.), Vol. 15A, pp. 330–350. Springer-Verlag, Berlin and New York.

Döbereiner, J. and Day, J. M. (1975). Nitrogen fixation in the rhizosphere of tropical grasses. *In* "Nitrogen Fixation by Free-living Micro-organisms" (W. D. P. Stewart, ed.), pp. 39–56. Cambridge Univ. Press, Cambridge.

Dodge, C. D. and Hiatt, A. J. (1972). Relationship of pH to ion uptake imbalance by varieties of wheat (*Triticum vulgare* L.). *Agron. J.* **64**, 476–477.

Dogar, M. A. and van Hai, T. (1980). Effect of P, N and HCO_3^- levels in the nutrient solution on rate of Zn absorption by rice roots and Zn content in plants. *Z. Pflanzenphysiol.* **98**, 203–212.

Doll, S., Rodier, F. and Willenbrink, J. (1979). Accumulation of sucrose in vacuoles isolated from red beet tissue. *Planta* **144**, 407–411.

Doman, D. C. and Geiger, D. R. (1979). Effect of exogenously supplied foliar potassium on phloem loading in *Beta vulgaris* L. *Plant Physiol.* **64**, 528–533.

Domsch, K. H. (1969). Microbial stimulation and inhibition of plant growth. *Trans. Int. Congr. Soil Sci., 9th, 1968,* Vol. 3, pp. 455–463.

Douglas, T. J. and Walker, R. R. (1983). 4-Desmethylsterol composition of citrus root-stocks of different salt exclusion capacity. *Physiol. Plant.* **58**, 69–74.

Douglas, T. J. and Walker, R. R. (1984). Phospholipids, free sterols and adenosine triphosphatase of plasma membrane-enriched preparations from roots of citrus genotypes differing in chloride exclusion ability. *Physiol. Plant.* **62**, 51–58.

Downton, W. J. S. (1977a). Photosynthesis in salt-stressed grapevines. *Aust. J. Plant Physiol.* **4**, 183–192.

Downton, W. J. S. (1977b). Chloride accumulation in different species of grapevine. *Sci. Hortic. (Amsterdam)* **7**, 249–253.

Downton, W. J. S. (1984). Salt tolerance of food crops: prospectives for improvements. *In* "CRC Critical Reviews in Plant Sciences." Vol. 1, pp. 183–201.

Dracup, M. N. H., Barrett-Lennard, E. C., Greenway, H. and Robson, A. D. (1984). Effect of phosphorus deficiency on phosphatase activity of cell walls from roots of subterranean clover. *J. Exp. Bot.* **35**, 466–480.

Dressel, J. and Jung, J. (1979). Gehaltsniveau an Vitaminen des B-Komplexes in Abhängigkeit von Stickstoffzufuhr und Standort. *Landwirtsch. Forsch., Sonderh.* **35**, 261–270.

Drew, M. C. (1979). Plant responses to anaerobic conditions in soil and solution culture. *Curr. Adv. Plant Sci.* **11**, No. 9, 36.1–36.14.

Drew, M. C. and Saker, L. R. (1975). Nutrient supply and the growth of the seminal root system in barley. II. Localized compensatory increases in lateral root growth and rates of nitrate uptake when nitrate supply is restricted to only part of the root system. *J. Exp. Bot.* **26**, 79–90.

Drew, M. C. and Saker, L. R. (1978). Nutrient supply and the growth of the seminal root system in barley. III. Compensatory increase in growth of lateral roots, and in rates of phosphate uptake, in response to a localized supply of phosphate. *J. Exp. Bot.* **29**, 435–451.

Drew, M. C. and Sisworo, E. J. (1979). The development of waterlogging damage in young barley plants in relation to plant nutrient status and changes in soil properties. *New Phytol.* **82**, 301–314.

Drew, M. C., Jackson, M. B. and Giffard, S. (1979). Ethylene-promoted adventitious rooting and development of cortical air spaces (aerenchyma) in roots may be adaptive responses to flooding in *Zea mays* L. *Planta* **147**, 83–88.

Drew, M. C., Chamel, A., Garrec, J.-P. and Fourcy, A. (1980). Cortical air spaces (aerenchyma) in roots of corn subjected to oxygen stress. Structure and influence on uptake and translocation of ^{86}rubidium. *Plant Physiol.* **65**, 506–511.

Drew, M. C., Saker, L. R., Barber, S. A. and Jenkins, W. (1984). Changes in the kinetics of phosphate and potassium absorption in nutrient-deficient barley roots measured by a solution-depletion technique. *Planta* **160**, 490–499.

Driessche, van den R. (1978). Response of Douglas fir seedlings to nitrate and ammonium nitrogen sources at different levels of pH and iron supply. *Plant Soil* **49**, 607–623.

Drissche, van den T. (1978). The molecular mechanism of Mimosa leaf seismonastic movement. A re-evaluation. *Arch. Biol.* **89**, 435–449.

Droppa, M., Terry, N. and Horvath, G. (1984). Effects of Cu deficiency on photosynthetic electron transport. *Proc. Natl. Acad. Sci.* **81**, 2369–2373.

Dudel, G. and Kohl, G. (1974). Über die Verteilung der Nitratreduktaseaktivität in Wurzel und Blatt bei *Hordeum vulgare* L. und ihre Abhängigkeit vom exogenen Nitratangebot. *Arch. Acker- Pflanzenbau Bodenkd.* **18**, 233–242.

Dugger, W. M. (1983). Boron in plant metabolism. *In* "Encyclopedia of Plant Physiology, New Series" (A. Läuchli and R. L. Bieleski, eds.), Vol. 15B, pp. 626–650. Springer-Verlag, Berlin and New York.

Dugger, W. M. and Palmer, R. L. (1980). The effect of boron in incorporation of glucose from UDPG into cotton fibers grown in vitro. *Plant Physiol.* **65**, 266–273.

Dunlop, J. (1974). The transport of potasssium to the xylem exudate of ryegrass. *J. Exp. Bot.* **25**, 1–10.

Dunlop, J. and Bowling, D. J. F. (1978). Uptake of phosphate by white clover. II. The effect of pH on the electrogenic phosphate pump. *J. Exp. Bot.* **29**, 1147–1153.

Düring, H. and Alleweldt, G. (1980). Effects of plant hormones on phloem transport in grapevines. *Ber. Dtsch. Bot. Ges.* **93**, 339–347.

Durrant, M. J., Draycott, A. P. and Milford, G. F. J. (1978). Effect of sodium fertilizer on water status and yield of sugar beet. *Ann. Appl. Biol.* **88**, 321–328.

Duvick, D. N., Kleese, R. A. and Frey, N. M. (1981). Breeding for tolerance of nutrient imbalance and constraints to growth in acid, alkaline and saline soils. *J. Plant. Nutr.* **4**, 111–129.

Dwelle, R. B., Kleinkopf, G. E., Steinhorst, R. K., Pavek, J. J. and Hurley, P. J. (1981). The influence of physiological processes on tuber yield of potato clones (*Solanum tuberosum* L.). Stomatal diffusive resistance, stomatal conductance, gross photosynthetic rate, leaf canopy, tissue nutrient levels, and tuber enzyme activities. *Potato Res.* **24**, 33–47.

Dwivedi, R. W. and Takkar, P. N. (1974). Ribonuclease activity as an index of hidden hunger of zinc in crops. *Plant Soil.* **40**, 173–181.
Ebeid, M. M. and Kutacek, M. (1979). Effects of EDTA on transport forms of manganese in maize xylem exudate. *Biol. Plant.* **21**, 178–182.
Eder, A. and Huber, W. (1977). Zur Wirkung von Abscisinsäure und Kinetin auf biochemische Veränderungen in *Pennisetum typhoides* unter Stresseinwirkungen. *Z. Pflanzenphysiol.* **84**, 303–311.
Edwards, D. G. and Asher, C. J. (1982). Tolerance of crop and pasture species to manganese toxicity. *In* "Proceedings of the Ninth Plant Nutrition Colloquium, Warwick, England" (A. Scaife, ed.), pp. 145–150. Commonw. Agric. Bur., Farnham Royal, Bucks.
Edwards, G. and Walker, D. (1983). "C_3, C_4: Mechanisms, and Cellular and Environmental Regulation, of Photosynthesis." Blackwell, Oxford.
Egli, S. (1983). Ectomykorrhiza bei Eiche und Fichte. *Allg. Forst Z.* **9/10**, 234–236.
Egmond, F. van and Atkas, M. (1977). Iron nutritional aspects of the ionic balance of plants. *Plant Soil* **48**, 685–703.
Egmond, F. van and Breteler, H. (1972). Nitrate reductase activity and oxalate content of sugar-beet leaves. *Neth. J. Agric. Sci.* **20**, 193–198.
Ehleringer, J. R. (1978). Implications of quantitative yield differences on the distributions of C_3 and C_4 grasses. *Oecologia* **31**, 255–267.
Eibach, H. (1982). Die vesikulare-arbuskulare Mycorrhiza der Rebe. Dissertation, Universität Hohenheim.
Eichhorn, M. and Augsten, H. (1974). Der Einfluss des Bors auf verschiedenartige Populationen von *Wolffia arrhiza* (L.) Wimm. in Chemostaten-Kultur. *Biochem. Physiol. Pflanz.* **165**, 371–385.
Ela, W. W., Anderson, M. A. and Brill, W. J. (1982). Screening and selection of maize to enhance associative bacterial nitrogen fixation. *Plant Physiol.* **70**, 1564–1567.
Elawad, S. H., Gascho, G. J. and Street, J. J. (1982a). Response of sugarcane to silicate source and rate. I. Growth and yield. *Agron. J.* **74**, 481–484.
Elawad, S. H., Street, J. J. and Gascho, G. J. (1982b). Response of sugarcane to silicate source and rate. II. Leaf freckling and nutrient content. *Agron. J.* **74**, 484–487.
Elgawhary, S. M., Malzer, G. L. and Barber, S. A. (1972). Calcium and strontium transport to plant roots. *Soil Sci. Soc. Am. Proc.* **36**, 794–799.
Elliott, G. C. and Läuchli, A. (1985). Phosphorus efficiency and phosphate–iron interaction in maize. *Agron. J.* **77**, 399–403.
Ellis, R. C. (1971). The mobilization of iron by extracts of Eucalyptus leaf litter. *J. Soil Sci.* **22**, 8–22.
El-Sheikh, A. M. and Ulrich, A. (1970). Interactions of rubidium, sodium, and potassium on the nutrition of sugar beet plants. *Plant Physiol.* **46**, 645–649.
El-Sheikh, A. M., Ulrich, A., Awad, S. K. and Mawardy, A. E. (1971). Boron tolerance of squash, melon, cucumber and corn. *J. Am. Soc. Hortic. Sci.* **96**, 536–537.
Elstner, E. F. (1982). Oxygen activation asnd oxygen toxicity. *Annu. Rev. Plant Physiol.* **33**, 73–96.
Elwali, A. M. O. and Gascho, G. J. (1984). Soil testing, foliar analysis, and DRIS as guides for sugarcane fertilization. *Agron. J.* **76**, 466–470.
Elzam, O. E. and Epstein, E. (1965). Absorption of chloride by barley roots: Kinetics and selectivity. *Plant Physiol.* **40**, 620–624.

Emmert, F. H. (1972). Effect of time, water flow, and pH on centripetal passage of radio-phosphorus across roots of intact plants. *Plant Physiol.* **50**, 332–335.

Ende, J. van den, Koornneef, P. and Sonneveld, C. (1975). Osmotic pressure of the soil solution: Determination and effects on some glasshouse crops. *Neth. J. Agric. Sci.* **23**, 181–190.

Engelbrecht, L., Orban, K. and Heese, W. (1969). Leaf-miner caterpillars and cytokinins in the "green islands" of autumn leaves. *Nature (London)* **223**, 319–321.

Engels, C. (1983). Wachstumsrate der Knollen von *Solanum tuberosum* L. var. Ostara in Abhängigkeit von exogenen und endogenen Faktoren—Konkurrenz zwischen Einzelknollen um Assimilate. Dissertation, Universität Hohenheim.

English, J. E. and Maynard, D. N. (1981). Calcium efficiency among tomato strains. *J. Am. Soc. Hortic. Sci.* **106**, 552–557.

Enoch, S. and Glinka, Z. (1981). Changes in potassium fluxes in cells of carrot storage tissue related to turgor pressure. *Physiol. Plant.* **53**, 548–552.

Epstein, E. (1965). Mineral metabolism. *In* "Plant Biochemistry" (J. Bonner and J. E. Varner, eds.), pp. 438–466. Academic Press, London and Orlando.

Epstein, E. (1972). "Mineral Nutrition of Plants: Principles and Perspectives." Wiley, New York.

Epstein, E. and Hagen, C. E. (1952). A kinetic study of the absorption of alkali cations by barley roots. *Plant Physiol.* **27**, 457–474.

Epstein, E., Rains, D. W. and Elzam, O. E. (1963). Resolution of dual mechanisms of potassium absorption by barley roots. *Proc. Natl. Acad. Sci. U.S.A.* **49**, 684–692.

Epstein, E., Norlyn, J. D., Rush, D. W., Kingsbury, R. W., Kelley, D. B., Cunningham, G. A. and Wrona, A. F. (1980). Saline culture of crops: A genetic approach. *Science* **210**, 399–404.

Erdei, L., Stuiver, B. and Kuiper, P. J. C. (1980). The effect of salinity on lipid composition and on activity of Ca^{2+} and Mg^{2+}-stimulated ATPases in salt-sensitive and salt-tolerant Plantago species. *Physiol. Plant.* **49**, 315–319.

Ergle, D. R. and Eaton, F. M. (1951). Sulfur nutrition of cotton. *Plant Physiol.* **26**, 639–654.

Eriksson, M. (1979). The effect of boron on nectar production and seed setting of red clover (*Trifolium pratense* L.). *Swed. J. Agric. Res.* **9**, 37–41.

Ernst, W. H. O. (1982). Schwermetallpflanzen. *In* "Pflanzenökologie und Mineralstoffwechsel" (H. Kinzel, ed.), pp. 472–506. Ulmer, Stuttgart.

Ernst, W. H. O. and Joosse-van Damme, E. N. G. (1983). "Umweltbelastung durch Mineralstoffe—Biologische Effekte." Fischer, Stuttgart.

Eschrich, W. (1976). "Strasburger's Kleines Botanisches Praktikum für Anfänger." Fischer, Stuttgart.

Eschrich, W. (1980). Free space invertase, its possible role in phloem unloading. *Ber. Dtsch. Bot. Ges.* **93**, 363–378.

Eschrich, W. (1984). Untersuchungen zur Regulation des Assimilattransports. *Ber. Dtsch. Bot. Ges.* **97**, 5–14.

Eshel, A. (1985). Response of *Sueda aegyptiaca* to KCl, NaCl and Na_2SO_4 treatments. *Physiol. Plant.* **64**, 308–315.

Eskew, D. L., Welch, R. M. and Norvell, W. A. (1984). Nickel in higher plants. Further evidence for an essential role. *Plant Physiol.* **76**, 691–693.

Evans, H. J. and Barber, L. E. (1977). Biological nitrogen fixation for food and fiber production. *Science* **197**, 332–339.

Evans, H. J. and Wildes, R. A. (1971). Potassium and its role in enzyme activation. *Proc. 8th Colloq. Int. Potash Inst. Bern,* pp. 13–39.

Evans, L. T., Wardlaw, I. F. and Fischer, R. A. (1975). Wheat. *In* "Crop Physiology" (L. T. Evans, ed.), pp. 101–109. Cambridge University Press, Cambridge.

Ewald, E. (1964). Die Wirkung unterschiedlicher Stickstoffdüngung auf Sommerweizen unter besonderer Berücksichtigung der Kornproteine und der Backqualität. Dissertation, Universität Hohenheim.

Ewart, J. A. D. (1978). Glutamin and dough tenacity. *J. Sci. Food Agric.* **29**, 551–556.

Eyster, C., Brown, T. E., Tanner, H. A. and Hood, S. L. (1958). Manganese requirement with respect to growth, Hill reaction and photosynthesis. *Plant Physiol.* **33**, 235–241.

Ezeta, F. N. and Jackson, W. A. (1975). Nitrate translocation by detopped corn seedlings. *Plant Physiol.* **56**, 148–156.

Fackler, U., Goldbach, H., Weiler, E. W. and Amberger, A. (1985). Influence of boron-deficiency on indol-3yl-acetic acid and abscisic acid levels in root and shoot tips. *J. Plant Physiol.* **119**, 295–299.

Faiz, S. M. A. and Weatherley, P. E. (1982). Root contraction in transpiring plants. *New Phytol.* **92**, 333–344.

Falchuk, K. H., Ulpino, L., Mazus, B. and Valee, B. L. (1977). E. gracilis RNA polymerase. I. A zinc metalloenzyme. *Biochem. Biophys. Res. Commun.* **74**, 1206–1212..

Faller, N. (1972). Schwefeldioxid, Schwefelwasserstoff, nitrose Gase und Ammoniak als ausschliessliche S- bzw. N-Quellen der höheren Pflanzen. *Z. Pflanzenernaehr. Bodenkd.* **131**, 120–130.

Fankhauser, H. and Brunold, C. (1978). Localization of adenosine 5′-phosphosulfate sulfotransferase in spinach leaves. *Planta* **143**, 285–289.

Fankhauser, H. and Brunold, C. (1979). Localization of O-acetyl-L-serine sulfhydrylase in *Spinacia oleracea* L. *Plant Sci. Lett.* **14**, 185–192.

Farley, R. F. and Draycott, A. P. (1973). Manganese deficiency of sugar beet in organic soil. *Plant Soil.* **38**, 235–241.

Farley, R. F. and Draycott, A. P. (1978). Manganese deficiency in sugar beet and the incorporation of manganese in the coating of pelleted seed. *Plant Soil* **49**, 71–83.

Farquhar, G. D., Wetselaar, R. and Firth, P. M. (1979). Ammonia volatization from senescing leaves of maize. *Science* **203**, 1257–1258.

Farrahi-Aschtiani, S. (1972). Einfluss von Ammonium- und Nitratstickstoff, Eisenchelaten und CCC auf den Chlorophyll- und Gesamtzuckergehalt der Blätter chlorotischer immergrüner Pflanzen auf alkalischen Böden Isfahans. *Z. Pflanzenernaehr. Bodenkd.* **131**, 190–196.

Farrow, R. P., Johnson, J. H., Gould, W. A. and Charbonneau, J. E. (1971). Detinning in canned tomatoes caused by accumulations of nitrate in the fruit. *J. Food Sci.* **36**, 341–345.

Fassbender, H. W. (1977). Modellversuch mit jungen Fichten zur Erfassung des internen Nährstoffumsatzes. *Oecol. Plant.* **12**, 263–272.

Faust, M. and Klein, J. D. (1974). Levels and sites of metabolically active calcium in apple fruit. *J. Am. Soc. Hortic. Sci.* **99**, 93–94.

Feldman, L. J. (1984). Regulation of root development. *Annu. Rev. Plant Physiol.* **35**, 223–242.

Fensom, D. S., Thompson, R. G. and Alexander, K. G. (1984). Stem anoxia temporarily interrupts translocation of ^{11}C-photosynthate in sunflower. *J. Exp. Bot.* **35**, 1582–1594.

Fentem, P. A., Lea, P. J. and Stewart, G. R. (1983). Ammonia assimilation in the roots of nitrate- and ammonia-grown *Hordeum vulgare* (cv. Golden promise). *Plant Physiol.* **71**, 496–501.

Ferguson, I. B. and Bollard, E. G. (1976). The movement of calcium in woody stems. *Ann. Bot (London)* [N.S.] **40**, 1057–1065.

Ferguson, I. B. and Clarkson, D. T. (1975). Ion transport and endothermal suberization in the roots of *Zea mays. New Phytol.* **75**, 69–79.

Ferguson, I. B. and Clarkson, D. T. (1976). Simultaneous uptake and translocation of magnesium and calcium in barley (*Hordeum vulgare* L.) roots. *Planta* **128**, 267–269.

Ferrari, G. and Renosto, F. (1972). Regulation of sulfate uptake by excised barley roots in the presence of selenate. *Plant Physiol.* **49**, 114–116.

Fido, R. J., Gundry, C. S., Hewitt, E. J. and Notton, B. A. (1977). Ultrastructural features of molybdenum deficiency and whiptail of cauliflower leaves. Effect of nitrogen source and tungsten substitution for molybdenum. *Aust. J. Plant Physiol.* **4**, 675–689.

Findenegg, G. R., Salihu, M. and Ali, N. A. (1982). Internal self-regulation of H^+-ion concentration in acid damaged and healthy plants of *Sorghum bicolor* (L.) Moench. *In* "Proceedings of the Ninth International Plant Nutrition Colloquium, Warwick, England" (A. Scaife, ed.), pp. 174–179. Commonw. Agric. Bur., Farnham Royal, Bucks.

Finn, B. J., Bourget, S. J., Nielson, K. F. and Dow, B. K. (1961). Effects of different soil moisture tensions on grass and legume species. *Can. J. Soil Sci.* **41**, 16–23.

Fiscus, E. L. and Kramer, P. J. (1970). Radial movement of oxygen in plant roots. *Plant Physiol.* **45**, 667–669.

Fisher, D. (1978). An evaluation of the Münch hypothesis for phloem transport in soybean. *Planta* **139**, 25–28.

Fisher, J. D. and Hodges, T. K. (1969). Monovalent ion stimulated adenosine triphosphatase from oat roots. *Plant Physiol.* **44**, 385–395.

Fisher, J. D., Hausen, D. and Hodges, T. K. (1970). Correlation between ion fluxes and ion stimulated adenosine triphosphatase activity of plant roots. *Plant Physiol.* **46**, 812–814.

Fleige, H., Strebel, O., Renger, M. and Grimme, H. (1981). Die potentielle P-Anlieferung durch Diffusion als Funktion von Tiefe, Zeit und Durchwurzelung bei einer Parabraunerde aus Löss. *Mitt. Dtsch. Bodenkd. Ges.* **32**, 305–310.

Fleige, H., Grimme, H., Renger, M. and Strebel, O. (1983). Zur Erfassung der Nährstoffanlieferung durch Diffusion im effektiven Wurzelraum. *Mitt. Dtsch. Bodenkd. Ges.* **38**, 381–386.

Flowers, T. J. and Läuchli, A. (1983). Sodium versus potassium: Substitution and compartmentation. *In* "Encyclopedia of Plant Physiology, New Series" (A. Läuchli and R. L. Bieleski, eds.), Vol. 15B, pp. 651–681. Springer-Verlag, Berlin and New York.

Flowers, T. J., Troke, P. F. and Yeo, A. R. (1977). The mechanism of salt tolerance in halophytes. *Annu. Rev. Plant Physiol.* **28**, 89–121.

Flügge, U. I., Freisl, M. and Heldt, H. W. (1980). Balance between metabolite accumulation and transport in relation to photosynthesis by isolated spinach chloroplasts. *Plant Physiol.* **65**, 574–577.

Foehse, D. and Jungk, A. (1983). Influence of phosphate and nitrate supply on root hair formation of rape, spinach and tomato plants. *Plant Soil* **74**, 359–368.

Forno, D. A., Yoshida, S. and Asher, C. J. (1975). Zinc deficiency in rice. I. Soil factors associated with the deficiency. *Plant Soil* **42**, 537–550.

Foroughi, M., Marschner, H. and Döring, H.-W. (1973). Auftreten von Bormangel bei *Citrus aurantium* L. (Bitterorangen) am Kaspischen Meer (Iran). *Z. Pflanzenernaehr. Bodenkd.* **136**, 220–228.

Forster, H. (1970). Der Einfluss einiger Ernährungsunterbrechungen auf die Ausbildung von Ertrags- und Qualitätsmerkmalen der Zuckerrübe. *Landwirtsch. Forsch. Sonderh.* **25**(II), 99–105.

Forster, H. (1980). Einfluss von unterschiedlich starkem Magnesiummangel bei Gerste auf den Kornertrag und seine Komponenten. *Z. Pflanzenernaehr. Bodenkd.* **143**, 627–637.

Forster, H. (1981). K-Bedarf und K-Versorgung von Kartoffeln. *Kali-Briefe* **15**, 745–760.

Forsyth, C. and Van Staden, J. (1981). The effect of root decapitation on lateral root formation and cytokinin production in *Pisum sativum*. *Physiol. Plant.* **51**, 375–379.

Fox, R. H. (1978). Selection for phosphorus efficiency in corn. *Commun. Soil Sci. Plant Anal.* **9**, 13–37.

Fox, R. L. and Lipps, R. C. (1961). Distribution and activity of roots in relation to soil properties. *Trans. Int. Congr. Soil Sci., 7th, 1960,* pp. 260–267.

Foy, C. D. (1974). Effect of aluminium on plant growth. *In* "The Plant Root and its Environment" (E. W. Carson, ed.), pp. 601–642. Univ. Press of Virginia, Charlottesville.

Foy, C. D. (1983). The physiology of plant adaptation to mineral stress. *Iowa State J. Res.* **57**, 355–391.

Foy, C. D. and Fleming, A. L. (1982). Aluminium tolerances of two wheat genotypes related to nitrate reductase activities. *J. Plant Nutr.* **5**, 1313–1333.

Foy, C. D., Fleming, A. L., Burns, G. R. and Arminger, W. H. (1967). Characterization of differential aluminium tolerance among varieties of wheat and barley. *Soil Sci. Soc. Am. Proc.* **31**, 513–521.

Foy, C. D., Fleming, A. L. and Arminger, W. H. (1969). Aluminium tolerance of soybean varieties in relation to calcium nutrition. *Agron. J.* **61**, 505–511.

Foy, C. D., Fleming, A. L. and Gerloff, G. C. (1972). Differential aluminium tolerance in two snapbean varieties. *Agron. J.* **64**, 815–818.

Foy, C. D., Fleming, A. L. and Schwartz, J. W. (1973). Opposite aluminium and manganese tolerances of two wheat varieties. *Agron. J.* **65**, 123–126.

Foy, C. D., Lafever, H. N., Schwartz, J. W. and Fleming, A. L. (1974). Aluminium tolerance of wheat cultivars related to region of origin. *Agron. J.* **66**, 751–758.

Foy, C. D., Chaney, R. L. and White, M. C. (1978). The physiology of metal toxicity in plants. *Annu. Rev. Plant Physiol.* **29**, 511–566.

Foy, C. D., Webb, H. W. and Jones, J. E. (1981). Adaptation of cotton genotypes to an acid, manganese toxic soil. *Agron. J.* **73**, 107–111.

Franck, E. von and Finck, A. (1980). Ermittlung von Zink-Ertragsgrenzwerten für Hafer und Weizen. *Z. Pflanzenernaehr. Bodenkd.* **143**, 38–46.

Franco, A. A. and Munns, D. N. (1981). Response of *Phaseolus vulgaris* L. to molybdenum under acid conditions. *Soil Sci. Soc. Am. J.* **45**, 1144–1148.

Franco, A. A. and Munns, D. N. (1982a). Acidity and aluminium restraints on nodulation, nitrogen fixation, and growth of *Phaseolus vulgaris* in solution culture. *Soil Sci. Soc. Am. J.* **46**, 296–301.

Franco, A. A. and Munns, D. N. (1982b). Nodulation and growth of *Phaseolus vulgaris* in solution culture. *Plant Soil* **66**, 149–160.
François, L. E. and Clark, R. A. (1979). Accumulation of sodium and chloride in leaves of sprinkler-irrigated grapes. *J. Am. Soc. Hortic. Sci.* **104**, 11–13.
Franke, W. (1967). Mechanism of foliar penetration of solutions. *Annu. Rev. Plant Physiol.* **18**, 281–300.
Franke, W. (1975). Stoffaufnahme durch das Blatt unter besonderer Berücksichtigung der Ektodesmen. *Bodenkultur* **26**, 331–341.
Freney, J. R., Delwiche, C. C. and Johnson, C. M. (1959). The effect of chloride on the free amino acids of cabbage and cauliflower plants. *Aust. J. Soil Sci.* **12**, 160–167.
Freney, J. R., Spencer, K. and Jones, M. B. (1978). The diagnosis of sulphur deficiency in wheat. *Aust. J. Agric. Res.* **29**, 727–738.
Fridovich, I. (1983). Superoxide radical: an endogenous toxicant. *Annu. Rev. Pharmacol. Toxicol.* **23**, 239–257.
Froehlich, D. M. and Fehr, W. R. (1981). Agronomic performance of soybeans with differing levels of iron deficiency chlorosis on calcareous soil. *Crop Sci.* **21**, 438–440.
Frota, J. N. E. and Tucker, T. C. (1978). Salt and water stress influences nitrogen metabolism in red kidney beans. *Soil Sci. Soc. Am. J.* **42**, 743–746.
Fußeder, A. (1984). Der Einfluß von Bodenart, Durchlüftung des Bodens, N-Ernährung und Rhizosphärenflora auf die Morphologie des seminalen Wurzelsystems von Mais. *Z. Pflanzenernaehr. Bodenkd.* **147**, 553–564.
Fuchs, W. H. and Grossman, F. (1972). Ernährung und Resistenz von Kulturpflanzen gegenüber Krankheitserregern und Schädlingen. *In* "Handbuch der Pflanzenernährung und Düngung" (H. Linser, ed.), Vol. 1, part 2, pp. 1007–1107. Springer-Verlag, Berlin and New York.
Fulton, J. M. and Erickson, A. E. (1964). Relation between soil aeration and ethyl alcohol accumulation in xylem exudate of tomatoes. *Soil Sci. Soc. Am. Proc.* **28**, 610–614.
Funkhouser, E. A. and Price, C. A. (1974). Chloroplast RNA: Possible site of an early lesion in iron deficiency. *Plant Cell Physiol.* **15**, 883–889.
Fushiya, S., Takahashi, K., Nakatsuyama, S., Sato, Y., Nozoe, S. and Takagi, S.-J. (1982). Co-occurrence of nicotianamine and avenic acids in *Avena sativa* and *Oryza sativa*. *Phytochemistry* **21**, 1907–1908.
Gabelman, W. H. and Gerloff, G. C. (1983). The search for and interpretation of genetic controls that enhance plant growth under deficiency levels of a macronutrient. *Plant Soil* **72**, 335–350.
Galling, G. (1963). Analyse des Magnesium-Mangels bei synchronisierten Chlorellen. *Arch. Mikrobiol.* **46**, 150–184.
Garcia, R. L. and Hanway, J. J. (1976). Foliar fertilization of soybeans during the seed filling period. *Agron. J.* **68**, 653–657.
Gardner, W. K., Parbery, D. G. and Barber, D. A. (1981). Proteoid root morphology and function in *Lupinus albus*. *Plant Soil* **60**, 143–147.
Gardner, W. K., Barber, D. A. and Parbery, D. C. (1982a). Effects of microorganisms on the formation and activity of proteoid roots of *Lupinus albus* L. *Aust. J. Bot.* **30**, 303–309.
Gardner, W. K., Parbery, D. G. and Barber, D. A. (1982b). The acquisition of phosphorus by *Lupinus albus* L. I. Some characteristics of the soil/root interface. *Plant Soil* **68**, 19–32.

Gardner, W. K., Parbery, D. G. and Barber, D. A. (1982c). The acquisition of phosphorus by *Lupinus albus* L. II. The effect of varying phosphorus supply and soil type on some characteristics of the soil/root interface. *Plant Soil* **68**, 33–41.
Gardner, W. K., Barber, D. A. and Parbery, D. G. (1983a). The acquisition of phosphorus by *Lupinus albus* L. III. The probable mechanism by which phosphorus movement in the soil/root interface is enhanced. *Plant Soil* **70**, 107–114.
Gardner, W. K., Barber, D. A. and Parbery, D. G. (1983b). Non-infecting rhizosphere micro-organisms and the mineral nutrition of temperate cereals. *J. Plant Nutr.* **6**, 185–199.
Gardner, W. K., Parbery, D. G., Barber, D. A. and Swinden, L. (1983c). The acquisition of phosphorus by *Lupinus albus* L. V. The diffusion of exudates away from roots: a computer simulation. *Plant Soil* **72**, 13–29.
Garg, O. K., Sharma, A. N. and Kona, G. R. S. S. (1979). Effect of boron on the pollen vitality and yield of rice plant (*Oryza sativa* L. var. Jaya). *Plant Soil* **52**, 591–594.
Gärtel, W. (1974). Die Mikronährstoffe—ihre Bedeutung für die Rebenernährung unter besonderer Berücksichtigung der Mangel- und Überschußerscheinungen. *Weinberg Keller* **21**, 435–507.
Garwood, E. A. and Williams, T. E. (1967). Growth, water use and nutrient uptake from the subsoil by grass swards. *J. Agric. Sci.* **69**, 125–130.
Garz, J. (1966). Menge, Verteilung und Bindungsform der Mineralstoffe (P, K, Mg und Ca) in den Leguminosensamen in Abhängigkeit von der Mineralstoffumlagerung innerhalb der Pflanze und den Ernährungsbedingungen. *Kuehn-Arch.* **80**, 137–194.
Gascho, G. J. (1977). Response of sugarcane to calcium silicate slag. I. Mechanisms of response in Florida. *Proc. Soil Crop Sci. Soc. Fla.* **37**, 55–58.
Gashaw, L. and Mugwira, L. M. (1981). Ammonium-N and nitrate-N effects on the growth and mineral compositions of triticale, wheat and rye. *Agron. J.* **73**, 47–51.
Gavalas, N. A. and Manetas, Y. (1980). Calcium inhibition of phosphoenolpyruvate carboxylase. Possible physiological consequences for C_4-photosynthesis. *Z. Pflanzenphysiol.* **100**, 179–184.
Geering, H. R., Hodgson, J. F. and Sdano, C. (1969). Micronutrient cation complexes in soil solution. IV. The chemical state of manganese in soil solution. *Soil Sci. Soc. Am. Proc.* **33**, 81–85.
Geiger, D. R. (1975). Phloem loading. *In* "Encyclopedia of Plant Physiology. New Series" (M. H. Zimmermann and J. A. Milburn, eds.), Vol. 1, pp. 396–431. Springer-Verlag, Berlin and New York.
Geijn, S. C. van de and Petit, C. M. (1979). Transport of divalent cations. Cation exchange capacity of intact xylem vessels. *Plant Physiol.* **64**, 954–958.
Geijn, S. C. van de and Smeulders, F. (1981). Diurnal changes in the flux of calcium towards meristems and transpiring leaves in tomato and maize plants. *Planta* **151**, 265–271.
Geisler, G. (1967). Interactive effects of CO_2 and O_2 in soil on root and top growth of barley and peas. *Plant Physiol.* **42**, 305–307.
Geisler, G. (1968). Über den Einfluss von Unterbodenverdichtungen auf den Luft- und Wasserhaushalt des Bodens und das Wurzelwachstum. *Landwirtsch. Forsch., Sonderh.* **22**, 61–69.
Geisler, G. and Krützfeldt, B. (1983). Untersuchungen zur Wirkung von "Stickstoff" auf die Morphologie, die Trockensubstanzbildung und die Aufnahmeleistung der Wurzelsysteme von Mais-, Sommergersten- und Ackerbohnensorten

unter Berücksichtigung der Temperatur. I. Wurzelmorphologie. *Z. Acker Pflanzenbau* **152**, 336–353.
Gepstein, S. (1982). Light-induced H^+ secretion and the relation to senescence of oat leaves. *Plant Physiol.* **70**, 1120–1124.
Gerath, H., Borchmann, W. and Zajonc, I. (1975). Zur Wirkung des Mikronährstoffs Bor auf die Ertragsbildung von Winterraps (*Brassica napus* L. ssp. *oleifera*). *Arch. Acker- Pflanzenbau Bodenkd.* **19**, 781–792.
Gerloff, G. C. and Gabelman, W. H. (1983). Genetic basis of inorganic plant nutrition. *In* "Encyclopedia of Plant Physiology, New Series" (A. Läuchli and R. L. Bieleski, eds.), Vol. 15B, pp. 453–480. Springer-Verlag, Berlin and New York.
Gerwick, B. C. and Black, C. C., Jr. (1979). Sulfur assimilation in C_4 plants. Intercellular compartmentation of adenosine 5′-triphosphate sulfurylase in crab grass leaves. *Plant Physiol.* **64**, 590–593.
Ghoneim, M. F. and Bussler, W. (1980). Diagnosis of zinc deficiency in cotton. *Z. Pflanzenernaehr. Bodenkd.* **143**, 377–384.
Gianinazzi-Pearson, V. and Diem, H. G. (1982). Endomycorrhizae in the tropics. *In* "Microbiology of Tropical Soils: Implications in Soil Management" (Y. R. Dommergues and H. G. Diem, eds.), pp. 209–251. Martinus Nijhoff, The Hague.
Gianinazzi-Pearson, V. and Gianinazzi, S. (1983). The physiology of vesicular–arbuscular mycorrhizal roots. *Plant Soil* **71**, 197–209.
Giaquinta, R. T. (1977). Phloem loading of sucrose. *Plant Physiol.* **59**, 750–755.
Giaquinta, R. T. (1978). Source and sink leaf metabolism in relation to phloem translocation. *Plant Physiol.* **61**, 380–385.
Giaquinta, R. T. (1979). Sucrose translocation and storage in the sugar beet. *Plant Physiol.* **63**, 828–832.
Giaquinta, R. T. (1980). Mechanism and control of phloem loading of sucrose. *Ber. Dtsch. Bot. Ges.* **93**, 187–201.
Giaquinta, R. T. (1983). Phloem loading of sucrose. *Annu. Rev. Plant Physiol.* **34**, 347–387.
Giaquinta, R. T. and Geiger, D. R. (1977). Mechanism of cyanide inhibition of phloem translocation. *Plant Physiol.* **59**, 178–180.
Giaquinta, R. T. and Quebedeaux, B. (1980). Phosphate-induced changes in assimilate partitioning in soybean leaves during pod filling. *Plant Physiol.* **65**, Suppl., 119.
Gibson, T. S., Speirs, J. and Brady, C. J. (1984). Salt tolerance in plants. II. In vitro translation of m-RNA from salt-tolerant and salt-sensitive plants on wheat germ ribosomes. Responses to ions and compatible organic solutes. *Plant, Cell Environ.* **7**, 579–587.
Gigon, A. and Rorison, I. H. (1972). The response of some ecologically distinct plant species to nitrate- and to ammonium-nitrogen. *J. Ecol.* **60**, 93–102.
Gilbert, W. A., Ludwick, A. E. and Westfall, D. G. (1983). Predicting in-season N requirements of sugar beets based on soil and petiole nitrate. *Agron. J.* **73**, 1018–1022.
Gildon, A. and Tinker, P. B. (1983a). Interactions of vesicular–arbuscular mycorrhizal infection and heavy metals in plants. I. The effects of heavy metals on the development of vesicular–arbuscular mycorrhizas. *New Phytol.* **94**, 247–261.
Gildon, A. and Tinker, P. B. (1983b). Interactions of vesicular–arbuscular mycorrhizal infections and heavy metals in plants. II. The effects of infection on uptake of copper. *New Phytol.* **95**, 263–268.

Gilfillan, I. M. and Jones, W. W. (1968). Effect of iron and manganese deficiency on the chlorophyll, amino acid and organic acid status of leaves of Macadamia. *Proc. Am. Soc. Hortic. Sci.* **93**, 210–214.

Gilmour, J. T. (1977). Micronutrient status of the rice plant. I. Plant and soil solution concentrations as function of time. *Plant Soil* **46**, 549–557.

Ginzburg, B. Z. (1961). Evidence for a protein gel structure cross-linked by metal cations in the intercellular cement of plant tissue. *J. Exp. Bot.* **12**, 85–107.

Gladstones, J. S., Loneragan, J. F. and Goodchild, N. A. (1977). Field responses to cobalt and molybdenum by different legume species, with interferences on the role of cobalt in legume growth. *Aust. J. Agric. Res.* **28**, 619–628.

Glass, A. D. M. (1983). Regulation of ion transport. *Annu. Rev. Plant Physiol.* **34**, 311–326.

Glass, A. D. M. and Dunlop, J. (1979). The regulation of K^+ influx in excised barley roots. Relationship between K^+ influx and electrochemical potential differences. *Planta* **145**, 395–397.

Glass, A. D. M. and Siddiqi, M. Y. (1982). Cation-stimulated H^+ efflux by intact roots of barley. *Plant, Cell Environ.* **5**, 385–393.

Gliemeroth, G. (1953). Bearbeitung und Düngung des Unterbodens in ihrer Wirkung auf Wurzelentwicklung, Stoffaufnahme und Pflanzenwachstum. *Z. Acker-Pflanzenbau.* **96**, 1–44.

Glinka, Z. and Reinhold, L. (1971). Abscisic acid raises the permeability of plant cells to water. *Plant Physiol.* **48**, 103–105.

Glusker, J. B. (1968). Mechanism of aconitase action deduced from crystallographic studies of its substrates. *J. Mol. Biol.* **38**, 149–162.

Godbold, D. L., Horst, W. J., Marschner, H., Collins, J. C. and Thurman, D. A. (1983). Root growth and Zn uptake by two ecotypes of *Deschampsia caespitosa* as affected by high Zn concentrations. *Z. Pflanzenphysiol.* **112**, 315–324.

Godo, G. H. and Reisenauer, H. M. (1980). Plant effects on soil manganese availability. *Soil Sci. Soc. Am. J.* **44**, 993–995.

Goeschl, J. D., Magnuson, C. E., Fares, Y., Jaeger, C. H., Nelson, C. E. and Strain, B. R. (1984). Spontaneous and induced blocking and unblocking of phloem transport. *Plant, Cell Environ.* **7**, 607–613.

Goldbach, E., Goldbach, H., Wagner, H. and Michael, G. (1975). Influence of N-deficiency on the abscisic acid content of sunflower plants. *Physiol. Plant.* **34**, 138–140.

Goldbach, H. and Goldbach, E. (1977). Abscisic acid translocation and influence of water stress on grain abscisic acid content. *J. Exp. Bot.* **28**, 1342–1350.

Goldbach, H. and Michael, G. (1976). Abscisic acid content of barley grains during ripening as affected by temperature and variety. *Crop Sci.* **16**, 797–799.

Goldbach, H., Goldbach, E. and Michael, G. (1977). Transport of abscisic acid from leaves to grains in wheat and barley plants. *Naturwissenschaften* **64**, 488.

Goor, B. J. van (1966). The role of calcium and cell permeability in the disease blossom end rot of tomatoes. *Physiol. Plant* **21**, 1110–1121.

Goor, B. J. van and Lune, P. van (1980). Redistribution of potassium, boron, iron, magnesium and calcium in apple trees determined by an indirect method. *Physiol. Plant.* **48**, 21–26.

Gorham, J., Hughes, L. and Wyn Jones, R. G. (1980). Chemical composition of salt-marsh plants from Ynys Mon (Anglesey): The concept of physiotypes. *Plant, Cell Environ.* **3**, 309–318.

Gorham, J., Hughes, L. and Wyn Jones, R. G. (1981). Low-molecular-weight carbohydrates in some salt-stressed plants. *Physiol. Plant.* **53**, 27–33.

Goss, M. J. (1977). Effects of mechanical impedance on root growth in barley (*Hordeum vulgare* L.). I. Effects on the elongation and branching of seminal root axes. *J. Exp. Bot.* **28**, 96–111.

Goss, M. J. and Scott-Russell, R. (1980). Effects of mechanical impedance on root growth in barley (*Hordeum vulgare* L.). III. Observations on the mechanism of response. *J. Exp. Bot.* **31**, 577–588.

Graham, J. H., Leonard, R. T. and Menge, J. A. (1981). Membrane-mediated decrease in root exudation responsible for phosphorus inhibition of vesicular–arbuscular mycorrhiza formation. *Plant Physiol.* **68**, 548–552.

Graham, R. D. (1975). Male sterility in wheat plants deficient in copper. *Nature (London)* **254**, 514–515.

Graham, R. D. (1976). Anomalous water relations in copper-deficient wheat plants. *Aust. J. Plant. Physiol.* **3**, 229–236.

Graham, R. D. (1979). Transport of copper and manganese to the xylem exudate of sunflower. *Plant, Cell Environ.* **2**, 139–143.

Graham, R. D. (1980a). The distribution of copper and soluble carbohydrates in wheat plants grown at high and low levels of copper supply. *Z. Pflanzenernaehr. Bodenkd.* **143**, 161–169.

Graham, R. D. (1980b). Susceptibility to powdery mildew of wheat plants deficient in copper. *Plant Soil* **56**, 181–185.

Graham, R. D. (1981). Absorption of copper by plant roots. *In* "Copper in Soils and Plants" (J. F. Loneragan, A. D. Robson and R. D. Graham, eds.), pp. 141–163. Academic Press, London and Orlando.

Graham, R. D. (1984). Breeding for nutritional characteristics in cereals. *In* "Advances in Plant Nutrition." (P. B. Tinker and A. Läuchli, eds.), Vol. 1, pp. 57–102. Praeger, New York.

Graham, R. D. and Pearce, D. T. (1979). The sensitivity of hexaploid and octaploid triticales and their parent species to copper deficiency. *Aust. J. Agric. Res.* **30**, 791–799.

Graham, R. D., Anderson, G. D. and Ascher, J. S. (1981). Absorption of copper by wheat, rye and some hybrid genotypes. *J. Plant Nutr.* **3**, 679–686.

Graham, R. D., Davies, W. J., Sparrow, D. H. B. and Ascher, J. S. (1982). Tolerance of barley and other cereals to manganese-deficient calcareous soils of South Australia. *In* "Genetic Specificity of Mineral Nutrition of Plants" (M. R. Saric, ed.), Vol. 13, pp. 277–283. Serb. Acad. Sci. Arts, Beograd.

Graham, J. H. and Syvertsen, J. P. (1984). Influence of vesicular–arbuscular mycorrhiza on the hydraulic conductivity of roots of two citrus root-stocks. *New Phytol.* **97**, 277–284.

Grasmanis, V. O. and Edwards, G. E. (1974). Promotion of flower initiation in apple trees by short exposure to the ammonium ion. *Aust. J. Plant. Physiol.* **1**, 99–105.

Grattan, S. R. and Maas, E. V. (1984). Interactive effects of salinity and substrate phosphate on soybean. *Agron. J.* **76**, 668–676.

Graven, E. H., Attoe, O. J. and Smith, D. (1965). Effect of liming and flooding on manganese toxicity in alfalfa. *Soil Sci. Soc. Am. Proc.* **29**, 702–706.

Graves, C. J., Adams, P. and Winsor, G. W. (1977). Some effects of indol-3-ylacetic acid on the rate of initiation and development of flower buds of *Chrysanthemum moriflorum. Ann. Bot. (London)* [N.S.] **41**, 747–753.

Graw, D., Moawad, M. and Rehm, S. (1979). Untersuchungen zur Wirts- und Wirkungsspezifität der VA-Mycorrhiza. *Z. Acker- Pflanzenbau* **148**, 85–98.
Greacen, E. L. and Oh, J. S. (1972). Physics of root growth. *Nature (London), New Biol.* **235**, 24–25.
Green, D. G. and Warder, F. G. (1973). Accumulation of damaging concentrations of phosphorus by leaves of Selkirk wheat. *Plant Soil* **38**, 567–572.
Green, J. F. and Muir, R. M. (1979). Analysis of the role of potassium in the growth effects of cytokinin, light and abscisic acid on cotyledon expansion. *Physiol. Plant.* **46**, 19–24.
Greenleaf, W. H. and Adams, F. (1969). Genetic control of blossom-end rot disease in tomatoes through calcium metabolism. *J. Am. Soc. Hortic. Sci.* **94**, 248–251.
Greenway, H. (1962). Plant response to saline substrates. I. Growth and ion uptake of several varieties of Hordeum during and after sodium chlorine treatment. *Aust. J. Biol. Sci.* **15**, 16–38.
Greenway, H. and Gunn, A. (1966). Phosphorus retranslocation in *Hordeum vulgare* during early tillering. *Planta* **71**, 43–67.
Greenway, H. and Munns, R. (1980). Mechanism of salt tolerance in nonhalophytes. *Annu. Rev. Plant Physiol.* **31**, 149–190.
Greenway, H. and Osmond, C. B. (1972). Salt responses of enzymes from species differing in salt tolerance. *Plant Physiol.* **49**, 256–259.
Greenway, H. and Pitman, M. G. (1965). Potassium retranslocation in seedlings of *Hordeum vulgare. Aust J. Biol. Sci.* **18**, 235–247.
Greenwood, D. J. (1967). Studies on the transport of oxygen through the stems and roots of vegetable seedlings. *New Phytol.* **66**, 337–347.
Greenwood, D. J. (1983). Quantitative theory and the control of soil fertility. *New Phytol.* **94**, 1–18.
Greenwood, D. J., Gerwitz, A., Stone, D. A. and Barnes, A. (1982). Root development of vegetable crops. *Plant Soil* **68**, 75–96.
Grewal, J. S. and Singh, S. N. (1980). Effect of potassium nutrition on frost damage and yield of potato plants on alluvial soils of the Punjab (India). *Plant Soil* **57**, 105–110.
Grieve, C. M. and Maas, E. V. (1981). Betaine accumulation in salt stressed sorghum. *Physiol. Plant.* **61**, 167–171.
Griffiths, D. W. and Thomas, T. A. (1981). Phytate and total phosphorus content of field beans (*Vicia faba* L.). *J. Sci. Food Agric.* **32**, 187–192.
Grimme, H. (1973). Magnesium diffusion in soils at different water and magnesium contents. *Z. Pflanzenernaehr. Bodenkd.* **134**, 9–19.
Grimme, H. (1982). The effect of Al on Mg uptake and yield of oats. *In* "Proceedings of the Ninth International Plant Nutrition Colloquium, Warwick, England. (A. Scaife, ed.), pp. 198–203. Commonw. Agric. Bur., Farnham Royal, Bucks.
Grimme, H. (1983). Aluminium induced magnesium deficiency in oats. *Z. Pflanzenernaehr. Bodenkd.* **146**, 666–676.
Grimme, H., Strebel, O., Renger, M. and Fleige, H. (1981). Die potentielle K-Anlieferung an die Pflanzenwurzel durch Diffusion. *Mitt. Dtsch. Bodenkd. Ges.* **36**, 367–374.
Grinsted, M. J., Hedley, M. J., White, R. E. and Nye, P. H. (1982). Plant-induced changes in the rhizosphere of rape (*Brassica napus* var. Emerald) seedlings. I. pH change and the increase in P concentration in the soil solution. *New Phytol.* **91**, 19–29.

Grosse-Brauckmann, E. (1957). Über den Einfluss der Kieselsäure auf den Mehltaubefall von Getreide bei unterschiedlicher Stickstoffdüngung. *Phytopathol. Z* **30**, 112–115.
Grosse-Brauckmann, E. (1979). Zur Mehltaubekämpfung in Gefässversuchen mittels Kieselsäuredüngung und Fungicidanwendung. *Landwirtsch. Forsch.* **32**, 150–156.
Grossmann, F. (1976). Outlines of host-parasite interactions in bacterial diseases in relation to plant nutrition. *Proc. 12th Colloq. Int. Potash Inst. Bern,* pp. 221–224.
Gruhn, K. (1961). Einfluss einer Molybdän-Düngung auf einige Stickstoff-Fraktionen von Luzerne und Rotklee. *Z. Pflanzenernaehr., Dueng., Bodenkd.* **95**, 110–118.
Grundon, N. J. (1980). Effectiveness of soil-dressing and foliar sprays of copper sulphate in correcting copper deficiency of wheat *(Triticum aestivum)* in Queensland. *Aust. J. Exp. Agric. Anim. Husb.* **20**, 717–723.
Grunes, D. L., Stout, P. R. and Brownell, J. R. (1970). Grass tetany of ruminants. *Adv. Agron.* **22**, 332–374.
Grunwald, G., Ehwald, R., Pietzsch, W. and Göring, H. (1979). A special role of the rhizodermis in nutrient uptake by plant roots. *Biochem. Physiol. Pflanz.* **174**, 831–837.
Guardia, M. D. de la and Benlloch, M. (1980). Effects of potassium and gibberellic acid on stem growth of whole sunflower plants. *Physiol. Plant.* **49**, 443–448.
Guardiola, J. L., Monerri, C. and Agusti, M. (1982). The inhibitory effect of gibberellic acid on flowering of citrus. *Physiol. Plant.* **55**, 136–142.
Gubler, W. D., Gorgan, R. G. and Osterli, P. P. (1982). Yellows of melons caused by molybdenum deficiency in acid soil. *Plant Dis.* **66**, 449–451.
Guerrero, M. G., Vega, J. M. and Losada, M. (1981). The assimilatory nitrate-reducing system and its regulation. *Annu. Rev. Plant Physiol.* **32**, 169–204.
Gunasena, H. P. M. and Harris, P. M. (1971). The effect of CCC, nitrogen and potassium on the growth and yield of two varieties of potatoes. *J. Agric. Sci.* **76**, 33–52.
Gupta, U. C. (1979). Boron nutrition of crops. *Adv. Agron.* **31**, 273–307.
Gupta, U. C. and Lipsett, J. (1981). Molybdenum in soils, plants and animals. *Adv. Agron.* **34**, 73–115.
Gurley, W. H. and Giddens, J. (1969). Factors affecting uptake, yield response, and carry over of molybdenum in soybean seed. *Agron. J.* **61**, 7–9.
Guttridge, C. G., Bradfield, E. G. and Holder, R. (1981). Dependence of calcium transport into strawberry leaves on positive pressure in the xylem. *Ann. Bot. (London)* [N.S.] **48**, 473–480.
Guzman, C. C. de and Dela Fuente, R. K. (1984). Polar calcium flux in sunflower hypocotyl segments. I. The effect of auxin. *Plant Physiol.* **76**, 347–352.
Gzik, A. and Günther, G. (1984). Einfluß von Cytokininen auf die Nitratreduktaseaktivität in Blättern von *Beta vulgaris* (Zuckerrüben) und *Chenopodium album* (weißer Gänsefuß). *Biochem. Physiol. Pflanz.* **179**, 295–301.
Haeder, H.-E. and Beringer, H. (1981). Influence of potassium nutrition and water stress on the abscisic acid content in grains and flag leaves during grain development. *J. Sci. Food Agric.* **32**.
Haeder, H.-E. and Beringer, H. (1984a). Long distance transport of potassium in cereals during grain filling in detached ears. *Physiol. Plant.* **62**, 433–438.
Haeder, H.-E. and Beringer, H. (1984b). Long distance transport of potassium in cereals during grain filling in intact plants. *Physiol. Plant.* **62**, 439–444.

Haeder, H. E., Mengel, K. and Forster, H. (1973). The effect of potassium on translocation of photosynthates and yield pattern of potato plants. *J. Sci. Food. Agric.* **24**, 1479–1487.

Hager, A. and Helmle, M. (1981). Properties of an ATP-fueled, Cl^--dependent proton pump localized in membranes of microsomal vesicles from maize coleoptiles. *Z. Naturforsch., Biosci. C:* **36**, 997–1008.

Hajibagheri, M. A., Hall, J. L. and Flowers, T. J. (1984). Stereological analysis of leaf cells of the halophyte *Sudea maritima* (L.) Dum. *J. Exp. Bot.* **35**, 1547–1557.

Hajrasuliha, S. (1980). Accumulation and toxicity of chloride in bean plants. *Plant Soil* **55**, 133–138.

Hale, K. A. and Sanders, F. E. (1982). Effects of benomyl on vesicular–arbuscular mycorrhizal infection of red clover (*Trifolium pratense* L.) and consequences for phosphorus inflow. *J. Plant Nutr.* **5**, 1355–1367.

Hall, A. J. and Milthorpe, F. L. (1978). Assimilation source–sink relationship in *Capsicum annuum* L. III. The effect of fruit excision on photosynthesis and leaf and stem carbohydrates. *Aust. J. Plant Physiol.* **5**, 1–13.

Hall, S. M. and Baker, D. A. (1972). The chemical composition of *Ricinus* phloem exudate. *Planta* **106**, 131–140.

Haller, T. and Stolp, H. (1985). Quantitative estimation of root exudation of maize plant. *Plant Soil* **86**, 207–216.

Halliwell, B. (1978). Biochemical mechanisms accounting for the toxic action of oxygen on living organisms: The key role of superoxide dismutase. *Cell Biol. Int. Rep.* **2**, 113–128.

Hallock, D. L. and Garren, K. H. (1968). Pod breakdown, yield and grade of Virginia type peanuts as affected by Ca, Mg, and K sulfates. *Agron. J.* **60**, 253–357.

Hallsworth, E. G., Greenwood, E. A. N. and Yates, M. G. (1964). Studies on the nutrition of forage legumes. III. The effect of copper on nodulation of *Trifolium subterraneum* L. and *Trifolium repens. Plant Soil* **20**, 17–33.

Hambidge, K. M. and Walravens, P. A. (1976). Zinc deficiency in infants and preadolescent children. *In* "Trace Elements in Human Health and Disease" (A. S. Prasad, and D. Overleas, eds.), Vol. 1, Chapter 2, pp. 21–32. Academic Press, New York.

Hamilton, J. L. and Lowe, R. H. (1981). Organic matter and N effects on soil nitrite accumulation and resultant nitrite toxicity to tobacco transplants. *Agron. J.* **73**, 787–790.

Hammes, P. S. and Beyers, E. A. (1973). Localization of the photoperiodic perception in potatoes. *Potato Res.* **16**, 68–72.

Hampe, T. and Marschner, H. (1982). Effect of sodium on morphology, water relations and net photosynthesis in sugar beet leaves. *Z. Pflanzenphysiol.* **108**, 151–162.

Hamze, M. and Nimah, M. (1982). Iron content during lime-induced chlorosis with two citrus rootstocks. *J. Plant Nutr.* **5**, 797–804.

Hancock, J. G. and Huisman, O. C. (1981). Nutrient movement in host-pathogen systems. *Annu. Rev. Phytopathol.* **19**, 309–331.

Handreck, K. K. and Riceman, D. S. (1969). Cobalt distribution in several pasture species grown in culture solutions. *Aust. J. Agric. Res.* **20**, 213–226.

Hanisch, H.-C. (1980). Zum Einfluss der Stickstoffdüngung und vorbeugender Spritzung von Natronwasserglas zu Weizenpflanzen auf deren Widerstandsfähigkeit gegen Getreideblattläuse. *Kali-Briefe* **15**, 287–296.

Hänisch ten Cate, C. H. and Breteler, H. (1982). Effect of plant growth regulators on nitrate utilization by roots of nitrogen-depleted dwarf bean. *J. Exp. Bot.* **33**, 37–46.

Hanson, J. B. (1984). The function of calcium in plant nutrition. *In* "Advances in Plant Nutrition" (P. B. Tinker and A. Läuchli, eds.), pp. 149–208. Praeger, New York.

Hardiman, R. T. and Banin, A. (1982). Cadmium absorption and translocation by bush beans at low external concentrations. *In* "Proceedings of the Ninth International Plant Nutrition Colloquium, Warwick, England" (A. Scaife, ed.), pp. 209–214. Commonw. Agric. Bur., Farnham Royal, Bucks.

Harper, S. H. T. and Lynch, J. M. (1980). Microbial effects on the germination and seedling growth of barley. *New Phytol.* **84**, 473–481.

Harper, S. H. T. and Lynch, J. M. (1982). The role of water-soluble components in phytotoxicity from decomposing straw. *Plant Soil* **65**, 11–17.

Harrison, S. J., Lepp, N. W. and Phipps, D. A. (1979). Uptake of copper by excised roots. II. Copper desorption from the free space. *Z. Pflanzenphysiol.* **94**, 27–34.

Harper, J. E. and Gibson, A. H. (1984). Differential nodulation tolerance to nitrate among legume species. *Crop Sci.* **24**, 797–801.

Hartel, H. (1977). Wirkung einer Harnstoffernährung auf Harnstoffumsatz und N-Stoffwechsel von Mais und Sojabohnen. Dissertation, Technische Universität, München.

Hartt, C. E. (1969). Effect of potassium deficiency upon translocation of ^{14}C in attached blades and entire plants of sugarcane. *Plant Physiol.* **44**, 1461–1469.

Hartung, W., Heilmann, B. and Gimmler, H. (1981). Do chloroplasts play a role in abscisic acid synthesis? *Plant Sci. Lett.* **22**, 235–242.

Hartzook, A., Karstadt, D., Naveh, M. and Feldman, S. (1974). Differential iron absorption efficiency of peanut (*Arachis hypogaea* L.) cultivars grown on calcareous soils. *Agron. J.* **66**, 114–115.

Harvey, D. M. R., Hall, J. L., Flowers, T. J. and Kent, B. (1981). Quantitative ion localization within *Suaeda maritima* leaf mesophyll cells. *Planta* **151**, 555–560.

Haschke, H. P. and Lüttge, K. (1975). Interactions between IAA, potassium, and malate accumulation and growth in Avena coleoptile segments. *Z. Pflanzenphysiol.* **76**, 450–455.

Hatch, M. D. and Slack, C. R. (1970). Photosynthetic CO_2-fixation pathways. *Annu. Rev. Plant Physiol.* **21**, 141–162.

Häussling, M., Leisen, E., Marschner, H. and Römheld, V. (1985). An improved method for non-destructive measurements of the pH at the root-soil interface (rhizosphere). *J. Plant Physiol.* **117**, 371–375.

Hawker, J. S. and Hatch, M. D. (1965). Mechanism of sugar storage by mature stem tissue of sugarcane. *Plant Physiol.* **18**, 444–453.

Hawker, J. S. and Smith, G. M. (1982). Salt tolerance and regulation of enzymes of starch synthesis in cassava (*Manihot esculenta* Crantz). *Aust. J. Plant Physiol.* **9**, 509–518.

Hawker, J. S., Marschner, H. and Downton, W. J. S. (1974). Effect of sodium and potassium on starch synthesis in leaves. *Aust. J. Plant Physiol.* **1**, 491–501.

Haynes, R. J. (1983). Soil acidification induced by leguminous crops. *Grass Forage Sci.*, (Oxford) **38**, 1–11.

Haynes, R. J. and Goh, K. M. (1978). Ammonium and nitrate nutrition of plants. *Biol. Rev. Cambridge Philos. Soc.* **53**, 465–510.

Haynes, R. J. and Ludecke, T. E. (1981). Yield, root morphology and chemical composition of two pasture legumes as affected by lime and phosphorus applications to an acid soil. *Plant Soil* **62**, 241–254.
Haystead, A. and Sprent, J. I. (1981). Symbiotic nitrogen fixation. *In* "Physiological Processes Limiting Plant Productivity" (C. B. Johnson, ed). pp. 345–364. Butterworth, London.
Heath, M. C. (1979). Partial characterization of the electron-opaque deposits formed in the non-host plant, French bean, after cowpea rust infection. *Physiol. Plant Pathol.* **15**, 141–148.
Heber, U., Kirk, M. R., Gimmler, H. and Schäfer, G. (1974). Uptake and reduction of glycerate by isolated chloroplasts. *Planta* **120**, 32–46.
Hecht-Buchholz, C. (1967). Über die Dunkelfärbung des Blattgrüns bei Phosphormangel. *Z. Pflanzenernaehr. Bodenkd.* **118**, 12–22.
Hecht-Buchholz, C. (1972). Wirkung der Mineralstoffernährung auf die Feinstruktur der Pflanzenzelle. *Z. Pflanzenernaehr. Bodenkd.* **132**, 45–68.
Hecht-Buchholz, C. (1973). Molybdänverteilung und -verträglichkeit bei Tomate, Sonnenblume und Bohne. *Z. Pflanzenernaehr. Bodenkd.* **136**, 110–119.
Hecht-Buchholz, C. (1979). Calcium deficiency and plant ultrastructure. *Commun. Soil. Sci. Plant Anal.* **10**, 67–81.
Hecht-Buchholz, C., Pflüger, R. and Marschner, H. (1971). Einfluss von Natriumchlorid auf Mitochondrienzahl und Atmung von Maiswurzelspitzen. *Z. Pflanzenphysiol.* **65**, 410–417.
Hedley, M. J., Nye, P. H. and White, R. E. (1982). Plant-induced changes in the rhizosphere of rape (*Brassica napus* var. Emerald) seedlings. II. Origin of the pH change. *New Phytol.* **91**, 31–44.
Heenan, D. P. and Campbell, L. C. (1981). Influence of potassium and manganese on growth and uptake of magnesium by soybeans (*Glycine max* (L.) Merr. cv Bragg). *Plant Soil* **61**, 447–456.
Heenan, D. P. and Carter, O. G. (1977). Influence of temperature on the expression of manganese toxicity by two soybean varieties. *Plant Soil* **47**, 219–227.
Heenan, D. P., Campbell, L. C. and Carter, O. G. (1981). Inheritance of tolerance to high manganese supply in soybean. *Crop Sci.* **21**, 625–627.
Hehl, G. and Mengel, K. (1972). Der Einfluss einer variierten Kalium- und Stickstoffdüngung auf den Kohlenhydratgehalt verschiedener Futterpflanzen. *Landwirtsch. Forsch., Sonderh.* **27**(2), 117–129.
Hein, M. B., Brenner, M. L. and Brun, W. A. (1979). Source/sink interaction in soybeans. II. A possible role of IAA. *Plant Physiol., Suppl.* **63**, 43.
Hein, M. B., Brenner, M. L. and Brun, W. A. (1984a). Concentrations of abscisic acid and indole-3-acetic acid in soybean seeds during development. *Plant Physiol.* **76**, 951–954.
Hein, M. B., Brenner, M. L. and Brun, W. A. (1984b). Effects of pod removal on the transport and accumulation of abscisic acid and indole-3-acetic acid in soybean leaves. *Plant Physiol.* **76**, 955–958.
Heise, K.-P. and Jacobi, G. (1973). Vergleichende Untersuchungen über die Lipidzusammensetzung von Etioplasten und Chloroplasten aus einer Mutante von Nicotiana. *Planta* **111**, 137–148.
Helal, H. M. and Mengel, K. (1979). Nitrogen metabolism of young barley plants as affected by NaCl-salinity and potassium. *Plant Soil* **51**, 457–462.

Helal, H. M. and Mengel, K. (1981). Interaction between light intensity and NaCl salinity and their effects on growth, CO_2 assimilation, and photosynthate conversion in young broad beans. *Plant Physiol.* **67**, 999–1002.
Helal, H. M. and Sauerbeck, D. R. (1981). Phosphatumsetzungen im Wurzelraum von Pflanzen. *Mitt. Dtsch. Bodenkd. Ges.* **32**, 295–304.
Helal, H. M. and Sauerbeck, D. R. (1984a). Influence of plant roots on C and P metabolism in soil. *Plant Soil* **76**, 175–182.
Helal, H. M and Sauerbeck, D. (1984b). Zur Veränderung organischer Fraktionen des Bodenphosphors in Wurzelnähe. *Mitt. Dtsch. Bodenkd. Ges.* **39**, 29–33.
Helder, R. J. and Boerma, J. (1969). An electron-microscopical study of the plasmodesmata in the roots of young barley seedlings. *Acta Bot. Neerl.* **18**, 99–107.
Heldt, H. W., Chon, C. J., Maronde, D., Herold, A., Stankovic, Z. S., Walker, D. A., Kraminer, A., Kirk, M. R. and Heber, U. (1977). Role of orthophosphate and other factors in the regulation of starch formation in leaves and isolated chloroplasts. *Plant Physiol.* **59**, 1146–1155.
Helms, K. and David, D. J. (1973). Far red and white light-promoted utilization of calcium by seedlings of *Phaseolus vulgaris* L. *Plant Physiol.* **51**, 37–42.
Hendriks, L., Claassen, N. and Jungk, A. (1981). Phosphatverarmung des wurzelnahen Bodens und Phosphataufnahme von Mais und Raps. *Z. Pflanzenernaehr. Bodenkd.* **144**, 486–499.
Hendrix, J. E. (1967). The effect of pH on the uptake and accumulation of phosphate and sulfate ions by bean plants. *Am. J. Bot.* **54**, 560–564.
Henis, Y. (1976). Effect of mineral nutrients on soil-borne pathogens and host resistance. *Proc. 12th Colloq. Int. Potash Inst. Bern,* pp. 101–112.
Hepper, C. M. (1983). The effect of nitrate and phosphate on the vesicular–arbuscular mycorrhizal infection of lettuce. *New Phytol.* **93**, 389–399.
Herms, U. and Brümmer, G. (1979). Einfluss der Redoxbedingungen auf die Löslichkeit von Schwermetallen in Böden und Sedimenten. *Mitt. Dtsch. Bodenkd. Ges.* **29**, 533–544.
Herms, U. and Brümmer, G. (1980). Einfluß der Bodenreaktion auf Löslichkeit und tolerierbare Gesamtgehalte an Nickel, Kupfer, Zink, Cadmium und Blei in Böden und kompostierten Siedlungsabfällen. *Landwirtsch. Forsch.* **33**, 408–423.
Herridge, D. F. and Pate, J. S. (1977). Utilization of net photosynthate for nitrogen fixation and protein production in an annual legume. *Plant Physiol.* **60**, 759–764.
Hertel, R. (1983). The mechanism of auxin transport as a model for auxin action. *Z. Pflanzenphysiol.* **112**, 53–67.
Herzog, H. (1981). Wirkung von zeitlich begrenzten Stickstoff- und Cytokiningaben auf die Fahnenblatt- und Kornentwicklung von Weizen. *Z. Pflanzenernaehr. Bodenkd.* **144**, 241–253.
Herzog, H. and Geisler, G. (1977). Der Einfluss von Cytokininapplikation auf die Assimilateinlagerung und die endogene Cytokininaktivität der Karyopsen bei zwei Sommerweizensorten. *Z. Acker- Pflanzenbau* **144**, 230–242.
Hess, D. (1981). In vitro associations between non-legumes and Rhizobium. *In* "Biology of Inorganic Nitrogen and Sulfur" (H. Bothe and A. Trebst, eds.), pp. 287–298. Springer-Verlag, Berlin and New York.
Hewitt, E. J. and Gundry, C. S. (1970). The molybdenum requirement of plants in relation to nitrogen supply. *J. Hortic. Sci.* **45**, 351–358.
Hewitt, E. J. and McCready, C. C. (1956). Molybdenum as a plant nutrient. VII. The effects of different molybdenum and nitrogen supplies on yields and composition of tomato plants grown in sand culture. *J. Hortic. Sci.* **31**, 284–290.

Heyns, K. (1979). Über die endogene Nitrosamin-Entstehung beim Menschen. *Landwirtsch. Forsch., Sonderh.* **39**, 145–162.
Heyser, W., Evert, F. R., Fritz, E. and Eschrich, W. (1978). Sucrose in the free space of translocating maize leaf bundles. *Plant Physiol.* **62**, 491–494.
Hiatt, A. J. (1967a). Relationship of cell sap pH to organic acid change during ion uptake. *Plant Physiol.* **42**, 294–298.
Hiatt, A. J. (1967b). Reactions in vitro of enzymes involved in CO_2-fixation accompanying salt uptake by barley roots. *Z. Pflanzenphysiol.* **56**, 233–245.
Hiatt, A. J. and Hendricks, S. B. (1967). The role of CO_2-fixation in accumulation of ions by barley roots. *Z. Pflanzenphysiol.* **56**, 220–232.
Higinbotham, N., Etherton, B. and Foster, R. J. (1967). Mineral ion contents and cell transmembrane electropotentials of pea and oat seedling tissue. *Plant Physiol.* **42**, 37–46.
Hill, J. (1980). The remobilization of nutrients from leaves. *J. Plant Nutr.* **2**, 407–444.
Hill, J., Robson, A. D. and Loneragan, J. F. (1978). The effect of copper and nitrogen supply on the retranslocation of copper in four cultivars of wheat. *Aust. J. Agric. Res.* **29**, 925–939.
Hill, J., Robson, A. D. and Loneragan, J. F. (1979a). The effects of Cu supply and shading on Cu retranslocation from old wheat leaves. *Ann. Bot. (London)* [N.S.] **43**, 449–457.
Hill, J., Robson, A. D. and Loneragan, J. F. (1979b). The effect of copper supply on the senescence and the retranslocation of nutrients of the oldest leaf of wheat. *Ann. Bot. (London)* [N.S.] **44**, 279–287.
Hill, J., Robson, A. D. and Loneragan, J. F. (1979c). The effect of copper and nitrogen supply on the distribution of copper in dissected wheat grain. *Aust. J. Agric. Res.* **30**, 233–237.
Hirsch, A. M. and Torrey, J. G. (1980). Ultrastructural changes in sunflower root cells in relation to boron deficiency and added auxin. *Can. J. Bot.* **58**, 856–866.
Hirsch, A. M., Pengelly, W. L. and Torrey, J. G. (1982). Endogenous IAA levels in boron-deficient and control root tips of sunflower. *Bot. Gaz. (Chicago)* **143**, 15–19.
Ho, L. C. and Baker, D. A. (1982). Regulation of loading and unloading in long distance transport systems. *Physiol. Plant.* **56**, 225–230.
Hoagland, D. R. (1948). "Lectures on the Inorganic Nutrition of Plants," pp. 48–71. Chronica Botanica, Waltham, Massachusetts.
Hocking, P. J. (1980). The composition of phloem exudate and xylem sap from tree tobacco (*Nicotiana glauca* Groh). *Ann. Bot. (London)* [N.S.] **45**, 633–643.
Hocking, P. J. and Plate, J. S. (1978). Accumulation and distribution of mineral elements in annual lupins *Lupinus albus* and *Lupinus angustifolius* L. *Aust. J. Agric. Res.* **29**, 267–280.
Hocking, P. J., Pate, J. S., Wee, S. C. and McComb, A. J. (1977). Manganese nutrition of *Lupinus* ssp. especially in relation to developing seeds. *Ann. Bot. (London)* [N.S.] **41**, 677–688.
Hodenberg, A. von and Finck, A. (1975). Ermittlung von Toxizitäts-Grenzwerten für Zink, Kupfer und Blei in Hafer und Rotklee. *Z. Pflanzenernaehr. Bodenkd.* **138**, 489–503.
Hodgson, J. F., Lindsay, W. L. and Trierweiler, J. T. (1966). Micronutrient cation complexing in soil solution. II. Complexing of zinc and copper in displaced solution from calcareous soils. *Soil Sci. Soc. Am. Proc.* **30**, 723–726.

Hodson, M. J. and Parry, W. D. (1982). The ultrastructure and analytical microscopy of silicon deposition in the aleurone layer of the caryopsis of *Setaria italica* (L.) Beauv. *Ann. Bot. (London)* [N.S.] **50**, 221–228.

Hoffman, G. J. (1981). Alleviating salinity stress. *ASAE Monogr.* **4**, 305–346.

Hoffman, G. J. and Jobes, J. A. (1978). Growth and water relations of cereal crops as influenced by salinity and relative humidity. *Agron. J.* **70**, 765–769.

Hoffman, G. J. and Phene, C. J. (1971). Effect of constant salinity levels on water use efficiency of bean and cotton. *Trans. ASAE* **14**, 1103–1106.

Höfner, W. (1970). Eisen- und manganhaltige Verbindungen im Blutungssaft von *Helianthus annuus. Physiol. Plant.* **23**, 673–677.

Höfner, W. and Grieb, R. (1979). Einfluss von Fe- and Mo-Mangel auf den Ionengehalt mono- und dikotyler Pflanzen unterschiedlicher Chloroseanfälligkeit. *Z. Pflanzenernaehr. Bodenkd.* **142**, 626–638.

Högberg, P. and Nylund, J.-E. (1981). Ectomycorrhizae in coastal miombo woodland of Tanzania. *Plant Soil* **63**, 283–289.

Hooymans, J. J. M. (1976). Competition between vacuolar accumulation and upward translocation of K^+ ions in barley plants. *Z. Pflanzenphysiol.* **79**, 182–186.

Hope, A. B. and Stevens, P. G. (1952). Electrical potential differences in bean roots on their relation to salt uptake. *Aust. J. Sci. Res., Ser. B* **5**, 335–343.

Hopkins, H. T., Specht, A. W. and Hendricks, S. B. (1950). Growth and nutrient accumulation as controlled by oxygen supply to plant roots. *Plant Physiol.* **25**, 193–208.

Horesh, I. and Levy, Y. (1981). Response of iron-deficient citrus trees to foliar iron sprays with a low-surface-tension surfactant. *Sci. Hortic. (Amsterdam)* **15**, 227–233.

Horgan, J. M. and Wareing, P. F. (1980). Cytokinins and the growth response of seedlings of *Betula pendula* Roth. and *Acer pseudoplatanus* L. to nitrogen and phosphorus deficiency. *J. Exp. Bot.* **31**, 525–532.

Horst, W. J. (1982). Quick screening of cowpea genotypes for manganese tolerance during vegetative and reproductive growth. *Z. Pflanzenernaehr. Bodenkd.* **145**, 423–435.

Horst, W. J. (1983). Factors responsible for genotypic manganese tolerance in cowpea *(Vigna unguiculata). Plant Soil* **72**, 213–218.

Horst, W. J. and Marschner, H. (1978a). Effect of silicon on manganese tolerance of bean plants (*Phaseolus vulgaris* L.). *Plant Soil* **50**, 287–303.

Horst, W. J. and Marschner, H. (1978b). Symptome von Mangan-Überschuß bei Bohnen (*Phaseolus vulgaris* L.). *Z. Pflanzenernaehr. Bodenkd.* **141**, 129–142.

Horst, W. J. and Marschner, H. (1978c). Effect of excessive manganese supply on uptake and translocation of calcium in bean plants (*Phaseolus vulgaris* L.). *Z. Pflanzenphysiol.* **87**, 137–148.

Horst, W. J., Wagner, A. and Marschner, H. (1982). Mucilage protects root meristems from aluminium injury. *Z. Pflanzenphysiol.* **105**, 435–444.

Horst, W. J., Wagner, A. and Marschner, H. (1983). Effect of aluminium on root growth, cell division rate and mineral element contents in roots of *Vigna unguiculata* genotypes. *Z. Pflanzenphysiol.* **109**, 95–103.

Howard, D. D. and Adams, R. (1965). Calcium requirement for penetration of subsoil by primary cotton roots. *Soil Sci. Soc. Am. Proc.* **29**, 558–562.

Howeler, R. H. (1973). Iron-induced oranging disease of rice in relation to physicochemical changes in a flooded oxisol. *Soil Sci. Soc. Am. Proc.* **37**, 898–903.

Howeler, R. H. and Cadavid, L. F. (1976). Screening of rice cultivars for tolerance to Al-toxicity in nutrient solutions as compared with field screening method. *Agron. J.* **68**, 554–555.

Howeler, R. H., Cadavid, L. F. and Burckhardt, E. (1982a). Response of cassava to VA mycorrhizal inoculation and phosphorus application in greenhouse and field experiments. *Plant Soil* **69**, 327–339.

Howeler, R. H., Edwards, D. G. and Asher, C. J. (1982b). Micronutrient deficiencies and toxicities of cassava plants grown in nutrient solutions. I. Critical tissue concentrations. *J. Plant Nutr.* **5**, 1059–1076.

Hsu, W. and Miller, G. W. (1968). Iron in relation to aconitate hydratase activity in *Glycine max* Merr. *Biochim. Biophys. Acta* **151**, 711–713.

Huber, S. C. (1984). Biochemical basis for effects of K-deficiency on assimilate export rate and accumulation of soluble sugars in soybean leaves. *Plant Physiol.* **76**, 424–430.

Huber, S. C. and Moreland, D. E. (1981). Ca-transport of potassium and sugars across the plasmalemma of mesophyll protoplasts. *Plant Physiol.* **67**, 163–169.

Hughes, J. C. and Evans, J. L. (1969). Studies on after-cooking blackening. V. Changes in after-cooking blackening and the chemistry of Magestic and Ulster Beacon tubers during the growing season. *Eur. Potato J.* **12**, 26–40.

Humble, G. D. and Hsiao, T. C. (1970). Light-dependent influx and efflux of potassium of guard cells during stomatal opening and closing. *Plant Physiol.* **46**, 483–487.

Humble, G. D. and Raschke, K. (1971). Stomatal opening quantitatively related to potassium transport. *Plant Physiol.* **48**, 447–453.

Hundt, I., Bergmann, W., Fischer, F. and Schilling, G. (1970a). Untersuchungen über den Einfluss des Mikronährstoffes Bor auf den N-Umsatz von *Helianthus annuus* L. *Albrecht-Thaer-Arch.* **14**, 713–724.

Hundt, J., Schilling, G., Fischer, F. and Bergmann, W. (1970b). Untersuchungen über den Einfluss des Mikronährstoffes Bor auf den Nukleinsäurestoffwechsel und die Gewebestruktur von *Helianthus annuus* L. *Albrecht-Thaer-Arch.* **14**, 725–737.

Hunt, P. G., Campbell, R. B., Sojka, R. E. and Parsons, J. E. (1981). Flooding-induced soil and plant ethylene accumulation and water status response of field-grown tobacco. *Plant Soil* **59**, 427–439.

Hunter, W. J., Fahring, C. J., Olsen, S. R. and Porter, L. K. (1982). Location of nitrate reduction in different soybean cultivars. *Crop Sci.* **22**, 944–948.

Hutton, J. T. and Norrish, K. (1974). Silicon content of wheat husks in relation to water transpired. *Aust. J. Agric. Res.* **25**, 203–212.

Hwang, B. K., Ibenthal, W.-D. and Heitefuss, R. (1983). Age, rate of growth, carbohydrate and amino acid content of spring barley plants in relation to their resistance to powdery mildew *(Erysiphe graminis* f. sp. *hordei). Physiol. Plant Pathol.* **22**, 1–14.

Hylton, L. O., Jr., Ulrich, A. and Cornelius, D. R. (1965). Comparison of nitrogen constituents as indicators of the nitrogen status of Italian ryegrass and relation of top to root growth. *Crop Sci.* **5**, 21–22.

Hylton, L.-O., Jr., Ulrich, A. and Cornelius, D. R. (1967). Potassium and sodium interrelations in growth and mineral content of Italian ryegrass. *Agron. J.* **59**, 311–314.

Ichioka, P. S. and Arnon, D. I. (1955). Molybdenum in relation to nitrogen metabolism. II. Assimilation of ammonia and urea without molybdenum by Scenedesmus. *Plant Physiol.* **69**, 1040–1045.

Idris, M., Hossain, M. M. and Choudhury, F. A. (1975). The effect of silicon on lodging of rice in presence of added nitrogen. *Plant Soil* **43**, 691–695.

Idris, M., Vinther, F. P. and Jensen, V. (1981). Biological nitrogen fixation associated with roots of field-grown barley (*Hordeum vulgare* L.). *Z. Pflanzenernaehr. Bodenkd.* **144**, 385–394.

Iglesias, A. and Satter, R. L. (1983). H^+ fluxes in excised Samanea motor tissue. I. Promotion by light. *Plant Physiol.* **72**, 564–569.

Ikehashi, H. and Ponnamperuma, F. N. (1978). Varietal tolerance or rice for adverse soils. *In* "Soils and Rice," pp. 801–823, Int. Rice Res. Inst., Los Baños, Philippines.

Inbal, E. and Feldman, M. (1982). The response of a hormonal mutant of common wheat to bacteria of the genus *Azospirillum. Isr. J. Bot.* **31**, 257–263.

Iruthayathas, E. E., Vlassak, K. and Gunasekaran, S. (1983). The influence of *Azospirillum* cell content and some amino-acids on winged bean *(Psophocarpus tetragonolobus)* Rhizobium symbiosis. *Z. Pflanzenernaehr. Bodenkd.* **146**, 405–408.

Isegawa, Y., Nakano, Y. and Kitaoka, S. (1984). Conversion and distribution of cobalamin in *Euglena gracilis* Z, with special reference to its location and probable function within chloroplasts. *Plant Physiol.* **76**, 814–818.

Isermann, K. (1975). Mögliche Ursachen der Mangan-Toleranz bestimmter Reis-Sorten. *Z. Pflanzenernaehr. Bodenkd.* **138**, 235–247.

Isermann, K. (1978). Einfluss von Chelatoren auf die Calcium-Verlagerung im Sproß höherer Pflanzen. *Z. Pflanzenernaehr. Bodenkd.* **141**, 285–298.

Ishizuka, J. (1982). Characterization of molybdenum absorption and translocation in soybean plants. *Soil Sci. Plant Nutr. (Tokyo)* **28**, 63–78.

Islam, A. K. M. S., Edwards, G. and Asher, C. J. (1980). pH optima for crop growth. Results of a flowing solution culture experiment with six species. *Plant Soil* **54**, 339–337.

Islam, M. M. and Ponnamperuma, F. N. (1982). Soil and plant tests for available sulfur in wetland rice soils. *Plant Soil* **68**, 97–113.

Islam, R. and Ayanaba, A. (1981). Effect of seed inoculation and preinfecting cowpea *(Vigna unguiculata)* with *Glomus mosseae* on growth and seed yield of the plants under field conditions. *Plant Soil* **61**, 341–350.

Ismunadji, M. (1976). Rice diseases and physiological disorders related to potassium deficiency. *Proc. 12th Colloq. Int. Potash Inst. Bern,* 47–60.

Ismunadji, M. and Dijkshoorn, W. (1971). Nitrogen nutrition of rice plants measured by growth and nutrient content in pot experiments. Ionic balance and selective uptake. *Neth. J. Agric. Sci.* **19**, 223–236.

Itoh, S. and Barber, S. A. (1983a). Phosphorus uptake by six plant species as related to root hairs. *Agron. J.* **75**, 457–461.

Itoh, S. and Barber, S. A. (1983b). A numerical solution of whole plant nutrient uptake for soil-root systems with root hairs. *Plant Soil* **70**, 403–413.

Ivins, J. D. and Bremner, P. M. (1964). Growth, development and yield in the potato. *Outlook Agric.* **4**, 211–217.

Jackson, C., Dench, J., Moore, A. L., Halliwell, B., Foyer, C. H. and Hall, D. O. (1978). Subcellular localization and identification of superoxide dismutase in the leaves of higher plants. *Eur. J. Biochem.* **91**, 339–344.

Jackson, M. B. and Campbell, D. L. (1979). Effects of benzyladenine and gibberellic acid on the responses of tomato plants to anaerobic root environments and to ethylene. *New Phytol.* **82**, 331–340.

Jackson, M. B. and Kowalewska, A. K. B. (1983). Positive and negative messages from roots induce foliar desiccation and stomatal closure in flooded pea plants. *J. Exp. Bot.* **34**, 493–506.

Jackson, M. B., Drew, M. C. and Giffard, S. C. (1981). Effects of spraying ethylene to the root system of *Zea mays* on growth and nutrient concentration in relation to flooding tolerance. *Physiol. Plant.* **52**, 23–28.

Jackson, M. B., Herman, B. and Goodenough, A. (1982). An examination of the importance of ethanol in causing injury to flooded plants. *Plant, Cell Environ.* **5**, 163–172.

Jackson, P. C. and St. John, J. B. (1980). Changes in membrane lipids of roots associated with changes in permeability. I. Effect of undissociated organic acids. *Plant Physiol.* **66**, 801–804.

Jackson, W. A., Volk, R. J. and Tucker, T. C. (1972). Apparent induction of nitrate uptake in nitrate depleted plants. *Agron. J.* **64**, 518–521.

Jacobson, L. (1955). CO_2-fixation and ion absorption in barley roots. *Plant Physiol.* **30**, 264–269.

Jacobson, L., Moore, D. P. and Hannapel, R. J. (1960). Role of calcium in absorption of monovalent cations. *Plant Physiol.* **35**, 352–358.

Jacoby, B. (1967). The effect of the roots on calcium ascent in bean stems. *Ann. Bot. (London)* [N.S.] **31**, 725–730.

Jacoby, B. and Rudich, B. (1980). Proton-chloride symport in barley roots. *Ann. Bot. (London)* [N.S.] **46**, 493–498.

Jaffe, M. J., Huberman, M., Johnson, J. and Telewski, F. W. (1985). Thigmomorphogenesis: The induction of callose formation and ethylene evolution by mechanical perturbation in bean stem. *Physiol. Plant.* **64**, 271–279.

Jager, A. de (1979). Localized stimulation of root growth and phosphate uptake in *Zea mays* L. resulting from restricted phosphate supply. *In* "The Soil-Root Interface" (J. L. Harley and R. Scott-Russell, eds.), pp. 391–403. Academic Press, London and Orlando.

Jager, A. de (1984). Effects of localized supply of H_2PO_4, NO_3, Ca and K on the concentration of that nutrient in the plant and the rate of uptake by the roots in young maize plants in solution culture. *Neth. J. Agric. Sci.* **32**, 43–56.

Jagnow, G. (1979). Nitrogen-fixing bacteria associated with graminaceous roots with special reference to *Spirillum lipoferum* Beyerinck. *Z. Pflanzenernaehr. Bodenkd.* **142**, 399–410.

Jagnow, G. (1983). Nitrogenase (C_2H_4) activity in roots of non-cultivated and cereal plants: Influence of nitrogen fertilizer on populations and activity of nitrogen-fixing bacteria. *Z. Pflanzenernaehr. Bodenkd.* **146**, 217–227.

Jameson, P. E., McWha, J. A. and Wright, G. J. (1982). Cytokinins and changes in their activity during development of grains of wheat (*Triticum aestivum* L.). *Z. Pflanzenphysiol.* **106**, 27–36.

Jarrell, W. M. and Beverly, R. B. (1981). The dilution effect in plant nutrition studies. *Adv. Agron.* **34**, 197–224.

Jarvis, S. C. (1981). Copper concentrations in plants and their relationship to soil properties. *In* "Copper in Soils and Plants" (J. F. Loneragan, A. D. Robson and R. D. Graham, eds.), pp. 265–285. Academic Press, London and Orlando.

Jarvis, S. C. (1982). Sodium absorption and distribution in forage grasses of different potassium status. *Ann. Bot. (London)* [N.S.] **49**, 199–206.

Jarvis, S. C. and Robson, A. D. (1982). Absorption and distribution of copper in plants with sufficient or deficient supplies. *Ann. Bot. (London)* [N.S.] **50**, 151–160.

Jasper, D. A., Robson, A. D. and Abbott, L. K. (1979). Phosphorus and the formation of vesicular–arbuscular mycorrhizas. *Soil Biol. Biochem.* **11**, 501–505.

Jauregui, M. A. and Reisenauer, H. M. (1982). Dissolution of oxides of manganese and iron by root exudate components. *Soil Sci. Soc. Am. J.* **46**, 314–317.

Jayman, T. C. Z. and Sivasubramaniam, S. (1975). Release of bound iron and aluminium from soils by the root exudates of tea *(Camellia sinensis)* plants. *J. Sci. Food Agric.* **26**, 1895–1898.

Jeffrey, D. W. (1968). Phosphate nutrition of Australian heath plants. II. The formation of polyphosphate by five heath species. *Aust. J. Bot.* **16**, 603–613.

Jenkyn, J. F. (1976). Nitrogen and leaf diseases of spring barley. *Proc. 12th Colloq. Int. Potash Inst. Bern,* pp. 119–128.

Jenner, C. F. (1980a). The conversion of sucrose to starch in development fruits. *Ber. Dtsch. Bot. Ges.* **93**, 289–298.

Jenner, C. F. (1980b). Effects of shading or removing spikelets in wheat: Testing assumptions. *Aust. J. Plant Physiol.* **7**, 113–121.

Jennings, D. H. (1968). Halophytes, succulence and sodium in plants—a unified theory. *New Phytol.* **67**, 899–911.

Jennings, D. H. (1976). The effect of sodium chloride on higher plants. *Biol. Rev. Cambridge Philos. Soc.* **51**, 453–486.

Jenny, H. (1965). Die Grenzzone von Wurzel und Boden in ihrer Bedeutung für die Aufnahme von Eisen aus kalkhaltigen Böden. *Wein-Wiss.* **20**, 49–61

Jenny, H. and Grossenbacher, K. (1963). Root-soil boundary zones as seen in the electron microscope. *Soil Sci. Soc. Am. Proc.* **27**, 273–277.

Jenny, H. and Overstreet, R. (1939). Cation interchange between plant roots and soil colloids. *Soil Sci.* **47**, 257–272.

Jensen, C. R., Stolzy, L. H. and Letey, J. (1967). Tracer studies of oxygen diffusion through roots of barley corn, and rice. *Soil Sci.* **103**, 23–29.

Jensen, P. (1982). Effects of interrupted K^+ supply on growth and uptake of K^+, Ca^{2+}, Mg^{2+} and Na^+ in spring wheat. *Physiol. Plant.* **56**, 259–265.

Jeschke, W. D. (1972). Über den lichtgeförderten Influx von Ionen in Blättern von *Elodea densa.* Vergleich der Influxe von K^+- und Cl^--Ionen. *Planta* **103**, 164–180.

Jeschke, W. D. (1977). K^+-Na^+-exchange and selectivity in barley root cells: Effect of Na^+ on the Na^+ fluxes. *J. Exp. Bot.* **28**, 1289–1305.

Jeschke, W. D. (1982). Shoot-dependent regulation of sodium and potassium fluxes in roots of whole barley seedlings. *J. Exp. Bot.* **33**, 601–618.

Jeschke, W. D. and Jambor, W. (1981). Determination of unidirectional sodium fluxes in roots of intact sunflower seedlings. *J. Exp. Bot.* **32**, 1257–1272.

Jeschke, W. D. and Stelter, W. (1976). Measurement of longitudinal ion profiles in single roots of Hordeum and Atriplex by use of flameless atomic absorption spectroscopy. *Planta* **128**, 107–112.

Jeschke, W. D., Atkins, C. A. and Pate, J. S. (1985). Ion circulation via phloem and xylem between root and shoot of nodulated white lupin. *J. Plant Physiol.* **117,** 319–330.

Jochem, P., Rona, J.-P., Smith, J. A. C. and Lüttge, U. (1984). Anion-sensitive ATPase activity and proton transport in isolated vacuoles of species of the CAM genus *Kalanchoë. Physiol. Plant.* **62**, 410–415.

Johansen, C., Edwards, D. G. and Loneragan, J. F. (1968). Interaction between potassium and calcium in their absorption by intact barley plants. II. Effects of calcium and potassium concentration on potassium absorption. *Plant Physiol.* **43**, 1722–1726.

Johansen, C., Edwards, D. G. and Loneragan, J. F. (1970). Potassium fluxes during potassium absorption by intact barley plants of increasing potassium content. *Plant Physiol.* **45**, 601–603.

Johnson, A. D. and Simons, J. G. (1979). Diagnostic indices of zinc deficiency in tropical legumes. *J. Plant Nutr.* **1**, 123–149.

Johnson, C. M., Stout, P. R., Broyer, T. C. and Carlton, A. B. (1957). Comparative chlorine requirements of different plant species. *Plant Soil* **8**, 337–353.

Johnson, C. R., Jarrell, W. M. and Menge, J. A. (1984). Influence of ammonium: nitrate ratio and solution pH on mycorrhizal infection, growth and nutrient composition of *Chrysanthemum moriflorum* var. Circus. *Plant Soil* **77**, 151–157.

Johnson, W. C. and Wear, J. I. (1967). Effect of boron on white clover (*Trifolium repens* L.) seed production. *Agron. J.* **59**, 205–206.

Johnston, M., Grof, C. P. L. and Brownell, P. F. (1984). Responses to ambient CO_2 concentrations by sodium-deficient C_4 plants. *Aust. J. Plant Physiol.* **11**, 137–141.

Jolivet, Y., Larher, F. and Hamelin, J. (1982). Osmoregulation in halophytic higher plants: The protective effect of glycine betaine against the heat destabilization of membranes. *Plant Sci. Lett.* **25**, 193–201.

Jones, F. G. W. (1976). Pests, resistance and fertilizers. *Proc. 12th Colloq. Int. Potash Inst. Bern,* pp. 233–258.

Jones, J. B. Jr. (1967). Interpretation of plant analysis for several agronomic crops. *In* "Soil Testing and Plant Analysis" (L. M. Walsh and J. D. Beaton, eds.), 1st edn. Vol. 2, pp. 49–58. Soil Sci. Soc. Am., Madison, Wisconsin.

Jones, J. B. Jr. (1972). Plant tissue analysis for micronutrients. *In* "Micronutrients in Agriculture" (J. J. Mortvedt, P. M. Giordano and W. L. Lindsay, eds.), pp. 319–346. Soil Sci. Soc. Am., Madison, Wisconsin.

Jones, L. H.-P. (1978). Mineral components of plant cell walls. *Am. J. Clin. Nutr.* **31**, 94–98.

Jones, L. H.-P. and Handreck, K. A. (1965). Studies of silica in the oat plant. III. Uptake of silica from soils by the plant. *Plant Soil* **23**, 79–96.

Jones, L. H.-P. and Handreck, K. H. (1967). Silica in soils, plants and animals. *Adv. Agron.* **19**, 107–149.

Jones, L. H.-P. and Handreck, K. A. (1969). Uptake of silica by *Trifolium incarnatum* in relation to the concentration in the external solution and to transpiration. *Plant Soil* **30**, 71–80.

Jones, L. H.-P, Hartley, R. D. and Jarvis, S. C. (1978). Mineral content of forage plants in relation to nutritional quality. Silicon. *In* "Annual Report of Grassland Research Institute, Hurley," pp. 25–26.

Jones, R. L. and Carbonell, J. (1984). Regulation of the synthesis of barley aleurone α-amylase by gibberellic acid and calcium ions. *Plant Physiol.* **76**, 213–218.

Judel, G. K. (1972). Änderungen in der Aktivität der Peroxidase und der Katalase und im Gehalt an Gesamtphenolen in den Blättern der Sonnenblume unter dem Einfluss von Kupfer- und Stickstoffmangel. *Z. Pflanzenernaehr. Bodenkd.* **133**, 81–92.

Judy, W., Melton, J., Lessman, G., Ellis, B. and Davies, J. (1965). Zinc fertilization of pea, beans, corn, and sugar beet in 1964. *Mich., Agric. Exp. Stn., Farm Sci. Res. Rep.* **33**, 1–8.

Jung, J. (1980). Zur praktischen Anwendung pflanzlicher Bioregulatoren. *Arzneim.-Forsch.* **30**, 1974–1980.

Jung, J. and Sturm, H. (1966). "Der Wachstumsregulator CCC," Rep. pp. 257–280. Landwirtsch. Vers. Stn. Limburgerhof der BASF.

Jungk, A. (1970). Wechselwirkungen zwischen Stickstoffkonzentration (NH_4, NH_4NO_3 und NO_3) und pH der Nährlösung auf Wuchs und Ionenhaushalt von Tomatenpflanzen. *Gartenbauwissenschaft* **35**, 13–28.

Jungk, A. (1984). Phosphatdynamik in der Rhizosphäre und Phosphatverfügbarkeit für Pflanzen. *Die Bodenkultur (Wien)* **35**, 99–107.

Jungk, A., Claassen, N. and Kuchenbuch, R. (1982). Potassium depletion of the soil-root interface in relation to soil parameters and root properties. *In* "Proceedings of the Ninth International Plant Nutrition Colloquium, Warwick, England" (A. Scaife, ed.), pp. 250–255. Commonw. Agric. Bur., Farnham Royal, Bucks.

Jurinak, J. J. and Inouye, T. S. (1962). Some aspects of zinc and copper phosphate formation in aqueous systems. *Soil Sci. Soc. Am. Proc.* **26**, 144–147.

Jyung, W. H., Ehmann, A., Schlender, K. K. and Scala, J. (1975). Zinc nutrition and starch metabolism in *Phaseolus vulgaris* L. *Plant Physiol.* **55**, 414–420.

Kaiser, W. M. and Hartung, W. (1981). Uptake and release of abscisic acid by isolated photoautotrophic mesophyll cells, depending on pH gradients. *Plant Physiol.* **68**, 202–206.

Kakie, T. (1969). Phosphorus fractions in tobacco plants as affected by phosphate application. *Soil. Sci. Plant Nutr. (Tokyo)* **15**, 81–85.

Kang, B. T. and Fox, R. L. (1980). A methodology for evaluating the manganese tolerance of cowpea *(Vigna unguiculata)* and some preliminary results of field trials. *Field Crops Res.* **3**, 199–210.

Kang, B. T., Islam, R., Sanders, F. E. and Aynanaba, A. (1980). Effect of phosphate fertilization and inoculation with VA-mycorrhizal fungi on performance of cassava (*Manihot esculenta* Crantz) grown on an alfisol. *Field Crops Res.* **3**, 83–94.

Kannan, S. and Ramani, S. (1978). Studies on molybdenum absorption and transport in bean and rice. *Plant Physiol.* **62**, 179–181.

Kapoor, A. C. and Li, P. H. (1982). Effects of age and variety on nitrate reductase and nitrogen fractions in potato plants. *J. Sci. Food Agric.* **33**, 401–406.

Kapulnik, Y., Kigel, J., Okon, Y., Nur, I. and Henis, Y. (1981a). Effect of Azospirillum inoculation and some growth parameters and N-content of wheat, sorghum and panicum. *Plant Soil* **61**, 65–70.

Kapulnik, Y., Okon, Y., Kigel, J., Nur, I. and Henis, Y. (1981b). Effect of temperature, nitrogen fertilization, and plant age on nitrogen fixation by *Setaria italica* inoculated with *Azospirillum brasilense* (strain cd). *Plant Physiol.* **68**, 340–343.

Karmoker, J. L. and Van Steveninck, R. F. M. (1979). The effect of abscisic acid on sugar levels in seedlings of *Phaseolus vulgaris* L. cv. Redland Pioneer. *Planta* **146**, 25–30.

Katouli, M. and Marchant, R. (1981). Effect of phytotoxic metabolites of *Fusarium culmorum* on barley root and root-hair development. *Plant Soil* **60**, 385–397.

Katyal, J. C. and Sharma, B. D. (1980). A new technique of plant analysis to resolve iron chlorosis. *Plant Soil* **55**, 105–119.

Katyal, J. C. and Sharma, B. D. (1984). Association of soil properties and soil and plant iron to iron deficiency response in rice (*Oriza sativa* L.). *Commun. Soil Sci. Plant Qual.* **15**, 1065–1081.

Katz, A., Dehan, K. and Itai, C. (1978). Kinetin reversal of NaCl effects. *Plant Physiol.* **62**, 836–837.

Kawase, M. (1981). Anatomical and morphological adaptation of plants to waterlogging. *HortScience* **16**, 30–34.

Kelday, L. S. and Bowling, D. J. F. (1980). Profiles of chloride concentration and PD in the root of *Commelina communis* L. *J. Exp. Bot.* **31**, 1347–1355.
Keller, P. and Deuel, H. (1957). Kationenaustauschkapazität und Pektingehalt von Pflanzenwurzeln. *Z. Pflanzenernaehr., Dueng., Bodenkd.* **79**, 119–131.
Keller, T. (1981). Auswirkungen von Luftverunreinigungen auf Pflanzen. *HLH, Heiz., Lueft. Klimatech., Haustech.* **48**, 22–24.
Kelley, P. M. and Izawa, S. (1978). The role of chloride ion in photosystem II. I. Effects of chloride ion on photosystem II electron transport and on hydroxylamine inhibition. *Biochim. Biophys. Acta* **502**, 198–210.
Keren, R., Gast, R. G. and Bar-Josef, B. (1981). pH dependent boron adsorption by Na-montmorillonite. *Soil Sci. Soc. Am. J.* **45**, 45–48.
Kerridge, P. C., Cook, B. G. and Everett, M. L. (1973). Application of molybdenum trioxide in the seed pellet for sub-tropical pasture legumes. *Trop. Grassl.* **7**, 229–232.
Kessler, E. (1955). On the role of manganese in the oxygen evolving system of photosynthesis. *Arch. Biochem. Biophys.* **59**, 527–529.
Keyser, H. H. and Munns, D. N. (1979). Effects of calcium, manganese, and aluminium on growth of rhizobia and acid media. *Soil Sci. Soc. Am. J.* **43**, 500–503.
Khan, A. H. and Marshall, C. (1981). Salt tolerance within populations of chewing fescue (*Festuca rubra* L.). *Commun. Soil Sci. Plant Anal.* **12**, 1271–1281.
Kianmehr, H. (1978). The response of *Helianthemum chamaecistus* Mill. to mycorrhizal infection in two different types of soil. *Plant Soil* **50**, 719–722.
Killham, K. and Firestone, M. K. (1983). Vesicular–arbuscular mycorrhizal mediation of grass response to acidic and heavy metal depositions. *Plant Soil* **72**, 39–48.
King, R. W. and Patrick, J. W. (1982). Control of assimilate movement in wheat. Is abscisic acid involved? *Z. Pflanzenphysiol.* **106**, 375–380.
Kingsbury, R. W., Epstein, E. and Pearcy, R. W. (1984). Physiological responses to salinity in selected lines of wheat. *Plant Physiol.* **74**, 417–423.
Kinzel, H. (1982). "Pflanzenökologie und Mineralstoffwechsel." Ulmer, Stuttgart.
Kinzel, H. (1983). Influence of limestone, silicates and soil pH on vegetation. *In* "Encyclopedia of Plant Physiology, New Series" (O. L. Lange, P. S. Nobel, C. B. Osmond and H. Ziegler, eds.), Vol. 12C, pp. 201–244. Springer-Verlag, Berlin and New York.
Kiraly, Z. (1964). Effect of nitrogen fertilization on phenol metabolism and stem rust susceptibility of wheat. *Phytopathol. Z.* **51**, 252–261.
Kiraly, Z. (1976). Plant disease resistance as influenced by biochemical effects of nutrients in fertilizers. *Proc. 12th Colloq. Int. Potash Inst. Bern,* pp. 33–46.
Kirkby, E. A. (1967). A note on the utilization of nitrate, urea, and ammonium nitrogen by *Chenopodium album. Z. Pflanzenernaehr. Bodenkd.* **117**, 204–209.
Kirkby, E. A. (1968). Influence of ammonium and nitrate nutrition on the cation-anion balance and nitrogen and carbohydrate metabolism of white mustard plants grown in dilute nutrient solutions. *Soil Sci.* **105**, 133–141.
Kirkby, E. A. (1979). Maximizing calcium uptake by plants. *Commun. Soil. Sci. Plant Anal.* **10**, 89–113.
Kirkby, E. A. (1981). Plant growth in relation to nitrogen supply. *In* "Terrestrial Nitrogen Cycles, Processes, Ecosystem Strategies and Management Impacts" (F. E. Clarke and T. Rosswall, eds.), pp. 249–267. Ecol. Bull., Stockholm.
Kirkby, E. A. and Armstrong, M. J. (1980). Nitrate uptake by roots as regulated by nitrate assimilation in the shoot of castor oil plants. *Plant Physiol.* **65**, 286–290.

Kirkby, E. A. and Knight, A. H. (1977). Influence of the level of nitrate nutrition on ion uptake and assimilation, organic acid accumulation, and cation-anion balance in whole tomato plants. *Plant Physiol.* **60**, 349–353.

Kirkby, E. A. and Mengel, K. (1967). Ionic balance in different tissues of the tomato plant in relation to nitrate, urea, or ammonium nutrition. *Plant Physiol.* **42**, 6–14.

Kirkby, E. A. and Mengel, K. (1970). Preliminary observations on the effect of urea nutrition on the growth and nitrogen metabolism of sunflower plants. *In* "Nitrogen Nutrition of the Plant" (E. A. Kirkby, ed.), pp. 35–38. The University of Leeds.

Kirkby, E. A. and Mengel, K. (1976). The role of magnesium in plant nutrition. *Z. Pflanzenernaehr. Bodenkd.* **139**, 209–222.

Kirkby, E. A. and Pilbeam, D. J. (1984). Calcium as a plant nutrient. *Plant Cell Environ.* **7**, 397–405.

Kirkham, D. S. (1954). Significance of the ratio of the water soluble aromatic and nitrogen constituents of apple and pear in the host–parasite relationship of *Venturia* sp. *Nature (London)* **173**, 690–691.

Kivilaan, A. and Scheffer, R. P. (1958). Factors affecting development of bacterial stem rot of *Pelargonium. Phytopathology* **48**, 185–191.

Kiyosawa, K. (1979). Unequal distribution of potassium and anions within the *Phaseolus* pulvinus during circadian leaf movement. *Plant Cell Physiol.* **20**, 1621–1634.

Kleeberger, A. and Klingmüller, W. (1980). Plasmid-mediated transfer of nitrogen-fixing capability to bacteria from the rhizosphere of grasses. *Mol. Gen. Genet.* **180**, 621–627.

Klein, H., Priebe, A. and Jäger, H.-J. (1979). Putrescine and spermidine in peas: Effects of nitrogen source and potassium supply. *Physiol. Plant.* **45**, 497–499.

Kleinkopf, G. E., Westermann, D. T. and Dwelle, R. B. (1981). Dry matter production and nitrogen utilization by six potato cultivars. *Agron. J.* **73**, 799–802.

Klemm, K. (1966). Der Einfluss der N-Form auf die Ertragsbildung verschiedener Kulturpflanzen. *Bodenkultur* **17**, 265–284.

Klepper, B. and Kaufmann, M. R. (1966). Removal of salt from xylem sap by leaves and stems of guttating plants. *Plant Physiol.* **41**, 1743–1747.

Kliewer, M. and Evans, H. J. (1963a). Identification of cobamide coenzyme in nodules of symbionts and isolation of the B_{12} coenzyme from *Rhizobium meliloti. Plant Physiol.* **38**, 55–59.

Kliewer, M. and Evans, H. J. (1963b). Cobamide coenzyme contents of soybean nodules and nitrogen fixing bacteria in relation to physiological conditions. *Plant Physiol.* **38**, 99–104.

Klucas, R. V., Hanus, F. J. and Russell, S. A. (1983). Nickel. A micronutrient element for hydrogen-dependent growth of *Rhizobium japonicum* and for expression of urease activity in soybean leaves. *Proc. Natl. Acad. Sci. USA* **90**, 2253–2257.

Kluge, R. (1971). Beitrag zum Problem des B-Mangels bei landwirtschaftlichen Kulturen als Folge der Bodentrockenheit. *Arch. Acker Pflanzenbau Bodenkd.* **15**, 749–754.

Kluge, R. and Beer, K. H. (1979). Einfluss des pH-Wertes auf die B-Adsorption von Aluminiumhydroxidgel, Tonmineralen und Böden. *Arch. Acker Pflanzenbau Bodenkd.* **23**, 279–287.

Knight, A. H., Crooke, W. M. and Burridge, J. C. (1973). Cation exchange capacity, chemical composition and the balance of carboxylic acids in the floral parts of various plant species. *Ann. Bot. (London)* [N.S.] **37**, 159–166.

Koch, K. E. and Johnson, C. R. (1984). Photosynthate partitioning in split-root citrus seedlings with mycorrhizal and nonmycorrhizal root systems. *Plant Physiol.* **75**, 26–30.

Koch, K. and Mengel, K. (1974). The influence of the level of potassium supply to young tobacco plants (*Nicotiana tabacum* L.) on short-term uptake and utilisation of nitrate nitrogen. *J. Sci. Food Agric.* **25**, 465–471.

Koda, Y. (1982). Changes in levels of butanol- and water-soluble cytokinins during the life cycyle of potato tubers. *Plant Cell Physiol.* **23**, 843–850.

Kojima, M. and Conn, E. E. (1982). Tissue distribution of chlorogenic acid and of enzymes involved in its metabolism in leaves of *Sorghum bicolor. Plant Physiol.* **70**, 922–925.

Kolesch, H., Oktay, M. and Höfner, W. (1984). Effect of iron chlorosis-inducing factors on the pH of the cytoplasm of sunflower *(Helianthus annuus). Plant Soil* **82**, 215–221.

Komor, E., Rotter, M. and Tanner, W. (1977). A proton-cotransport system in a higher plant: Sucrose transport in *Ricinus communis. Plant Sci. Lett.* **9**, 153–162.

Komor, E., Rotter, M., Waldhauser, J., Martin, E. and Cho, B. H. (1980). Sucrose proton symport for phloem loading in the Ricinus seedling. *Ber. Dtsch. Bot. Ges.* **93**, 211–219.

Komor, E., Thom, M. and Maretzki, A. (1982). Vacuoles from sugarcane suspension cultures. III. Protonmotive potential differences. *Plant Physiol.* **69**, 1326–1330.

Kondo, T. (1982). Correlation between potassium uptake rhythm and nitrate uptake rhythm in *Lemna gibba* G 3. *Plant Cell Physiol.* **23**, 909–916.

Konings, H. and Verschuren, G. (1980). Formation of aerenchyma in roots of *Zea mays* in aerated solutions and its relation to nutrient supply. *Physiol. Plant.* **49**, 265–270.

Konno, H., Yamaya, T., Yamasaki, Y. and Matsumoto, H. (1984). Pectic polysaccharide break-down of cell walls in cucumber roots grown with calcium starvation. *Plant Physiol.* **76**, 633–637.

Koontz, H. V. and Foote, R. E. (1966). Transpiration and calcium deposition by unifoliate leaves of *Phaseolus vulgaris* differing in maturity. *Physiol. Plant.* **19**, 313–321.

Kouchi, H. (1977). Rapid cessation of mitosis and elongation of root tip cells of *Vicia faba* as affected by boron deficiency. *Soil Sci. Plant Nutr.* **23**, 113.

Kovanci, I. and Colakoglu, H. (1976). The effect of varying K level on yield components and susceptibility of young wheat plants to attack by *Puccinia striiformis* West. *Proc. 12th Colloq. Int. Potash Inst. Bern,* pp. 177–182.

Kovanci, I. Hakerlerler, H. and Höfner W. (1978). Ursachen der Chlorosen an Mandarinen (*Citrus reticulata* Blanco) der ägäischen Region. *Plant Soil* **50**, 193–205.

Kraffczyk, I., Trolldenier, G. and Beringer, H. (1984). Soluble root exudates of maize: Influence of potassium supply and rhizosphere microorganisms. *Soil Biol. Biochem.* **16**, 315–322.

Kramer, D. (1983). The possible role of transfer cells in the adaptation of plants to salinity. *Physiol. Plant.* **58**, 549–555.

Kramer, D., Läuchli, A., Yeo, A. R. and Gullasch, J. (1977). Transfer cells in roots of *Phaseolus coccineus:* Ultrastructure and possible function in exclusion of sodium from the shoot. *Ann. Bot. (London).* [N.S.] **41**, 1031–1040.

Kramer, D., Römheld, V., Landsberg, E. and Marschner, H. (1980). Induction of transfer-cell formation by iron deficiency in the root epidermis of *Helianthus annuus*. *Planta* **147**, 335–339.
Kramer, P. J. (1983). "Water Relations in Plants." Academic Press, New York.
Krauss, A. (1971). Einfluss der Ernährung des Salats mit Massennährstoffen auf den Befall mit *Botrytis cinera* Pers. *Z. Pflanzenernaehr. Bodenkd.* **128**, 12–23.
Krauss, A. (1978a). Tuberization and abscisic acid content in *Solanum tuberosum* as affected by nitrogen nutrition. *Potato Res.* **21**, 183–193.
Krauss, A. (1978b). Endogenous regulation mechanisms in tuberization of potato plants in relation to environmental factors. *EAPR Abstr. Conf. Pap.* **7**, 47–48.
Krauss, A. (1980). Influence of nitrogen nutrition on tuber initiation of potatoes. *Proc. 15th Colloq. Int. Potash Inst. Bern,* pp. 175–184.
Krauss, A. and Marschner, H. (1971). Einfluss der Stickstoffernährung der Kartoffeln auf Induktion und Wachstumsrate der Knolle. *Z. Pflanzenernaehr. Bodenkd.* **128**, 153–168.
Krauss, A. and Marschner, H. (1975). Einfluss des Calcium-Angebotes auf Wachstumsrate und Calcium-Gehalt von Kartoffelknollen. *Z. Pflanzenernaehr. Bodenkd.* **138**, 317–326.
Krauss, and Marschner, H. (1976). Einfluss von Stickstoffernährung und Wuchsstoffapplikation auf die Knolleninduktion bei Kartoffelpflanzen. *Z. Pflanzenernaehr. Bodenkd.* **139**, 143–155.
Krauss, A. and Marschner, H. (1982). Influence of nitrogen nutrition, daylength and temperature on contents of gibberellic and abscisic acid and on tuberization in potato plants. *Potato Res.* **25**, 13–21.
Kriedemann, P. E. and Sands, R. (1984). Salt resistance and adaptation to root-zone hypoxia in sunflower. *Aust. J. Plant Physiol.* **11**, 287–301.
Krogmann, D. W., Jagendorf, A. T. and Avron, M. (1959). Uncouplers of spinach chloroplast photosynthetic phosphorylation. *Plant Physiol.* **34**, 272–277.
Krosing, M. (1978). Der Einfluss von Bormangel und von mechanischer Zerstörung des Spitzenmeristems auf die Zellteilung bei Sonnenblumen. *Z. Pflanzenernaehr. Bodenkd.* **141**, 641–654.
Krotzky, A., Berggold, R., Jaeger, D., Dart, P. J. and Werner, D. (1983). Enhancement of aerobic nitrogenase activity (acetylene reduction assay) by phenol in soils and the rhizosphere of cereals. *Z. Pflanzenernaehr. Bodenkd.* **146**, 634–642.
Krug, H., Wiebe, H.-J. and Jungk, A. (1972). Calciummangel an Blumenkohl unter konstanten Klimabedingungen. *Z. Pflanzenernaehr. Bodenkd.* **133**, 213–226.
Kubota, J. and Allaway, W. H. (1972). Geographic distribution of trace element problems. *In* "Micronutrients in Agriculture" (J. J. Mortvedt, P. M. Giordano and W. L. Lindsay. eds.), pp. 525–554. Soil Sci. Soc. Am., Madison, Wisconsin.
Kucey, R. M. N. and Paul, E. A. (1982). Biomass of mycorrhizal fungi associated with bean roots. *Soil Biol. Biochem.* **14**, 413–414.
Kuchenbuch, R. and Jungk, A. (1984). Wirkung der Kaliumdüngung auf die Kaliumverfügbarkeit in der Rhizosphäre von Raps. *Z. Pflanzenernaehr. Bodenkd.* **147**, 435–448.
Kühn, H., Schuster, W. and Linser, H. (1977). Starke Halmverkürzung bei Winterroggen durch kombinierte Anwendung von CCC und Ethephon im Feldversuch. *Z. Acker- Pflanzenbau* **145**, 22–30.
Kuiper, P. J. C. (1968). Lipids in grape roots in relation to chloride transport. *Plant Physiol.* **43**, 1367–1371.

Kuiper, P. J. C. (1980). Lipid metabolism as a factor in environmental adaptation. *In* "Biogenesis and Function of Plant Lipids" (P. Mezliok *et al.*, eds.), pp. 169–196. Elsevier/North-Holland Biomedical Press, Amsterdam.

Kuiper, P. J. C., Kähr, M., Stuiver, C. E. E. and Kylin, A. (1974). Lipid composition of whole roots and of Ca^{2+}, Mg^{2+}-activated adenosine triphosphatases from wheat and oat as related to mineral nutrition. *Physiol. Plant.* **32**, 33–36.

Kumon, K. and Tsurumi, S. (1984). Ion efflux from pulvinar cells during slow downward movement of the petiole of *Mimosa pudica* L. induced by photostimulation. *J. Plant Physiol.* **115**, 439–443.

Kuo, J., Pate, J. S., Rainbird, R. M. and Atkins, C. A. (1980). Internodes of grain legumes—New location of xylem parenchyma transfer cells. *Protoplasma* **104**, 181–185.

Kurvits, A. and Kirkby, E. A. (1980). The uptake of nutrients by sunflower plants *(Helianthus annuus)* growing in a continuous flowing culture system, supplied with nitrate or ammonium as nitrogen source. *Z. Pflanzenernaehr. Bodenkd.* **143**, 140–149.

Kuznetsova, G. A., Kuznetsova, M. G. and Grineva, G. M. (1981). Characteristics of water exchange and anatomical-morphological structure in corn plants under conditions of flooding. *Sov. Plant Physiol. (Engl. Transl.)* **28**, 241–248.

Kylin, A. and Hansson, G. (1971). Transport of sodium and potassium, and properties of (sodium + potassium) activated adenosine triphosphatase: Possible connection with salt tolerance in plants. *Proc. 8th Colloq. Int. Potash Inst. Bern,* pp. 64–68.

Lachno, D. R., Harrison-Murray, R. S. and Audus, L. J. (1982). The effect of mechanical impedance to growth on the levels of ABA and IAA in root tips of *Zea mays* L. *J. Exp. Bot.* **33**, 943–951.

Lafever, H. N., Campbell, L. G. and Foy, C. D. (1977). Differential response of wheat cultivars to Al. *Agron. J.* **69**, 563–568.

LaHaye, P. A. and Epstein, E. (1971). Calcium and salt tolerance by bean plants. *Physiol. Plant.* **25**, 213–218.

Lambers, H. (1982). Cyanide-resistant respiration: A non-phosphorylating electron transport pathway acting as an energy overflow. *Physiol. Plant.* **55**, 478–485.

Lambers, H., Blacquiere, T. and Stuiver, B. (C. E. E.) (1981a). Interactions between osmoregulation and the alternative respiratory pathway in *Plantago coronopus* as affected by salinity. *Physiol. Plant.* **51**, 63–68.

Lambers, H., Posthumus, F., Stulen, I., Lantin, L., Dijk, S. J. van de and Hostra, R. (1981b). Energy metabolism of *Plantago lanceolata* as dependent on the supply of mineral nutrients. *Physiol. Plant.* **51**, 85–92.

Lambers, H., Day, D. A. and Azcón-Bieto, J. (1983). Cyanide-resistant respiration in roots and leaves. Measurements with intact tissues and isolated mitochondria. *Physiol. Plant.* **58**, 148–154.

Lambert, D. H., Baker, D. E. and Cole, H. Jr. (1979). The role of mycorrhizae in the interactions of phosphorus with zinc, copper and other elements. *Soil Sci. Soc. Am. J.* **43**, 976–980.

Lambert, D. H., Cole, H. Jr. and Baker, D. E. (1980). Variation in the response of alfalfa clones and cultivars to mycorrhizae and phosphorus. *Crop Sci.* **20**, 615–618.

Lameta D'Arcy, A. (1982). Etude des exsudats racinaíres de Soja et de Lentille. I. Cinétique d'exsudation des composes phénologiques des amino acides et des sucres, au cours de premiers jours de la vie des plantules. *Plant Soil* **68**, 399–403.

Lamont, B. (1982). Mechanisms for enhancing nutrient uptake, with particular reference to Mediterranean South Africa and Western Australia. *Bot. Rev.* **48**, 597–689.

Landsberg, E.-C. (1979). Einfluss des Säurestoffwechsels und der Nitratreduktion auf Eisenmangel-bedingte Veränderungen des Substrat-pH-Wertes bei mono- und dikotylen Pflanzenarten. Dissertation, D 83 No. 84. Technische Universität, Berlin.

Landsberg, E.-C. (1981). Organic acid synthesis and release of hydrogen ions in response to Fe deficiency stress of mono- and dicotyledonous plant species. *J. Plant Nutr.* **3**, 579–591.

Lantzsch, H. J., Marschner, H., Wilberg, E. and Scheuermann, S. (1980). The improvement of the bioavailability of zinc in wheat and barley grains following application of zinc fertilizer. *Proc. Miner. Elements, Helsinki 1980,* Part I, pp. 323–328.

Larcher, W. (1980). "Ökologie der Pflanzen." Ulmer, Stuttgart.

Larkum, A. W. D. (1968). Ionic relations of chloroplasts in vivo. *Nature (London)* **218**, 447–449.

Lass, B. and Ullrich-Eberius, C. I. (1984). Evidence for proton/sulfate co-transport and its kinetics in *Lemna gibba* G 1. *Planta* **161**, 53–60.

Laties, G. G. and Budd, K. (1964). The developments of differential permeability in isolated stele of corn roots. *Proc. Natl. Acad. Sci. U.S.A.* **52**, 462–469.

Läuchli, A. (1976a). Symplasmic transport and ion release to the xylem. *In* "Transport and Transfer Processes in Plants" (I. F. Wardlaw and J. B. Passioura, eds.), Chapter 9, pp. 101–112. Academic Press, New York.

Läuchli, A. (1976b). Genotypic variation in transport. *In* "Transport in Plants 2, Part A" (U. Lüttge and M. G. Pitman, eds.), pp. 372–393. Springer-Verlag, Berlin and New York.

Läuchli, A. and Pflüger, R. (1978). Potassium transport through plant cell membranes and metabolic role of potassium in plants. *Proc. 11th Congr. Int. Potash Inst. Bern,* pp. 111–163.

Läuchli, A. and Wieneke, J. (1979). Studies on growth and distribution of Na^+, K^+ and Cl^- in soybean varieties differing in salt tolerance. *Z. Pflanzenernaehr. Bodenkd.* **142**, 3–13.

Läuchli, A., Spurr, A. R. and Epstein, E. (1971). Lateral transport of ions into the xylem of corn roots. II. Evaluation of a stelar pump. *Plant Physiol.* **48**, 118–124.

Läuchli, A., Lüttge, U. and Pitman, M. G. (1973). Ion uptake and transport through barley seedlings: Differential effect of cycloheximide. *Z. Naturforsch.* **28**C, 431–434.

Läuchli, A., Pitman, M. G., Kramer, D. and Ball, E. (1978). Are developing xylem vessels the sites of ion exudation from root to shoot? *Plant, Cell Environ.* **1**, 217–223.

Lawlor, D. W. and Milford, G. F. J. (1973). The effect of sodium on growth of water-stressed sugar-beet. *Ann. Bot. (London)* [N.S.] **37**, 597–604.

Layzell, D. B. and LaRue, T. A. (1982). Modeling C and N transport to developing soybean fruits. *Plant Physiol.* **70**, 1290–1298.

Lee, J. A. and Woolhouse, H. W. (1969a). A comparative study of bicarbonate inhibitions of root growth in calcicole and calcifuge grasses. *New Phytol.* **68**, 1–11.

Lee, J. A. and Woolhouse, H. W. (1969b). Root growth and dark fixation of carbon dioxide in calcicoles and calcifuges. *New Phytol.* **68**, 247–255.

Lee, J. S., Mulkey, T. J. and Evans, M. L. (1984). Inhibition of polar calcium movement and gravitropism in roots treated with auxin-transport inhibitors. *Planta* **160**, 536–543.

Lee, R. B. (1977). Effects of organic acids on the loss of ions from barley roots. *J. Exp. Bot.* **28**, 578–587.

Lee, R. B. (1982). Selectivity and kinetics of ion uptake of barley plants following nutrient deficiency. *Ann. Bot. (London)* [N.S.] **50**, 429–449.

Lee, R. B. and Ratcliffe, R. G. (1983). Phosphorus nutrition and the intracellular distribution of inorganic phosphate in pea root tips: A quantitative study using ^{31}P–NMR. *J. Exp. Bot.* **34**, 1222–1244.

Lee, S. G. and Aronoff, S. (1967). Boron in plants: A biochemical role. *Science* **158**, 798–799.

Lefebvre, D. D. and Glass, A. D. M. (1982). Regulation of phosphate influx in barley roots; effects of phosphate deprivation and reduction of influx with provision of orthophosphate. *Physiol. Plant.* **54**, 199–206.

Le Gales, Y., Lamant, A. and Heller, R. (1980). Fixation du calcium par des fractions macromoleculaires solubles isolées a partir de végétaux supérieurs. *Physiol. Veg.* **18**, 431–441.

Legge, R. L., Thompson, E., Baker, J. E. and Lieberman, M. (1982). The effect of calcium on the fluidity and phase properties of microsomal membranes isolated from postclimacteric Golden Delicious apples. *Plant Cell Physiol.* **23**, 161–169.

Legget, I. E. and Epstein, E. (1956). Kinetics of sulfate absorption by barley roots. *Plant Physiol.* **31**, 222–226.

Lehr, J. J. (1953). Sodium as a plant nutrient, *J. Sci. Food Agric.* **4**, 460–468.

Leigh, R. A. and Tomos, A. D. (1983). An attempt to use isolated vacuoles to determine the distribution of sodium and potassium in cells of storage roots of red beet (*Beta vulgaris* L.). *Planta* **159**, 469–475.

Leigh, R. A. and Wyn Jones, R. G. (1984). A hypothesis relating critical potassium concentrations for growth to the distribution and functions of this ion in the plant cell. *New Phytol.* **97**, 1–13.

Leigh, R. A., Ahmad, N. and Wyn Jones, R. G. (1981). Assessment of glycine-betaine and proline compartmentation by analysis of isolated beet vacuoles. *Planta* **153**, 34–41.

Leigh, R. A., Stribley, D. P. and Johnston, A. E. (1982). How should tissue nutrient concentrations be expressed? *In* "Proceedings of the Ninth International Plant Nutrition Colloquium, Warwick, England" (A. Scaife, ed.), pp. 39–44. Commonw. Agric. Bur., Farnham Royal, Bucks.

Lemon, E. and Houtte, R. van (1980). Ammonia exchange at the land surface. *Agron. J.* **72**, 876–883.

Lenz, F. (1970). Einfluss der Früchte auf das Wachstum, den Wasserverbrauch und die Nährstoffaufnahme von Auberginen. *Gartenbauwissenschaft* **35**, 281–292.

Lenz, F. and Döring, H. W. (1975). Fruit effects on growth and water consumption in Citrus. *Gartenbauwissenschaft* **6**, 257–260.

Leonard, R. T. and Hotchkiss, C. W. (1976). Cation-stimulated adenosine triphosphatase activity and cation transport in corn roots. *Plant Physiol.* **58**, 331–335.

Lerer, M and Bar-Akiva, A. (1976). Nitrogen constituents in manganese-deficient lemon leaves. *Physiol. Plant.* **38**, 13–18.

Lerner, H. R., Reinhold, L. Guy, R., Braun, Y., Hasidim, M. and Poljakoff-Mayber, A. (1983). Salt activation and inhibition of membrane ATPase from roots of the halophyte *Atriplex nummularia*. *Plant, Cell Environ.* **6**, 501–506.

Leshem, Y. Y., Sridhara, S. and Thompson, J. E. (1984). Involvement of calcium and calmodulin in membrane deterioration during senescence of pea foliage. *Plant Physiol.* **75**, 329–335.

Lessani, H. and Marschner, H. (1978). Relation between salt tolerance and long distance transport of sodium and chloride in various crop species. *Aust. J. Plant. Physiol.* **5**, 27–37.

Leuchs, F. (1959). Über Beziehungen zwischen Fäulniserscheinungen, Wundheilung und Kaliversorgung an Rosenkohl. *Z. Pflanzenkr. (Pflanzenpathol.) Pflanzenschutz* **66**, 499–508.

Levitt, J. (1980). "Responses of Plants to Environmental Stresses," 2nd ed., Vol. 2. Academic Press, New York.

Lewin, J. and Reimann, B. E. F. (1969). Silicon and plant growth. *Annu. Rev. Plant Physiol.* **20**, 289–304.

Levy, Y. and Horesh, I. (1984). Importance of penetration through stomata in the correction of chlorosis with iron salts and low-surface-tension surfactants. *J. Plant Nutr.* **7**, 279–281.

Lewis, D. G. and Quirk, J. P. (1967). Phosphate diffusion in soil and uptake by plants. III. ^{31}P-movement and uptake by plants as indicated by ^{32}P autoradiography. *Plant Soil* **27**, 445–453.

Lewis, D. H. (1980a). Boron, lignification and the origin of vascular plants—a unified hypothesis. *New Phytol.* **84**, 209–229.

Lewis, D. H. (1980b). Are there inter-relations between the metabolic role of boron, synthesis of phenolic phytoalexins and the germination of pollen? *New Phytol.* **84**, 261–270.

Lexmond, T. M. and Vorm, P. D. J. van der (1981). The effect of pH on copper toxicity to hydroponically grown maize. *Neth. J. Agric. Sci.* **29**, 217–238.

Liegel, W. (1970). Calciumoxalat-Abscheidung in Fruchtstielen einiger Apfelvarietäten. *Angew. Bot.* **44**, 223–232.

Lillo, C. and Henriksen, A. (1984). Comparative studies of diurnal variations of nitrate reductase activity in wheat, oat and barley. *Physiol. Plant.* **62**, 89–94.

Lin, C. H. and Stocking, C. R. (1978). Influence of leaf age, light, dark and iron deficiency on polyribosome levels in maize leaves. *Plant Cell Physiol.* **19**, 461–470.

Lin, D. C. and Nobel, P. S. (1971). Control of photosynthesis by Mg^{2+}. *Arch. Biochem. Biophys.* **145**, 622–632.

Lin. P. P. C., Egli, D. B., Li, G. M. and Meckel, L. (1984). Polyamine titer in the embryonic axis and the cotyledons of *Glycine max* (L.) during seed growth and maturation. *Plant Physiol.* **76**, 366–371.

Lindberg, S. (1980). Kinetic studies of a ($Na^+ + K^+ + Mg^{2+}$) ATPase in sugar beet roots. III. A proposed model for the ($Na^+ + K^+$) activation and its significance for field properties. *Physiol. Plant.* **48**, 65–70.

Lindsay, W. L. (1974). Role of chelations in micronutrient availability. *In* "The Plant Root and its Environment" (E. W. Carson, ed.), pp. 507–524. Univ. Press of Virginia, Charlottesville.

Lindsay, W. L. and Norvell, W. A. (1978). Development of a DTPA soil test for zinc, iron, manganese and copper. *Soil Sci. Soc. Am. J.* **42**, 421–428.

Lindsay, W. L. and Schwab, A. P. (1982). The chemistry of iron in soils and its availability to plants. *J. Plant Nutr.* **5**, 821–840.

Lingle, J. C. and Lorenz, O. A. (1969). Potassium nutrition of tomatoes. *J. Am. Soc. Hortic. Sci.* **94**, 679–683.

Linser, H., Raafat, A. and Zeid, F. A. (1974). Reinprotein und Chlorophyll bei *Daucus carota* im Verlauf der Vegetationsperiode des ersten Jahres unter dem Einfluss von Wachstumsregulatoren. *Z. Pflanzenernaehr. Bodenkd.* **137**, 36–48.

Liu, W. (1979). Potassium and phosphate uptake in corn roots. Further evidence for an electrogenic H^+/K^+ exchanger and an OH^-/P_i antiporter. *Plant Physiol.* **63**, 952–955.

Loescher, W. H., Marlow, G. C. and Kennedy, R. A. (1982). Sorbitol metabolism and sink-source interconversion in developing apple leaves. *Plant Physiol.* **701,** 335–339.

Löhnis, M. P. (1960). Effect of magnesium on calcium supply on the uptake of manganese by various crop plants. *Plant Soil* **12**, 339–376.

Lolas, G. M., Palamidis, N. and Markakis, P. (1976). The phytic-acid total phosphorus relationship in barley, oats, soybeans and wheat. *Cereal Chem.* **53**, 867–870.

Loneragan, J. F. (1975). The availability and absorption of trace elements in soil-plant systems and their relation to movement and concentrations of trace elements in plants. *In* "Trace Elements in Soil-Plant-Animal Systems" (D. J. D. Nicholas and A. R. Egan, eds.), pp. 109–134. Academic Press, London and Orlando.

Loneragan, J. F. and Asher, C. H. (1967). Response of plants to phosphate concentration in solution culture. II. Role of phosphate absorption and its relation to growth. *Soil Sci.* **103**, 311–318.

Loneragan, J. F. and Snowball, K. (1969). Calcium requirements of plants. *Aust. J. Agric. Res.* **20**, 465–478.

Loneragan, J. F., Snowball, K. and Simmons, W. J. (1968). Response of plants to calcium concentration in solution culture. *Aust. J. Agric. Res.* **19**, 845–857.

Loneragan, J. F., Snowball, K. and Robson, A. D. (1976). Remobilization of nutrients and its significance in plant nutrition. *In* "Transport and Transfer Process in Plants" (I. F. Wardlaw and J. B. Passioura, eds.), pp. 463–469. Academic Press, London and Orlando.

Loneragan, J. F., Grove, T. S., Robson, A. D. and Snowball, K. (1979). Phosphorus toxicity as a factor in zinc-phosphorus interactions in plants. *Soil Sci. Soc. Am. J.* **43**, 966–972.

Loneragan, J. F., Delhaize, E. and Webb, J. (1982a) Enzymic diagnosis of copper deficiency in subterranean clover. I. Relationship of ascorbate oxidase activity in leaves to plant copper status. *Aust. J. Agric. Res.* **33**, 967–979.

Loneragan, J. F., Grunes, D. L., Welch, R. M., Aduayi, E. A., Tengah, A., Lazar, V. A. and Cary, E. E. (1982b). Phosphorus accumulation and toxicity in leaves in relation to zinc supply. *Soil Sci. Soc. Am. J.* **46**, 345–352.

Longstreth, D. J. and Nobel, P. S. (1979). Salinity effects on leaf anatomy. Consequences for photosynthesis. *Plant Physiol.* **63**, 700–703.

Longstreth, D. J., Bolaños, J. A. and Smith, J. E. (1984). Salinity effects on photosynthesis and growth of *Alternanthera philoxeroides* (Mart.) Giseb[1]. *Plant Physiol.* **75**, 1044–1047.

Lookeren-Campagne, R. N. van (1957). Light-dependent chloride absorption in *Vallisneria* leaves. *Acta Bot. Neerl.* **6**, 543–582.

Lott, J. N. A. and Buttrose, M. S. (1978). Globoids in protein bodies of legume seed cotyledons. *Aust. J. Plant Physiol.* **5**, 89–111.

Lott, J. N. A. and Vollmer, C. M. (1973). Changes in the cotyledons of *Cucurbita maxima* during germination. IV. Protein bodies. *Protoplasma* **78**, 255–271.

Loughman, B. C. (1966). The mechanism of absorption and utilization of phosphate by barley plants in relation to subsequent transport to the shoot. *New Phytol.* **65**, 388–397.

Loughman, B. C., Webb, M. J. and Loneragan, J. F. (1982). Zinc and the utilization of phosphate in wheat plants. *In* "Proceedings of the Ninth International Plant Nutrition Colloquium, Warwick, England" (A. Scaife, ed), pp. 335–340. Commonw. Agric. Bur., Farnham Royal, Bucks.

Louwerse, W. (1967). The influence of the plant nutrition status on bleeding and salt uptake. *Acta Bot. Neerl.* **16**, 42–55.

Lovatt, C. J., Albert, L. S. and Tremblay, G. C. (1981). Synthesis, salvage and catabolism of uridine nucleotides in boron-deficient squash roots. *Plant Physiol.* **68**, 1389–1394.

Lowther, W. L. and Loneragan, J. F. (1968). Calcium and nodulation in subterranean clover. (*Trifolium subterraneum* L.). *Plant Physiol.* **43**, 1362–1366.

Lüdders, P. and Bünemann, G. (1970). Die Wirkung des Zeitpunktes von Harnstoffspritzungen auf Apfelbäume. *Z. Pflanzenernaehr. Bodenkd.* **125**, 144–155.

Lüdders, P. and Fischer-Bölükbasi, T. (1980). Einfluss von Alar und TIBA auf den Mineralstoffgehalt der Früchte bei unterschiedlichem Fruchtbehang. *Gartenbauwissenschaft* **45**, 235–240.

Lund, Z. F. (1970). The effect of calcium and its relation to several cations in soybean root growth. *Soil Sci. Soc. Am. Proc.* **34**, 456–459.

Lüttge, U. and Laties, G. G. (1966). Dual mechanism of ion absorption in relation to long distance transport in plants. *Plant Physiol.* **41**, 1531–1539.

Lune, P. van and Goor, B. J. van (1977). Ripening disorders of tomato as affected by the K/Ca ratio in the culture solution. *J. Hortic. Sci.* **52**, 173–180.

Luxmoore, R. J., Fischer, R. A. and Stolzy, L. H. (1973). Flooding and soil temperature effects on wheat during grain filling. *Agron. J.* **65**, 361–364.

Lynch, J., Epstein, E. and Läuchli, A. (1982). Na^+-K^+ relationship in salt-stressed barley. *In* "Proceedings of the Ninth International Plant Nutrition Colloquium, Warwick, England" (A. Scaife, ed.), pp. 347–352. Commonw. Agric. Bur., Farnham Royal, Bucks.

Lynch, J. M. (1978). Production and phytotoxicity of acetic acid in anaerobic soils containing plant residues. *Soil Biol. Biochem.* **10**, 131–135.

Lyttleton, J. W. (1960). Stabilization by manganese ions of ribosomes from embryonic plant tissue. *Nature (London)* **187**, 1026–1027.

Maas, E. V. and Hoffman, G. J. (1977). Crop salt tolerance—current assessment. *J. Irrig. Drain. Div. Am. Soc. Civ. Eng.* **103**, 115–134.

Maas, E. V., Hoffman, G. J., Chaba, G. D., Poss, J. A. and Shannon, M. C. (1983). Salt sensitivity of corn at various growth stages. *Irrigation Sci.* **4**, 45–57.

McCaslin, B. D., Samson, R. F. and Baltensperger, A. A. (1981). Selection for turf-type bermudagrass genotypes with reduced iron chlorosis. *Commun. Soil. Sci. Plant Anal.* **12**, 189–204.

McClendon, J. H. (1976). Elemental abundance as a factor on the origins of mineral nutrient requirements. *J. Mol. Evol.* **8**, 175–195.

McClure, J. M. (1976). Physiology and functions of flavanoids. *In* "The Flavanoids" (J. B. Harborne, T. Mabry and H. Mabry, eds.), pp. 970–1055. Chapman and Hall, London.

McDaniel, M. E. and Dunphy, D. J. (1978). Differential iron chlorosis of oat cultivars. *Crop Sci.* **18**, 136–138.

MacDonald, I. R., Macklon, A. E. S. and MacLeod, R. W. G., (1975). Energy supply and light-enhanced chloride uptake in wheat laminae. *Plant Physiol.* **56**, 699–702.
McGrath, J. F. and Robson, A. D. (1984). The movement of zinc through excised stems of seedlings of *Pinus radiata* D. Don. *Ann. Bot.* **54**, 231–242.
McGregor, A. J. and Wilson, G. C. S. (1964). The effect of applications of manganese sulphate to a neutral soil upon the yield of tubers and the incidence of common scab in potatoes. *Plant Soil* **20**, 59–64.
Machold, O. (1967). Untersuchungen an stoffwechseldefekten Mutanten der Kulturtomate. III. Die Wirkung von Ammonium- und Nitratstickstoff auf den Chlorophyllgehalt. *Flora (Jena), Abt. A.* **157**, 536–551.
Machold, O. (1968). Einfluss der Ernährungsbedingungen auf den Zustand des Eisens in der Blätter, den Chlorophyllgehalt und die Katalase- sowie Peroxydaseaktivität. *Flora (Jena), Abt. A* **159**, 1–25.
Machold, O. (1972). Lamellarproteine grüner und chlorotischer Chloroplasten. *Biochem. Physiol. Pflanz.* **163**, 30–41.
Machold, O. and Stephan, U. W. (1969). The function of iron in porphyrin and chlorophyll biosynthesis. *Phytochemistry* **8**, 2189–2192.
Machold, O., Meisel, W. and Schnorr, H. (1968). Bestimmung der Bindungsformen des Eisens in Blättern durch Mössbauer-Spektrometrie. *Naturwissenschaften* **55**, 499–500.
McHughen, A. and Swartz, M. (1984). A tissue-culture derived salt-tolerant line of flax *(Linum usitatissimum). J. Plant Physiol.* **117**, 109–117.
MacInnes, C. B. and Albert, L. S. (1969). Effect of light intensity and plant size on rate of development of early boron deficiency symptoms in tomato root tips. *Plant Physiol.* **44**, 965–967.
MacLeod, L. B. (1969). Effects of N, P and K and their interactions on the yield and kernel weight of barley in hydroponic culture. *Agron. J.* **61**, 26–29.
Mackay, A. D. and Barber, S. A. (1984). Soil temperature effects on root growth and phosphorus uptake by corn. *Soil Sci. Soc. Am. J.* **48**, 818–823.
McManmon, M. and Crawford, R. M. M. (1971). A metabolic theory of flooding tolerance: The significance of enzyme distribution and behaviour. *New Phytol.* **70**, 299–306.
McNeil, D. L. (1980). The role of the stem in phloem loading of minerals in *Lupinus albus* L. cv. Ultra *Ann. Bot. (London)* [N.S.] **45**, 329–338.
MacRobbie, E. A. C. (1981). Effects of ABA in "isolated" guard cells of *Commelina communis* L. *J. Exp. Bot.* **32**, 563–572.
McSwain, B. D., Tsujimoto, H. Y. and Arnon, D. I. (1976). Effects of magnesium and chloride ions on light-induced electron transport in membrane fragments from a blue-green alga. *Biochim. Biophys. Acta* **423**, 313–322.
Mäder, M. (1977). Die Lokalisation der Peroxidase-Isoenzymgruppe G_I in der Zellwand von Tabakgeweben. *Planta* **131**, 11–15.
Mäder, M. and Füssl, R. (1982). Role of peroxidase in lignification of tobacco cells. II. Regulation by phenolic compounds. *Plant Physiol.* **70**, 1132–1134.
Mahmoud, E. A. and Hill, M. J. (1981). Salt tolerance of sugar beet at various temperatures. *N.Z. J. Agric. Res.* **24**, 67–71.
Maier-Maercker, U. (1979). "Peristomatal transpiration" and stomatal movement: A controversial view. I. Additional proof of peristomatal transpiration by photography and a comprehensive discussion in the light of recent results. *Z. Pflanzenphysiol.* **91**, 25–43.

Maizlich, N. A., Fritton, D. D. and Kendall, W. A. (1980). Root morphology and early development of maize at varying levels of nitrogen., *Agron. J.* **72**, 25–31.

Makmur, A., Gerloff, G. C. and Gabelman, W. H. (1978). Physiology and inheritance of efficiency in potassium utilization in tomatoes. (*Lycopersicon esculentum* Mill.) grown under potassium stress. *J. Am. Soc. Hortic. Sci.* **103**, 545–549.

Malash, N. M. A. R. and Flowers, T. J. (1984). The effect of phenylmercuric acetate on salt tolerance in wheat. *Plant Soil* **81**, 269–279.

Malek, F. and Baker, D. A. (1977). Proton cotransport of sugars in phloem loading. *Planta* **135**, 297–299.

Malzer, G. L. and Barber, S. A. (1975). Precipitation of calcium and strontium sulfates around plant roots and its evaluation. *Soil Sci. Soc. Am. Proc.* **39**, 492–495.

Manolakis, E. and Lüdders, P. (1977). Die Wirkung gleichmäßiger und jahreszeitlich abwechselnder Ammonium- und Nitraternährung auf Apfelbäume. I. Einfluss auf das vegetative Wachstum. *Gartenbauwissenschaft* **42**, 1–7.

Marion, G. M., Hendrix, D. M., Dutt, G. R. and Fuller, W. H. (1976). Aluminium and silica solubility in soils. *Soil Sci.* **121**, 76–85.

Mark, F. van der., Lange, T. de and Bienfait, H. F. (1981). The role of ferritin in developing primary bean leaves under various light conditions. *Planta* **153**, 338–342.

Marmé, D., (1983). Calcium transport and function. *In* "Encyclopedia of Plant Physiology, New Series" (A. Läuchli and R. L. Bieleski, eds.), Vol. 15B, pp. 599–625. Springer-Verlag, Berlin and New York.

Marquard, R., Kühn, H. and Linser, H. (1968). Der Einfluss der Schwefelernährung auf die Senfölbildung. *Z. Pflanzenernaehr. Bodenkd.* **121**, 221–230.

Marschner, H. (1971). Why can sodium replace potassium in plants? *Proc. 8th Colloq. Int. Potash Inst. Bern,* pp. 50–63.

Marschner, H. (1983). General introduction to the mineral nutrition of plants. *In* "Encyclopedia of Plant Physiology, New Series" (A. Läuchli and R. L. Bieleski, eds.), Vol. 15A, pp. 5–60. Springer-Verlag, Berlin and New York.

Marschner, H. and Döring, H. W. (1977). Effects of K^+ and Na^+ on ADPG-starch synthetase. *Proc. 13th Colloq. Int. Potash Inst. Bern,* pp. 101–113.

Marschner, H. and Michael, G. (1960). Untersuchungen über Schwefelabscheidung und Sulfataustausch an Weizenwurzeln. *Z. Pflanzenernaehr., Dueng., Bodenkd.* **91**, 24–44.

Marschner, H. and Mix, G. (1974). Einfluss von Natrium und Mycostatin auf den Mineralstoffgehalt im Blattgewebe und die Feinstruktur der Chloroplasten. *Z. Pflanzenernaehr. Bodenkd.* **136**, 203–219.

Marschner, H, and Ossenberg-Neuhaus, H. (1977). Wirkung von 2,3,5-Trijodbenzoesäure (TIBA) auf den Calciumtransport und die Kationenaustauschkapazität in Sonnenblumen. *Z. Pflanzenphysiol.* **85**, 29–44.

Marschner, H. and Possingham, J. V. (1975). Effect of K^+ and Na^+ on growth of leaf discs of sugar beet and spinach. *Z. Pflanzenphysiol.* **75**, 6–16.

Marschner, H. and Richter, C. (1973). Akkumulation und Translokation von K^+, Na^+ und Ca^{2+} bei Angebot zu einzelnen Wurzelzonen von Maiskeimpflanzen. *Z. Pflanzenernaehr. Bodenkd.* **135**, 1–15.

Marschner, H. and Richter, C. (1974) Calcium-Transport in Wurzeln von Mais- und Bohnenkeimpflanzen. *Plant Soil* **40**, 193–210.

Marschner, H. and Römheld, V. (1983). In-vivo measurement of root-induced pH changes at the soil-root interface: Effect of plant species and nitrogen source. *Z. Pflanzenphysiol.* **111**, 241–251.
Marschner, H. and Schafarczyk, W. (1967). Vergleich der Nettoaufnahme von Natrium und Kalium bei Mais- und Zuckerrübenpflanzen. *Z. Pflanzenernaehr. Bodenkd.* **118**, 172–187.
Marschner, H. and Schropp, A. (1977). Vergleichende Untersuchungen über die Empfindlichkeit von 6 Unterlagensorten der Weinrebe gegenüber Phosphat-induziertem Zink-Mangel. *Vitis* **16**, 79–88.
Marschner, H., Kalisch, A. and Römheld, V. (1974). Mechanism of iron uptake in different plant species. *In* "Proceedings of the Seventh International Colloquium on Plant Analysis and Fertilizer Problems" (J. Wehrmann, ed.), pp. 273–281. Hannover.
Marschner, H., Römheld, V. and Azarabadi, S. (1978). Iron stress response of efficient and inefficient plant species. *In* "Proceedings of the Eighth International Colloquium on Plant Analysis and Fertilizer Problems" (A. R. Ferguson, R. L. Bieleski, I. B. Ferguson, eds.), pp. 319–327. *Inf. Ser. N.Z. Dep. Sci. Ind. Res.* **134**.
Marschner, H., Kylin, A. and Kuiper, P. J. C. (1981a). Differences in salt tolerance of three sugar beet genotypes. *Physiol. Plant.* **51**, 234–238.
Marschner, H., Kuiper, P. J. C. and Kylin, A. (1981b). Genotypic differences in the response of sugar beet plants to replacement of potassium by sodium. *Physiol. Plant.* **51**, 239–244.
Marschner, H., Römheld, V. and Ossenberg-Neuhaus, H. (1982). Rapid method for measuring changes in pH and reducing processes along roots of intact plants. *Z. Pflanzenphysiol.* **105**, 407–416.
Martin, C. (1976). Nutrition and virus diseases of plants. *Proc. 12th Colloq. Int. Potash Inst. Bern,* pp. 193–200.
Martin, F., Canet, D., Rolin, D., Marchal, J. P. and Larher, F. (1983). Phosphorus-31 nuclear magnetic resonance study of polyphosphate metabolism in intact ectomycorrhizal fungi. *Plant Soil* **71**, 469–476.
Martin, J.-B., Bligney, R., Rebeille, F., Douce, R. Leguay, J.-J., Mathieu, Y. and Guern, J. (1982). A ^{31}P nuclear magnetic resonance study of intracellular pH of plant cells cultivated in liquid medium. *Plant Physiol.* **70**, 1156–1161.
Martin, J. K. (1977). Factors influencing the loss of organic carbon from wheat roots. *Soil Biol. Biochem.* **9**, 1–7.
Martin, P. (1970). Pathway of translocation of ^{15}N from labelled nitrate or ammonium in kidney bean plants. *In* "Nitrogen Nutrition of the Plant" (E. A. Kirkby, ed.), pp. 104–112. University of Leeds.
Martin, P. (1971). Wanderwege des Stickstoffs in Buschbohnenpflanzen beim Aufwärtstransport nach der Aufnahme durch die Wurzel. *Z. Pflanzenphysiol.* **64**, 206–222.
Martin, P. (1973). Nitratstickstoff in Buschbohnenblättern unter dem Gesichtspunkt der Kompartimentierung der Zellen. *Z. Pflanzenphysiol.* **70**, 158–165.
Martin, P. (1982). Stem xylem as a possible pathway for mineral retranslocation from senescing leaves to the ear in wheat. *Aust. J. Plant Physiol.* **9**, 197–207.
Martin, P. and Glatzle, A. (1982). Mutual influences of *Azospirillum* spp. and grass seedlings. *In "Azospirillum,* Genetics, Physiology, Ecology Workshop, 1981 Bayreuth" (W. Klingmüller, ed.), pp. 108–120. Birkhäuser-Verlag.

Martin, P. and Platz, V. (1982). Retranslocation of nitrogen and potassium from leaves to grains in wheat. *In* "Proceedings of the Ninth International Plant Nutrition Colloquium Warwick, England" (A. Scaife, ed.), pp. 360–365. Commonw. Agric. Bur., Farnham Royal, Bucks.

Martin, P., Glatzle, A. and Kolb, W. (1984). Möglicher Beitrag N_2-bindender Bakterien in der Rhizosphäre zur Nährstoffversorgung von Pflanzen. *Landwirtsch. Forsch. Sonderh.* **40**, 241–249.

Martinez-Carrasco, R. and Thorne, G. N. (1979). Physiological factors limiting grain size in wheat. *J. Exp. Bot.* **30**, 669–679.

Martini, F. and Thellier, M. (1980). Use of an (n, α) nuclear reaction to study the long distance transport of boron in *Trifolium repens* after foliar application. *Planta* **150**, 197–205.

Martinoia, E., Heck, U and Wienecken, A. (1981). Vacuoles as storage compartments for nitrate in barley leaves. *Nature (London)* **289**, 292–294.

Marx, C., Dexheimer, J., Gianinazzi-Pearson, V. and Gianinazzi, S. (1982). Enzymatic studies on the metabolism of vesicular–arbuscular mycorrhizas. IV. Ultracytoenzymological evidence (ATPase) for active transfer processes in the host-arbuscle interface. *New Phytol.* **90**, 37–43.

Mascarenhas, J. P. and Machlis, L. (1964). Chemotropic response of the pollen of *Antirrhinum majus* to calcium. *Plant Physiol.* **39**, 70–77.

Mashhady, A. S. and Rowell, D. L. (1978). Soil alkalinity. II. The effect of Na_2CO_3 on iron and manganese supply to tomatoes. *J. Soil Sci.* **29**, 367–372.

Massey, H. F. and Loeffel, A. (1967). Species specific variations in zinc content of corn kernels. *Agron. J.* **59**, 214–217.

Matar, A. E., Paul, J. L. and Jenny, H. (1967). Two phase experiments with plants growing in phosphate-treated soil. *Soil Sci. Soc. Am. Proc.* **31**, 235–237.

Mathers, A. C., Thomas, J. D., Steward, B. A. and Herring, J. E. (1980). Manure and inorganic fertilizer effects on sorghum and sunflower growth on iron-deficient soil. *Agron. J.* **72**, 1025–1029.

Mathys, W. (1977). The role of malate, oxalate, and mustard oil glucosides in the evolution of zinc-resistance in herbage plants. *Physiol. Plant.* **40**, 130–136.

Matocha, J. E. and Smith, L. (1980). Influence of potassium on *Helminthosporium cynodontis* and dry matter yields of coastal Bermudagrass. *Agron. J.* **72**, 565–567.

Matsumoto, H. and Tamura, K. (1981). Respiratory stress in cucumber roots treated with ammonium or nitrate nitrogen. *Plant Soil* **60**, 195–204.

Matsumoto, H., Hirasawa, F., Torikai, H. and Takahashi, E. (1976a). Localization of absorbed aluminium in pea root and its binding to nucleic acid. *Plant Cell Physiol.* **17**, 127–137.

Matsumoto, H., Hirasawa, E., Morimura, S. and Takahashi, E. (1976b). Localization of aluminium in tea leaves. *Plant Cell Physiol.* **17**, 627–631.

Matsuyama, N. (1975). The effect of ample nitrogen fertilizer on cell wall materials and its significance to rice blast disease. *Ann. Phytopathol. Soc. Jpn.* **4**, 56–61.

Matsuyama, N. and Dimond, A. E. (1973). Effect of nitrogenous fertilizer on biochemical processes that could affect lesion size of rice blast. *Phytopathology* **63**, 1202–1203.

Mattoo, A. K. and Lieberman, M. (1977). Localization of the ethylene-synthesizing system in apple tissue. *Plant Physiol.* **60**, 794–799.

Mayr, H. H. (1968). Anwendung und Bedeutung von Chlorcholinchlorid (CCC) in der Landwirtschaft. *Landwirtsch. Forsch.* **31**, 195–202.

Means, A. R. and Dedman, J. R. (1980). Calmodulin—an intracellular calcium receptor. *Nature (London)* **285**, 73–77.
Meeks, J. C., Wolk, C. P., Schilling, N., Schaffer, P. W., Avissar, Y. and Chien, W.-S. (1978). Initial organic products of fixation of ^{13}N-dinitrogen by root nodules of soybean *(Glycine max). Plant Physiol.* **61**, 980–983.
Meiri, A., Hofman, G. J., Shannon, M. C. and Poss, J. A. (1982). Salt tolerance of 2 muskmelon cultivars under 2 radiation levels. *J. Am. Soc. Hortic. Sci.* **107**, 1168–1172.
Mejstřik, V. K. and Cudlin, P. (1983). Mycorrhiza in some plant desert species in Algeria. *Plant Soil* **71**, 363–366.
Memon, A. R. and Yatazawa, M. (1984). Nature of manganese complexes in manganese accumulator plant—*Acanthopanax sciadophylloides. J. Plant Nutr.* **7**, 961–974.
Menary, R. C. and Van Staden, J. (1976). Effect of phosphorus nutrient and cytokinins on flowering in the tomato, *Lycopersicon esculentum* Mill. *Aust. J. Plant Physiol.* **3**, 201–205.
Menge, J. A. (1983). Utilization of vesicular–arbuscular mycorrhizal fungi in agriculture. *Can. J. Bot.* **61**, 1015–1024,
Menge, J. A., Labanauskas, C. K., Johnson, E. L. V. and Pratt, R. G. (1978). Partial substitution of mycorrhizal fungi for phosphorus fertilization in the greenhouse culture of citrus. *Soil Sci. Soc. Am. J.* **42**, 926–930.
Menge, J. A., LaRue, J., Labanauskas, C. K. and Johnson, E. L. (1980). The effect of two mycorrhizal fungi upon growth and nutrition of avocado seedlings grown with six fertilizer treatments. *J. Am. Soc. Hortic. Sci.* **105**, 400–404.
Mengel, K. (1962). Die K- und Ca-Aufnahme der Pflanze in Abhängigkeit vom Kohlenhydratgehalt ihrer Wurzel. *Z. Pflanzenernaehr., Dueng., Bodenkd.* **98**, 44–54.
Mengel, K. and Bübl, W. (1983). Verteilung von Eisen in Blättern von Weinreben mit HCO_3^- induzierter Chlorose. *Z. Pflanzenernaehr. Bodenkd.* **146**, 560–571.
Mengel, K. and Haeder, H. E. (1977). Effect of potassium supply on the rate of phloem sap exudation and the composition of phloem sap of *Rizinus communis. Plant Physiol.* **59**, 282–284.
Mengel, K. and Helal, M. (1967). Der Einfluss des austauschbaren Ca^{2+} junger Gerstenwurzeln auf den Flux von K^+ und Phosphat- eine Interpretation des Viets-Effektes. *Z. Pflanzenphysiol.* **57**, 223–234.
Mengel, K. and Helal, M. (1968). Der Einfluss einer variierten N- und K-Ernährung auf den Gehalt an löslichen Aminoverbindungen in der oberirdischen Pflanzenmasse von Hafer. *Z. Pflanzenernaehr. Bodenkd.* **120**, 12–20.
Mengel, K. and Kirkby, E. A. (1982). "Principles of Plant Nutrition." 3rd ed. Int. Potash Inst. Bern, Switzerland.
Mengel, K. and Malissiovas, N. (1981). Bicarbonat als auslösender Faktor der Eisenchlorose bei der Weinrebe *(Vitis vinifera). Vitis* **20**, 235–243.
Mengel, K. and Schneider, B. (1965). Die K-Aufnahme als Funktion der Influxrate und der Zellpermeabilität, mathematisch und experimentell an der K-Aufnahme junger Gerstenwurzeln dargestellt. *Physiol. Plant.* **18**, 1105–1114.
Mengel, K. and Steffens, D. (1982). Beziehung zwischen Kationen/Anionen-Aufnahme von Rotklee und Protonenabscheidung der Wurzeln. *Z. Pflanzenernaehr. Bodenkd.* **145**, 229–236.
Mengel, K., Grimme, H. and Nemeth, K. (1969). Potentielle und effektive Verfügbarkeit von Pflanzennährstoffen im Boden. *Landwirtsch. Forsch. Sonderh.* **16**, 79–91.

Mengel, K., Haghparast, M. and Koch, K. (1974). The effect of potassium on the fixation of molecular nitrogen by root nodules of *Vicia faba*. *Plant Physiol.* **54**, 535–538.

Mengel, K., Viro, M. and Hehl, G. (1976). Effect of potassium on uptake and incorporation of ammonium-nitrogen of rice plants. *Plant Soil* **44**, 547–558.

Mengel, K., Scherer, H. W. and Malissiovas, N. (1979). Die Chlorose aus der Sicht der Bodenchemie und Rebenernährung. *Mitt. Klosterneuburg* **29**, 151–156.

Mengel, K., Robin, P. and Salsac, L. (1983). Nitrate reductase activity in shoots and roots of maize seedlings as affected by form of nitrogen nutrition and the pH of the nutrient solution. *Plant Physiol.* **71**, 618–622.

Mengel, K., Breininger, M. T. and Bübl, W. (1984). Bicarbonate, the most important factor inducing iron chlorosis in vine grapes on calcareous soil. *Plant Soil* **81**, 333–344.

Merker, E. (1961). Welche Ursachen hat die Schädigung der Insekten durch die Düngung im Walde? *Allg. Forst- Jagdztg.* **132**, 73–82.

Mertens, T. and Hess, D. (1984). Yield increases in spring wheat (*Triticum aestivum* L.) inoculated with *Azospirillum lipoferum* under greenhouse and field conditions of a temperate region. *Plant Soil* **82**, 87–99.

Meshram, S. U. and Shende, S. T. (1982). Response of maize to *Azotobacter chroococcum*. *Plant Soil* **69**, 265–273.

Mettler, I. J., Mandala, S. and Taiz, L. (1982). Characterization of in vitro proton pumping by microsomal vesicles isolated from corn coleoptiles. *Plant Physiol.* **70**, 1738–1742.

Michael, B., Zink, F. and Lantzsch, H. J. (1980). Effect of phosphate application on phytin-P and other phosphate fractions in developing wheat grains. *Z. Pflanzenernaehr. Bodenkd.* **143**, 369–376.

Michael, G. (1941). Über die Aufnahme und Verteilung des Magnesiums und dessen Rolle in der höheren grünen Pflanze. *Z. Pflanzenernaehr., Dueng., Bodenkd.* **25**, 65–120.

Michael, G. and Beringer, H. (1980). The role of hormones in yield formation. *Proc. 15th Colloq. Int. Potash Inst. Bern,* pp. 85–116.

Michael, G., Faust, H. and Blume, B. (1960). Die Verteilung von spät gedüngtem ^{15}N in der reifenden Gerstenpflanze unter besonderer Berücksichtigung der Korneiweisse. *Z. Pflanzenernaehr., Dueng., Bodenkd.* **91**, 158–169.

Michael, G., Schumacher, H. and Marschner, H. (1965). Aufnahme von Ammonium- und Nitratstickstoff als markierte Ammoniumnitrate und deren Verteilung in der Pflanze. *Z. Pflanzenernaehr., Dueng., Bodenkd.* **110**, 225–238.

Michael, G., Wilberg, E. and Kouhsiahi-Tork, K. (1969). Durch hohe Luftfeuchtigkeit induzierter Bormangel. *Z. Pflanzenernaehr. Bodenkd.* **122**, 1–3.

Middleton, K. R. and Smith, G. S. (1979). A comparison of ammoniacal and nitrate nutrition of perennial ryegrass throughout a thermodynamical model. *Plant Soil* **53**, 487–504.

Milford, G. F. J., Cormack, W. F. and Durrant, M. J. (1977). Effects of sodium chloride on water status and growth of sugar beet. *J. Exp. Bot.* **28**, 1380–1388.

Miller, G. W., Denney, A., Pushnik, J. and Yu, M. H. (1982). The transformation of delta-amino-levulinate a precursor for chlorophyll, in barley and the role of iron. *J. Plant Nutr.* **5**, 289–300.

Millet, E. and Feldman, M. (1984). Yield response of a common spring wheat cultivar to inoculation with *Azospirillum brasilense* at various levels of nitrogen fertilization. *Plant Soil* **80**, 255–259.

Millet, E., Avavi, Y. and Feldman, M. (1984). Yield response of various wheat genotypes to inoculation with *Azospirillum brasilense. Plant Soil* **80**, 261–266.
Millikan, C. R. (1963). Effects of different levels of zinc and phosphorus on the growth of subterranean clover (*Trifolium subterraneum* L). *Aust. J. Agric. Res.* **14**, 180–205.
Minchin, F. R. and Pate, J. S. (1974). Diurnal fluctioning of the legume root nodule. *J. Exp. Bot.* **25**, 295–308.
Minchin, P. E. H. and Thorpe, M. R. (1982). Evidence of a flow of water into sieve tubes associated with phloem loading. *J. Exp. Bot.* **33**, 233–240.
Mino, Y., Ishida, T., Ota, N., Inoue, M., Nomoto, K., Takemoto, T., Tanaka, H. and Sugiura, Y. (1983). Mugineic acid—iron (III) complex and its structurally analogous cobalt (III) complex: Characterization and implication for absorption and transport of iron in gramineous plants. *J. Am. Chem. Soc.* **105**, 4671–4676.
Minotti, P. L. and Jackson, W. A. (1970). Nitrate reduction in the roots and shoots of wheat seedlings. *Planta* **95**, 36–44.
Mirswa, W. and Ansorge, H. (1981). Einfluss der K-Düngung auf Ertrag und Qualität der Kartoffel. *Arch. Acker- Pflanzenbau Bodenkd.* **25**, 165–171.
Mishra, D. and Kar, M. (1974). Nickel in plant growth and metabolism. *Bot. Rev.* **40**, 395–452.
Mitchell, P. (1966). Chemiosmotic coupling in oxidative and photosynthetic phosphorylation. *Biol. Rev. Cambridge Philos. Soc.* **41**, 445–502.
Mitscherlich, E. A. (1954). "Bodenkunde für Landwirte, Förster und Gärtner," 7th ed. Parey, Berlin.
Mitsui, T., Christeller, J. T., Hara-Nishimura, I. and Akazawa, T. (1984). Possible roles of calcium and calmodulin in the biosynthesis and secretion of α-amylase in rice seed scutellar epithelium. *Plant Physiol.* **75**, 21–25.
Mittelheuser, C. J. and Van Steveninck, R. F. M. (1971). Rapid action of abscisic acid on photosynthesis and stomatal resistance. *Planta* **97**, 83–86.
Mix, G. and Marschner, H. (1974). Mineralstoffverteilung zwischen Chloroplasten und übrigem Blattgewebe. *Z. Pflanzenphysiol.* **73**, 307–312.
Mix, G. P. and Marschner, H. (1976a). Calciumgehalte in Früchten von Paprika, Bohnen, Quitte und Hagebutte im Verlauf des Fruchtwachstums. *Z. Pflanzenernaehr. Bodenkd.* **139**, 537–549.
Mix, G. P. and Marschner, H. (1976b). Einfluss exogener und endogener Faktoren auf den Calciumgehalt von Paprika- und Bohnenfrüchten. *Z. Pflanzenernaehr. Bodenkd.* **139**, 551–563.
Mix, G. P. and Marschner, H. (1976c). Calcium-Umlagerung in Bohnenfrüchten während des Samenwachstums. *Z. Pflanzenphysiol.* **80**, 354–366.
Miyake, Y. and Takahashi, E. (1978). Silicon deficiency of tomato plant. *Soil Sci. Plant Nutr. (Tokyo)* **24**, 175–189.
Miyake, Y. and Takahashi, E. (1983). Effect of silicon on the growth of solution-cultured cucumber plant. *Soil Sci. Plant Nutr. (Tokyo)* **29**, 71–83.
Mizuno, N., Inazu, O. and Kamada, K. (1982). Characteristics of concentrations of copper, iron and carbohydrates in copper deficient wheat plants. *In* "Proceedings of the Ninth International Plant Nutrition Colloquium, Warwick, England" (A. Scaife, ed.), pp. 396–399. Commonw. Agric. Bur., Farnham Royal, Bucks.
Moghimi, A., Tate, M. E. and Oades, J. M. (1978). Characterization of rhizosphere products especially 2-ketogluconic acid. *Soil Biol. Biochem.* **10**, 283–287.
Mohr, H. (1983). Zur Faktorenanalyse des Baumsterbens. *Biol. Unserer Zeit* **13**, 74–78.

Monestiez, M., Lamant, A. and Heller, R. (1982). Endocellular distribution of calcium and Ca-ATPases in horse-bean roots: Possible relation to the ecological status of the plant. *Physiol. Plant.* **55**, 445–452.

Monselise, S. P., Varga, A., Knegt, E. and Bruinsma, J. (1978). Course of the zeatin content in tomato fruits and seeds developing in intact or partially defoliated plants. *Z. Pflanzenphysiol.* **90**, 451–460.

Moore, H. M. and Hirsch, A. M. (1983). Effects of boron deficiency on mitosis and incorporation of tritiated thymidine into nuclei of sunflower root tips. *Am. J. Bot.* **70**, 165–172.

Moore, R. and Black, C. C., Jr. (1979). Nitrogen assimilation pathways in leaf mesophyll and bundle sheath cells of C_4 photosynthetic plants formulated from comparative studies with *Digitaria sanguinalis* (L.). Scap. *Plant Physiol.* **64**, 309–313.

Moraghan, J. T. (1979). Manganese toxicity in flax growing on certain calcareous soils low in available iron. *Soil Sci. Soc. Am. J.* **43**, 1177–1180.

Moraghan, J. T. and Freeman, T. J. (1978). Influence of Fe EDDHA on growth and manganese accumulation in flax. *Soil Sci. Soc. Am. J.* **42**, 455–460.

Morgan, J. M. (1980). Possible role of abscisic acid in reducing seed set in water stressed plants. *Nature (London)* **285**, 655–657.

Morgan, M. A., Volk, R. J. and Jackson, W. A. (1973). Simultaneous influx and efflux of nitrate during uptake by perennial ryegrass. *Plant Physiol.* **51**, 267–272.

Morgan, P. W., Joham, H. E. and Amin, J. U. (1966). Effect of manganese toxicity on indoleacetic and oxidase system of cotton. *Plant Physiol.* **41**, 718–724.

Morgan, P. W., Taylor, D. M. and Joham, H. E. (1976). Manipulation of IAA-oxidase activity and auxine-deficiency symptoms in intact cotton plants with manganese nutrition. *Plant Physiol.* **37**, 149–156.

Morimura, S. and Matsumoto, H. (1978). Effect of aluminium on some properties and template activity of purified pea DNA. *Plant Cell Physiol.* **19**, 429–436.

Morimura, S., Takahashi, E. and Matsumoto, H. (1978). Association of aluminium with nuclei and inhibition of cell division in onion *(Allium cepa)* roots. *Z. Pflanzenphysiol.* **88**, 395–401.

Morré, D. J. and Bracker, C. E. (1976). Ultrastructural alteration of plant plasma membranes induced by auxin and calcium ions. *Plant Physiol.* **58**, 544–547.

Moorby, H. and Nye, P. H. (1984). The effect of temperature variation over the root system on root extension and phosphate uptake by rape. *Plant Soil* **78**, 283–293.

Morrison, R. S., Brooks, R. D., Reeves, R. D., Malaise, F., Horowitz, P., Aronson, M. and Merriam, G. R. (1981). The diverse chemical forms of heavy-metals in tissue extracts of some metallophytes from Shaba province, Zaïre. *Phytochemistry* **20**, 455–458.

Mortvedt, J. J. (1981). Nitrogen and molybdenum uptake and dry matter relationship in soybeans and forage legumes in response to applied molybdenum on acid soil. *J. Plant Nutr.* **3**, 245–256.

Mortvedt, J. J., Fleischfresser, M. H., Berger, K. C. and Darling, H. M. (1961). The relation of some soluble manganese to the incidence of common scab in potatoes. *Am. Potato J.* **38**, 95–100.

Mortvedt, J. J., Berger, K. C. and Darling, H. M. (1963). Effects of manganese and copper on the growth of *Streptomyces scabies* and the incidence of potato scab. *Am. Potato J.* **40**, 96–102.

Mostafa, M. A. E. and Ulrich, A. (1976). Absorption, distribution and form of Ca in relation to Ca deficiency (tip burn) of sugarbeets. *Crop Sci.* **16**, 27–30.
Mothes, K. (1939). Über den Schwefelstoffwechsel der Pflanzen. *Planta* **29**, 67–109.
Mothes, K. (1960). Über das Altern der Blätter und die Möglichkeit ihrer Wiederverjüngung. *Naturwissenschaften* **47**, 337–351.
Mounla, M. A. K., Bangerth, F. and Stoy, V. (1980). Gibberellin-like substances and indole type auxins in developing grains of normal- and high-lysine genotypes of barley. *Physiol. Plant.* **48**, 568–573.
Muchovej, R. M. C. and Muchovej, J. J. (1982). Calcium suppression of *Sclerotium*-induced twin stem abnormality of soybean. *Soil Sci.* **134**, 181–184.
Mueller, P. and Rudin, D. O. (1967). Development of $K^+ - Na^+$ discrimination in experimental bimolecular lipid membranes by macrophilic antibiotics. *Biochem. Biophys. Res. Commun.* **26**, 398–405.
Mueller, W. C. and Beckman, C. H. (1978). Ultrastructural localization of polyphenoloxidase and peroxidase in roots and hypocotyls of cotton seedlings. *Can. J. Bot.* **56**, 1579–1587.
Mugwira, L. M. and Patel, S. U. (1977). Root zone pH changes and ion uptake imbalances by triticale, wheat and rye. *Agron. J.* **69**, 719–722.
Mugwira, L. M., Sapra, V. T., Patel, S. U. and Choudry, M. A. (1981). Aluminium tolerance of triticale and wheat cultivars developed in different regions. *Agron. J.* **73**, 470–475.
Mukherjee. I. (1974). Effect of potassium on proline accumulation in maize during wilting. *Physiol. Plant.* **31**, 288–291.
Mukherji, S., Dey, B., Paul, A. K. and Sircar, S. M. (1971). Changes in phosphorus fractions and phytase activity of rice seeds during germination. *Physiol. Plant.* **25**, 94–97.
Mulder, E. G. (1975). Physiology and ecology of free-living, nitrogen-fixing bacteria. *In* "Nitrogen Fixation by Free-living Micro-Organisms" (W. D. P. Stewart, ed.), pp. 3–28. Cambridge University Press, Cambridge.
Mulette, K. L., Hannon, N. J. and Elliott, A. G. L. (1974). Insoluble phosphorus usage by *Eucalyptus*. *Plant Soil* **41**, 199–205.
Münch, E. (1930). "Die Stoffbewegungen in der Pflanze." Fischer, Jena.
Munk, H. (1982). Zur Bedeutung silikatischer Stoffe bei der Düngung landwirtschaftlicher Kulturpflanzen. *Landwirtsch. Forsch., Sonderh.* **38**, 264–277.
Munns, D. N. (1968a). Nodulation of *Medicago sativa* in solution culture. III. Effects of nitrate on root hairs and infection. *Plant Soil* **29**, 33–47.
Munns, D. N. (1968b). Nodulation of *Medicago sativa* in solution culture. IV. Effects of indole-3-acetate in relation to activity and nitrate. *Plant Soil* **29**, 257–262.
Munns, D. N. (1970). Nodulation of *Medicago sativa* in solution culture. V. Calcium and pH requirements during infection. *Plant Soil* **32**, 90–102.
Munns, D. N., Fox, R. L. and Koch, B. L. (1977). Influence of lime on nitrogen fixation by tropical and temperate legumes. *Plant Soil* **46**, 591–601.
Munns, R., Greenway, H., Delane, R. and Gibbs, J. (1982). Ion concentration and carbohydrate status of the elongating leaf tissue of *Hordeum vulgare* growing at high external NaCl. II. Cause of the growth reduction. *J. Exp. Bot.* **33**, 574–583.
Munns, R., Greenway, H. and Kirst, G. O. (1983). Halotolerant Eukaryotes. *In* "Encyclopedia of Plant Physiology" (O. L. Lange, P. S. Nobel, C. B. Osmond and H. Ziegler, eds.), Vol. 12C, pp. 59–135. Springer-Verlag, Berlin and New York.

Murty, K. S., Smith, T. A. and Bould, C. (1971). The relation between the putrescine content and potassium status of black current leaves. *Ann. Bot. (London)* [N.S.] **35**, 687–695.

Musick, H. B. (1978). Phosphorus toxicity in seedlings of *Larrea divaricata* grown in solution culture. *Bot Gaz. (Chicago)* **139**, 108–111.

Mylonas. V. A. and McCants, C. B. (1980). Effects of humic and fulvic acids on growth of tobacco. I. Root initiation and elongation. *Plant Soil* **54**, 485–490.

Nable, R. O. and Brownell, P. F. (1984). Effect of sodium and light upon the concentrations of alanine in leaves of C_4 plants. *Aust. J. Plant Physiol.* **11**, 319–324.

Nable, R. O. and Loneragan, J. F. (1984). Translocation of manganese in subterranean clover (*Trifolium subterraneum* L. cv. Seaton Park). II. Effects of leaf senescence and of restricting supply of manganese to part of a split root system. *Aust. J. Plant Physiol.* **11**, 113–118.

Nable, R. O., Bar-Akiva, A. and Loneragan, J. F. (1984). Functional manganese requirement and its use as a critical value for diagnosis of manganese deficiency in subterranean clover (*Trifolium subterraneum* L. cv. Seaton Park). *Ann. Bot.* **54**, 39–49.

Nabors, M. W., Gibbs, S.-E., Bernstein, C. S. and Mais, M. E. (1980). NaCl-tolerant tobacco plants from cultured cells. *Z. Pflanzenphysiol.* **97**, 13–17.

Nagarajah, S. and Ulrich, A. (1966). Iron nutrition of the sugar beet plant in relation to growth, mineral balance and riboflavin formation. *Soil Sci.* **102**, 399–407.

Nagarajah, S., Posner, A. M. and Quirk, J. P. (1970). Competitive adsorptions of phosphate with polygalacturonate and other organic anions on kaolinite and oxide surfaces. *Nature (London)* **228**, 83–84.

Naidoo, G., Steward, J. McD. and Lewis, R. J. (1978). Accumulation sites of Al in snapbean and cotton roots. *Agron. J.* **70**, 489–492.

Nair, V. D. and Prenzel, J. (1978). Calculations of equilibrium concentration of mono- and polynuclear hydroxyaluminium species of different pH and total aluminium concentrations. *Z. Pflanzenernaehr. Bodenkd.* **141**, 741–751.

Naito, K., Nagumo, S., Furuya, K. and Suzuki, H. (1981). Effect of benzyladenine on RNA and protein synthesis in intact bean leaves at various stages of ageing. *Physiol. Plant.* **52**, 343–348.

Nambiar, E. K. S. (1976a). Uptake of Zn^{65} from dry soil by plants. *Plant Soil* **44**, 267–271.

Nambiar, E. K. E. (1976b). The uptake of zinc-65 by roots in relation to soil water content and root growth. *Aust. J. Soil Res.* **14**, 67–74.

Nambiar, E. K. S. (1976c). Genetic differences in the copper nutrition of cereals. I. Differential responses of genotypes to copper. *Aust. J. Agric. Res.* **27**, 453–463.

Nambiar, P. T. C., Nigain, S. N., Dart, P. J. and Gibbons, R. W. (1983). Absence of root hairs in non-nodulating groundnut, *Arachis hypogaea* L. *J. Exp. Bot.* **34**, 484–488.

Nandi, A. S. and Sen, S. P. (1981). Utility of some nitrogen fixing microorganisms in the phyllosphere of crop plants. *Plant Soil* **63**, 465–476.

Nason, A., Kaplan, O. and Colowick, S. P. (1951). Changes in enzymatic constitution in zinc-deficient *Neurospora. J. Biol. Chem.* **188**, 397–406.

Nátr, L. (1975). Influence of mineral nutrition on photosynthesis and the use of assimilates. *Photosynth. Prod. Differ. Environ. [Proc. IBP Synth. Meet.], 1973*, Vol. 3, pp. 537–555.

Nayyar, V. K. and Takkar, P. N. (1980). Evaluation of various zinc sources for rice grown on alkali soil. *Z. Pflanzenernaehr. Bodenkd.* **143**, 489–493.

Nelson, L. E. (1983). Tolerance of 20 rice cultivars to excess Al and Mn. *Agron. J.* **75**, 134–138.

Nemeth, K. (1982). Electro-ultrafiltration of aqueous soil suspension with simultaneously varying temperature and voltage. *Plant Soil* **64**, 7–23.

Nemeth, K. (1985). Recent advances in EUF research (1980–1983). *Plant Soil* **83**, 1–19.

Nemeth, K., Makhdhum, I. Q., Koch, K. and Beringer, H. (1979). Determination of categories of soil nitrogen by electro-ultrafiltration (EUF). *Plant Soil* **53**, 445–453.

Ness, P. J. and Woolhouse, H. W. (1980). RNA synthesis in *Phaseolus* chloroplasts. I. Ribonucleic acid synthesis and senescing leaves. *J. Exp. Bot.* **31**, 223–233.

Neumann, K. H. and Steward, F. C. (1968). Investigations on the growth and metabolism of cultured explants of *Daucus carota* I. Effects of iron, molybdenum and manganese on growth. *Planta* **81**, 333–350.

Neumann, P. M. (1982). Late-season foliar fertilisation with macronutrients—Is there a theoretical basis for increased seed yields? *J. Plant Nutr.* **5**, 1209–1215.

Neumann, P. M. and Stein, Z. (1984). Relative rates of delivery of xylem solute to shoot tissues: Possible relationship to sequential leaf senescence. *Physiol. Plant.* **62**, 390–397.

Neumann, P. M., Ehrenreich, Y. and Golab, Z. (1983). Foliar fertilizer damage to corn leaves: Relation to cuticular penetration. *Agron. J.* **73**, 979–982.

Nevins, D. J. and Loomis, R. S. (1970). Nitrogen nutrition and photosynthesis in sugar beet (*Beta vulgaris* L.). *Crop Sci.* **10**, 21–25.

Neyra, C. A. and Hageman, R. H. (1978). Pathway for nitrate assimilation in corn (*Zea mays* L.) leaves. Cellular distribution of enzyme and energy sources for nitrate reduction. *Plant Physiol.* **62**, 618–621.

Nguyen, J. and Feierabend, J. (1978). Some properties and subcellular localization of xanthine dehydrogenase in pea leaves. *Plant Sci. Lett.* **13**, 125–132.

Nicholas, D. J. D., Wilson, P. R., Heinen, W., Palmer, G. and Beinert, H. (1962). Use of electron paramagnetic resonance spectroscopy in investigations of function of metal components in microorganisms. *Nature (London)* **196**, 433–436.

Nicholson, C., III, Stein, J. and Wilson, K. A. (1980). Identification of the low molecular weight copper protein from copper-intoxicated mung bean plants. *Plant Physiol.* **66**, 272–275.

Nishio, J. N. and Terry, N. (1983). Iron nutrition-mediated chloroplast development. *Plant Physiol.* **71**, 688–691.

Nissen, P., Fageria, N. K., Rayar, A. J., Hassan, M. M. and van Hai, T. (1980). Multiphasic accumulation of nutrients by plants. *Physiol. Plant.* **49**, 222–240.

Nitsos, R. E. and Evans, H. J. (1969). Effects of univalent cations on the activity of particulate starch synthetase. *Plant Physiol.* **44**, 1260–1266.

Nösberger, J. and Humphries, E. C. (1965). The influence of removing tubers on drymatter production and net assimilation rate of potato plants. *Ann. Bot. (London)* [N.S.] **29**, 579–588.

Notton, B. A. and Hewitt, E. J. (1979). Structure and properties of higher plant nitrate reductase especially *Spinacia oleracea*. *in* "Nitrogen Assimilation of Plants" (E. J. Hewitt and C. V. Cutting, eds.), pp. 227–244. Academic Press, London and Orlando.

Nyatsanaga, T. and Pierre, W. H. (1973). Effect of nitrogen fixation by legumes on soil acidity. *Agron. J.* **65**, 936–940.

Nye, P. H. (1981). Changes in pH across the rhizosphere induced by roots. *Plant Soil* **61**, 7–26.

Nye, P. H. and Greenland, D. J. (1960). "The Soil under Shifting Cultivation." Commonw. Agric. Bur., Farnham Royal, Bucks.

Nye, P. H. and Tinker, P. B. (1977). "Solute Movements in the Root-Soil System." Blackwell, Oxford.

Oades, J. M. (1978). Mucilages at the root surface. *J. Soil Sci.* **29**, 1–16.

Oaks, A., Wallace, W. and Stevens, D. (1972). Synthesis and turnover of nitrate reductase in corn roots. *Plant Physiol.* **50**, 649–654.

Oaks, A., Aslam, M. and Boesel, I. (1977). Ammonium and amino acids as regulators of nitrate reductase in corn roots. *Plant Physiol.* **59**, 391–394.

O'Conner, G. A., Lindsay, W. L. and Olsen, S. R. (1971). Diffusion of iron and iron chelates in soil. *Soil Sci. Soc. Am. Proc.* **35**, 407–410.

Oertli, J. J. (1962). Loss of boron from plants through guttation. *Soil Sci.* **94**, 214–219.

Oertli, J. J. and Grgurevic, E. (1975). Effect of pH on the absorption of boron by excised barley roots. *Agron. J.* **67**, 278–280.

Oertli, J. J. and Richardson, W. F. (1970). The mechanism of boron immobility in plants. *Physiol. Plant.* **23**, 108–116.

Oertli, J. J. and Roth, J. A. (1969). Boron supply of sugar beet, cotton and soybean. *Agron. J.* **61**, 191–195.

Ogawa, M., Tanaka, K. and Kasai, Z. (1979a). Accumulation of phosphorus, magnesium and potassium in developing rice grains: followed by electron microprobe X-ray analysis focusing on the aleurone layer. *Plant Cell Physiol.* **20**, 19–27.

Ogawa, M., Tanaka, K. and Kasai, Z. (1979b). Energy-dispersive X-ray analysis of phytin globoids in aleurone particles of developing rice grains. *Soil Sci. Plant Nutr. (Tokyo)* **25**, 437–448.

Ohki, K. (1976). Effect of zinc nutrition on photosynthesis and carbonic anhydrase activity in cotton. *Physiol. Plant.* **38**, 300–304.

Ohki, K., Wilson, D. O. and Anderson, O. E. (1981). Manganese deficiency and toxicity sensitivities of soybean cultivar. *Agron. J.* **72**, 713–716.

Ohki, K., Boswell, F. C., Parker, M. B., Shuman, L. M. and Wilson, D. O. (1979). Critical manganese deficiency level of soybean related to leaf position. *Agron. J.* **71**, 233–234.

O'Kelley, J. C. (1969). Mineral nutrition of algae. *Annu. Rev. Plant Physiol.* **19**, 89–112.

Okuda, A. and Takahashi, E. (1965). The role of silicon. *Min. Nutr. Rice Plant, Proc. Symp. Int. Rice Res. Inst., 1964,* pp. 123–146.

Oldenkamp. L. and Smilde, K. W. (1966). Copper deficiency in douglas fir (*Pseudotsuga menziesii* Mirb. Franco). *Plant Soil* **25**, 150–152.

Ollagnier, M. and Renard, J.-L. (1976). The influence of potassium on the resistance of oil palms to *Fusarium. Proc. 12th Colloq. Int. Potash Inst. Bern,* pp. 157–166.

Olsen, R. A. and Brown, J. C. (1980). Factors related to iron uptake by dicotyledonous and monocotyledonous plants. I. pH and reductant. *J. Plant Nutr.* **2**, 629–645.

Olsen, R. A., Bennett, J. H., Blume, D. and Brown, J. C. (1981). Chemical aspects of the Fe stress response mechanism in tomatoes. *J. Plant Nutr.* **3**, 905–921.

O'Neal, D. and Joy, K. W. (1974). Glutamine synthetase of pea leaves. Divalent cation effects, substrate specificity, and other properties. *Plant Physiol.* **54**, 775–779.

O'Neill, S. D. and Spanwick, R. M. (1984). Characterization of native and reconstituted plasma membrane H^+-ATPase from the plasma membrane of *Beta vulgaris*. *J. Membr. Biol.* **79**, 245–256.

O'Neill, S. D., Bennett, A. B. and Spanswick, R. M. (1983). Characterization of a NO_3^- sensitive H^+-ATPase from corn roots. *Plant Physiol.* **72**, 837–846.

Orton, T. J. (1980). Comparison of salt tolerance between *Hordeum vulgare* and *Hordeum jubatum* in whole plants and callus cultures. *Z. Pflanzenphysiol.* **98**, 105–118.

Osmond, C. B. (1967). Acid metabolism in *Atriplex*. I. Regulation in oxalate synthesis by the apparent excess cation absorption. *Aust. J. Biol. Sci.* **20**, 575–587.

O'Sullivan, M. (1971). Aldolase activity in plants as an indicator of zinc deficiency. *J. Sci. Food Agric.* **21**, 607–609.

Oteifa, B. A. and Elgindi, A. Y. (1976). Potassium nutrition of cotton, *Gossypium barbadense,* in relation to nematode infection by *Meliodogyne incognita* and *Rotylenchulus reniformis. Proc. 12th Colloq. Int. Potash Inst. Bern,* pp. 301–306.

Ottow, J. C. G., Benckiser, G., Santiago, S. and Watanabe, I. (1982). Iron toxicity of wetland rice (*Oriza sativa* L.) as a multiple nutritional stress. *In* "Proceedings of the Ninth International Plant Nutrition Colloquium, Warwick, England" (A. Scaife, ed.), pp. 454–460. Commonw. Agric. Bur., Farnham Royal, Bucks.

Overleas, D. (1973). Phytates. *In* "Toxicants Occurring Naturally in Foods" 2nd ed., Chapter 17, pp. 363–371. Natl. Acad. Sci., Washington, D.C.

Ozanne, P. G. (1958). Chlorine deficiency in soils. *Nature (London)* **182**, 1172–1173.

Ozanne, P. G., Woolley, J. T. and Broyer, T. C. (1957). Chlorine and bromine in the nutrition of higher plants. *Aust. J. Biol. Sci.* **10**, 66–79.

Ozanne, P. G., Greenwood, E. A. N. and Shaw, T. C. (1963). The cobalt requirement of subterranean clover in the field. *Aust. J. Agric. Res.* **14**, 39–50.

Ozanne, P. G., Keay, J. and Biddiscombe, E. F. (1969). The comparative applied phosphate requirement of eight annual pasture species. *Aust. J. Biol. Sci.* **20**, 809–818.

Pairunan, A. K., Robson, A. D. and Abbott, L. K. (1980). The effectiveness of vesicular–arbuscular mycorrhiza in increasing growth and phosphorus uptake of subterranean clover from phosphorus sources of different solubilities. *New Phytol.* **84**, 327–338.

Pais, I. (1983). The biological importance of titanium. *J. Plant Nutr.* **6**, 3–131.

Pal, U. R. and Malik, H. S. (1981). Contribution of *Azospirillum brasilense* to the nitrogen needs of grain (*Sorghum bicolor* (L.) Moench) in humid subtropics. *Plant Soil* **63**, 501–504.

Palavan, N. and Galston, A. W. (1982). Polyamine biosynthesis and titer during various development stages of *Phaseolus vulgaris. Physiol. Plant.* **55**, 438–444.

Palmieri, S. and Giovinazzi, F. (1982). Ascorbic acid as negative effector of the peroxidase-catalyzed degradation of indole-3-acetic acid. *Physiol. Plant.* **56**, 1–5.

Palzkill, D. A., Tibbitts, T. W. and Williams, P. H. (1976). Enhancement of calcium transport to inner leaves of cabbage for prevention of tipburn. *J. Am. Soc. Hortic. Sci* **101**, 645–648.

Pandita, M. L. and Andrew, W. T. (1967). A correlation between phosphorus content of leaf tissue and days to maturity in tomato and lettuce. *Proc. Am. Soc. Hortic. Sci.* **91**, 544–549.

Papastylianou, I., Graham, R. D. and Puckridge, D. W. (1982). The diagnosis in wheat by means of a critical nitrate concentration in stem bases. *Commun. Soil Sci. Plant Anal.* **13**, 473–485.

Pardee, A. B. (1967). Crystallization of a sulfate binding protein (permease) from *Salmonella typhimurium. Science* **156**, 1627–1628.

Parfitt, R. L. (1979). The availability of P from phosphate-goethite bridging complexes. Description and uptake by ryegrass. *Plant Soil* **53**, 55–65.

Parker, M. B. and Harris, H. B. (1977). Yield and leaf nitrogen of nodulating soybeans as affected by nitrogen and molybdenum. *Agron. J.* **69**, 551–554.

Parker, M. B., Gascho, G. J. and Gaines, T. P. (1983). Chloride toxicity of soybeans grown on Atlantic coast flatwoods soils. *Agron. J.* **75**, 439–443.

Parry, D. W. and Hodson, M. J. (1982). Silica distribution in the caryopsis and inflorescence bracts of foxtail millet (*Setaria italica* L. Beauv.) and its possible significance in carcinogenesis. *Ann. Bot. (London)* [N.S.] **49**, 531–540.

Parry, D. W. and Kelso, M. (1975). The distribution of silicon deposits in the root of *Molinia caerulea* (L.) Moench and *Sorghum bicolor* (L.) Moench. *Ann. Bot. (London)* [N.S.] **39**, 995–1001.

Parry, D. W. and Smithson, F. (1964). Types of opaline silica deposition in the leaves of British grasses. *Ann. Bot. (London)* [N.S.] **28**, 169–185.

Parthier, B. (1979). The role of phytohormones (cytokinin) in chloroplast development. *Biochem. Physiol. Pflanz.* **144**, 173–214.

Pasricha, N. S., Nayyar, V. K., Randhawa, N. S. and Sinha, M. K. (1977). Influence of sulphur fertilization on suppression of molybdenum uptake by berseem *(Trifolium alexandrinum)* and oats *(Avena sativa)* grown on a molybdenum-toxic soil. *Plant Soil* **46**, 245–250.

Passioura, J. B. and Leeper, G. W. (1963). Soil compaction and manganese deficiency. *Nature (London)* **200**, 29–30.

Pate, J. S. (1973). Uptake, assimilation and transport of nitrogen compounds by plants. *Soil Biol. Biochem.* **1**, 109–119.

Pate, J. S. (1975). Exchange of solutes between phloem and xylem and circulation in the whole plant. *In* "Encyclopedia of Plant Physiology, New Series" (M. H. Zimmermann and J. A. Milburn, eds.), Vol. 1, pp. 451–468. Springer-Verlag, Berlin and New York.

Pate, J. S. and Atkins, C. A. (1983). Xylem and phloem transport and the functional economy of carbon and nitrogen of a legume leaf. *Plant Physiol.* **71**, 835–840.

Pate, J. S. and Gunning, B. E. S. (1972). Transfer cells. *Annu. Rev. Plant Physiol.* **23**, 173–196.

Pate, J. S. and Herridge, D. F. (1978). Partitioning and utilization of net photosynthate in nodulated annual legumes. *J. Exp. Bot.* **29**, 401–412.

Pate, J. S., Wallace, W. and Die, J. van (1964). Petiole bleeding sap in the examination of the circulation of nitrogenous substances in plants. *Nature (London)* **204**, 1073–1074.

Pate, J. S., Sharkey, P. J. and Lewis, O. A. M. (1974). Xylem to phloem transfer of solutes in fruiting shoots of legumes, studied by a phloem bleeding technique. *Planta* **122**, 11–26.

Pate, J. S., Kuo, J. and Hocking, P. J. (1978). Functioning of conducting elements of phloem and xylem in the stalk of the developing fruit of *Lupinus albus* L. *Aust. J. Plant Physiol.* **5**, 321–326.

Pate, J. S., Layzell, D. B. and Atkins, C. A. (1979). Economy of carbon and nitrogen in a nodulated and nonnodulated (NO_3-grown) legume. *Plant Physiol.* **64**, 1083–1088.
Pate, J. S., Atkins, C. A., White, S. T., Rainbird, R. M. and Woo, K. C. (1980). Nitrogen nutrition and xylem transport of nitrogen in ureide-producing grain legumes. *Plant Physiol.* **65**, 961–965.
Patrick, J. W. (1984). Photosynthate unloading from seed coats of *Phaseolus vulgaris* L. Control by tissue water relations. *J. Plant Physiol.* **115**, 297–310.
Patrick, Z. A. (1971). Phytotoxic substances associated with the decomposition in soil of plant residues. *Soil Sci.* **111**, 13–18.
Pauls, K. P. and Thompson, J. E. (1984). Evidence for the accumulation of peroxidized lipids in membranes of senescing cotyledons. *Plant Physiol.* **75**, 1152–1157.
Pearson, C. J. and Steer, B. T. (1977). Daily changes in nitrate uptake and metabolism in *Capsicum annuum. Planta* **137**, 107–112.
Pearson, C. J., Volk, R. J. and Jackson, W. A. (1981). Daily changes in nitrate influx, efflux and metabolism in maize and pearl millet. *Planta* **152**, 319–324.
Pearson, R. W., Radcliffe, L. F. and Taylor, H. M. (1970). Effect of soil temperature, strength and pH on cotton seedlings root elongation. *Agron. J.* **62**, 243–246.
Pearson, R. W., Childs, J. and Lund, Z. F. (1973). Uniformity of limestone mixing in acid subsoil as a factor in cotton root penetration. *Soil Sci. Soc. Am. Proc.* **37**, 727–732.
Peck, N. H., Grunes, D. L., Welch, R. M. and MacDonald, G. E. (1980). Nutritional quality of vegetable crops as affected by phosphorus and zinc fertilizer. ¼Agron. J. **72**, 528–534.
Peel, A. J. and Rogers, S. (1982). Stimulation of sugar loading into sieve elements of willow by potassium and sodium salts. *Planta* **154**, 94–96.
Peirson, D. R. and Elliot, J. R. (1981). In vivo nitrite reduction in leaf tissue of *Phaseolus vulgaris. Plant Physiol.* **68**, 1068–1072.
Peoples. M. B., Pate, J. S. and Atkins, C. A. (1983). Mobilization of nitrogen in fruiting plants of a cultivar of cowpea. *J. Exp. Bot.* **34**, 563–578.
Peoples, T. R. and Koch, D. W. (1979). Role of potassium in carbon dioxide assimilation in *Medicago sativa* L. *Plant Physiol.* **63**, 878–881.
Perkins, H. J. and Aronoff, S. (1956). Identification of the blue fluorescent compounds in boron deficient plants. *Arch. Biochem. Biophys.* **64**, 506–507.
Perrenoud, S. (1977). Potassium and plant health. *In* "Research Topics", No.3, pp. 1–118. Int. Potash Inst., Bern, Switzerland.
Perret, P. and Koblet, W. (1981). Nachweis höherer Äthylengehalte in der Bodenluft eines von der Verdichtungschlorose befallenen Rebberges. *Vitis* **20**, 320–328.
Persson, H. (1979). Fine root production, mortality, and decomposition in forest ecosystems. *Vegetatio* **41**, 101–109.
Persson, L. (1969). Labile-bound sulfate in wheat-roots: Localization, nature and possible connection to active absorption mechanism. *Physiol. Plant.* **22**, 959–977.
Perur, N. G., Smith R. L. and Wiebe, H. H. (1961). Effect of iron chlorosis on protein fraction on corn leaf tissue. *Plant Physiol.* **36**, 736–739.
Peterson, F. J. and Butler, G. W. (1967). Significance of selenocystathionine in an Australian selenium-accumulating plant, *Neptunia amplexicaulis. Nature (London)* **213**, 599–600.
Petit, C. M. and Geijn, S. C. van de (1978). In vivo measurement of cadmium (^{115m}Cd) transport and accumulation in the stems of intact tomato plants

(*Lycopersicon esculentum,* Mill.). I. Long distance transport and local accumulation. *Planta* **138**, 137–143.

Petit, C. M., Ringoet, A. and Myttenaere, C. (1978). Stimulation of cadmium uptake in relation to the cadmium content of plants. *Plant Physiol.* **62**, 554–557.

Peverley, J. H., Adamec, J. and Parthasarathy, M. V. (1978). Association of potassium and some other monovalent cations with occurrence of polyphosphate. *Plant Physiol.* **62**, 120–126.

Pflüger, R. and Cassier, A. (1977). Influence of monovalent cations on photosynthetic CO_2 fixation. *Proc. 13th Colloq. Int. Potash Inst. Bern.* pp. 95–100.

Pflüger, R. and Wiedemann, R. (1977). Der Einfluss monovalenter Kationen auf die Nitratreduktion von *Spinacia oleracea* L. *Z. Pflanzenphysiol.* **85**, 125–133.

Pilbeam, D. J. and Kirkby, E. A. (1983). The physiological role of boron in plants. *J. Plant Nutr.* **6**, 563–582.

Pilet, P. E. (1981). Root growth and gravireaction: Endogenous hormone balance. *In* "Structure and Function of Plant Roots" (R. Brouwer, O. Gasparikova, J. Kolek and B. E. Loughman, eds.), pp. 89–93. Martin Nijhoff/Junk Publ., The Hague.

Pill, W. G., Lambeth, V. N. and Hinckley, T. M. (1978). Effects of nitrogen forms and level on ion concentrations, water stress, and blossom-end rot incidence in tomato. *J. Am. Soc. Hortic. Sci.* **1031, 265–268.**

Pingel, U. (1976). Der Einfluss phenolischer Aktivatoren und Inhibitoren der IES-Oxidase-Aktivität auf die Adventivbewurzelung bei *Tradescantia albiflora. Z. Pflanzenphysiol.* **79**, 109–120.

Pirson, A. (1937). Ernaehrungs- und stoffwechselphysiologische Untersuchungen an *Frontalis* und *Chlorella. Z. Bot.* **31**, 193–267.

Pissarek, H. P. (1973). Zur Entwicklung der Kalium-Mangelsymptome von Sommerraps. *Z. Pflanzenernaehr. Bodenkd.* **136**, 1–19.

Pissarek, H. P. (1974). Untersuchungen der durch Kupfermangel bedingten anatomischen Veränderungen bei Hafer- und Sonnenblumen. *Z. Pflanzenernaehr. Bodenkd.* **137**, 224–234.

Pissarek, H. P. (1979). Der Einfluss von Grad und Dauer des Mg-Mangels auf den Kornertrag von Hafer. *Z. Acker- Pflanzenbau* **148**, 62–71.

Pissarek, H. P. (1980). Makro- und Mikrosymptome des Bormangels bei Sonnenblumen, Chinakohl und Mais. *Z. Pflanzenernaehr. Bodenkd.* **143**, 150–160.

Pitman, M. G. (1972a). Uptake and transport of ions in barley seedlings. II. Evidence for two active stages in transport to the shoot. *Aust. J. Biol. Sci.* **25**, 243–257.

Pitman, M. G. (1972b). Uptake and transport of ions in barley seedlings. III. Correlation between transport to the shoot and relative growth rate. *Aust. J. Biol. Sci.* **25**, 905–919.

Pitman, M. G., Mowat, J. and Nair, H. (1971). Interactions of processes for accumulation of salt and sugar in barley plants. *Aust. J. Biol. Sci.* **24**, 619–631.

Pitman, M. G., Lüttge, U., Läuchli, A. and Ball, E. (1974). Ion uptake to slices of barley leaves, and regulation of K content in cells of the leaves. *Z. Pflanzenphysiol.* **72**, 75–88.

Pitman, M. G., Schaefer, N. and Wildes, R. A. (1975). Relation between permeability to potassium and sodium ions and fusicoccin-stimulated hydrogen ion efflux in barley roots. *Planta* **126**, 61–73.

Platero, M. and Tejerina, G. (1976). Calcium nutrition in *Phaseolus vulgaris* in relation to its resistance to *Erwinia carotavora. Phytopathol. Z.* **85**, 314–319.

Platt-Aloia, K. A., Thomson, W. W. and Terry, N. (1983). Changes in plastid ultrastructure during iron nutrition-mediated chloroplast development. *Protoplasma* **114**, 85–92.

Plenchette, C., Fortin, J. A. and Furlan, V. (1983). Growth response of several plant species to mycorrhizae in a soil of moderate P-fertility. II. Soil fumigation induced stunting of plants corrected by reintroduction of the wild endomycorrhizal flora. *Plant Soil* **70**, 211–217.

Pohlman, A. A. and McColl, J. G. (1982). Nitrogen fixation in the rhizosphere and rhizoplane of barley. *Plant Soil* **69**, 341–352.

Polacco, J. C. (1977). Nitrogen metabolism in soybean tissue culture. II. Urea utilization and urease synthesis require Ni^{2+}. *Plant Physiol.* **59**, 827–830.

Polar, E. (1975). Zinc in pollen and its incorporation into seeds. *Planta* **123**, 97–103.

Poljakoff-Mayber, A. (1975). Morphological and anatomical changes in plants as a response to salinity stress. *In* "Plants in Saline Environments" (A. Poljakoff-Mayber and J. Gale, eds.), pp. 97–117. Springer-Verlag, Berlin and New York.

Pollard, A. S. and Wyn Jones, R. G. (1979). Enzyme activities in concentrated solutions of glycinebetaine and other solutes. *Planta* **144**, 291–298.

Pollard, A. S., Parr, A. J. and Loughman, B. C. (1977). Boron in relation to membrane function in higher plants. *J. Exp. Bot.* **28**, 831–841.

Ponnamperuma, F. N. (1972). The chemistry of submerged soils. *Adv. Agron.* **24**, 29–96.

Ponnamperuma, F. N. (1982). Genotypic adaptability as a substitute for amendments on toxic and nutrient-deficient soils. *In* "Proceedings of the Ninth International Plant Nutrition Colloquium, Warwick, England" (A. Scaife, ed.), pp. 467–473. Commonw. Agric. Bur., Farnham Royal, Bucks.

Poole, R. J. (1978). Energy coupling for membrane transport. *Annu. Rev. Plant Physiol.* **29**, 437–460.

Poovaiah, B. W. (1979). Role of calcium in ripening and senescence. *Commun. Soil Sci. Plant Anal.* **10**, 83–88.

Poovaiah, B. W. and Leopold, A. C. (1973). Deferral of leaf senescence with calcium. *Plant Physiol.* **52**, 236–239.

Portis, A. R., Jr. (1981). Evidence of a low stromal Mg^{2+} concentration in intact chloroplasts in the dark. I. Studies with the ionophore A 23187. *Plant Physiol.* **67**, 985–989.

Portis, A. R., Jr. (1982). Effects of the relative extra-chloroplastic concentrations of inorganic phosphate, 3-phosphoglycerate and dihydroxyacetone phosphate on the rate of starch synthesis in isolated spinach chloroplasts. *Plant Physiol.* **70**, 393–396.

Portis, A. R., Jr. and Heldt, H. W. (1976). Light-dependent changes of the Mg^{2+} concentration in the stroma in relation to the Mg^{2+} depending of CO_2 fixation in intact chloroplasts. *Biochim. Biophys. Acta.* **449**, 434–446.

Possingham, J. V. (1954). The effect of molybdenum on the organic acids and inorganic phosphorus of plants. *Aust. J. Biol. Sci.* **7**, 221–224.

Possingham, J. V. (1957). The effect of mineral nutrition on the content of free amino acids and amides in tomato plants. I. A comparison of the effects of deficiencies of copper, zinc, manganese, iron, and molybdenum. *Aust. J. Biol. Sci.* **10**, 539–551.

Possingham, J. V., Vesk, M. and Mercer, F. V. (1964). The fine structure of leaf cells of manganese deficient spinach. *J. Ultrastruct. Res.* **11**, 68–83.

Poston, J. M. (1978). Coenzyme B_{12}-dependent enzymes in potato: leucine 2,3-aminomutase and methylmalomyl-CoA mutase. *Phytochemistry* **17**, 401–402.

Powell, C. L. (1980). Effect of phosphate fertilizers on the production of mycorrhizal inoculum in soil. *N.Z. J. Agric. Res.* **23**, 219–233.

Powell, P. E., Szaniszlo, P. J., Cline, G. R. and Reid, C. P. P. (1982). Hydroxamate siderophores in the iron nutrition of plants. *J. Plant Nutr.* **5**, 653–673.

Powrie, J. K. (1964). The effect of young lucerne on a siliceous sand. *Plant Soil* **21**, 81–93.

Pradet, A. and Raymond, P. (1983). Adenine nucleotide ratios and adenylate energy charge in energy metabolism. *Annu. Rev. Plant Physiol.* **34**, 199–224.

Prask, J. A. and Plocke, D. J. (1971). A role of zinc in the structural integrity of the cytoplasmic ribosomes of *Euglena gracilis. Plant Physiol.* **48**, 150–155.

Preusser, E., Khalil, F. A. and Göring. H. (1981). Regulation of activity of the granule-bound starch synthetase by monovalent cations. *Biochem. Physiol. Pflanz.* **176**, 744–752.

Prins, W. H. (1983). Effect of a wide range of nitrogen applications on the herbage nitrate content in long-term fertilizer trials on all-grass swards. *Fert. Res.* **4**, 101–113.

Prohaska, K. R. and Fehr, W. R. (1981). Recurrent selection for resistance to iron deficiency in soybean. *Crop Sci. 321*, 524–526.

Prummel, J. (1975). Effect of soil structure on phosphate nutrition of crop plants. *Neth. J. Agric. Sci.* **23**, 62–68.

Purves, D. and Mackenzie, E. J. (1974). Phytotoxicity due to boron in municipal compost. *Plant Soil* **40**, 231–235.

Queen, W. H. (1967). Water and ^{32}P entrance through grape roots. *Plant Physiol.* **42**, 5–18.

Quick, W. A. and Li, P. L. (1976). Phosphate balance in potato tubers. *Potato Res.* **19**, 305–312.

Quispel, A. (1983). Dinitrogen-fixing symbioses with legumes, non-legume angiosperms and associative symbioses. *In* "Encyclopedia of Plant Physiology, New Series" (A. Läuchli and R. L. Bieleski, eds.), Vol. 15A, pp. 286–329. Springer-Verlag, Berlin and New York.

Rademacher, W. (1978). Gaschromatographische Analyse der Veränderungen im Hormongehalt des wachsenden Weizenkorns. Dissertation, Universtität Göttingen.

Radin, J. W. (1978). A physiological basis for the division of nitrate assimilation between roots and leaves. *Plant Sci. Lett.* **13**, 21–25.

Radin, J. W. (1983). Control of plant growth by nitrogen: Differences between cereals and broadleaf species. *Plant, Cell Environ.* **6**, 65–68.

Radin, J. W. (1984). Stomatal responses to water stress and to abscisic acid in phosphorus-deficient cotton plants. *Plant Physiol.* **76**, 392–394.

Radin, J. W. and Ackerson, R. C. (1981). Water relations of cotton plants under nitrogen deficiency. III. Stomatal conductance. *Plant Physiol.* **67**, 115–119.

Radin, J. W. and Boyer, J. S. (1982). Control of leaf expansion by nitrogen nutrition in sunflower plants: Role of hydraulic conductivity and turgor. *Plant Physiol.* **69**, 771–775.

Radin, J. W. and Eidenbock, M. P. (1984). Hydraulic conductance as a factor limiting leaf expansion of phosphorus-deficient cotton plants. *Plant Physiol.* **75**, 372–377.

Radin, J. W. and Parker, L. L. (1978). Water relation of cotton plants under nitrogen deficiency. I. Dependence upon leaf structure. *Plant Physiol.* **64**, 495–498.

Radin, J. W. and Parker, L. L. (1979). Water relations of cotton plants under nitrogen deficiency. *Plant Physiol.* **64**, 495–498.

Radin, J. W., Parker, L. L. and Guinn, G. (1982). Water relations of cotton plants under nitrogen deficiency. V. Environmental control of abscisic acid accumulation and stomatal sensitivity to abscisic acid. *Plant Physiol.* **70**, 1066–1070.

Radley, M. (1978). Factors affecting grain enlargement in wheat. *J. Exp. Bot.* **29**, 919–934.

Raghavendra, A. A., Rao, J. M. and Das, V. S. R. (1976). Replaceability of potassium by sodium for stomatal opening in epidermal strips of *Commelina benghalensis*. *Z. Pflanzenphysiol.* **80**, 36–42.

Rahimi, A. (1970). Kupfermangel bei höheren Pflanzen. *Landwirtsch. Forsch., Sonderh.* **25**(I), 42–47.

Rahimi, A. and Bussler, W. (1973a). Die Diagnose des Kupfermangels mittels sichtbarer Symptome an höheren Pflanzen. *Z. Pflanzenernaehr. Bodenkd.* **135**, 267–283.

Rahimi, A. and Bussler, W. (1973b). Physiologische Voraussetzungen für die Bildung der Kupfermangelsymptome. *Z. Pflanzenernaehr. Bodenkd.* **136**, 25–32.

Rahimi, A. and Bussler, W. (1974). Kuperfermangel bei höheren Pflanzen und sein histochemischer Nachweis. *Landwirtsch. Forsch., Sonderh.* **30**(II), 101–111.

Rahimi, A. and Bussler, W. (1978). Makro- und Mikrosymptome des Zinkmangels bei höheren Pflanzen. *Z. Pflanzenernaehr. Bodenkd.* **141**, 567–581.

Rahimi, A. and Bussler, W. (1979). Die Entwicklung und der Zn-, Fe- und P-Gehalt höherer Pflanzen in Abhängigkeit vom Zinkangebot. *Z. Pflanzenernaehr. Bodenkd.* **142**, 15–27.

Rahimi, A. and Schropp, A. (1984). Carboanhydraseaktivität und extrahierbares Zink als Maßstab für die Zink-Versorgung von Pflanzen. *Z. Pflanzenernaehr. Bodenkd.* **147**, 572–583.

Rahman, M. S. and Wilson, J. H. (1977). Effect of phosphorus applied as superphosphate on rate of development and spikelet number per ear in different cultivars of wheat. *Aust. J. Agric. Res.* **28**, 183–186.

Raij, B. van and Diest, A. van (1979). Utilization of phosphate from different sources by six plant species. *Plant Soil* **51**, 577–589.

Railton, I. D. and Wareing, P. F. (1973). Effects of daylength on endogenous gibberellins in leaves of *Solanum andigenum*. I. Changes in levels of free acidic gibberellin-like substances. *Physiol. Plant.* **28**, 88–94.

Rainbird, R. M., Atkins, C. A. and Pate, J. S. (1983). Diurnal variation in the functioning of cowpea nodules. *Plant Physiol.* **72**, 308–312.

Rains, D. W. (1968). Kinetics and energetics of light-enhanced potassium absorption by corn leaf tissue. *Plant Physiol.* **431, 394–400.**

Rains, D. W. (1969). Cation absorption by slices of stem tissues of bean and cotton. *Experientia* **25**, 215–216.

Rajaratnam, J. A. and Hock, L. I. (1975). Effect of boron nutrition on intensity of red spider mite attack on oil palm seedlings. *Exp. Agric.* **11**, 59–63.

Rajaratnam, J. A. and Lowry, J. B. (1974). The role of boron in the oil-palm *(Elaeis guineensis). Ann. Bot. (London)* [N.S.] **38**, 193–200.

Ramshorn, K. (1958). Zur partiellen "aeroben" Gärung in den Wurzeln von *Vicia faba*. II. *Flora (Jena), Abt. A* **146**, 178–211.

Randall, P. J. (1969). Changes in nitrate and nitrate reductase levels on restoration of molybdenum to molybdenum-deficient plants. *Aust. J. Agric. Res.* **20**, 635–642.

Randall, P. J. and Bouma, D. (1973). Zinc deficiency, carbonic anhydrase, and photosynthesis in leaves of spinach. *Plant Physiol.* **52**, 229–232.
Rao, N. R., Naithani, S. C., Jasdanwala, R. T. and Singh, Y. D. (1982). Changes in indoleacetic acid oxidase and peroxidase activities during cotton fibre development. *Z. Pflanzenphysiol.* **106**, 157–166.
Rao, V. R. and Rao, J. L. N. (1984). Nitrogen fixation (C_2H_4 reduction) in soil samples from rhizosphere of rice grown under alternate flooded and nonflooded conditions. *Plant Soil* **81**, 111–118.
Raschke, K. and Schnabl, H. (1978). Availability of chloride affects the balance between potassium chloride cells of *Vicia faba* L. *Plant Physiol.* **62**, 84–87.
Raskin, J. and Kende, H. (1983). How does deep water rice solve its aeration problem? *Plant Physiol.* **72**, 447–454.
Rasmussen, H. P. (1968). Entry and distribution of aluminium in *Zea mays*. The mode of entry and distribution of aluminium in *Zea mays*: Electron microprobe-X-ray analysis. *Planta* **81**, 28–37.
Rasmussen, P. E., Ramig, R. E., Ekin, L. G. and Rhode, C. R. (1977). Tissue analyses guidelines for diagnosing sulfur deficiency in white wheat. *Plant Soil* **46**, 153–163.
Ratnayake, M., Leonard, R. T. and Menge, A. (1978). Root exudation in relation to supply of phosphorus and its possible relevance to mycorrhizal infection. *New Phytol.* **81**, 543–552.
Rauser, W. E. (1983). Estimating thiol-rich copper-binding protein in small root samples. *Z. Pflanzenphysiol.* **112**, 69–77.
Rauser, W. E. (1984). Isolation and partial purification of cadmium-binding protein from roots of the grass *Agrostis gigantea. Plant Physiol.* **74**, 1025–1029.
Rauser, W. E. and Curvetto, N. R. (1980). Metallothionein occurs in roots of *Agrostis* tolerant to excess copper. *Nature (London)* **287**, 563–564.
Raven, J. A. (1977). H^+ and Ca^{2+} in phloem and symplast: Relation of relative immobility of the ions to the cytoplasmic nature of the transport path. *New Phytol.* **79**, 465–480.
Raven, J. A. (1980). Short- and long-distance transport of boric acid in plants. *New Phytol.* **84**, 231–249.
Raven, J. A. (1983). The transport and function of silicon in plants. *Biol. Rev. Cambridge Philos. Soc.* **58**(2), 179–207.
Raven, J. A. and Smith, F. A. (1976). Nitrogen assimilation and transport in vascular land plants in relation to intracellular pH regulation. *New Phytol.* **76**, 415–431.
Ray, T. B. and Black, C. C. (1979). The C4 pathway and its regulation. *In* "Encyclopedia of Plant Physiology, New Series" (M. Gibbs and E. Latzko, eds.), Vol. 6, pp. 77–101. Springer-Verlag, Berlin and New York.
Rebeille, F., Bligny, R. and Douce, R. (1984). Is the cytosolic P_i concentration a limiting factor for plant cell respiration? *Plant Physiol.* **74**, 355–359.
Reddy, D. T. and Raj, A. S. (1975). Cobalt nutrition of groundnut in relation to growth and yield. *Plant Soil* **42**, 145–152.
Reddy, S. V. K. and Venkaiah, B. (1984). Subcellular localization and identification of superoxide dismutase isoenzymes from *Pennisetum typhoideum* seedlings. *J. Plant Physiol.* **116**, 81–85.
Reid, C. P. P., Kidd, F. A. and Ekwebelam, S. A. (1983). Nitrogen nutrition, photosynthesis and carbon allocation in ectomycorrhizal pine. *Plant Soil* **71**, 415–432.

Reid, D. M. and Railton, I. D. (1974). Effect of flooding on the growth of tomato plants: Involvement of cytokinins and gibberellins. *In* "Mechanisms of Regulation of Plant Growth" (R. L. Bieleski *et al.*, eds.), Cresswell Bull. No. 12, pp. 789–792. R. Soc. N.Z., Wellington.

Reid, D. M., Crozier, A. and Harvey, B. M. R. (1969). The effects of flooding on the export of gibberellins from the root to the shoot. *Planta* **89**, 376–379.

Reid, C. P. P., Crowley, D. E., Kim, H. J., Powell, P. E. and Szaniszlo, P. J. (1984a). Utilization of iron by oat when supplied as ferrated synthetic chelate or as ferrated hydroxamate siderophore. *J. Plant Nutr.* **7**, 437–447.

Reid, R. K., Reid, C. P. P., Powell, P. E. and Szaniszlo, P. J. (1984b). Comparison of siderophore concentrations in aqueous extracts of rhizosphere and adjacent bulk soils. *Pedobiologia* **26**, 263–266.

Reinhold, J. G., Nasr, K., Lahimgarzadeh, A. and Hedayati, H. (1973). Effects of purified phytate and phytate-rich bread upon metabolism of zinc, calcium, phosphorus and nitrogen in man. *Lancet* **I**, 283–291.

Reiss, H. D. and Herth, W. (1978). Visualization of the Ca^{2+}-gradient in growing pollen tubers of *Lilium longiflorum* with chlorotetracycline fluorescence. *Protoplasma* **97**, 373–378.

Reiss, H. D. and Herth, W. (1979). Calcium ionophore A 23187 affects localized wall secretion in tip region of pollen tubers of *Lilium longiflorum. Planta* **145**, 225–232.

Renger, M., Strebel, O., Grimme, H. and Fleige, H. (1981). Nährstoffanlieferung an die Pflanzenwurzel durch Massenfluss. *Mitt. Dtsch. Bodenkd. Ges.* **30**, 63–70.

Rennenberg, H. (1982). Glutathione metabolism and possible biological roles in higher plants. *Phytochemistry* **21**, 2771–2781.

Rennenberg, H., Schmitz, K. and Bergmann, L. (1979). Long-distance transport of sulfur in *Nicotiana tabacum. Planta* **147**, 57–62.

Rensing, L. and Cornelius, G. (1980). Biologische Membranen als Komponenten oszillierender Systeme. *Biol. Rundsch.* **18**, 197–209.

Reuter, D. J., Robson, A. D., Loneragan, J. F. and Tranthim-Fryer, D. J. (1981). Copper nutrition of subterranean clover (*Trifolium subterraneum* L. cv. Seaton Park). II. Effects of copper supply on distribution of copper and the diagnosis of copper deficiency by plant analysis. *Aust. J. Agric. Res.* **32**, 267–282.

Rheinbaben, W. von and Trolldenier, G. (1983). Influence of mineral nutrition on denitrification on plants in nutrient solution culture. *Z. Pflanzenernaehr. Bodenkd.* **146**, 180–187.

Rheinbaben, W. von and Trolldenier, G. (1984). Influence of plant growth on denitrification in relation to soil moisture and potassium nutrition. *Z. Pflanzenernaehr. Bodenkd.* **147**, 730–738.

Rhodes, L. H. and Gerdemann, J. W. (1975). Phosphate uptake zones of mycorrhizal and non-mycorrhizal onions. *New Phytol.* **75**, 555–561.

Rhue, R. D. and Grogan, C. O. (1977). Screening corn for Al tolerance using different Ca and Mg concentrations. *Agron. J.* **69**, 755–760.

Richards, B. N. and Bevege, D. I. (1969). Critical foliage concentrations of nitrogen and phosphorus as a guide to the nutrient status of *Araucaria* underplanted to *Pinus. Plant Soil* **31**, 328–336.

Richards, D. (1981). Root-shoot interactions in fruiting tomato plants. *In* "Structure and Function of Plant Roots" (P. Brouwer, O. Gasparikova, J. Kolek and B. C. Loughman, eds.), pp. 373–380. Martinus Nijhoff/Junk, The Hague.

Richards, S. J., Hagan, R. M. and McCalla, T. M. (1952). Soil temperature and plant growth. *In* "Soil Physical Conditions and Plant Growth" (B. T. Shaw, ed.), pp. 303–480. Academic Press, London and Orlando.

Richmond, E. A. and Lang, A. (1957). Effect of kinetin on protein content and survival of detached *Xanthium* leaves. *Science* **125**, 650–651.

Rigney, C. J. and Wills, R. B. H. (1981). Calcium movement, a regulating factor in the initiation of tomato fruit ripening. *HortScience* **16**, 550–551.

Rijven, A. H. G. C. and Gifford, R. M. (1983). Accumulation and conversion of sugars by developing wheat grains. IV. Effects of phosphate and potassium ions in endosperm slices. *Plant, Cell Environ.* **6**, 625–631.

Riley, D. and Barber, S. A. (1971). Effect of ammonium and nitrate fertilization on phosphorus uptake as related to root-induced pH changes at the root-soil interface. *Soil Sci. Soc. Am. Proc.* **35**, 301–306.

Rivera, C. M. and Penner, D. (1978). Effect of calcium and nitrogen on soybean *(Glycine max)* root fatty acid composition and uptake of linuron. *Weed Sci.* **26**, 647–650.

Robards, A. W., Jackson, S. M., Clarkson, D. T. and Sanderson, J. (1973). The structure of barley roots in relation to the transport of ions into the stele. *Protoplasma* **77**, 291–311.

Roberts, J. K. M., Linker, C. S., Benoit, A. G., Jardetzky, O. and Nieman, R. H. (1984). Salt stimulation of phosphate uptake in maize root tips studied by ^{31}P nuclear magnetic resonance. *Plant Physiol.* **75**, 947–950.

Robertson, G. A. and Loughman, B. C. (1974a). Reversible effects of boron on the absorption and incorporation of phosphate in *Vicia faba* L. *New Phytol.* **73**, 291–298.

Robertson, G. A. and Loughman, B. C. (1974b). Response to boron deficiency: A comparison with responses produced by chemical methods of retarding root elongation. *New Phytol.* **73**, 821–832.

Robinson, P. W. and Hodges, C. F. (1981). Nitrogen-induced changes in the sugars and amino acids of sequentially senescing leaves of *Poa pratensis* and phathogenesis by *Drechslera sorokiniana. Phytopathol. Z.* **101**, 348–361.

Robinson, S. P. and Downton, W. J. S. (1984). Potassium, sodium, and chloride content of isolated intact chloroplasts in relation to ionic compartmentation in leaves. *Arch. Biochem. Biophys.* **228**, 197–206.

Robinson, S. P., Downton, W. J. S. and Millhouse, J. A. (1983). Photosynthesis and ion content of leaves and isolated chloroplasts of salt-stressed spinach. *Plant Physiol.* **73**, 238–242.

Robson, A. D. and Loneragan, J. F. (1970a). Nodulation and growth of *Medicago truncatula* on acid soils. I. Effect of $CaCO_3$ and inoculation level on the nodulation of *M. truncatula* on a moderately acid soil. *J. Agric. Res.* **21**, 427–434.

Robson, A. D. and Loneragan, J. F. (1970b). Nodulation and growth of *Medicago truncatula* on acid soils. II. Colonization of acid soils by *Rhizobium meliloti. Aust. J. Agric. Res.* **21**, 435–445.

Robson, A. D. and Mead, G. R. (1980). Seed cobalt in *Lupinus angustifolius. Aust. J. Agric. Res.* **31**, 109–116.

Robson, A. D. and Pitman, M. G. (1983). Interactions between nutrients in higher plants. *In* "Encyclopedia of Plant Physiology, New Series" (A. Läuchli and R. L. Bieleski, eds.), Vol 15A, pp. 147–180. Springer-Verlag, Berlin and New York.

Robson, A. D. and Reuter, D. J. (1981). Diagnosis of copper deficiency and toxicity. *In* "Copper in Soils and Plants" (J. F. Loneragan, A. D. Robson and

R. D. Graham, eds.), pp. 287–312. Academic Press, London and Orlando.
Robson, A. D., Dilworth, M. J. and Chatel, D. L. (1979). Cobalt and nitrogen fixation in *Lupinus angustifolius* L. I. Growth nitrogen concentrations and cobalt distribution. *New Phytol.* **83**, 53–62.
Robson, A. D., O'Hara, A. O. and Abbott, L. K. (1981a). Involvement of phosphorus in nitrogen fixation by subterranean clover *Trifolium subterraneum* L.). *Aust. J. Plant Physiol.* **8**, 427–436.
Robson, A. D., Hartley, R. D. and Jarvis, S. C. (1981b). Effect of copper deficiency on phenolic and other constituents of wheat cell walls. *New Phytol.* **89**, 361–373.
Roeb, G. W., Wieneke, J. and Führ, F. (1982). Auswirkungen hoher NaCl-Konzentrationen im Nährmedium auf die Transpiration, den Abscisinsäure-, Cytokinin- und Prolingehalt zweier Sojabohnensorten. *Z. Pflanzenernaehr. Bodenkd.* **145,** 103–116.
Rognes, S. E. (1980). Anion regulation of lupin asparagine synthetase: Chloride activation of the glutamine-utilizing reaction. *Phytochemistry* **19**, 2287–2293.
Römer, W. (1971). Untersuchungen über die Auslastung des Photosyntheseapparates bei Gerste (*Hordeum distichon* L.) und weissem Senf (*Sinapis alba* L.) in Abhängigkeit von den Umweltbedingungen. *Arch. Bodenfruchtbarkeit Pflanzenprod.* **15**, 414–423.
Römheld, V. (1984). pH-Veränderungen in der Rhizosphäre in Abhängigkeit vom Nährstoffangebot. *Landwirtsch. Forsch. Sonderh.* **40**, 226–230.
Römheld, V. and Kramer, D. (1983). Relationship between proton efflux and rhizodermal transfer cells induced by iron deficiency. *Z. Pflanzenphysiol.* **113**, 73–83.
Römheld, V. and Marschner, H. (1981a). Rhythmic iron stress reactions in sunflower at suboptimal iron supply. *Physiol. Plant.* **53**, 347–353.
Römheld, V. and Marschner, H. (1981b). Iron deficiency stress induced morphological and physiological changes in root tips of sunflower. *Physiol. Plant.* **53**, 354–360.
Römheld, V. and Marschner, H. (1983). Mechanism of iron uptake by peanut plants. I. Fe^{III} reduction, chelate splitting, and release of phenolics. *Plant Physiol.* **71**, 949–954.
Römheld, V., Marschner, H. and Kramer, D. (1982). Response to Fe deficiency in roots of "Fe efficient" plant species. *J. Plant Nutr.* **5**, 489–498.
Ronen, M. and Mayak, S. (1981). Interrelationship between abscisic acid and ethylene in the control of senescence processes in carnation flowers. *J. Exp. Bot.* **32**, 759–765.
Rossiter, R. C. (1978). Phosphorus deficiency and flowering in subterranean clover (*Tr. subterraneum* L.). *Ann. Bot. (London)* [N.S.] **42**, 325–329.
Roth-Bejerano, N. and Itai, C. (1981). Effect of boron on stomatal opening of epidermal strips of *Commelina communis*. *Physiol. Plant.* **52**, 302–304.
Rovira, A. D., Bowen, G. D. and Foster, R. C. (1983). The significance of rhizosphere microflora and mycorrhizas in plant nutrition. *In* "Encyclopedia of Plant Physiology, New Series" (A. Läuchli and R. L. Bieleski, eds.), Vol. 15A, pp. 61–89. Springer-Verlag, Berlin and New York.
Rovira, A. P., Foster, R. C. and Martin, J. K. (1979). Origin, nature and nomenclature of the organic materials in the rhizosphere. *In* "The Soil-Root Interface" (J. L. Harley and R. Scott-Russell, eds.), pp. 1–4. Academic Press, London.
Rufty, T. W., Jr., Miner, W. S. and Raper, C. D., Jr. (1979). Temperature effects on growth and manganese tolerance in tobacco. *Agron. J.* **71**, 638–644.

Rufty, T. W., Jr., Jackson, W. A. and Raper, C. D., Jr. (1981). Nitrate reduction in roots as affected by presence of potassium and by flux of nitrate through the roots. *Plant Physiol.* **68**, 605–609.

Rufty, T. W., Jr., Jackson, W. A. and Raper, C. D., Jr. (1982a). Inhibition of nitrate assimilation in roots in the presence of ammonium: The moderating influence of potassium. *J. Exp. Bot.* **33**, 1122–1137.

Rufty, T. W., Jr., Raper, C. D. and Jackson, W. A. (1982b). Nitrate uptake, root and shoot growth, and ion balance of soybean plants during acclimation to root-zone acidity. *Bot. Gaz. (Chicago)* **143**, 5–14.

Rufty, T. W., Jr., Volk, R. J., McClure, R. R., Israel, D. W. and Raper, C. D., Jr. (1982c). Relative content of NO_3^- and reduced N in xylem exudate as an indicator of root reduction of concurrently absorbed $^{15}NO_3$. *Plant Physiol.* **69**, 166–170.

Ruinen, J. (1975). Nitrogen fixation in the phyllosphere. *In* "Nitrogen Fixation by Free-living Micro-organisms" (W. D. P. Stewart, ed.), pp. 85–100. Cambridge University Press, Cambridge.

Rupp, H. and Weser, U. (1978). Circular dichroism of metallothioneins, a structural approach. *Biochim. Biophys. Acta* **533**, 209–226.

Rush, D. W. and Epstein, E. (1976). Genotypic responses to salinity. Differences between salt-sensitive and salt-tolerant genotypes of the tomato. *Plant Physiol.* **162**, 162–166.

Rush, D. W. and Epstein, E. (1981). Comparative studies on the sodium, potassium, and chloride relations of a wild halophilic and a domestic salt-sensitive tomato species. *Plant Physiol.* **68**, 1308–1313.

Rutland, R. B. (1971). Radioisotopic evidence of immobilization of iron in *Azalea* by excess calcium carbonate. *J. Am. Soc. Hortic. Sci.* **96**, 653–655.

Rutland, R. B. and Bukovac, M. J. (1971). The effect of calcium bicarbonate on iron absorption and distribution by *Chrysanthemum morifolium* (Ram.). *Plant Soil* **35**, 225–236.

Rygiewicz, P. T. and Bledsoe, C. S. (1984). Mycorrhizal effects on potassium fluxes by northwest coniferous seedlings. *Plant Physiol.* **76**, 918–923.

Rygiewicz, P. T., Bledsoe, C. S. and Zasoski, R. J. (1984). Effects of ectomycorrhizae and solution pH on (^{15}N)ammonium uptake by coniferous seedlings. *Can. J. For. Res.* **14**, 885–892.

Ryle, G. J. A., Powell, C. E. and Gordon, A. J. (1979). The respiratory costs of nitrogen fixations in the soybean, cowpea, and white clover. II. Comparisons of the costs of nitrogen fixation and the utilization of combined nitrogen. *J. Exp. Bot.* **30**, 145–153.

Rynders, L. and Vlassak, K. (1982). Use of *Azospirillum brasilense* as biofertilizer in intensive wheat cropping. *Plant Soil* **66**, 217–223.

Saalbach, E. and Aigner, H. (1970). Über die Wirkung einer Natriumdüngung auf Natriumgehalt, Ertrag und Trockensubstanzgehalt einiger Gras- und Kleearten. *Landwirtsch. Forsch.* **23**, 264–274.

Saftner, R. A. and Wyse, R. E. (1980). Alkali cation/sucrose co-transport in the root sink of sugar beet. *Plant Physiol.* **66**, 884–889.

Saftner, R. A., Daie, J. and Wyse, R. E. (1983). Sucrose uptake and compartmentation in sugar beet transport tissue. *Plant Physiol.* **72**, 1–6.

Saglio, P. H., Rancillac, M., Bruzan, F. and Pradet, A. (1984). Critical oxygen pressure for growth and respiration of excised and intact roots. *Plant Physiol.* **76**, 151–154.

Sahrawat, K. L. (1980). Control of urea hydrolysis and nitrification in soil by chemicals—prospects and problems. *Plant Soil* **57**, 335–352.

Saini, H. S. and Aspinall, D. (1982). Sterility in wheat (*Triticum aestivum* L.) induced by water deficit or high temperature: Possible mediation by abscisic acid. *Aust. J. Plant Physiol.* **9**, 529–537.

Salama, A. M. S. El-D. and Wareing, P. F. (1979). Effects of mineral nutrition on endogenous cytokinins in plants of sunflower (*Helianthus annuus* L.). *J. Exp. Bot.* **30**, 971–981.

Salami, A. U. and Kenefick, D. G. (1970). Stimulation of growth in zinc-deficient corn seedlings by the addition of tryptophan. *Crop Sci.* **10**, 291–294.

Same, B. I., Robson, A. D. and Abbott, L. K. (1983). Phosphorus, soluble carbohydrates and endomycorrhizal infection. *Soil Biol. Biochem.* **15**, 593–597.

Samimy, C. (1978). Influence of cobalt on soybean hypocotyl growth and its ethylene evolution. *Plant Physiol.* **62**, 1005–1006.

Sanchez, P. A. and Salinas, G. (1981). Low input technology for managing Oxisols and Ultisols in tropical America. *Adv. Agron.* **34**, 280–406.

Sanders, F. E. (1975). The effect of foliar-applied phosphate on mycorrhizal infections of onion roots. *In* "Endomycorrhizas" (F. E. Sanders, B. Mosse and P. B. Tinker, eds.), pp. 261–287. Academic Press, London and Orlando.

Sanders, F. E. and Sheikh, N. A. (1983). The development of vesicular–arbuscular mycorrhizal infection in plant root system. *Plant Soil* **71**, 223–246.

Sanders, J. R. (1983). The effect of pH on the total and free ionic concentrations of manganese, zinc and cobalt in soil solutions. *J. Soil Sci.* **34**, 315–323.

Sanderson, J. (1983). Water uptake by different regions of barley root. Pathway of radial flow in relation to development of the endodermis. *J. Exp. Bot.* **34**, 240–253.

Sandhu, G. R., Aslam, Z., Salim, M., Sattor, A., Qureshi, R. H., Ahmad, N. and Wyn Jones, R. G. (1981). The effect of salinity on the yield and composition of *Diplachne fusca* (Kallar grass). *Plant, Cell Environ.* **4**, 177–181.

Sandmann, G. and Böger, P. (1983). The enzymatological function of heavy metals and their role in electron transfer processes of plants. *In* "Encyclopedia of Plant Physiology, New Series" (A. Läuchli and R. L. Bieleski, eds.), Vol. 15A, pp. 563–596. Springer-Verlag, Berlin and New York.

Sangster, A. G. (1970). Intracellular silica deposition in immature leaves in three species of the *Gramineae. Ann. Bot. (London)* [N.S.] **34**, 245–257.

Sangster, A. G., Hodson, M. J. and Wynn Parry, D. (1983). Silicon deposition and anatomical studies in the inflorescence with their possible relevance to carcinogenesis. *New Phytol.* **93**, 105–122.

Sano, Y., Fujii, T., Iyama, S., Hirota, Y. and Komagata, K. (1981). Nitrogen fixation in the rhizosphere of cultivated and wild rice strains. *Crop Sci.* **21**, 758–761.

Santoro, L. G. and Magalhaes, A. C. N. (1983). Changes in nitrate reductase activity during development of soybean leaf. *Z. Pflanzenphysiol.* **112**, 113–121.

Sarkar, A. N., Jenkins, D. A. and Wyn Jones, R. G. (1979). Modification to mechanical and mineralogical composition of soil within the rhizosphere. *In* "The Soil-Root Interface" (J. L. Harley and R. Scott-Russell eds.), pp. 125–136. Academic Press, London and Orlando.

Sartain, J. B. and Kamprath, E. J. (1975). Effect of liming a highly Al-saturated soil on the top and root growtl d soybean nodulation. *Agron. J.* **67**, 507–510.

Sartirana, M. L. and Bianchetti, R. (1967). The effect of phosphate on the development of phytase in the wheat embryo. *Physiol. Plant.* **20**, 1066–1075.

Sattelmacher, B. and Marschner. H. (1978a). Nitrogen nutrition and cytokinin activity in *Solanum tuberosum. Physiol. Plant.* **42**, 185–189.

Sattelmacher, B. and Marschner, H. (1978b). Relation between nitrogen nutrition, cytokinin and tuberization in *Solanum tuberosum. Physiol. Plant.* **44**, 65–68.

Sattelmacher, B. and Marschner, H. (1979). Tuberization in potato plants as affected by application of nitrogen to the roots and leaves. *Potato Res.* **22**, 39–47.

Satter, R. L., Applewhite, P. B. and Galston, A. W. (1974). Rhythmic potassium flux in *Albizza*. Effect of aminophylline, cations and inhibitors of respiration and protein synthesis. *Plant Physiol.* **54**, 280–285.

Sauerbeck, D. and Johnen, B. (1976). Der Umsatz von Pflanzenwurzeln im Laufe der Vegetationsperiode und dessen Beitrag zur "Bodenatmung." *Z. Pflanzenernaehr. Bodenkd.* **139**, 315–328.

Sauerbeck, D., Nonnen, S. and Allard, J. L. (1981). Assimilateverbrauch und -umsatz im Wurzelraum in Abhängigkeit von Pflanzenart und -anzucht. *Landwirtsch. Forsch., Sonderh.* **37**, 207–216.

Savage, W., Berry, W. L. and Reed, C. A. (1981). Effects of trace element stress on the morphology of developing seedlings of lettuce (*Lactuca sativa* L. Grand Rapids) as shown by scanning electron microscopy. *J. Plant Nutr.* **3**, 129–138.

Saxena, N. P. and Sheldrake, A. R. (1980). Iron chlorosis in chickpea (*Cicer arietinum* L.) grown on high pH calcareous vertisol. *Field Crops Res.* **3**, 211–214.

Schacherer, A. and Beringer, H. (1984). Zahl und Größe von Endospermzellen im wachsenden Getreidekorn als Indikator der Speicherkapazität. *Ber. Dtsch. Bot. Ges.* **97**, 183–195.

Schachtschabel, P. and Beyme, B. (1980). Löslichkeit des anorganischen Bodenphosphors und Phosphatdüngung. *Z. Pflanzenernaehr. Bodenkd.* **143**, 306–316.

Schaff, B. E. and Skogley, E. D. (1982). Diffusion of potassium, calcium, and magnesium in Bozeman silt loam as influenced by temperature and moisture. *Soil Sci. Soc. Am. J.* **46**, 521–524.

Scheffer, K., Schreiber, A. and Kickuth, R. (1982). Die sorptive Bindung von Düngerphosphaten im Boden und die phosphatmobilisierende Wirkung der Kieselsäure. 2. Mitt. Die phosphatmobilisierende Wirkung der Kieselsäure. *Arch. Acker- Pflanzenbau Bodenkd.* **26**, 143–152.

Schenk, M. K. and Barber, S. A. (1979a). Phosphate uptake by corn as affected by soil characteristics and root morphology. *Soil Sci. Soc. Am. J.* **43**, 880–883.

Schenk, M. K. and Barber, S. A. (1979b). Root characteristics of corn genotypes as related to phosphorus uptake. *Agron. J.* **71**, 921–924.

Schenk, M. K. and Wehrmann, J. (1979). The influence of ammonia in nutrient solution on growth and metabolism of cucumber plants. *Plant Soil* **52**, 403–414.

Schiff, J. A. (1983). Reduction and other metabolic reactions of sulfate. *In* "Encyclopedia of Plant Physiology, New Series" (A. Läuchli, and R. L. Bieleski, eds.), Vol. 15A, pp. 401–421. Springer-Verlag, Berlin and New York.

Schilling, G. (1983). Genetic specificity of nitrogen nutrition in leguminous plants. *Plant Soil* **72**, 321–334.

Schilling, G. and Trobisch, S. (1970). Einfluss zusätzlich später Stickstoffgaben auf die Ertragsbildung von Kruziferen in Gefäß- und Feldversuchen. *Albrecht-Thaer-Arch.* **14**, 739–750.

Schilling, G. and Trobisch, S. (1971). Untersuchungen über die Verlagerung ^{15}N-markierter Stickstoffverbindungen in Abhängigkeit von der Proteinsynthese am Zielort bei *Sinapis alba. Arch. Acker- Pflanzenbau Bodenkd.* **15**, 671–682.

Schimansky, C. (1981). Der Einfluss einiger Versuchsparameter auf das Fluxverhalten von ^{28}Mg bei Gerstenkeimpflanzen in Hydrokulturversuchen. *Landwirtsch. Forsch.* **34**, 154–165.
Schinas, S. and Rowell, D. L. (1977). Lime-induced chlorosis. *J. Soil Sci.* **28**, 351–368.
Schjørring, J. K. and Jensen, P. (1984). Phosphorus nutrition of barley, buckwheat and rape seedlings. II. Influx and efflux of phosphorus by intact roots of different P status. *Physiol. Plant.* **61**, 584–590.
Schlee, D., Reinbothe, D. and Fritsche, W. (1968). Der Einfluss von Eisen auf den Purinstoffwechsel und die Riboflavinbildung von *Candida guilliermondii* (Cast.) Lang et G. *Allg. Mikrobiol.* **8**, 127–138.
Schleiff, U. (1980–1981). Osmotic potentials of roots of onions and their rhizospheric soil solution when irrigated with saline drainage waters. *Agric. Water Manage.* **3**, 317–323.
Schlichting, E. (1976). Pflanzen- und Bodenanalysen zur Charakterisierung des Nährstoffzustandes von Standorten. *Landwirtsch. Forsch.* **29**, 317–321.
Schmid, K. (1967). Zur Stickstoffdüngung im Tabakbau. *Dtsch. Tabakbau* **14**, 129–133.
Schmidt, H. E., Wrazidlo, W., Bergmann, W. and Schmelzer, K. (1972). Nachweis von Zinkmangel als Ursache der Kräuselkrankheit des Hopfens. *Biol. Zentralbl.* **91**, 729–742.
Schmidt, S. and Buban, T. (1971). Beziehungen zwischen dem ^{32}P-Einbau in anorganische Polyphosphate der Blätter und dem Beginn der Blütendifferenzierung bei *Malus domestica. Biochem. Physiol. Pflanze* **162**, 265–271.
Schmit. J.-N. (1981). Le calcium dans le cellule génératrice en mitose. Etude dans le tube pollinique en germitation de *Clivia nobilis* Lindl. *(Amaryllidacee) C.R. Seances Acad. Sci., Ser. III.* **293**, 755–760.
Schmutz, D. and Brunold, C. (1982). Regulation of sulfate assimilation in plants. XIII. Assimilatory sulfate reduction during ontogenesis of primary leaves of *Phaseolus vulgaris* L. *Plant Physiol.* **70**, 524–527.
Schmutz, D. and Brunold, C. (1984). Intercellular localization of assimilatory sulfate reduction in leaves of *Zea mays* and *Triticum aestivum. Plant Physiol.* **74**, 866–870.
Schnabl, H. (1980). Der Anionenmetabolismus in stärkehaltigen und stärkefreien Schließzellenprotoplasten. *Ber. Dtsch. Bot. Ges.* **93**, 595–605.
Schnabl, H. and Ziegler, H. (1977). The mechanism of stomatal movement in *Allium cepa* L. *Planta* **136**, 37–43.
Schofield, R. K. (1955). Can a precise meaning be given to "available" soil phosphorus? *Soils Fert.* **28**, 373–375.
Scholz, G. (1958). Über die Bedeutung des Bors für die Alkaloidproduktion von *Nicotiana rustica. Z. Pflanzenernaehr. Dueng., Bodenkd.* **80**, 149–155.
Scholz, G. and Böhme, H. (1980). Biochemical mutants in higher plants as tools for chemical and physiological investigations—a survey. *Kulturpflanze* **28**, 11–32.
Schönherr, H. and Bukovac, M. J. (1970). Preferential polar pathways in the cuticle and their relationship to ectodesmata. *Planta* **92**, 189–201.
Schönwitz, R. and Ziegler, H. (1982). Exudation of water-soluble vitamins and of some carbohydrates by intact roots of maize seedlings (*Zea mays* L.) into a mineral nutrient solution. *Z. Pflanzenphysiol.* **107**, 7–14.
Schropp, A. and Marschner, H. (1977). Wirkung hoher Phosphatdüngung auf die Wachstumsrate, den Zn-Gehalt und das P/Zn-Verhältnis in Weinreben *(Vitis vinifera). Z. Pflanzenernaehr. Bodenkd.* **140**, 525–529.

Schubert, K. R., Jennings, N. T. and Evans. H. J. (1978). Hydrogen reactions of nodulated leguminous plants. *Plant Physiol.* **61**, 398–401.
Schultz, R. C., Kormanik, P. P. and Bryan, W. C. (1981). Effects of fertilization and vesicular–arbuscular mycorrhizal inoculation on growth of hardwood seedlings. *Soil Sci. Soc. Am. J.* **45**, 961–965.
Schulze, E. D., Turner, N. C. and Glatzle, G. (1984). Carbon, water and nutrient relations of two mistletoes and their hosts: A hypothesis. *Plant, Cell Environ.* **7**, 293–299.
Schulze, W. (1957). Über den Einfluss der Düngung auf die Bildung der Chloroplastenpigmente. *Z. Pflanzenernaehr., Dueng., Boedenkd.* **76**, 1–19.
Schumacher, R. and Frankenhauser, F. (1968). Fight against bitter pit. *Schweiz. Z. Obst- Weinbau.* **104**, 424.
Schussler, J. R., Brenner, M. I. and Brun, W. A. (1984). Abscisic acid and its relationship to seed filling in soybeans. *Plant Physiol.* **76**, 301–306.
Schütte, K. H. (1967). The influence of boron and copper deficiency upon infection by *Erysiphe graminis* D. C. the powdery mildew in wheat var. Kenya. *Plant Soil* **27**, 450–452.
Schwarz, M. and Gale, J. (1981). Maintenance respiration and carbon balance of plants at low levels of sodium chloride salinity. *J. Exp. Bot.* **32**, 933–941.
Scott, H. D., Beasley, S. D. and Thompson, L. F. (1976). Effect of lime on boron transport to and uptake by cotton. *Soil Sci. Soc. Am. Proc.* **39**, 1116–1121.
Scott-Russell, R. (1977). "Plant Root Systems: Their Function and Interaction with the Soil." McGraw-Hill, New York.
Scott-Russell, R. and Goss, M. J. (1974). Physical aspects of soil fertility—The response of roots to mechanical impedance. *Neth. J. Agric. Sci.* **22**, 305–318.
Seckback, J. (1982). Ferreting out the secrets of plant ferritin—a review. *J. Plant Nutr.* **5**, 369–394.
Sekiya, J., Schmidt, A., Wilson, L. G. and Filner, P. (1982a). Emission of hydrogen sulfide by leaf-tissue in response to L-cysteine. *Plant Physiol.* **70**, 430–436.
Sekiya, J., Wilson, L. G. and Filner, P. (1982b). Resistance to injury by sulfur dioxide. Correlation with its reduction to and emission of hydrogen sulfide in *Cucurbitaceae. Plant Physiol.* **70**, 437–441.
Sen Gupta, B., Nandi, A. S. and Sen, S. P. (1982). Utility of phyllosphere N_2-fixing micro-organisms in the improvement of crop growth. I. Rice. *Plant Soil* **68**, 55–67.
Servaites, J. C., Schrader, L. E. and Jung, D. M. (1979). Energy-dependent loading of amino acids and sucrose into the phloem of soybean. *Plant Physiol.* **64**, 546–550.
Seth, A. K. and Wareing, P. F. (1967). Hormone-directed transport of metabolites and its possible role in plant senescence. *J. Exp. Bot.* **18**, 65–77.
Setter, T. L. and Meller, V. H. (1984). Reserve carbohydrate in maize stem. ^{14}C glucose and ^{14}C sucrose uptake characteristics. *Plant Physiol.* **75**, 617–622.
Setter, T. L., Brun, W. A. and Brenner, M. L. (1980). Effect of obstructed translocation on leaf abscisic acid, and associated stomatal closure and photosynthesis decline. *Plant Physiol.* **65**, 1111–1115.
Sevilla, F., Lopez-Gorge, J., Gomez, M. and Del Rio, L. A. (1980). Manganese superoxide dismutase from higher plant. Purification of a new Mn-containing enzyme. *Planta* **150**, 153–157.
Shaked, A. and Bar-Akiva, A. (1967). Nitrate reductase activity as an indication of molybdenum level and requirement of citrus plants. *Phytochemistry* **6**, 347–350.
Shanmugam, K. T., O'Gara, F., Andersen, K. and Valentine, R. C. (1978). Biological nitrogen fixation. *Annu. Rev. Plant Physiol.* **29**, 263–276.

Sharma, B. D., Takkar, P. N. and Sadana, U. S. (1982). Evaluation of levels and methods of zinc application to rice in sodic soils. *Fert. Res.* **3**, 161–167.

Sharma, C. P., Sharma, P. N., Bisht, S. S. and Nautiyal, B. D. (1982). Zinc deficiency induced changes in cabbage. *In* "Proceedings of the Ninth Plant Nutrition Colloquium, Warwick, England" (A. Scaife, ed.), pp. 601–606. Commonw. Agric. Bur., Farnham Royal, Bucks.

Sharpless, R. O. and Johnson, D. S. (1977). The influence of calcium on senescense changes in apple. *Ann. Appl. Biol.* **85**, 450–453.

Shea, P. F., Gabelman, W. H. and Gerloff, G. C. (1967). The inheritance of efficiency in potassium utilization in snap beans (*Phaseolus vulgaris* L.) *Proc. Am. Soc. Hortic. Sci.* **91**, 286–293.

Shear, C. B. (1975). Calcium related disorders of fruits and vegetables. *HortScience.* **10**, 361–365.

Sherwood, R. T. and Vance, C. P. (1980). Resistance to fungal penetration in *Gramineae. Phytopathology* **70**, 273–279.

Shetty, A. S. and Miller, G. W. (1966). Influence of iron chlorosis on pigment and protein metabolism in leaves of *Nicotiana tabacum* L. *Plant Physiol.* **41**, 415–421.

Shierlaw, J. and Alston, A. M. (1984). Effect of soil compaction on root growth and uptake of phosphorus. *Plant Soil* **77**, 15–28.

Shih, L.-M, Kaur-Sawhney, R., Führer, J., Samata, S. and Galston, A. W. (1982). Effect of exogenous 1,3-diaminopropane and spermidine on senescence of oat leaves. I. Inhibition of protease activity, ethylene production and chlorophyll loss as related to polyamine content. *Plant Physiol.* **70**, 1592–1596.

Shimshi, D. (1969). Interaction between irrigation and plant nutrition. *Proc. 7th Colloq. Int. Potash Inst. Bern.*, pp. 111–120.

Shkol'nik, M. Y. (1974). General conception of the physiological role of boron in plants. *Sov. Plant Physiol. (Engl. Transl.)* **21**, 140–150.

Shkol-nik, M. Y., Krupnikova, T. A. and Smirnov, Y. S. (1981). Activity of polyphenol oxidase and sensitivity to boron deficiency in monocots and dicots. *Sov. Plant Physiol. (Engl. Transl.)* **28**, 279–283.

Shone, M. G. T., Clarkson, D. T., Sanderson, J. and Wood, A. V. (1973). A comparison of the uptake and translocation of some organic molecules and ions in higher plants. *In* "Ion Transport in Plants" (W. P. Anderson, ed.), pp. 571–582. Academic Press, London and Orlando.

Shrift, A. (1969). Aspects of selenium metabolism in higher plants. *Annu. Rev. Plant Physiol.* **20**, 475–494.

Shrotri, C. K., Mohanty, P., Rathore, V. C. and Tewari, M. N. (1983). Zinc deficiency limits the photosynthetic enzyme activation in *Zea mays* L. *Biochem. Physiol. Pflanz.* **178**, 213–217.

Shukla, U. C., and Raj, H. (1980). Zinc response in pigeon pea as influenced by genotypic variability. *Plant Soil* **57**, 323–333.

Siegel, N. and Haug, A. (1983). Calmodulin-dependent formation of membrane potential in barley root plasma membrane vesicles: A biochemical model of aluminium toxicity in plants. *Physiol. Plant.* **59**, 285–291.

Sieverding, E. (1984). Influence of soil water regimes on VA mycorrhiza: III. Comparison of 3 mycorrhizal fungi and their influences on transpiration. *Z. Acker Pflanzenbau* **153**, 52–61.

Sieverding, E. and Leihner, D. E. (1984). Effect of herbicides on population dynamics of VA-mycorrhiza with cassava. *Angew. Bot.* **58**, 283–294.

Sijmons, P. C., Kolattukudy, P. E. and Bienfait, H. F. (1985). Iron deficiency decreases suberization in bean roots through a decrease in suberin-specific peroxidase activity. *Plant Physiol.* **78**, 115–120.

Silberbush, M. and Barber, S. A. (1984). Phosphorus and potassium uptake of field-grown soybean cultivars predicted by a simulation model. *Soil Sci. Soc. Am. J.* **48**, 592–596.

Silberbush, M., Hallmark, W. B. and Barber, S. A. (1983). Simulation of effects of soil bulk density and P addition on K uptake of soybean. *Comm. Soil Sci. Plant Anal.* **14**, 287–296.

Silva, P. R. F. da and Stutte, C. A. (1981). Nitrogen loss in conjunction with transpiration from rice leaves as influenced by growth stage, leaf position, and N supply. *Agron. J.* **73**, 38–42.

Silvius, J. E., Kremer, D. F. and Lee, D. R. (1978). Carbon assimilation and translocation in soybean leaves at different stages of development. *Plant Physiol.* **62**, 54–58.

Simojoki, P. (1972). Results of boron fertilizer experiments in barley. *Ann. Agric. Fenn.* **11**, 333–341.

Simpson, R. J., Lambers, H. and Dalling, M. J. (1982). Translocation of nitrogen in a vegetative wheat plant *(Triticum aestivum). Physiol. Plant.* **56**, 11–17.

Sims, G. K. and Dunigan, E. P. (1984). Diurnal and seasonal variation in nitrogenase activity (C_2H_2 reduction) of rice roots. *Soil Biol. Biochem.* **16**, 15–18.

Sims, J. L. and Patrick, W. H., Jr. (1978). The distribution of micronutrient cations in soil under conditions of varying redox potential and pH. *Soil Sci. Soc. Am. J.* **42**, 258–262.

Sinclair, T. R. and de Wit, C. T. (1976). Analysis of the carbon and nitrogen limitations to soybean yield. *Agron. J.* **68**, 319–324.

Singer, S. J. (1972). A fluid lipid-globular mosaic model of membrane structure. *Ann. N.Y. Acad. Sci.* **195**, 16–23.

Singh, B. K. and Jenner, C. F. (1982). Association between concentrations of organic nutrients in the grain, endosperm cell number and grain dry weight within the ear of wheat. *Aust. J. Plant Physiol.* **9**, 83–95.

Singh, M. (1981). Effect of zinc, phosphorus and nitrogen on tryptophan concentration in rice grains grown on limed and unlimed soils. *Plant Soil* **62**, 305–308.

Singh, S. P. and Paleg, L. G. (1984). Low-temperature-induced GA_3 sensitivity of wheat. II. Changes in lipids associated with the low temperature-induced GA_3 sensitivity of isolated aleurone of kite. *Plant Physiol.* **76**, 143–147.

Singleton, P. W., El Swaify, S. A. and Bohlool, B. B. (1982). Effect of salinity on *Rhizobium* growth and survival. *Appl. Environ. Microbiol.* **44**, 884–890.

Sinha, B. K. and Singh, N. T. (1974). Effect of transpiration rate on salt accumulation around corn roots in a saline soil. *Agron. J.* **66**, 557–560.

Skokut, T. A., Wolk, C. P., Thomas, J., Meeks, J. C., Shaffer, P. W. and Chien, W. S. (1978). Initial organic products of assimilation of ^{13}N-ammonium and ^{13}N nitrate by tobacco cell cultured on different sources of nitrogen. *Plant Physiol.* **62**, 299–304.

Skoog, F. (1940). Relationship between zinc and auxin in the growth of higher plants. *Am. J. Bot.* **27**, 939–950.

Slankis, V. (1973). Hormonal relationship in mycorrhizal development. *In* "Ectomycorrhizae: Their Ecology and Physiology" (G. C. Marks and T. T. Kozlowski, eds.), pp. 231–298. Academic Press, London and Orlando.

Slocum, R. C. and Roux, S. J. (1983). Cellular and subcellular localization of calcium in gravistimulated oat coleoptiles and its possible significance in the establishment of tropic curvature. *Planta* **157**, 481–492.

Smeulders, F. and Geijn, S. C. van de (1983). In situ immobilization of heavy metals with tetraethylenepentamine (tetren) in natural soils and its effect on toxicity and plant growth. III. Uptake and mobility of copper and its tetren-complex in corn plants. *Plant Soil* **70**, 59–68.

Smiley, R. W. (1974). Rhizosphere pH as influenced by plants, soils, and nitrogen fertilizers. *Soil Sci. Soc. Am. Proc.* **38**, 795–799.

Smiley, R. W. (1978). Colonization of wheat roots by *Gaeumannomyces graminis* inhibited by specific soils, microorganisms, and ammonium nitrogen. *Soil Biol. Biochem.* **10**, 175–179.

Smiley, R. W. and Cook, R. J. (1973). Relationship between take-all of wheat and rhizosphere pH in soils fertilized with ammonium-*vs.* nitrate-nitrogen. *Phytopathology* **63**, 882–890.

Smirnoff, N. and Stewart, G. R. (1985). Nitrate assimilation and translocation by higher plants: Comparative physiology and ecological consequences. *Physiol. Plant.* **64**, 133–140.

Smirnov, Y. S., Krupnikova, T. A. and Shkol'nik, M. Y. (1977). Content of IAA in plants with different sensitivity to boron deficits. *Sov. Plant Physiol. (Engl. Transl.)* **24**, 270–276.

Smith, A. F. and Raven, J. A. (1979). Intracellular pH and its regulation. *Annu. Rev. Plant Physiol.* **30**, 289–311.

Smith, B. N. (1984). Iron in higher plants: Storage and metabolic role. *J. Plant Nutr.* **7**, 759–766.

Smith, D. and Struckmeyer, B. E. (1977). Effects of high-levels of chlorine in alfalfa shoots. *Can. J. Plant Sci.* **57**, 293–296.

Smith, F. A. (1973). The internal control of nitrate uptake into excised barley roots with differing salt contents. *New Phytol.* **72**, 769–782.

Smith, F. A. and Walker, N. A. (1976). Chloride transport in *Chara corallina* and the electrochemical potential difference for hydrogen ions. *J. Exp. Bot.* **27**, 451–459.

Smith, F. W. (1974). The effect of sodium on potassium nutrition and ionic relations in Rhodes grass. *Aust. J. Agric. Res.* **25**, 407–414.

Smith, G. S., Middleton, K. R. and Edmonds, A. S. (1978). Sodium and potassium contents of top-dressed pastures in New Zealand in relation to plant and animal nutrition. *N.Z. J. Exp. Agric.* **6**, 217–225.

Smith, G. S., Middleton, K. R. and Edmonds, A. S. (1980). Sodium nutrition of pasture plants. II. Effects of sodium chloride on growth, chemical composition and reduction of nitrate nitrogen. *New Phytol.* **84**, 613–622.

Smith, G. S., Lauren, D. R., Cornforth, I. S. and Agnew, M. P. (1982). Evaluation of putrescine as a biochemical indicator of the potassium requirements of lucerne. *New Phytol.* **91**, 419–428.

Smith, J. A. C. and Milburn, J. A. (1980). Water stress and phloem loading. *Ber. Dtsch. Bot. Ges.* **93**, 269–280.

Smith, M. K. and McComb, J. A. (1981). Effect of NaCl on the growth of whole plants and their corresponding callus cultures. *Aust. J. Plant Physiol.* **8**, 267–275.

Smith, R. H. and Johnson, W. C. (1969). Effect of boron on white clover nectar production. *Crop Sci.* **9**, 75–76.

Smith, R. L., Bouton, J. H., Schank, S. C., Quesenberry, V. H., Tyler, M. E., Milam, J. R., Gaskin, M. H. and Littell, R. S. (1976). Nitrogen fixation inoculated with *Spirillum lipoferum*. *Science* **193**, 1003–1005.

Smith, T. A. (1973). Amine levels in mineral-deficient *Hordeum vulgare* leaves. *Phytochemistry* **12**, 2091–2100.

Smith, T. A. and Sinclair, C. (1967). The effect of acid feeding on amine formation in barley. *Ann. Bot. (London)* [N.S.] **31**, 103–111.

Smith, T. A. and Wilshire, G. (1975). Distribution of cadaverine and other amines in higher plants. *Phytochemistry* **14**, 2341–2346.

Smyth, D. A. and Chevalier, P. (1984). Increases in phosphatase and β-glucosidase activities in wheat seedlings in response to phosphorus-deficient growth. *J. Plant Nutr.* **7**, 1221–1231.

Snellgrove, R. C., Splitstoesser, W. E., Stribley, D. P. and Tinker, P. B. (1982). The distribution of carbon and the demand of the fungal symbiont in leek plants with vesicular–arbuscular mycorrhizas. *New Phytol.* **92**, 75–88.

Snowball, K., Robson, A. D. and Loneragan, J. F. (1980). The effect of copper on nitrogen fixation in subterranean clover *(Trifolium subterraneum)*. *New Phytol.* **85**, 63–72.

Sogawa, K. (1982). The rice brown plant hopper: Feeding physiology and host plant interactions. *Annu. Rev. Entomol.* **27**, 49–73.

Sommer, K. and Six, R. (1982a). Längenwachstum und Assimilateinlagerung bei Wintergerste in Abhängigkeit von der Stickstoffversorgung und möglichen Witterungseinflüssen. *Landwirtsch. Forsch.* **35**, 14–25.

Sommer, K. and Six, R. (1982b). Ammonium als Stickstoffquelle beim Anbau von Futtergerste. *Landwirtsch. Forsch., Sonderh.* **38**, 151–161.

Soni, S. L. and Parry, D. W. (1973). Electron probe microanalysis of silicon deposition in the inflorescence bracts of the rice plant *(Oryza sativa)*. *Am. J. Bot.* **60**, 111–116.

Sonneveld, C. and Ende, J. van den (1975). The effect of some salts on head weight and tipburn of lettuce and on fruit production and blossom-end rot of tomatoes. *Neth. J. Agric. Sci.* **23**, 192–201.

Sonneveld, C. and Voogt, S. J. (1975). Studies on the manganese uptake of lettuce on steam-sterilized glasshouse soils. *Plant Soil* **42**, 49–62.

Soon, Y. K. and Miller, M. H. (1977). Changes in the rhizosphere due to NH_4^+ and NO_3^- fertilization and phosphorus uptake by corn seedlings (*Zea mays* L.). *Soil Sci. Soc. Am. Proc.* **41**, 77–80.

Sovonick, S. A., Geiger, D. R. and Fellows, R. J. (1974). Evidence for active phloem loading in the minor veins of sugar beet. *Plant Physiol.* **54**, 886–891.

Sowokinos, J. R. (1981). Pyrophosphorylases in *Solanum tuberosum*. II. Catalytic properties and regulation of ADP-Glucose and UDP-Glucose pyrophosphorylase activities in potatoes. *Plant Physiol.* **68**, 924–929.

Spaeth, S. C. and Sinclair, T. R. (1983). Variation in nitrogen accumulation among soybean cultivars. *Field Crops Res.* **7**, 1–12.

Spain, J. M., Francis, C. A., Howeler, R. H. and Calvo, F. (1975). Differential species and varietal tolerances to soil acidity in tropical crops and pastures. *In* "Soil Management in Tropical America" (E. Bornemisza and A. Alvarado, eds.), pp. 308–329. North Carolina State University, Raleigh.

Spanner, D. C. (1975). Electroosmotic flow. *In*, "Encyclopedia of Plant Physiology, New Series" (M. H. Zimmermann, and J. A. Milburn, eds.), Vol. 1, pp. 301–327. Springer-Verlag, Berlin and New York.

Spanswick, R. M. (1981). Electronic ion pumps. *Annu. Rev. Plant Physiol.* **32**, 267–289.

Spencer, D. and Possingham, J. V. (1960). The effect of nutrient deficiencies on the Hill reaction of isolated chloroplasts from tomato. *Aust. J. Biol. Sci.* **13**, 441–445.

Spencer, D. and Possingham, J. V. (1961). The effect of manganese deficiency on photophosphorylation and the oxygen-evolving sequence in spinach chloroplasts. *Biochim. Biophys. Acta* **52**, 379–381.

Sperrazza, J. M. and Spremulli, L. L. (1983). Quantitation of cation binding to wheat germ ribosomes: influences on subunit association equilibria and ribosome activity. *Nucleic Acid Res.* **11**, 2665–2679.

Spiertz, J. H. J. and Ellen, J. (1978). Effects of nitrogen on crop development and grain growth of winter wheat in relation to assimilation and utilization of assimilates and nutrients. *Neth. J. Agric. Sci.* **26**, 210–231.

Spiller, S. C., Castelfranco, A. M. and Castelfranco, P. A. (1982). Effects of iron and oxygen on chlorophyll biosynthesis. I. In vivo observations on iron and oxygen-deficient plants. *Plant Physiol.* **69**, 107–111.

Srivastava, O. P. and Sethi, B. C. (1981). Contribution of farm yard manure on the build up of available zinc in an aridisol. *Commun. Soil Sci. Plant Anal.* **12**, 355–361.

Starck, Z. and Karwowska, R. (1978). Effect of salt-stresses on the hormonal regulation of growth, photosynthesis and distribution of ^{14}C-assimilates in bean plants. *Acta Soc. Bot. Pol.* **47**, 245–267.

Starck, Z., Choluj, D. and Szczepanska, B. (1980). Photosynthesis and photosynthates distribution in potassium-deficient radish plants treated with indolyl-3-acetic acid or gibberellic acid. *Photosynthetica* **14**, 497–505.

Steer, B. T., Hocking, P. J., Kortt, A. A. and Roxburgh, C. M. (1984). Nitrogen nutrition of sunflower (*Helianthus annuus* L.) yield components, the timing of their establishment and seed characteristics in response to nitrogen supply. *Field Crops Res* **9**, 219–236.

Steffens, D. (1984). Wurzelstudien und Phosphat-Aufnahme von Weidelgras und Rotklee unter Feldbedingungen. *Z. Pflanzenernaehr. Bodenkd.* **147**, 85–97.

Steffens, D. and Mengel, K. (1980). Das Aneignungsvermögen von *Lolium perenne* im Vergleich zu *Trifolium pratense* für Zwischenschicht-Kalium der Tonminerale. *Landwirtsch. Forsch.* **36**, 120–127.

Steffens, D. and Mengel, K. (1981). Vergleichende Untersuchungen zwischen *Lolium perenne* und *Trifolium pratense* über das Aneignungsvermögen von Kalium. *Mitt. Dtsch. Bodenkdl. Gesellsch.* **32**, 375–386.

Stein, M. and Willenbrink, J. (1976). Zur Speicherung von Saccharose in der wachsenden Zuckerrübe. *Z. Pflanzenphysiol.* **79**, 310–322.

Steingröver, E. (1981). The relationship between cyanide-resistant root respiration and the storage of sugars in the transport in *Daucus carota* L. *J. Exp. Bot.* **32**, 911–919.

Steingröver, E. (1983). Storage of osmotically active compounds in the taproot of *Daucus carota* L. *J. Exp. Bot.* **34**, 425–433.

Steingröver, E., Oosterhuis, R. and Wieringa, F. (1982). Effect of light treatment and nutrition on nitrate accumulation in spinach (*Spinacea oleracea* L.). *Z. Pflanzenphysiol.* **107**, 97–102.

Stelzer, R. and Läuchli, A. (1977). Salz- und Überflutungstoleranz von *Puccinellia peisonis*. II. Strukturelle Differenzierung der Wurzel in Beziehung zur Funktion. *Z. Pflanzenphysiol.* **84**, 95–108.

Stenlid, G. (1982). Cytokinins as inhibitors of root growth. *Physiol. Plant.* **56**, 500–506.

Stevenson, F. J. and Fitch, A. (1981). Reactions with organic matter. *In* "Copper in Soils and Plants" (J. F. Loneragan, A. D. Robson and R. D. Graham, eds.), pp. 69–95. Academic Press, New York.

Stewart, W. D. P., Haystead, A. and Dharmawardene, M. W. N. (1975). Nitrogen assimilation and metabolism in blue-green algae. *In* "Nitrogen Fixation by Free-living Micro-organisms" (D. P. Stewart, ed), pp. 129–158. Cambridge University Press. Cambridge.

Stockman, Y. M., Fischer, R. A. and Brittain, E. G. (1983). Assimilate supply and floret development within the spike of wheat (*Triticum aestivum* L.). *Aust. J. Plant Physiol.* **10**, 585–594.

Stocking, C. R. (1975). Iron deficiency and the structure and physiology of maize chloroplasts. *Plant Physiol.* **55**, 626–631.

Storey, R., Pitman, M. G., Stelzer, R. and Carter, C. (1983). X-ray micro-anaylsis of cells and cell compartments of *Atriplex spongiosa*. I. Leaves. *J. Exp. Bot.* **34**, 778–794.

Stout, P. R., Meager, W. R., Pearson, G. A. and Johnson, C. M. (1951). Molybdenum nutrition of crop plants. I. The influence of phosphate and sulfate on the absorption of molybdenum from soils and solution cultures. *Plant Soil* **3**, 51–87.

Stout, R. G., Johnson, K. D. and Rayle, D. L. (1978). Rapid auxin- and fusicoccin-enhanced Rb^+ uptake and malate synthesis in *Avena* coleoptile sections. *Planta* **139**, 35–41.

Strebel, O., Grimme, H., Renger, M. and Fleige, H. (1980). A field study with nitrogen-15 of soil and fertilizer nitrate uptake and of water withdrawal by spring wheat. *Soil Sci.* **130**, 205–210.

Strebel, O., Duynisveld, W. H. M., Grimme, H., Renger, M. and Fleige, H. (1983). Wasserentzug durch Wurzeln und Nitratanlieferung (Massenfluss, Diffusion) als Funktion von Bodentiefe und Zeit bei einem Zuckerrübenbestand. *Mitt. Dtsch. Bodenkd. Ges.* **38**, 153–158.

Streeter, J. G. (1978). Effect of N starvation on soybean plants at various stages of growth on seed yield on N-concentration in plant parts at maturity. *Agron. J.* **70**, 74–76.

Streeter, J. G. (1979). Allantoin and allantoic acid in tissues and stem exudate from field-grown soybean plants. *Plant Physiol.* **63**, 478–480.

Streeter, J. G. (1982). Synthesis and accumulation of nitrite in soybean nodules supplied with nitrate. *Plant Physiol.* **69**, 1429–1434.

Stribley, D. P., Tinker, P. B. and Snellgrove, R. C. (1981). Effect of vesicular–arbuscular mycorrhizal fungi on the relations of plant growth, internal phosphorus concentration and soil phosphate analyses. *J. Soil Sci.* **31**, 655–672.

Strullu, D. G., Harley, L., Gourret, J. P. and Garrec, J. P. (1982). Ultra-structure and microanalysis of the polyphosphate granules of the endomycorrhizas of *Fagus sylvatica*. *New Phytol.* **92**, 417–424.

Struve, I., Weber, A., Lüttge, U., Ball, E. and Smith, J. A. C. (1985). Increased vacuolar ATPase activity correlated with CAM induction in *Mesembryanthemum crystallinum* and *Kalanchoë blossfeldiana* cv. Tom Thumb. *J. Plant Physiol.* **117**, 451–468.

Stryker, R. B., Gilliam, J. W. and Jackson, W. A. (1974). Nonuniform transport of phosphorus from single roots to the leaves of *Zea mays*. *Physiol. Plant.* **30**, 231–239.

Strzelczyk, E. and Pokojska-Burdziej, A. (1984). Production of auxins and gibberellin-like substances by mycorrhizal fungi, bacteria and actinomycetes isolated from soil and the mycorrhizosphere of pine (*Pinus silvestris* L.). *Plant Soil* **81**, 185–194.

Stuiver, C. E. E., Kuiper, P. J. C. and Marschner, H. (1978). Lipids from bean, barley and sugar beet in relation to salt resistance. *Physiol. Plant.* **42**, 124–128.

Stuiver, C. E. E., Kuiper, P. J. C., Marschner, H. and Kylin, A. (1981). Effects of salinity and replacement of K^+ by Na^+ on lipid composition in two sugar beet inbred lines. *Physiol. Plant.* **52**, 77–82.

Suelter, C. H. (1970). Enzymes activated by monovalent cations. *Science* **168**, 789–795.

Sugiura, Y., Tanaka, H., Mino, Y., Ishida, T., Ota, N., Inoue, M., Nomoto, K., Yoshioka, H. and Takemoto, T. (1981). Structure, properties, and transport mechanism of iron (III) complex of mugineic acid, a possible phytosiderophore. *J. Am. Chem. Soc.* **103**, 6979–6982.

Sugiyama, T., Nakyama, N. and Akazawa, T. (1968). Structure and function of chloroplast proteins. V. Homotropic effect of bicarbonate in RuBP carboxylase relation and the mechanism of activation by magnesium ions. *Arch. Biochem. Biophys.* **126**, 734–745.

Sugiyama, T., Matsumoto, C., Akazawa, T. and Miyachi, S. (1969). Structure and function of chloroplast proteins. VII. Ribulose-1,5-diphosphate carboxylase of *Chlorella ellipsoida. Arch. Biochem. Biophys.* **129**, 597–602.

Sumner, M. E. (1977). Application of Beaufils' diagnostic indices to maize data published in the literature irrespectively of age and conditions. *Plant Soil* **46**, 359–369.

Sundstrom, F. J., Morse, R. D. and Neal, J. L. (1982). Nodulation and nitrogen fixation of *Phaseolus vulgaris* L. grown in minesoil as affected by soil compaction and N fertilization. *Commun. Soil Sci. Plant Anal.* **13**, 231–242.

Svensson, S.-B. (1971). The effect of coumarin on root growth and root histology. *Physiol. Plant.* **24**, 446–470.

Svensson, S.-B. (1972). The effect of coumarin on growth, production of dry matter, protein and nucleic acids in roots of maize and wheat and the interaction of coumarin with metabolic inhibitors. *Physiol. Plant.* **27**, 13–24.

Sweatt, M. R. and Davies, F. T. Jr. (1984). Mycorrhizae, water relations, growth, and nutrient uptake of geranium grown under moderately high phosphorus regimes. *J. Am. Soc. Hortic. Sci.* **109**, 210–213.

Sze, H. (1984). H^+-translocating ATPases on the plasma membrane and tonoplast of plant cells. *Physiol. Plant.* **61**, 683–691.

Tabor, J. A., Pennington, D. A. and Warwick, A. W. (1984). Sampling variability of petiole nitrate in irrigated cotton. *Commun. Soil Sci. Plant Anal.* **15**, 573–585.

Takagi, S. (1976). Naturally occurring iron-chelating compounds in oat- and rice-root washings. I. Activity measurements and preliminary characterization. *Soil Sci. Plant Nutr.* **22**, 423–433.

Takagi, S., Nomoto, K. and Takemoto, T. (1984). Physiological aspect of mugineic acid, a possible phytosiderophore of graminaceous plants. *J. Plant Nutr.* **7**, 469–477.

Takahashi, E. and Miyake, Y. (1977). Silica and plant growth. *Proc. Int. Semin. Soil Environ. Fert. Manage. Intensive Agric., 1977,* 603–611.

Takaki, H. and Kushizaki, M. (1970). Accumulation of free tryptophane and tryptamine in zinc deficient maize seedlings. *Plant Cell Physiol.* **2**, 793–804.

Takeoka, Y., Kondo, K. and Kaufman, P. B. (1983). Leaf surface fine-structures in rice plants cultured under shaded, and non-shaded conditions. *Jpn. J. Crop Sci.* **52**, 534–543.

Takeoka, Y., Wada, T., Naito, K. and Kaufman, P. B. (1984). Studies on silification of epidermal tissues of grasses as investigated by soft X-ray image analysis. II. Differences in frequency of silica bodies in bulliform cells at different positions in the leaves of rice plants. *Jpn. J. Crop Sci.* **53**, 197–203.

Tal, M. and Shannon, M. C. (1983). Salt tolerance in the wild relations of the cultivated tomato: Responses of *Lycopersicon esculentum, L. cheesmanii, L. peruvianum, Solanum pennellii* and F_1 hybrids to high salinity. *Aust. J. Plant Physiol.* **10**, 109–117.

Talha, M., Amberger, A. and Burkart, N. (1979). Effect of soil compaction and soil moisture level on plant growth and potassium uptake. *Z. Acker- Pflanzenbau* **148**, 156–164.

Talouizte, A., Champigny, M. L., Bismuth, E. and Moyse, A. (1984). Root carbohydrate metabolism associated with nitrate assimilation in wheat previously deprived of nitrogen. *Physiol. Veg.* **22**, 19–27.

Tanada, T. (1978). Boron–key element in the actions of phytochrome and gravity. *Planta* **143**, 109–111.

Tanada, T. (1982). Role of boron in the far-red delay of nyctinastic closure of *Albizzia* pinnules. *Plant Physiol.* **70**, 320–321.

Tanaka, A. and Navasero, S. A. (1964). Loss of nitrogen from the rice plant through rain or dew. *Soil Sci. Plant. Nutr. (Tokyo)* **10**, 36–39.

Tanaka, H. (1966). Response of *Lemna pausicostata* to boron as affected by light intensity. *Plant Soil* **25**, 425–434.

Tanaka, H. (1967). Boron adsorption by plant roots. *Plant Soil* **27**, 300–302.

Tannenbaum, S. R., Fett, D., Young, V. R., Lan, P. D. and Bruce, W. R. (1978). Nitrite and nitrate are formed by endogenous synthesis in the human intestine. *Science* **200**, 1487–1488.

Tanner, P. D. (1978). A relationship between premature sprouting on the cob and the molybdenum and nitrogen status of maize grain. *Plant Soil* **49**, 427–432.

Tanner, P. D. (1982). The molybdenum requirements of maize in Zimbabwe. *Zimbabwe Agric. J.* **79**, 61–64.

Tanner, W. (1980). On the possible role of ABA on phloem unloading. *Ber. Dtsch. Bot. Ges.* **93**, 349–351.

Taylor, H. M. and Ratliff, L. F. (1969). Root elongation rates of cotton and peanuts as a function of soil strength and soil water content. *Soil Sci.* **108**, 113–119.

Terry, N. (1977). Photosynthesis, growth, and the role of chloride. *Plant Physiol.* **60**, 69–75.

Terry, N. (1980). Limiting factors in photosynthesis. I. Use of iron stress to control photochemical capacity in vivo. *Plant Physiol.* **65**, 114–120.

Terry, N. and Low, G. (1982). Leaf chlorophyll content and its relation to the intracellular location of iron. *J. Plant Nutr.* **5**, 301–310.

Terry, N. and Ulrich, A. (1973). Effects of phosphorus deficiency on the photosynthesis and respiration of leaves in sugar beet. *Plant Physiol.* **51**, 43–47.

Terry, N. and Ulrich, A. (1974). Effects of magnesium deficiency on the photosynthesis and respiration of leaves of sugar beet. *Plant Physiol.* **54**, 379–381.

Theg, S. M. and Homann, P. H. (1982). Light-, pH- and uncoupler-dependent association of chloride with chloroplast thylakoids. *Biochim. Biophys. Acta* **679**, 221–234.

Thellier, M., Duval, Y. and Demarty, M. (1979). Borate exchanges of *Lemna minor* L. as studied with the help of the enriched stable isotope and of a (n, α) nuclear reaction. *Plant Physiol.* **63**, 283–288.

Theodorides, T. N. and Pearson, C. J. (1982). Effect of temperature on nitrate uptake, translocation and metabolism in *Pennisetum americanum. Aust. J. Plant Physiol.* **9**, 309–320.

Thiel, H. and Finck, A. (1973). Ermittlung von Grenzwerten optimaler Kupfer-Versorgung für Hafer und Sommergerste. *Z. Pflanzenernaehr. Bodenkd.* **134**, 107–125.

Thomas, J. and Prasad, R. (1983). Mineralization of urea, coated urea and nitrification inhibitor treated urea in different rice growing soils. *Z. Pflanzenernaehr. Bodenkd.* **146**, 341–347.

Thomas, W. A. (1967). Dye and calcium ascent in dogwood trees. *Plant Physiol.* **42**, 1800–1802.

Tien, T. M., Gaskins, M. H. and Hubbell, D. H. (1979). Plant growth substances produced by *Azospirillum brasilense* and their effect on the growth of pearl millet (*Pennisetum americanum* L.). *Appl. Environ. Microbiol.* **37**, 1016–1024.

Tietz, A., Ludewig, M., Dingkuhn, M. and Dörffling, K. (1981). Effect of abscisic acid on the transport of assimilates in barley. *Planta* **152**, 557–561.

Tiffin, L. O. (1970). Translocation of iron citrate and phosphorus in xylem exudate of soybean. *Plant Physiol.* **45**, 280–283.

Tiller, K. and Merry, R. H. (1981). Copper pollution of agricultural soils. *In* "Copper in Soils and Plants" (J. F. Loneragan, A. D. Robson and R. D. Graham, eds.), pp 119–137. Academic Press, London and Orlando.

Timpo, E. E. and Neyra, C. A. (1983). Expression of nitrate and nitrite reductase activities under various forms of nitrogen nutrition in *Phaseolus vulgaris* L. *Plant Physiol.* **72**, 71–75.

Tinker, P. B. (1980). Role of rhizosphere microorganisms in phosphorus uptake by plants. *Role Phosphorus Agric., Proc. Symp., 1976.* Chapter 22, pp. 617–654.

Tinker, P. B. (1984). The role of microorganisms in mediating and facilitating the uptake of plant nutrients from soil. *Plant Soil* **76**, 77–91.

Tomati, U. and Galli, E. (1979). Water stress and -SH-dependent physiological activities in young maize plants. *J. Exp. Bot.* **30**, 557–563.

Tombesi, L., Calé, M. T. and Tiborne, B. (1969). Effects of nitrogen, phosphorus and potassium fertilizers on assimilation capacity of *Beta vulgaris* chloroplasts. *Plant Soil* **31**, 65–76.

Torii, K. and Laties, G. G. (1966). Dual mechanism of ion uptake in relation to vacuolation in corn roots. *Plant Physiol.* **41**, 863–870.

Toriyama, H. and Jaffe, M. J. (1972). Migration of calcium and its role in the regulation of seismonasty in the motor cell of *Mimosa pudica* L. *Plant Physiol.* **49**, 72–81.

Tran Dang Hong, Minchin, F. R. and Summerfield, R. J. (1977). Recovery of nodulated cowpea plants (*Vigna unguiculata* (L.) Walp.) from waterlogging during vegetative growth. *Plant Soil* **48**, 661–672.

Trehan, S. P. and Sekhon, G. S. (1977). Effect of clay, organic matter and $CaCO_3$ content of zinc adsorption by soils. *Plant Soil* **46**, 329–336.

Treharne, K. J. and Cooper, J. P. (1969). Effect of temperature on the activity of carboxylase in tropical and temperate *Graminaea. J. Exp. Bot.* **20**, 170–175.

Trewavas, A. (1981). How do plant growth substances work? *Plant, Cell Environ.* **4**, 203–208.

Trier, K. and Bergmann, W. (1974). Ergebnisse zur wechselseitigen Beeinflussung der Zink- und Phosphorsäureernährung von Mais (*Zea mays* L.). *Arch. Acker-Pflanzenbau Bodenkd.* **18**, 65–75.

Tripepi, R. R. and Mitchell, C. A. (1984). Metabolic response of river birch and European birch roots to hypoxia. *Plant Physiol.* **76**, 31–35.

Triplett, E. W., Barnett, N. M. and Blevins, D. G. (1980). Organic acids and ionic balance in xylem exudate of wheat during nitrate or sulfate absorption. *Plant Physiol.* **65**, 610–613.

Trobisch, S. (1966). Ein Beitrag zur Aufklärung der pH- und Düngungsabhängigkeit der Mo-Aufnahme. *Albrecht-Thaer-Arch.* **10**, 1097–1099.

Trobisch, S. and Germar, R. (1959). Ergebnisse eines Molybdän-Düngungsversuches zu Blumenkohl. *Dtsch. Landwirtsch.* **10**, 189–191.

Trobisch, S. and Schilling, G. (1969). Untersuchungen über Zusammenhänge zwischen Massenentwicklung und N-Umsatz während der generativen Phase bei *Sinapis alba* L. *Albrecht-Thaer-Arch.* **13**, 867–878.

Trobisch, S. and Schilling, G. (1970). Beitrag zur Klärung der physiologischen Grundlage der Samenbildung bei einjährigen Pflanzen und zur Wirkung später zusätzlicher N-Gaben auf diesen Prozess am Beispiel von *Sinapsis alba* L. *Albrecht-Thaer-Arch.* **14**, 253–265.

Trolldenier, G. (1973). Secondary effects of potassium and nitrogen nutrition of rice: Change in microbial activity and iron reduction in the rhizosphere. *Plant Soil* **38**, 267–279.

Trolldenier, G. (1977). Influence of some environmental factors on nitrogen fixation in the rhizosphere of rice. *Plant Soil* **47**, 203–217.

Trolldenier, G. (1981). Influence of soil moisture, soil acidity and nitrogen source on take-all of wheat. *Phytopathol. Z.* **102**, 163–177.

Trolldenier, G. (1982). Effect of soil temperature on nitrogen fixation on roots of rice and reed. *Plant Soil* **68**, 217–221.

Trolldenier, G. and Hecht-Buchholz, C. (1984). Effect of aeration status of nutrient solution on microorganisms, mucilage and ultrastructure of wheat roots. *Plant Soil* **80**, 381–390.

Trolldenier, G. and Rheinbaben, W. von (1981a). Root respiration and bacterial population of roots. I. Effect of nitrogen source, potassium nutrition and aeration of roots. *Z. Pflanzenernaehr. Bodenkd.* **144**, 366–371.

Trolldenier, G. and Rheinbaben, W. von (1981b). Root respiration and bacterial population of roots. II. Effects of nutrient deficiency. *Z. Pflanzenernaehr. Bodenkd.* **144**, 378–384.

Tromp. J. (1979). The intake curve for calcium into apple fruits under various environmental conditions. *Commun. Soil Sci. Plant Anal.* **10**, 325–335.

Trought, M. C. T. and Drew, M. C. (1980a). The development of waterlogging damage in wheat seedlings (*Triticum aestivum* L.). I. Shoot and root growth in relation to changes in the concentration of dissolved gases and solutes in the soil solution. *Plant Soil* **54**, 77–94.

Trought, M. C. T. and Drew, M. C. (1980b). The development of waterlogging damage in wheat seedlings (*Triticum aestivum* L.). II. Accumulation and redistribution of nutrients by the shoot. *Plant Soil* **56**, 187–199.

Trought, M. C. T. and Drew, M. C. (1982). Effects of waterlogging on young wheat plants (*Triticum aestivum* L.) and on soil solutes at different soil temperatures. *Plant Soil* **69**, 311–326.

Truelsen, T. A. and Wyndaele, R. (1984). Recycling efficiency in hydrogenase uptake positive strains of *Rhizobium leguminosarum*. *Physiol. Plant.* **62**, 45–60.
Tsui, C. (1948). The role of zinc in auxin synthesis in the tomato plant. *Am. J. Bot.* **35**, 172–179.
Tu, J. C. (1981). Effect of salinity on rhizobium-root-hair interaction, nodulation and growth of soybean. *Can. J. Plant Sci.* **61**, 231–239.
Tukendorf, A., Lyszcz, S. and Baszynski, T. (1984). Copper binding proteins in spinach tolerant to excess copper. *J. Plant Physiol.* **115**, 351–360.
Tukey, H. B., Jr. (1970). The leaching of substances from plants. *Annu. Rev. Plant Physiol.* **21**, 305–324.
Tukey, H. B., Jr. and Morgan, J. V. (1963). Injury to foliage and its effect upon the leaching of nutrients from above-ground plant parts. *Physiol. Plant.* **16**, 557–564.
Turner, R. G. (1970). The subcellular distribution of zinc and copper within the roots of metal-tolerant clones of *Agrostis tenuis* Sibth. *New Phytol.* **69**, 725–731.
Tyagi, V. K. and Chauhan, S. K. (1982). The effect of leaf exudates on the spore germination of phylloplane mycoflora of chilli (*Capsicum annuum* L.) cultivars. *Plant Soil* **65**, 249–256.
Tyree, M. T. (1970). The symplast concept. A general theory of symplastic transport according to the thermodynamics of irreversible processes. *J. Theor. Biol.* **26**, 181–124.
Uehara, K., Fujimoto, S. and Taniguchi, T. (1974a). Studies on violet-colored acid phosphatase of sweet potato. I. Purification and some physical properties. *J. Biochem. (Tokyo)* **75**, 627–638.
Uehara, K., Fujimoto, S. and Taniguchi, T. (1974b). Studies on violet-colored acid phosphatase of sweet potato. II. Enzymatic properties and amino acid composition. *J. Biochem. (Tokyo)* **75**, 639–649.
Uemura, M. and Yoshida, S. (1984). Involvement of plasma membrane alterations in cold acclimation of winter rye seedlings (*Secale cereale* L. cv. Puma). *Plant Physiol.* **75**, 818–826.
Ullrich, W. R. (1983). Uptake and reduction of nitrate: Algae and fungi. *In* "Encyclopedia of Plant Physiology, New Series" (A. Läuchli and R. L. Bieleski, eds.), Vol. 15A, pp. 376–397. Springer-Verlag, Berlin and New York.
Ullrich, W. R. and Novacky, A. (1981). Nitrate-dependent membrane potential changes and their induction in *Lemna gibba* G1. *Plant Sci. Lett.* **22**, 211–217.
Ullrich-Eberius, C. I., Novacky, A., Fischer, E. and Lüttge, U. (1981). Relationship between energy-dependent phosphate uptake and the electrical membrane potential in *Lemna gibba* G1. *Plant Physiol.* **67**, 797–801.
Ulrich, A. (1941). Metabolism of non-volatile organic acids in excised barley roots as related to cation-anion balance during salt accumulation. *Am. J. Bot.* **28**, 526–537.
Ulrich, A. and Hylton, L. O., Jr. (1968). Sulfur nutrition of Italian ryegrass measured by growth and mineral content. *Plant Soil* **29**, 274–284.
Ulrich, A. and Ohki, K. (1956). Chlorine, bromine and sodium as nutrients for sugar beet plants. *Plant Physiol.* **313**, 171–181.
U.S. Salinity Laboratory Staff (1954). Diagnosis and improvement of saline and alkali soils. *U.S., Dept. Agric., Agric. Handb.* **60**.
Uren, N. C. (1982). Chemical reduction at the root surface. *J. Plant Nutr.* **5**, 515–520.
Vakhmistrov, D. B. (1967). On the function of the apparent free space in plant roots. A comparative study of the absorption power of epidermal and cortex cells in barley roots. *Sov. Plant Physiol. (Engl. Transl.)* **14**, 123–129.

Vakhmistrov, D. B. (1981). Specialization of root tissues in ion transport. *Plant Soil* **63**, 33–38.
Vakhmistrov, D. B. and Ali-Zade, V. M. (1974). Physiological analysis of transitional processes in exudation of bleeding sap by the root system of sunflower. *Sov. Plant Physiol. (Engl. Transl.)* **20**, 654–663.
Vance, C. P. and Stade, S. (1984). Alfalfa root nodule carbon dioxide fixation. II. Partial purification and characterization of root nodule phosphoenolpyruvate carboxylase. *Plant Physiol.* **75**, 261–264.
Vance, C. P., Kirk, T. K. and Sherwood, R. T. (1980). Lignification as a mechanism of disease resistance. *Annu. Rev. Phytopathol.* **18**, 259–288.
Vance, C. P., Stade, S. and Maxwell, C. A. (1983). Alfalfa root nodule carbon dioxide fixation. I. Association with nitrogen fixation and incorporation into amino acids. *Plant Physiol.* **72**, 469–473.
Van Kirk, C. A. and Raschke, K. (1978). Presence of chloride reduces malate production in epidermis during stomatal opening. *Plant Physiol.* **61**, 361–364.
Van Miegroet, H. and Cole, D. W. (1984). The impact of nitrification on soil acidification and cation leaching in a red alder ecosystem. *J. Environ. Qual.* **13**, 586–590.
Van Staden, J. and Davey, J. E. (1979). The synthesis, transport and metabolism of endogenous cytokinins. *Plant, Cell Environ.* **2**, 93–106.
Van Steveninck, R. F. M. (1965). The significance of calcium on the apparent permeability of cell membranes and the effects of substitution with other divalent ions. *Physiol. Plant.* **18**, 54–69.
Van Steveninck, R. F. M., Van Steveninck, M. E., Stelzer, R. and Läuchli, A. (1982). Studies on the distribution of Na and Cl in two species of lupin (*Lupinus luteus* and *Lupinus angustifolius*) differing in salt tolerance. *Physiol. Plant.* **56**, 465–473.
Varga, A. and Bruinsma, J. (1974). The growth and ripening of tomato fruits at different levels of endogenous cytokinins. *J. Hortic. Sci.* **49**, 135–142.
Vaughan, A. K. F. (1977). The relation between the concentration of boron in the reproductive and vegetative organs of maize plants and their development. *Rhod. J. Agric. Res.* **15**, 163–170.
Vaughan, D., DeKock, P. C. and Ord, B. G. (1982). The nature and localization of superoxide dismutase in fronds of *Lemma gibba* L. and the effect of copper and zinc deficiency on its activity. *Physiol. Plant.* **54**, 253–257.
Venkatarayappa, T., Tsujita, M. J. and Murr, D. P. (1980). Influence of cobaltous ion (Co^{2+}) on the postharvest behaviour of "Samantha" roses. *J. Am. Soc. Hortic. Sci.* **105**, 148–151.
Venkat-Raju, K. and Marschner, H. (1981). Inhibition of iron stress reactions in sunflower by bicarbonate. *Z. Pflanzenernaehr. Bodenkd.* **144**, 339–355.
Venkat-Raju, K., Marschner, H. and Römheld, V. (1972). Effect of iron nutritional status on ion uptake, substrate pH and production and release of organic acids and riboflavine by sunflower plants. *Z. Pflanzenernaehr. Bodenkd.* **132**, 177–190.
Venter, H. A. van de and Currier, H. B. (1977). The effect of boron deficiency on callose formation and ^{14}C translocation in bean *(Phaseolus vulgaris)* and cotton (*Gossypium hirsutum* L.). *Am. J. Bot.* **64**, 861–865.
Vermeer, J. and McCully, M. E. (1981). Fucose in the surface deposits of axenic and field grown roots of *Zea mays* L. *Protoplasma* **109**, 233–248.
Vertregt, N. (1968). Relation between black spot and composition of the potato tuber. *Eur. Potato J.* **11**, 34–44.

Vesk, M., Possingham, V. and Mercer, F. V. (1966). The effect of mineral nutrient deficiency on the structure of the leaf cells of tomato, spinach and maize. *Aust. J. Bot.* **14**, 1–18.

Vetter, H. and Früchtenicht, K. (1979). Berücksichtigung von Standort, Bodenzustand und Bewirtschaftungsbedingungen bei der Auswertung der Bodenuntersuchung. *Landwirtsch. Forsch., Sonderh.* **36**, 438–451.

Vetter, H. and Teichmann, W. (1968). Feldversuche mit gestaffelten Kupfer- und Stickstoff- Düngergaben in Weser–Ems. *Z. Pflanzenernaehr. Bodenkd.* **121**, 97–111.

Vetter, H., Früchtenicht, K. and Mählhop, R. (1978). Untersuchungen über den Aussagewert verschiedener Bodenuntersuchungsmethoden für die Ermittlung des Phosphatdüngerbedarfs. *Landwirtsch. Forsch., Sonderh.* **34**, 121–132.

Vielemeyer, H. P., Fischer, F. and Bergmann, W. (1966). Über den Einfluss der Eisen- und Manganernährung auf die Peroxidase- und Katalaseaktivität sowie den Gehalt an löslichen Kohlenhydraten in den Blättern einiger landwirtschaftlicher Kulturpflanzen. *Albrecht-Thaer-Arch.* **10**, 727–745.

Vielemeyer, H. P., Fischer, F. and Bergmann, W. (1969). Untersuchungen über den Einfluss der Mikronährstoffe Eisen und Mangan auf den Stickstoff-Stoffwechsel landwirtschaftlicher Kulturpflanzen. 2. Mitt.: Untersuchungen über die Wirkung des Mangans auf die Nitratreduktion und den Gehalt an freien Aminosäuren in jungen Buschbohnenpflanzen. *Albrecht-Thaer-Arch.* **13**, 393–404.

Viets, F. G., Jr. (1944). Calcium and other polyvalent cations as accelerators of ion accumulation by excised barley roots. *Plant Physiol.* **19**, 466–480.

Viets, F. G., Jr., Boawn, L. C. and Crawford, C. L. (1954). Zinc contents and deficiency symptoms of 26 crops grown on a zinc-deficient soil. *Soil Sci.* **78**, 305–316.

Vinther, F. P. (1982). Nitrogenase activity (acetylene reduction) during the growth cycle of spring barley (*Hordeum vulgare* L.). *Z. Pflanzenernaehr. Bodenkd.* **145**, 356–362.

Vlamis, J. and Williams, E. E. (1964). Iron and manganese relations in rice and barley. *Plant Soil* **20**, 221–231.

Vlamis, J. and Williams, D. E. (1967). Manganese and silicon interaction in the Gramineae. *Plant Soil* **27**, 131–140.

Vlcek, L. M. and Gassman, M. L. (1979). Reversal of α,α'dipyridyl-induced porphyrin synthesis in etiolated and greening red kidney bean leaves. *Plant Physiol.* **64**, 393–397.

Voigt, P. W., Dewald, C. L., Matocha, J. E. and Foy, C. D. (1982). Adaptation of iron-efficient and inefficient lovegrass strains to calcareous soils. *Crop Sci.* **22**, 672–676.

Volk, R. J., Kahn, R. P. and Weintraub, R. L. (1958). Silicon content of the rice plant as a factor influencing its resistance to infection by the blast fungus, *Piricularia oryzae*. *Phytopathology* **48**, 179–184.

Volz, M. G. and Jacobson, L. (1974). A specific calcium requirement for potassium uptake by excised vetch roots. *Plant Soil* **41**, 647–659.

Vorm, P. D. J. van der (1980). Uptake of Si by five plant species as influenced by variations in Si-supply. *Plant Soil* **56**, 153–156.

Vorm, P. D. J. van der and Diest, A. van (1979). Aspects of the Fe and Mn nutrition of rice plants. II. Iron and manganese uptake by rice plants, grown on aerobic water cultures. *Plant Soil* **52**, 19–29.

Vose, P. B. (1982). Iron nutrition in plants: A world overview. *J. Plant Nutr.* **5**, 233–249.

Vose, P. B. (1983). Rationale of selection for specific nutritional characters in crop improvement with *Phaseolus vulgaris* L. as a case of study. *Plant Soil* **72**, 351–364.

Vreugdenhil, D. (1983). Abscisic acid inhibits phloem loading of sucrose. *Physiol. Plant.* **57**, 463–467.

Wagatsuma, T. (1983). Characterization of absorption sites for aluminium in the roots. *Soil Sci. Plant Nutr.* **29**, 499–515.

Wagner, G. J. and Trotter, M. M. (1982). Inducible Cd binding complexes of cabbage and tobacco. *Plant Physiol.* **69**, 804–809.

Wagner, H. and Michael, G. (1971). Der Einfluss unterschiedlicher Stickstoffversorgung auf die Cytokininbildung in Wurzeln von Sonnenblumenpflanzen. *Biochem. Physiol. Pflanz.* **162**, 147–158.

Wahle, K. W. J. and Davies, N. T. (1977). Involvement of copper in microsomal mixed-function oxidase reactions: A review. *J. Sci. Food Agric.* **28**, 93–97.

Wainwright, S. J. and Woolhouse, H. W. (1975). Physiological mechanisms of heavy metal tolerance in plants. *Symp. Br. Ecol. Soc.* **15**, 231–257.

Wainwright, S. J. and Woolhouse, H. W. (1977). Some physiological aspects of copper and zinc tolerance in *Agrostis tenuis* Sibth: Cell elongation and membrane damage. *J. Exp. Bot.* **28**, 1029–1036.

Wakhloo, J. L. (1975a). Studies on the growth, flowering, and production of female sterile flowers as affected by different levels of foliar potassium in *Solanum sisymbrifolium* Lam. I. Effect of potassium content of the plant on vegetative growth and flowering. *J. Exp. Bot.* **26**, 425–433.

Wakhloo, J. L. (1975b). Studies on the growth, flowering and production of female sterile flowers as affected by different levels of foliar potassium in *Solanum sisymbrifolium* Lam. II. Interaction between foliar potassium and applied gibberellic acid and 6-furfuylaminopurine. *J. Exp. Bot.* **26**, 433–440.

Walker, C. D. and Webb, J. (1981). Copper in plants, forms and behaviour. *In* "Copper in Soils and Plants" (J. F. Loneragan, A. D. Robson and R. D. Graham, eds.), pp. 189–212. Academic Press, London and Orlando.

Walker, D. A. (1980). Regulation of starch synthesis in leaves—the role of orthophosphate. *Proc. 15th Colloq. Int. Potash Inst. Bern,* pp. 195–207.

Walker, J. M. (1969). One-degree increments in soil temperature affect maize seedling behaviour. *Soil Sci. Soc. Am. Proc.* **33**, 729–736.

Walker, R. R., Tötökfalvy, E., Scott, N. S. and Kriedemann, P. E. (1981). An analysis of photosynthetic response to salt treatment in *Vitis vinifera*. *Aust. J. Plant Physiol.* **8**, 359–374.

Wallace, A. (1980a). Effect of excess chelating agent on micronutrient concentrations in bush beans grown in solution culture. *J. Plant Nutr.* **2**, 163–170.

Wallace, A. (1980b). Effect of chelating agents on uptake of trace metals when chelating agents are applied to soil in contrast to when they are applied to solution culture. *J. Plant Nutr.* **2**, 171–175.

Wallace, A. (1982). Effect of nitrogen fertilizer and nodulation on lime-induced chlorosis in soybean. *J. Plant Nutr.* **5**, 363–368.

Wallace, A., Frolich, E. and Lunt, O. R. (1966). Calcium requirements of higher plants. *Nature (London)* **209**, 634.

Wallace, A., Abou-Zamzam, A. M. and Motoyama, E. (1971). Cation and anion balance in the xylem exudate of tobacco roots. *Plant Soil* **35**, 433–438.

Wallace, A., Abou-Zamzam, A. M. and Mueller, R. T. (1972). Transport of sodium into the xylem exudate of tobacco. *Plant Physiol.* **50**, 388–390.

Wallace, T. (1961). "The Diagnosis of Mineral Deficiencies in Plants by Visual Symptoms. A Colour Atlas and Guide." H.M. Stationery Office, London.

Wallace, W. and Pate, J. S. (1965). Nitrate reductase in the field pea (*Pisum arvense* L.). *Ann. Bot. (London)* [N.S.] **29**, 655–671.

Wallsgrove, R. M., Keys, A. J., Lea, P. J. and Miflin, B. J. (1983). Photosynthesis, photorespiration and nitrogen metabolism. *Plant, Cell Environ.* **6**, 301–309.

Wang, C. H., Liem, T. H. and Mikkelsen, D. S. (1976). Sulfur deficiency—a limiting factor in rice production in the lower Amazon basin. II. Sulfur requirement for rice production. *IRI Res. Inst. [Rep.]* **48**, 9–30.

Wang, T. S. C., Yang, T. K. and Chuang, Z. T. (1967). Soil phenolic acids as plant growth inhibitors. *Soil Sci.* **103**, 239–246.

Warembourg, F. R. and Billes, G. (1979). Estimation carbon transfers in the plant rhizosphere. *In* "The Soil-Root Interface" (J. L. Harley and R. Scott-Russell, eds.), pp. 183–196. Academic Press, London and Orlando.

Waters, S. P., Martin, P. and Lee, B. T. (1984). The influence of sucrose and abscisic acid on the determination of grain number in wheat. *J. Exp. Bot.* **35**, 829–840.

Watson, D. J. (1952). The physiological basis of variation in yield. *Adv. Agron.* **4**, 101–144.

Watson, E. R., Lapins, P. and Barron, R. J. W. (1976). Effect of waterlogging on the growth, grain and straw yield of wheat, barley and oats. *Aust. J. Exp. Agric. Anim. Husb.* **16**, 114–122.

Webb, T. and Armstrong, W. (1983). The effect of anoxia and carbohydrates on the growth and viability of rice, pea and pumpkin roots. *J. Exp. Bot.* **34**, 579–603.

Weckenmann, D. and Martin, P. (1981). Changes in the pattern of endopeptidases during senescence of bush bean leaves (*Phaseolus vulgaris* L.). *Z. Pflanzenphysiol.* **104**, 103–108.

Wedler, A. (1980). Untersuchungen über Nitratgehalte in einigen ausgewählten Gemüsearten. *Landwirtsch. Forsch., Sonderh.* **36**, 128–137.

Wehrmann, J. and Scharpf, H. C. (1979). Der Mineralstickstoffgehalt des Bodens als Maßstab für den Stickstoffdüngerbedarf. *Plant Soil* **52**, 109–126.

Wehrmann, J., Scharpf, H. C., Böhmer, M. and Wollring, J. (1982). Determination of nitrogen fertilizer requirements by nitrate analysis of the soil and of the plant. *In* "Proceedings of the Ninth International Plant Nutrition Colloquium, Warwick, England" (A. Scaife, ed.), pp. 702–709. Commonw. Agric. Bur., Farnham Royal, Bucks.

Weigel, R. J., Jr., Schillinger, J. A., McCaw, B. A., Gauch, H. G. and Hsiao, E. (1974). Nutrient-nitrate levels and the accumulation of chloride in leaves of snap beans and roots of soybeans. *Crop Sci.* **13**, 411–412.

Weigl, H. and Ziegler, H. (1962). Die räumliche Verteilung von ^{35}S und die Art der markierten Verbindungen in Spinatblättern nach Begasung mit $^{35}SO_2$. *Planta* **58**, 435–447.

Weiler, E. W. and Ziegler, H. (1981). Determination of phytohormones in phloem exudate from species by radioimunoassay. *Planta* **152**, 168–170.

Weir, R. C. and Hudson, A. (1966). Molybdenum deficiency in maize in relation to seed reserves. *Aust. J. Exp. Agric. Anim. Husb.* **6**, 35–41.

Weir, R. G., Nagle, R. K., Noonan, J. B. and Towner, A. G. W. (1976). Effect of foliar and soil applied molybdenum treatments on molybdenum concentration of maize grain. *Aust. J. Exp. Agric. Anim. Husb.* **16**, 761–764.

Weiser, C. J., Blaney, L. T. and Li, P. (1964). The question of boron and sugar translocation. *Physiol. Plant.* **17**, 589–599.

Weiss, A. and Herzog, A. (1978). Isolation and characterization of a silicon-organic complex from plants. *In* "Biochemistry of Silicon and Related Problems" (G. Bendz and I. Lindqvist, eds.), pp. 109–127. Plenum, New York.

Weiss, M. G. (1943). Inheritance and physiology of efficiency in iron utilization in soybeans. *Genetics* **28**, 253–268.

Welch, R. M. (1981). The biological significance of nickel. *J. Plant Nutr.* **3**, 345–356.

Welch, R. M., House, W. A. and Allaway, W. H. (1974). Availability of zinc from pea seed to rats. *J. Nutr.* **104**, 733–740.

Welch, R. M., Webb, M. J. and Loneragan, J. F. (1982). Zinc in membrane function and its role in phosphorus toxicity. *In* "Proceedings of the Ninth Plant Nutrition Colloquium, Warwick, England" (A. Scaife, ed.), pp. 710–715. Commonw. Agric. Bur., Farnham Royal, Bucks.

Welp, G., Herms, U. and Brümmer, G. (1983). Einfluss von Bodenreaktion, Redoxbedingungen und organischer Substanz auf die Phosphatgehalte der Bodenlösung. *Z. Pflanzenernaehr. Bodenkd.* **146**, 38–52.

Welte, E. and Müller, K. (1966). Über den Einfluss der Kalidüngung auf die Dunkelung von rohem Kartoffelbrei. *Eur. Potato J.* **9**, 36–45.

Wenzel, G. and Kreutzer, K. (1971). Der Einfluss des Manganmangels auf die Resistenz der Fichten (*Picea abies* Karst.) gegen *Fomes annosus* (Fr) Cooke. *Z. Pflanzenernaehr. Bodenkd.* **128**, 123–129.

Werner, D. (1967). Untersuchungen über die Rolle der Kieselsäure in der Entwicklung höherer Pflanzen. I. Analyse der Hemmung durch Germaniumsäure. *Planta* **76**, 25–36.

Werner, D. (1980). Stickstoff (N_2)-Fixierung und Produktionsbiologie. *Angew. Bot.* **54**, 67–75.

Werner, D. and Roth, R. (1983). Silica metabolism. *In* "Encyclopedia of Plant Physiology, New Series" (A. Läuchli and R. L. Bieleski, eds.), Vol. 15B, pp. 682–694. Springer-Verlag, Berlin and New York.

Werner, D., Mörsch, E., Stripf, R. and Winchenbach, W. (1980). Development of nodules of *Glycine max* infected with an ineffective strain of *Rhizobium japonicum. Planta* **147**, 320–329.

Werner, D., Wilcockson, J., Tripf, R., Mörschel, E. and Papen, H. (1981). Limitations of symbiotic and associated nitrogen fixation by developmental stages in the system *Rhizobium japonicum* with *Glycine max* and *Azospirillum brasilense* with grasses, e.g. *Triticum aestivum*. *In* "Biology of Inorganic Nitrogen and Sulfur" (H. Bothe and A. Trebst, eds.), pp. 299–308. Springer-Verlag, Berlin and New York.

Werner, W. (1959). Die Wirkung einer Magnesiumdüngung zu Kartoffeln in Abhängigkeit von Bodenreaktion und Stickstofform. *Kartoffelbau* **10**, 13–14.

Weser, U., Rupp, H., Donay, F., Linnemann, F., Voelter, W., Voetsch, W. and Jung, G. (1973). Characterization of Cd, Zn-thionein (metallothionein) isolated from rat and chicken liver. *Eur. J. Biochem.* **39**, 127–140.

West, D. W. (1978). Water use and sodium chloride uptake by apple trees. II. The response to soil oxygen deficiency. *Plant Soil* **50**, 51–65.

West, D. W. and Taylor, J. A. (1980). The effect of temperature on salt uptake by tomato plants with diurnal and nocturnal waterlogging of salinized rootzones. *Plant Soil* **56**, 113–121.

Westerman, L. and Roddick, J. G. (1981). Annual variation in sterol levels in leaves of *Taraxacum officinale* Weber. *Plant Physiol.* **68**, 872–875.

Wetselaar, R. and Farquhar, G. D. (1980). Nitrogen losses from tops of plants. *Adv. Agron.* **33**, 263–302.

White, M. C., Decker, A. M. and Chaney, R. L. (1979). Differential cultivar tolerance in soybean to phytotoxic levels of soil Zn. I. Range of cultivar response. *Agron. J.* **71**, 121–126.

White, M. C., Decker, A. M. and Chaney, R. L. (1981a). Metal complexation in xylem fluid. I. Chemical composition of tomato and soybean stem exudate. *Plant Physiol.* **67**, 292–300.

White, M. C., Decker, A. M. and Chaney, R. L. (1981b). Metal complexation in xylem fluid. II. Theoretical equilibrium model and computational computer program. *Plant Physiol.* **67**, 301–310.

White, R. E. (1977). Studies on mineral absorption by plants. III. The interaction of aluminium, phosphate and pH on the growth of *Medicago sativa. Plant Soil* **46**, 195–208.

Whiteaker, G., Gerloff, G. C., Gabelman, W. H. and Lindgren, D. (1976). Intraspecific differences in growth of beans at stress levels of phosphorus. *J. Am. Soc. Hortic. Sci.* **101**, 472–475.

Wiebe, H. J., Schätzler, H. P. and Kühn, W. (1977). On the movement and distribution of calcium in white cabbage in dependence of the water status. *Plant Soil* **48**, 409–416.

Wieneke, J. and Führ, F. (1973). Untersuchungen zur Translokation von ^{45}Ca im Apfelbaum. I. Transport und Verteilung in Abhängigkeit von Aufnahme-Zeitpunkt. *Gartenbauwissenschaft* **38**, 91–108.

Wiersma, D. and van Goor, B. J. (1979). Chemical forms of nickel and cobalt in phloem of *Ricinus communis. Physiol. Plant.* **45**, 440–442.

Wiersum, L. K. (1957). The relationship of the size and structural rigidity of pores to their penetration by roots. *Plant Soil* **9**, 75–85.

Wiersum, L. K. (1966). Calcium content of fruits and storage tissue in relation to the mode of water supply. *Acta Bot. Neerl.* **15**, 406–418.

Wiersum, L. K. (1967). Potential subsoil utilization by roots. *Plant Soil* **27**, 383–400.

Wiese, G. and Veith, J. A. (1975). Komplexbildung zwischen Zitronensäure und Aluminium. *Z. Naturforsch., B: Anorg. Chem., Org. Chem.* **303**, 446–453.

Wightman, F., Schneider, E. A. and Thimann, K. V. (1980). Hormonal factors controlling the initiation and development of lateral roots. II. Effects of exogenous growth factors on lateral root formation in pea roots. *Physiol. Plant.* **49**, 304–314.

Wignarajah, K., Greenway, H. and John, C. D. (1976). Effect of waterlogging on growth and activity of alcohol dehydrogenase in barley and rice. *New Phytol.* **77**, 585–592.

Wild, A., Woodhouse, P. J. and Hopper, M. J. (1979). A comparison between uptake of potassium by plants from solutions of constant potassium concentration and during depletion. *J. Exp. Bot.* **30**, 697–704.

Willenbrink, J. (1964). Lichtabhängiger ^{35}S-Einbau in organische Bindung in Tomatenpflanzen. *Z. Naturforsch.* **19**, 356–357.

Willenbrink, J. (1967). Über Beziehungen zwischen Proteinumsatz und Schwefelversorgung der Chloroplasten. *Z. Pflanzenphysiol.* **56**, 427–438.

Willenbrink, J. (1983). Mechanismen des Zuckertransports durch Membranen bei *Beta vulgaris. Kali-Briefe* **16**, 585–594.

Willenbrink, J., Doll, S., Getz, H.-P. and Meyer, S. (1984). Zuckeraufnahme in isolierten Vakuolen und Protoplasten aus dem Speichergewebe von Beta-Rüben. *Ber. Dtsch. Bot. Ges.* **97**, 27–39.

Williams, D. E. and Vlamis, J. (1957). The effect of silicon on yield and manganese-54 uptake and distribution in the leaves of barley plants grown in culture solutions. *Plant Physiol.* **32**, 404–409.

Williams, E. G. and Knight, A. H. (1963). Evaluations of soil phosphate status by pot experiments, conventional extraction methods and labile phosphate values estimated with the aid of phosphorus 32. *J. Sci. Food Agric.* **14**, 555–563.

Willmer, C. M. and Mansfield, T. A. (1970). Active cation transport and stomatal opening: A possible physiological role of sodium ions. *Z. Pflanzenphysiol.* **61**, 398–400.

Wills, R. B. H., Tirmazi, S. I. H. and Scott, K. J. (1977). Use of calcium to delay ripening of tomatoes. *HortScience* **12**, 551–552.

Wilson, D. O. and Reisenauer, H. M. (1963). Cobalt requirement of symbiotically grown alfalfa. *Plant Soil* **19**, 364–373.

Wilson, D. O., Boswell, F. C., Ohki, K., Parker, M. B., Shuman, L. M. and Jellum, M. D. (1982). Changes in soybean seed oil and protein as influenced by manganese nutrition. *Crop Sci.* **22**, 948–952.

Wilson, S. B. and Hallsworth, E. G. (1965). Studies on the nutrition of the forage legumes. IV. The effect of cobalt on the growth of nodulated and non-nodulated *Trifolium subterraneum* L. *Plant Soil* **22**, 260–279.

Wilson, S. B. and Nicholas, D. J. D. (1967). A cobalt requirement for non-nodulated legumes and for wheat. *Phytochemistry* **6**, 1057–1066.

Winkler, R. G., Polacco, J. C., Eskew, D. L. and Welch, R. M. (1983). Nickel is not required for apo-urease synthesis in soybean seeds. *Plant Physiol.* **72**, 262–263.

Winter, E. (1982). Salt tolerance of *Trifolium alexandrinum* L. III. Effects of salt on ultrastructure of phloem and xylem transfer cells in petioles and leaves. *Aust. J. Plant Physiol.* **9**, 239–250.

Wirth, E., Kelly, G. J., Fischbeck, G. and Latzko, E. (1977). Enzyme activities and products of CO_2-fixation in various photosynthetic organs of wheat and oat. *Z. Pflanzenphysiol.* **82**, 78–87.

Witt, H. H. and Jungk, A. (1974). The nitrate inducible nitrate reductase activity in relation to nitrogen nutritional status of plants. *Proc. 7th Int. Colloq. Plant Anal. Fert. Probl.,* pp. 519–527. Hannover.

Witt, H. H. and Jungk, A. (1977). Beurteilung der Molybdänversorgung von Pflanzen mit Hilfe der Mo-induzierbaren Nitratreduktase-Aktivität *Z. Pflanzenernaehr. Bodenkd.* **140**, 209–222.

Witty, J. F., Keay, P. J., Frogatt, P. J. and Dart, P. J. (1979). Algal nitrogen fixation on temperate arable fields. The Broadbalk experiment. *Plant Soil* **52**, 151–164.

Wollring, J. and Wehrmann, J. (1981). Der Nitrat-Schnelltest—Entscheidungshilfe für die N-Spätdüngung. *Mitt. DLG* **8**, 448–449.

Wolswinkel, P. (1978). Accumulation of phloem-mobile mineral elements at the site of attachment of *Cuscuta europea* L. *Z. Pflanzenphysiol.* **86**, 77–84.

Wolswinkel, P., Ammerlaan, A. and Peters, F. C. (1984). Phloem unloading of amino acids at the site of attachment of *Cuscuta europaea. Plant Physiol.* **75**, 13–20.

Wolterbeek, H. T., Luipen, J. van and Bruin, M. de (1984). Non-steady state xylem transport of fifteen elements into the tomato leaf as measured by gamma-ray spectroscopy: A model. *Physiol. Plant.* **61**, 599–606.

Wong, M. H. and Bradshaw, A. D. (1982). A comparison of the toxicity of heavy metals, using root elongation of rye grass, *Lolium perenne. New Phytol.* **91**, 255–261.

Wong You Cheong, Y. and Chan, P. Y. (1973). Incorporation of ^{32}P in phosphate esters of the sugar cane plant and the effect of Si and Al on the distribution of these esters. *Plant Soil* **38**, 113–123.

Woodrow, I. E. and Rowan, K. S. (1979). Change of flux of orthophosphate between cellular compartments in ripening tomato fruits in relation to the climacteric rise in respiration. *Aust. J. Plant Physiol.* **6**, 39–46.

Woolhouse, H. W. (1978). Light gathering and carbon assimilation processes in photosynthesis their adaptive modification and significance for agriculture. *Endeavour* [N.S.] **2**, 35–46.

Woolhouse, H. W. (1983). Toxicity and tolerance in response of plants to metals. *In* "Encyclopedia of Plant Physiology, New Series" (O. L. Lange *et al.,* eds.), Vol. 12C, pp. 246–300. Springer-Verlag, Berlin and New York.

Wright, J. P. and Fisher, D. B. (1981). Measurement of the sieve tube membrane potential. *Plant Physiol.* **67**, 845–848.

Wrigley, C. W., du Cros, D. L., Archer, M. J., Downie, P. G. and Roxburgh, C. M. (1980). The sulfur content of wheat endosperm and its relevance to grain quality. *Aust. J. Plant Physiol.* **7**, 755–766.

Wu, L., Thurman, D. A. and Bradshaw, A. D. (1975). The uptake of copper and its effect upon respiratory processes of roots of copper-tolerant and non-tolerant clones of *Agrostis stolonifera. New Phytol.* **75**, 225–229.

Wunderlich, F. (1978). Die Kernmatrix: Dynamisches Protein-Gerüst in Zellkernen. *Naturwiss. Rundsch.* **31**, 282–288.

Wutscher, H. K., Olsen, E. O., Shuli, A. V. and Plynado, A. (1970). Leaf nutrient levels, chlorosis, and growth of young grapefruit trees on 16 rootstocks grown on calcareous soil. *J. Am. Soc. Hortic. Sci.* **95**, 259–261.

Wyn Jones, R. G. (1981). Salt tolerance. *In* "Physiological Processes Limiting Plant Productivity" (C. B. Johnson, ed.), pp. 271–292. Butterworth, London.

Wyn Jones, R. G. and Lunt, O. R. (1967). The function of calcium in plants. *Bot. Rev.* **33**, 407–426.

Wyn Jones, R. G. and Pollard, A. (1983). Proteins, enzymes and inorganic ions. *In* "Encyclopedia of Plant Physiology, New Series" (A. Läuchli and R. L. Bieleski, eds.), Vol. 15B, pp. 528–562. Springer-Verlag, Berlin and New York.

Wyn Jones, R. G., Brady, C. J. and Speirs, J. (1979). Ionic and osmotic relations in plant cells. *In* "Recent Advances in the Biochemistry of Cereals" (D. L. Laidman and R. G. Wyn Jones, eds.), pp. 63–103. Academic Press, London and Orlando.

Yahalom, E., Kapulnik, Y. and Okon, Y. (1984). Response of *Setaria italica* to inoculation with *Azospirillum brasilense* as compared to *Acotobacter chroococcum. Plant Soil* **82**, 77–85.

Yamada, Y., Bukovac, M. J. and Wittwer, S. H. (1964). Ion binding by surfaces of isolated cuticular membranes. *Plant Physiol.* **39**, 978–982.

Yeo, A. R. (1983). Salinity resistance: Physiologies and prices. *Physiol. Plant.* **58**, 214–222.

Yeo, A. R., Läuchli, A., Kramer, D. and Gullasch, J. (1977). Ion measurements by X-ray microanalysis in unfixed, frozen, hydrated plant cells of species differing in salt tolerance. *Planta* **134**, 35–38.

Yih, R. Y. and Clark, H. E. (1965). Carbohydrate and protein content of boron-deficient tomato root tips in relation to anatomy and growth. *Plant Physiol.* **40**, 312–315.
Yoshida, S. (1984). Chemical and biochemical changes in the plasma membrane during cold acclimation of mulberry bark cells (*Morus bombycis* Koidz, cv. Goroji). *Plant Physiol.* **76**, 257–265.
Yoshida, S. and Tadano, T. (1978). Adaptation of plants to submerged soils. *ASA Spec. Publ.* **32**, 233–256.
Yoshida, S., Navasero, S. A. and Ramirez, E. A. (1969). Effects of silica and nitrogen supply on some leaf characters of the rice plant. *Plant Soil* **31**, 48–56.
Yoshida, T. and Ancajas, R. R. (1973). Nitrogen-fixing activity in upland and flooded rice fields. *Soil Sci. Soc. Am. Proc.* **37**, 42–46.
Yoshida, T. and Yoneyama, T. (1980). Atmospheric dinitrogen fixation in the flooded rice rhizosphere as determined by the N-15 isotope technique. *Soil Sci. Plant Nutr. (Tokyo)* **26**, 551–559.
Yu, P. T., Stolzy, L. H. and Letey, J. (1969). Survival of plants under prolonged flooded conditions. *Agron. J.* **61**, 844–847.
Zeevaart, J. A. D. and Boyer, G. L. (1984). Accumulation and transport of abscisic acid and its metabolites in *Ricinnus* and *Xanthium. Plant Physiol.* **74**, 934–939.
Zehler, E. (1981). Die Natrium-Versorgung von Mensch, Tier und Pflanze. *Kali-Briefe* **15**, 773–792.
Zelleke, A. and Kliewer, W. M. (1981). Factors affecting the qualitative and quantitative levels of cytokinins in xylem sap of grapevine. *Vitis* **20**, 93–104.
Ziegler, H. (1975). Nature of transported substances. *In* "Encyclopedia of Plant Physiology, New Series" (M. H. Zimmermann and J. A. Milburn, eds.), Vol. 1, pp. 59–100. Springer-Verlag, Berlin and New York.
Zimmermann, U. and Steudle, E. (1970). Bestimmung von Reflexionskoeffizienten an der Membran der Alge *Valonia utricularis. Z. Naturforsch., B: Anorg. Chem., Org. Chem., Biochem., Biophys., Biol.* **25B**, 500–504.
Zsoldos, F. and Haunold, E. (1982). Influence of 2,4-D and low pH on potassium, ammonium and nitrate uptake by rice roots. *Physiol. Plant.* **54**, 63–68.
Zsoldos, F. and Karvaly, B. (1978). Effects of Ca^{2+} and temperature on potassium uptake along roots of wheat, rice and cucumber. *Physiol. Plant.* **43**, 326–330.
Zur, B., Jones, J. W., Boote, K. J. and Hammond, L. C. (1982). Total resistance to water flow in field soybeans. II. Limiting soil moisture. *Agron. J.* **74**, 99–105.

Index